The nth-Order Comprehensive Adjoint Sensitivity Analysis Methodology, Volume I

Dan Gabriel Cacuci

The nth-Order Comprehensive Adjoint Sensitivity Analysis Methodology, Volume I

Overcoming the Curse of Dimensionality: Linear Systems

Springer

Dan Gabriel Cacuci
University of South Carolina
Columbia, SC, USA

ISBN 978-3-030-96363-7 ISBN 978-3-030-96364-4 (eBook)
https://doi.org/10.1007/978-3-030-96364-4

This Springer imprint is published by the registered company Springer Nature Switzerland AG
The registered company address is: Gewerbestrasse 11, 6330 Cham, Switzerland

Preface

The computational models of physical systems comprise parameters, independent and dependent variables. Since the physical processes themselves are seldom known precisely and since most of the model's parameters stem from experimental procedures that are also subject to imprecisions and/or uncertainties, the results predicted by computational models are also imprecise, being affected by the uncertainties underlying the respective model. Analyzing the reliability of the model results (called model "responses") requires the computation and use of the functional derivatives (called "sensitivities") of the model's response to the model's imprecisely known parameters. Model parameters include imprecisely known internal and external boundaries that characterize the physical system represented by the respective model. Sensitivities are needed for many purposes: eliminating unimportant parameters and/or processes, quantifying the uncertainties induced in a model response due to model parameter uncertainties, validating the model by comparing computations to experiments, optimizing the model, assimilating additional data into the model, reducing uncertainties in the results predicted by the model, and so on.

It is well known that the number of sensitivities increases exponentially with the order of the respective sensitivities. Hence, the computation of sensitivities by conventional deterministic and/or statistical methods, which require re-computations using the original model, is limited in practice by the well-known "curse of dimensionality." Furthermore, conventional methods can provide only approximate, rather than exact, values for the sensitivities.

The material presented in the three volumes of this book stems from the author's quest to overcome, to the largest possible extent, the curse of dimensionality while providing exact expressions for computing efficiently model response sensitivities of high order for large-scale models comprising many parameters. Volume 1 of this book presents the mathematical framework of the "nth-Order Comprehensive Adjoint Sensitivity Analysis Methodology for Response-Coupled Forward and Adjoint Linear Systems" (abbreviated as "nth-CASAM-L"), along with didactic applications thereof. The need for determining not only first-order but also higher-order sensitivities of model responses to model parameters is underscored in Chap. 1

of this volume. The "nth-CASAM-L" mathematical framework, which is presented in Chap. 2 of this volume, was developed specifically for linear systems because the most important model responses produced by such systems can depend simultaneously on both the forward and adjoint state functions governing the respective linear system. Nonlinear operators do not admit adjoint operators (only linearized versions of nonlinear operators can admit a bona fide adjoint operator), so responses in nonlinear systems can only depend on the system's forward state functions. Therefore, the sensitivity analysis of responses that simultaneously involve both forward and adjoint state functions makes it necessary to treat linear models/systems in their own right, rather than treating them as particular cases of nonlinear systems.

The nth-CASAM-L is formulated in linearly increasing higher-dimensional Hilbert spaces, as opposed to exponentially increasing parameter-dimensional spaces. Consequently, the nth-CASAM-L enables the most efficient computation of exactly determined expressions of arbitrarily high-order sensitivities of a generic system response which depends on both the forward and adjoint state functions. The integer "n" denotes an arbitrarily high order (for the respective sensitivity) and the qualifier "comprehensive" is employed in order to highlight the fact that the framework of the nth-CASAM-L comprises not only internal model parameters but also includes the system's uncertain boundaries and internal interfaces in the model's phase-space.

The validity of the mathematical framework of the nth-CASAM-L is established using mathematical induction. The nth-CASAM-L can be applied to any model (deterministic, statistical, etc.). The higher-level adjoint functions are computed using the same forward and adjoint solvers (i.e., computer codes) as used for solving the original forward and adjoint systems, thereby requiring relatively minor additional software development for computing the various-order sensitivities. The nth-CASAM-L also enables the direct and exact derivation (first on paper, then programing the exact expression thus obtained for numerical evaluation) of the expression of a specific high-order sensitivity of interest, which can subsequently be computed directly; such a direct computation is not possible with any statistical method. The larger the number of model parameters, the more efficient the nth-CASAM-L becomes for computing arbitrarily high-order response sensitivities.

The application of the nth-CASAM-L is illustrated in Chap. 3 at a graduate-student's level by considering two paradigm physical models. One model describes particle transport in a homogenized medium characterized by imprecisely known parameters, internal and external boundaries in a spatial domain. The other model describes the time-evolution—in a temporal domain having imprecisely known starting and final-time instances—of the concentration of a substance obeying a conservation law in a medium characterized by imprecisely known parameters and initial state. These models have been selected for illustrating the application of the nth-CASAM-L because they involve many uncertain parameters, internal and external interfaces and imprecisely known initial- and final-time instances, yet they admit exact closed-form expressions for all sensitivities, for convenient verification of all results. The paradigm models analyzed in Chap. 3 admit many relative sensitivities that have identical values. The precise and efficient computation of many relative sensitivities that have identical values does not pose any difficulties to the nth-

CASAM-L methodology but is known to pose significant challenges to conventional statistical methods which rely on sensitivities being readily "sortable" by having relative values that clearly differ from each other. Since the models analyzed in Chap. 3 involve numerous parameters yet are sufficiently simple to admit closed-form exact solutions for all quantities of interest (forward and adjoint state functions, responses, sensitivities, etc.), these paradigm models can be used as benchmarks for comparing the results produced by the nth-CASAM-L to the results produced by conventional statistical and/or finite-difference methods.

Chapter 4 illustrates the application of the nth-CASAM-L to perform sensitivity analysis of ratios of generic functionals of the forward and adjoint neutron flux distributions in a multiplying subcritical system which comprises a non-fission neutron source. The forward and adjoint flux distributions in such a system are modeled by the forward and, respectively, adjoint time-independent integro-differential neutron transport Boltzmann equations. This physical system has been chosen as a paradigm application of the nth-CASAM-L because the Boltzmann equation comprises both differential and integral operators, in six independent variables, as well as a large number of model parameters. Using the integro-differential Boltzmann equation makes it possible to illustrate simultaneously the construction of adjoint operators both for differential and for integral operators. Solving numerically the integro-differential Boltzmann equation epitomizes "large-scale computations" involving many parameters (including isotopic number densities, microscopic cross sections, fission spectrum parameters, neutron sources, and detector response functions).

The Roussopoulos and Schwinger functionals are particularly important model responses that depend on both the forward and adjoint state functions, playing prominent roles in the construction of variational principles for system control, optimization and solving operator-equations. Therefore, Chap. 4 also presents the application of the nth-CASAM-L to these functionals. The material presented in Chap. 4 will be used to perform a fourth-order sensitivity and uncertainty analysis of an OECD/NEA reactor physics benchmark modeled by the Boltzmann transport equation involving many (over twenty thousand) uncertain model parameters. This analysis, which also quantifies the potentially large errors incurred when neglecting higher-order sensitivities, will be presented in Volume 2 of this book.

Volume 3 of this book will present the "nth-Order Comprehensive Adjoint Sensitivity Analysis Methodology for Nonlinear Systems" (abbreviation: nth-CASAM-N) for the practical, efficient, and exact computation of arbitrarily high-order sensitivities of responses to model parameters for nonlinear systems.

The material presented in this book is aimed at graduate students and researchers in the fields of uncertainty quantification, model validation, optimization, data assimilation, model calibration, sensor fusion, reduced-order modeling, inverse problems, and predictive modeling.

The author gratefully acknowledges the constructive reviewing and recommendations from Dr. Mihaela Ionescu-Bujor, scientist at the Karlsruhe Institute of Technology, Germany. Her observations and suggestions have significantly streamlined the didactical presentation of the contents of this book. The Author is also grateful to his colleague, Dr. Ruixian Fang, Research Associate Professor at the University of South Carolina, for many fruitful discussions, for his thorough reviewing and final proofreading of the manuscript.

Columbia, SC, USA Dan Gabriel Cacuci

Contents

Chapter 1
Motivation: Overcoming the Curse of Dimensionality in Sensitivity Analysis, Uncertainty Quantification, and Predictive Modeling

1.1 Introduction

The mathematical/computational models of physical systems comprise parameters, independent and dependent variables. Since the physical processes themselves are seldom known precisely and since most of the model's parameters stem from experimental procedures that are also subject to imprecisions and/or uncertainties, the results predicted by these models are also imprecise, being affected by the uncertainties underlying the respective model. The functional derivatives (also called "sensitivities") of results (also called "responses") produced by mathematical/computational models are needed for many purposes, including: (1) understanding the model by ranking the importance of the various parameters; (2) performing "reduced-order modeling" by eliminating unimportant parameters and/or processes; (3) quantifying the uncertainties induced in a model response due to model parameter uncertainties; (4) performing "model validation," by comparing computations to experiments to address the question "does the model represent reality?" (5) prioritizing improvements in the model; (6) performing data assimilation and model calibration as part of forward "predictive modeling" to obtain best-estimate predicted results with reduced predicted uncertainties; (7) performing inverse "predictive modeling"; and (8) designing and optimizing the system.

The importance of computing accurately not only the first-order sensitivities (i.e., functional derivatives) but also the higher-order sensitivities of model responses (i.e., results of interest) with respect to the model's parameters is underscored in Sect. 1.2. By considering a simple model of neutron scattering and absorption in a large medium, chosen because it admits exact closed-form solutions for all of the response sensitivities and moments (expectation, variance, skewness, kurtosis) in the parameters' phase-space, it is shown that the neglect of higher-order response sensitivities leads to substantial errors in predicting the model's response and accompanying uncertainties. In particular, if only first-order sensitivities are used, the third moment of the response distribution is predicted to be zero, which is

D. G. Cacuci, *The nth-Order Comprehensive Adjoint Sensitivity Analysis Methodology, Volume I*, https://doi.org/10.1007/978-3-030-96364-4_1

erroneous unless the unknown response distribution happens to be symmetrical in the phase-space of model parameters. At least second-order sensitivities must be determined and employed in order to estimate the skewness of the response distribution.

Section 1.3 highlights the exponential increase—with the order of response sensitivities—of the number of large-scale computations needed to determine the higher-order sensitivities. For a model comprising a total number of *TP* parameters, the computation by conventional methods of the nth-order sensitivities of a model response with respect to the model parameters requires at least $O(TP^n)$ large-scale computations. This exponential increase of the number of large-scale computations needed to determine the higher-order sensitivities is the manifestation of the "curse of dimensionality in sensitivity analysis," by analogy to the expression coined by Bellman (1957) to express the difficulty of using "brute-force" grid search when optimizing a function with many input variables.

Section 1.4 presents the general formulas for computing the first four moments of model responses in the phase-space of model parameters. These formulas are consistent up to and including the sixth-order standard deviations of the parameters, thus generalizing the formulas originally provided by Tukey (1957). These formulas clearly illustrate that the curse of dimensionality also affects the computation of the uncertainties induced in the response distribution by uncertainties in the distribution of parameters.

1.2 Need for Computation of High-Order Response Sensitivities: An Illustrative Example

This section illustrates the need to compute high-order sensitivities for predicting variations in the model's response caused by variations in the model's parameters (which is the scope of "sensitivity analysis"), as well as predicting the impact of higher-order sensitivities on the uncertainties induced in responses by uncertain model parameters (which is the scope of "uncertainty quantification"). The illustrative model used for this purpose describes the energy-dependent distribution of neutrons in a medium of large ("infinite") extent, in which neutrons emitted by a source (which may be an external source or an internal fission source) will lose energy ("slow down") following scattering collisions and will disappear following absorption collisions with the nuclei in the medium. A widely used neutron slowing down model (Weiberg and Wigner 1958; Amaldi 1959) is described by the following integral equation:

$$\int_E^{\infty} \Phi(E')\Sigma_s(E')\frac{dE'}{E'} + Q(E) = \Phi(E)\Sigma(E). \tag{1.1}$$

The integral (1.1) for the energy-dependent neutron flux, denoted as $\Phi(E)$, is derived under the assumptions that: (1) the absorbing isotope does not contribute to scattering; (2) $\Sigma_s(E)$ represents the macroscopic scattering cross section of hydrogen, since scattering by the heavy nuclide(s) is negligible; and (3) the macroscopic total cross section is denoted as $\Sigma(E) \triangleq \Sigma_s(E) + \Sigma_a(E)$, where the absorption cross section $\Sigma_a(E)$ stems mostly from the heavy nuclide (in the hydrogen-moderated system), since the absorption in hydrogen is negligible by comparison to its scattering cross section. The quantity $Q(E)$ in (1.1) denotes the energy-dependent source of neutrons in the medium. The macroscopic scattering cross section of hydrogen is defined as $\Sigma_s(E) = N_H\sigma_s(E)$, where N_H and $\sigma_s(E)$ denote the atomic number density and, respectively, the microscopic scattering cross section of hydrogen. The macroscopic absorption cross section of the heavy nuclide is defined as $\Sigma_a(E) = N_a\sigma_a(E)$, where N_a and $\sigma_a(E)$ denote the atomic number density and, respectively, the microscopic absorption cross section of the absorbing heavy nuclide. Under the additional assumption that the (resonance) absorption sets in below the low-energy limit, E_s, of the fission spectrum (an assumption which also does not affect the sensitivity analysis to follow), the solution of (1.1) has the following form:

$$\begin{aligned} \Phi(\boldsymbol{\alpha}; E) &= \frac{p(\boldsymbol{\alpha}; E)\int_{E_s(\boldsymbol{\alpha})}^{\infty} Q(\boldsymbol{\alpha}; E')dE'}{E\Sigma(\boldsymbol{\alpha}; E)}; \\ p(\boldsymbol{\alpha}; E) &\triangleq \exp\left[-\int_{E}^{E_s(\boldsymbol{\alpha})} \frac{\Sigma_a(\boldsymbol{\alpha}; E')}{\Sigma(\boldsymbol{\alpha}; E')}\frac{dE'}{E'}\right]. \end{aligned} \tag{1.2}$$

Evidently, the flux $\Phi(\boldsymbol{\alpha}; E)$ depends on the fission spectrum and the various number densities and microscopic cross sections of the isotopes contained in the medium under consideration. Therefore, it is convenient to use the generic notation introduced in Sect. 1.2 and thus to consider all of the uncertain model parameters which influence the flux $\Phi(\boldsymbol{\alpha}; E)$ to be the components of a "vector of parameters" denoted as $\boldsymbol{\alpha} \triangleq (\alpha_1, \ldots, \alpha_{TP})^{\dagger} \in \mathbb{R}^{TP}$, where TP denotes the "total number of model parameters." Variations in the model's parameters (i.e., cross sections, number densities) will induce variations in the flux $\Phi(\boldsymbol{\alpha}; E)$, the quantification of which is the scope of sensitivity analysis. Uncertainties in the model parameters will induce uncertainties in the flux, which can be quantified by using the expressions provided in Sect. 1.3, below.

As indicated by (1.2), the neutron flux $\Phi(\boldsymbol{\alpha}; E)$ varies as $1/E$ and undergoes sharp dips/depressions at resonance absorption energies, where $\Sigma(\boldsymbol{\alpha}; E)$ becomes very large. In addition, the flux $\Phi(\boldsymbol{\alpha}; E)$ is attenuated by the resonance escape probability, $p(\boldsymbol{\alpha}; E)$, as the neutrons undergo moderating scattering collisions. The qualitative form of $\Phi(\boldsymbol{\alpha}; E)$ shown in (1.2) is also representative of the neutron slowing down flux in different mixtures of materials/isotopes (Weiberg and Wigner 1958; Amaldi 1959).

The impact of the scattering process can be qualitatively determined by considering a system in which neutron absorption is negligible, which would be the case of neutron slowing down in pure hydrogen. In such a system, the energy-dependent flux would take on the following particular form of (1.2):

$$\Phi_H(\boldsymbol{\alpha};E) = \frac{Q(\boldsymbol{\alpha})}{E\Sigma_s(\boldsymbol{\alpha};E)};\ Q(\boldsymbol{\alpha}) \triangleq \int_{E_s(\boldsymbol{\alpha})}^{\infty} Q(\boldsymbol{\alpha};E')dE'\,(E). \tag{1.3}$$

Although (1.3) is strictly correct only for an infinite hydrogen medium, it is often used to estimate the energy-dependent flux in water-moderated reactors (Weiberg and Wigner 1958; Amaldi 1959).

The neutron flux $\Phi_H(\boldsymbol{\alpha};E)$ will serve as the "model response," i.e., the result of interest produced by the computational model represented by (1.3). The effects of the scattering cross sections can be quantified in isolation from the effects of the other model parameters by considering that the source does not vary from its nominal value denoted as $Q(\boldsymbol{\alpha}^0)$ and that the number density of hydrogen, N_H, remains unchanged, as well. Consider that the energy-dependent microscopic total cross section of hydrogen, $\sigma_s(E)$, was measured in a series of measurements at an arbitrary but fixed energy value E_m, $0 < E_m < \infty$, and was found to have a mean value denoted as σ_m. Under these assumptions, the flux $\Phi_H(\boldsymbol{\alpha};E)$ in (1.3) takes on the following expression:

$$\Phi_H(\sigma;E) = \frac{Q(\boldsymbol{\alpha}^0)}{EN_H\sigma_s(E)}. \tag{1.4}$$

The Taylor series expansion of $\Phi_H(\sigma;E)$ around the nominal parameter value σ_m has the following form:

$$\Phi_H(\sigma_s;E) = \Phi_H^0(\sigma_m;E) + \sum_{n=1}^{\infty} S_n(E_m)[\sigma_s(E) - \sigma_m]^n, \tag{1.5}$$

where

$$\Phi_H^0(\sigma_m;E) \triangleq \frac{Q(\boldsymbol{\alpha}^0)}{EN_H\sigma_m}, \tag{1.6}$$

and where the quantities $S_n(E)$, $n = 1, 2, \ldots$, denote the *absolute* sensitivities of $\Phi_H(\sigma;E)$ with respect to the parameter $\sigma_s(E_m) = \sigma_m$ at energy $E = E_m$ and have the following expressions:

$$S_n(E_m) \triangleq \frac{Q(\boldsymbol{\alpha}^0)}{EN_H}\left\{\frac{1}{n!}\frac{d^n}{d[\sigma_s(E)]^n}[\sigma_s(E)]^{-1}\right\}_{(E_m)} = \Phi_H^0(\sigma_m;E)(-1)^n\left[\frac{1}{\sigma_m}\right]^n. \tag{1.7}$$

The expression of $S_n(E)$ shown in (1.7) has been obtained by using the following result:

$$\frac{d^n}{dx^n}\left(x^{-1}\right) = (-1)^n n! x^{-(n+1)}, \quad n = 0, 1, \ldots . \tag{1.8}$$

It follows from the expression given in (1.5) that the *relative* sensitivities of $\Phi_H(\sigma; E)$ with respect to the parameter $\sigma_s(E)$, which will be denoted as $R_n(E)$, $n = 1, 2, \ldots$, have the following expressions:

$$R_n(E) = S_n(E)\frac{(\sigma_m)^n}{\Phi_H(\sigma_m; E)} = (-1)^n, \quad n = 1, 2, \ldots . \tag{1.9}$$

It is evident from (1.9) that the *relative values of the sensitivities of all orders are all equal to unity*; their values do *not* decrease with increasing order $n = 1, 2, \ldots$.

Using the result from (1.9) in (1.5) yields the following Taylor series for $\Phi_H(\sigma; E)$:

$$\Phi_H(\sigma; E) = \lim_{n \to \infty}\left[\Phi_H^{(n)}(\sigma; E)\right], \tag{1.10}$$

where the quantity $\Phi_H^{(n)}(\sigma; E)$ denotes the "*nth-order approximation of the value of the actual flux* $\Phi_H(\sigma; E)$," and is defined as follows:

$$\Phi_H^{(n)}(\sigma; E) \triangleq \Phi_H^0(\sigma_m; E)\sum_{k=0}^{n} R_k t^k; \quad t \triangleq \frac{\sigma_s(E) - \sigma_m}{\sigma_m}. \tag{1.11}$$

The Taylor series in (1.10) converges only for values of $\sigma_s(E)$ confined within the following interval:

$$\left(1 - \frac{\beta}{|R_n|}\right)\sigma_m \le \sigma_s(E) \le \left(1 + \frac{\beta}{|R_n|}\right)\sigma_m; \quad 0 < \beta < 1. \tag{1.12}$$

The results obtained in (1.10) and (1.12) are essential for using the probability distribution of $\sigma_s(E)$, which is seldom known completely in practice, to construct statistical indicators (i.e., mean value, standard deviation) for the distribution of $\Phi_H(\sigma; E)$. It is especially important *to ensure that the values of* $\sigma_s(E)$ *used for constructing statistical properties for* $\Phi_H(\sigma; E)$ *do not fall outside the interval of convergence shown in (1.12).*

In view of the relation shown in (1.12), it follows that if (in addition to the measured mean value σ_m) a standard deviation, $SD[\sigma_s(E)]$, is also provided, then it must be ensured that the values $\sigma_s(E) \pm SD[\sigma_s(E)]$ fall within the interval of convergence shown in (1.12) for the Taylor series shown in (1.10). If a standard deviation is not provided for the measurement of $\sigma_s(E_m)$, then the most conservative assumption is to consider that $\sigma_s(E)$ is uniformly distributed within the interval provided in (1.12), in which the Taylor series of $\Phi_H(\sigma; E)$ convergences, since the uniform distribution is the least informative (hence, the least biased) distribution. It will henceforth be assumed that the distribution for $\sigma_s(E)$ is uniform in the interval given in (1.12), in which case the uniform probability distribution, $P[\sigma_s(E)]$, the expectation, $\mathbf{E}[\sigma_s(E)]$, the variance, $Var[\sigma_s(E)]$, and the standard deviation, $SD[\sigma_s(E)]$, of $\sigma_s(E)$ have the following expressions:

$$\begin{aligned} &P[\sigma_s(E)] = \frac{1}{2\beta\sigma_m}; \ \mathbf{E}[\sigma_s(E)] = \sigma_m; \ Var[\sigma_s(E)] = \frac{(\beta\sigma_m)^2}{3}; \\ &SD[\sigma_s(E)] = \frac{1}{\sqrt{3}}\beta\sigma_m; \ 0 < \beta < 1. \end{aligned} \tag{1.13}$$

The third and fourth moments, $\mu_3[\sigma_s(E)]$ and $\mu_4[\sigma_s(E)]$, of the uniform distribution of $\sigma_s(E)$ have the following expressions:

$$\begin{aligned} &\mu_3[\sigma_s(E)] \triangleq \int_{(1-\beta)\sigma_m}^{(1+\beta)\sigma_m} [\sigma_s(E) - \sigma_m]^3 \, d\sigma_s(E) = 0, \\ &\mu_4[\sigma_s(E)] \triangleq \int_{(1-\beta)\sigma_m}^{(1+\beta)\sigma_m} [\sigma_s(E) - \sigma_m]^4 \, d\sigma_s(E) = \frac{(\beta\sigma_m)^4}{5}. \end{aligned} \tag{1.14}$$

The uniform distribution defined in (1.13) is symmetrical with respect to σ_m and can be visualized as a rectangle of height $P[\sigma_s(E)] = [2\beta\sigma_m]^{-1}$ and width extending from $(1 - \beta)\sigma_m$ to $(1 + \beta)\sigma_m$ for $0 < \beta < 1$. When β increases toward unity, the uniform distribution $P[\sigma_s(E)] = [2\beta\sigma_m]^{-1}$ widens and its height decreases toward $0.5/\sigma_m$. At the other extreme, when β decreases toward zero, the distribution $P[\sigma_s(E)] = [2\beta\sigma_m]^{-1}$ becomes very narrow and peaked, tending toward a Dirac-delta distribution centered at σ_m. It follows from the expressions provided in (1.13) and (1.12) that, if $\sigma_s(E)$ is distributed uniformly as shown in (1.13), then the Taylor series shown in (1.10) for $\Phi_H(\sigma; E)$ converges only in the following interval expressed in terms of "standard deviations of $\sigma_s(E)$":

$$\sigma_m - \sqrt{3} \cdot SD[\sigma_s(E)] \le \sigma_s(E) \le \sigma_m + \sqrt{3} \cdot SD[\sigma_s(E)]; \ 0 < \beta < 1. \tag{1.15}$$

1.2.1 Sensitivity Analysis

The relative error, $\varepsilon_{\Phi}^{(n)}$, between the exact value of $\Phi_H(\sigma; E)$ provided in (1.4) and the Taylor series in (1.10) truncated to the *nth*-order is given by the following expression:

$$\varepsilon_{\Phi}^{(n)} \triangleq \frac{1}{\Phi_H(\sigma; E)} \left[\Phi_H(\sigma; E) - \Phi_H^0(\sigma_m; E) \sum_{k=0}^{n} R_n(E) t^k \right] = R_{n+1}(E) t^{n+1}. \quad (1.16)$$

The relation provided in (1.16) can be used to estimate the smallest order, n, which would be required to ensure that the Taylor series expansion shown in (1.10) represents the exact flux $\Phi_H(\sigma; E)$ given in (1.4) within a target error which would not exceed $\left| \varepsilon_{\Phi}^{(n)} \right|$.

As an example of the use of the Taylor series expansion in *sensitivity analysis*, consider that it is desired to predict the value of $\Phi_H(\sigma; E)$ when $\sigma_s(E)$ is "one standard deviation away from the mean," i.e., when:

$$\sigma_s^{1SD}(E) = \sigma_m + SD[\sigma_s(E)] = \sigma_m \left(1 + \frac{1}{\sqrt{3}} \beta \right). \quad (1.17)$$

Replacing the result obtained in (1.17) into the definition provided in (1.11) and taking $\beta = 1 - 10^{-4}$ (to be extremely close to the outer boundaries of the "one-standard deviation interval") yields the following result when keeping the first four decimal digits:

$$t = \beta / \sqrt{3} = \left(1 - 10^{-4}\right) / \sqrt{3} \cong 0.5773. \quad (1.18)$$

In sensitivity analysis, the value provided in (1.18) is used in (1.11) to investigate how accurately the various orders of approximations, $\Phi_H^{(n)}(\sigma; E)$, can predict the actual value of the flux at $\Phi_H(\sigma; E)$. The following results, to four significant decimals, are obtained from (1.4), (1.11), and (1.17) for $\beta = 1 - 10^{-4}$ and $t = 0.5773$:

1. The exact value of the flux is:

$$\Phi_H\left(\sigma_s^{1SD}; E\right) = \Phi_H^0(\sigma_m; E) \left(1 + \frac{1 - 10^{-4}}{\sqrt{3}} \right)^{-1} \cong 0.6340\, \Phi_H^0(\sigma_m; E). \quad (1.19)$$

2. If only the *first-order sensitivities* of $\Phi_H(\sigma; E)$ with respect to $\sigma_s(E)$ *are available*, then the *first-order expansion* ($n = 1$) in (1.11) yields the following result:

$$\Phi_H^{(1)}(\sigma; E) \triangleq \Phi_H^0(\sigma_m; E)(1 + R_1 t) = 0.4227\, \Phi_H^0(\sigma_m; E). \quad (1.20)$$

It follows from (1.19) and (1.20) that the error of the first-order approximation is:

$$\varepsilon_{\Phi}^{(1)} \triangleq \frac{\Phi_H\left(\sigma_s^{1SD}; E\right) - \Phi_H^{(1)}(\sigma; E)}{\Phi_H\left(\sigma_s^{1SD}; E\right)} = 33.33\%. \quad (1.21)$$

Thus, the result obtained by keeping only the first-order sensitivities in the Taylor expansion *underestimates* the exact flux by 33.33%.

3. If *first-order and second-order sensitivities* of $\Phi_H(\sigma; E)$ with respect to $\sigma_s(E)$ *are available*, then the *second-order expansion* ($n = 2$) in (1.11) yields the following result:

$$\Phi_H^{(2)}(\sigma; E) \triangleq \Phi_H^0(\sigma_m; E)\left(\sum_{k=1}^{2} R_k t^k\right) = 0.7560\,\Phi_H^0(\sigma_m; E). \qquad (1.22)$$

It follows from (1.22) and (1.19) that the error of the second-order approximation is as follows:

$$\varepsilon_\Phi^{(2)} \triangleq \frac{\Phi_H\left(\sigma_s^{1SD}; E\right) - \Phi_H^{(2)}(\sigma; E)}{\Phi_H\left(\sigma_s^{1SD}; E\right)} = -19.24\%. \qquad (1.23)$$

Thus, the result obtained by keeping the first- and second-order sensitivities in the Taylor expansion *overestimates* the exact flux by 19.24%.

4. If all sensitivities up to and including the *third-order sensitivities* of $\Phi_H(\sigma; E)$ with respect to $\sigma_s(E)$ *are available*, then the *third-order expansion* ($n = 3$) in (1.11) yields the following result:

$$\Phi_H^{(3)}(\sigma; E) \triangleq \Phi_H^0(\sigma_m; E)\left(\sum_{k=1}^{3} R_k t^k\right) = 0.5636\,\Phi_H^0(\sigma_m; E). \qquad (1.24)$$

It follows from (1.24) and (1.19) that the error of the third-order approximation is as follows:

$$\varepsilon_\Phi^{(3)} \triangleq \frac{\Phi_H\left(\sigma_s^{1SD}; E\right) - \Phi_H^{(3)}(\sigma; E)}{\Phi_H\left(\sigma_s^{1SD}; E\right)} = 11.11\%. \qquad (1.25)$$

Thus, the result obtained by keeping the first-, second-, and third-order sensitivities in the Taylor expansion *underestimates* the exact flux by 11.11%.

5. If all sensitivities up to and including the *fourth-order sensitivities* of $\Phi_H(\sigma; E)$ with respect to $\sigma_s(E)$ *are available*, then the *fourth-order expansion* ($n = 4$) in (1.11) yields the following result:

$$\Phi_H^{(4)}(\sigma; E) \triangleq \Phi_H^0(\sigma_m; E)\left(\sum_{k=1}^{4} R_k t^k\right) = 0.6747\Phi_H^0(\sigma_m; E). \qquad (1.26)$$

It follows from (1.26) and (1.19) that the error of the fourth-order approximation is as follows:

$$\varepsilon_\Phi^{(4)} \triangleq \frac{\Phi_H\left(\sigma_s^{1SD}; E\right) - \Phi_H^{(4)}(\sigma; E)}{\Phi_H\left(\sigma_s^{1SD}; E\right)} = -6.42\%.. \qquad (1.27)$$

Thus, the result obtained by keeping the first-, second-, third-, and fourth-order sensitivities in the Taylor expansion *overestimates* the exact flux by 6.42%.

As indicated by the results in (1.21), (1.23), (1.25), and (1.27), the odd-order approximations underestimate the true value, and the even-order approximations overestimate the true value of the flux, while converging to the actual, true value, of the flux $\Phi_H(\sigma; E)$. This behavior is expected since the flux is represented by an alternating series, cf. (1.11). The results obtained in (1.21), (1.23), (1.25), and (1.27) are consistent with the error estimate provided in (1.16). The relation provided in (1.16) can be used to determine which order of approximation would need to be used to obtain an approximate result which would be within a predetermined error by comparison to the exact result. For example, over 46,000 terms in the Taylor series expansion provided in (1.11) would be required to represent the flux $\Phi_H(\sigma; E)$ within an accuracy of 1% at a distance of $\sigma_s(E) = [1 + (1 - 10^{-4})/\sqrt{3}]\sigma_m$ away from σ_m; the expected extremely slow convergence of the Taylor series expansion very close to the end of the interval of convergence is evident.

1.2.2 Uncertainty Quantification: Moments of the Response Distribution

The *expectation*, $\mathbf{E}[\Phi_H]$, of the flux $\Phi_H(\sigma; E)$, is obtained by integrating either (1.4) or (1.10) over the uniform probability distribution for $\sigma_s(E)$ shown in (1.13). The result of the respective integration is:

$$\begin{aligned}
\mathbf{E}[\Phi_H] &= \frac{Q(\boldsymbol{\alpha}^0)}{EN_H}\frac{1}{2\beta\sigma_m}\int\limits_{(1-\beta)\sigma_m}^{(1+\beta)\sigma_m}\frac{d\sigma_s(E)}{\sigma_s(E)} = \frac{\Phi_H^0(\sigma_m; E)}{2\beta}\ln\frac{1+\beta}{1-\beta} \\
&= \Phi_H^0(\sigma_m; E)\sum_{k=0}^{\infty}\frac{R_k}{2\beta\sigma_m}\int\limits_{(1-\beta)\sigma_m}^{(1+\beta)\sigma_m}\left[\frac{\sigma_s(E)-\sigma_m}{\sigma_m}\right]^k d\sigma_s(E) \\
&= \Phi_H^0(\sigma_m; E)\sum_{k=0}^{\infty}\frac{\beta^{2k}}{2k+1}; \quad 0 < \beta < 1.
\end{aligned} \tag{1.28}$$

The last expression on the right side of (1.28) has been obtained by noting that $R_{2k} \equiv 1$. Furthermore, the result obtained in (1.28) indicates that $\mathbf{E}[\Phi_H]$ is an even function of β, i.e., $\mathbf{E}[\Phi_H, \beta] = \mathbf{E}[\Phi_H, -\beta]$.

The *nth*-order approximation, denoted as $\mathbf{E}^{(n)}[\Phi_H]$, of the expectation $\mathbf{E}[\Phi_H]$, is provided by the following truncated Taylor series:

$$\mathbf{E}^{(n)}[\Phi_H] = \Phi_H^0(\sigma_m; E)\sum_{k=0}^{n}\frac{\beta^{2k}}{2k+1}; \quad n = 0, 1, \ldots. \tag{1.29}$$

Notably, the convergence of the infinite series representation of $\mathbf{E}[\Phi_H]$ obtained in (1.28) depends solely on the chosen value for the positive parameter β: the smaller the chosen value for β, the faster the convergence of the series representation for $\mathbf{E}[\Phi_H]$.

The *variance*, denoted as $V[\Phi_H]$, of the flux $\Phi_H(\sigma; E)$, is obtained by recalling the following relation:

$$V[\Phi_H] = \mathbf{E}\left[(\Phi_H)^2\right] - \mathbf{E}^2[\Phi_H]. \tag{1.30}$$

Using (1.4) and (1.28) in (1.30) yields the following result:

$$\begin{aligned} V[\Phi_H] &= \left[\frac{Q(\boldsymbol{\alpha}^0)}{EN_H}\right]^2 \frac{1}{2\beta\sigma_m} \int_{(1-\beta)\sigma_m}^{(1+\beta)\sigma_m} \frac{d\sigma_s(E)}{[\sigma_s(E)]^2} - \left[\frac{Q(\boldsymbol{\alpha}^0)}{EN_H}\frac{1}{2\beta\sigma_m}\ln\frac{1+\beta}{1-\beta}\right]^2 \\ &= \left[\Phi_H^0(\sigma_m; E)\right]^2 \left[\frac{1}{1-\beta^2} - \frac{1}{4\beta^2}\left(\ln\frac{1+\beta}{1-\beta}\right)^2\right]. \end{aligned} \tag{1.31}$$

The result obtained in (1.31) indicates that $V[\Phi_H]$ is an even function of β, i.e.,

$$V[\Phi_H, \beta] = V[\Phi_H, -\beta] = \left[\Phi_H^0(\sigma_m; E)\right]^2 \left[\frac{1}{1-\beta^2} - \frac{1}{4\beta^2}\left(\ln\frac{1+\beta}{1-\beta}\right)^2\right]. \tag{1.32}$$

As indicated by the result obtained in (1.31), the convergence of the series representation of $V[\Phi_H]$ would also depend solely on the chosen value for the positive parameter β: the smaller the chosen value for β, the faster would be the convergence of the series representation for $V[\Phi_H]$.

The *nth*-order approximation, denoted as $V^{(n)}[\Phi_H]$, of the variance $V[\Phi_H]$, is also obtained by using the relation provided in (1.30), which takes on the following form:

$$V^{(n)}[\Phi_H] = \mathbf{E}^{(n)}\left[\left(\Phi_H^{(n)}\right)^2\right] - \left(\mathbf{E}^{(n)}\right)^2\left[\Phi_H^{(n)}\right]; \quad n = 1, 2, \ldots. \tag{1.33}$$

The *third-order moment*, denoted as $\mu_3[\Phi_H]$, of the flux $\Phi_H(\sigma; E)$, is obtained using (1.4) and (1.28) in the definition of $\mu_3[\Phi_H]$ and performing the ensuing integrations, to obtain the following result:

$$\begin{aligned} \mu_3[\Phi_H] &\triangleq \frac{1}{2\beta\sigma_m} \int_{(1-\beta)\sigma_m}^{(1+\beta)\sigma_m} \{\Phi_H(\sigma; E) - \mathbf{E}[\Phi_H]\}^3 d\sigma_s(E) \\ &= \left[\Phi_H^0(\sigma_m; E)\right]^3 \frac{1}{2\beta}\left[\frac{2\beta}{\left(1-\beta^2\right)^2} - \frac{3}{1-\beta^2}\ln\frac{1+\beta}{1-\beta} + \frac{1}{2\beta^2}\left(\ln\frac{1+\beta}{1-\beta}\right)^3\right]. \end{aligned} \tag{1.34}$$

Note that $\mu_3[\Phi_H]$ is an even function of β, i.e.,

$$\mu_3[\Phi_H, \beta] = \mu_3[\Phi_H, -\beta]$$
$$= \left[\Phi_H^0(\sigma_m; E)\right]^3 \frac{1}{2\beta}\left[\frac{2\beta}{\left(1-\beta^2\right)^2} - \frac{3}{1-\beta^2}\ln\frac{1+\beta}{1-\beta} + \frac{1}{2\beta^2}\left(\ln\frac{1+\beta}{1-\beta}\right)^3\right]. \tag{1.35}$$

The *skewness* of the distribution of the response $\Phi_H(\sigma; E)$ will be denoted as *Skew* $[\Phi_H]$ and is defined as follows:

$$Skew[\Phi_H] \triangleq \frac{\mu_3[\Phi_H]}{\{V[\Phi_H]\}^{3/2}}$$
$$= \frac{\frac{1}{2\beta}\left[\frac{2\beta}{\left(1-\beta^2\right)^2} - \frac{3}{1-\beta^2}\ln\frac{1+\beta}{1-\beta} + \frac{1}{2\beta^2}\left(\ln\frac{1+\beta}{1-\beta}\right)^3\right]}{\left[\frac{1}{1-\beta^2} - \frac{1}{4\beta^2}\left(\ln\frac{1+\beta}{1-\beta}\right)^2\right]^{\frac{3}{2}}}. \tag{1.36}$$

Skewness indicates the direction and relative magnitude of a distribution's deviation from the normal distribution. Note that, for this illustrative model, *Skew* $[\Phi_H, \beta] = Skew[\Phi_H, -\beta]$.

The *nth-order* approximation of the third-order moment, denoted as $\mu_3^{(n)}[\Phi_H]$, is defined below:

$$\mu_3^{(n)}[\Phi_H] \triangleq \frac{1}{2\beta\sigma_m}\int_{(1-\beta)\sigma_m}^{(1+\beta)\sigma_m}\left\{\Phi_H^{(n)}(\sigma; E) - \mathbf{E}^{(n)}[\Phi_H]\right\}^3 d\sigma_s(E). \tag{1.37}$$

It follows from (1.37), (1.11), and (1.29) that the first-order approximation of the third-order moment, denoted as $\mu_3^{(1)}[\Phi_H]$, has the following expression:

$$\mu_3^{(1)}[\Phi_H] \triangleq \frac{1}{2\beta\sigma_m}\int_{(1-\beta)\sigma_m}^{(1+\beta)\sigma_m}\left\{\Phi_H^{(1)}(\sigma; E) - \mathbf{E}^{(1)}[\Phi_H]\right\}^3 d\sigma_s(E)$$
$$= \frac{1}{2\beta\sigma_m}\int_{(1-\beta)\sigma_m}^{(1+\beta)\sigma_m}\left\{\Phi_H^0(\sigma_m; E)[1 + R_1 t] - \Phi_H^0(\sigma_m; E)\right\}^3 d\sigma_s(E) \equiv 0. \tag{1.38}$$

It also follows from (1.37), (1.11), and (1.29) that the second-order approximation of the third-order moment, denoted as $\mu_3^{(2)}[\Phi_H]$, has the following expression:

$$\begin{aligned}\mu_3^{(2)}[\Phi_H] &\triangleq \frac{1}{2\beta\sigma_m}\int_{(1-\beta)\sigma_m}^{(1+\beta)\sigma_m}\left\{\Phi_H^{(2)}(\sigma;E)-\mathbf{E}^{(2)}[\Phi_H]\right\}^3 d\sigma_s(E)\\ &= \frac{\Phi_H^0(\sigma_m;E)}{2\beta\sigma_m}\int_{(1-\beta)\sigma_m}^{(1+\beta)\sigma_m}\left[\left(\sum_{k=0}^{2}R_k t^k\right)-\frac{Q(\boldsymbol{\alpha}^0)}{EN_H\sigma_m}\left(1+\frac{\beta^2}{3}\right)\right]^3 d\sigma_s(E)\\ &= \left[\Phi_H^0(\sigma_m;E)\right]^3\frac{4}{15}\beta^4+O\left(\beta^6\right).\end{aligned} \tag{1.39}$$

It is evident that high-order approximations for the third-order moment are very laborious to obtain. On the other hand, since the definition of skewness involves not only the response's third moment but also the response's variance, it is possible to compute approximate values of the skewness, denoted as $Skew^{(m,\,n)}[\Phi_H]$, by using approximations of different orders for the response's variance and third moment, respectively, as follows:

$$Skew^{(m,n)}[\Phi_H] \triangleq \frac{\mu_3^{(m)}[\Phi_H]}{\left\{V^{(n)}[\Phi_H]\right\}^{3/2}};\quad m,n=1,2,\ldots. \tag{1.40}$$

The flexible definition for the approximate skewness $Skew^{(m,\,n)}[\Phi_H]$ provided in (1.40) enables the use of higher-order approximations for the response variance, which require less computational effort to determine by comparison to the same order of approximation for the third moment. The flexible definition of $Skew^{(m,\,n)}[\Phi_H]$ could provide higher accuracy for the skewness with less computational effort by using a lower-order approximation for the third moment than the order of approximation used for computing the variance.

The *fourth-order moment*, denoted as $\mu_4[\Phi_H]$, of the flux $\Phi_H(\sigma; E)$, is defined as follows:

$$\begin{aligned}\mu_4[\Phi_H] &\triangleq \frac{1}{2\beta\sigma_m}\int_{(1-\beta)\sigma_m}^{(1+\beta)\sigma_m}\left\{\Phi_H(\sigma;E)-\mathbf{E}[\Phi_H]\right\}^4 d\sigma_s(E)\\ &= \left[\Phi_H^0(\sigma_m;E)\right]^4\left[\frac{1+\beta^2/3}{\left(1-\beta^2\right)^3}-\frac{2}{\beta\left(1-\beta^2\right)^2}\ln\frac{1+\beta}{1-\beta}\right.\\ &\quad\left.+\frac{3}{2\beta^2\left(1-\beta^2\right)}\left(\ln\frac{1+\beta}{1-\beta}\right)^2-\frac{3}{16\beta^4}\left(\ln\frac{1+\beta}{1-\beta}\right)^4\right].\end{aligned} \tag{1.41}$$

Note that $\mu_4[\Phi_H]$ is an even function of β, i.e., $\mu_4[\Phi_H, \beta] = \mu_4[\Phi_H, -\beta]$.

The *nth-order* approximation of the fourth-order moment, denoted as $\mu_4^{(n)}[\Phi_H]$, is defined below:

$$\mu_4^{(n)}[\Phi_H] \triangleq \frac{1}{2\beta\sigma_m} \int_{(1-\beta)\sigma_m}^{(1+\beta)\sigma_m} \left\{ \Phi_H^{(n)}(\sigma;E) - \mathbf{E}^{(n)}[\Phi_H] \right\}^4 d\sigma_s(E). \tag{1.42}$$

It follows from (1.42), (1.11), and (1.29) that the first-order approximation of the fourth-order moment, denoted as $\mu_4^{(1)}[\Phi_H]$, has the following expression:

$$\begin{aligned} \mu_4^{(1)}[\Phi_H] &\triangleq \frac{1}{2\beta\sigma_m} \int_{(1-\beta)\sigma_m}^{(1+\beta)\sigma_m} \left\{ \Phi_H^{(1)}(\sigma;E) - \mathbf{E}^{(1)}[\Phi_H] \right\}^4 d\sigma_s(E) \\ &= \frac{\Phi_H^0(\sigma_m;E)}{2\beta\sigma_m} \int_{(1-\beta)\sigma_m}^{(1+\beta)\sigma_m} \left[\left(\sum_{k=0}^{1} R_k t^k \right) - 1 \right]^4 d\sigma_s(E) = \left[\Phi_H^0(\sigma_m;E)\right]^4 \frac{\beta^4}{5}. \end{aligned} \tag{1.43}$$

It also follows from (1.42), (1.11), and (1.29) that the second-order approximation of the fourth-order moment, denoted as $\mu_4^{(2)}[\Phi_H]$, has the following expression:

$$\begin{aligned} \mu_4^{(2)}[\Phi_H] &\triangleq \frac{1}{2\beta\sigma_m} \int_{(1-\beta)\sigma_m}^{(1+\beta)\sigma_m} \left\{ \Phi_H^{(2)}(\sigma;E) - \mathbf{E}^{(2)}[\Phi_H] \right\}^4 d\sigma_s(E) \\ &= \frac{\Phi_H^0(\sigma_m;E)}{2\beta\sigma_m} \left[\left(\sum_{k=0}^{2} R_k t^k \right) - \left(1 + \frac{\beta^2}{3} \right) \right]^4 \int_{(1-\beta)\sigma_m}^{(1+\beta)\sigma_m} d\sigma_s(E) \\ &= \left[\Phi_H^0(\sigma_m;E)\right]^4 \left[\frac{\beta^4}{5} + \frac{6}{7}\beta^6 + O(\beta^8) \right]. \end{aligned} \tag{1.44}$$

Evidently, high-order approximations for the fourth-order moment $\mu_4[\Phi_H]$ are very laborious to obtain.

The *kurtosis* of the distribution of the response $\Phi_H(\sigma;E)$ will be denoted as *Kurt* $[\Phi_H]$ and is defined as follows:

$$\begin{aligned} Kurt[\Phi_H] &\triangleq \frac{\mu_4[\Phi_H]}{\{V[\Phi_H]\}^2} = \left[\frac{1}{1-\beta^2} - \frac{1}{4\beta^2} \left(\ln \frac{1+\beta}{1-\beta} \right)^2 \right]^{-2} \\ &\times \left[\frac{1+\beta^2/3}{(1-\beta^2)^3} - \frac{2}{\beta(1-\beta^2)^2} \ln \frac{1+\beta}{1-\beta} \right. \\ &\left. + \frac{3}{2\beta^2(1-\beta^2)} \left(\ln \frac{1+\beta}{1-\beta} \right)^2 - \frac{3}{16\beta^4} \left(\ln \frac{1+\beta}{1-\beta} \right)^4 \right]. \end{aligned} \tag{1.45}$$

The kurtosis of a probability distribution indicates the propensity of the respective distribution to produce outliers or heavy tails.

The approximate kurtosis, denoted as $Kurt^{(m,n)}[\Phi_H]$, can be computed by using approximations of different orders for the response's variance and fourth moment, respectively, as follows:

$$Kurt^{(m,n)}[\Phi_H] = \frac{\mu_4^{(m)}[\Phi_H]}{\left\{V^{(n)}[\Phi_H]\right\}^2}; \quad m,n = 1,2,\ldots. \tag{1.46}$$

The flexible definition for the approximate kurtosis $Kurt^{(m,n)}[\Phi_H]$ provided in (1.46) aims at obtaining higher accuracy for the kurtosis with less computational effort than would be required if the fourth moment were computed at the same high order $m = n$ as the variance. Thus, the variance would be computed relatively inexpensively at an order of approximation $n > m$, while the fourth moment would be computed only up to the order of approximation $m < n$, expecting that $Kurt^{(m,n)}[\Phi_H] \simeq Kurt^{(n,n)}[\Phi_H]$ for a sufficiently high order m.

Section 1.2.1 has presented the investigation of the order of sensitivities that would be needed to achieve a target accuracy for predicting changes in the flux due to changes in the cross section $\sigma_s(E)$ by using the Taylor series expansion given in (1.11). Analogously, it is of interest to investigate the accuracy provided by the approximations $\mathbf{E}^{(n)}[\Phi_H]$, $V^{(n)}[\Phi_H]$, $Skew^{(m,n)}[\Phi_H]$, and $Kurt^{(m,n)}[\Phi_H]$, $n = 1, 2, \ldots$; $m = 1, 2, \ldots$, for the expectation $\mathbf{E}[\Phi_H]$, variance $V[\Phi_H]$, skewness $Skew[\Phi_H]$, and kurtosis $Kurt[\Phi_H]$, respectively.

The expressions of the various orders of approximation for $\mathbf{E}^{(n)}[\Phi_H]$ and $V^{(n)}[\Phi_H]$, $n = 1, 2, \ldots$, will be computed by using the expression of the approximations $\Phi_H^{(n)}$ in (1.33), in order to verify independently the partial series that would be obtained by expanding (in truncated Taylor series in β) the exact results obtained in (1.28) and (1.31). The following results are thus obtained:

1. If only the *first-order sensitivities* of $\Phi_H(\sigma; E)$ with respect to $\sigma_s(E)$ *are available*, then the *first-order approximation* $\Phi_H^{(1)}(\sigma; E)$ of $\Phi_H(\sigma; E)$ takes on the following particular form of (1.11):

$$\Phi_H^{(1)}(\sigma; E) \triangleq \Phi_H^0(\sigma_m; E)(1 + R_1 t); \quad t \triangleq \frac{\sigma_s(E) - \sigma_m}{\sigma_m}. \tag{1.47}$$

The first-order approximate expectation $\mathbf{E}^{(1)}[\Phi_H]$ has the following expression:

$$\mathbf{E}^{(1)}[\Phi_H] = \frac{\Phi_H^0(\sigma_m; E)}{2\beta} \int_{-\beta}^{\beta} (1 - t)\, dt = \Phi_H^0(\sigma_m; E) = \mathbf{E}^{(0)}[\Phi_H]. \tag{1.48}$$

The expectation $\mathbf{E}^{(1)}[\Phi_H]$ is accurate only to the *zeroth-order* term since the contribution from the first-order sensitivities vanishes after the integration of the respective first-order (odd) term over the symmetric interval around the expectation of the parameter $\sigma_s(E)$.

Replacing the expression given in (1.47) into (1.33) yields:

$$V^{(1)}[\Phi_H] = \left[\Phi_H^0(\sigma_m; E)\right]^2 \left[\frac{1}{2\beta}\int_{-\beta}^{\beta}(1-t)^2 dt\right] - \left[\Phi_H^0(\sigma_m; E)\right]^2 = \left[\Phi_H^0(\sigma_m; E)\right]^2 \frac{\beta^2}{3}. \tag{1.49}$$

The variance $V^{(1)}[\Phi_H]$ is accurate up to and including the contributions stemming from the *first-order sensitivities.*

2. If *first-order and second-order sensitivities* of $\Phi_H(\sigma; E)$ with respect to $\sigma_s(E)$ *are available*, then the *second-order expansion* $\Phi_H^{(2)}(\sigma; E)$ of $\Phi_H(\sigma; E)$ takes on the following particular form of (1.11):

$$\Phi_H^{(2)}(\sigma; E) \triangleq \Phi_H^0(\sigma_m; E)\left(1 - t + t^2\right); \; t \triangleq \frac{\sigma_s(E) - \sigma_m}{\sigma_m}. \tag{1.50}$$

The expectation $\mathbf{E}^{(2)}[\Phi_H]$ has the following expression:

$$\mathbf{E}^{(2)}[\Phi_H] = \frac{\Phi_H^0(\sigma_m; E)}{2\beta}\int_{-\beta}^{\beta}\left(1 - t + t^2\right) dt = \Phi_H^0(\sigma_m; E)\left(1 + \frac{\beta^2}{3}\right). \tag{1.51}$$

The expectation $\mathbf{E}^{(2)}[\Phi_H]$ is accurate up to and including the contributions stemming from the *second-order sensitivities.*

Replacing the expression given in (1.50) into (1.33) and using (1.51) yields:

$$V^{(2)}[\Phi_H] = \left[\Phi_H^0(\sigma_m; E)\right]^2 \left[\frac{1}{2\beta}\int_{-\beta}^{\beta}\left(1 - t + t^2\right)^2 dt\right] - \left[\Phi_H^0(\sigma_m; E)\right]^2 \left(1 + \frac{\beta^2}{3}\right)^2 = \left[\Phi_H^0(\sigma_m; E)\right]^2 \left(\frac{\beta^2}{3} + \frac{4}{45}\beta^4\right). \tag{1.52}$$

The variance $V^{(2)}[\Phi_H]$ is accurate up to and including the contributions stemming from the *first-order sensitivities but the contributions from the second-order sensitivities are incomplete* because the term $4\beta^4/45$ is incomplete. Additional contributions to the term $4\beta^4/45$ will arise from the third-order sensitivities, as will be shown below.

3. If all sensitivities up to and including the *third-order sensitivities* of $\Phi_H(\sigma; E)$ with respect to $\sigma_s(E)$ *are available*, then the *third-order expansion* $\Phi_H^{(3)}(\sigma; E)$ of $\Phi_H(\sigma; E)$ takes on the following particular form of (1.11):

$$\Phi_H^{(3)}(\sigma; E) \triangleq \Phi_H^0(\sigma_m; E)\left(1 - t + t^2 - t^3\right); \quad t \triangleq \frac{\sigma_s(E) - \sigma_m}{\sigma_m}. \tag{1.53}$$

The expectation $\mathbf{E}^{(3)}[\Phi_H]$ has therefore the following expression:

$$\begin{aligned} \mathbf{E}^{(3)}[\Phi_H] &= \Phi_H^0(\sigma_m; E)\frac{1}{2\beta}\int_{-\beta}^{\beta}\left(1 - t + t^2 - t^3\right)dt \\ &= \Phi_H^0(\sigma_m; E)\left(1 + \frac{\beta^2}{3}\right) = \mathbf{E}^{(2)}[\Phi_H]. \end{aligned} \tag{1.54}$$

The expectation $\mathbf{E}^{(3)}[\Phi_H]$ is only as accurate as the lower-order approximation $\mathbf{E}^{(2)}[\Phi_H]$, i.e., it is accurate up to and including the contributions stemming from the *second-order sensitivities*, because the contribution from the third-order sensitivities vanishes after the integration of the respective third-order (odd) term over the symmetric interval around the expectation of the parameter $\sigma_s(E)$.

Replacing the expression given in (1.53) into (1.33) and using (1.54) yields:

$$\begin{aligned} V^{(3)}[\Phi_H] &= \left[\Phi_H^0(\sigma_m; E)\right]^2\left[\frac{1}{2\beta}\int_{-\beta}^{\beta}\left(1 - t + t^2 - t^3\right)^2 dt\right] \\ &- \left[\Phi_H^0(\sigma_m; E)\right]^2\left(1 + \frac{\beta^2}{3}\right)^2 = \left[\Phi_H^0(\sigma_m; E)\right]^2\left(\frac{\beta^2}{3} + \frac{22}{45}\beta^4 + \frac{\beta^6}{7}\right). \end{aligned} \tag{1.55}$$

The variance $V^{(3)}[\Phi_H]$ is accurate up to and including the contributions stemming from the *second-order sensitivities but the contributions from the third-order sensitivities are incomplete* because the term $\beta^6/7$ is incomplete. Additional contributions to the term $\beta^6/7$ will arise from the fourth-order sensitivities, as will be shown below.

4. If all sensitivities up to and including the *fourth-order sensitivities* of $\Phi_H(\sigma; E)$ with respect to $\sigma_s(E)$ *are available*, then the *fourth-order expansion* $\Phi_H^{(4)}(\sigma; E)$ of $\Phi_H(\sigma; E)$ takes on the following particular form of (1.11):

$$\Phi_H^{(4)}(\sigma; E) \triangleq \Phi_H^0(\sigma_m; E)\left(1 - t + t^2 - t^3 + t^4\right); \quad t \triangleq \frac{\sigma_s(E) - \sigma_m}{\sigma_m}. \tag{1.56}$$

The expectation $\mathbf{E}^{(4)}[\Phi_H]$ has the following expression:

$$\mathbf{E}^{(4)}[\Phi_H] = \Phi_H^0(\sigma_m; E)\frac{1}{2\beta}\int_{-\beta}^{\beta}\left(1 - t + t^2 - t^3 + t^4\right)dt$$
$$= \Phi_H^0(\sigma_m; E)\left(1 + \frac{\beta^2}{3} + \frac{\beta^4}{5}\right). \tag{1.57}$$

The expectation $\mathbf{E}^{(4)}[\Phi_H]$ is accurate up to and including the contributions stemming from the *fourth-order sensitivities.*

Replacing the expression given in (1.57) into (1.33) and using (1.57) yields:

$$V^{(4)}[\Phi_H] = \left[\Phi_H^0(\sigma_m; E)\right]^2\left[\frac{1}{2\beta}\int_{-\beta}^{\beta}\left(1 - t + t^2 - t^3 + t^4\right)^2 dt\right]$$
$$-\left[\Phi_H^0(\sigma_m; E)\right]^2\left(1 + \frac{\beta^2}{3} + \frac{\beta^4}{5}\right)^2 \tag{1.58}$$
$$= \left[\Phi_H^0(\sigma_m; E)\right]^2\left(\frac{\beta^2}{3} + \frac{22}{45}\beta^4 + \frac{31}{105}\beta^6 + \frac{16}{225}\beta^8\right).$$

The variance $V^{(4)}[\Phi_H]$ is accurate up to and including the contributions stemming from the *third-order sensitivities but the contributions from the fourth-order sensitivities are incomplete* because the term $16\beta^8/225$ is incomplete. Additional contributions to the term $16\beta^8/225$ will arise from the fifth-order sensitivities.

As has been mentioned in the foregoing, and as indicated by the results obtained thus far, the convergence of the infinite series representations of the expectation and variance, $\mathbf{E}[\Phi_H]$ and, respectively, $V[\Phi_H]$ depends solely on the chosen value for the positive parameter β: the smaller the chosen value for β, the faster the convergence of the series representation for $\mathbf{E}[\Phi_H]$ and for $V[\Phi_H]$, respectively. It is therefore of interest to investigate numerically the errors associated with the first four orders of approximations $\mathbf{E}^{(n)}[\Phi_H]$ and $V^{(n)}[\Phi_H]$ for $n = 1, 2, 3, 4$, for several representative values for $\beta, 0 < \beta < 1$. Three values will be considered for the parameter β, as follows:

(a) $\beta = 0.2$, which corresponds to a uniform distribution of height $P[\sigma_s(E)] = 2.5/\sigma_m$ and (very narrow) width confined to the interval $0.8\,\sigma_m \leq \sigma_s(E) \leq 1.2\,\sigma_m$, with a standard deviation $SD[\sigma_s(E)] = \frac{1}{\sqrt{3}}\beta\sigma_m = (11.55\%)\sigma_m$.
(b) $\beta = 0.5$, which corresponds to a uniform distribution of height $P[\sigma_s(E)] = 1/\sigma_m$ and width confined to the interval $0.5\,\sigma_m \leq \sigma_s(E) \leq 1.5\,\sigma_m$, with a standard deviation $SD[\sigma_s(E)] = (28.87\%)\sigma_m$.
(c) $\beta = 0.95$, which corresponds to a uniform distribution of height $P[\sigma_s(E)] = 0.5263/\sigma_m$ and width confined to the interval $0.05\,\sigma_m \leq \sigma_s(E) \leq 1.95\,\sigma_m$, with a standard deviation $SD[\sigma_s(E)] = (54.85\%)\sigma_m$.

The exact normalized expectation $\mathbf{E}_N[\Phi_H]$ and normalized variance $V_N[\Phi_H]$ of the flux $\Phi_H(\sigma; E)$ are defined by using (1.28) and (1.31), as follows:

$$\mathbf{E}_N[\Phi_H] \triangleq \frac{\mathbf{E}[\Phi_H]}{\Phi_H^0(\sigma_m; E)} = \frac{1}{2\beta} \ln \frac{1+\beta}{1-\beta}, \tag{1.59}$$

$$V_N[\Phi_H] \triangleq \frac{V[\Phi_H]}{\left[\Phi_H^0(\sigma_m; E)\right]^2} = \left[\frac{1}{1-\beta^2} - \frac{1}{4\beta^2}\left(\ln \frac{1+\beta}{1-\beta}\right)^2\right]. \tag{1.60}$$

The normalized exact standard deviation is defined as follows: $SD_N[\Phi_H] = \sqrt{V_N[\Phi_H]}$. The normalized *nth-order* approximations $\mathbf{E}_N^{(n)}[\Phi_H]$, $V_N^{(n)}[\Phi_H]$, and $SD_N^{(n)}[\Phi_H]$ of the expectation, variance, and standard deviation of the flux response $\Phi_H(\sigma; E)$, respectively, are defined as follows:

$$\begin{aligned} &\mathbf{E}_N^{(n)}[\Phi_H] \triangleq \mathbf{E}^{(n)}[\Phi_H]\left[\Phi_H^0(\sigma_m; E)\right]^{-1}; V_N^{(n)}[\Phi_H] \triangleq V^{(n)}[\Phi_H]\left[\Phi_H^0(\sigma_m; E)\right]^{-2}; \\ &SD_N^{(n)}[\Phi_H] \triangleq \sqrt{V_N^{(n)}[\Phi_H]}; \quad n = 1, 2, 3, 4. \end{aligned} \tag{1.61}$$

The results obtained from using the exact expression provided in (1.59) and (1.60) are compared to the approximate results obtained by using the expressions provided in (1.61), for $n = 1, 2, 3, 4$, and for $\beta = 0.2$, $\beta = 0.5$, and $\beta = 0.95$, respectively. Since all four moments (i.e., expectation, variance, skewness, and kurtosis) of the distribution of the response $\Phi_H(\sigma; E)$ are even functions of β, it suffices to consider only positive values for β. The results obtained for the expectation and variance of the predicted distribution of the response $\Phi_H(\sigma; E)$ are summarized in Tables 1.1, 1.2, and 1.3, below.

The results presented in Tables 1.1, 1.2, and 1.3 indicate that:

1. For very small values ($\beta = 0.2$) of the parameter β, the expansions (in powers of β) that represent the various orders ($n = 1, 2, 3, 4$) of approximations of the normalized expectation $\mathbf{E}_N^{(n)}[\Phi_H]$ and normalized variance $V_N^{(n)}[\Phi_H]$ of the

Table 1.1 Exact and approximate values for the expectation, variance, and standard deviation of $\Phi_H(\sigma; E)$ for $\beta = 0.2$

Moment	$\mathbf{E}_N[\Phi_H]$	$V_N[\Phi_H]$	$SD_N[\Phi_H]$
Exact value	1.014	0.0142	0.1190
First-order value	1.000	0.0133	0.1155
([a]Error)	(1.4%)	(5%)	(3%)
Second-order value	1.000	0.0134	0.1158
([a]Error)	(1.4%)	(3.88%)	(2.7%)
Third-order value	1.013	0.0142	0.0190
([a]Error)	(0.07%)	([b]Exact)	([b]Exact)
Fourth-order value	1.013	0.0142	0.0190
([a]Error)	(0.07%)	([b]Exact)	([b]Exact)

[a]Error = (exact−approximate)/exact
[b]Exact to at least four decimal digits

Table 1.2 Exact and approximate values for the expectation, variance, and standard deviation of $\Phi_H(\sigma; E)$ for $\beta = 0.5$

Moment	$\mathbf{E}_N[\Phi_H]$	$V_N[\Phi_H]$	$SD_N[\Phi_H]$
Exact value	1.0990	0.1264	0.3555
First-order value	1.000	0.0833	0.2887
([a]Error)	(9.01%)	(34.05%)	(18.8%)
Second-order value	1.083	0.0889	0.2981
([a]Error)	(1.43%)	(29.67%)	(16.2%)
Third-order value	1.083	0.1161	0.3375
([a]Error)	(1.43%)	(8.13%)	(5.06%)
Fourth-order value	1.096	0.1188	0.3447
([a]Error)	(0.27%)	(6.03%)	(3.05%)

[a]Error = (exact−approximate)/exact

Table 1.3 Exact and approximate values for the expectation, variance, and standard deviation of $\Phi_H(\sigma; E)$ for $\beta = 0.95$

Moment	$\mathbf{E}_N[\Phi_H]$	$V_N[\Phi_H]$	$SD_N[\Phi_H]$
Exact value	1.928	6.5385	2.5570
First-order value	1.000	0.3008	0.5485
([a]Error)	(48.14%)	(95.40%)	(78.55%)
Second-order value	1.301	0.3732	0.6109)
([a]Error)	(32.53%)	(94.29%)	(76.11%)
Third-order value	1.301	0.8040	0.8967
([a]Error)	(32.53%)	(87.70%)	(64.93%)
Fourth-order value	1.464	0.9632	0.9814
([a]Error)	(24.1%)	(85.27%)	(62.62%)

[a]Error = (exact−approximate)/exact

unknown distribution of the flux $\Phi_H(\sigma; E)$ converge very quickly to the respective exact values $\mathbf{E}_N[\Phi_H]$ and $V_N[\Phi_H]$, as indicated by the results presented in Table 1.1. Already the first-order approximations $\mathbf{E}_N^{(1)}[\Phi_H]$ and $V_N^{(1)}[\Phi_H]$ yield results that are within 5% of the exact values of the expectation and, respectively, variance of the flux $\Phi_H(\sigma; E)$. The third-order approximation $\mathbf{E}_N^{(3)}[\Phi_H]$ is within 0.07% of the exact value for the expectation of the flux $\Phi_H(\sigma; E)$ while the third-order approximation $SD_N^{(3)}[\Phi_H]$ of the standard deviation of the flux $\Phi_H(\sigma; E)$ is exact up to at least four decimals. Recall that a very small value of β implies that the measurement $\sigma_m \triangleq \sigma_s(E_m)$ is extremely accurate, i.e., the uniform distribution of $\sigma_s(E_m)$ around σ_m is extremely narrow.

2. The value $\beta = 0.5$ represents a measurement of moderate precision, so it is representative of the majority of evaluated cross sections such as the scattering cross section $\sigma_s(E_m)$. The results in Table 1.2 indicate that the first-order approximations are not satisfactory: the first-order approximation $\mathbf{E}_N^{(1)}[\Phi_H]$ of the exact expectation $\mathbf{E}_N[\Phi_H]$ is in error by 9%, while the first-order approximation $SD_N^{(1)}[\Phi_H]$ of the exact standard deviation $SD_N[\Phi_H]$ is in error by 19%. Only the third-order (or higher-order) approximations, $\mathbf{E}_N^{(3)}[\Phi_H]$ and $SD_N^{(3)}[\Phi_H]$, would

provide values for the expectation and, respectively, standard deviation of the flux $\Phi_H(\sigma; E)$ which would be within 5% of the respective exact values.

3. The value $\beta = 0.95$ characterizes either an imprecise measurement or the need to construct "tolerance intervals" that cover a large segment of the unknown distribution of the quantity under investigation, i.e., the flux $\Phi_H(\sigma; E)$, in this illustrative example. As the results in Table 1.3 indicate, the convergence of the various approximations is extremely slow. Even the fourth-order approximations $\mathbf{E}_N^{(4)}[\Phi_H]$ and $SD_N^{(4)}[\Phi_H]$ are very far off the exact values $\mathbf{E}_N[\Phi_H]$ and $SD_N[\Phi_H]$, respectively, of the expectation and the standard deviation of $\Phi_H(\sigma; E)$.

The accuracies of the first-order and second-order approximations for the third-order moment, $\mu_3[\Phi_H]$, and the skewness, $Skew[\Phi_H]$, of $\Phi_H(\sigma; E)$ are illustrated by the results presented in Table 1.4. The exact moment, $\mu_3[\Phi_H]$, of the predicted distribution of the response $\Phi_H(\sigma; E)$ is positive, indicating a distribution which would be skewed toward values more positive than the distribution's expected value $\mathbf{E}_N[\Phi_H]$. As the positive values of the parameter β increase toward unity, which would represent an increasingly broader distribution for $\Phi_H(\sigma; E)$, $\mu_3[\Phi_H]$ also increases; hence, the skewness of the distribution for $\Phi_H(\sigma; E)$ also increases. *With pronounced skewness, standard statistical inference procedures such as a confidence interval for a mean will be not only incorrect, in the sense that the true coverage level will differ from the nominal* (e.g., *95%) level, but they will also result in unequal error probabilities on each side.*

Notably, *the results presented in bold in* Table 1.4 *indicate that the first-order approximation for the third moment* $\mu_3^{(1)}[\Phi_H]$ *of the predicted response distribution is always zero. Thus, using just the first-order response sensitivities will always be 100% incorrect unless the distribution of the response under consideration happens to be symmetric.* Thus, the response distribution is always predicted to be symmetrical if only first-order sensitivities are considered. Hence, a "first-order sensitivity

Table 1.4 Exact and approximate values for the third-order moment and skewness of $\Phi_H(\sigma; E)$

Third-order moment	$\beta = 0.2$	$\beta = 0.5$	$\beta = 0.95$
Exact: $\mu_3[\Phi_H]$	0.0005	0.0353	60.20
Exact: $Skew[\Phi_H]$	0.2955	0.7857	3.6
First-order: $\mu_3^{(1)}[\Phi_H]$ ([a]Error)	**0.0000 (100%)**	**0.0000 (100%)**	**0.0000 (100%)**
First-order: $Skew^{(1,1)}[\Phi_H]$ ([a]Error)	**0.0000 (100%)**	**0.0000 (100%)**	**0.0000 (100%)**
Second-order: $\mu_3^{(2)}[\Phi_H]$ ([a]Error)	0.0004 (20%)	0.0267 (32.2%)	0.2172 (99.6%)
Second-order: $Skew^{(2,2)}[\Phi_H]$	0.2576	1.0073	0.9527
([a]Error)	(13.2%)	(−28.3%)	(73.6%)
$Skew^{(2,4)}[\Phi_H]$	0.2364	0.6521	0.2298
([a]Error)	(20%)	(17.0%)	(93.6%)

[a]Error = (exact−approximate)/exact

and uncertainty quantification" will always produce an erroneous third moment (and hence skewness) of the predicted response distribution, unless the distribution happens to be symmetrical. *At least second-order sensitivities must be used in order to estimate the third-order moment (and hence the skewness) of the response distribution.* The overall conclusion that follows from the results presented in Table 1.4 is that low-order (i.e., first- and second-order) approximations for the third-order moment, $\mu_3[\Phi_H]$, of the response distribution can provide the correct sign (i.e., positive or negative) of the response skewness, but cannot provide reliably correct values for the skewness. Significantly higher-order approximations, which would require the computation of high-order response sensitivities to model parameters, are needed in order to obtain reliable values for the third-order response moment $\mu_3[\Phi_H]$ and, hence, for the skewness of the response distribution. Comparing the results in Table 1.4 for increasing (positive) values for the parameter β, the low-order (first- and second-order) approximations become increasingly unreliable as β increases toward unity. This trend indicates that, as the response distribution becomes wider, sensitivities of increasingly higher order would be needed in order to improve the predicted values for the response skewness. Using, e.g., a second-order approximation for the third moment $\mu_3[\Phi_H]$ but a fourth-order approximation for the variance (so that a comparable amount of computational effort would be needed for computing the respective approximations) does not significantly improve the final results. If accurate values for the response skewness are required, it is not possible to circumvent the need for computing high-order sensitivities, which in turn enable the computation of high-order approximations to the third-order moment and, hence, skewness.

Illustrative results for the first-order and second-order approximations for the fourth-order moment $\mu_4[\Phi_H]$ of $\Phi_H(\sigma; E)$ are presented in Table 1.5, below.

Table 1.5 Exact and approximate values for the fourth-order moment of $\Phi_H(\sigma; E)$

Fourth-order moment	$\beta = 0.2$	$\beta = 0.5$	$\beta = 0.95$
Exact: $\mu_4[\Phi_H]$	0.0004	0.0411	779.47
Exact: $Kurt[\Phi_H]$	1.9837	2.5725	18.23
First-order: $\mu_4^{(1)}[\Phi_H]$ ([a]Error)	0.0003 (25%)	0.0125 (69.6%)	0.1629 (100%)
First-order: $Kurt^{(1,1)}[\Phi_H]$ ([a]Error)	1.6960 (14.5%)	1.8014 (29.97%)	1.8004 (90%)
Second-order: $\mu_4^{(2)}[\Phi_H]$ ([a]Error)	0.0004 ([b]Exact)	0.0259 (37%)	0.7930 (99.99%)
Second-order: $Kurt^{(2,2)}[\Phi_H]$	2.2277	3.2771	5.694
([a]Error)	(−12.3%)	(−27.39)	(68.8%)
$Kurt^{(2,4)}[\Phi_H]$	1.9837	1.8351	0.8548
([a]Error)	([b]Exact)	(28.6%)	(95.3%)

[a]Error = (exact−approximate)/exact
[b]Exact to at least four decimal digits

The main conclusions that follow from the results presented in Table 1.5 are similar to those that were drawn from the results presented in Table 1.4. Thus, the results presented in Table 1.5 indicate that for small values of the parameter β, the first-order approximation, $\mu_4^{(1)}[\Phi_H]$, provides an approximate value which differs by 25% from the exact value $\mu_4[\Phi_H]$, while the second-order approximation $\mu_4^{(2)}[\Phi_H]$ provides the exact first four significant digits of the exact value of $\mu_4[\Phi_H]$. Low-order (i.e., first- and second-order) approximations for the fourth-order moment, $\mu_4[\Phi_H]$, of the response distribution can provide the correct sign of the response kurtosis but cannot provide reliably correct values for the kurtosis. Significantly higher-order approximations, which would require the computation of high-order response sensitivities to model parameters, are needed in order to obtain reliable values for the fourth-order response moment $\mu_4[\Phi_H]$ and, hence, for the kurtosis of the response distribution. The results in Table 1.5 indicate that the low-order (first- and second-order) approximations become increasingly unreliable as β increases (with positive values) toward unity, which means that, as the response distribution becomes wider, sensitivities of increasingly higher order would be needed in order to improve the predicted values for the response kurtosis. If accurate values for the response kurtosis are required, it is not possible to circumvent the need for computing high-order sensitivities.

It is known (Weiberg and Wigner 1958; Amaldi 1959) that the form showed in (1.3) of the slowing down neutron flux in a hydrogenous medium also holds for the asymptotic (below the cut-off energy of the fission source) slowing down flux in an infinite *non*-hydrogenous medium, where it has the following expression:

$$\Phi_m(\boldsymbol{\alpha}; E) = \frac{Q(\boldsymbol{\alpha})}{\bar{\xi}(E) E \Sigma_s(\boldsymbol{\alpha}; E)}. \tag{1.62}$$

In (1.62), the quantity $\bar{\xi}(E)$ denotes the average increase in lethargy per collision and depends on the atomic/molecular masses and the scattering cross sections of the nuclides in the medium. The analysis of the dependence of the flux $\Phi_m(\boldsymbol{\alpha}; E)$ on the scattering cross section of the mixture, $\Sigma_s(\boldsymbol{\alpha}; E)$, is similar to the analysis of the slowing-down flux $\Phi_H(\sigma; E)$ in hydrogen (Weiberg and Wigner 1958; Amaldi 1959).

The expression of the slowing down neutron flux distribution, $\Phi(\boldsymbol{\alpha}; E)$, in a hydrogenous medium in which both scattering and absorption are important was provided in (1.2). The Taylor series expansion of $\Phi(\boldsymbol{\alpha}; E)$ at the nominal parameter value σ_m has the following form:

$$\Phi\left(\sigma^H; E\right) = \sum_{n=0}^{\infty} \frac{1}{n!} \left\{ \frac{d^n \Phi\left(\sigma^H; E\right)}{d\left(\sigma^H\right)^n} \right\}_{\left(E_m; \sigma_m; \boldsymbol{\alpha}^0\right)} \left(\sigma^H - \sigma_m\right)^n, \tag{1.63}$$

where:

$$
\begin{aligned}
\frac{d^n\Phi(\sigma^H;E)}{d(\sigma^H)^n} &= \sum_{k=0}^{n}\binom{n}{k}\left\{\frac{d^k}{d(\sigma^H)^k}\left[\frac{1}{\Sigma(\sigma^H;E)}\right]\right\}\frac{d^{n-k}G(\boldsymbol{\alpha};E)}{d(\sigma^H)^{n-k}} \\
&= G(\boldsymbol{\alpha};E)\frac{d^n}{d(\sigma^H)^n}\left[\frac{1}{\Sigma(\sigma^H;E)}\right] \\
&+\binom{n}{1}\frac{dG(\boldsymbol{\alpha};E)}{d(\sigma^H)}\frac{d^{n-1}}{d(\sigma^H)^{n-1}}\left[\frac{1}{\Sigma(\sigma^H;E)}\right]+\ldots+\frac{1}{\Sigma(\sigma^H;E)}\frac{d^n G(\boldsymbol{\alpha};E)}{d(\sigma^H)^n}.
\end{aligned}
\tag{1.64}
$$

It follows from (1.63) and (1.64) that the first term, denoted as $\Phi_1(\sigma^H;E)$, in the Taylor series of $\Phi(\sigma^H;E)$ is the following infinite series:

$$
\begin{aligned}
\Phi_1\left(\sigma^H;E\right) &= G\left(\sigma_m;\boldsymbol{\alpha}^0;E_m\right) \\
&\times\sum_{n=0}^{\infty}\frac{1}{n!}\left\{\frac{d^n}{d(\sigma^H)^n}\left[\frac{1}{\Sigma(\sigma^H;E)}\right]\right\}_{(E_m;\,\sigma_m;\,\boldsymbol{\alpha}^0)}\left(\sigma^H-\sigma_m\right)^n,
\end{aligned}
\tag{1.65}
$$

which has the same functional form as the series for $\Phi_H(\sigma;E)$ in (1.10). In fact, each successive term in the series representation of $\Phi(\boldsymbol{\alpha};E)$ shown in (1.63) has the same functional form as the series for $\Phi_H(\sigma;E)$ in (1.10). Therefore, the analysis that has been performed in the foregoing for $\Phi_H(\sigma;E)$ is also applicable to perform sensitivity and uncertainty analyses aimed at quantifying the impact of the uncertainties afflicting the other parameters (cross sections, number densities, source parameters, etc.) on the distribution of the slowing down flux $\Phi(\boldsymbol{\alpha};E)$ in a scattering and absorbing hydrogenous medium.

1.3 The Curse of Dimensionality in Sensitivity Analysis: Computation of High-Order Response Sensitivities to Model Parameters

As the simple illustrative system considered in Sect. 1.2 has indicated, the system's response is, ultimately, a function of the system's parameters $\boldsymbol{\alpha}\triangleq(\alpha_1,\ldots,\alpha_{TP})^\dagger\in\mathbb{R}^{TP}$. The Taylor series of a system response, denoted as $R_k(\boldsymbol{\alpha})$, around the expected (or nominal) parameter values $\boldsymbol{\alpha}^0$ has the following well-known formal expression:

$$
R_k^{(0)}(\boldsymbol{\alpha}) = R_k\left(\boldsymbol{\alpha}^0\right),
\tag{1.66}
$$

$$
R_k^{(1)}(\boldsymbol{\alpha}) = R_k^{(0)}(\boldsymbol{\alpha})+\sum_{j_1=1}^{TP}\frac{\partial R_k(\boldsymbol{\alpha}^0)}{\partial\alpha_{j_1}}\left(\alpha_{j_1}-\alpha_{j_1}^0\right),
\tag{1.67}
$$

$$
R_k^{(2)}(\boldsymbol{\alpha}) = R_k^{(1)}(\boldsymbol{\alpha})+\frac{1}{2}\sum_{j_1=1}^{TP}\sum_{j_2=1}^{TP}\frac{\partial^2 R_k(\boldsymbol{\alpha}^0)}{\partial\alpha_{j_1}\partial\alpha_{j_2}}\left(\alpha_{j_1}-\alpha_{j_1}^0\right)\left(\alpha_{j_2}-\alpha_{j_2}^0\right),
\tag{1.68}
$$

$$R_k^{(3)}(\boldsymbol{\alpha}) = R_k^{(2)}(\boldsymbol{\alpha}) + \frac{1}{3!}\sum_{j_1=1}^{TP}\sum_{j_2=1}^{TP}\sum_{j_3=1}^{TP}\frac{\partial^3 R_k(\boldsymbol{\alpha}^0)}{\partial\alpha_{j_1}\partial\alpha_{j_2}\partial\alpha_{j_3}}\left(\alpha_{j_1}-\alpha_{j_1}^0\right)\left(\alpha_{j_2}-\alpha_{j_2}^0\right) \times \left(\alpha_{j_3}-\alpha_{j_3}^0\right), \tag{1.69}$$

$$R_k^{(4)}(\boldsymbol{\alpha}) = R_k^{(3)}(\boldsymbol{\alpha}) + \frac{1}{4!}\sum_{j_1=1}^{TP}\sum_{j_2=1}^{TP}\sum_{j_3=1}^{TP}\sum_{j_4=1}^{TP}\frac{\partial^4 R_k(\boldsymbol{\alpha}^0)}{\partial\alpha_{j_1}\partial\alpha_{j_2}\partial\alpha_{j_3}\partial\alpha_{j_4}} \times\left[\left(\alpha_{j_1}-\alpha_{j_1}^0\right)\left(\alpha_{j_2}-\alpha_{j_2}^0\right)\left(\alpha_{j_3}-\alpha_{j_3}^0\right)\left(\alpha_{j_4}-\alpha_{j_4}^0\right)\right], \tag{1.70}$$

$$R_k^{(5)}(\boldsymbol{\alpha}) = R_k^{(4)}(\boldsymbol{\alpha}) + \frac{1}{5!}\sum_{j_1=1}^{TP}\sum_{j_2=1}^{TP}\sum_{j_3=1}^{TP}\sum_{j_4=1}^{TP}\sum_{j_5=1}^{TP}\frac{\partial^5 R_k(\boldsymbol{\alpha}^0)}{\partial\alpha_{j_1}\partial\alpha_{j_2}\partial\alpha_{j_3}\partial\alpha_{j_4}\partial\alpha_{j_5}} \times\left[\left(\alpha_{j_1}-\alpha_{j_1}^0\right)\left(\alpha_{j_2}-\alpha_{j_2}^0\right)\left(\alpha_{j_3}-\alpha_{j_3}^0\right)\left(\alpha_{j_4}-\alpha_{j_4}^0\right)\left(\alpha_{j_5}-\alpha_{j_5}^0\right)\right], \tag{1.71}$$

$$R_k^{(6)}(\boldsymbol{\alpha}) = R_k^{(5)}(\boldsymbol{\alpha}) + \frac{1}{6!}\sum_{j_1=1}^{TP}\sum_{j_2=1}^{TP}\sum_{j_3=1}^{TP}\sum_{j_4=1}^{TP}\sum_{j_5=1}^{TP}\sum_{j_6=1}^{TP}\frac{\partial^6 R_k(\boldsymbol{\alpha}^0)}{\partial\alpha_{j_1}\partial\alpha_{j_2}\partial\alpha_{j_3}\partial\alpha_{j_4}\partial\alpha_{j_5}\partial\alpha_{j_6}} \times\left[\left(\alpha_{j_1}-\alpha_{j_1}^0\right)\left(\alpha_{j_2}-\alpha_{j_2}^0\right)\left(\alpha_{j_3}-\alpha_{j_3}^0\right)\left(\alpha_{j_4}-\alpha_{j_4}^0\right)\left(\alpha_{j_5}-\alpha_{j_5}^0\right)\left(\alpha_{j_6}-\alpha_{j_6}^0\right)\right]. \tag{1.72}$$

As is well known, and as indicated by (1.66)–(1.72), the Taylor series of a function of TP-variables [e.g., $R_k(\boldsymbol{\alpha})$] comprises TP first-order derivatives, TP(-TP + 1)/2 distinct second-order derivatives, $TP(TP + 1)(TP + 2)/6$ distinct third-order derivatives, and so on. The computation by conventional methods of the nth-order functional derivatives (called "sensitivities" in the field of sensitivity analysis) of a response with respect to the TP-parameters it depends on would require at least $O(TP^n)$ large-scale computations. This exponential increase—with the order of response sensitivities—of the number of large-scale computations needed to determine higher-order sensitivities is the manifestation of the "curse of dimensionality in sensitivity analysis," by analogy to expression coined by Bellman (1957) to express the difficulty of using "brute-force" grid search when optimizing a function with many input variables.

1.4 The Curse of Dimensionality in Uncertainty Quantification: Moments of the Response Distribution in Parameter Phase-Space

In practice, the model parameters are not known exactly, even though they are not bona fide random quantities. For practical purposes, however, these model parameters are considered to be variates that obey a multivariate probability distribution function, denoted as $p_\alpha(\boldsymbol{\alpha})$, which is seldom known, particularly for large-scale systems involving many parameters. Considering that the multivariate distribution

$p_\alpha(\boldsymbol{\alpha})$ is formally defined on a domain D_α, the various moments of $p_\alpha(\boldsymbol{\alpha})$ can be defined in a standard manner by using the following notation

$$\langle u(\boldsymbol{\alpha})\rangle_\alpha \triangleq \int_{D_\alpha} u(\boldsymbol{\alpha}) p_\alpha(\boldsymbol{\alpha}) d\boldsymbol{\alpha}, \tag{1.73}$$

where $u(\boldsymbol{\alpha})$ is a continuous function of the parameters $\boldsymbol{\alpha}$. Using the notation defined in (1.73), the moments of $p_\alpha(\boldsymbol{\alpha})$ are defined as follows:

1. The expected (or mean) value of a model parameter α_j, denoted as α_j^0, is defined as follows:

$$\alpha_j^0 \triangleq \langle \alpha_j \rangle_\alpha \triangleq \int_{D_\alpha} \alpha_j p_\alpha(\boldsymbol{\alpha}) d\boldsymbol{\alpha}, \quad j = 1, \ldots, TP. \tag{1.74}$$

The expected values α_j^0 are considered to be the components of the following vector of mean (expected) values:

$$\boldsymbol{\alpha}^0 \triangleq \left(\alpha_1^0, \ldots, \alpha_{TP}^0\right)^\dagger. \tag{1.75}$$

2. The covariance, $\text{cov}(\alpha_{j_1}, \alpha_{j_2})$, of two parameters, α_{j_1} and α_{j_2}, is defined as follows:

$$\mu_2^{j_1 j_2}(\boldsymbol{\alpha}) \triangleq \text{cov}\left(\alpha_{j_1}, \alpha_{j_2}\right) \triangleq \left\langle \left(\alpha_{j_1} - \alpha_{j_1}^0\right)\left(\alpha_{j_2} - \alpha_{j_2}^0\right)\right\rangle_\alpha, \quad j_1, j_2 = 1, \ldots, TP. \tag{1.76}$$

The variance, $\text{var}(\alpha_j)$, of a parameter α_j, is defined as follows:

$$\text{var}\left(\alpha_j\right) \triangleq \left\langle \left(\alpha_j - \alpha_j^0\right)^2 \right\rangle_\alpha, \quad j = 1, \ldots, TP. \tag{1.77}$$

The standard deviation, σ_j, of α_j is defined as follows: $\sigma_j \triangleq \sqrt{\text{var}\left(\alpha_j\right)}$. The correlation, $\rho_{j_1 j_2}$, between two parameters α_{j_1} and α_{j_2} is defined as follows:

$$\rho_{j_1 j_2} \triangleq \text{cov}\left(\alpha_{j_1}, \alpha_{j_2}\right) / \left(\sigma_{j_1} \sigma_{j_2}\right); \quad j_1, j_2 = 1, \ldots, TP. \tag{1.78}$$

3. The third-order parameter moment, $\mu_3^{j_1 j_2 j_3}$, and the associated third-order correlation $t_{j_1 j_2 j_3}$ among three parameters are defined as follows:

$$\mu_3^{j_1 j_2 j_3} \triangleq \left\langle \left(\alpha_{j_1} - \alpha_{j_1}^0\right)\left(\alpha_{j_2} - \alpha_{j_2}^0\right)\left(\alpha_{j_3} - \alpha_{j_3}^0\right)\right\rangle_\alpha$$
$$\triangleq t_{j_1 j_2 j_3}\sigma_{j_1}\sigma_{j_2}\sigma_{j_3};\ j_1, j_2, j_3 = 1, \ldots, TP. \tag{1.79}$$

4. The fourth-order parameter moment, $\mu_4^{j_1 j_2 j_3 j_4}$, and the associated fourth-order correlation $q_{j_1 j_2 j_3 j_4}$ among four parameters are defined as follows:

$$\mu_4^{j_1 j_2 j_3 j_4} \triangleq \left\langle \left(\alpha_{j_1} - \alpha_{j_1}^0\right)\left(\alpha_{j_2} - \alpha_{j_2}^0\right)\left(\alpha_{j_3} - \alpha_{j_3}^0\right)\left(\alpha_{j_4} - \alpha_{j_4}^0\right)\right\rangle_\alpha$$
$$\triangleq q_{j_1 j_2 j_3 j_4}\sigma_{j_1}\sigma_{j_2}\sigma_{j_3}\sigma_{j_4};\ j_1, j_2, j_3, j_4 = 1, \ldots, TP. \tag{1.80}$$

5. The fifth-order parameter moment, $\mu_5^{j_1 j_2 j_3 j_4 j_5}$, and the associated fifth-order correlation $p_{j_1 j_2 j_3 j_4 j_5}$ among five parameters are defined as follows:

$$\mu_5^{j_1 j_2 j_3 j_4 j_5} \triangleq \left\langle \left(\alpha_{j_1} - \alpha_{j_1}^0\right)\left(\alpha_{j_2} - \alpha_{j_2}^0\right)\left(\alpha_{j_3} - \alpha_{j_3}^0\right)\left(\alpha_{j_4} - \alpha_{j_4}^0\right)\left(\alpha_{j_5} - \alpha_{j_5}^0\right)\right\rangle_\alpha$$
$$\triangleq p_{j_1 j_2 j_3 j_4 j_5}\sigma_{j_1}\sigma_{j_2}\sigma_{j_3}\sigma_{j_4}\sigma_{j_5};\ j_1, j_2, j_3, j_4, j_5 = 1, \ldots, TP. \tag{1.81}$$

6. The sixth-order parameter moment, $\mu_6^{j_1 j_2 j_3 j_4 j_5 j_6}$, and the associated sixth-order correlation $s_{j_1 j_2 j_3 j_4 j_5 j_6}$ among six parameters are defined as follows:

$$\mu_6^{j_1 j_2 j_3 j_4 j_5 j_6} \triangleq \left\langle \left(\alpha_{j_1} - \alpha_{j_1}^0\right)\left(\alpha_{j_2} - \alpha_{j_2}^0\right) \times \ldots \times \left(\alpha_{j_5} - \alpha_{j_5}^0\right)\left(\alpha_{j_6} - \alpha_{j_6}^0\right)\right\rangle_\alpha$$
$$\triangleq s_{j_1 j_2 j_3 j_4 j_5 j_6}\sigma_{j_1}\sigma_{j_2}\sigma_{j_3}\sigma_{j_4}\sigma_{j_5}\sigma_{j_6};\quad j_1, j_2, j_3, j_4, j_5, j_6 = 1, \ldots, TP. \tag{1.82}$$

Higher-order moments of the parameter distribution are defined similarly to the first six moments defined in (1.74)–(1.82). Uncertainties in the model's parameters will evidently give rise to uncertainties in the model responses $R_k(\boldsymbol{\alpha})$. The approximate moments of the unknown distribution of $R_k(\boldsymbol{\alpha})$ are obtained by using the so-called "propagation of errors" methodology, integrating formally, over the parameter distribution, various expressions involving the truncated Taylor series expansion of the response provided in (1.72). Using this concept of integrating the truncated Taylor series expansion of the response, Tukey (1957) has presented the most extensive formulas to date for the mean, variance, skewness, and kurtosis of the response distribution, which include terms up to fourth-order response sensitivities. The remainder of this section presents formulas which are consistent up to and including the sixth-order standard deviations, thus generalizing the formulas presented by Tukey (1957).

1.4.1 Expectation Value of a Response

The expectation (value), $E(R_k)$, of a response $R_k(\boldsymbol{\alpha})$ is defined using (1.73) as follows:

$$E(R_k) \triangleq \langle R_k(\boldsymbol{\alpha}) \rangle_{\alpha} \triangleq \int\limits_{D_\alpha} R_k(\boldsymbol{\alpha}) p_\alpha(\boldsymbol{\alpha}) d\boldsymbol{\alpha}. \tag{1.83}$$

Using (1.66)–(1.72) in the definition provided in (1.83) yields the following expressions for the *nth*-order approximate expectation, $E^{(n)}(R_k)$, $n = 1, \ldots, 6$, of the response $R_k(\boldsymbol{\alpha})$:

$$E^{(1)}(R_k) = E^{(0)}(R_k) \triangleq R_k\left(\boldsymbol{\alpha}^0\right), \tag{1.84}$$

$$E^{(2)}(R_k) = E^{(1)}(R_k) + \frac{1}{2} \sum_{j_1=1}^{TP} \sum_{j_2=1}^{TP} \frac{\partial^2 R_k(\boldsymbol{\alpha}^0)}{\partial \alpha_{j_1} \partial \alpha_{j_2}} \rho_{j_1 j_2} \sigma_{j_1} \sigma_{j_2}, \tag{1.85}$$

$$E^{(3)}(R_k) = E^{(2)}(R_k) + \frac{1}{6} \sum_{j_1=1}^{TP} \sum_{j_2=1}^{TP} \sum_{j_3=1}^{TP} \frac{\partial^3 R_k(\boldsymbol{\alpha}^0)}{\partial \alpha_{j_1} \partial \alpha_{j_2} \partial \alpha_{j_3}} t_{j_1 j_2 j_3} \sigma_{j_1} \sigma_{j_2} \sigma_{j_3}, \tag{1.86}$$

$$\begin{aligned} E^{(4)}(R_k) &= E^{(3)}(R_k) \\ &+ \frac{1}{4!} \sum_{j_1=1}^{TP} \sum_{j_2=1}^{TP} \sum_{j_3=1}^{TP} \sum_{j_4=1}^{TP} \frac{\partial^4 R_k(\boldsymbol{\alpha}^0)}{\partial \alpha_{j_1} \partial \alpha_{j_2} \partial \alpha_{j_3} \partial \alpha_{j_4}} q_{j_1 j_2 j_3 j_4} \sigma_{j_1} \sigma_{j_2} \sigma_{j_3} \sigma_{j_4}, \end{aligned} \tag{1.87}$$

$$\begin{aligned} E^{(5)}(R_k) &= E^{(4)}(R_k) + \frac{1}{5!} \sum_{j_1=1}^{TP} \sum_{j_2=1}^{TP} \sum_{j_3=1}^{TP} \sum_{j_4=1}^{TP} \sum_{j_5=1}^{TP} \frac{\partial^5 R_k(\boldsymbol{\alpha}^0)}{\partial \alpha_{j_1} \partial \alpha_{j_2} \partial \alpha_{j_3} \partial \alpha_{j_4} \partial \alpha_{j_5}} \\ &\times p_{j_1 j_2 j_3 j_4 j_5} \sigma_{j_1} \sigma_{j_2} \sigma_{j_3} \sigma_{j_4} \sigma_{j_5}, \end{aligned} \tag{1.88}$$

$$\begin{aligned} E^{(6)}(R_k) = E^{(5)}(R_k) &+ \frac{1}{6!} \sum_{j_1=1}^{TP} \sum_{j_2=1}^{TP} \sum_{j_3=1}^{TP} \sum_{j_4=1}^{TP} \sum_{j_5=1}^{TP} \sum_{j_6=1}^{TP} \frac{\partial^6 R_k(\boldsymbol{\alpha}^0)}{\partial \alpha_{j_1} \partial \alpha_{j_2} \partial \alpha_{j_3} \partial \alpha_{j_4} \partial \alpha_{j_5} \partial \alpha_{j_6}} \\ &\times s_{j_1 j_2 j_3 j_4 j_5 j_6} \sigma_{j_1} \sigma_{j_2} \sigma_{j_3} \sigma_{j_4} \sigma_{j_5} \sigma_{j_6}. \end{aligned} \tag{1.89}$$

The radius/domain of convergence of the series in (1.89) must also be established before using it for any subsequent purpose, since the radius/domain of convergence of the series in (1.89) may differ from the radius/domain of convergence of the Taylor series in (1.72). As indicated in (1.85), the second-order sensitivities contribute the leading correction terms to the response's expected value, causing it to differ from the response's computed value, $R_k(\boldsymbol{\alpha}^0)$.

1.4.2 Response-Parameter Covariances

For $n = 1, \ldots, 6$, the nth-order approximation of the covariance, $\mathrm{cov}(\alpha_i, R_k) \triangleq \left\langle \left(\alpha_i - \alpha_i^0\right)\left[R_k(\boldsymbol{\alpha}) - E(R_k)\right] \right\rangle_\alpha$, between a computed responses and a parameter is defined as follows:

$$\begin{aligned} cov\left(\alpha_i, R_k^{(n)}\right) &\triangleq \left\langle \left(\alpha_i - \alpha_i^0\right)\left[R_k^{(n)}(\boldsymbol{\alpha}) - E^{(n)}(R_k)\right] \right\rangle_\alpha \\ &= \left\langle \left(\alpha_i - \alpha_i^0\right) R_k^{(n)}(\boldsymbol{\alpha}) \right\rangle_\alpha, \quad n = 1, \ldots, 6; i = 1, \ldots, TP. \end{aligned} \tag{1.90}$$

Using (1.66)–(1.72) in (1.90) yields the following expressions for $i = 1, \ldots, TP$:

$$cov\left(\alpha_i, R_k^{(1)}\right) \triangleq \left\langle \left(\alpha_i - \alpha_i^0\right) R_k^{(1)}(\boldsymbol{\alpha}) \right\rangle_\alpha = \sum_{j_1=1}^{TP} \frac{\partial R_k(\boldsymbol{\alpha}^0)}{\partial \alpha_{j_1}} \rho_{i,j_1} \sigma_i \sigma_{j_1}; \tag{1.91}$$

$$\begin{aligned} cov\left(\alpha_i, R_k^{(2)}\right) &\triangleq \left\langle \left(\alpha_i - \alpha_i^0\right) R_k^{(2)}(\boldsymbol{\alpha}) \right\rangle_\alpha \\ &= cov\left(\alpha_i, R_k^{(1)}\right) + \frac{1}{2} \sum_{j_1=1}^{TP} \sum_{j_2=1}^{TP} \frac{\partial^2 R_k(\boldsymbol{\alpha}^0)}{\partial \alpha_{j_1} \partial \alpha_{j_2}} t_{i,j_1 j_2} \sigma_i \sigma_{j_1} \sigma_{j_2}; \end{aligned} \tag{1.92}$$

$$\begin{aligned} cov\left(\alpha_i, R_k^{(3)}\right) &\triangleq \left\langle \left(\alpha_i - \alpha_i^0\right) R_k^{(3)}(\boldsymbol{\alpha}) \right\rangle_\alpha = cov\left(\alpha_i, R_k^{(2)}\right) \\ &+ \frac{1}{6} \sum_{j_1=1}^{TP} \sum_{j_2=1}^{TP} \sum_{j_3=1}^{TP} \frac{\partial^3 R_k(\boldsymbol{\alpha}^0)}{\partial \alpha_{j_1} \partial \alpha_{j_2} \partial \alpha_{j_3}} q_{i,j_1 j_2 j_3} \sigma_i \sigma_{j_1} \sigma_{j_2} \sigma_{j_3}; \end{aligned} \tag{1.93}$$

$$\begin{aligned} cov\left(\alpha_i, R_k^{(4)}\right) &\triangleq \left\langle \left(\alpha_i - \alpha_i^0\right) R_k^{(4)}(\boldsymbol{\alpha}) \right\rangle_\alpha = cov\left(\alpha_i, R_k^{(3)}\right) \\ &+ \frac{1}{4!} \sum_{j_1=1}^{TP} \sum_{j_2=1}^{TP} \sum_{j_3=1}^{TP} \sum_{j_4=1}^{TP} \frac{\partial^4 R_k(\boldsymbol{\alpha}^0)}{\partial \alpha_{j_1} \partial \alpha_{j_2} \partial \alpha_{j_3} \partial \alpha_{j_4}} p_{i,j_1 j_2 j_3 j_4} \sigma_i \sigma_{j_1} \sigma_{j_2} \sigma_{j_3} \sigma_{j_4}; \end{aligned} \tag{1.94}$$

$$\begin{aligned} cov\left(\alpha_i, R_k^{(5)}\right) &\triangleq \left\langle \left(\alpha_i - \alpha_i^0\right) R_k^{(5)}(\boldsymbol{\alpha}) \right\rangle_\alpha = cov\left(\alpha_i, R_k^{(4)}\right) \\ &+ \frac{1}{5!} \sum_{j_1=1}^{TP} \sum_{j_2=1}^{TP} \sum_{j_3=1}^{TP} \sum_{j_4=1}^{TP} \sum_{j_5=1}^{TP} \frac{\partial^5 R_k(\boldsymbol{\alpha}^0)}{\partial \alpha_{j_1} \partial \alpha_{j_2} \partial \alpha_{j_3} \partial \alpha_{j_4} \partial \alpha_{j_5}} s_{i,j_1 j_2 j_3 j_4 j_5} \sigma_i \sigma_{j_1} \sigma_{j_2} \sigma_{j_3} \sigma_{j_4} \sigma_{j_5}. \end{aligned} \tag{1.95}$$

1.4.3 Covariance of Two Responses

For $n = 1, \ldots, 6$, the nth-order approximation, $cov\left(R_k^{(n)}, R_\ell^{(n)}\right)$, of the covariance, $cov(R_k, R_\ell) \triangleq \langle [R_k(\boldsymbol{\alpha}) - E(R_k)][R_\ell(\boldsymbol{\alpha}) - E(R_\ell)] \rangle_\alpha$, between two responses R_k and R_ℓ is defined as follows:

$$\left\{cov\left(R_k^{(n)},R_\ell^{(n)}\right)\right\}_{inc} \triangleq \left\langle \left[R_k^{(n)}(\boldsymbol{\alpha}) - E^{(n)}(R_k)\right]\left[R_\ell^{(n)}(\boldsymbol{\alpha}) - E^{(n)}(R_\ell)\right]\right\rangle_\alpha. \quad (1.96)$$

It will be seen that the result, denoted as $\left\{cov\left(R_k^{(n)},R_\ell^{(n)}\right)\right\}_{inc}$, which is provided by the right side of (1.96), is inconsistent (i.e., incomplete) in the approximation of the highest-order term, which contains the product of standard deviations $\sigma_{j_1} \times \sigma_{j_2} \times \ldots \times \sigma_{j_n}$. The alternative definition in (1.97) provides an approximation for a "consistent" covariance, which is denoted as $\left\{cov\left(R_k^{(n)},R_\ell^{(n)}\right)\right\}_{con}$ and which comprises the consistent (i.e., complete) expression for the term containing $\sigma_{j_1} \times \sigma_{j_2} \times \ldots \times \sigma_{j_n}$, when the terms of $O(\sigma^{n+1})$ are discarded, i.e.:

$$\left\{cov\left(R_k^{(n)},R_\ell^{(n)}\right)\right\}_{con} \triangleq cov\left(R_k^{(n+1)},R_\ell^{(n+1)}\right) - O\left(\sigma^{n+1}\right), \quad n = 1, \ldots, 5. \quad (1.97)$$

The standard deviation, $SD\left[R_k^{(n)}(\boldsymbol{\alpha})\right]$, of $R_k^{(n)}(\boldsymbol{\alpha})$ is obtained by setting $k = \ell$ in either (1.96) or (1.97) and taking the square root of the resulting expression, respectively, to obtain:

$$\left\{SD\left(R_k^{(n)}\right)\right\}_{inc} \triangleq \sqrt{\left\{\text{var}\left(R_k^{(n)},R_k^{(n)}\right)\right\}_{inc}} + O\left(\sigma^{n/2}\right), \quad n = 1, 2, \ldots. \quad (1.98)$$

$$\left\{SD\left(R_k^{(n)}\right)\right\}_{con} \triangleq \sqrt{\left\{\text{var}\left(R_k^{(n)},R_k^{(n)}\right)\right\}_{con}}, \quad n = 1, 2, \ldots. \quad (1.99)$$

Thus, as indicated by the right side of (1.98), the standard deviation $\left\{SD\left(R_k^{(n)}\right)\right\}_{inc}$ corresponding to the inconsistent variance $\left\{\text{var}\left(R_k^{(n)},R_k^{(n)}\right)\right\}_{inc}$ will *not* comprise the complete number of terms that contain the *nth*-order parameter standard deviations. In contradistinction to $\left\{SD\left(R_k^{(n)}\right)\right\}_{inc}$, the consistent standard deviation, $\left\{SD\left(R_k^{(n)}\right)\right\}_{con}$, which corresponds to the consistent variance $\left\{\text{var}\left(R_k^{(n)},R_k^{(n)}\right)\right\}_{con}$, will comprise all of the terms that contain the *nth*-order parameter standard deviations.

Using (1.66)–(1.89) in the definition provided in (1.96) yields the following expressions for the covariances and standard deviations of various orders of approximation $n = 1, \ldots, 5$:

1. The first-order approximation of the covariance $cov\left(R_k^{(1)},R_\ell^{(1)}\right)$

$$cov\left(R_k^{(1)},R_\ell^{(1)}\right) \triangleq \left\langle \left[R_k^{(1)}(\boldsymbol{\alpha}) - E^{(1)}(R_k)\right]\left[R_\ell^{(1)}(\boldsymbol{\alpha}) - E^{(1)}(R_\ell)\right]\right\rangle_\alpha$$
$$= \sum_{j_1=1}^{TP}\sum_{j_2=1}^{TP} \frac{\partial R_k(\boldsymbol{\alpha}^0)}{\partial \alpha_{j_1}}\frac{\partial R_\ell(\boldsymbol{\alpha}^0)}{\partial \alpha_{j_2}}\rho_{j_1j_2}\sigma_{j_1}\sigma_{j_2}. \quad (1.100)$$

Notably, the expression of $cov\left(R_k^{(1)}, R_\ell^{(1)}\right)$ provided in (1.100) is consistent both in the highest order (in this case: first order) of sensitivities and also in the highest order of parameter standard deviation (in this case, second order) since all of the terms involving the product $\sigma_{j_1}\sigma_{j_2}$ are consistently included (i.e., none are missing) in the expression provided. Therefore, the standard deviation will also be correct to first order in the standard deviations of the parameters, i.e.,

$$SD\left(R_k^{(1)}\right) = \left\{\sum_{j_1=1}^{TP}\sum_{j_2=1}^{TP} \frac{\partial R_k(\boldsymbol{\alpha}^0)}{\partial \alpha_{j_1}} \frac{\partial R_k(\boldsymbol{\alpha}^0)}{\partial \alpha_{j_2}} \rho_{j_1 j_2}\sigma_{j_1}\sigma_{j_2}\right\}^{\frac{1}{2}}. \tag{1.101}$$

2. The second-order approximation of the covariance $cov\left(R_k^{(2)}, R_\ell^{(2)}\right)$:

(a) Using (1.96) yields the following expression:

$$\begin{aligned}\left\{cov\left(R_k^{(2)}, R_\ell^{(2)}\right)\right\}_{inc} &= cov\left(R_k^{(1)}, R_\ell^{(1)}\right) \\ &+\frac{1}{2}\sum_{j_1=1}^{TP}\sum_{j_2=1}^{TP}\sum_{j_3=1}^{TP}\left(\frac{\partial^2 R_k(\boldsymbol{\alpha}^0)}{\partial \alpha_{j_1}\partial \alpha_{j_2}}\frac{\partial R_\ell(\boldsymbol{\alpha}^0)}{\partial \alpha_{j_3}} + \frac{\partial R_k(\boldsymbol{\alpha}^0)}{\partial \alpha_{j_1}}\frac{\partial^2 R_\ell(\boldsymbol{\alpha}^0)}{\partial \alpha_{j_2}\partial \alpha_{j_3}}\right) t_{j_1 j_2 j_3}\sigma_{j_1}\sigma_{j_2}\sigma_{j_3} \\ &+\frac{1}{4}\sum_{j_1=1}^{TP}\sum_{j_2=1}^{TP}\sum_{j_3=1}^{TP}\sum_{j_4=1}^{TP}\frac{\partial^2 R_k(\boldsymbol{\alpha}^0)}{\partial \alpha_{j_1}\partial \alpha_{j_2}}\frac{\partial^2 R_\ell(\boldsymbol{\alpha}^0)}{\partial \alpha_{j_3}\partial \alpha_{j_4}}\left(q_{j_1 j_2 j_3 j_4} - \rho_{j_1 j_2}\rho_{j_3 j_4}\right)\sigma_{j_1}\sigma_{j_2}\sigma_{j_3}\sigma_{j_4}.\end{aligned} \tag{1.102}$$

(b) Using (1.97) yields the following expression:

$$\begin{aligned}\left\{cov\left(R_k^{(2)}, R_\ell^{(2)}\right)\right\}_{con} &= \left\{cov\left(R_k^{(2)}, R_\ell^{(2)}\right)\right\}_{inc} \\ &+\frac{1}{6}\sum_{j_1=1}^{TP}\sum_{j_2=1}^{TP}\sum_{j_3=1}^{TP}\sum_{j_4=1}^{TP}\left(\frac{\partial^3 R_k(\boldsymbol{\alpha}^0)}{\partial \alpha_{j_1}\partial \alpha_{j_2}\partial \alpha_{j_3}}\frac{\partial R_\ell(\boldsymbol{\alpha}^0)}{\partial \alpha_{j_4}} + \frac{\partial R_k(\boldsymbol{\alpha}^0)}{\partial \alpha_{j_1}}\frac{\partial^3 R_\ell(\boldsymbol{\alpha}^0)}{\partial \alpha_{j_2}\partial \alpha_{j_3}\partial \alpha_{j_4}}\right) \\ &\times q_{j_1 j_2 j_3 j_4}\sigma_{j_1}\sigma_{j_2}\sigma_{j_3}\sigma_{j_4}.\end{aligned} \tag{1.103}$$

Notably, the expression of $\left\{cov\left(R_k^{(2)}, R_\ell^{(2)}\right)\right\}_{inc}$ provided in (1.102) is consistent in the second order of sensitivities but is inconsistent in the fourth order of standard deviations of parameters, i.e., $O(\sigma_{j_1}\sigma_{j_2}\sigma_{j_3}\sigma_{j_4})$. On the other hand, the expression $\left\{cov\left(R_k^{(2)}, R_\ell^{(2)}\right)\right\}_{con}$ provided in (1.103) is consistent in the fourth order (i.e., $\sigma_{j_1}\sigma_{j_2}\sigma_{j_3}\sigma_{j_4}$) of parameter standard deviations. Therefore, if the third-order sensitivities are available, then the expression provided in (1.103) should be used, since it is correct up to, and including, the fourth-order terms containing the products $\sigma_{j_1}\sigma_{j_2}\sigma_{j_3}\sigma_{j_4}$. Consequently, the consistent second-order standard deviation $\left\{SD\left(R_k^{(2)}\right)\right\}_{con} \triangleq \sqrt{\left\{cov\left(R_k^{(2)}, R_\ell^{(2)}\right)\right\}_{con}}$ will *not* have any second-order errors in the parameter standard deviations. In contradistinction, the second-order *inconsistent standard deviation* $\left\{SD\left(R_k^{(2)}\right)\right\}_{inc} \triangleq \sqrt{\left\{cov\left(R_k^{(2)}, R_\ell^{(2)}\right)\right\}_{inc}}$ *will have second-order errors in the parameter standard deviations.*

3. The third-order approximation of the covariance $cov\left(R_k^{(3)}, R_\ell^{(3)}\right)$:

    (a) Using (1.96) yields the following expression:

$$\left\{cov\left(R_k^{(3)}, R_\ell^{(3)}\right)\right\}_{inc} = \frac{1}{12}\sum_{j_1=1}^{TP}\sum_{j_2=1}^{TP}\sum_{j_3=1}^{TP}\sum_{j_4=1}^{TP}\sum_{j_5=1}^{TP}\left[\frac{\partial^3 R_k(\boldsymbol{\alpha}^0)}{\partial\alpha_{j_1}\partial\alpha_{j_2}\partial\alpha_{j_3}}\frac{\partial^2 R_\ell(\boldsymbol{\alpha}^0)}{\partial\alpha_{j_4}\partial\alpha_{j_5}}\right.$$
$$\times\left(p_{j_1j_2j_3j_4j_5} - t_{j_1j_2j_3}\rho_{j_4j_5}\right) + \frac{\partial^2 R_k(\boldsymbol{\alpha}^0)}{\partial\alpha_{j_1}\partial\alpha_{j_2}}\frac{\partial^3 R_\ell(\boldsymbol{\alpha}^0)}{\partial\alpha_{j_3}\partial\alpha_{j_4}\partial\alpha_{j_5}}$$
$$\left.\times\left(p_{j_1j_2j_3j_4j_5} - \rho_{j_1j_2}t_{j_3j_4j_5}\right)\right]\sigma_{j_1}\sigma_{j_2}\sigma_{j_3}\sigma_{j_4}\sigma_{j_5} + \left\{cov\left(R_k^{(2)}, R_\ell^{(2)}\right)\right\}_{con}. \tag{1.104}$$

    (b) Using (1.97) yields the following expression:

$$\left\{cov\left(R_k^{(3)}, R_\ell^{(3)}\right)\right\}_{con} = \left\{cov\left(R_k^{(3)}, R_\ell^{(3)}\right)\right\}_{inc}$$
$$+\frac{1}{24}\sum_{j_1=1}^{TP}\sum_{j_2=1}^{TP}\sum_{j_3=1}^{TP}\sum_{j_4=1}^{TP}\sum_{j_5=1}^{TP}\left[\frac{\partial^4 R_k(\boldsymbol{\alpha}^0)}{\partial\alpha_{j_1}\partial\alpha_{j_2}\partial\alpha_{j_3}\partial\alpha_{j_4}}\frac{\partial R_\ell(\boldsymbol{\alpha}^0)}{\partial\alpha_{j_5}}\right. \tag{1.105}$$
$$\left.+\frac{\partial R_k(\boldsymbol{\alpha}^0)}{\partial\alpha_{j_1}}\frac{\partial^4 R_\ell(\boldsymbol{\alpha}^0)}{\partial\alpha_{j_2}\partial\alpha_{j_3}\partial\alpha_{j_4}\partial\alpha_{j_5}}\right]p_{j_1j_2j_3j_4j_5}\sigma_{j_1}\sigma_{j_2}\sigma_{j_3}\sigma_{j_4}\sigma_{j_5}.$$

The expression of $\left\{cov\left(R_k^{(3)}, R_\ell^{(3)}\right)\right\}_{inc}$ provided in (1.104) is consistent in the third-order sensitivities but is inconsistent in the highest order of parameter standard deviations, i.e., $O(\sigma_{j_1}\sigma_{j_2}\sigma_{j_3}\sigma_{j_4}\sigma_{j_5})$, while the expression $\left\{cov\left(R_k^{(3)}, R_\ell^{(3)}\right)\right\}_{con}$ provided in (1.105) is consistent in the highest order of parameter standard deviations. Therefore, if the fourth-order sensitivities are available, then the expression provided in (1.105) should be used, since it is correct up to, and including, the fifth-order terms containing the products $\sigma_{j_1}\sigma_{j_2}\sigma_{j_3}\sigma_{j_4}\sigma_{j_5}$. Also noteworthy is the fact that neither the consistent nor the inconsistent standard deviations computed using the third-order variance approximations provided in (1.104) or (1.105), respectively, are correct to *third-order* (i.e., $\sqrt{\sigma_{j_1}\sigma_{j_2}\sigma_{j_3}\sigma_{j_4}\sigma_{j_5}\sigma_{j_6}}$) parameter standard deviations. In order to obtain standard deviations which are correct up to and including the third-order terms containing $O\left(\sqrt{\sigma_{j_1}\sigma_{j_2}\sigma_{j_3}\sigma_{j_4}\sigma_{j_5}\sigma_{j_6}}\right)$, it is necessary to use the fourth-order variance approximations to be provided in (1.106) and (1.107), below.

4. The fourth-order approximation of the covariance $cov\left(R_k^{(4)}, R_\ell^{(4)}\right)$:

(a) Using (1.96) yields the following expression:

$$\begin{aligned}
\left\{cov\left(R_k^{(4)}, R_\ell^{(4)}\right)\right\}_{inc} &= \left\{cov\left(R_k^{(3)}, R_\ell^{(3)}\right)\right\}_{con} \\
&+ \frac{1}{36}\sum_{j_1=1}^{TP}\sum_{j_2=1}^{TP}\sum_{j_3=1}^{TP}\sum_{j_4=1}^{TP}\sum_{j_5=1}^{TP}\sum_{j_6=1}^{TP}\frac{\partial^3 R_k(\boldsymbol{\alpha}^0)}{\partial\alpha_{j_1}\partial\alpha_{j_2}\partial\alpha_{j_3}}\frac{\partial^3 R_l(\boldsymbol{\alpha}^0)}{\partial\alpha_{j_4}\partial\alpha_{j_5}\partial\alpha_{j_6}} \\
&\times \left(s_{j_1j_2j_3j_4j_5j_6} - t_{j_1j_2j_3}t_{j_4j_5j_6}\right)\sigma_{j_1}\sigma_{j_2}\sigma_{j_3}\sigma_{j_4}\sigma_{j_5}\sigma_{j_6} \\
&+ \frac{1}{48}\sum_{j_1=1}^{TP}\sum_{j_2=1}^{TP}\sum_{j_3=1}^{TP}\sum_{j_4=1}^{TP}\sum_{j_5=1}^{TP}\sum_{j_6=1}^{TP}\frac{\partial^4 R_k(\boldsymbol{\alpha}^0)}{\partial\alpha_{j_1}\partial\alpha_{j_2}\partial\alpha_{j_3}\partial\alpha_{j_4}}\frac{\partial^2 R_l(\boldsymbol{\alpha}^0)}{\partial\alpha_{j_5}\partial\alpha_{j_6}} \\
&\times \left(s_{j_1j_2j_3j_4j_5j_6} - q_{j_1j_2j_3j_4}\rho_{j_5j_6}\right)\sigma_{j_1}\sigma_{j_2}\sigma_{j_3}\sigma_{j_4}\sigma_{j_5}\sigma_{j_6} \\
&+ \frac{1}{48}\sum_{j_1=1}^{TP}\sum_{j_2=1}^{TP}\sum_{j_3=1}^{TP}\sum_{j_4=1}^{TP}\sum_{j_5=1}^{TP}\sum_{j_6=1}^{TP}\frac{\partial^2 R_k(\boldsymbol{\alpha}^0)}{\partial\alpha_{j_1}\partial\alpha_{j_2}}\frac{\partial^4 R_l(\boldsymbol{\alpha}^0)}{\partial\alpha_{j_3}\partial\alpha_{j_4}\partial\alpha_{j_5}\partial\alpha_{j_6}} \\
&\times \left(s_{j_1j_2j_3j_4j_5j_6} - \rho_{j_1j_2}q_{j_3j_4j_5j_6}\right)\sigma_{j_1}\sigma_{j_2}\sigma_{j_3}\sigma_{j_4}\sigma_{j_5}\sigma_{j_6}.
\end{aligned} \tag{1.106}$$

(b) Using (1.97) yields the following expression:

$$\begin{aligned}
\left\{cov\left(R_k^{(4)}, R_\ell^{(4)}\right)\right\}_{con} &= \left\{cov\left(R_k^{(4)}, R_\ell^{(4)}\right)\right\}_{inc} + \\
&+ \frac{1}{120}\sum_{j_1=1}^{TP}\sum_{j_2=1}^{TP}\sum_{j_3=1}^{TP}\sum_{j_4=1}^{TP}\sum_{j_5=1}^{TP}\sum_{j_6=1}^{TP}\left[\frac{\partial^5 R_k(\boldsymbol{\alpha}^0)}{\partial\alpha_{j_1}\partial\alpha_{j_2}\partial\alpha_{j_3}\partial\alpha_{j_4}\partial\alpha_{j_5}}\frac{\partial R_l(\boldsymbol{\alpha}^0)}{\partial\alpha_{j_6}}\right. \\
&\left. + \frac{\partial R_k(\boldsymbol{\alpha}^0)}{\partial\alpha_{j_1}}\frac{\partial^5 R_l(\boldsymbol{\alpha}^0)}{\partial\alpha_{j_2}\partial\alpha_{j_3}\partial\alpha_{j_4}\partial\alpha_{j_5}\partial\alpha_{j_6}}\right] s_{j_1j_2j_3j_4j_5j_6}\sigma_{j_1}\sigma_{j_2}\sigma_{j_3}\sigma_{j_4}\sigma_{j_5}\sigma_{j_6}.
\end{aligned} \tag{1.107}$$

The expression of $\left\{cov\left(R_k^{(4)}, R_\ell^{(4)}\right)\right\}_{inc}$ provided in (1.106) is consistent in the fourth order of sensitivities but is inconsistent in the highest order (sixth order) of parameter standard deviations, while the expression $\left\{cov\left(R_k^{(4)}, R_\ell^{(4)}\right)\right\}_{con}$ provided in (1.107) is consistent in the sixth order of parameter standard deviations $\sigma_{j_1}\sigma_{j_2}\sigma_{j_3}\sigma_{j_4}\sigma_{j_5}\sigma_{j_6}$. Therefore, if the fifth-order sensitivities are available, then the expression provided in (1.107) should be used, since it is correct up to, and including, the fifth-order terms containing the products $\sigma_{j_1}\sigma_{j_2}\sigma_{j_3}\sigma_{j_4}\sigma_{j_5}\sigma_{j_6}$. Notably, the standard deviation $\left\{SD\left(R_k^{(4)}\right)\right\}_{con} \triangleq \sqrt{\left\{cov\left(R_k^{(4)}, R_\ell^{(4)}\right)\right\}_{con}}$ would comprise all of the third-order standard deviations of the form $\sqrt{\sigma_{j_1}\sigma_{j_2}\sigma_{j_3}\sigma_{j_4}\sigma_{j_5}\sigma_{j_6}}$.

1.4.4 Triple Correlations Among Responses and Parameters

1. The ***triple-correlations*** (or third-order moment of the distribution of responses), denoted as $\mu_3(R_k, R_\ell, R_m)$, ***among three responses***, $R_k(\boldsymbol{\alpha})$, $R_\ell(\boldsymbol{\alpha})$, and $R_m(\boldsymbol{\alpha})$, are defined as follows:

$$\mu_3(R_k, R_\ell, R_m) \triangleq \langle [R_k(\boldsymbol{\alpha}) - E(R_k)][R_\ell(\boldsymbol{\alpha}) - E(R_\ell)][R_m(\boldsymbol{\alpha}) - E(R_m)] \rangle_\alpha. \tag{1.108}$$

(a) If only *first-order response sensitivities are available*, the following first-order approximate expression, denoted as $\mu_3^{(1)}(R_k, R_\ell, R_m)$, can be obtained:

$$\mu_3^{(1)}(R_k, R_\ell, R_m) = \sum_{j_1=1}^{TP} \sum_{j_2=1}^{TP} \sum_{j_3=1}^{TP} \times \frac{\partial R_k(\boldsymbol{\alpha}^0)}{\partial \alpha_{j_1}} \frac{\partial R_\ell(\boldsymbol{\alpha}^0)}{\partial \alpha_{j_2}} \frac{\partial R_m(\boldsymbol{\alpha}^0)}{\partial \alpha_{j_3}} t_{j_1 j_2 j_3} \sigma_{j_1} \sigma_{j_2} \sigma_{j_3}. \tag{1.109}$$

(b) If *first- and second-order response sensitivities are available*, the following second-order approximate expression, denoted as $\mu_3^{(2)}(R_k, R_\ell, R_m)$, can be obtained:

$$\begin{aligned} \mu_3^{(2)}(R_k, R_\ell, R_m) &= \mu_3^{(1)}(R_k, R_\ell, R_m) + \mu_3^{(2)}(\sigma^5) + \mu_3^{(2)}(\sigma^6) \\ &+ \frac{1}{2} \sum_{j_1=1}^{TP} \sum_{j_2=1}^{TP} \sum_{j_3=1}^{TP} \sum_{j_4=1}^{TP} \left[\frac{\partial R_k(\boldsymbol{\alpha}^0)}{\partial \alpha_{j_1}} \frac{\partial R_\ell(\boldsymbol{\alpha}^0)}{\partial \alpha_{j_2}} \frac{\partial^2 R_m(\boldsymbol{\alpha}^0)}{\partial \alpha_{j_3} \partial \alpha_{j_4}} \right. \\ &\times \left(q_{j_1 j_2 j_3 j_4} - \rho_{j_1 j_2} \rho_{j_3 j_4} \right) + \frac{\partial R_k(\boldsymbol{\alpha}^0)}{\partial \alpha_{j_1}} \frac{\partial^2 R_\ell(\boldsymbol{\alpha}^0)}{\partial \alpha_{j_2} \partial \alpha_{j_3}} \frac{\partial R_m(\boldsymbol{\alpha}^0)}{\partial \alpha_{j_4}} \\ &\times \left. \left(q_{j_1 j_2 j_3 j_4} - \rho_{j_1 j_4} \rho_{j_2 j_3} \right) \right] \sigma_{j_1} \sigma_{j_2} \sigma_{j_3} \sigma_{j_4}, \end{aligned} \tag{1.110}$$

where the quantity $\mu_3^{(2)}(\sigma^5)$ contains products of the form $\sigma_{j_1}\sigma_{j_2}\sigma_{j_3}\sigma_{j_4}\sigma_{j_5}$ and is defined as follows:

$$\begin{aligned} \mu_3^{(2)}(\sigma^5) &\triangleq \frac{1}{4} \sum_{j_1=1}^{TP} \sum_{j_2=1}^{TP} \sum_{j_3=1}^{TP} \sum_{j_4=1}^{TP} \sum_{j_5=1}^{TP} \left[\frac{\partial R_k(\boldsymbol{\alpha}^0)}{\partial \alpha_{j_1}} \frac{\partial^2 R_\ell(\boldsymbol{\alpha}^0)}{\partial \alpha_{j_2} \partial \alpha_{j_3}} \frac{\partial^2 R_m(\boldsymbol{\alpha}^0)}{\partial \alpha_{j_4} \partial \alpha_{j_5}} \right. \\ &\times \left(p_{j_1 j_2 j_3 j_4 j_5} - t_{j_1 j_2 j_3} \rho_{j_4 j_5} - t_{j_1 j_4 j_5} \rho_{j_2 j_3} \right) + \frac{\partial^2 R_k(\boldsymbol{\alpha}^0)}{\partial \alpha_{j_1} \partial \alpha_{j_2}} \frac{\partial R_\ell(\boldsymbol{\alpha}^0)}{\partial \alpha_{j_3}} \frac{\partial^2 R_m(\boldsymbol{\alpha}^0)}{\partial \alpha_{j_4} \partial \alpha_{j_5}} \\ &\times \left(p_{j_1 j_2 j_3 j_4 j_5} - \rho_{j_1 j_2} t_{j_3 j_4 j_5} - t_{j_1 j_2 j_3} \rho_{j_4 j_5} \right) + \frac{\partial^2 R_k(\boldsymbol{\alpha}^0)}{\partial \alpha_{j_1} \partial \alpha_{j_2}} \frac{\partial^2 R_\ell(\boldsymbol{\alpha}^0)}{\partial \alpha_{j_3} \partial \alpha_{j_4}} \frac{\partial R_m(\boldsymbol{\alpha}^0)}{\partial \alpha_{j_5}} \\ &\times \left. \left(p_{j_1 j_2 j_3 j_4 j_5} - \rho_{j_1 j_2} t_{j_3 j_4 j_5} - t_{j_1 j_2 j_5} \rho_{j_3 j_4} \right) \right] \sigma_{j_1} \sigma_{j_2} \sigma_{j_3} \sigma_{j_4} \sigma_{j_5}, \end{aligned} \tag{1.111}$$

while the quantity $\mu_3^{(2)}\left(\sigma^6\right)$ contains products of the form $\sigma_{j_1}\sigma_{j_2}\sigma_{j_3}\sigma_{j_4}\sigma_{j_5}\sigma_{j_6}$ and is defined as follows:

$$\begin{aligned}
\mu_3^{(2)}\left(\sigma^6\right) \triangleq \frac{1}{8} \sum_{j_1=1}^{TP} \sum_{j_2=1}^{TP} \sum_{j_3=1}^{TP} \sum_{j_4=1}^{TP} \sum_{j_5=1}^{TP} \sum_{j_6=1}^{TP} & \frac{\partial^2 R_k(\boldsymbol{\alpha}^0)}{\partial \alpha_{j_1} \partial \alpha_{j_2}} \frac{\partial^2 R_\ell(\boldsymbol{\alpha}^0)}{\partial \alpha_{j_3} \partial \alpha_{j_4}} \frac{\partial^2 R_m(\boldsymbol{\alpha}^0)}{\partial \alpha_{j_5} \partial \alpha_{j_6}} \\
\times \left(s_{j_1 j_2 j_3 j_4 j_5 j_6} + 2\rho_{j_1 j_2}\rho_{j_3 j_4}\rho_{j_5 j_6} - \rho_{j_1 j_2} q_{j_3 j_4 j_5 j_6} \right. & \\
\left. - q_{j_1 j_2 j_5 j_6}\rho_{j_3 j_4} - q_{j_1 j_2 j_3 j_4}\rho_{j_5 j_6} \right) & \sigma_{j_1}\sigma_{j_2}\sigma_{j_3}\sigma_{j_4}\sigma_{j_5}\sigma_{j_6}.
\end{aligned} \tag{1.112}$$

As can be observed from (1.110), the expression of $\mu_3^{(2)}(R_k, R_\ell, R_m)$ consistently contains all of the first- and second-order response sensitivities and all of the fifth-order terms in standard deviations (of the form $\sigma_{j_1}\sigma_{j_2}\sigma_{j_3}\sigma_{j_4}\sigma_{j_5}$) but does not contain all of the sixth-order terms in standard deviations (of the form $\sigma_{j_1}\sigma_{j_2}\sigma_{j_3}\sigma_{j_4}\sigma_{j_5}\sigma_{j_6}$). The sixth-order terms in standard deviations, containing the remaining terms of the form $\sigma_{j_1}\sigma_{j_2}\sigma_{j_3}\sigma_{j_4}\sigma_{j_5}\sigma_{j_6}$, are provided in (1.114), below.

2. The availability of the *third-order response sensitivities* enables the computation of additional terms that comprise sixth-order terms containing the products $\sigma_{j_1}\sigma_{j_2}\sigma_{j_3}\sigma_{j_4}\sigma_{j_5}\sigma_{j_6}$. The following third-order (in sensitivities) approximate expression, denoted as $\mu_3^{(3)}(R_k, R_\ell, R_m)$, can be obtained:

$$\begin{aligned}
\mu_3^{(3)}(R_k, R_l, R_m) \triangleq & \mu_3^{(2)}(R_k, R_\ell, R_m) \\
& + \frac{1}{12} \sum_{j_1=1}^{TP} \sum_{j_2=1}^{TP} \sum_{j_3=1}^{TP} \sum_{j_4=1}^{TP} \sum_{j_5=1}^{TP} \sum_{j_6=1}^{TP} \left[\frac{\partial R_k(\boldsymbol{\alpha}^0)}{\partial \alpha_{j_1}} \frac{\partial^2 R_l(\boldsymbol{\alpha}^0)}{\partial \alpha_{j_2} \partial \alpha_{j_3}} \frac{\partial^3 R_m(\boldsymbol{\alpha}^0)}{\partial \alpha_{j_4} \partial \alpha_{j_5} \partial \alpha_{j_6}} \right. \\
& \times \left(s_{j_1 j_2 j_3 j_4 j_5 j_6} - t_{j_1 j_2 j_3} t_{j_4 j_5 j_6} - q_{j_1 j_4 j_5 j_6}\rho_{j_2 j_3} \right) + \frac{\partial R_k(\boldsymbol{\alpha}^0)}{\partial \alpha_{j_1}} \frac{\partial^3 R_l(\boldsymbol{\alpha}^0)}{\partial \alpha_{j_2} \partial \alpha_{j_3} \partial \alpha_{j_4}} \frac{\partial^2 R_m(\boldsymbol{\alpha}^0)}{\partial \alpha_{j_5} \partial \alpha_{j_6}} \\
& \times \left(s_{j_1 j_2 j_3 j_4 j_5 j_6} - t_{j_1 j_5 j_6} t_{j_2 j_3 j_4} - q_{j_1 j_2 j_3 j_4}\rho_{j_5 j_6} \right) + \frac{\partial^2 R_k(\boldsymbol{\alpha}^0)}{\partial \alpha_{j_1} \partial \alpha_{j_2}} \frac{\partial R_l(\boldsymbol{\alpha}^0)}{\partial \alpha_{j_3}} \frac{\partial^3 R_m(\boldsymbol{\alpha}^0)}{\partial \alpha_{j_4} \partial \alpha_{j_5} \partial \alpha_{j_6}} \\
& \times \left(s_{j_1 j_2 j_3 j_4 j_5 j_6} - t_{j_1 j_2 j_3} t_{j_4 j_5 j_6} - \rho_{j_1 j_2} q_{j_3 j_4 j_5 j_6} \right) + \frac{\partial^2 R_k(\boldsymbol{\alpha}^0)}{\partial \alpha_{j_1} \partial \alpha_{j_2}} \frac{\partial^3 R_l(\boldsymbol{\alpha}^0)}{\partial \alpha_{j_3} \partial \alpha_{j_4} \partial \alpha_{j_5}} \frac{\partial R_m(\boldsymbol{\alpha}^0)}{\partial \alpha_{j_6}} \\
& \times \left(s_{j_1 j_2 j_3 j_4 j_5 j_6} - t_{j_1 j_2 j_6} t_{j_3 j_4 j_5} - \rho_{j_1 j_2} q_{j_3 j_4 j_5 j_6} \right) + \frac{\partial^3 R_l(\boldsymbol{\alpha}^0)}{\partial \alpha_{j_1} \partial \alpha_{j_2} \partial \alpha_{j_3}} \frac{\partial R_l(\boldsymbol{\alpha}^0)}{\partial \alpha_{j_4}} \frac{\partial^2 R_m(\boldsymbol{\alpha}^0)}{\partial \alpha_{j_5} \partial \alpha_{j_6}} \\
& \times \left(s_{j_1 j_2 j_3 j_4 j_5 j_6} - t_{j_1 j_2 j_3} t_{j_4 j_5 j_6} - q_{j_1 j_2 j_3 j_4}\rho_{j_5 j_6} \right) + \frac{\partial^3 R_l(\boldsymbol{\alpha}^0)}{\partial \alpha_{j_1} \partial \alpha_{j_2} \partial \alpha_{j_3}} \frac{\partial^2 R_l(\boldsymbol{\alpha}^0)}{\partial \alpha_{j_4} \partial \alpha_{j_5}} \frac{\partial R_m(\boldsymbol{\alpha}^0)}{\partial \alpha_{j_6}} \\
& \left. \times \left(s_{j_1 j_2 j_3 j_4 j_5 j_6} - t_{j_1 j_2 j_3} t_{j_4 j_5 j_6} - q_{j_1 j_2 j_3 j_6}\rho_{j_4 j_5} \right) \right] \sigma_{j_1}\sigma_{j_2}\sigma_{j_3}\sigma_{j_4}\sigma_{j_5}\sigma_{j_6}.
\end{aligned} \tag{1.113}$$

3. If *fourth-order response sensitivities* are also available, the following fourth-order (in sensitivities) approximate expression $\mu_3^{(4)}(R_k, R_l, R_m)$ can be obtained:

$$\mu_3^{(4)}(R_k, R_l, R_m) \triangleq \mu_3^{(3)}(R_k, R_l, R_m) + \frac{1}{24}\sum_{j_1=1}^{TP}\sum_{j_2=1}^{TP}\sum_{j_3=1}^{TP}\sum_{j_4=1}^{TP}\sum_{j_5=1}^{TP}\sum_{j_6=1}^{TP}\left[\frac{\partial R_k(\boldsymbol{\alpha}^0)}{\partial \alpha_{j_1}}\frac{\partial R_l(\boldsymbol{\alpha}^0)}{\partial \alpha_{j_2}}\frac{\partial^4 R_m(\boldsymbol{\alpha}^0)}{\partial \alpha_{j_3}\partial \alpha_{j_4}\partial \alpha_{j_5}\partial \alpha_{j_6}}\right.$$
$$\times\left(s_{j_1j_2j_3j_4j_5j_6} - \rho_{j_1j_2}q_{j_3j_4j_5j_6}\right) + \frac{\partial R_k(\boldsymbol{\alpha}^0)}{\partial \alpha_{j_1}}\frac{\partial^4 R_l(\boldsymbol{\alpha}^0)}{\partial \alpha_{j_2}\partial \alpha_{j_3}\partial \alpha_{j_4}\partial \alpha_{j_5}}\frac{\partial R_m(\boldsymbol{\alpha}^0)}{\partial \alpha_{j_6}} \quad (1.114)$$
$$\times\left(s_{j_1j_2j_3j_4j_5j_6} - \rho_{j_1j_6}q_{j_2j_3j_4j_5}\right) + \frac{\partial^4 R_k(\boldsymbol{\alpha}^0)}{\partial \alpha_{j_1}\partial \alpha_{j_2}\partial \alpha_{j_3}\partial \alpha_{j_4}}\frac{\partial R_l(\boldsymbol{\alpha}^0)}{\partial \alpha_{j_5}}\frac{\partial R_m(\boldsymbol{\alpha}^0)}{\partial \alpha_{j_6}}$$
$$\left.\times\left(s_{j_1j_2j_3j_4j_5j_6} - q_{j_1j_2j_3j_4}\rho_{j_5j_6}\right)\right]\sigma_{j_1}\sigma_{j_2}\sigma_{j_3}\sigma_{j_4}\sigma_{j_5}\sigma_{j_6}.$$

As can be observed from (1.114), the quantity $\mu_3^{(4)}(R_k, R_l, R_m)$ contains all of the terms involving sixth-order products $\sigma_{j_1}\sigma_{j_2}\sigma_{j_3}\sigma_{j_4}\sigma_{j_5}\sigma_{j_6}$ of standard deviations.

4. The ***triple-correlations among one parameter***, α_i, ***and two responses***, $R_k(\boldsymbol{\alpha})$ and $R_\ell(\boldsymbol{\alpha})$, are defined as follows:

$$\mu_3(\alpha_i, R_k, R_\ell) \triangleq \left\langle \left(\alpha_i - \alpha_i^0\right)[R_k(\boldsymbol{\alpha}) - E(R_k)][R_\ell(\boldsymbol{\alpha}) - E(R_\ell)]\right\rangle_\alpha. \quad (1.115)$$

Using the Taylor series in (1.72) up to and including the sixth order in standard deviations, the expression of $\mu_3(\alpha_i, R_k, R_\ell)$ is as follows:

$$\mu_3(\alpha_i, R_k, R_\ell) = \sum_{j_1=1}^{TP}\sum_{j_2=1}^{TP}\frac{\partial R_j(\boldsymbol{\alpha}^0)}{\partial \alpha_{j_1}}\frac{\partial R_k(\boldsymbol{\alpha}^0)}{\partial \alpha_{j_2}}t_{j_1j_2i}\sigma_{j_1}\sigma_{j_2}\sigma_i$$
$$+\frac{1}{2}\sum_{j_1=1}^{TP}\sum_{j_2=1}^{TP}\sum_{j_3=1}^{TP}\left[\frac{\partial R_j(\boldsymbol{\alpha}^0)}{\partial \alpha_{j_1}}\frac{\partial^2 R_k(\boldsymbol{\alpha}^0)}{\partial \alpha_{j_2}\partial \alpha_{j_3}}\left(q_{j_1j_2j_3i} - \rho_{j_1i}\rho_{j_2j_3}\right)\right. \quad (1.116)$$
$$\left.+\frac{\partial^2 R_j(\boldsymbol{\alpha}^0)}{\partial \alpha_{j_1}\partial \alpha_{j_2}}\frac{\partial R_k(\boldsymbol{\alpha}^0)}{\partial \alpha_{j_3}}\left(q_{j_1j_2j_3i} - \rho_{j_1j_2}\rho_{j_3i}\right)\right]\sigma_{j_1}\sigma_{j_2}\sigma_{j_3}\sigma_i$$
$$+\mu_3^{(5)}(\alpha_i, R_k, R_\ell) + \mu_3^{(6)}(\alpha_i, R_k, R_\ell),$$

where the quantity $\mu_3^{(5)}(\alpha_i, R_k, R_\ell)$ comprises terms in fifth-order standard deviations involving products of the form $\sigma_{j_1}\sigma_{j_2}\sigma_{j_3}\sigma_{j_4}\sigma_i$, and is defined as follows:

$$
\begin{aligned}
\mu_3^{(5)}(\alpha_i, R_k, R_\ell) &\triangleq \frac{1}{4}\sum_{j_1=1}^{TP}\sum_{j_2=1}^{TP}\sum_{j_3=1}^{TP}\sum_{j_4=1}^{TP}\frac{\partial^2 R_j(\boldsymbol{\alpha}^0)}{\partial\alpha_{j_1}\partial\alpha_{j_2}}\frac{\partial^2 R_k(\boldsymbol{\alpha}^0)}{\partial\alpha_{j_3}\partial\alpha_{j_4}} \\
&\times\left(p_{j_1j_2j_3j_4i} - t_{j_1j_2i}\rho_{j_3j_4} - \rho_{j_1j_2}t_{j_3j_4i}\right)\sigma_{j_1}\sigma_{j_2}\sigma_{j_3}\sigma_{j_4}\sigma_i \\
&+\frac{1}{6}\sum_{j_1=1}^{TP}\sum_{j_2=1}^{TP}\sum_{j_3=1}^{TP}\sum_{j_4=1}^{TP}\left[\frac{\partial R_j(\boldsymbol{\alpha}^0)}{\partial\alpha_{j_1}}\frac{\partial^3 R_k(\boldsymbol{\alpha}^0)}{\partial\alpha_{j_2}\partial\alpha_{j_3}\partial\alpha_{j_4}}\right. \\
&\times\left(p_{j_1j_2j_3j_4i} - \rho_{j_1i}t_{j_2j_3j_4}\right) + \frac{\partial^3 R_j(\boldsymbol{\alpha}^0)}{\partial\alpha_{j_1}\partial\alpha_{j_2}\partial\alpha_{j_3}}\frac{\partial R_k(\boldsymbol{\alpha}^0)}{\partial\alpha_{j_4}} \\
&\left.\times\left(p_{j_1j_2j_3j_4i} - t_{j_1j_2j_3}\rho_{j_4i}\right)\right]\sigma_{j_1}\sigma_{j_2}\sigma_{j_3}\sigma_{j_4}\sigma_i,
\end{aligned}
\tag{1.117}
$$

while the quantity $\mu_3^{(6)}(\alpha_i, R_k, R_\ell)$ comprises terms in sixth-order standard deviations involving products of the form $\sigma_{j_1}\sigma_{j_2}\sigma_{j_3}\sigma_{j_4}\sigma_{j_5}\sigma_i$, and is defined as follows:

$$
\begin{aligned}
\mu_3^{(6)}(\alpha_i, R_k, R_\ell) &\triangleq \frac{1}{12}\sum_{j_1=1}^{TP}\sum_{j_2=1}^{TP}\sum_{j_3=1}^{TP}\sum_{j_4=1}^{TP}\sum_{j_5=1}^{TP}\left[\frac{\partial^2 R_j(\boldsymbol{\alpha}^0)}{\partial\alpha_{j_1}\partial\alpha_{j_2}}\frac{\partial^3 R_k(\boldsymbol{\alpha}^0)}{\partial\alpha_{j_3}\partial\alpha_{j_4}\partial\alpha_{j_5}}\right. \\
&\times\left(s_{j_1j_2j_3j_4j_5i} - t_{j_1j_2i}t_{j_3j_4j_5} - \rho_{j_1j_2}q_{j_3j_4j_5i}\right) + \frac{\partial^3 R_j(\boldsymbol{\alpha}^0)}{\partial\alpha_{j_1}\partial\alpha_{j_2}\partial\alpha_{j_3}}\frac{\partial^2 R_k(\boldsymbol{\alpha}^0)}{\partial\alpha_{j_4}\partial\alpha_{j_5}} \\
&\left.\times\left(s_{j_1j_2j_3j_4j_5i} - t_{j_1j_2j_3}t_{j_4j_5i} - q_{j_1j_2j_3i}\rho_{j_4j_5}\right)\right]\sigma_{j_1}\sigma_{j_2}\sigma_{j_3}\sigma_{j_4}\sigma_{j_5}\sigma_i \\
&+\frac{1}{24}\sum_{j_1=1}^{TP}\sum_{j_2=1}^{TP}\sum_{j_3=1}^{TP}\sum_{j_4=1}^{TP}\sum_{j_5=1}^{TP}\left[\frac{\partial R_j(\boldsymbol{\alpha}^0)}{\partial\alpha_{j_1}}\frac{\partial^4 R_k(\boldsymbol{\alpha}^0)}{\partial\alpha_{j_2}\partial\alpha_{j_3}\partial\alpha_{j_4}\partial\alpha_{j_5}}\right. \\
&\times\left(s_{j_1j_2j_3j_4j_5i} - \rho_{j_1i}q_{j_2j_3j_4j_5}\right) + \frac{\partial^4 R_j(\boldsymbol{\alpha}^0)}{\partial\alpha_{j_1}\partial\alpha_{j_2}\partial\alpha_{j_3}\partial\alpha_{j_4}}\frac{\partial R_k(\boldsymbol{\alpha}^0)}{\partial\alpha_{j_5}} \\
&\left.\times\left(s_{j_1j_2j_3j_4j_5i} - q_{j_1j_2j_3j_4}\rho_{j_5i}\right)\right]\sigma_{j_1}\sigma_{j_2}\sigma_{j_3}\sigma_{j_4}\sigma_{j_5}\sigma_i.
\end{aligned}
\tag{1.118}
$$

5. The ***triple-correlations among two parameters*** α_i and α_j ***and one response*** $R_k(\boldsymbol{\alpha})$ are defined as follows:

$$
\begin{aligned}
&\mu_3(\alpha_i, \alpha_j, R_k) \\
&\triangleq\left\langle\left(\alpha_i - \alpha_i^0\right)\left(\alpha_j - \alpha_j^0\right)\left[R_k(\boldsymbol{\alpha}) - E(R_k)\right]\right\rangle_\alpha.
\end{aligned}
\tag{1.119}
$$

Using the Taylor series in (1.72) up to and including the sixth order in standard deviations, the expression of $\mu_3(\alpha_i, \alpha_j, R_k)$ is as follows:

$$
\begin{aligned}
\mu_3\left(\alpha_i, \alpha_j, R_k\right) &= \sum_{j_1=1}^{TP} \frac{\partial R_k(\boldsymbol{\alpha}^0)}{\partial \alpha_{j_1}} t_{j_1 ij} \sigma_{j_1} \sigma_i \sigma_j \\
&+ \frac{1}{2} \sum_{j_1=1}^{TP} \sum_{j_2=1}^{TP} \frac{\partial^2 R_k(\boldsymbol{\alpha}^0)}{\partial \alpha_{j_1} \partial \alpha_{j_2}} \left(q_{j_1 j_2 ij} - \rho_{j_1 j_2} \rho_{ij} \right) \sigma_{j_1} \sigma_{j_2} \sigma_i \sigma_j \\
&+ \frac{1}{6} \sum_{j_1=1}^{TP} \sum_{j_2=1}^{TP} \sum_{j_3=1}^{TP} \frac{\partial^3 R_k(\boldsymbol{\alpha}^0)}{\partial \alpha_{j_1} \partial \alpha_{j_2} \partial \alpha_{j_3}} \left(p_{j_1 j_2 j_3 ij} - t_{j_1 j_2 j_3} \rho_{ij} \right) \sigma_{j_1} \sigma_{j_2} \sigma_{j_3} \sigma_i \sigma_j \\
&+ \frac{1}{24} \sum_{j_1=1}^{TP} \sum_{j_2=1}^{TP} \sum_{j_3=1}^{TP} \sum_{j_4=1}^{TP} \frac{\partial^4 R_k(\boldsymbol{\alpha}^0)}{\partial \alpha_{j_1} \partial \alpha_{j_2} \partial \alpha_{j_3} \partial \alpha_{j_4}} \\
&\times \left(s_{j_1 j_2 j_3 j_4 ij} - q_{j_1 j_2 j_3 j_4} \rho_{ij} \right) \sigma_{j_1} \sigma_{j_2} \sigma_{j_3} \sigma_{j_4} \sigma_i \sigma_j .
\end{aligned}
\tag{1.120}
$$

1.4.5 Quadruple Correlations Among Responses and Parameters

1. ***The quadruple-correlations*** (or fourth-order moment of the distribution of responses), denoted as $\mu_4(R_k, R_\ell, R_m, R_n)$, ***among four responses,*** $R_k(\boldsymbol{\alpha})$, $R_\ell(\boldsymbol{\alpha})$, $R_m(\boldsymbol{\alpha})$, and $R_n(\boldsymbol{\alpha})$, are defined as follows:

$$
\begin{aligned}
\mu_4(R_k, R_\ell, R_m, R_n) &\triangleq \langle [R_k(\boldsymbol{\alpha}) - E(R_k)][R_\ell(\boldsymbol{\alpha}) - E(R_\ell)] \\
&\times [R_m(\boldsymbol{\alpha}) - E(R_m)][R_n(\boldsymbol{\alpha}) - E(R_n)] \rangle_\alpha .
\end{aligned}
\tag{1.121}
$$

   (a) If only *first-order response sensitivities are available*, the following first-order approximate expression, denoted as $\mu_4^{(1)}(R_k, R_l, R_m, R_n)$, can be obtained:

$$
\begin{aligned}
&\mu_4^{(1)}(R_k, R_l, R_m, R_n) \triangleq \left\langle \left[R_k^{(1)}(\boldsymbol{\alpha}) - E^{(1)}(R_k) \right] \left[R_\ell^{(1)}(\boldsymbol{\alpha}) - E^{(1)}(R_\ell) \right] \right. \\
&\left. \times \left[R_m^{(1)}(\boldsymbol{\alpha}) - E^{(1)}(R_m) \right] \left[R_n^{(1)}(\boldsymbol{\alpha}) - E^{(1)}(R_n) \right] \right\rangle_\alpha \\
&= \sum_{j_1=1}^{TP} \sum_{j_2=1}^{TP} \sum_{j_3=1}^{TP} \sum_{j_4=1}^{TP} \frac{\partial R_k(\boldsymbol{\alpha}^0)}{\partial \alpha_{j_1}} \frac{\partial R_l(\boldsymbol{\alpha}^0)}{\partial \alpha_{j_2}} \frac{\partial R_m(\boldsymbol{\alpha}^0)}{\partial \alpha_{j_3}} \frac{\partial R_n(\boldsymbol{\alpha}^0)}{\partial \alpha_{j_4}} q_{j_1 j_2 j_3 j_4} \sigma_{j_1} \sigma_{j_2} \sigma_{j_3} \sigma_{j_4} .
\end{aligned}
\tag{1.122}
$$

   The expression of $\mu_4^{(1)}(R_k, R_l, R_m, R_n)$ is consistent in that it comprises all of the first-order sensitivities and all of the terms involving fourth-order products of standard deviations of the form $\sigma_{j_1} \sigma_{j_2} \sigma_{j_3} \sigma_{j_4}$.

(b) If all *first- and second-order response sensitivities* are available, the following approximation of the quadruple-correlations among four responses $R_k(\boldsymbol{\alpha})$, $R_\ell(\boldsymbol{\alpha})$, $R_m(\boldsymbol{\alpha})$, and $R_n(\boldsymbol{\alpha})$ can be obtained:

$$\begin{aligned}
&\mu_4^{(2)}(R_k,R_\ell,R_m,R_n) \triangleq \mu_4^{(1)}(R_k,R_\ell,R_m,R_n) \\
&+\frac{1}{2}\sum_{j_1=1}^{TP}\sum_{j_2=1}^{TP}\sum_{j_3=1}^{TP}\sum_{j_4=1}^{TP}\sum_{j_5=1}^{TP}\left[\frac{\partial R_k(\boldsymbol{\alpha}^0)}{\partial \alpha_{j_1}}\frac{\partial R_\ell(\boldsymbol{\alpha}^0)}{\partial \alpha_{j_2}}\frac{\partial R_m(\boldsymbol{\alpha}^0)}{\partial \alpha_{j_3}}\frac{\partial^2 R_n(\boldsymbol{\alpha}^0)}{\partial \alpha_{j_4}\partial \alpha_{j_5}}\right. \\
&\times\left(p_{j_1j_2j_3j_4j_5}-t_{j_1j_2j_3}\rho_{j_4j_5}\right)+\frac{\partial R_k(\boldsymbol{\alpha}^0)}{\partial \alpha_{j_1}}\frac{\partial R_\ell(\boldsymbol{\alpha}^0)}{\partial \alpha_{j_2}}\frac{\partial^2 R_m(\boldsymbol{\alpha}^0)}{\partial \alpha_{j_3}\partial \alpha_{j_4}}\frac{\partial R_n(\boldsymbol{\alpha}^0)}{\partial \alpha_{j_5}} \\
&\times\left(p_{j_1j_2j_3j_4j_5}-t_{j_1j_2j_5}\rho_{j_3j_4}\right)+\frac{\partial R_k(\boldsymbol{\alpha}^0)}{\partial \alpha_{j_1}}\frac{\partial^2 R_\ell(\boldsymbol{\alpha}^0)}{\partial \alpha_{j_2}\partial \alpha_{j_3}}\frac{\partial R_m(\boldsymbol{\alpha}^0)}{\partial \alpha_{j_4}}\frac{\partial R_n(\boldsymbol{\alpha}^0)}{\partial \alpha_{j_5}} \\
&\times\left(p_{j_1j_2j_3j_4j_5}-t_{j_1j_4j_5}\rho_{j_2j_3}\right)+\frac{\partial^2 R_k(\boldsymbol{\alpha}^0)}{\partial \alpha_{j_1}\partial \alpha_{j_2}}\frac{\partial R_\ell(\boldsymbol{\alpha}^0)}{\partial \alpha_{j_3}}\frac{\partial R_m(\boldsymbol{\alpha}^0)}{\partial \alpha_{j_4}}\frac{\partial R_n(\boldsymbol{\alpha}^0)}{\partial \alpha_{j_5}} \\
&\left.\times\left(p_{j_1j_2j_3j_4j_5}-\rho_{j_1j_2}t_{j_3j_4j_5}\right)\right]\sigma_{j_1}\sigma_{j_2}\sigma_{j_3}\sigma_{j_4}\sigma_{j_5}+\mu_4^{(2,a)}\left(\sigma^6\right)+\mu_4^{(2,b)}\left(\sigma^6\right),
\end{aligned} \tag{1.123}$$

where the quantities $\mu_4^{(2,a)}\left(\sigma^6\right)$ and $\mu_4^{(2,b)}\left(\sigma^6\right)$ contain the terms involving first-order and second-order sensitivities but sixth-order products of standard deviations of the form $\sigma_{j_1}\sigma_{j_2}\sigma_{j_3}\sigma_{j_4}\sigma_{j_5}\sigma_{j_6}$, and are defined as follows:

$$\begin{aligned}
&\mu_4^{(2,a)}\left(\sigma^6\right) \triangleq \frac{1}{4}\sum_{j_1=1}^{TP}\sum_{j_2=1}^{TP}\sum_{j_3=1}^{TP}\sum_{j_4=1}^{TP}\sum_{j_5=1}^{TP}\sum_{j_6=1}^{TP}\left[\frac{\partial R_k(\boldsymbol{\alpha}^0)}{\partial \alpha_{j_1}}\frac{\partial R_\ell(\boldsymbol{\alpha}^0)}{\partial \alpha_{j_2}}\frac{\partial^2 R_m(\boldsymbol{\alpha}^0)}{\partial \alpha_{j_3}\partial \alpha_{j_4}}\frac{\partial^2 R_n(\boldsymbol{\alpha}^0)}{\partial \alpha_{j_5}\partial \alpha_{j_6}}\right. \\
&\times\left(s_{j_1j_2j_3j_4j_5j_6}+\rho_{j_1j_2}\rho_{j_3j_4}\rho_{j_5j_6}-q_{j_1j_2j_5j_6}\rho_{j_3j_4}-q_{j_1j_2j_3j_4}\rho_{j_5j_6}\right) \\
&+\frac{\partial R_k(\boldsymbol{\alpha}^0)}{\partial \alpha_{j_1}}\frac{\partial^2 R_\ell(\boldsymbol{\alpha}^0)}{\partial \alpha_{j_2}\partial \alpha_{j_3}}\frac{\partial R_m(\boldsymbol{\alpha}^0)}{\partial \alpha_{j_4}}\frac{\partial^2 R_n(\boldsymbol{\alpha}^0)}{\partial \alpha_{j_5}\partial \alpha_{j_6}} \\
&\times\left(s_{j_1j_2j_3j_4j_5j_6}+\rho_{j_1j_4}\rho_{j_2j_3}\rho_{j_5j_6}-q_{j_1j_4j_5j_6}\rho_{j_2j_3}-q_{j_1j_2j_3j_4}\rho_{j_5j_6}\right) \\
&+\frac{\partial R_k(\boldsymbol{\alpha}^0)}{\partial \alpha_{j_1}}\frac{\partial^2 R_\ell(\boldsymbol{\alpha}^0)}{\partial \alpha_{j_2}\partial \alpha_{j_3}}\frac{\partial^2 R_m(\boldsymbol{\alpha}^0)}{\partial \alpha_{j_4}\partial \alpha_{j_5}}\frac{\partial R_n(\boldsymbol{\alpha}^0)}{\partial \alpha_{j_6}} \\
&\left.\times\left(s_{j_1j_2j_3j_4j_5j_6}+\rho_{j_1j_6}\rho_{j_2j_3}\rho_{j_4j_5}-q_{j_1j_4j_5j_6}\rho_{j_2j_3}-q_{j_1j_2j_3j_6}\rho_{j_4j_5}\right)\right] \\
&\times\sigma_{j_1}\sigma_{j_2}\sigma_{j_3}\sigma_{j_4}\sigma_{j_5}\sigma_{j_6},
\end{aligned} \tag{1.124}$$

$$\begin{aligned}
\mu_4^{(2,b)}\left(\sigma^6\right) &\triangleq \frac{1}{4}\sum_{j_1=1}^{TP}\sum_{j_2=1}^{TP}\sum_{j_3=1}^{TP}\sum_{j_4=1}^{TP}\sum_{j_5=1}^{TP}\sum_{j_6=1}^{TP}\left[\frac{\partial^2 R_k(\boldsymbol{\alpha}^0)}{\partial\alpha_{j_1}\partial\alpha_{j_2}}\frac{\partial R_\ell(\boldsymbol{\alpha}^0)}{\partial\alpha_{j_3}}\frac{\partial R_m(\boldsymbol{\alpha}^0)}{\partial\alpha_{j_4}}\frac{\partial^2 R_n(\boldsymbol{\alpha}^0)}{\partial\alpha_{j_5}\partial\alpha_{j_6}}\right.\\
&\times\left(s_{j_1j_2j_3j_4j_5j_6}+\rho_{j_1j_2}\rho_{j_3j_4}\rho_{j_5j_6}-\rho_{j_1j_2}q_{j_3j_4j_5j_6}-q_{j_1j_2j_3j_4}\rho_{j_5j_6}\right)\\
&+\frac{\partial^2 R_k(\boldsymbol{\alpha}^0)}{\partial\alpha_{j_1}\partial\alpha_{j_2}}\frac{\partial R_\ell(\boldsymbol{\alpha}^0)}{\partial\alpha_{j_3}}\frac{\partial^2 R_m(\boldsymbol{\alpha}^0)}{\partial\alpha_{j_4}\partial\alpha_{j_5}}\frac{\partial R_n(\boldsymbol{\alpha}^0)}{\partial\alpha_{j_6}}\\
&\times\left(s_{j_1j_2j_3j_4j_5j_6}+\rho_{j_1j_2}\rho_{j_3j_6}\rho_{j_4j_5}-\rho_{j_1j_2}q_{j_3j_4j_5j_6}-q_{j_1j_2j_3j_6}\rho_{j_4j_5}\right)\\
&+\frac{\partial^2 R_k(\boldsymbol{\alpha}^0)}{\partial\alpha_{j_1}\partial\alpha_{j_2}}\frac{\partial^2 R_l(\boldsymbol{\alpha}^0)}{\partial\alpha_{j_3}\partial\alpha_{j_4}}\frac{\partial R_m(\boldsymbol{\alpha}^0)}{\partial\alpha_{j_5}}\frac{\partial R_n(\boldsymbol{\alpha}^0)}{\partial\alpha_{j_6}}\\
&\left.\times\left(s_{j_1j_2j_3j_4j_5j_6}+\rho_{j_1j_2}\rho_{j_3j_4}\rho_{j_5j_6}-\rho_{j_1j_2}q_{j_3j_4j_5j_6}-q_{j_1j_2j_5j_6}\rho_{j_3j_4}\right)\right]\\
&\times\sigma_{j_1}\sigma_{j_2}\sigma_{j_3}\sigma_{j_4}\sigma_{j_5}\sigma_{j_6}.
\end{aligned}\tag{1.125}$$

(c) If *third-order response sensitivities are also available*, then the remaining terms containing sixth-order products of standard deviations of the form $\sigma_{j_1}\sigma_{j_2}\sigma_{j_3}\sigma_{j_4}\sigma_{j_5}\sigma_{j_6}$ can also be computed, to obtain the following approximation consistent in sixth-order products of standard deviations, denoted as $\mu_4^{(3)}(R_k, R_l, R_m, R_n)$:

$$\begin{aligned}
&\mu_4^{(3)}(R_k, R_\ell, R_m, R_n)\triangleq\mu_4^{(2)}(R_k, R_\ell, R_m, R_n)\\
&+\frac{1}{6}\sum_{j_1=1}^{TP}\sum_{j_2=1}^{TP}\sum_{j_3=1}^{TP}\sum_{j_4=1}^{TP}\sum_{j_5=1}^{TP}\sum_{j_6=1}^{TP}\left[\frac{\partial R_k(\boldsymbol{\alpha}^0)}{\partial\alpha_{j_1}}\frac{\partial R_l(\boldsymbol{\alpha}^0)}{\partial\alpha_{j_2}}\frac{\partial R_m(\boldsymbol{\alpha}^0)}{\partial\alpha_{j_3}}\frac{\partial^3 R_n(\boldsymbol{\alpha}^0)}{\partial\alpha_{j_4}\partial\alpha_{j_5}\partial\alpha_{j_6}}\right.\\
&\times\left(s_{j_1j_2j_3j_4j_5j_6}-t_{j_1j_2j_3}t_{j_4j_5j_6}\right)+\frac{\partial R_k(\boldsymbol{\alpha}^0)}{\partial\alpha_{j_1}}\frac{\partial R_l(\boldsymbol{\alpha}^0)}{\partial\alpha_{j_2}}\frac{\partial^3 R_m(\boldsymbol{\alpha}^0)}{\partial\alpha_{j_3}\partial\alpha_{j_4}\partial\alpha_{j_5}}\frac{\partial R_n(\boldsymbol{\alpha}^0)}{\partial\alpha_{j_6}}\\
&\times\left(s_{j_1j_2j_3j_4j_5j_6}-t_{j_1j_2j_6}t_{j_3j_4j_5}\right)+\frac{\partial R_k(\boldsymbol{\alpha}^0)}{\partial\alpha_{j_1}}\frac{\partial^3 R_l(\boldsymbol{\alpha}^0)}{\partial\alpha_{j_2}\partial\alpha_{j_3}\partial\alpha_{j_4}}\frac{\partial R_m(\boldsymbol{\alpha}^0)}{\partial\alpha_{j_5}}\frac{\partial R_n(\boldsymbol{\alpha}^0)}{\partial\alpha_{j_6}}\\
&\times\left(s_{j_1j_2j_3j_4j_5j_6}-t_{j_1j_5j_6}t_{j_2j_3j_4}\right)+\frac{\partial^3 R_k(\boldsymbol{\alpha}^0)}{\partial\alpha_{j_1}\partial\alpha_{j_2}\partial\alpha_{j_3}}\frac{\partial R_l(\boldsymbol{\alpha}^0)}{\partial\alpha_{j_4}}\frac{\partial R_m(\boldsymbol{\alpha}^0)}{\partial\alpha_{j_5}}\frac{\partial R_n(\boldsymbol{\alpha}^0)}{\partial\alpha_{j_6}}\\
&\left.\times\left(s_{j_1j_2j_3j_4j_5j_6}-t_{j_1j_2j_3}t_{j_4j_5j_6}\right)\right]\sigma_{j_1}\sigma_{j_2}\sigma_{j_3}\sigma_{j_4}\sigma_{j_5}\sigma_{j_6}.
\end{aligned}\tag{1.126}$$

2. The ***quadruple-correlations*** denoted as $\mu_4(\alpha_i, R_j, R_k, R_\ell)$, ***among one parameter***, α_i, ***and three responses,*** $R_j(\boldsymbol{\alpha})$, $R_k(\boldsymbol{\alpha})$, and $R_\ell(\boldsymbol{\alpha})$, are defined as follows:

$$\begin{aligned}
\mu_4\left(\alpha_i, R_j, R_k, R_\ell\right)\triangleq&\left\langle\left(\alpha_i-\alpha_i^0\right)\left[R_j(\boldsymbol{\alpha})-E\left(R_j\right)\right]\right.\\
&\left.\times\left[R_k(\boldsymbol{\alpha})-E(R_k)\right]\left[R_\ell(\boldsymbol{\alpha})-E(R_\ell)\right]\right\rangle_\alpha.
\end{aligned}\tag{1.127}$$

(a) If *first- and second-order response sensitivities are available*, the following second-order approximate expression, denoted as $\mu_4^{(2)}(\alpha_i, R_j, R_k, R_\ell)$, can be obtained:

$$
\begin{aligned}
\mu_4^{(2)}(\alpha_i, R_j, R_k, R_\ell) &= \sum_{j_1=1}^{TP}\sum_{j_2=1}^{TP}\sum_{j_3=1}^{TP} \frac{\partial R_j(\boldsymbol{\alpha}^0)}{\partial \alpha_{j_1}} \frac{\partial R_k(\boldsymbol{\alpha}^0)}{\partial \alpha_{j_2}} \frac{\partial R_l(\boldsymbol{\alpha}^0)}{\partial \alpha_{j_3}} q_{j_1 j_2 j_3 i}\sigma_{j_1}\sigma_{j_2}\sigma_{j_3}\sigma_i \\
&+ \frac{1}{2}\sum_{j_1=1}^{TP}\sum_{j_2=1}^{TP}\sum_{j_3=1}^{TP}\sum_{j_4=1}^{TP} \left[\frac{\partial R_j(\boldsymbol{\alpha}^0)}{\partial \alpha_{j_1}} \frac{\partial R_k(\boldsymbol{\alpha}^0)}{\partial \alpha_{j_2}} \frac{\partial^2 R_l(\boldsymbol{\alpha}^0)}{\partial \alpha_{j_3}\partial \alpha_{j_4}} \left(p_{j_1 j_2 j_3 j_4 i} - t_{j_1 j_2 i}\rho_{j_3 j_4}\right) \right. \\
&+ \frac{\partial R_j(\boldsymbol{\alpha}^0)}{\partial \alpha_{j_1}} \frac{\partial^2 R_k(\boldsymbol{\alpha}^0)}{\partial \alpha_{j_2}\partial \alpha_{j_3}} \frac{\partial R_l(\boldsymbol{\alpha}^0)}{\partial \alpha_{j_4}} \left(p_{j_1 j_2 j_3 j_4 i} - \rho_{j_2 j_3} t_{j_1 j_4 i}\right) \\
&+ \left. \frac{\partial^2 R_j(\boldsymbol{\alpha}^0)}{\partial \alpha_{j_1}\partial \alpha_{j_2}} \frac{\partial R_l(\boldsymbol{\alpha}^0)}{\partial \alpha_{j_3}} \frac{\partial R_l(\boldsymbol{\alpha}^0)}{\partial \alpha_{j_4}} \left(p_{j_1 j_2 j_3 j_4 i} - \rho_{j_1 j_2 i} t_{j_3 j_4 i}\right) \right] \sigma_{j_1}\sigma_{j_2}\sigma_{j_3}\sigma_{j_4}\sigma_i \\
&+ \frac{1}{4}\sum_{j_1=1}^{TP}\sum_{j_2=1}^{TP}\sum_{j_3=1}^{TP}\sum_{j_4=1}^{TP}\sum_{j_5=1}^{TP} \left[\frac{\partial R_j(\boldsymbol{\alpha}^0)}{\partial \alpha_{j_1}} \frac{\partial^2 R_k(\boldsymbol{\alpha}^0)}{\partial \alpha_{j_2}\partial \alpha_{j_3}} \frac{\partial^2 R_l(\boldsymbol{\alpha}^0)}{\partial \alpha_{j_4}\partial \alpha_{j_5}} \right. \\
&\times \left(s_{j_1 j_2 j_3 j_4 j_5 i} - \rho_{j_2 j_3} q_{j_1 j_4 j_5 i} - q_{j_1 j_2 j_3 i}\rho_{j_4 j_5} + \rho_{j_1 i}\rho_{j_2 j_3}\rho_{j_4 j_5}\right) \\
&+ \frac{\partial^2 R_j(\boldsymbol{\alpha}^0)}{\partial \alpha_{j_1}\partial \alpha_{j_2}} \frac{\partial R_l(\boldsymbol{\alpha}^0)}{\partial \alpha_{j_3}} \frac{\partial^2 R_l(\boldsymbol{\alpha}^0)}{\partial \alpha_{j_4}\partial \alpha_{j_5}} \left(s_{j_1 j_2 j_3 j_4 j_5 i} - \rho_{j_1 j_2} q_{j_3 j_4 j_5 i}\right. \\
&\left. - q_{j_1 j_2 j_3 i}\rho_{j_4 j_5} + \rho_{j_1 j_2}\rho_{j_3 i}\rho_{j_4 j_5}\right) + \frac{\partial^2 R_j(\boldsymbol{\alpha}^0)}{\partial \alpha_{j_1}\partial \alpha_{j_2}} \frac{\partial^2 R_k(\boldsymbol{\alpha}^0)}{\partial \alpha_{j_3}\partial \alpha_{j_4}} \frac{\partial R_l(\boldsymbol{\alpha}^0)}{\partial \alpha_{j_5}} \\
&\times \left. \left(s_{j_1 j_2 j_3 j_4 j_5 i} - \rho_{j_1 j_2} q_{j_3 j_4 j_5 i} - q_{j_1 j_2 j_5 i}\rho_{j_3 j_4} + \rho_{j_1 j_2}\rho_{j_3 j_4}\rho_{j_5 i}\right) \right] \sigma_{j_1}\sigma_{j_2}\sigma_{j_3}\sigma_{j_4}\sigma_{j_5}\sigma_i .
\end{aligned}
\tag{1.128}
$$

(b) If *third-order response sensitivities are also available*, the following third-order approximate expression, denoted as $\mu_4^{(3)}(\alpha_i, R_j, R_k, R_\ell)$, which contains all of the sixth-order terms in the standard deviations, can be obtained:

$$
\begin{aligned}
&\mu_4^{(3)}(\alpha_i, R_j, R_k, R_\ell) = \mu_4^{(2)}(\alpha_i, R_j, R_k, R_\ell) \\
&+ \frac{1}{6}\sum_{j_1=1}^{TP}\sum_{j_2=1}^{TP}\sum_{j_3=1}^{TP}\sum_{j_4=1}^{TP}\sum_{j_5=1}^{TP} \left[\frac{\partial R_j(\boldsymbol{\alpha}^0)}{\partial \alpha_{j_1}} \frac{\partial R_k(\boldsymbol{\alpha}^0)}{\partial \alpha_{j_2}} \frac{\partial^3 R_l(\boldsymbol{\alpha}^0)}{\partial \alpha_{j_3}\partial \alpha_{j_4}\partial \alpha_{j_5}} \right. \\
&\times \left(s_{j_1 j_2 j_3 j_4 j_5 i} - t_{j_1 j_2 i} t_{j_3 j_4 j_5}\right) + \frac{\partial R_j(\boldsymbol{\alpha}^0)}{\partial \alpha_{j_1}} \frac{\partial^3 R_k(\boldsymbol{\alpha}^0)}{\partial \alpha_{j_2}\partial \alpha_{j_3}\partial \alpha_{j_4}} \frac{\partial R_l(\boldsymbol{\alpha}^0)}{\partial \alpha_{j_5}} \\
&\times \left(s_{j_1 j_2 j_3 j_4 j_5 i} - t_{j_1 j_5 i} t_{j_2 j_3 j_4}\right) + \frac{\partial^3 R_j(\boldsymbol{\alpha}^0)}{\partial \alpha_{j_1}\partial \alpha_{j_2}\partial \alpha_{j_3}} \frac{\partial R_k(\boldsymbol{\alpha}^0)}{\partial \alpha_{j_4}} \frac{\partial R_l(\boldsymbol{\alpha}^0)}{\partial \alpha_{j_5}} \\
&\times \left. \left(s_{j_1 j_2 j_3 j_4 j_5 i} - t_{j_1 j_2 j_3} t_{j_4 j_5 i}\right)\right] \sigma_{j_1}\sigma_{j_2}\sigma_{j_3}\sigma_{j_4}\sigma_{j_5}\sigma_i .
\end{aligned}
\tag{1.129}
$$

3. The ***quadruple-correlations***, denoted as $\mu_4(\alpha_i, \alpha_j, R_k, R_\ell)$, ***among two parameters*** α_i and α_j ***and two responses*** $R_k(\boldsymbol{\alpha})$ and $R_\ell(\boldsymbol{\alpha})$ are defined as follows:

$$
\begin{aligned}
&\mu_4(\alpha_i, \alpha_j, R_k, R_\ell) \triangleq \left\langle \left(\alpha_i - \alpha_i^0\right)\left(\alpha_j - \alpha_j^0\right) \right.\\
&\quad \times [R_k(\boldsymbol{\alpha}) - E(R_k)][R_\ell(\boldsymbol{\alpha}) - E(R_\ell)]\rangle_\alpha \\
&= \sum_{j_1=1}^{TP}\sum_{j_2=1}^{TP} \frac{\partial R_k(\boldsymbol{\alpha}^0)}{\partial \alpha_{j_1}} \frac{\partial R_l(\boldsymbol{\alpha}^0)}{\partial \alpha_{j_2}} q_{j_1 j_2 ij}\sigma_{j_1}\sigma_{j_2}\sigma_i\sigma_j \\
&+ \frac{1}{2}\sum_{j_1=1}^{TP}\sum_{j_2=1}^{TP}\sum_{j_3=1}^{TP}\left[\frac{\partial R_k(\boldsymbol{\alpha}^0)}{\partial \alpha_{j_1}}\frac{\partial^2 R_l(\boldsymbol{\alpha}^0)}{\partial \alpha_{j_2}\partial \alpha_{j_3}}\left(p_{j_1 j_2 j_3 ij} - t_{j_1 ij}\rho_{j_2 j_3}\right)\right. \\
&+ \left.\frac{\partial^2 R_k(\boldsymbol{\alpha}^0)}{\partial \alpha_{j_1}\partial \alpha_{j_2}}\frac{\partial R_l(\boldsymbol{\alpha}^0)}{\partial \alpha_{j_3}}\left(p_{j_1 j_2 j_3 ij} - \rho_{j_1 j_2}t_{j_3 ij}\right)\right]\sigma_{j_1}\sigma_{j_2}\sigma_{j_3}\sigma_i\sigma_j \\
&+ \frac{1}{4}\sum_{j_1=1}^{TP}\sum_{j_2=1}^{TP}\sum_{j_3=1}^{TP}\sum_{j_4=1}^{TP}\left[\frac{\partial^2 R_k(\boldsymbol{\alpha}^0)}{\partial \alpha_{j_1}\partial \alpha_{j_2}}\frac{\partial^2 R_l(\boldsymbol{\alpha}^0)}{\partial \alpha_{j_3}\partial \alpha_{j_4}}\left(s_{j_1 j_2 j_3 j_4 ij} + \rho_{j_1 j_2}\rho_{j_3 j_4}\rho_{ij}\right.\right. \\
&\left.\left. -\rho_{j_1 j_2}q_{j_3 j_4 ij} - q_{j_1 j_2 ij}\rho_{j_3 j_4}\right)\right]\sigma_{j_1}\sigma_{j_2}\sigma_{j_3}\sigma_{j_4}\sigma_i\sigma_j \\
&+ \frac{1}{6}\sum_{j_1=1}^{TP}\sum_{j_2=1}^{TP}\sum_{j_3=1}^{TP}\sum_{j_4=1}^{TP}\left[\frac{\partial R_k(\boldsymbol{\alpha}^0)}{\partial \alpha_{j_1}}\frac{\partial^3 R_l(\boldsymbol{\alpha}^0)}{\partial \alpha_{j_2}\partial \alpha_{j_3}\partial \alpha_{j_4}}\left(s_{j_1 j_2 j_3 j_4 ij} - t_{j_1 ij}t_{j_2 j_3 j_4}\right)\right. \\
&+ \left.\frac{\partial^3 R_k(\boldsymbol{\alpha}^0)}{\partial \alpha_{j_1}\partial \alpha_{j_2}\partial \alpha_{j_3}}\frac{\partial R_l(\boldsymbol{\alpha}^0)}{\partial \alpha_{j_4}}\left(s_{j_1 j_2 j_3 j_4 ij} - t_{j_1 j_2 j_3}t_{j_4 ij}\right)\right]\sigma_{j_1}\sigma_{j_2}\sigma_{j_3}\sigma_{j_4}\sigma_i\sigma_j.
\end{aligned}
\tag{1.130}
$$

4. The ***quadruple-correlations***, denoted as $\mu_4(\alpha_i, \alpha_j, \alpha_k, R_\ell)$, ***among three parameters*** α_i, α_j, and α_k ***and one response***, $R_\ell(\boldsymbol{\alpha})$, are defined as follows:

$$
\begin{aligned}
&\mu_4(\alpha_i, \alpha_j, \alpha_k, R_\ell) \triangleq \left\langle \left(\alpha_i - \alpha_i^0\right)\left(\alpha_j - \alpha_j^0\right)\left(\alpha_k - \alpha_k^0\right)[R_\ell(\boldsymbol{\alpha}) - E(R_\ell)]\right\rangle_\alpha \\
&= \sum_{j_1=1}^{TP}\frac{\partial R_\ell(\boldsymbol{\alpha}^0)}{\partial \alpha_{j_1}} q_{j_1 ijk}\sigma_{j_1}\sigma_i\sigma_j\sigma_k \\
&+ \frac{1}{2}\sum_{j_1=1}^{TP}\sum_{j_2=1}^{TP}\frac{\partial^2 R_\ell(\boldsymbol{\alpha}^0)}{\partial \alpha_{j_1}\partial \alpha_{j_2}}\left(p_{j_1 j_2 ijk} - \rho_{j_1 j_2}t_{ijk}\right)\sigma_{j_1}\sigma_{j_2}\sigma_i\sigma_j\sigma_k \\
&+ \frac{1}{6}\sum_{j_1=1}^{TP}\sum_{j_2=1}^{TP}\sum_{j_3=1}^{TP}\frac{\partial^3 R_\ell(\boldsymbol{\alpha}^0)}{\partial \alpha_{j_1}\partial \alpha_{j_2}\partial \alpha_{j_3}}\left(s_{j_1 j_2 j_3 ijk} - t_{j_1 j_2 j_3}t_{ijk}\right)\sigma_{j_1}\sigma_{j_2}\sigma_{j_3}\sigma_i\sigma_j\sigma_k.
\end{aligned}
\tag{1.131}
$$

This section has presented formulas for the moments of model responses in the phase-space of model parameters. These formulas are consistent up to and including the sixth-order standard deviations, thus generalizing the formulas presented by

Tukey (1957). It has been shown that the first-order sensitivities contribute the leading terms to the second-, third-, and fourth-order moments of the response distribution, thus providing the leading contributions to the responses variance/ covariances, skewness, and kurtosis. Obtaining the exact and complete set of first-order sensitivities of responses to model parameters is of paramount importance for any analysis of a computational model.

The second-order sensitivities contribute the leading correction terms to the response's expected value, causing it to differ from the response's computed value. The second-order sensitivities also contribute to the response variances and covariances. If the parameters follow a normal (Gaussian) multivariate distribution, the second-order sensitivities contribute the leading terms to the response's third-order moment. Thus, neglecting the second-order response sensitivities to normally distributed parameters would nullify the third-order response correlations and hence would nullify the skewness of a response.

In particular, the skewness of a single response, R, is customarily denoted as $\gamma_1(R)$, and is defined as follows: $\gamma_1(R) = \mu_3(R)/[\mathrm{var}(R)]^{3/2}$, where $\mu_3(R)$ denotes the third central moment of the response distribution. As is well known, the skewness provides a quantitative measure of the asymmetries in the respective distribution. Hence, a "first-order sensitivity and uncertainty quantification" will always produce an erroneous third moment (and hence skewness) of the predicted response distribution, unless the unknown response distribution happens to be symmetrical. At least second-order sensitivities must be used in order to estimate the third-order moment (and hence the skewness) of the response distribution. Skewness indicates the direction and relative magnitude of a distribution's deviation from the normal distribution while kurtosis indicates the propensity of the predicted response distribution to have heavy tails and/or outliers. With pronounced skewness, standard statistical inference procedures such as constructing a confidence interval for the mean (expectation) of a computed/predicted model response will be not only incorrect, in the sense that the true coverage level will differ from the nominal (e.g., 95%) level, but the error probabilities will be unequal on each side of the predicted mean.

1.5 Chapter Summary

This chapter has underscored the need for computing not only the first-order but also the higher-order sensitivities of model responses with respect to the model's parameters. The contributions of the higher-order sensitivities to predicted model responses and the moments of the responses' distribution in the parameters' phase-space has also been highlighted by presenting general formulas for computing the first four moments of model responses in the phase-space of model parameters. These formulas are consistent up to and including the sixth-order standard deviations of the parameters and corresponding parameter correlations, thus generalizing the formulas originally provided by Tukey (1957).

Computing just the first-order sensitivities cannot provide any assurance that the higher-order sensitivities are negligible. For example, using just the first-order sensitivities will always produce a zero-valued third moment (and hence skewness) of the predicted response distribution, which is erroneous unless the unknown response distribution happens to be symmetrical. At least second-order sensitivities must be used in order to estimate the third-order moment (and hence the skewness) of the response distribution. Skewness indicates the direction and relative magnitude of a distribution's deviation from the normal distribution. With pronounced skewness, standard statistical inference procedures such as constructing a confidence interval for the mean (expectation) of a computed/predicted model response will be not only incorrect, in the sense that the true coverage level will differ from the nominal (e.g., 95%) level, but the error probabilities will be unequal on each side of the predicted mean. The fourth-order moment of the response distribution yields the response's kurtosis, which indicates the propensity of the predicted response distribution to have heavy tails and/or outliers.

As a matter of course, therefore, at least the second-order response sensitivities should be computed, to prove quantitively if they are indeed unimportant. And if any of the second-order sensitivities turn out to be important, then the corresponding third-order sensitivities should be computed and investigated appropriately. As is well known, for a model comprising a total number of TP-parameters, for each model response there are TP first-order sensitivities, $TP(TP + 1)/2$ distinct second-order sensitivities, $TP(TP + 1)(TP + 2)/6$ distinct third-order sensitivities, and so on. Thus, the number of sensitivities increases exponentially with the order of the respective sensitivities. Hence, their computation by conventional methods (e.g., finite difference schemes, statistical assessments) which require re-computations using the original model (with suitably altered parameter values) is hampered by the curse of dimensionality while providing only approximate (often inaccurate), rather than exact, values for the sensitivities. The only method which overcomes, to the largest possible extent, the curse of dimensionality while providing exact expressions for computing efficiently sensitivities of all orders is the comprehensive adjoint sensitivity analysis methodology, conceived and developed by Cacuci (1981a, b, 2015, 2016, 2018, 2019a, b, c, d, 2021a, b) and presented in this 3-volume book. The qualifier "comprehensive" is employed in order to highlight that the model parameters considered within the framework of this methodology includes the system's uncertain boundaries and internal interfaces in phase-space.

Since linear operators admit bona fide adjoint operators, responses of models that are linear in the state functions (i.e., dependent variables) can depend simultaneously on both the forward and the adjoint state functions. Hence, the sensitivity analysis of such responses makes it necessary to treat linear models/systems in their own right, rather than treating them as particular cases of nonlinear systems. This is in contradistinction to responses for nonlinear systems, which can depend only on the forward state functions, since nonlinear operators do not admit bona fide adjoint operators (only a linearized form of a nonlinear operator admits an adjoint operator).

Chapter 2 of this book presents the mathematical framework of the *nth-Order Comprehensive Adjoint Sensitivity Analysis Methodology for Response-Coupled Forward/Adjoint Linear Systems* (abbreviated as "nth-CASAM-L"), which is conceived for the most efficient computation of *exactly obtained* mathematical expressions of arbitrarily high-order (nth-order) sensitivities of a generic system response with respect to *all* of the parameters (including boundary and initial conditions) underlying the respective forward/adjoint systems. Volume 2 of this book will present an application of the nth-CASAM-L to perform a fourth-order sensitivity and uncertainty analysis of an OECD/NEA reactor physics benchmark (Valentine 2006) which comprises many uncertain parameters (including 180 group-averaged total microscopic cross sections, 21,600 group-averaged scattering microscopic cross sections, 120 fission process parameters, 60 fission spectrum parameters, 10 parameters describing the experiment's nuclear sources, and 6 isotopic number densities). Volume 3 of this book will present the "nth-Order Comprehensive Adjoint Sensitivity Analysis Methodology for Nonlinear Systems" (abbreviation: nth-CASAM-N) for the practical, efficient, and exact computation of arbitrarily high-order sensitivities of responses to model parameters for nonlinear systems. The application of the nth-CASAM-L and the nth-CASAM-N is expected to revolutionize all of the fields of activities which require response sensitivities, including the fields of uncertainty quantification, model validation, optimization, data assimilation, model calibration, sensor fusion, reduced-order modeling, inverse problems, and predictive modeling.

Chapter 2
The nth-Order Comprehensive Adjoint Sensitivity Analysis Methodology for Response-Coupled Forward/Adjoint Linear Systems (nth-CASAM-L)

2.1 Introduction

For each response of a model comprising TP-parameters, there will exist TP first-order sensitivities, $TP(TP+1)/2$ distinct second-order sensitivities, $TP(TP+1)(TP+2)/6$ distinct third-order sensitivities, and so on. The number of sensitivities increases exponentially with the order of the respective sensitivities. Hence, their computation by conventional deterministic methods (e.g., finite difference methods, Green's function methods, the forward sensitivity analysis procedure), which require re-computations using the original model, is limited in practice by the so-called "curse of dimensionality" (Bellman 1957). Furthermore, such conventional methods cannot provide exact values for the sensitivities, but only approximate values which are often inaccurate.

On the other hand, statistical methods construct an approximate response distribution (often called "response surface") in the parameters space, and subsequently use scatter plots, regression, rank transformation, correlation, and partial correlation analyses, to identify approximate expectation values, variances, and covariances for the responses. These statistical quantities are subsequently used to construct quantities that play the role of (approximate) first-order response sensitivities. Thus, statistical methods commence with "uncertainty analysis" and subsequently attempt an approximate "sensitivity analysis" of the computed model response under consideration. The currently popular statistical methods for uncertainty and sensitivity analysis are broadly categorized as sampling-based methods, variance-based methods, and spectral methods. The main advantage of using statistical methods for uncertainty and sensitivity analysis is that they are conceptually easy to implement. On the other hand, statistical methods for uncertainty and sensitivity analysis have three major inherent practical drawbacks, as follows:

1. Since many thousands of simulations are needed to obtain reliable results, statistical methods are at best expensive (for small systems), or, at worst, impracticable (e.g., for large time-dependent systems).

D. G. Cacuci, *The nth-Order Comprehensive Adjoint Sensitivity Analysis Methodology, Volume I*, https://doi.org/10.1007/978-3-030-96364-4_2

2. Response sensitivities cannot be computed exactly.
3. Since the response sensitivities and parameter uncertainties are amalgamated, inherently and inseparably, within the results produced by statistical methods, improvements in parameter uncertainties cannot be directly propagated to improve response uncertainties; rather, the entire set of simulations and statistical post-processing must be repeated anew. In particular, a "fool-proof" statistical method for analyzing correctly models involving highly correlated parameters does not seem to exist currently, so that particular care must be used when interpreting regression results for such models.
4. Evidently, statistical methods are also limited by the curse of dimensionality.

The pioneering use of adjoint operators, which enables the most efficient computation of exactly obtained expressions for the first-order sensitivities of responses with respect to model parameters, is attributed to Wigner (1945), who applied first-order perturbation theory to obtain the first-order sensitivities of the eigenvalue (multiplication factor) of the linear neutron transport equation. Many works using adjoint operators in conjunction with the application of first- and second-order perturbation theory to the linear neutron transport and/or diffusion equations were published in the nuclear science and engineering field during the 1960s and 1970s. Cacuci (1981a, b) is credited (see, e.g., Práger and Kelemen, 2014; Luo et al. 2020) with the conception of the rigorous first-order adjoint sensitivity analysis methodology for generic large-scale *nonlinear* (as opposed to linearized) systems involving generic operator responses, including scalar-valued responses as particular cases. The use of adjoint operators for the computation of first- and second-order response sensitivities to particular problems spread to other fields in the 1980s, too numerous to mention here. Application to large-scale systems was presented in the book by Cacuci et al. (2005). The general mathematical framework for the second-order adjoint sensitivity analysis methodology was conceived by Cacuci (2015, 2016, 2018). However, the vast majority of the applications to date of the first- and second-order adjoint sensitivity analysis methodology considered that the phase-space location of the physical model's boundary is perfectly well known and did not investigate the consequences of imprecisely known (i.e., uncertain) model boundaries and/or internal interfaces, which will be addressed within the theoretical framework to be presented in this chapter.

Striving to overcome, to the largest possible extent, the curse of dimensionality while providing exact expressions for computing efficiently model response sensitivities for large-scale models comprising many parameters, Cacuci (2019a, b, c, 2020, 2021a, b) has extended the framework of the second-order adjoint sensitivity analysis methodology to third- and fourth-order, for both linear and nonlinear systems comprising not only uncertain model parameters but also imprecisely known internal and external boundaries. The generic mathematical framework for computing third- and fourth-order sensitivities of responses that depend simultaneously on both the forward and the adjoint state functions was conceived and recently presented (Cacuci, 2019d, 2021a, b).

The need for developing a separate methodology for linear systems arose because the most important model responses produced by such systems can depend *simultaneously* on both the forward and adjoint state functions governing the respective linear system. Nonlinear operators do not admit adjoint operators (only linearized versions of nonlinear operators can admit a bona fide adjoint operator), so responses in nonlinear systems can only depend on the system's forward state functions. Therefore, the sensitivity analysis of responses that simultaneously involve both forward and adjoint state functions makes it necessary to treat linear models/systems in their own right, rather than treating them as particular cases of nonlinear systems.

Among the most important responses of linear systems that involve both the forward and adjoint functions are various Lagrangian functionals, the Raleigh quotient for computing eigenvalues and/or separation constants when solving partial differential equations, and the Schwinger functional for first-order "normalization-free" solutions (see, e.g., Lewins 1965; Stacey 1974). These functionals play a fundamental role in optimization and control procedures, derivation of numerical methods for solving equations (differential, integral, integro-differential), etc.

This chapter presents the mathematical framework of the *nth-Order Comprehensive Adjoint Sensitivity Analysis Methodology for Response-Coupled Forward/Adjoint Linear Systems* (abbreviated as "nth-CASAM-L"). The nth-CASAM-L enables the efficient computation of exactly determined expressions of arbitrarily high-order sensitivities of a generic system response—which can depend on both the forward and adjoint state functions—with respect to all of the parameters that characterize the physical system. The qualifier "comprehensive" is employed in order to highlight that the model parameters considered within the framework of the nth-CASAM-L include the system's uncertain boundaries and internal interfaces in phase-space. The mathematical framework nth-CASAM-L presented in this chapter is based on the work by Cacuci (2019a, b, c, d, 2021a, b) and is structured as follows: Section 2.2 presents the general mathematical modeling of response-coupled linear forward and adjoint systems. Sections 2.3–2.6 present the mathematical frameworks of the nth-CASAM-L for the particular cases $n = 1, 2, 3$, and 4, respectively. Section 2.7 presents the proof by "mathematical induction" that the mathematical framework underlying the nth-CASAM-L is correct, as follows: (1) establish the mathematical framework underlying the nth-CASAM-L for $n = 1$; (2) assume that the mathematical framework is valid for an arbitrarily high-order, n; and (3) prove that the mathematical framework is valid for $n + 1$.

As an additional application of the general mathematical framework of the nth-CASAM-L, Section 2.8 presents the "5th-Order Comprehensive Adjoint Sensitivity Analysis Methodology for Coupled Forward/Adjoint Linear Systems" (5th-CASAM-L). The derivation of the 5th-CASAM-L is performed explicitly, by starting with the results produced by the 4th-CASAM-L. The mathematical framework thus derived for the 5th-CASAM-L is subsequently shown to coincide with the particular formulas obtained by setting $n = 5$ in the general nth-CASAM-L framework. Section 2.9 summarizes the main characteristics and significance of the nth-CASAM-L in the quest to overcome the curse of dimensionality in sensitivity analysis, uncertainty quantification, and predictive modeling.

2.2 Mathematical Modeling of Response-Coupled Linear Forward and Adjoint Systems

The mathematical model of a physical system comprises independent variables (e.g., space, time), dependent variables (aka "state functions"; e.g., temperature, mass, momentum), and various parameters (appearing in correlations, coordinates of physical boundaries, etc.), which are all interrelated by equations that usually represent conservation laws. The model parameters usually stem from processes that are external to the system under consideration and are seldom, if ever, known precisely. Without loss of generality, the model parameters can be considered to be real scalar quantities, having known nominal (or mean) values and, possibly, known higher-order moments or cumulants (i.e., variance/covariances, skewness, kurtosis), which are usually determined from experimental data and/or processes external to the physical system under consideration. These imprecisely known model parameters will be denoted as $\alpha_1,\ldots, \alpha_{TP}$, where the subscript "*TP*" indicates "*Total* (number of) *Parameters*." It will be convenient to consider that these parameters are components of a "vector of parameters" denoted as $\boldsymbol{\alpha} \triangleq (\alpha_1, \ldots, \alpha_{TP})^\dagger \in \mathbb{R}^{TP}$, where $\mathbb{R}^{TP}$ denotes the TP-dimensional subset of the set of real scalars. These model parameters are considered to include imprecisely known geometrical parameters that characterize the physical system's boundaries in the phase-space of the model's independent variables. Thus, $\boldsymbol{\alpha} \in \mathbb{R}^{TP}$ is a TP-dimensional column vector and its components are considered to include all of the uncertain model parameters, including those that may enter into defining the system's boundary in the phase-space of independent variables. For subsequent developments, it is convenient to denote matrices and vectors using capital and, respectively, lower case bold letters. The symbol "$\triangleq$" will be used to denote "is defined as" or "is by definition equal to." Transposition will be indicated by a dagger (†) superscript.

The model is considered to comprise *TI* independent variables which will be denoted as x_i, $i = 1, \ldots, TI$, and are considered to be the components of a *TI*-dimensional column vector denoted as $\mathbf{x} \triangleq (x_1, \ldots, x_{TI})^\dagger \in \mathbb{R}^{TI}$, where the sub/superscript "*TI*" denotes the "*Total* (number of) *Independent* variables." The vector $\mathbf{x} \in \mathbb{R}^{TI}$ of independent variables is considered to be defined on a phase-space domain, denoted as $\Omega(\boldsymbol{\alpha})$, $\Omega(\boldsymbol{\alpha}) \triangleq \{-\infty \leq \lambda_i(\boldsymbol{\alpha}) \leq x_i \leq \omega_i(\boldsymbol{\alpha}) \leq \infty; i = 1, \ldots, TI\}$, the boundaries of which may depend on (some of) the model parameters $\boldsymbol{\alpha}$. The lower boundary-point of an independent variable is denoted as $\lambda_i(\boldsymbol{\alpha})$ (e.g., the inner radius of a sphere or cylinder, the lower range of an energy-variable, the initial time-value), while the corresponding upper boundary-point is denoted as $\omega_i(\boldsymbol{\alpha})$ (e.g., the outer radius of a sphere or cylinder, the upper range of an energy-variable, the final time-value). A typical example of boundary conditions that depend on imprecisely known model and geometrical parameters occurs when modeling particle diffusion within a medium, the boundaries of which are facing vacuum. For such models, the boundary conditions for the respective state (dependent) variables (i.e., particle flux and/or current) are imposed not on the physical boundary but on the "extrapolated boundary" of the respective spatial domain. The "extrapolated boundary" depends both on the imprecisely known physical dimensions of the medium's domain/extent and also on the medium's properties, i.e., atomic number densities and microscopic transport cross sections. The boundary of $\Omega(\boldsymbol{\alpha})$, which will be denoted as $\partial\Omega(\boldsymbol{\alpha})$, comprises the set of all of the endpoints $\lambda_i(\boldsymbol{\alpha})$, $\omega_i(\boldsymbol{\alpha})$, $i = 1, \ldots, TI$, of the respective intervals on which the components of $\mathbf{x}$ are defined, i.e., $\partial\Omega(\boldsymbol{\alpha}) \triangleq \{\lambda_i(\boldsymbol{\alpha}) \cup \omega_i(\boldsymbol{\alpha}),\ i = 1, \ldots, TI\}$.

A linear physical system is generally modeled by a system of coupled linear operator-equations which can be generally represented as follows:

$$\mathbf{L}(\mathbf{x};\boldsymbol{\alpha})\boldsymbol{\varphi}(\mathbf{x}) = \mathbf{q}_{\varphi}(\mathbf{x};\boldsymbol{\alpha}), \quad \mathbf{x} \in \Omega(\boldsymbol{\alpha}), \tag{2.1}$$

where $\boldsymbol{\varphi}(\mathbf{x}) \triangleq [\varphi_1(\mathbf{x}), \ldots, \varphi_{TD}(\mathbf{x})]^{\dagger}$ is a TD-dimensional column vector of dependent variables and where the sub/superscript "*TD*" denotes the "*Total* (number of) *Dependent* variables." The functions $\varphi_i(\mathbf{x})$, $i = 1, \ldots, TD$, denote the system's "dependent variables" (also called "state functions"). The matrix $\mathbf{L}(\mathbf{x};\boldsymbol{\alpha}) \triangleq [L_{ij}(\mathbf{x};\boldsymbol{\alpha})\ ], i, j = 1, \ldots, TD$, has dimensions $TD \times TD$. The components $\mathbf{L}_{ij}(\mathbf{x};\boldsymbol{\alpha})$ are operators that act linearly on the dependent variables $\varphi_j(\mathbf{x})$ and also depend (in general, nonlinearly) on the uncertain model parameters $\boldsymbol{\alpha}$. The components of the TD-dimensional column vector $\mathbf{q}_{\varphi}(\mathbf{x};\boldsymbol{\alpha}) \triangleq [q_{\varphi,\ 1}(\mathbf{x};\boldsymbol{\alpha}), \ldots, q_{\varphi,\ TD}(\mathbf{x};\boldsymbol{\alpha})]^{\dagger}$ represent inhomogeneous source terms. The components $q_{\varphi,\ i}(\mathbf{x};\boldsymbol{\alpha})$ of $\mathbf{q}_{\varphi}(\mathbf{x};\boldsymbol{\alpha})$, where the subscript "$\varphi$" indicates sources associated with the "forward" system of equations, are also nonlinear functions of $\boldsymbol{\alpha}$. Since the right side of (2.1) may contain distributions, the equality in this equation is considered to hold in the weak (i.e., "distributional") sense. Similarly, all of the equalities that involve differential equations in this work will be considered to hold in the distributional sense.

When $\mathbf{L}(\mathbf{x};\boldsymbol{\alpha})$ contains differential operators, a set of boundary and initial conditions which define the domain of $\mathbf{L}(\mathbf{x};\boldsymbol{\alpha})$ must also be given. Since the complete mathematical model is considered to be linear in $\boldsymbol{\varphi}(\mathbf{x})$, the boundary and/or initial conditions needed to define the domain of $\mathbf{L}(\mathbf{x};\boldsymbol{\alpha})$ must also be linear in $\boldsymbol{\varphi}(\mathbf{x})$. Such linear boundary and initial conditions are represented in the following operator form:

$$\mathbf{B}_{\varphi}(\mathbf{x};\boldsymbol{\alpha})\ \boldsymbol{\varphi}(\mathbf{x}) - \mathbf{c}_{\varphi}(\mathbf{x};\boldsymbol{\alpha}) = \mathbf{0}, \ \ \mathbf{x} \in \partial\Omega_x(\boldsymbol{\alpha}). \tag{2.2}$$

In (2.2), the components $B_{ij}(\mathbf{x};\boldsymbol{\alpha})$; $i = 1, \ldots, N_B$; $j = 1, \ldots, TD$, of the $N_B \times TD$-dimensional matrix-operator $\mathbf{B}_{\varphi}(\mathbf{x};\boldsymbol{\alpha})$ are operators that act linearly on $\boldsymbol{\varphi}(\mathbf{x})$ and nonlinearly on $\boldsymbol{\alpha}$; the quantity N_B denotes the total number of boundary and initial conditions. The operator $\mathbf{c}_{\varphi}(\mathbf{x};\boldsymbol{\alpha}) \triangleq \left[c_{\varphi,1}(\mathbf{x};\boldsymbol{\alpha}), \ldots, c_{\varphi,N_B}(\mathbf{x};\boldsymbol{\alpha})\right]^{\dagger}$ is a N_B-dimensional vector comprising components that are operators which, in general, act nonlinearly on $\boldsymbol{\alpha}$. The subscript "φ" in (2.2) indicates boundary conditions associated with the forward state function $\boldsymbol{\varphi}(\mathbf{x})$.

Physical problems modeled by linear systems and/or operators are naturally defined in Hilbert spaces. The dependent variables $\varphi_i(\mathbf{x})$, $i = 1, \ldots, TD$, for the physical system represented by (2.1) and (2.2) are considered to be square-integrable functions of the independent variables and are considered to belong to a Hilbert space which will be denoted as H_0, where the subscript "zero" denotes "zeroth-level" or "original." Higher-level Hilbert spaces, which will be denoted as H_1, H_2, etc., will also be introduced and used in this chapter. The Hilbert space H_0 is considered to be endowed with the following inner product, denoted as $\langle \boldsymbol{\varphi}(\mathbf{x}), \boldsymbol{\psi}(\mathbf{x}) \rangle_0$, between two elements $\boldsymbol{\varphi}(\mathbf{x}) \in H_0$ and $\boldsymbol{\psi}(\mathbf{x}) \in H_0$:

$$\begin{aligned} \langle \boldsymbol{\varphi}(\mathbf{x}), \boldsymbol{\psi}(\mathbf{x}) \rangle_0 &\triangleq \int_{\Omega_x} \boldsymbol{\varphi}(\mathbf{x}) \cdot \boldsymbol{\psi}(\mathbf{x}) d\mathbf{x} \triangleq \prod_{i=1}^{TI} \int_{\lambda_i(\boldsymbol{\alpha})}^{\omega_i(\boldsymbol{\alpha})} \boldsymbol{\varphi}(\mathbf{x}) \cdot \boldsymbol{\psi}(\mathbf{x})\, d\mathbf{x} \\ &= \sum_{i=1}^{TD} \int_{\lambda_1(\boldsymbol{\alpha})}^{\omega_1(\boldsymbol{\alpha})} \cdots \int_{\lambda_{TI}(\boldsymbol{\alpha})}^{\omega_{TI}(\boldsymbol{\alpha})} \varphi_i(\mathbf{x}) \psi_i(\mathbf{x})\, dx_1 \ldots dx_{TI}. \end{aligned} \tag{2.3}$$

The dot in (2.3) indicates the "scalar product of two vectors," which is defined as follows:

$$\boldsymbol{\varphi}(\mathbf{x}) \cdot \boldsymbol{\psi}(\mathbf{x}) \triangleq \sum_{i=1}^{TD} \varphi_i(\mathbf{x}) \psi_i(\mathbf{x}). \tag{2.4}$$

The product-notation $\prod_{i=1}^{TI} \int_{\lambda_i(\boldsymbol{\alpha})}^{\omega_i(\boldsymbol{\alpha})} [\,]\, dx_i$ in (2.3) denotes the respective multiple integrals. The linear operator $\mathbf{L}(\mathbf{x}; \boldsymbol{\alpha})$ admits an adjoint operator, which will be denoted as $\mathbf{L}^*(\mathbf{x}; \boldsymbol{\alpha})$ and which is defined through the following relation for a vector $\boldsymbol{\psi}(\mathbf{x}) \in H_0$:

$$\langle \boldsymbol{\psi}(\mathbf{x}), \mathbf{L}(\mathbf{x}; \boldsymbol{\alpha})\ \boldsymbol{\varphi}(\mathbf{x}) \rangle_0 = \langle \mathbf{L}^*(\mathbf{x}; \boldsymbol{\alpha})\ \boldsymbol{\psi}(\mathbf{x}), \boldsymbol{\varphi}(\mathbf{x}) \rangle_0. \tag{2.5}$$

In (2.5), the formal adjoint operator $\mathbf{L}^*(\mathbf{x}; \boldsymbol{\alpha})$ is the $TD \times TD$ matrix comprising elements $L_{ji}^*(\mathbf{x}; \boldsymbol{\alpha})$ obtained by transposing the formal adjoints of the forward operators $L_{ij}(\mathbf{x}; \boldsymbol{\alpha})$, i.e.,

$$\mathbf{L}^*(\mathbf{x}; \boldsymbol{\alpha}) \triangleq \left[L_{ji}^*(\mathbf{x}; \boldsymbol{\alpha}) \right], \quad i,j = 1, \ldots, TD. \tag{2.6}$$

Hence, the system adjoint to the linear system represented by (2.1) and (2.2) can generally be represented as follows:

$$\mathbf{L}^*(\mathbf{x}; \boldsymbol{\alpha})\ \boldsymbol{\psi}(\mathbf{x}) = \mathbf{q}_\psi(\mathbf{x}; \boldsymbol{\alpha}), \quad \mathbf{x} \in \Omega_x(\boldsymbol{\alpha}), \tag{2.7}$$

$$\mathbf{B}_\psi(\mathbf{x}; \boldsymbol{\alpha})\ \boldsymbol{\psi}(\mathbf{x})\ - \mathbf{c}_\psi(\mathbf{x}; \boldsymbol{\alpha}) = \mathbf{0}, \quad \mathbf{x} \in \partial\Omega_x(\boldsymbol{\alpha}). \tag{2.8}$$

When $\mathbf{L}(\mathbf{x}; \boldsymbol{\alpha})$ comprises differential operators, the operations (e.g., integration by parts) that implement the transition from the left side to the right side of (2.5) give rise to boundary terms which are collectively called the "bilinear concomitant." The domain of $\mathbf{L}^*(\mathbf{x};\boldsymbol{\alpha})$ is determined by selecting adjoint boundary and/or initial conditions so as to ensure that the bilinear concomitant vanishes when the selected adjoint boundary conditions are implemented together with the forward boundary conditions given in (2.2). The adjoint boundary conditions thus selected are represented in operator form by (2.8), where the subscript "ψ" indicates adjoint boundary and/or initial conditions associated with the adjoint state function $\boldsymbol{\psi}(\mathbf{x})$. In most practical situations, the Hilbert space H_0 is self-dual.

The nominal solution of (2.1) and (2.2), which will be denoted as $\boldsymbol{\varphi}^0(\mathbf{x})$, and the nominal solution of (2.7) and (2.8), which will be denoted as $\boldsymbol{\psi}^0(\mathbf{x})$, are obtained by solving these equations at the nominal (or mean) values, denoted as $\boldsymbol{\alpha}^0$, of the model parameters. The superscript "zero" will henceforth be used to denote "nominal" (or, equivalently, "expected" or "mean" values). Thus, the vectors $\boldsymbol{\varphi}^0(\mathbf{x})$, $\boldsymbol{\psi}^0(\mathbf{x})$, and $\boldsymbol{\alpha}^0$ satisfy the following equations:

$$\mathbf{L}\left(\mathbf{x};\boldsymbol{\alpha}^0\right)\boldsymbol{\varphi}^0(\mathbf{x}) = \mathbf{q}_\varphi\left(\mathbf{x};\boldsymbol{\alpha}^0\right), \quad \mathbf{x}\in\Omega_x\left(\boldsymbol{\alpha}^0\right), \tag{2.9}$$

$$\mathbf{B}_\varphi\left(\mathbf{x};\boldsymbol{\alpha}^0\right)\ \boldsymbol{\varphi}^0(\mathbf{x}) - \mathbf{c}_\varphi\left(\mathbf{x};\boldsymbol{\alpha}^0\right) = \mathbf{0}, \quad \mathbf{x}\in\partial\Omega_x\left(\boldsymbol{\alpha}^0\right), \tag{2.10}$$

$$\mathbf{L}^*\left(\mathbf{x};\boldsymbol{\alpha}^0\right)\ \boldsymbol{\psi}^0(\mathbf{x}) = \mathbf{q}_\psi\left(\mathbf{x};\boldsymbol{\alpha}^0\right), \quad \mathbf{x}\in\Omega_x\left(\boldsymbol{\alpha}^0\right), \tag{2.11}$$

$$\mathbf{B}_\psi\left(\mathbf{x};\boldsymbol{\alpha}^0\right)\boldsymbol{\psi}^0\left(\mathbf{x};\boldsymbol{\alpha}^0\right) - \mathbf{c}_\psi\left(\mathbf{x};\boldsymbol{\alpha}^0\right) = \mathbf{0}, \quad \mathbf{x}\in\partial\Omega_x\left(\boldsymbol{\alpha}^0\right). \tag{2.12}$$

The relationship shown in (2.5), which is the basis for defining the adjoint operator, also provides the following fundamental "reciprocity-like" relation between the sources of the forward and the adjoint equations, cf. (2.1) and (2.7), respectively:

$$\left\langle\boldsymbol{\psi}(\mathbf{x}),\mathbf{q}_\varphi(\mathbf{x};\boldsymbol{\alpha})\right\rangle_0 = \left\langle\mathbf{q}_\psi(\mathbf{x};\boldsymbol{\alpha}),\boldsymbol{\varphi}(\mathbf{x})\right\rangle_0. \tag{2.13}$$

The functional on the right side of (2.13) represents a "detector response," i.e., a reaction-rate between the particles and the medium represented by $\mathbf{q}_\psi(\mathbf{x};\boldsymbol{\alpha})$ which is equivalent to the "number of counts" of particles incident on a detector of particles that "measures" the particle flux $\boldsymbol{\varphi}(\mathbf{x})$. In view of the relation provided in (2.13), the source term $\mathbf{q}_\psi(\mathbf{x};\boldsymbol{\alpha}) \triangleq [q_{\psi,1}(\mathbf{x};\boldsymbol{\alpha}),\ldots,q_{\psi,TD}(\mathbf{x};\boldsymbol{\alpha})]^\dagger$ in the adjoint Eq. (2.7) is usually associated with the "result of interest" to be measured and/or computed, which is customarily called the system's "response." In particular, if $\mathbf{q}_\psi(\mathbf{x};\boldsymbol{\alpha}) = \delta(-\mathbf{x} - \mathbf{x}_d)$, then $\langle\mathbf{q}_\psi(\mathbf{x};\boldsymbol{\alpha}),\boldsymbol{\varphi}(\mathbf{x})\rangle_0 = \boldsymbol{\varphi}(\mathbf{x}_d)$, which means that, in such a case, the right side of (2.13) provides the value of the dependent variable (particle flux, temperature, velocity, etc.) at the point in phase-space where the respective measurement is performed.

The relation given in (2.13) indicates that for linear physical systems, the system's response may depend not only on the model's state-functions and on the system parameters, but also on the adjoint state function. The system's response will be denoted as $R[\boldsymbol{\varphi}(\mathbf{x}),\boldsymbol{\psi}(\mathbf{x});\boldsymbol{\alpha}]$ and, in the most general case, is a nonlinear operator that acts on the model's forward and adjoint state functions, as well as on imprecisely known model parameters, both directly and indirectly through the state functions.

The nominal value of the response, $R[\boldsymbol{\varphi}^0(\mathbf{x}), \boldsymbol{\psi}^0(\mathbf{x}); \boldsymbol{\alpha}^0]$, is determined by using the nominal parameter values $\boldsymbol{\alpha}^0$, the nominal value $\boldsymbol{\varphi}^0(\mathbf{x})$ of the forward state function, and the nominal value $\boldsymbol{\psi}^0(\mathbf{x})$ of the adjoint state function.

In general, a function-valued (operator) response $R_{op}[\boldsymbol{\varphi}(\mathbf{x}), \boldsymbol{\psi}(\mathbf{x}); \boldsymbol{\alpha}]$ can be represented by a spectral expansion in multidimensional orthogonal polynomials or Fourier series of the form:

$$R_{op}[\boldsymbol{\varphi}(\mathbf{x}), \boldsymbol{\psi}(\mathbf{x}); \boldsymbol{\alpha}] = \sum_{m_1} \dots \sum_{m_{TI}} c_{m_1 \dots m_{TI}} P_{m_1}(x_1) P_{m_2}(x_2) \dots P_{m_{TI}}(x_{TI}), \tag{2.14}$$

where the quantities $P_{m_i}(x_i)$, $i = 1, \dots, TI$, denote the corresponding spectral functions (e.g., orthogonal polynomials or Fourier exponential/trigonometric functions) and where the spectral Fourier coefficients $c_{m_1 \dots m_{TI}}$ are defined as follows:

$$c_{m_1 \dots m_{TI}} \triangleq \prod_{i=1}^{TI} \int_{\lambda_i(\boldsymbol{\alpha})}^{\omega_i(\boldsymbol{\alpha})} R_{op}[\boldsymbol{\varphi}(\mathbf{x}), \boldsymbol{\psi}(\mathbf{x}); \boldsymbol{\alpha}] P_{m_1}(x_1) \dots P_{m_i}(x_i) \dots P_{m_{TI}}(x_{N_x}) dx_i. \tag{2.15}$$

The coefficients $c_{m_1 \dots m_{TI}}$ can themselves be considered as system responses since the spectral polynomials $P_{m_i}(x_i)$ are perfectly well known while the expansion coefficients will contain all of the dependencies of the respective response on the imprecisely known model and response parameters. This way, the sensitivity analysis of operator-valued responses can be reduced to the sensitivity analysis of scalar-valued responses.

A measurement of a physical quantity can be represented as a response $R_p[\boldsymbol{\varphi}(\mathbf{x}_p), \boldsymbol{\psi}(\mathbf{x}_p); \boldsymbol{\alpha}]$ located at a specific point, $\mathbf{x}_p$, in phase-space, having the following form:

$$R_p[\boldsymbol{\varphi}(\mathbf{x}_p), \boldsymbol{\psi}(\mathbf{x}_p); \boldsymbol{\alpha}] \triangleq \prod_{i=1}^{TI} \int_{\lambda_i(\boldsymbol{\alpha})}^{\omega_i(\boldsymbol{\alpha})} R[\boldsymbol{\varphi}(\mathbf{x}), \boldsymbol{\psi}(\mathbf{x}); \boldsymbol{\alpha}] \delta(\mathbf{x} - \mathbf{x}_p) dx_i, \tag{2.16}$$

where $\delta(\mathbf{x} - \mathbf{x}_p)$ denotes the multidimensional Dirac-delta functional. Responses which occur in many fields (e.g., optimization, control, model verification, data assimilation, model validation, model calibration, predictive modeling) are Lagrangians having the following functional form:

$$\begin{aligned} R[\boldsymbol{\varphi}(\mathbf{x}), \boldsymbol{\psi}(\mathbf{x}); \boldsymbol{\alpha}] \triangleq \prod_{i=1}^{TI} \int_{\lambda_i(\boldsymbol{\alpha})}^{\omega_i(\boldsymbol{\alpha})} & \{\boldsymbol{\varphi}(\mathbf{x}) \mathbf{q}_{\psi}(\mathbf{x}; \boldsymbol{\alpha}) \\ & + \boldsymbol{\psi}(\mathbf{x}) [\mathbf{L}(\boldsymbol{\alpha}) \boldsymbol{\varphi}(\mathbf{x}) - \mathbf{q}_{\varphi}(\mathbf{x}; \boldsymbol{\alpha})] \} dx_i. \end{aligned} \tag{2.17}$$

A particular form of the response defined in (2.17) is the Levine and Schwinger (1949) "normalization-free Lagrangian," which has the following form:

$$
\begin{aligned}
R[\boldsymbol{\varphi}(\mathbf{x}),\boldsymbol{\psi}(\mathbf{x});\boldsymbol{\alpha}] &\triangleq \prod_{i=1}^{TI}\int_{\lambda_i(\boldsymbol{\alpha})}^{\omega_i(\boldsymbol{\alpha})} \boldsymbol{\varphi}(\mathbf{x})\mathbf{q}_{\psi}(\mathbf{x};\boldsymbol{\alpha})dx_i \\
&\times \prod_{i=1}^{TI}\int_{\lambda_i(\boldsymbol{\alpha})}^{\omega_i(\boldsymbol{\alpha})} \boldsymbol{\psi}(\mathbf{x})\mathbf{q}_{\varphi}(\mathbf{x};\boldsymbol{\alpha})\,dx_i \Big/ \prod_{i=1}^{TI}\int_{\lambda_i(\boldsymbol{\alpha})}^{\omega_i(\boldsymbol{\alpha})} \boldsymbol{\psi}(\mathbf{x})\mathbf{L}(\mathbf{x};\boldsymbol{\alpha})\boldsymbol{\varphi}(\mathbf{x};\mathbf{x})\,dx_i.
\end{aligned}
\tag{2.18}
$$

The system responses defined in (2.14) through (2.18) indicate that their sensitivities (i.e., functional derivatives) with respect to the model parameters can be studied by considering a generic response of the following form:

$$
R[\boldsymbol{\varphi}(\mathbf{x}),\boldsymbol{\psi}(\mathbf{x});\boldsymbol{\alpha}] \triangleq \prod_{i=1}^{TI}\int_{\lambda_i(\boldsymbol{\alpha})}^{\omega_i(\boldsymbol{\alpha})} S[\boldsymbol{\varphi}(\mathbf{x}),\boldsymbol{\psi}(\mathbf{x});\boldsymbol{\alpha}]dx_i, \tag{2.19}
$$

where $S[\boldsymbol{\varphi}(\mathbf{x}),\boldsymbol{\psi}(\mathbf{x});\boldsymbol{\alpha}]$ denotes a suitably Gateaux- (G-) differentiable function of the indicated arguments. In general, $S[\boldsymbol{\varphi}(\mathbf{x}),\boldsymbol{\psi}(\mathbf{x});\boldsymbol{\alpha}]$ is nonlinear in $\boldsymbol{\varphi}(\mathbf{x})$, $\boldsymbol{\psi}(\mathbf{x})$, and $\boldsymbol{\alpha}$, and the components of $\boldsymbol{\alpha}$ are considered to also include parameters that may specifically appear only in the definition of the response under consideration (but which might not appear in the definition of the model). Thus, the (physical) "system" is defined and understood to comprise both the system's computational model and the system's response.

It is important to note that the response of a linear system is almost always a nonlinear function of the model's parameters, even if it is a linear function of the forward or adjoint state functions. Only in particularly trivial situations is the system response a linear function of the system parameters. As a simple example, consider a linear system modeled by the following linear evolution equation for the quantity $\varphi(t)$ in the (independent) time variable t:

$$
\frac{d\varphi(t)}{dt} + \mu\varphi(t) = q, \quad 0 < t < t_f, \tag{2.20}
$$

$$
\varphi(0) = 0. \tag{2.21}
$$

Consider that the system response of interest is the value $\varphi(t_0)$ of the system's state function at a time-instance $t = t_0 < t_f$. This response is represented mathematically as follows:

$$
\varphi(t_0) = \int_0^{t_f} \varphi(t)\delta(t - t_0)\,dt. \tag{2.22}
$$

The quantities μ, q, t_0, and t_f are model parameters. Note that both the physical system, modeled by (2.20) and (2.21), as well as the system response, i.e., (2.22), are linear in the state function $\varphi(t)$. Nevertheless, the response $\varphi(t_0)$ is nonlinear in the model parameters, as can be readily seen by deriving its explicit expression, which is provided below:

$$\varphi(t_0) = \frac{q}{\mu}[1 - \exp(-\mu t_0)]. \tag{2.23}$$

Of course, if the response were ab initio a nonlinear functional of $\varphi(t)$, it would evidently be a nonlinear function of the model parameters, as well.

The generic response defined in (2.19) provides the basis for constructing any other responses of specific interest and will therefore be used for the generic "*nth-Order Comprehensive Adjoint Sensitivity Analysis Methodology for Response-Coupled Forward/Adjoint Linear Systems,*" which will be abbreviated as "*nth-CASAM-L*" (where "*n*" is a finite number that indicates any desired order) and which will be developed in the remainder of this chapter. Note that the generic response defined in (2.19) is, in general, a nonlinear function of all of its arguments, i.e., $S[\boldsymbol{\varphi}(\mathbf{x}), \boldsymbol{\psi}(\mathbf{x}); \boldsymbol{\alpha}]$ is nonlinear in $\boldsymbol{\varphi}(\mathbf{x})$, $\boldsymbol{\psi}(\mathbf{x})$, and $\boldsymbol{\alpha}$.

2.3 The First-Order Comprehensive Sensitivity Analysis Methodology for Response-Coupled Linear Forward and Adjoint Systems (1st-CASAM-L)

The model parameters α_i are imprecisely known quantities, so their actual values may differ from their nominal values by quantities denoted as $\delta\alpha_i \triangleq \alpha_i - \alpha_i^0, i = 1, \ldots, TP$. Since the model parameters $\boldsymbol{\alpha}$ and the state functions are related to each other through the forward and adjoint systems, it follows that variations $\delta\boldsymbol{\alpha} \triangleq (\delta\alpha_1, \ldots, \delta\alpha_{TP})$ in the model parameters will cause corresponding variations $\delta\boldsymbol{\varphi} \triangleq (\delta\varphi_1, \ldots, \delta\varphi_{TD})$, $\delta\varphi_i \triangleq \varphi_i - \varphi_i^0$, $i = 1, \ldots, TD$ and $\delta\boldsymbol{\psi} \triangleq (\delta\psi_1, \ldots, \delta\psi_{TD})$, $\delta\psi_i \triangleq \psi_i - \psi_i^0$, $i = 1, \ldots, TD$ in the forward and, respectively, adjoint state functions. In turn, the variations $\delta\boldsymbol{\alpha}$, $\delta\boldsymbol{\varphi}$, and $\delta\boldsymbol{\psi}$ will cause a response variation $R(\boldsymbol{\varphi}^0 + \delta\boldsymbol{\varphi}; \boldsymbol{\psi}^0 + \delta\boldsymbol{\psi}; \boldsymbol{\alpha}^0 + \delta\boldsymbol{\alpha})$ around the nominal response value $R[\boldsymbol{\varphi}^0(\mathbf{x}), \boldsymbol{\psi}^0(\mathbf{x}); \boldsymbol{\alpha}^0]$.

The first-order sensitivities of the response, $R[\boldsymbol{\varphi}(\mathbf{x}), \boldsymbol{\psi}(\mathbf{x}); \boldsymbol{\alpha}]$, can be computed *approximately* using the well-known finite-difference formula presented below:

$$\begin{aligned}
&\frac{\partial R(\boldsymbol{\alpha})}{\partial \alpha_j} \approx \frac{1}{2h_j}\left(R_{j+1} - R_{j-1}\right) + O\left(h_j^2\right), \\
&R_{j+1} \triangleq R\left[\boldsymbol{\varphi}\left(\alpha_j + h_j; \mathbf{x}\right), \boldsymbol{\psi}\left(\alpha_j + h_j; \mathbf{x}\right); \alpha_j + h_j\right], \\
&R_{j-1} \triangleq R\left[\boldsymbol{\varphi}\left(\alpha_j - h_j; \mathbf{x}\right), \boldsymbol{\psi}\left(\alpha_j - h_j; \mathbf{x}\right); \alpha_j - h_j\right],
\end{aligned} \tag{2.24}$$

where h_j denotes a "judiciously chosen" variation in the parameter α_j around its nominal value α_j^0. The values R_{j+1} and R_{j-1} are obtained by re-solving the original forward and adjoint systems of equations repeatedly, using the changed parameter values $(\alpha_j \pm h_j)$. The value of the variation h_j is chosen by "trial and error" for each parameter α_j, and varies from parameter to parameter. If the value of h_j is too large or too small, the result produced by (2.24) will be in considerable error from the exact value of the derivative $\partial R(\boldsymbol{\alpha})/\partial \alpha_j$, and this negative outcome may be exacerbated by the fact that the user may not be aware of this error since the exact result for the respective sensitivity can be computed only by using the "forward or the adjoint sensitivity analysis methodologies" to be presented in this section. It is important to note that finite difference formulas introduce their intrinsic "methodological errors," such as the error $O(h_j^2)$ indicated in (2.24), which are in addition to, and independent of, the errors that might be incurred in the computation of $R[\boldsymbol{\varphi}^0(\mathbf{x}), \boldsymbol{\psi}^0(\mathbf{x}); \boldsymbol{\alpha}^0]$. In other words, even if the computation of $R[\boldsymbol{\varphi}^0(\mathbf{x}), \boldsymbol{\psi}^0(\mathbf{x}); \boldsymbol{\alpha}^0]$ were perfect (error-free), the finite-difference formulas nevertheless introduce their own, intrinsic, numerical errors into the computation of the sensitivity $\{dR(\boldsymbol{\alpha})/d\alpha_j\}_{\boldsymbol{\alpha}^0}$. This statement is also valid for any of the statistical methods (e.g., Latin-hypercubes) that might be used in conjunction with re-computations.

The aim of the "*Forward Sensitivity Analysis Methodology*" (FSAM) is to compute the response sensitivities without introducing any methodological approximations and/or errors, as is the case when using statistical methods or finite-difference schemes. The mathematical framework of the FSAM is set in the vector-space of the model parameters $\boldsymbol{\alpha}$. In this vector-space, the first-order partial sensitivity of $R[\boldsymbol{\varphi}(\mathbf{x}), \boldsymbol{\psi}(\mathbf{x}); \boldsymbol{\alpha}]$ to a generic model parameter, α_i, can in principle be obtained by determining the first-order functional derivative of the response and equations underlying the model with respect to a generic model/respond parameter α_i, evaluated at the nominal parameter values $\boldsymbol{\alpha}^0$. Recall that in a linear vector space comprising function-valued elements of the form $\boldsymbol{\eta}(\mathbf{x}) \triangleq [\eta_1(\mathbf{x}), \ldots, \eta_K(\mathbf{x})]^\dagger$, the most comprehensive—and practically computable—concept of a "partial derivative" of a nonlinear operator $F[\eta_1(\mathbf{x}), \ldots, \eta_K(\mathbf{x})]$ with respect to any one of its arguments, η_i, stems from the properties of the first-order *partial Gateaux- (G-) variation*, $\delta F_i[\eta_1(\mathbf{x}), \ldots, \eta_K(\mathbf{x}); h_i(\mathbf{x})]$, of $F[\eta_1(\mathbf{x}), \ldots, \eta_K(\mathbf{x})]$, at a point η_i^0 which is defined as follows:

$$\begin{aligned} &\delta F_i\left[\eta_1(\mathbf{x}), \ldots, \eta_i^0(\mathbf{x}), \ldots, \eta_K(\mathbf{x}); h_i(\mathbf{x})\right] \triangleq \{\delta F_i[\ldots, \eta_i(\mathbf{x}), \ldots; h_i(\mathbf{x})]\}_{\eta_i^0} \\ &\triangleq \left\{\frac{d}{d\varepsilon} F\left[\eta_1(\mathbf{x}), \ldots, \eta_i^0(\mathbf{x}) + \varepsilon h_i(\mathbf{x}), \ldots, \eta_K(\mathbf{x})\right]\right\}_{\varepsilon=0}, \end{aligned} \tag{2.25}$$

where ε is a scalar, i.e., $\varepsilon \in \mathrm{F}$, where F denotes the underlying field of real scalars, and where $h_i(\mathbf{x})$ denotes an arbitrary variation in $\eta_i(\mathbf{x})$. The convenient notation $\{G(x)\}_y$ will be used to denote that the quantity within the braces [in this example: $G(x)$] is to be evaluated at the quantity that appears as a subscript to the right side of the respective braces [in this example: at $x = y$].

The first-order *total* G-variation $\delta F_i[\eta_1(\mathbf{x}),\ldots,\eta_K(\mathbf{x});h_1(\mathbf{x}),\ldots,h_K(\mathbf{x})]$ of $F[\eta_1(\mathbf{x}),\ldots,\eta_K(\mathbf{x})]$ for arbitrary variations $\mathbf{h}\triangleq(h_1,\ldots,h_K)^\dagger$ is defined as follows:

$$\begin{aligned}&\{\delta F_i[\eta_1(\mathbf{x}),\ldots,\eta_K(\mathbf{x});h_1(\mathbf{x}),\ldots,h_K(\mathbf{x})]\}_{\boldsymbol{\eta}^0}\\&\triangleq\left\{\frac{d}{d\varepsilon}F\left[\eta_1^0(\mathbf{x})+\varepsilon h_1(\mathbf{x}),\ldots,\eta_K^0(\mathbf{x})+\varepsilon h_K(\mathbf{x})\right]\right\}_{\varepsilon=0}.\end{aligned}\tag{2.26}$$

The first-order G-variation $\delta F(\boldsymbol{\eta};\mathbf{h})$ is an operator defined on the same domain as $F(\boldsymbol{\eta})$, and has the same range as $F(\boldsymbol{\eta})$. The G-variation $\delta F(\boldsymbol{\eta};\mathbf{h})$ satisfies the relation $F(\boldsymbol{\eta}+\varepsilon\mathbf{h})-F(\boldsymbol{\eta})=\delta F(\boldsymbol{\eta};\mathbf{h})+\Delta(\mathbf{h})$, with $\lim_{\varepsilon\to 0}[\Delta(\varepsilon\mathbf{h})]/\varepsilon=0$. The existence of the G-variation $\delta F(\boldsymbol{\eta};\mathbf{h})$ does not guarantee its numerical computability. Numerical methods (e.g., Newton's method and variants thereof) most often require that $\delta F(\boldsymbol{\eta};\mathbf{h})$ be linear in the variations $\mathbf{h}$ in a neighborhood $(\boldsymbol{\eta}^0+\varepsilon\mathbf{h})$ around $\boldsymbol{\eta}^0$. The necessary and sufficient conditions for the G-differential $\delta F(\boldsymbol{\eta};\mathbf{h})$ of a nonlinear operator $F(\boldsymbol{\eta})$ to be linear in the variations $\mathbf{h}$ in a neighborhood $(\boldsymbol{\eta}+\varepsilon\mathbf{h})$ around $\boldsymbol{\eta}$ are as follows:

1. $F(\boldsymbol{\eta})$ satisfies a weak Lipschitz condition at $\boldsymbol{\eta}^0$; i.e.,

$$\left\|F\left(\boldsymbol{\eta}^0+\varepsilon\mathbf{h}\right)-F\left(\boldsymbol{\eta}^0\right)\right\|\le k\left\|\varepsilon\boldsymbol{\eta}^0\right\|,\ k<\infty.\tag{2.27}$$

2. For two arbitrary vectors of variations $\mathbf{h}_1$ and $\mathbf{h}_2$, the operator $F(\boldsymbol{\eta})$ satisfies the relation

$$F\left(\boldsymbol{\eta}^0+\varepsilon\mathbf{h}_1+\varepsilon\mathbf{h}_2\right)-F\left(\boldsymbol{\eta}^0+\varepsilon\mathbf{h}_1\right)-F\left(\boldsymbol{\eta}^0+\varepsilon\mathbf{h}_2\right)+F\left(\boldsymbol{\eta}^0\right)=o(\varepsilon).\tag{2.28}$$

In practice, it is not necessary to investigate if $F(\boldsymbol{\eta})$ satisfies the conditions presented in (2.27) and (2.28) since it is usually evident if the right side of (2.26) is linear (or not) in the variations $\mathbf{h}$. *A total variation* $\delta F(\boldsymbol{\eta};\mathbf{h})$ *which is linear in* $\mathbf{h}$ *is called the "total differential of* $F(\boldsymbol{\eta})$" and is usually denoted as $DF(\boldsymbol{\eta};\mathbf{h})$. In this case, the partial variation $\delta F_i(\eta_1,\ldots,\eta_K;h_i)$ is called the "partial differential $\delta F_i(\eta_1,\ldots,\eta_K;h_i)$ of $F(\boldsymbol{\eta})$ with respect to η_i." Thus, the total differential $DF(\boldsymbol{\eta};\mathbf{h})$ is linear in the variations h_i and the following representation holds:

$$DF(\eta_1,\ldots,\eta_K;h_1,\ldots,h_K)=\sum_{i=1}^{K}\frac{\partial F(\eta_1,\ldots,\eta_K)}{\partial\eta_i}h_i,\tag{2.29}$$

where the quantities $\partial F(\eta_1,\ldots,\eta_K)/\partial\eta_i$ denote the partial derivatives of $F(\boldsymbol{\eta})$ with respect to its arguments η_i. It will henceforth be assumed that all of the operators considered in this chapter satisfy the conditions presented in (2.27) and (2.28), and therefore admit partial G-derivatives such that the representation shown in (2.29) exists.

The sensitivity $\{\partial R[\boldsymbol{\varphi}(\mathbf{x}),\boldsymbol{\psi}(\mathbf{x});\boldsymbol{\alpha}]/\partial\alpha_j\}_{(\boldsymbol{\alpha}^0)}$ of $R[\boldsymbol{\varphi}(\mathbf{x}),\boldsymbol{\psi}(\mathbf{x});\boldsymbol{\alpha}]$ to a model parameter α_j, evaluated at the nominal parameter values $\boldsymbol{\alpha}^0$, is obtained by determining the partial G-differential $\delta R[\boldsymbol{\varphi}^0(\mathbf{x}),\boldsymbol{\psi}^0(\mathbf{x});\boldsymbol{\alpha}^0;\delta\alpha_{j_1}] = \{\partial R[\boldsymbol{\varphi}(\mathbf{x}),\boldsymbol{\psi}(\mathbf{x});\boldsymbol{\alpha}]/\partial\alpha_{j_1}\}_{(\boldsymbol{\alpha}^0)}\delta\alpha_{j_1}$ of $R[\boldsymbol{\varphi}(\mathbf{x}),\boldsymbol{\psi}(\mathbf{x});\boldsymbol{\alpha}]$ with respect to the respective parameter α_{j_1}, for each $j_1 = 1, \ldots, TP$. Since the forward state function $\boldsymbol{\varphi}(\mathbf{x})$ and the adjoint state function $\boldsymbol{\psi}(\mathbf{x})$ are the solutions of (2.9)–(2.12), it follows that both $\boldsymbol{\varphi}(\mathbf{x})$ and $\boldsymbol{\psi}(\mathbf{x})$ are implicit functions of the model parameters $\boldsymbol{\alpha} \triangleq (\alpha_1,\ldots,\alpha_{TP})^\dagger$. Hence, considering that $\boldsymbol{\varphi}(\mathbf{x}) = \boldsymbol{\varphi}(\mathbf{x};\boldsymbol{\alpha})$, $\boldsymbol{\psi}(\mathbf{x}) = \boldsymbol{\psi}(\mathbf{x};\boldsymbol{\alpha})$, and applying the definition provided in (2.25) to the response defined in (2.19) yields the following expression for the first-order partial sensitivity $\{\partial R[\boldsymbol{\varphi}(\mathbf{x}),\boldsymbol{\psi}(\mathbf{x});\boldsymbol{\alpha}]/\partial\alpha_{j_1}\}_{(\boldsymbol{\alpha}^0)}$, for each $j_1 = 1, \ldots, TP$:

$$
\begin{aligned}
&\left\{\frac{\partial R[\boldsymbol{\varphi}(\mathbf{x}),\boldsymbol{\psi}(\mathbf{x});\boldsymbol{\alpha}]}{\partial\alpha_{j_1}}\right\}_{\boldsymbol{\alpha}^0} \\
&\triangleq \frac{1}{\delta\alpha_{j_1}}\left\{\frac{d}{d\varepsilon}R\left[\boldsymbol{\varphi}\left(\alpha_{j_1}^0+\varepsilon\delta\alpha_{j_1}\right);\boldsymbol{\psi}\left(\alpha_{j_1}^0+\varepsilon\delta\alpha_{j_1}\right);\left(\alpha_{j_1}^0+\varepsilon\delta\alpha_{j_1}\right)\right]\right\}_{\varepsilon=0} \\
&=\prod_{k=1}^{TI}\left\{\int\limits_{\lambda_k(\boldsymbol{\alpha})}^{\omega_k(\boldsymbol{\alpha})} dx_k\,\frac{\partial S[\boldsymbol{\varphi}(\mathbf{x}),\boldsymbol{\psi}(\mathbf{x});\boldsymbol{\alpha}]}{\partial\alpha_{j_1}}\right\}_{\boldsymbol{\alpha}^0} \\
&+\sum_{k=1}^{TI}\prod_{m=1,k\neq m}^{TI}\left\{\int\limits_{\lambda_m(\boldsymbol{\alpha})}^{\omega_m(\boldsymbol{\alpha})} dx_m S[\boldsymbol{\varphi}(\ldots,\omega_k,\ldots),\boldsymbol{\psi}(\ldots,\omega_k,\ldots);\boldsymbol{\alpha}]\frac{\partial\omega_k(\boldsymbol{\alpha})}{\partial\alpha_{j_1}}\right. \\
&\left.-S[\boldsymbol{\varphi}(\ldots,\lambda_k,\ldots),\boldsymbol{\psi}(\ldots,\lambda_k,\ldots,);\boldsymbol{\alpha}]\frac{\partial\lambda_k(\boldsymbol{\alpha})}{\partial\alpha_{j_1}}\right\}_{\boldsymbol{\alpha}^0} \\
&+\prod_{k=1}^{TI}\left\{\int\limits_{\lambda_k(\boldsymbol{\alpha})}^{\omega_k(\boldsymbol{\alpha})} dx_k\,\frac{\partial S[\boldsymbol{\varphi}(\mathbf{x}),\boldsymbol{\psi}(\mathbf{x});\boldsymbol{\alpha}]}{\partial\boldsymbol{\varphi}}\frac{\partial\boldsymbol{\varphi}}{\partial\alpha_{j_1}}+\frac{\partial S[\boldsymbol{\varphi}(\mathbf{x}),\boldsymbol{\psi}(\mathbf{x});\boldsymbol{\alpha}]}{\partial\boldsymbol{\psi}}\frac{\partial\boldsymbol{\psi}}{\partial\alpha_{j_1}}\right\}_{\boldsymbol{\alpha}^0}.
\end{aligned}
\tag{2.30}
$$

The functions $\partial\boldsymbol{\varphi}(\mathbf{x})/\partial\alpha_{j_1}$ and $\partial\boldsymbol{\psi}(\mathbf{x})/\partial\alpha_{j_1}$ are obtained by solving the system of equations obtained by determining the first partial G-derivatives of (2.1), (2.2), (2.7), and (2.8) with respect to the model parameter α_{j_1}. Applying the definition provided in (2.25) to (2.1), (2.2), (2.7), and (2.8) yields, after canceling the scalar parameter variation $\delta\alpha_{j_1}$ on both sides of the equal signs in resulting expressions, the following "*First-Order Forward Sensitivity System*" (1st-OFSS):

$$\begin{aligned} &\mathbf{L}\left(\mathbf{x};\boldsymbol{\alpha}^0\right)\frac{\partial\boldsymbol{\varphi}}{\partial\alpha_{j_1}}=\left\{\mathbf{f}^{(1)}(j_1;\boldsymbol{\varphi};\boldsymbol{\alpha})\right\}_{(\boldsymbol{\alpha}^0)};\mathbf{x}\in\Omega_x;j_1=1,\ldots,TP;\\ &\mathbf{f}^{(1)}(j_1;\boldsymbol{\varphi};\boldsymbol{\alpha})\triangleq\frac{\partial\mathbf{q}_\varphi(\mathbf{x};\boldsymbol{\alpha})}{\partial\alpha_{j_1}}-\frac{\partial\mathbf{L}(\mathbf{x};\boldsymbol{\alpha})}{\partial\alpha_{j_1}}\boldsymbol{\varphi}(\mathbf{x});\end{aligned}\tag{2.31}$$

$$\begin{aligned} &\left\{\mathbf{B}_\varphi(\mathbf{x};\boldsymbol{\alpha})\right\}_{(\boldsymbol{\alpha}^0)}\frac{\partial\boldsymbol{\varphi}}{\partial\alpha_{j_1}}=\left\{\mathbf{c}_{\boldsymbol{\varphi}}^{(1)}(j_1;\boldsymbol{\varphi};\boldsymbol{\alpha})\right\}_{(\boldsymbol{\alpha}^0)};\mathbf{x}\in\partial\Omega_x\left(\boldsymbol{\alpha}^0\right);\\ &\mathbf{c}_\varphi^{(1)}(j_1;\boldsymbol{\varphi};\boldsymbol{\alpha})\triangleq\frac{\partial\mathbf{c}_\varphi(\mathbf{x};\boldsymbol{\alpha})}{\partial\alpha_{j_1}}-\frac{\partial \boldsymbol{B}_\varphi(\mathbf{x};\boldsymbol{\alpha})}{\partial\alpha_{j_1}}\boldsymbol{\varphi}(\mathbf{x});\end{aligned}\tag{2.32}$$

$$\begin{aligned} &\mathbf{L}^*\left(\mathbf{x};\boldsymbol{\alpha}^0\right)\frac{\partial\boldsymbol{\psi}(\mathbf{x})}{\partial\alpha_{j_1}}=\left\{\mathbf{g}^{(1)}(j_1;\boldsymbol{\psi};\boldsymbol{\alpha})\right\}_{(\boldsymbol{\alpha}^0)};\mathbf{x}\in\Omega_x;\ j_1=1,\ldots,TP;\\ &\mathbf{g}^{(1)}(j_1;\boldsymbol{\psi};\boldsymbol{\alpha})\triangleq\frac{\partial\mathbf{q}_\psi(\mathbf{x};\boldsymbol{\alpha})}{\partial\alpha_{j_1}}-\frac{\partial\mathbf{L}^*(\boldsymbol{\alpha})}{\partial\alpha_{j_1}}\boldsymbol{\psi}(\mathbf{x});\end{aligned}\tag{2.33}$$

$$\begin{aligned} &\left\{\mathbf{B}_\psi(\mathbf{x};\boldsymbol{\alpha})\right\}_{(\boldsymbol{\alpha}^0)}\frac{\partial\boldsymbol{\psi}(\mathbf{x})}{\partial\alpha_{j_1}}=\left\{\mathbf{c}_\psi^{(1)}(j_1;\boldsymbol{\psi};\boldsymbol{\alpha})\right\}_{(\boldsymbol{\alpha}^0)};\mathbf{x}\in\partial\Omega_x\left(\boldsymbol{\alpha}^0\right);\\ &\mathbf{c}_\psi^{(1)}(j_1;\boldsymbol{\psi};\boldsymbol{\alpha})\triangleq-\frac{\partial\mathbf{B}_\psi(\mathbf{x};\boldsymbol{\alpha})}{\partial\alpha_{j_1}}\boldsymbol{\psi}(\mathbf{x})+\frac{\partial\mathbf{c}_\psi(\mathbf{x};\boldsymbol{\alpha})}{\partial\alpha_{j_1}}.\end{aligned}\tag{2.34}$$

The system of equations comprising (2.31)–(2.34) is called the "*First-Order Forward Sensitivity System*" (1st-OFSS) because its solutions are the first-order derivatives of the forward and adjoint state functions $\partial\boldsymbol{\varphi}(\mathbf{x})/\partial\alpha_{j_1}$ and $\partial\boldsymbol{\psi}(\mathbf{x})/\partial\alpha_{j_1}$, respectively. Evidently, the 1st-OFSS would need to be solved 2*TP*-times, with different sources on their right sides, in order to compute the first-order derivatives $\partial\boldsymbol{\varphi}(\mathbf{x})/\partial\alpha_{j_1}$ and $\partial\boldsymbol{\psi}(\mathbf{x})/\partial\alpha_{j_1}$ of the forward and, respectively, adjoint state functions with respect to all model parameters $\alpha_{j_1}, j_1 = 1, \ldots, TP$. Subsequently, the solutions $\partial\boldsymbol{\varphi}(\mathbf{x})/\partial\alpha_{j_1}$ and $\partial\boldsymbol{\psi}(\mathbf{x})/\partial\alpha_{j_1}$ of the 1st-OFSS would be used in (2.30) to compute the first-order response sensitivity $\left\{\partial R[\boldsymbol{\varphi}(\mathbf{x}),\boldsymbol{\psi}(\mathbf{x});\boldsymbol{\alpha}]/\partial\alpha_{j_1}\right\}_{(\boldsymbol{\alpha}^0)}$, for each $j_1 = 1, \ldots, TP$, using quadrature formulas. The quadrature computations are small-scale inexpensive computations but solving the 1st-OFSS entails large-scale computations, even if the same linear operators, namely $\mathbf{L}(\mathbf{x};\boldsymbol{\alpha})$ and $\mathbf{L}^*(\mathbf{x};\boldsymbol{\alpha})$, would need to be inverted each time (which means that the same code/solvers would be used, without needing additional programming for inverting these operators). Thus, solving the 1st-OFSS is advantageous only if the model under consideration comprises just a few parameters, i.e., the number *TP* of model parameters is considerably less than the number of responses of interest. This is rarely the case in practice, however, since most problems of practical interest are characterized by many model parameters and comparatively few responses.

The aim of the *First-order* ***C****omprehensive* ***A****djoint* ***S****ensitivity* ***A****nalysis* ***M****ethodology for response-coupled* ***L****inear forward and adjoint systems* (acronym: 1st-CASAM-L) is to find an alternative way for expressing the contribution of the flux variation (the so-called "indirect-effect term" contribution) to the total response sensitivity, so as to avoid the need for having to compute the derivatives $\partial \boldsymbol{\varphi}/\partial \alpha_{j_1}$ and $\partial \boldsymbol{\psi}/\partial \alpha_{j_1}$ of the forward and adjoint state functions with respect to the model's parameters.

The first-order Gateaux- (G-) variation, denoted as $\delta R(\boldsymbol{\varphi}^0, \boldsymbol{\psi}^0, \boldsymbol{\alpha}^0; \delta\boldsymbol{\varphi}, \delta\boldsymbol{\psi}, \delta\boldsymbol{\alpha})$, of the response $\mathbf{R}[\boldsymbol{\varphi}(\mathbf{x}),\boldsymbol{\psi}(\mathbf{x}); \boldsymbol{\alpha}]$ for arbitrary variations $\delta\boldsymbol{\varphi}(\mathbf{x})$, $\delta\boldsymbol{\psi}(\mathbf{x})$, $\delta\boldsymbol{\alpha}$ in the model parameters and state functions, in a neighborhood $[\boldsymbol{\varphi}^0(\mathbf{x}) + \varepsilon\delta\boldsymbol{\varphi}(\mathbf{x}), \boldsymbol{\psi}^0(\mathbf{x}) + \varepsilon\delta\boldsymbol{\psi}(\mathbf{x}); \boldsymbol{\alpha}^0 + \varepsilon\delta\boldsymbol{\alpha}]$ around $(\boldsymbol{\varphi}^0, \boldsymbol{\psi}^0, \boldsymbol{\alpha}^0)$, where $\varepsilon \in$ F is a real scalar (F denotes the underlying field of scalars), is defined as follows:

$$\begin{aligned}
&\delta R\left(\boldsymbol{\varphi}^0, \boldsymbol{\psi}^0, \boldsymbol{\alpha}^0; \delta\boldsymbol{\varphi}, \delta\boldsymbol{\psi}, \delta\boldsymbol{\alpha}\right) \\
&\triangleq \left\{\frac{d}{d\varepsilon} R\left[\boldsymbol{\varphi}^0(\mathbf{x}) + \varepsilon\delta\boldsymbol{\varphi}(\mathbf{x}), \boldsymbol{\psi}^0(\mathbf{x}) + \varepsilon\delta\boldsymbol{\psi}(\mathbf{x}); \boldsymbol{\alpha}^0 + \varepsilon\delta\boldsymbol{\alpha}\right]\right\}_{\varepsilon=0} \\
&\triangleq \left\{\delta R\left(\boldsymbol{\varphi}^0, \boldsymbol{\psi}^0, \boldsymbol{\alpha}^0; \delta\boldsymbol{\varphi}, \delta\boldsymbol{\psi}\right)\right\}_{ind} + \left\{\delta R\left(\boldsymbol{\varphi}^0, \boldsymbol{\psi}^0, \boldsymbol{\alpha}^0; \delta\boldsymbol{\alpha}\right)\right\}_{dir},
\end{aligned} \tag{2.35}$$

where the "direct-effect" term $\{\delta R(\boldsymbol{\varphi}^0, \boldsymbol{\psi}^0, \boldsymbol{\alpha}^0; \delta\boldsymbol{\alpha})\}_{dir}$ depends only on the parameter variations $\delta\boldsymbol{\alpha}$ and is defined as follows:

$$\begin{aligned}
&\left\{\delta R\left(\boldsymbol{\varphi}^0, \boldsymbol{\psi}^0, \boldsymbol{\alpha}^0; \delta\boldsymbol{\alpha}\right)\right\}_{dir} \triangleq \left\{\frac{\partial}{\partial\boldsymbol{\alpha}} \prod_{k=1}^{TI} \int_{\lambda_k(\boldsymbol{\alpha})}^{\omega_k(\boldsymbol{\alpha})} dx_k S\left[\boldsymbol{\varphi}(\mathbf{x}), \boldsymbol{\psi}(\mathbf{x}); \boldsymbol{\alpha}\right]\right\}_{\boldsymbol{\alpha}^0} \delta\boldsymbol{\alpha} \\
&= \prod_{k=1}^{TI} \left\{\int_{\lambda_k(\boldsymbol{\alpha})}^{\omega_k(\boldsymbol{\alpha})} dx_k \frac{\partial S\left[\boldsymbol{\varphi}(\mathbf{x}), \boldsymbol{\psi}(\mathbf{x}); \boldsymbol{\alpha}\right]}{\partial\boldsymbol{\alpha}} \delta\boldsymbol{\alpha}\right\}_{\boldsymbol{\alpha}^0} \\
&+ \sum_{k=1}^{TI} \prod_{m=1, k\neq m}^{TI} \left\{\int_{\lambda_m(\boldsymbol{\alpha})}^{\omega_m(\boldsymbol{\alpha})} dx_m S\left[\boldsymbol{\varphi}(\ldots, \omega_k, \ldots), \boldsymbol{\psi}(\ldots, \omega_k, \ldots); \boldsymbol{\alpha}\right] \frac{\partial \omega_k(\boldsymbol{\alpha})}{\partial\boldsymbol{\alpha}}\right\}_{\boldsymbol{\alpha}^0} \delta\boldsymbol{\alpha} \\
&- \sum_{k=1}^{TI} \prod_{m=1, k\neq m}^{TI} \left\{\int_{\lambda_m(\boldsymbol{\alpha})}^{\omega_m(\boldsymbol{\alpha})} dx_m S\left[\boldsymbol{\varphi}(\ldots, \lambda_k, \ldots), \boldsymbol{\psi}(\ldots, \lambda_k, \ldots); \boldsymbol{\alpha}\right] \frac{\partial \lambda_k(\boldsymbol{\alpha})}{\partial\boldsymbol{\alpha}}\right\}_{\boldsymbol{\alpha}^0} \delta\boldsymbol{\alpha},
\end{aligned} \tag{2.36}$$

and where the "indirect-effect" term $\{\delta R(\boldsymbol{\varphi}^0, \boldsymbol{\psi}^0, \boldsymbol{\alpha}^0; \delta\boldsymbol{\varphi}, \delta\boldsymbol{\psi})\}_{ind}$ depends only on the variations $\delta\boldsymbol{\varphi}$ and $\delta\boldsymbol{\psi}$ in the state functions, and is defined as follows:

$$\begin{aligned}
&\left\{\delta R\left(\boldsymbol{\varphi}^0, \boldsymbol{\psi}^0, \boldsymbol{\alpha}^0; \delta\boldsymbol{\varphi}, \delta\boldsymbol{\psi}\right)\right\}_{ind} \\
&\triangleq \prod_{k=1}^{TI} \left\{\int_{\lambda_k(\boldsymbol{\alpha})}^{\omega_k(\boldsymbol{\alpha})} dx_k \left[\frac{\partial S\left[\boldsymbol{\varphi}(\mathbf{x}), \boldsymbol{\psi}(\mathbf{x}); \boldsymbol{\alpha}\right]}{\partial\boldsymbol{\varphi}} \delta\boldsymbol{\varphi} + \frac{\partial S\left[\boldsymbol{\varphi}(\mathbf{x}), \boldsymbol{\psi}(\mathbf{x}); \boldsymbol{\alpha}\right]}{\partial\boldsymbol{\psi}} \delta\boldsymbol{\psi}\right]\right\}_{\boldsymbol{\alpha}^0} .
\end{aligned} \tag{2.37}$$

In (2.36) and (2.37), the notation $\{\}_{\boldsymbol{\alpha}^0}$ has been used to indicate that the quantity within the brackets is to be evaluated at the nominal values of the parameters *and* state functions. This simplified notation is justified by the fact that when the parameters take on their nominal values, it implicitly means that the corresponding state functions also take on their corresponding nominal values, in view of (2.9)–(2.12). This simplified notation will be used throughout this book.

The "direct effect" term $\{\delta R(\boldsymbol{\varphi}^0, \boldsymbol{\psi}^0, \boldsymbol{\alpha}^0; \delta\boldsymbol{\alpha})\}_{dir}$ defined in (2.36) depends directly on the parameter variations $\delta\boldsymbol{\alpha}$, and can be computed immediately since it does not depend on the variations $\delta\boldsymbol{\varphi}(\mathbf{x})$ and $\delta\boldsymbol{\psi}(\mathbf{x})$. On the other hand, the "indirect effect" term $\{\delta R(\boldsymbol{\varphi}^0, \boldsymbol{\psi}^0, \boldsymbol{\alpha}^0; \delta\boldsymbol{\varphi}, \delta\boldsymbol{\psi})\}_{ind}$ defined in (2.37) depends indirectly on the parameter variations $\delta\boldsymbol{\alpha}$, through the variations $\delta\boldsymbol{\varphi}(\mathbf{x})$ and $\delta\boldsymbol{\psi}(\mathbf{x})$ in the forward and adjoint state functions $\boldsymbol{\varphi}(\mathbf{x})$ and $\boldsymbol{\psi}(\mathbf{x})$, and can be computed only after having computed the values of the variations $\delta\boldsymbol{\varphi}(\mathbf{x})$ and $\delta\boldsymbol{\psi}(\mathbf{x})$. Note that the expression in (2.35) is the total variation/differential that corresponds to the partial G-derivative expressed in (2.30).

The variations $\delta\boldsymbol{\varphi}(\mathbf{x})$ and $\delta\boldsymbol{\psi}(\mathbf{x})$ are the solutions of the system of equations obtained by taking the G-differentials of (2.1), (2.2), (2.7), and (2.8) which can be written in the following matrix-vector form:

$$\left\{\mathbf{V}^{(1)}(\boldsymbol{\alpha})\delta\mathbf{u}^{(1)}(\mathbf{x})\right\}_{\boldsymbol{\alpha}^0} = \left\{\mathbf{q}^{(1)}\left(\mathbf{u}^{(1)}; \boldsymbol{\alpha}; \delta\boldsymbol{\alpha}\right)\right\}_{\boldsymbol{\alpha}^0}, \quad \mathbf{x} \in \Omega_x, \tag{2.38}$$

$$\left\{\mathbf{b}_V^{(1)}\left(\mathbf{u}^{(1)}; \boldsymbol{\alpha}; \delta\mathbf{u}^{(1)}; \delta\boldsymbol{\alpha}\right)\right\}_{\boldsymbol{\alpha}^0} = \mathbf{0}, \quad \mathbf{x} \in \partial\Omega_x\left(\boldsymbol{\alpha}^0\right), \tag{2.39}$$

where

$$\mathbf{V}^{(1)}(\boldsymbol{\alpha}) \triangleq \begin{pmatrix} \mathbf{L}(\boldsymbol{\alpha}) & \mathbf{0} \\ \mathbf{0} & \mathbf{L}^*(\boldsymbol{\alpha}) \end{pmatrix}; \quad \mathbf{u}^{(1)}(\mathbf{x}) \triangleq \begin{pmatrix} \boldsymbol{\varphi}(\mathbf{x}) \\ \boldsymbol{\psi}(\mathbf{x}) \end{pmatrix}; \quad \delta\mathbf{u}^{(1)}(\mathbf{x}) \triangleq \begin{pmatrix} \delta\boldsymbol{\varphi}(\mathbf{x}) \\ \delta\boldsymbol{\psi}(\mathbf{x}) \end{pmatrix};$$
$$\mathbf{q}^{(1)}\left(\mathbf{u}^{(1)}; \boldsymbol{\alpha}; \delta\boldsymbol{\alpha}\right) \triangleq \begin{pmatrix} \mathbf{q}_1^{(1)}(\boldsymbol{\varphi}; \boldsymbol{\alpha}; \delta\boldsymbol{\alpha}) \\ \mathbf{q}_2^{(1)}(\boldsymbol{\psi}; \boldsymbol{\alpha}; \delta\boldsymbol{\alpha}) \end{pmatrix}; \quad \mathbf{b}_V^{(1)}\left(\mathbf{u}^{(1)}; \delta\mathbf{u}^{(1)}; \boldsymbol{\alpha}; \delta\boldsymbol{\alpha}\right) \triangleq \begin{pmatrix} \mathbf{b}_1^{(1)}(\boldsymbol{\varphi}; \boldsymbol{\alpha}; \delta\boldsymbol{\alpha}) \\ \mathbf{b}_2^{(1)}(\boldsymbol{\psi}; \boldsymbol{\alpha}; \delta\boldsymbol{\alpha}) \end{pmatrix}. \tag{2.40}$$

$$\left\{\mathbf{q}_1^{(1)}(\boldsymbol{\varphi}, \boldsymbol{\alpha}; \delta\boldsymbol{\alpha})\right\}_{\boldsymbol{\alpha}^0} \triangleq \left\{\frac{\partial\left[\mathbf{q}_\varphi(\boldsymbol{\alpha}) - \mathbf{L}(\boldsymbol{\alpha})\boldsymbol{\varphi}\right]}{\partial\boldsymbol{\alpha}}\right\}_{\boldsymbol{\alpha}^0} \delta\boldsymbol{\alpha}, \tag{2.41}$$

$$\left\{\mathbf{q}_2^{(1)}(\boldsymbol{\psi}, \boldsymbol{\alpha}; \delta\boldsymbol{\alpha})\right\}_{\boldsymbol{\alpha}^0} \triangleq \left\{\frac{\partial\left[\mathbf{q}_\psi(\boldsymbol{\alpha}) - \mathbf{L}^*(\boldsymbol{\alpha})\boldsymbol{\psi}(\mathbf{x})\right]}{\partial\boldsymbol{\alpha}}\right\}_{\boldsymbol{\alpha}^0} \delta\boldsymbol{\alpha}. \tag{2.42}$$

The matrices $\partial\mathbf{q}_\varphi(\boldsymbol{\alpha})/\partial\boldsymbol{\alpha}$ and $\partial[\mathbf{L}(\boldsymbol{\alpha})\boldsymbol{\varphi}]/\partial\boldsymbol{\alpha}$, which appear on the right side of (2.38), are defined as follows:

$$\frac{\partial \mathbf{q}_{\varphi}(\boldsymbol{\alpha})}{\partial \boldsymbol{\alpha}} \triangleq \begin{pmatrix} \frac{\partial q_{\varphi,1}}{\partial \alpha_1} & \cdots & \frac{\partial q_{\varphi,1}}{\partial \alpha_{TP}} \\ \vdots & \ddots & \vdots \\ \frac{\partial q_{\varphi,TD}}{\partial \alpha_1} & \cdots & \frac{\partial q_{\varphi,TD}}{\partial \alpha_{TP}} \end{pmatrix},$$

$$\frac{\partial [\mathbf{L}(\boldsymbol{\alpha})\boldsymbol{\varphi}]}{\partial \boldsymbol{\alpha}} \triangleq \begin{pmatrix} \frac{\partial \left[\sum_{j=1}^{TD} L_{1,j}(\boldsymbol{\alpha})\varphi_j\right]}{\partial \alpha_1} & \cdots & \frac{\partial \left[\sum_{j=1}^{TD} L_{1,j}(\boldsymbol{\alpha})\varphi_j\right]}{\partial \alpha_{TP}} \\ \vdots & \ddots & \vdots \\ \frac{\partial \left[\sum_{j=1}^{TD} L_{TD,j}(\boldsymbol{\alpha})\varphi_j\right]}{\partial \alpha_1} & \cdots & \frac{\partial \left[\sum_{j=1}^{TD} L_{TD,j}(\boldsymbol{\alpha})\varphi_j\right]}{\partial \alpha_{TP}} \end{pmatrix}. \tag{2.43}$$

The system of equations comprising Eqs. (2.38) and (2.39) will be called the "*1st-Level Variational Sensitivity System*" (1st-LVSS) and its solution, $\delta \mathbf{u}^{(1)}(\mathbf{x})$, will be called the "*1st-Level variational sensitivity function*." The superscript "(1)" used in (2.38) and (2.39) indicates "1st-Level" (as opposed to "first-order") because the 1st-LVSS differs from the 1st-OFSS in that the 1st-LVSS involves the total first-order differential $\delta \mathbf{u}^{(1)}(\mathbf{x})$ of the state functions whereas the 1st-OFSS involves the first-order partial derivatives of the state functions with respect to a single model parameter α_{j_1}. Of course, multiplying each equation of the 1st-OFSS by the corresponding parameter variation $\delta\alpha_{j_1}$ and summing the resulting equations over the index $j_1 = 1, \ldots, TP$ yields the 1st-LVSS. Therefore, solving the 1st-LVSS is equivalent to solving the 1st-OFSS for each possible component of the vectors of parameter variations $\delta\boldsymbol{\alpha}$.

The indirect-effect term $\{\delta R(\boldsymbol{\varphi}^0, \boldsymbol{\psi}^0, \boldsymbol{\alpha}^0; \delta\boldsymbol{\varphi}, \delta\boldsymbol{\psi})\}_{ind}$ defined in (2.37) can be computed by applying the First-Order Comprehensive Adjoint Sensitivity Analysis Methodology (1st-CASAM-L), which avoids the need for computing the functions $\delta\boldsymbol{\varphi}(\mathbf{x})$ and $\delta\boldsymbol{\psi}(\mathbf{x})$. The principles underlying the 1st-CASAM-L are as follows:

1. Introduce a Hilbert space, denoted as H_1, comprising square-integrable vector-valued elements of the form $\boldsymbol{\eta}^{(1)}(\mathbf{x}) \triangleq \left[\boldsymbol{\eta}_1^{(1)}(\mathbf{x}), \boldsymbol{\eta}_2^{(1)}(\mathbf{x})\right]^{\dagger}$, $\boldsymbol{\eta}_i^{(1)}(\mathbf{x}) \triangleq \left[\eta_{i,1}^{(1)}(\mathbf{x}), \ldots, \eta_{i,j}^{(1)}(\mathbf{x}), \ldots, \eta_{i,TD}^{(1)}(\mathbf{x})\right]^{\dagger}$, $i = 1, 2$, and endowed with an inner product between two elements, $\boldsymbol{\eta}^{(1)}(\mathbf{x}) \in H_1$, $\boldsymbol{\xi}^{(1)}(\mathbf{x}) \in H_1$, of this Hilbert space, which will be denoted as $\langle \boldsymbol{\eta}^{(1)}(\mathbf{x}), \boldsymbol{\xi}^{(1)}(\mathbf{x})\rangle_1$ and which is defined as follows:

$$\left\langle \boldsymbol{\eta}^{(1)}(\mathbf{x}), \boldsymbol{\xi}^{(1)}(\mathbf{x}) \right\rangle_1 \triangleq \sum_{i=1}^{2} \left\langle \boldsymbol{\eta}_i^{(1)}(\mathbf{x}), \boldsymbol{\xi}_i^{(1)}(\mathbf{x}) \right\rangle_0. \tag{2.44}$$

2. In the Hilbert H_1, form the inner product of (2.38) with a yet undefined vector-valued function $\mathbf{a}^{(1)}(\mathbf{x}) \triangleq \left[\mathbf{a}_1^{(1)}(\mathbf{x}), \mathbf{a}_2^{(1)}(\mathbf{x})\right]^{\dagger} \in H_1$ to obtain the following relation:

$$\left\{\left\langle \mathbf{a}^{(1)}(\mathbf{x}), \mathbf{V}^{(1)}(\boldsymbol{\alpha})\delta\mathbf{u}^{(1)}(\mathbf{x})\right\rangle_1\right\}_{\boldsymbol{\alpha}^0} = \left\{\left\langle \mathbf{a}^{(1)}(\mathbf{x}), \mathbf{q}^{(1)}(\boldsymbol{\varphi};\boldsymbol{\psi};\boldsymbol{\alpha};\delta\boldsymbol{\alpha})\right\rangle_1\right\}_{\boldsymbol{\alpha}^0}. \quad (2.45)$$

3. Using the definition of the adjoint operator in the Hilbert space H_1, recast the left side of (2.45) as follows:

$$\begin{aligned}\left\{\left\langle \mathbf{a}^{(1)}(\mathbf{x}), \mathbf{V}^{(1)}(\boldsymbol{\alpha})\delta\mathbf{u}^{(1)}(\mathbf{x})\right\rangle_1\right\}_{\boldsymbol{\alpha}^0} &= \left\{\left\langle \delta\mathbf{u}^{(1)}(\mathbf{x}), \mathbf{A}^{(1)}(\boldsymbol{\alpha})\mathbf{a}^{(1)}(\mathbf{x})\right\rangle_1\right\}_{\boldsymbol{\alpha}^0} \\ &+ \left\{P^{(1)}\left[\delta\mathbf{u}^{(1)}(\mathbf{x}); \mathbf{a}^{(1)}(\mathbf{x}); \boldsymbol{\alpha}; \delta\boldsymbol{\alpha}\right]\right\}_{\boldsymbol{\alpha}^0},\end{aligned} \quad (2.46)$$

where $\left\{P^{(1)}\left[\delta\mathbf{u}^{(1)}(\mathbf{x}); \mathbf{a}^{(1)}(\mathbf{x}); \boldsymbol{\alpha}; \delta\boldsymbol{\alpha}\right]\right\}_{\boldsymbol{\alpha}^0}$ denotes the bilinear concomitant defined on the phase-space boundary $\mathbf{x} \in \partial\Omega_x(\boldsymbol{\alpha}^0)$, and where $\mathbf{A}^{(1)}(\boldsymbol{\alpha})$ is the operator formally adjoint to $\mathbf{V}^{(1)}(\boldsymbol{\alpha})$, i.e.,

$$\mathbf{A}^{(1)}(\boldsymbol{\alpha}) \triangleq \left[\mathbf{V}^{(1)}(\boldsymbol{\alpha})\right]^* = \begin{pmatrix} \mathbf{L}^*(\boldsymbol{\alpha}) & \mathbf{0} \\ \mathbf{0} & \mathbf{L}(\boldsymbol{\alpha}) \end{pmatrix}. \quad (2.47)$$

4. Require the first term on right side of (2.46) to represent the indirect-effect term defined in (2.37), to obtain the following relation:

$$\mathbf{A}^{(1)}(\boldsymbol{\alpha})\mathbf{a}^{(1)}(\mathbf{x}) = \mathbf{s}^{(1)}\left(\mathbf{u}^{(1)}(\mathbf{x}); \boldsymbol{\alpha}\right), \quad (2.48)$$

where

$$\mathbf{s}^{(1)}\left[\mathbf{u}^{(1)}(\mathbf{x}); \boldsymbol{\alpha}\right] \triangleq \begin{pmatrix} \left[\mathbf{s}_1^{(1)}\left(\mathbf{u}^{(1)}; \boldsymbol{\alpha}\right)\right]^{\dagger} \\ \left[\mathbf{s}_2^{(1)}\left(\mathbf{u}^{(1)}; \boldsymbol{\alpha}\right)\right]^{\dagger} \end{pmatrix} \triangleq \begin{pmatrix} \left[\partial S\left(\mathbf{u}^{(1)}; \boldsymbol{\alpha}\right)/\partial\boldsymbol{\varphi}\right]^{\dagger} \\ \left[\partial S\left(\mathbf{u}^{(1)}; \boldsymbol{\alpha}\right)/\partial\boldsymbol{\psi}\right]^{\dagger} \end{pmatrix}. \quad (2.49)$$

5. Implement the boundary conditions given in (2.39) into (2.46) and eliminate the remaining unknown boundary-values of the functions $\delta\boldsymbol{\varphi}$ and $\delta\boldsymbol{\psi}$ from the expression of the bilinear concomitant $\left\{P^{(1)}\left[\delta\mathbf{u}^{(1)}(\mathbf{x}); \mathbf{a}^{(1)}(\mathbf{x}); \boldsymbol{\alpha}\right]\right\}_{\boldsymbol{\alpha}^0}$ by selecting appropriate boundary conditions for the function $\mathbf{a}^{(1)}(\mathbf{x}) \triangleq \left[\mathbf{a}_1^{(1)}(\mathbf{x}), \mathbf{a}_2^{(1)}(\mathbf{x})\right]^{\dagger}$, to ensure that (2.48) is well-posed while being independent of unknown values of $\delta\boldsymbol{\varphi}$, $\delta\boldsymbol{\psi}$, and $\delta\boldsymbol{\alpha}$. The boundary conditions thus chosen for the function $\mathbf{a}^{(1)}(\mathbf{x}) \triangleq \left[\mathbf{a}_1^{(1)}(\mathbf{x}), \mathbf{a}_2^{(1)}(\mathbf{x})\right]^{\dagger}$ can be represented in operator form as follows:

$$\left\{\mathbf{b}_A^{(1)}\left[\mathbf{u}^{(1)}(\mathbf{x}); \mathbf{a}^{(1)}(\mathbf{x}); \boldsymbol{\alpha}\right]\right\}_{\boldsymbol{\alpha}^0} = \mathbf{0}, \ \mathbf{x} \in \partial\Omega_x\left(\boldsymbol{\alpha}^0\right). \quad (2.50)$$

6. The selection of the boundary conditions for the adjoint function $\mathbf{a}^{(1)}(\mathbf{x}) \triangleq \left[\mathbf{a}_1^{(1)}(\mathbf{x}), \mathbf{a}_2^{(1)}(\mathbf{x})\right]^\dagger$ represented by (2.50) eliminates the appearance of the *unknown* values of $\delta\mathbf{u}^{(1)}(\mathbf{x})$ in $\left\{P^{(1)}\left[\delta\mathbf{u}^{(1)}(\mathbf{x}); \mathbf{a}^{(1)}(\mathbf{x}); \boldsymbol{\alpha}; \delta\boldsymbol{\alpha}\right]\right\}_{\boldsymbol{\alpha}^0}$ and reduces this bilinear concomitant to a residual quantity that contains boundary terms involving only known values of $\boldsymbol{\varphi}(\mathbf{x})$, $\boldsymbol{\psi}(\mathbf{x})$, $\mathbf{a}^{(1)}(\mathbf{x})$, $\boldsymbol{\alpha}$, and $\delta\boldsymbol{\alpha}$. This residual quantity will be denoted as $\left\{\hat{P}^{(1)}\left[\mathbf{u}^{(1)}(\mathbf{x}); \mathbf{a}^{(1)}(\mathbf{x}); \boldsymbol{\alpha}; \delta\boldsymbol{\alpha}\right]\right\}_{\boldsymbol{\alpha}^0}$. In general, this residual quantity does not automatically vanish, although it may do so occasionally.
7. The system of equations comprising (2.48) together with the boundary conditions represented by (2.50) constitute the *First-Level Adjoint Sensitivity System* (1st-LASS). The solution $\mathbf{a}^{(1)}(\mathbf{x}) \triangleq \left[\mathbf{a}_1^{(1)}(\mathbf{x}), \mathbf{a}_2^{(1)}(\mathbf{x})\right]^\dagger$ of the 1st-LASS will be called the *first-level adjoint sensitivity function.* The 1st-LASS is called "first-level" (as opposed to "first-order") because it does not contain any differential or functional-derivatives, but its solution, $\mathbf{a}^{(1)}(\mathbf{x})$, will be used below to compute the first-order sensitivities of the response with respect to the model parameters. This terminology will be also used subsequently in this chapter, when deriving the expressions for the second and third-order sensitivities.
8. It follows from (2.45) and (2.46) that the following relation holds:

$$
\begin{aligned}
\left\{\left\langle \mathbf{a}^{(1)}(\mathbf{x}), \mathbf{q}^{(1)}\left(\mathbf{u}^{(1)}; \boldsymbol{\alpha}; \delta\boldsymbol{\alpha}\right)\right\rangle_1\right\}_{\boldsymbol{\alpha}^0} &= \left\{\left\langle \delta\mathbf{u}^{(1)}(\mathbf{x}), \mathbf{A}^{(1)}(\boldsymbol{\alpha})\mathbf{a}^{(1)}(\mathbf{x})\right\rangle_1\right\}_{\boldsymbol{\alpha}^0} \\
&+ \left\{\hat{P}^{(1)}\left[\mathbf{u}^{(1)}(\mathbf{x}); \mathbf{a}^{(1)}(\mathbf{x}); \boldsymbol{\alpha}; \delta\boldsymbol{\alpha}\right]\right\}_{\boldsymbol{\alpha}^0}.
\end{aligned}
\tag{2.51}
$$

9. Recalling from (2.48) that the first term on the right side of (2.51) is the indirect-effect term $\{\delta R(\boldsymbol{\varphi}^0, \boldsymbol{\psi}^0, \boldsymbol{\alpha}^0; \delta\boldsymbol{\varphi}, \delta\boldsymbol{\psi})\}_{ind}$, it follows from (2.51) that the indirect-effect term can be expressed in terms of the first-level adjoint sensitivity function $\mathbf{a}^{(1)}(\mathbf{x}) \triangleq \left[\mathbf{a}_1^{(1)}(\mathbf{x}), \mathbf{a}_2^{(1)}(\mathbf{x})\right]^\dagger$ as follows:

$$
\begin{aligned}
&\left\{\delta R\left(\boldsymbol{\varphi}^0, \boldsymbol{\psi}^0, \boldsymbol{\alpha}^0; \delta\boldsymbol{\varphi}, \delta\boldsymbol{\psi}\right)\right\}_{ind} = \left\{\left\langle \mathbf{a}^{(1)}(\mathbf{x}), \mathbf{q}^{(1)}\left(\mathbf{u}^{(1)}; \boldsymbol{\alpha}; \delta\boldsymbol{\alpha}\right)\right\rangle_1\right\}_{\boldsymbol{\alpha}^0} \\
&- \left\{\hat{P}^{(1)}\left[\mathbf{u}^{(1)}; \mathbf{a}^{(1)}; \boldsymbol{\alpha}; \delta\boldsymbol{\alpha}\right]\right\}_{\boldsymbol{\alpha}^0} \\
&= \left\{\sum_{i=1}^{2}\left\langle \mathbf{a}_i^{(1)}(\mathbf{x}), \mathbf{q}_i^{(1)}\left(\mathbf{u}^{(1)}; \boldsymbol{\alpha}; \delta\boldsymbol{\alpha}\right)\right\rangle_0\right\}_{\boldsymbol{\alpha}^0} - \left\{\hat{P}^{(1)}\left[\mathbf{u}^{(1)}; \mathbf{a}^{(1)}; \boldsymbol{\alpha}; \delta\boldsymbol{\alpha}\right]\right\}_{\boldsymbol{\alpha}^0} \\
&\equiv \left\{\delta R\left(\mathbf{u}^{(1)}; \mathbf{a}^{(1)}; \boldsymbol{\alpha}; \delta\boldsymbol{\alpha}\right)\right\}_{ind}.
\end{aligned}
\tag{2.52}
$$

As indicated by the last equality in (2.52), the variations $\delta\boldsymbol{\varphi}$ and $\delta\boldsymbol{\psi}$ have been eliminated from the original expression of the indirect-effect term, which now instead depends on the first-level adjoint sensitivity function

$\mathbf{a}^{(1)}(\mathbf{x}) \triangleq \left[\mathbf{a}_1^{(1)}(\mathbf{x}), \mathbf{a}_2^{(1)}(\mathbf{x})\right]^\dagger$. As indicated in (2.48), solving the *First-Level Adjoint Sensitivity System* (1st-LASS) entails the following operations: (1) inverting (i.e., solving) the original left side of the adjoint equation with the source $[\partial S(\mathbf{u}^{(1)}; \boldsymbol{\alpha})/\partial \boldsymbol{\varphi}]^\dagger$ to obtain the first-level adjoint sensitivity function $\mathbf{a}_1^{(1)}(\mathbf{x})$; and (2) inverting the original left side of the forward equation with the source $[\partial S(\mathbf{u}^{(1)}; \boldsymbol{\alpha})/\partial \boldsymbol{\psi}]^\dagger$ to obtain the first-level adjoint sensitivity function $\mathbf{a}_2^{(1)}(\mathbf{x})$. It is very important to note that the *1st-LASS* is independent of parameter variations $\delta\boldsymbol{\alpha}$. Hence, the 1st-LASS needs to be solved only once (as opposed to the 1st-LVSS or the 1st-OFSS, which would need to be solved anew for each parameter variation) to determine the first-level adjoint sensitivity function $\mathbf{a}^{(1)}(\mathbf{x}) \triangleq \left[\mathbf{a}_1^{(1)}(\mathbf{x}), \mathbf{a}_2^{(1)}(\mathbf{x})\right]^\dagger$. Subsequently, the "indirect-effect term" $\{\delta R(\mathbf{u}^{(1)}; \mathbf{a}^{(1)}; \boldsymbol{\alpha}; \delta\boldsymbol{\alpha})\}_{ind}$ is computed efficiently and exactly by simply performing the integrations over the adjoint function $\mathbf{a}^{(1)}(\mathbf{x}) \triangleq \left[\mathbf{a}_1^{(1)}(\mathbf{x}), \mathbf{a}_2^{(1)}(\mathbf{x})\right]^\dagger$, as indicated on the right side of (2.52).

As indicated in (2.35), the total first-order sensitivity of the response $R[\boldsymbol{\varphi}(\mathbf{x}), \boldsymbol{\psi}(\mathbf{x}); \boldsymbol{\alpha}]$ to the model parameters is obtained by adding the expressions of the direct-effect term defined in (2.36) and indirect-effect term as obtained in (2.52), which yields the following expression:

$$
\begin{aligned}
&\left\{\delta R\left[\mathbf{u}^{(1)}(\mathbf{x}), \boldsymbol{\alpha}; \delta\boldsymbol{\varphi}, \delta\boldsymbol{\psi}, \delta\boldsymbol{\alpha}\right]\right\}_{\boldsymbol{\alpha}^0} \\
&= \prod_{k=1}^{TI}\left\{\int_{\lambda_k(\boldsymbol{\alpha})}^{\omega_k(\boldsymbol{\alpha})} dx_k \frac{\partial S[\boldsymbol{\varphi}(\mathbf{x}), \boldsymbol{\psi}(\mathbf{x}); \boldsymbol{\alpha}]}{\partial \boldsymbol{\alpha}} \delta\boldsymbol{\alpha}\right\}_{\boldsymbol{\alpha}^0} \\
&+ \sum_{k=1}^{TI} \prod_{m=1, k\neq m}^{TI} \left\{\int_{\lambda_m(\boldsymbol{\alpha})}^{\omega_m(\boldsymbol{\alpha})} dx_m S[\boldsymbol{\varphi}(\ldots, \omega_k, \ldots), \boldsymbol{\psi}(\ldots, \omega_k, \ldots); \boldsymbol{\alpha}] \frac{\partial \omega_k(\boldsymbol{\alpha})}{\partial \boldsymbol{\alpha}}\right\}_{\boldsymbol{\alpha}^0} \delta\boldsymbol{\alpha} \\
&- \sum_{k=1}^{TI} \prod_{m=1, k\neq m}^{TI} \left\{\int_{\lambda_m(\boldsymbol{\alpha})}^{\omega_m(\boldsymbol{\alpha})} dx_m S[\boldsymbol{\varphi}(\ldots, \lambda_k, \ldots), \boldsymbol{\psi}(\ldots, \lambda_k, \ldots); \boldsymbol{\alpha}] \frac{\partial \lambda_k(\boldsymbol{\alpha})}{\partial \boldsymbol{\alpha}}\right\}_{\boldsymbol{\alpha}^0} \delta\boldsymbol{\alpha}, \\
&+ \left\{\sum_{i=1}^{2} \left\langle \mathbf{a}_i^{(1)}(\mathbf{x}), \mathbf{q}_i^{(1)}\left(\mathbf{u}^{(1)}; \boldsymbol{\alpha}; \delta\boldsymbol{\alpha}\right)\right\rangle_0\right\}_{\boldsymbol{\alpha}^0} - \left\{\widehat{P}^{(1)}\left[\mathbf{u}^{(1)}; \mathbf{a}^{(1)}; \boldsymbol{\alpha}; \delta\boldsymbol{\alpha}\right]\right\}_{\boldsymbol{\alpha}^0} \\
&\equiv \left\{\delta R\left[\mathbf{u}^{(1)}(\mathbf{x}); \mathbf{a}^{(1)}(\mathbf{x}); \boldsymbol{\alpha}; \delta\boldsymbol{\alpha}\right]\right\}_{\boldsymbol{\alpha}^0}.
\end{aligned}
\tag{2.53}
$$

The expression in (2.53) no longer depends on the variations $\delta\boldsymbol{\varphi}$ and $\delta\boldsymbol{\psi}$, which are expensive to compute, but instead depends on the first-level adjoint sensitivity function $\mathbf{a}^{(1)}(\mathbf{x}) \triangleq \left[\mathbf{a}_1^{(1)}(\mathbf{x}), \mathbf{a}_2^{(1)}(\mathbf{x})\right]^\dagger$. In particular, this expression also reveals that the sensitivities of the response $R[\boldsymbol{\varphi}(\mathbf{x}), \boldsymbol{\psi}(\mathbf{x}); \boldsymbol{\alpha}]$ to parameters that characterize the system's boundary and/or internal interfaces can arise both from the direct-effect and indirect-effect terms. It also follows from (2.53) that the total first-order sensitivity $\delta R(\boldsymbol{\varphi}^0, \boldsymbol{\psi}^0; \mathbf{a}^{(1)}; \boldsymbol{\alpha}^0; \delta\boldsymbol{\alpha})$ can be expressed in terms of the first-level adjoint sensitivity functions as follows:

$$\left\{\delta R\left[\mathbf{u}^{(1)}(\mathbf{x}); \mathbf{a}^{(1)}(\mathbf{x}); \boldsymbol{\alpha}; \delta\boldsymbol{\alpha}\right]\right\}_{\boldsymbol{\alpha}^0} = \sum_{j_1=1}^{TP} \left\{\frac{\partial R\left[j_1; \mathbf{u}^{(1)}(\mathbf{x}); \mathbf{a}^{(1)}(\mathbf{x}); \boldsymbol{\alpha}\right]}{\partial \alpha_{j_1}}\right\}_{\boldsymbol{\alpha}^0} \delta\alpha_{j_1}, \tag{2.54}$$

where the quantities defined below

$$\begin{aligned}\frac{\partial R\left[j_1; \mathbf{u}^{(1)}(\mathbf{x}); \mathbf{a}^{(1)}(\mathbf{x}); \boldsymbol{\alpha}\right]}{\partial \alpha_{j_1}} &= \prod_{k=1}^{TI} \int_{\lambda_k(\boldsymbol{\alpha})}^{\omega_k(\boldsymbol{\alpha})} dx_k S^{(1)}\left[j_1; \mathbf{u}^{(1)}(\mathbf{x}); \mathbf{a}^{(1)}(\mathbf{x}); \boldsymbol{\alpha}\right] \\ &\triangleq R^{(1)}\left[j_1; \mathbf{u}^{(1)}(\mathbf{x}); \mathbf{a}^{(1)}(\mathbf{x}); \boldsymbol{\alpha}\right], \; j_1 = 1, \ldots, TP,\end{aligned} \tag{2.55}$$

represent the first-order response sensitivities, evaluated at the nominal values of the state functions and model parameters. In particular, if the residual bilinear concomitant is non-zero, the functions $S^{(1)}[j_1; \boldsymbol{\varphi}(\mathbf{x}), \boldsymbol{\psi}(\mathbf{x}); \mathbf{a}^{(1)}(\mathbf{x}); \boldsymbol{\alpha}]$ would contain suitably defined Dirac delta-functionals for expressing the respective non-zero boundary terms as volume-integrals over the phase-space of the independent variables. Dirac-delta functionals would also be needed to represent, within $S^{(1)}[j_1; \boldsymbol{\varphi}(\mathbf{x}), \boldsymbol{\psi}(\mathbf{x}); \mathbf{a}^{(1)}(\mathbf{x}); \boldsymbol{\alpha}]$, the terms containing the derivatives of the boundary endpoints with respect to the model and/or response parameters.

In the particular case in which the model's response depended only on the forward state function $\boldsymbol{\varphi}(\mathbf{x})$, the sensitivities $\partial R\left[\boldsymbol{\varphi}(\mathbf{x}); \mathbf{a}_1^{(1)}(\mathbf{x}); \boldsymbol{\alpha}\right]/\partial \alpha_{j_1}, j_1 = 1, \ldots, TP$, would depend only on the first-level adjoint sensitivity function $\mathbf{a}_1^{(1)}(\mathbf{x})$, which is the solution of the following adjoint-like 1st-LASS:

$$\mathbf{V}^{(1)}(\boldsymbol{\alpha}) = \mathbf{L}(\boldsymbol{\alpha}); \mathbf{A}^{(1)}(\boldsymbol{\alpha}) = \mathbf{L}^*(\boldsymbol{\alpha}); \mathbf{a}^{(1)}(\mathbf{x}) = \mathbf{a}_1^{(1)}(\mathbf{x}). \tag{2.56}$$

A similar simplification occurs in the boundary conditions represented by (2.50). Conversely, if the model's response depended only on the adjoint state function $\boldsymbol{\psi}(\mathbf{x})$, then the sensitivities $\partial R\left[\boldsymbol{\psi}(\mathbf{x}); \mathbf{a}_2^{(1)}(\mathbf{x}); \boldsymbol{\alpha}\right]/\partial \alpha_{j_1}, j_1 = 1, \ldots, TP$, would depend only on the first-level adjoint sensitivity function $\mathbf{a}_2^{(1)}(\mathbf{x})$, which is the solution of the following forward-like 1st-LASS:

$$\mathbf{V}^{(1)}(\boldsymbol{\alpha}) = \mathbf{L}^*(\boldsymbol{\alpha}); \mathbf{A}^{(1)}(\boldsymbol{\alpha}) = \mathbf{L}(\boldsymbol{\alpha}); \; \mathbf{a}^{(1)}(\mathbf{x}) = \mathbf{a}_2^{(1)}(\mathbf{x}). \tag{2.57}$$

A corresponding simplification would occur in the boundary conditions represented by (2.50). In summary, if the response depends only on the forward or only on the adjoint state functions, then the 1st-LASS becomes half as large by comparison to the case when the response depends on both the forward and the adjoint state functions.

The First-Order Comprehensive Adjoint Sensitivity Analysis Methodology (1st-CASAM-L) requires 2 large-scale computations: one large-scale computation for computing the first-level adjoint state function $\mathbf{a}_1^{(1)}(\mathbf{x})$ and one large-scale computation for computing the first-level adjoint state function $\mathbf{a}_2^{(1)}(\mathbf{x})$. The Forward Sensitivity Analysis Method requires $2TP$ large-scale computations: TP large-scale computations for computing the first-order derivatives of the forward state function $\boldsymbol{\varphi}(\boldsymbol{\alpha};\mathbf{x})$ and TP large-scale computations for computing the first-order derivatives of the adjoint state function $\boldsymbol{\psi}(\boldsymbol{\alpha};\mathbf{x})$. The finite-difference (FD) method requires $4\,TP$ large-scale computations: $2\,TP$ large-scale computations for computing the forward state functions $\boldsymbol{\varphi}(\alpha_j \pm h_j;\mathbf{x})$ and $2\,TP$ large-scale computations for computing the adjoint state functions $\boldsymbol{\psi}(\alpha_j \pm h_j;\mathbf{x})$.

2.4 The Second-Order Comprehensive Sensitivity Analysis Methodology for Response-Coupled Linear Forward and Adjoint Systems (2nd-CASAM-L)

There are $TP(TP+1)/2$ distinct second-order sensitivities of the response $R[\boldsymbol{\varphi}(\mathbf{x}),\boldsymbol{\psi}(\mathbf{x});\boldsymbol{\alpha}]$ with respect to the model and response parameters. These will be indexed using subscripts of the form $j_1 = 1,\ldots,TP$; $j_2 = 1,\ldots,j_1$. The second-order responses sensitivities could be computed *approximately* by "brute force" re-computations using standard forward or backward differences. Thus, the second-order unmixed sensitivities can be calculated using the following formula:

$$\frac{\partial^2 R(\boldsymbol{\alpha})}{\left(\partial \alpha_{j_1}\right)^2} \approx \frac{1}{{h_{j_1}}^2}\left(R_{j_1+1} - R_{j_1} + R_{j_1-1}\right) + O\left({h_{j_1}}^2\right),\ j_1 = 1,\ldots,TP, \tag{2.58}$$

where $R_{j_1+1} \triangleq R\left(\alpha_{j_1} + h_{j_1}\right)$, $R_{j_1} \triangleq R\left(\alpha_{j_1}\right)$, $R_{j1-1} \triangleq R(\alpha_{j1} - h_{j1})$ and h_{j1} denotes a "judiciously chosen" variation in the parameter α_{j_1} around its nominal value $\alpha_{j_1}^0$. The second-order mixed sensitivities, $\partial^2 R(\boldsymbol{\alpha})/\partial\alpha_{j_1}\partial\alpha_{j_2}$, can be calculated by using the following finite-difference formula:

$$\begin{aligned}\frac{\partial^2 R(\boldsymbol{\alpha})}{\partial\alpha_{j_1}\partial\alpha_{j_2}} &\approx \frac{1}{4h_{j_1}h_{j_2}}\left(R_{j_1+1,j_2+1} - R_{j_1-1,j_2+1} - R_{j_1+1,j_2-1} + R_{j_1-1,j_2-1}\right)\\ &+O\left({h_{j_1}}^2,{h_{j_2}}^2\right),\ for\ j_1 = 1,\ldots,TP;\ j_2 = 1,\ldots,j_1;\end{aligned} \tag{2.59}$$

where h_{j_1} and h_{j_2} denote the variations in the parameters α_{j1} and α_{j_2}, respectively. The values of the quantities $R_{j_1+1,j_2+1} \triangleq R\left(\alpha_{j_1} + h_{j_1}, \alpha_{j_2} + h_{j_2}\right)$, $R_{j_1-1,j_2+1} \triangleq R\left(\alpha_{j_1} - h_{j_1}, \alpha_{j_2} + h_{j_2}\right)$, $R_{j_1+1,j_2-1} \triangleq R\left(\alpha_{j_1} + h_{j_1}, \alpha_{j_2} - h_{j_2}\right)$, and $R_{j_1-1,j_2-1} \triangleq R\left(\alpha_{j_1} - h_{j_1}, \alpha_{j_2} - h_{j_2}\right)$ are obtained by repeatedly re-solving the original forward equations (2.1) and (2.2) as well as the original adjoint equations, (2.7) and (2.8), using the changed parameter values $(\alpha_{j_1} \pm h_{j_1})$ and $(\alpha_{j_2} \pm h_{j_2})$. The values of h_{j_1} and h_{j_2}, respectively, must be chosen "judiciously" by "trial and error" for each of the parameters α_{j_1} and α_{j_2}; otherwise, the result produced by (2.59) will be erroneous, significantly removed from the exact value of the derivative $\partial^2 R(\boldsymbol{\alpha})/\partial\alpha_{j_1}\partial\alpha_{j_2}$. These finite-difference formulas introduce their intrinsic "methodological errors" of order $O(h_{j_1}{}^2)$ and/or $O\left(h_{j_1}^2, h_{j_2}^2\right)$, as indicated in (2.58) and (2.59), which are in addition to, and independent of, the errors that might be incurred in the computation of $R(\boldsymbol{\alpha})$. In other words, even if the computation of $R(\boldsymbol{\alpha})$ were perfect (error-free), the finite-difference formulas nevertheless introduce their own, intrinsic, numerical errors into the computation of $\partial^2 R(\boldsymbol{\alpha})/\partial\alpha_{j_1}\partial\alpha_{j_2}$.

Within the framework of the Forward Sensitivity Analysis Methodology (FSAM), the second-order sensitivities $\partial^2 R(\boldsymbol{\alpha})/\partial\alpha_{j_1}\partial\alpha_{j_2}, j_1, j_2 = 1, \ldots, TP$, of the response to the model and response parameters are obtained by computing the partial G-derivatives of the first-order sensitivities represented by (2.30), which yields the following result:

$$\begin{aligned}\frac{\partial^2 R(\boldsymbol{\varphi};\boldsymbol{\psi};\boldsymbol{\alpha})}{\partial\alpha_{j_1}\partial\alpha_{j_2}} &\triangleq \frac{1}{\delta\alpha_{j_2}}\left\{\frac{d}{d\varepsilon}R^{(1)}\left[j_1;\boldsymbol{\varphi}\left(\alpha_{j_2}^0+\varepsilon\delta\alpha_{j_2}\right);\boldsymbol{\psi}\left(\alpha_{j_2}^0+\varepsilon\delta\alpha_{j_2}\right);\right.\right.\\ &\quad\left.\left.\alpha_{j_2}^0+\varepsilon\delta\alpha_{j_2};\frac{\partial\boldsymbol{\varphi}\left(\alpha_{j_2}^0+\varepsilon\delta\alpha_{j_2}\right)}{\partial\alpha_{j_1}};\frac{\partial\boldsymbol{\psi}\left(\alpha_{j_2}^0+\varepsilon\delta\alpha_{j_2}\right)}{\partial\alpha_{j_1}}\right]\right\}_{\varepsilon=0}\\ &= R^{(2)}\left[j_2;j_1;\boldsymbol{\varphi};\boldsymbol{\psi};\boldsymbol{\alpha};\frac{\partial\boldsymbol{\varphi}}{\partial\alpha_{j_2}};\frac{\partial\boldsymbol{\psi}}{\partial\alpha_{j_2}};\frac{\partial^2\boldsymbol{\varphi}}{\partial\alpha_{j_1}\partial\alpha_{j_2}};\frac{\partial^2\boldsymbol{\psi}}{\partial\alpha_{j_1}\partial\alpha_{j_2}}\right].\end{aligned} \tag{2.60}$$

As indicated by the functional dependence of the last term in (2.60), the evaluation of the second-order response sensitivities requires prior knowledge of the first- and second-order derivatives of the state functions with respect to the model parameters. The functions $\partial\boldsymbol{\varphi}(\mathbf{x})/\partial\alpha_{j_2}$ and $\partial\boldsymbol{\psi}(\mathbf{x})/\partial\alpha_{j_2}$ are obtained by solving the 1st-OFSS. The functions $\partial^2\boldsymbol{\varphi}/\partial\alpha_{j_1}\partial\alpha_{j_2}$ and $\partial^2\boldsymbol{\psi}/\partial\alpha_{j_1}\partial\alpha_{j_2}$ are the solutions of the following "*Second-Order Forward Sensitivity System*" (2nd-OFSS), which is obtained by taking the partial G-derivative of the 1st-OFSS with respect to a generic model parameter α_{j_2} to obtain the following equations for $j_1, j_2 = 1, \ldots, TP$:

$$\mathbf{L}(\mathbf{x};\boldsymbol{\alpha}^0)\frac{\partial^2\boldsymbol{\varphi}}{\partial\alpha_{j_2}\partial\alpha_{j_1}} = \left\{\mathbf{f}^{(2)}(j_2;j_1;\boldsymbol{\varphi};\boldsymbol{\alpha})\right\}_{\boldsymbol{\alpha}^0};\ \mathbf{x}\in\Omega_x; \tag{2.61}$$

$$\mathbf{B}_\varphi(\mathbf{x};\boldsymbol{\alpha}^0)\frac{\partial^2\boldsymbol{\varphi}}{\partial\alpha_{j_2}\partial\alpha_{j_1}} = \left\{\mathbf{c}_\varphi^{(2)}(j_2;j_1;\boldsymbol{\varphi};\boldsymbol{\alpha})\right\}_{\boldsymbol{\alpha}^0};\mathbf{x}\in\partial\Omega_x(\boldsymbol{\alpha}^0); \tag{2.62}$$

$$\mathbf{L}^*\left(\mathbf{x};\boldsymbol{\alpha}^0\right)\frac{\partial^2\boldsymbol{\psi}}{\partial\alpha_{j_2}\partial\alpha_{j_1}}=\left\{\mathbf{g}^{(2)}(j_2;j_1;\boldsymbol{\psi};\boldsymbol{\alpha})\right\}_{\boldsymbol{\alpha}^0};\ \mathbf{x}\in\Omega_x; \tag{2.63}$$

$$\mathbf{B}_\psi\left(\mathbf{x};\boldsymbol{\alpha}^0\right)\frac{\partial^2\boldsymbol{\psi}}{\partial\alpha_{j_2}\partial\alpha_{j_1}}=\left\{\mathbf{c}_\psi^{(2)}(j_2;j_1;\boldsymbol{\psi};\boldsymbol{\alpha})\right\}_{\boldsymbol{\alpha}^0};\mathbf{x}\in\partial\Omega_x\left(\boldsymbol{\alpha}^0\right); \tag{2.64}$$

where:

$$\mathbf{f}^{(2)}(j_2;j_1;\boldsymbol{\varphi};\boldsymbol{\alpha})\triangleq\frac{\partial\mathbf{f}^{(1)}(j_1;\boldsymbol{\varphi};\boldsymbol{\alpha})}{\partial\alpha_{j_2}}-\frac{\partial\mathbf{L}(\mathbf{x};\boldsymbol{\alpha})}{\partial\alpha_{j_2}}\frac{\partial\boldsymbol{\varphi}}{\partial\alpha_{j_1}}; \tag{2.65}$$

$$\mathbf{g}^{(2)}(j_2;j_1;\boldsymbol{\psi};\boldsymbol{\alpha})\triangleq\frac{\partial\mathbf{g}^{(1)}(j_1;\boldsymbol{\psi};\boldsymbol{\alpha})}{\partial\alpha_{j_2}}-\frac{\partial\mathbf{L}^*(\mathbf{x};\boldsymbol{\alpha})}{\partial\alpha_{j_2}}\frac{\partial\boldsymbol{\psi}}{\partial\alpha_{j_1}}; \tag{2.66}$$

$$\mathbf{c}_\varphi^{(2)}(j_2;j_1;\boldsymbol{\varphi};\boldsymbol{\alpha})\triangleq\frac{\partial\mathbf{c}_\varphi^{(1)}(j_1;\boldsymbol{\varphi};\boldsymbol{\alpha})}{\partial\alpha_{j_2}}-\frac{\partial\mathbf{B}_\varphi(\mathbf{x};\boldsymbol{\alpha})}{\partial\alpha_{j_2}}\frac{\partial\boldsymbol{\varphi}}{\partial\alpha_{j_1}};\mathbf{x}\in\partial\Omega_x\left(\boldsymbol{\alpha}^0\right); \tag{2.67}$$

$$\mathbf{c}_\psi^{(2)}(j_2;j_1;\boldsymbol{\psi};\boldsymbol{\alpha})\triangleq\frac{\partial\mathbf{c}_\psi^{(1)}(j_1;\boldsymbol{\psi};\boldsymbol{\alpha})}{\partial\alpha_{j_2}}-\frac{\partial\mathbf{B}_\psi(\mathbf{x};\boldsymbol{\alpha})}{\partial\alpha_{j_2}}\frac{\partial\boldsymbol{\psi}(\mathbf{x})}{\partial\alpha_{j_1}};\mathbf{x}\in\partial\Omega_x\left(\boldsymbol{\alpha}^0\right). \tag{2.68}$$

Evidently, the 1st-OFSS would need to be solved 2*TP*-times, with different right sides in order to compute the first-order derivatives $\partial\boldsymbol{\varphi}(\mathbf{x})/\partial\alpha_{j_2}$ and $\partial\boldsymbol{\psi}(\mathbf{x})/\partial\alpha_{j_2}$ with respect to all model parameters, which will entail large-scale computations. Furthermore, there are *TP*(*TP* + 1)/2 distinct second-order functions $\partial^2\boldsymbol{\varphi}/\partial\alpha_{j_1}\partial\alpha_{j_2}$ and just as many distinct second-order functions $\partial^2\boldsymbol{\psi}/\partial\alpha_{j_1}\partial\alpha_{j_2}$. This means that the 2nd-OFSS would need to be solved *TP*(*TP* + 1) times to compute these second-order functions. Hence, the computation of the *TP*(*TP* + 1)/2 distinct second-order response sensitivities to the model parameters using the FSAM would necessitate a total number of $TP^2 + 3TP$ large-scale computations, evidently highlighting the impact of the "curse of dimensionality" (Bellman 1957).

The starting point for the development of the 2nd-CASAM-L is provided by the expressions of the sensitivities provided in terms of the first-level adjoint sensitivity functions, namely (2.55), rather than the expression given in (2.30) for the first-order sensitivities produced by the FSAM. Thus, the second-order total sensitivity of the model response is obtained by applying the definition of the first-order G-differential to (2.55), which yields the following result:

$$\begin{aligned}
&\left\{\delta R^{(1)}\left[j_1;\mathbf{u}^{(1)}(\mathbf{x});\mathbf{a}^{(1)}(\mathbf{x});\boldsymbol{\alpha};\delta\mathbf{u}^{(1)}(\mathbf{x});\delta\mathbf{a}^{(1)}(\mathbf{x});\delta\boldsymbol{\alpha}\right]\right\}_{\boldsymbol{\alpha}^0} \\
&\triangleq\left\{\frac{d}{d\varepsilon}\left[\prod_{k=1}^{TI}\int_{\lambda_k(\boldsymbol{\alpha}^0+\varepsilon\delta\boldsymbol{\alpha})}^{\omega_k(\boldsymbol{\alpha}^0+\varepsilon\delta\boldsymbol{\alpha})} dx_k S^{(1)}\left[j_1;\boldsymbol{\varphi}^0+\varepsilon\delta\boldsymbol{\varphi};\boldsymbol{\psi}^0+\varepsilon\delta\boldsymbol{\psi};\mathbf{a}^{(1),0}(\mathbf{x})+\varepsilon\delta\mathbf{a}^{(1)};\right.\right.\right. \\
&\left.\left.\left.\boldsymbol{\alpha}^0+\varepsilon\delta\boldsymbol{\alpha}\right]\right\}_{\varepsilon=0}=\left\{\delta R^{(1)}\left[j_1;\mathbf{u}^{(1)}(\mathbf{x});\mathbf{a}^{(1)}(\mathbf{x});\boldsymbol{\alpha};\delta\boldsymbol{\alpha}\right]\right\}_{dir} \\
&+\left\{\delta R^{(1)}\left[j_1;\mathbf{u}^{(1)}(\mathbf{x});\mathbf{a}^{(1)}(\mathbf{x});\boldsymbol{\alpha};\delta\mathbf{u}^{(1)}(\mathbf{x});\delta\mathbf{a}^{(1)}(\mathbf{x})\right]\right\}_{ind}, j_1=1,\ldots,TP,
\end{aligned} \tag{2.69}$$

where the direct-effect term $\{\delta R^{(1)}[j_1;\mathbf{u}^{(1)}(\mathbf{x});\mathbf{a}^{(1)}(\mathbf{x});\boldsymbol{\alpha};\delta\boldsymbol{\alpha}]\}_{dir}$ is defined as follows:

$$\begin{aligned}
&\left\{\delta R^{(1)}\left[j_1;\mathbf{u}^{(1)}(\mathbf{x});\mathbf{a}^{(1)}(\mathbf{x});\boldsymbol{\alpha};\delta\boldsymbol{\alpha}\right]\right\}_{dir} \\
&\triangleq\left\{\frac{\partial}{\partial\boldsymbol{\alpha}}\prod_{k=1}^{TI}\int_{\lambda_k(\boldsymbol{\alpha})}^{\omega_k(\boldsymbol{\alpha})} dx_k S^{(1)}\left[j_1;\mathbf{u}^{(1)}(\mathbf{x});\mathbf{a}^{(1)}(\mathbf{x});\boldsymbol{\alpha}\right]\right\}_{\boldsymbol{\alpha}^0}\delta\boldsymbol{\alpha},
\end{aligned} \tag{2.70}$$

while $\{\delta R^{(1)}[j_1;\mathbf{u}^{(1)}(\mathbf{x});\mathbf{a}^{(1)}(\mathbf{x});\boldsymbol{\alpha};\delta\mathbf{u}^{(1)}(\mathbf{x});\delta\mathbf{a}^{(1)}(\mathbf{x})]\}_{ind}$ denotes the indirect-effect term and is defined as follows:

$$\begin{aligned}
&\left\{\delta R^{(1)}\left[j_1;\mathbf{u}^{(1)}(\mathbf{x});\mathbf{a}^{(1)}(\mathbf{x});\boldsymbol{\alpha};\delta\mathbf{u}^{(1)}(\mathbf{x});\delta\mathbf{a}^{(1)}(\mathbf{x})\right]\right\}_{ind} \\
&\triangleq\prod_{k=1}^{TI}\int_{\lambda_k(\boldsymbol{\alpha})}^{\omega_k(\boldsymbol{\alpha})}\left[\frac{\partial S^{(1)}\left(j_1;\mathbf{u}^{(1)};\mathbf{a}^{(1)};\boldsymbol{\alpha}\right)}{\partial\boldsymbol{\varphi}}\delta\boldsymbol{\varphi}(\mathbf{x})+\frac{\partial S^{(1)}\left(j_1;\mathbf{u}^{(1)};\mathbf{a}^{(1)};\boldsymbol{\alpha}\right)}{\partial\boldsymbol{\psi}}\delta\boldsymbol{\psi}(\mathbf{x})\right. \\
&\left.+\frac{\partial S^{(1)}\left(j_1;\mathbf{u}^{(1)};\mathbf{a}^{(1)};\boldsymbol{\alpha}\right)}{\partial\mathbf{a}_1^{(1)}}\delta\mathbf{a}_1^{(1)}(\mathbf{x})+\frac{\partial S^{(1)}\left(j_1;\mathbf{u}^{(1)};\mathbf{a}^{(1)};\boldsymbol{\alpha}\right)}{\partial\mathbf{a}_2^{(1)}}\delta\mathbf{a}_2^{(1)}(\mathbf{x})\right]dx_k.
\end{aligned} \tag{2.71}$$

The left side of (2.71) is to be computed at nominal parameter and state function values but the corresponding indication (namely, the superscript "zero") has been omitted in (2.71) in order to simplify the notation.

The direct-effect term $\{\delta R^{(1)}[j_1;\mathbf{u}^{(1)}(\mathbf{x});\mathbf{a}^{(1)}(\mathbf{x});\boldsymbol{\alpha};\delta\boldsymbol{\alpha}]\}_{dir}$ can be computed immediately. The term, $\{\delta R^{(1)}[j_1;\mathbf{u}^{(1)}(\mathbf{x});\mathbf{a}^{(1)}(\mathbf{x});\boldsymbol{\alpha};\delta\mathbf{u}^{(1)}(\mathbf{x});\delta\mathbf{a}^{(1)}(\mathbf{x})]\}_{ind}$, however, depends on the variations $\delta\boldsymbol{\varphi}(\mathbf{x})$, $\delta\boldsymbol{\psi}(\mathbf{x})$, and $\delta\mathbf{a}^{(1)}(\mathbf{x})$ in the forward state functions and the first-level adjoint sensitivity function. Recall that the functions $\delta\boldsymbol{\varphi}(\mathbf{x})$ and $\delta\boldsymbol{\psi}(\mathbf{x})$ are the solutions of the 1st-LVSS defined by (2.38) and (2.39). On the other hand, the function $\delta\mathbf{a}^{(1)}(\mathbf{x})$ is the solution of the system of equations obtained by G-differentiating the 1st-LASS. Applying the definition of the first G-differential to the 1st-LASS defined by (2.48) and (2.50) yields the following equations

$$\mathbf{L}^*\left(\boldsymbol{\alpha}^0\right)\delta\mathbf{a}_1^{(1)} - \left\{\frac{\partial^2 S\left(\mathbf{u}^{(1)};\boldsymbol{\alpha}\right)}{\partial\boldsymbol{\varphi}\partial\boldsymbol{\varphi}}\right\}_{\boldsymbol{\alpha}^0}\delta\boldsymbol{\varphi} - \left\{\frac{\partial^2 S\left(\mathbf{u}^{(1)};\boldsymbol{\alpha}\right)}{\partial\boldsymbol{\varphi}\partial\boldsymbol{\psi}}\right\}_{\boldsymbol{\alpha}^0}\delta\boldsymbol{\psi} = \left\{\mathbf{p}_1^{(2)}\left(\mathbf{u}^{(1)};\mathbf{a}_1^{(1)};\boldsymbol{\alpha};\delta\boldsymbol{\alpha}\right)\right\}_{\boldsymbol{\alpha}^0}, \tag{2.72}$$

$$\mathbf{L}\left(\boldsymbol{\alpha}^0\right)\delta\mathbf{a}_2^{(1)} - \left\{\frac{\partial^2 S\left(\mathbf{u}^{(1)};\boldsymbol{\alpha}\right)}{\partial\boldsymbol{\varphi}\partial\boldsymbol{\psi}}\right\}_{\boldsymbol{\alpha}^0}\delta\boldsymbol{\varphi} - \left\{\frac{\partial^2 S\left(\mathbf{u}^{(1)};\boldsymbol{\alpha}\right)}{\partial\boldsymbol{\psi}\partial\boldsymbol{\psi}}\right\}_{\boldsymbol{\alpha}^0}\delta\boldsymbol{\psi} = \left\{\mathbf{p}_2^{(2)}\left(\mathbf{u}^{(1)};\mathbf{a}_2^{(1)};\boldsymbol{\alpha};\delta\boldsymbol{\alpha}\right)\right\}_{\boldsymbol{\alpha}^0}, \tag{2.73}$$

subject to the following boundary conditions:

$$\begin{aligned}
&\left\{\delta\mathbf{b}_A^{(1)}\left(\mathbf{u}^{(1)};\mathbf{a}^{(1)};\boldsymbol{\alpha};\delta\boldsymbol{\varphi};\delta\boldsymbol{\psi};\delta\mathbf{a}_1^{(1)};\delta\mathbf{a}_2^{(1)};\delta\boldsymbol{\alpha}\right)\right\}_{\boldsymbol{\alpha}^0} \\
&\triangleq \left\{\frac{\partial\mathbf{b}_A^{(1)}\left(\mathbf{u}^{(1)};\mathbf{a}^{(1)};\boldsymbol{\alpha}\right)}{\partial\boldsymbol{\varphi}}\delta\boldsymbol{\varphi} + \frac{\partial\mathbf{b}_A^{(1)}\left(\mathbf{u}^{(1)};\mathbf{a}^{(1)};\boldsymbol{\alpha}\right)}{\partial\boldsymbol{\psi}}\delta\boldsymbol{\psi}\right\}_{\boldsymbol{\alpha}^0} \\
&+\left\{\frac{\partial\mathbf{b}_A^{(1)}\left(\mathbf{u}^{(1)};\mathbf{a}^{(1)};\boldsymbol{\alpha}\right)}{\partial\mathbf{a}_1^{(1)}}\delta\mathbf{a}_1^{(1)} + \frac{\partial\mathbf{b}_A^{(1)}\left(\mathbf{u}^{(1)};\mathbf{a}^{(1)};\boldsymbol{\alpha}\right)}{\partial\mathbf{a}_2^{(1)}}\delta\mathbf{a}_2^{(1)}\right\}_{\boldsymbol{\alpha}^0} \\
&+\left\{\frac{\partial\mathbf{b}_A^{(1)}\left(\mathbf{u}^{(1)};\mathbf{a}^{(1)};\boldsymbol{\alpha}\right)}{\partial\boldsymbol{\alpha}}\delta\boldsymbol{\alpha}\right\}_{\boldsymbol{\alpha}^0} = \mathbf{0}, \quad \mathbf{x}\in\partial\Omega_x\left(\boldsymbol{\alpha}^0\right).
\end{aligned} \tag{2.74}$$

where

$$\mathbf{p}_1^{(2)}\left(\boldsymbol{\varphi};\boldsymbol{\psi};\mathbf{a}_1^{(1)};\boldsymbol{\alpha};\delta\boldsymbol{\alpha}\right) \triangleq \frac{\partial^2 S\left(\mathbf{u}^{(1)};\boldsymbol{\alpha}\right)}{\partial\boldsymbol{\alpha}\partial\boldsymbol{\varphi}}\delta\boldsymbol{\alpha} - \frac{\partial\left[\mathbf{L}^*(\boldsymbol{\alpha})\mathbf{a}_1^{(1)}\right]}{\partial\boldsymbol{\alpha}}\delta\boldsymbol{\alpha}, \tag{2.75}$$

$$\mathbf{p}_2^{(2)}\left(\mathbf{u}^{(1)};\mathbf{a}_2^{(1)};\boldsymbol{\alpha};\delta\boldsymbol{\alpha}\right) \triangleq \frac{\partial^2 S\left(\mathbf{u}^{(1)};\boldsymbol{\alpha}\right)}{\partial\boldsymbol{\alpha}\partial\boldsymbol{\psi}}\delta\boldsymbol{\alpha} - \frac{\partial\left[\mathbf{L}(\boldsymbol{\alpha})\mathbf{a}_2^{(1)}\right]}{\partial\boldsymbol{\alpha}}\delta\boldsymbol{\alpha}. \tag{2.76}$$

Recalling the 1st-LVSS defined in (2.38) and (2.39), and considering (2.72) through (2.76), it follows that the functions $\delta\boldsymbol{\varphi}(\mathbf{x})$, $\delta\boldsymbol{\psi}(\mathbf{x})$, and $\delta\mathbf{a}^{(1)}(\mathbf{x}) \triangleq \left[\delta\mathbf{a}_1^{(1)}(\mathbf{x}), \delta\mathbf{a}_2^{(1)}(\mathbf{x})\right]^\dagger$ are the solution of the following system of equations:

$$\left\{\mathbf{V}^{(2)}\left(\mathbf{u}^{(1)};\boldsymbol{\alpha}\right)\delta\mathbf{u}^{(2)}(\mathbf{x})\right\}_{\boldsymbol{\alpha}^0} = \left\{\mathbf{q}^{(2)}\left(\mathbf{u}^{(2)};\boldsymbol{\alpha};\delta\boldsymbol{\alpha}\right)\right\}_{\boldsymbol{\alpha}^0}, \quad \mathbf{x}\in\Omega_x, \tag{2.77}$$

$$\left\{\mathbf{b}_{\nu}^{(2)}\left[\mathbf{u}^{(2)};\boldsymbol{\alpha};\delta\mathbf{u}^{(2)}(\mathbf{x});\delta\boldsymbol{\alpha}\right]\right\}_{\boldsymbol{\alpha}^0} \triangleq \left\{\begin{pmatrix}\mathbf{b}_{\nu}^{(1)}\left(\mathbf{u}^{(1)};\boldsymbol{\alpha};\delta\mathbf{u}^{(1)};\delta\boldsymbol{\alpha}\right)\\ \delta\mathbf{b}_{A}^{(1)}\left(\mathbf{u}^{(2)};\boldsymbol{\alpha};\delta\mathbf{u}^{(2)};\delta\boldsymbol{\alpha}\right)\end{pmatrix}\right\}_{\boldsymbol{\alpha}^0} = \begin{pmatrix}\mathbf{0}\\ \mathbf{0}\end{pmatrix}, \mathbf{x}\in\partial\Omega_x\left(\boldsymbol{\alpha}^0\right), \tag{2.78}$$

where

$$\mathbf{V}^{(2)}\left(\mathbf{u}^{(1)};\boldsymbol{\alpha}\right) \triangleq \begin{pmatrix} \mathbf{L}(\boldsymbol{\alpha}) & \mathbf{0} & \mathbf{0} & \mathbf{0}\\ \mathbf{0} & \mathbf{L}^*(\boldsymbol{\alpha}) & \mathbf{0} & \mathbf{0}\\ -\dfrac{\partial^2 S\left(\mathbf{u}^{(1)};\boldsymbol{\alpha}\right)}{\partial\boldsymbol{\varphi}\partial\boldsymbol{\varphi}} & -\dfrac{\partial^2 S\left(\mathbf{u}^{(1)};\boldsymbol{\alpha}\right)}{\partial\boldsymbol{\varphi}\partial\boldsymbol{\psi}} & \mathbf{L}^*(\boldsymbol{\alpha}) & \mathbf{0}\\ -\dfrac{\partial^2 S\left(\mathbf{u}^{(1)};\boldsymbol{\alpha}\right)}{\partial\boldsymbol{\psi}\partial\boldsymbol{\varphi}} & -\dfrac{\partial^2 S\left(\mathbf{u}^{(1)};\boldsymbol{\alpha}\right)}{\partial\boldsymbol{\psi}\partial\boldsymbol{\psi}} & \mathbf{0} & \mathbf{L}(\boldsymbol{\alpha}) \end{pmatrix}; \tag{2.79}$$

$$\mathbf{u}^{(2)}(\mathbf{x}) \triangleq \begin{pmatrix}\mathbf{u}^{(1)}(\mathbf{x})\\ \mathbf{a}^{(1)}(\mathbf{x})\end{pmatrix} = \begin{pmatrix}\boldsymbol{\varphi}(\mathbf{x})\\ \boldsymbol{\psi}(\mathbf{x})\\ \mathbf{a}_1^{(1)}(\mathbf{x})\\ \mathbf{a}_2^{(1)}(\mathbf{x})\end{pmatrix};$$
$$\delta\mathbf{u}^{(2)}(\mathbf{x}) \triangleq \begin{pmatrix}\delta\mathbf{u}^{(1)}(\mathbf{x})\\ \delta\mathbf{a}^{(1)}(\mathbf{x})\end{pmatrix} = \begin{pmatrix}\delta\boldsymbol{\varphi}(\mathbf{x})\\ \delta\boldsymbol{\psi}(\mathbf{x})\\ \delta\mathbf{a}_1^{(1)}(\mathbf{x})\\ \delta\mathbf{a}_2^{(1)}(\mathbf{x})\end{pmatrix}; \tag{2.80}$$

$$\mathbf{q}^{(2)}\left(\mathbf{u}^{(2)};\boldsymbol{\alpha};\delta\boldsymbol{\alpha}\right) \triangleq \begin{pmatrix}\mathbf{q}_1^{(2)}\left(\mathbf{u}^{(1)};\boldsymbol{\alpha};\delta\boldsymbol{\alpha}\right)\\ \mathbf{q}_2^{(2)}\left(\mathbf{u}^{(1)};\boldsymbol{\alpha};\delta\boldsymbol{\alpha}\right)\\ \mathbf{q}_3^{(2)}\left(\mathbf{u}^{(2)};\boldsymbol{\alpha};\delta\boldsymbol{\alpha}\right)\\ \mathbf{q}_4^{(2)}\left(\mathbf{u}^{(2)};\boldsymbol{\alpha};\delta\boldsymbol{\alpha}\right)\end{pmatrix} \triangleq \begin{pmatrix}\mathbf{q}_1^{(1)}\left(\boldsymbol{\varphi};\boldsymbol{\alpha};\delta\boldsymbol{\alpha}\right)\\ \mathbf{q}_2^{(1)}\left(\boldsymbol{\psi};\boldsymbol{\alpha};\delta\boldsymbol{\alpha}\right)\\ \mathbf{p}_1^{(2)}\left(\mathbf{u}^{(2)};\boldsymbol{\alpha};\delta\boldsymbol{\alpha}\right)\\ \mathbf{p}_2^{(2)}\left(\mathbf{u}^{(2)};\boldsymbol{\alpha};\delta\boldsymbol{\alpha}\right)\end{pmatrix}. \tag{2.81}$$

All of the components of the matrices and vectors defined in Eqs. (2.77) and (2.78) are to be computed at nominal parameter and state function values. The matrix $\mathbf{V}^{(2)}(\boldsymbol{\varphi};\boldsymbol{\psi};\boldsymbol{\alpha})$, which is defined in (2.79), has the following structure:

$$\mathbf{V}^{(2)}\left(\mathbf{u}^{(1)};\boldsymbol{\alpha}\right) = \begin{pmatrix}\mathbf{V}^{(1)} & [\mathbf{0}]_{2\times 2}\\ \mathbf{V}_{21}^{(2)} & \mathbf{V}_{22}^{(2)}\end{pmatrix};\ \mathbf{V}_{22}^{(2)} = \mathbf{A}^{(1)}(\boldsymbol{\alpha});$$
$$\mathbf{V}_{21}^{(2)} \triangleq \begin{pmatrix} -\dfrac{\partial^2 S\left(\mathbf{u}^{(1)};\boldsymbol{\alpha}\right)}{\partial\boldsymbol{\varphi}\partial\boldsymbol{\varphi}} & -\dfrac{\partial^2 S\left(\mathbf{u}^{(1)};\boldsymbol{\alpha}\right)}{\partial\boldsymbol{\varphi}\partial\boldsymbol{\psi}}\\ -\dfrac{\partial^2 S\left(\mathbf{u}^{(1)};\boldsymbol{\alpha}\right)}{\partial\boldsymbol{\psi}\partial\boldsymbol{\varphi}} & -\dfrac{\partial^2 S\left(\mathbf{u}^{(1)};\boldsymbol{\alpha}\right)}{\partial\boldsymbol{\psi}\partial\boldsymbol{\psi}} \end{pmatrix}. \tag{2.82}$$

The matrix $\mathbf{V}_{21}^{(2)}$ depends only on the system's response and is responsible for coupling the forward and adjoint systems. Although the forward and adjoint systems are coupled, they could nevertheless be solved successively rather than simultaneously, because the matrix $\mathbf{V}^{(2)}(\mathbf{u}^{(1)};\boldsymbol{\alpha})$ is block-diagonal.

The system of equations represented by (2.77) together with the boundary conditions represented by (2.78) will be called the "*2nd-Level Variational Sensitivity System*" (2nd-LVSS). The solution, $\delta\mathbf{u}^{(2)}(\mathbf{x})$, of the 2nd-LVSS will be called the "*second-level variational sensitivity function.*" The 2nd-LVSS differs from the 2nd-OFSS in that the 2nd-LVSS involves first-order differentials of the state functions whereas the 2nd-OFSS involves the second-order partial derivatives of the state functions with respect to model parameters α_{j_2} and $\alpha_{j_1}, j_1 = 1, \ldots, TP; j_2 = 1, \ldots, j_1$. In principle, the 2nd-LVSS could be solved for each possible component of the vectors of parameter variations $\delta\boldsymbol{\alpha}$ to obtain the second-level variational vector $\delta\mathbf{u}^{(2)}(\mathbf{x})$. Subsequently, $\delta\mathbf{u}^{(2)}(\mathbf{x})$ could be used together with the known parameter variations $\delta\boldsymbol{\alpha}$ in (2.71) to compute the indirect-effect term $\{\delta R^{(1)}[j_1;\mathbf{u}^{(2)}(\mathbf{x});\boldsymbol{\alpha};\delta\mathbf{u}^{(1)}(\mathbf{x});\delta\mathbf{a}^{(1)}(\mathbf{x})]\}_{ind}$. Computing $\{\delta R^{(1)}[j_1;\mathbf{u}^{(2)}(\mathbf{x});\boldsymbol{\alpha};\delta\mathbf{u}^{(1)}(\mathbf{x});\delta\mathbf{a}^{(1)}(\mathbf{x})]\}_{ind}$ by solving the 2nd-LVSS would require at least $2TP(TP + 1)$ large-scale computations (to solve the 2nd-LVSS) for every independent component of the vectors of parameter variations $\delta\boldsymbol{\alpha}$.

Following the principles introduced by Cacuci (2018), the application of the 2nd-CASAM-L avoids the need for solving the 2nd-LVSS, by expressing the indirect-effect term $\{\delta R^{(1)}[j_1;\mathbf{u}^{(2)}(\mathbf{x});\boldsymbol{\alpha};\delta\mathbf{u}^{(1)}(\mathbf{x});\delta\mathbf{a}^{(1)}(\mathbf{x})]\}_{ind}$ defined in (2.71) in an alternative manner, in terms of the solution of a *Second-Level Adjoint Sensitivity System* (2nd-LASS), so as to eliminate the appearance of the second-level variational vector $\delta\mathbf{u}^{(2)}(\mathbf{x})$ in the alternative expression of the indirect-effect term $\{\delta R^{(1)}[j_1;\mathbf{u}^{(2)}(\mathbf{x});\boldsymbol{\alpha};\delta\mathbf{u}^{(1)}(\mathbf{x});\delta\mathbf{a}^{(1)}(\mathbf{x})]\}_{ind}$. The construction of the requisite 2nd-LASS commences by considering a Hilbert space, denoted as H_2, comprising square-integrable vector-valued elements of the form $\boldsymbol{\eta}^{(2)}(\mathbf{x})\triangleq\left[\boldsymbol{\eta}_1^{(2)}(\mathbf{x}),\boldsymbol{\eta}_2^{(2)}(\mathbf{x}),\boldsymbol{\eta}_3^{(2)}(\mathbf{x}),\boldsymbol{\eta}_4^{(2)}(\mathbf{x})\right]^{\dagger}\in \mathrm{H}_2$, with $\boldsymbol{\eta}_i^{(2)}(\mathbf{x})\triangleq\left[\eta_{i,1}^{(2)}(\mathbf{x}),\ldots,\eta_{i,j}^{(2)}(\mathbf{x}),\ldots,\eta_{i,TD}^{(2)}(\mathbf{x})\right]^{\dagger}$, $i=1,2,3,4$. The inner product between two elements, $\boldsymbol{\eta}^{(2)}(\mathbf{x})\in \mathrm{H}_2$ and $\boldsymbol{\xi}^{(2)}(\mathbf{x})\in \mathrm{H}_2$, of this Hilbert space, will be denoted as $\langle\boldsymbol{\eta}^{(2)}(\mathbf{x}),\boldsymbol{\xi}^{(2)}(\mathbf{x})\rangle_2$ and is defined as follows:

$$\left\langle\boldsymbol{\eta}^{(2)}(\mathbf{x}),\boldsymbol{\xi}^{(2)}(\mathbf{x})\right\rangle_2\triangleq\sum_{i=1}^{4}\left\langle\boldsymbol{\eta}_i^{(2)}(\mathbf{x}),\boldsymbol{\xi}_i^{(2)}(\mathbf{x})\right\rangle_0. \tag{2.83}$$

In the Hilbert H_2, form the inner product of (2.77) with a set of yet undefined functions $\mathbf{a}^{(2)}(j_1;\mathbf{x})\triangleq\left[\mathbf{a}_1^{(2)}(j_1;\mathbf{x}),\mathbf{a}_2^{(2)}(j_1;\mathbf{x}),\mathbf{a}_3^{(2)}(j_1;\mathbf{x}),\mathbf{a}_4^{(2)}(j_1;\mathbf{x})\right]^{\dagger}\in \mathrm{H}_2$, $j_1 = 1, \ldots, TP$, to obtain the following relation:

$$\begin{aligned}&\left\{\left\langle\mathbf{a}^{(2)}(j_1;\mathbf{x}),\mathbf{V}^{(2)}\left(\mathbf{u}^{(1)};\boldsymbol{\alpha}\right)\delta\mathbf{u}^{(2)}(\mathbf{x})\right\rangle_2\right\}_{\boldsymbol{\alpha}^0}\\&=\left\{\left\langle\mathbf{a}^{(2)}(j_1;\mathbf{x}),\mathbf{q}^{(2)}\left(\mathbf{u}^{(2)};\boldsymbol{\alpha};\delta\boldsymbol{\alpha}\right)\right\rangle_2\right\}_{\boldsymbol{\alpha}^0}.\end{aligned} \tag{2.84}$$

Using the definition of the adjoint operator in the Hilbert space H_2, recast the left side of (2.84) as follows:

$$\begin{aligned}&\left\{\left\langle \mathbf{a}^{(2)}(j_1;\mathbf{x}),\mathbf{V}^{(2)}\left(\mathbf{u}^{(1)};\boldsymbol{\alpha}\right)\delta\mathbf{u}^{(2)}(\mathbf{x})\right\rangle_2\right\}_{\boldsymbol{\alpha}^0}\\&=\left\{\left\langle \delta\mathbf{u}^{(2)}(\mathbf{x}),\mathbf{A}^{(2)}\left(\mathbf{u}^{(1)};\boldsymbol{\alpha}\right)\mathbf{a}^{(2)}(j_1;\mathbf{x})\right\rangle_2\right\}_{\boldsymbol{\alpha}^0}\\&+\left\{P^{(2)}\left[\delta\mathbf{u}^{(2)}(\mathbf{x});\mathbf{u}^{(1)}(\mathbf{x});\mathbf{a}^{(2)}(j_1;\mathbf{x});\boldsymbol{\alpha};\delta\boldsymbol{\alpha}\right]\right\}_{\boldsymbol{\alpha}^0},\end{aligned}\tag{2.85}$$

where $\left\{P^{(2)}\left[\delta\mathbf{u}^{(2)}(\mathbf{x});\mathbf{u}^{(1)}(\mathbf{x});\mathbf{a}^{(2)}(j_1;\mathbf{x});\boldsymbol{\alpha};\delta\boldsymbol{\alpha}\right]\right\}_{\boldsymbol{\alpha}^0}$ denotes the bilinear concomitant defined on the phase-space boundary $\mathbf{x} \in \partial\Omega_x(\boldsymbol{\alpha}^0)$ and where $\mathbf{A}^{(2)}(\mathbf{u}^{(1)};\boldsymbol{\alpha}) \triangleq [\mathbf{V}^{(2)}(\mathbf{u}^{(1)};\boldsymbol{\alpha})]^*$ is the operator formally adjoint to $\mathbf{V}^{(2)}(\mathbf{u}^{(1)};\boldsymbol{\alpha})$ and has therefore the following form:

$$\begin{aligned}\mathbf{A}^{(2)}\left(\mathbf{u}^{(1)};\boldsymbol{\alpha}\right)&=\begin{pmatrix}\mathbf{A}^{(1)}(\boldsymbol{\alpha}) & \left[\mathbf{V}_{21}^{(2)}\left(\mathbf{u}^{(1)};\boldsymbol{\alpha}\right)\right]^*\\ [\mathbf{0}]_{(2\times2)} & \mathbf{V}^{(1)}(\boldsymbol{\alpha})\end{pmatrix}\\&=\begin{pmatrix}\mathbf{L}^*(\boldsymbol{\alpha}) & \mathbf{0} & -\dfrac{\partial^2 S(\mathbf{u}^{(1)};\boldsymbol{\alpha})}{\partial\boldsymbol{\varphi}\partial\boldsymbol{\varphi}} & -\dfrac{\partial^2 S(\mathbf{u}^{(1)};\boldsymbol{\alpha})}{\partial\boldsymbol{\psi}\partial\boldsymbol{\varphi}}\\ \mathbf{0} & \mathbf{L}(\boldsymbol{\alpha}) & -\dfrac{\partial^2 S(\mathbf{u}^{(1)};\boldsymbol{\alpha})}{\partial\boldsymbol{\varphi}\partial\boldsymbol{\psi}} & -\dfrac{\partial^2 S(\mathbf{u}^{(1)};\boldsymbol{\alpha})}{\partial\boldsymbol{\psi}\partial\boldsymbol{\psi}}\\ \mathbf{0} & \mathbf{0} & \mathbf{L}(\boldsymbol{\alpha}) & \mathbf{0}\\ \mathbf{0} & \mathbf{0} & \mathbf{0} & \mathbf{L}^*(\boldsymbol{\alpha})\end{pmatrix}.\end{aligned}\tag{2.86}$$

The first term on right side of (2.85) is now required to represent the indirect-effect term $\{\delta R^{(1)}[j_1;\mathbf{u}^{(2)}(\mathbf{x});\boldsymbol{\alpha};\delta\mathbf{u}^{(1)}(\mathbf{x});\delta\mathbf{a}^{(1)}(\mathbf{x})]\}_{ind}$ defined in (2.71). This requirement is satisfied by imposing the following relation on each element $\mathbf{a}^{(2)}(j_1;\mathbf{x})\triangleq\left[\mathbf{a}_1^{(2)}(j_1;\mathbf{x}),\mathbf{a}_2^{(2)}(j_1;\mathbf{x}),\mathbf{a}_3^{(2)}(j_1;\mathbf{x}),\mathbf{a}_4^{(2)}(j_1;\mathbf{x})\right]^\dagger \in H_2, j_1 = 1, \ldots, TP$:

$$\left\{\mathbf{A}^{(2)}\left(\mathbf{u}^{(1)};\boldsymbol{\alpha}\right)\mathbf{a}^{(2)}(j_1;\mathbf{x})\right\}_{\boldsymbol{\alpha}^0}=\left\{\mathbf{s}^{(2)}\left(j_1;\mathbf{u}^{(2)};\boldsymbol{\alpha}\right)\right\}_{\boldsymbol{\alpha}^0},\ j_1=1,\ldots,TP,\tag{2.87}$$

where the block-vector $\mathbf{s}^{(2)}(j_1;\mathbf{u}^{(2)};\boldsymbol{\alpha})$ comprises, for each $j_1 = 1, \ldots, TP$, four vector-components which are defined as follows:

$$\begin{aligned}&\mathbf{s}^{(2)}\left(j_1;\mathbf{u}^{(2)};\boldsymbol{\alpha}\right)\\&\triangleq\left[\mathbf{s}_1^{(2)}\left(j_1;\mathbf{u}^{(2)};\mathbf{x}\right),\mathbf{s}_2^{(2)}\left(j_1;\mathbf{u}^{(2)};\mathbf{x}\right),\mathbf{s}_3^{(2)}\left(j_1;\mathbf{u}^{(2)};\mathbf{x}\right),\mathbf{s}_4^{(2)}\left(j_1;\mathbf{u}^{(2)};\mathbf{x}\right)\right]^\dagger\\&\triangleq\left[\frac{\partial S^{(1)}(j_1;\mathbf{u}^{(2)};\boldsymbol{\alpha})}{\partial\boldsymbol{\varphi}},\frac{\partial S^{(1)}(j_1;\mathbf{u}^{(2)};\boldsymbol{\alpha})}{\partial\boldsymbol{\psi}},\frac{\partial S^{(1)}(j_1;\mathbf{u}^{(2)};\boldsymbol{\alpha})}{\partial\mathbf{a}_1^{(1)}},\frac{\partial S^{(1)}(j_1;\mathbf{u}^{(2)};\boldsymbol{\alpha})}{\partial\mathbf{a}_2^{(1)}}\right]^\dagger.\end{aligned}\tag{2.88}$$

The definition of the set of vectors $\mathbf{a}^{(2)}(j_1;\mathbf{x}) \triangleq \left[\mathbf{a}_1^{(2)}(j_1;\mathbf{x}), \mathbf{a}_2^{(2)}(j_1;\mathbf{x}), \mathbf{a}_3^{(2)}(j_1;\mathbf{x}), \mathbf{a}_4^{(2)}(j_1;\mathbf{x})\right]^\dagger$ will be completed by selecting boundary conditions for this set of vectors, which will be represented in operator form as follows:

$$\left\{\mathbf{b}_A^{(2)}\left[\mathbf{u}^{(2)}(\mathbf{x}); \mathbf{a}^{(2)}(j_1;\mathbf{x}); \boldsymbol{\alpha}\right]\right\}_{\boldsymbol{\alpha}^0} = \mathbf{0}, \ \mathbf{x} \in \partial\Omega_x(\boldsymbol{\alpha}^0), j_1 = 1, \ldots, TP. \tag{2.89}$$

The boundary conditions represented by (2.89) are selected so as to satisfy the following requirements:

1. The boundary conditions (2.89) together with the operator Eq. (2.87) constitute a well-posed problem for the functions $\mathbf{a}^{(2)}(j_1;\mathbf{x})$.
2. The implementation in (2.85) of the boundary conditions provided in (2.78) together with those provided in (2.89) eliminates all of the unknown values of the functions $\delta\mathbf{u}^{(2)}(\mathbf{x})$ and $\mathbf{a}^{(2)}(j_1;\mathbf{x})$ in the expression of the bilinear concomitant $\left\{P^{(2)}\left[\mathbf{u}^{(2)}(\mathbf{x}); \mathbf{a}^{(2)}(j_1;\mathbf{x}); \boldsymbol{\alpha}; \delta\mathbf{u}^{(2)}(\mathbf{x}); \delta\boldsymbol{\alpha}\right]\right\}_{\boldsymbol{\alpha}^0}$. The bilinear concomitant may vanish after these boundary conditions are implemented, but if it does not, it will be reduced to a residual quantity which will be denoted as $\widehat{P}^{(2)}\left[\mathbf{u}^{(2)}(\mathbf{x}); \mathbf{a}^{(2)}(j_1;\mathbf{x}); \boldsymbol{\alpha}; \delta\boldsymbol{\alpha}\right]$ and which will comprise only known values of $\mathbf{u}^{(2)}(\mathbf{x})$, $\mathbf{a}^{(2)}(j_1;\mathbf{x})$, $\boldsymbol{\alpha}$ and $\delta\boldsymbol{\alpha}$.

The system of equations represented by (2.87) together with the boundary conditions represented by (2.89) constitute the *Second-Level Adjoint Sensitivity System* (2nd-LASS). The solution of the 2nd-LASS, i.e., the four-component vector $\mathbf{a}^{(2)}(j_1;\mathbf{x}) \triangleq \left[\mathbf{a}_1^{(2)}(j_1;\mathbf{x}), \mathbf{a}_2^{(2)}(j_1;\mathbf{x}), \mathbf{a}_3^{(2)}(j_1;\mathbf{x}), \mathbf{a}_4^{(2)}(j_1;\mathbf{x})\right]^\dagger \in \mathrm{H}_2$, $j_1, \ldots, TP$, will be called the *second-level adjoint sensitivity function.* The results provided in (2.87), (2.85) and (2.84) are employed in (2.71) to obtain the following expression for the indirect-effect term $\{\delta R^{(1)}[j_1; \mathbf{u}^{(2)}(\mathbf{x}); \boldsymbol{\alpha}; \delta\mathbf{u}^{(1)}(\mathbf{x}); \delta\mathbf{a}^{(1)}(\mathbf{x})]\}_{ind}$ in terms of the second-level adjoint sensitivity functions $\mathbf{a}^{(2)}(j_1;\mathbf{x})$, for $j_1 = 1, \ldots, TP$:

$$\begin{aligned}
&\text{For } j_1 = 1, \ldots, TP: \ \left\{\delta R^{(1)}\left[j_1; \mathbf{u}^{(2)}(\mathbf{x}); \boldsymbol{\alpha}; \delta\mathbf{u}^{(1)}(\mathbf{x}); \delta\mathbf{a}^{(1)}(\mathbf{x})\right]\right\}_{ind} \\
&= \left\{\left\langle \mathbf{a}^{(2)}(j_1;\mathbf{x}), \mathbf{q}^{(2)}\left(\mathbf{u}^{(2)}; \boldsymbol{\alpha}; \delta\boldsymbol{\alpha}\right)\right\rangle_2\right\}_{\boldsymbol{\alpha}^0} \\
&\quad - \left\{\widehat{P}^{(2)}\left[\mathbf{u}^{(2)}(\mathbf{x}); \mathbf{a}^{(2)}(j_1;\mathbf{x}); \boldsymbol{\alpha}; \delta\boldsymbol{\alpha}\right]\right\}_{\boldsymbol{\alpha}^0}.
\end{aligned} \tag{2.90}$$

Inserting the expressions that define the vector $\mathbf{q}^{(2)}(\mathbf{u}^{(2)}; \boldsymbol{\alpha}; \delta\boldsymbol{\alpha})$ from (2.81) into (2.90) and adding the resulting expression for the indirect-effect term with the expression of the direct-effect term given in (2.70) yields the following expression for the total second-order G-differential of the response $R[\boldsymbol{\varphi}(\mathbf{x}), \boldsymbol{\psi}(\mathbf{x}); \boldsymbol{\alpha}]$:

$$
\begin{aligned}
&\left\{\delta R^{(1)}\left[j_1;\mathbf{u}^{(2)}(\mathbf{x});\boldsymbol{\alpha};\delta\mathbf{u}^{(2)}(\mathbf{x});\delta\boldsymbol{\alpha}\right]\right\}_{\boldsymbol{\alpha}^0} \\
&=\left\{\frac{\partial}{\partial\boldsymbol{\alpha}}\prod_{k=1}^{TI}\int_{\lambda_k(\boldsymbol{\alpha})}^{\omega_k(\boldsymbol{\alpha})}dx_k S^{(1)}\left(j_1;\mathbf{u}^{(2)};\boldsymbol{\alpha}\right)\right\}_{\boldsymbol{\alpha}^0}\delta\boldsymbol{\alpha} \\
&+\left\{\sum_{n=1}^{4}\left\langle\mathbf{a}_n^{(2)}(j_1;\mathbf{x}),\mathbf{q}_n^{(2)}\left(\mathbf{u}^{(2)};\boldsymbol{\alpha};\delta\boldsymbol{\alpha}\right)\right\rangle_0\right\}_{\boldsymbol{\alpha}^0}-\left\{\widehat{P}^{(2)}\left[\mathbf{u}^{(2)}(\mathbf{x});\mathbf{a}^{(2)}(j_1;\mathbf{x});\boldsymbol{\alpha};\delta\boldsymbol{\alpha}\right]\right\}_{\boldsymbol{\alpha}^0} \\
&\equiv\sum_{j_2=1}^{TP}\left\{R^{(2)}\left[j_2;j_1;\mathbf{u}^{(2)}(\mathbf{x});\mathbf{a}^{(2)}(j_1;\mathbf{x});\boldsymbol{\alpha}\right]\right\}_{\boldsymbol{\alpha}^0}\delta\alpha_{j_2},
\end{aligned}
\tag{2.91}
$$

where $R^{(2)}[j_2;j_1;\mathbf{u}^{(2)}(\mathbf{x});\mathbf{a}^{(2)}(j_1;\mathbf{x});\boldsymbol{\alpha}]$ denotes the *second-order partial sensitivity* of the response with respect to the model parameters, evaluated at the nominal parameter values $\boldsymbol{\alpha}^0$, and has the following expression for $j_1, j_2 = 1, \ldots, TP$:

$$
\begin{aligned}
&R^{(2)}\left[j_2;j_1;\mathbf{u}^{(2)}(\mathbf{x});\mathbf{a}^{(2)}(j_1;\mathbf{x});\boldsymbol{\alpha}\right]\equiv\frac{\partial R^{(1)}\left[j_1;\mathbf{u}^{(1)}(\mathbf{x});\mathbf{a}^{(1)};\boldsymbol{\alpha}\right]}{\partial\alpha_{j_2}} \\
&\equiv\frac{\partial^2 R\left[\mathbf{u}^{(1)}(\mathbf{x});\boldsymbol{\alpha}\right]}{\partial\alpha_{j_2}\partial\alpha_{j_1}}=\frac{\partial}{\partial\alpha_{j_2}}\left\{\prod_{k=1}^{TI}\int_{\lambda_k(\boldsymbol{\alpha})}^{\omega_k(\boldsymbol{\alpha})}dx_k S^{(1)}\left(j_1;\mathbf{u}^{(2)};\boldsymbol{\alpha}\right)\right\}_{\boldsymbol{\alpha}^0} \\
&+\left\{\left\langle\mathbf{a}_1^{(2)}(j_1;\mathbf{x}),\frac{\partial\left[\mathbf{q}_\varphi(\boldsymbol{\alpha})-\mathbf{L}(\boldsymbol{\alpha})\boldsymbol{\varphi}(\mathbf{x})\right]}{\partial\alpha_{j_2}}\right\rangle_0\right\}_{\boldsymbol{\alpha}^0} \\
&+\left\{\left\langle\mathbf{a}_2^{(2)}(j_1;\mathbf{x}),\frac{\partial\left[\mathbf{q}_\psi(\boldsymbol{\alpha})-\mathbf{L}^*(\boldsymbol{\alpha})\boldsymbol{\psi}(\mathbf{x})\right]}{\partial\alpha_{j_2}}\right\rangle_0\right\}_{\boldsymbol{\alpha}^0} \\
&+\left\{\left\langle\mathbf{a}_3^{(2)}(j_1;\mathbf{x}),\frac{\partial^2 S\left(\mathbf{u}^{(1)}(\mathbf{x});\boldsymbol{\alpha}\right)}{\partial\alpha_{j_2}\partial\boldsymbol{\varphi}}-\frac{\partial\left[\mathbf{L}^*(\boldsymbol{\alpha})\mathbf{a}_1^{(1)}\right]}{\partial\alpha_{j_2}}\right\rangle_0\right\}_{\boldsymbol{\alpha}^0} \\
&+\left\{\left\langle\mathbf{a}_4^{(2)}(j_1;\mathbf{x}),\frac{\partial^2 S\left(\mathbf{u}^{(1)}(\mathbf{x});\boldsymbol{\alpha}\right)}{\partial\alpha_{j_2}\partial\boldsymbol{\psi}}-\frac{\partial\left[\mathbf{L}(\boldsymbol{\alpha})\mathbf{a}_2^{(1)}\right]}{\partial\alpha_{j_2}}\right\rangle_0\right\}_{\boldsymbol{\alpha}^0} \\
&-\left\{\frac{\partial}{\partial\alpha_{j_2}}\widehat{P}^{(2)}\left[\mathbf{u}^{(2)}(\mathbf{x});\mathbf{a}^{(2)}(j_1;\mathbf{x});\boldsymbol{\alpha}\right]\right\}_{\boldsymbol{\alpha}^0}.
\end{aligned}
\tag{2.92}
$$

Since the *2nd-LASS* is independent of parameter variations $\delta\boldsymbol{\alpha}$, the exact computation of all of the partial second-order sensitivities $R^{(2)}[j_2;j_1;\mathbf{u}^{(2)}(\mathbf{x});\mathbf{a}^{(2)}(j_1;\mathbf{x});\boldsymbol{\alpha}]$ requires at most *TP* large-scale (adjoint) computations using the 2nd-LASS, rather than $O(TP^2)$ large-scale computations as would be required by forward methods. In component form, the equations comprising the 2nd-LASS are solved for each $j_1 = 1, \ldots, TP$ in the following order:

$$\mathbf{L}(\boldsymbol{\alpha})\mathbf{a}_3^{(2)}(j_1;\mathbf{x}) = \frac{\partial S^{(1)}\left(j_1;\mathbf{u}^{(2)};\boldsymbol{\alpha}\right)}{\partial \mathbf{a}_1^{(1)}}, \tag{2.93}$$

$$\mathbf{L}^*(\boldsymbol{\alpha})\mathbf{a}_4^{(2)}(j_1;\mathbf{x}) = \frac{\partial S^{(1)}\left(j_1;\mathbf{u}^{(2)};\boldsymbol{\alpha}\right)}{\partial \mathbf{a}_2^{(1)}}, \tag{2.94}$$

$$\begin{aligned} \mathbf{L}^*(\boldsymbol{\alpha})\mathbf{a}_1^{(2)}(j_1;\mathbf{x}) &= \frac{\partial^2 S\left(\mathbf{u}^{(1)};\boldsymbol{\alpha}\right)}{\partial\boldsymbol{\varphi}\partial\boldsymbol{\varphi}}\mathbf{a}_3^{(2)}(j_1;\mathbf{x}) \\ &+ \frac{\partial^2 S\left(\mathbf{u}^{(1)};\boldsymbol{\alpha}\right)}{\partial\boldsymbol{\psi}\partial\boldsymbol{\varphi}}\mathbf{a}_4^{(2)}(j_1;\mathbf{x}) + \frac{\partial S^{(1)}\left(j_1;\mathbf{u}^{(2)};\boldsymbol{\alpha}\right)}{\partial\boldsymbol{\varphi}}, \end{aligned} \tag{2.95}$$

$$\begin{aligned} \mathbf{L}(\boldsymbol{\alpha})\mathbf{a}_2^{(2)}(j_1;\mathbf{x}) &= \frac{\partial^2 S\left(\mathbf{u}^{(1)};\boldsymbol{\alpha}\right)}{\partial\boldsymbol{\varphi}\partial\boldsymbol{\psi}}\mathbf{a}_3^{(2)}(j_1;\mathbf{x}) \\ &+ \frac{\partial^2 S\left(\mathbf{u}^{(1)};\boldsymbol{\alpha}\right)}{\partial\boldsymbol{\psi}\partial\boldsymbol{\psi}}\mathbf{a}_4^{(2)}(j_1;\mathbf{x}) + \frac{\partial S^{(1)}\left(j_1;\mathbf{u}^{(2)};\boldsymbol{\alpha}\right)}{\partial\boldsymbol{\psi}}. \end{aligned} \tag{2.96}$$

It is important to note that *by solving the 2nd-LASS TP-times, the "off-diagonal" second-order mixed sensitivities* $\partial^2 R/\partial\alpha_{j_1}\partial\alpha_{j_2}$ *will be computed twice, in two different ways (i.e., using distinct second-level adjoint sensitivity functions), thereby providing an independent intrinsic (numerical) verification that the first- and second-order response sensitivities are computed accurately.* The information provided by the first-order sensitivities usually indicates which second-order sensitivities are important and which could be neglected. Therefore, it is useful to prioritize the computation of the second-order sensitivities by using the rankings of the relative magnitudes of the first-order sensitivities as a "priority indicator": the larger the magnitude of the relative first-order sensitivity, the higher the priority for computing the corresponding second-order sensitivities. Also, since vanishing first-order sensitivities may indicate critical points of the response in the phase-space of model parameters, it is also of interest to compute the second-order sensitivities that correspond to vanishing first-order sensitivities. In practice, only those second-order partial sensitivities which are deemed important would need to be computed.

Dirac delta-functionals in (2.92) may need to be used for expressing the non-zero residual terms in the respective residual bilinear concomitant and/or the terms containing derivatives with respect to the lower- and upper-boundary points, so that the expression of the partial second-order sensitivities $R^{(2)}[j_2;j_1;\mathbf{u}^{(2)}(\mathbf{x});\mathbf{a}^{(2)}(j_1;\mathbf{x});\boldsymbol{\alpha}]$ can be written in the following form, in preparation for computing the third-order sensitivities:

$$R^{(2)}\left[j_2;j_1;\mathbf{u}^{(2)}(\mathbf{x});\mathbf{a}^{(2)}(j_1;\mathbf{x});\boldsymbol{\alpha}\right] = \prod_{k=1}^{TI}\int_{\lambda_k(\boldsymbol{\alpha})}^{\omega_k(\boldsymbol{\alpha})} dx_k S^{(2)}\left[j_2;j_1;\mathbf{u}^{(2)}(\mathbf{x});\mathbf{a}^{(2)}(j_1;\mathbf{x});\boldsymbol{\alpha}\right]. \tag{2.97}$$

2.5 The Third-Order Comprehensive Sensitivity Analysis Methodology for Response-Coupled Linear Forward and Adjoint Systems (3rd-CASAM-L)

There are $TP(TP + 1)(TP + 2)/3!$ distinct third-order sensitivities of the response $R[\boldsymbol{\varphi}(\mathbf{x}), \boldsymbol{\psi}(\mathbf{x}); \boldsymbol{\alpha}]$ with respect to the model and response parameters. The third-order unmixed sensitivities, $\partial^3 R(\boldsymbol{\alpha})/\partial\alpha_{j_1}^3$, of the response can be approximately computed by using the well-known finite-difference formula presented below:

$$\frac{\partial^3 R(\boldsymbol{\alpha})}{\partial\alpha_{j_1}^3} \approx \frac{1}{2h_{j_1}^3}\left(R_{j_1+2} - 2R_{j_1+1} + 2R_{j_1-1} - R_{j_1-2}\right) + \mathrm{O}\left(h_{j_1}^2\right),\ j_1 = 1, \ldots, TP, \tag{2.98}$$

where $R_{j_1+2} \triangleq R\left(\alpha_{j_1} + 2h_{j_1}\right)$, $R_{j_1+1} \triangleq R\left(\alpha_{j_1} + h_{j_1}\right)$, $R_{j_1-1} \triangleq R\left(\alpha_{j_1} - h_{j_1}\right)$, $R_{j_1-2} \triangleq R\left(\alpha_{j_1} - 2h_{j_1}\right)$ and h_{j_1} denotes a "judiciously chosen" variation in the parameter α_{j_1} around its nominal value $\alpha_{j_1}^0$. The third-order mixed sensitivities, $\partial^3 R(\boldsymbol{\alpha})/\partial\alpha_{j_1}\partial\alpha_{j_2}\partial\alpha_{j_3}$, can be calculated by using the following finite-difference formula:

$$\begin{aligned}\frac{\partial^3 R(\boldsymbol{\alpha})}{\partial\alpha_{j_1}\partial\alpha_{j_2}\partial\alpha_{j_3}} &\approx \frac{1}{8h_{j_1}h_{j_2}h_{j_3}}\left(R_{j_1+1,j_2+1,j_3+1} - R_{j_1+1,j_2+1,j_3-1}\right. \\ &-R_{j_1+1,j_2-1,j_3+1} + R_{j_1+1,j_2-1,j_3-1} - R_{j_1-1,j_2+1,j_3+1} + R_{j_1-1,j_2+1,j_3-1} \\ &\left.+R_{j_1-1,j_2-1,j_3+1} - R_{j_1-1,j_2-1,j_3-1}\right) + \mathrm{O}\left(h_{j_1}^2, h_{j_2}^2, h_{j_3}^2\right),\end{aligned} \tag{2.99}$$

where $R_{j_1+1,j_2+1,j_3+1} = R\left(\alpha_{j_1} + h_{j_1}, \alpha_{j_2} + h_{j_2}, \alpha_{j_3} + h_{j_3}\right)$, $R_{j_1-1,j_2+1,j_3+1} = R\left(\alpha_{j_1} - h_{j_1}, \alpha_{j_2} + h_{j_2}, \alpha_{j_3} + h_{j_3}\right)$, etc. The values of the quantities on the rights sides of the expressions shown in (2.98) and (2.99) are obtained by re-solving the original forward and adjoint equations repeatedly, using the changed parameter values $\left(\alpha_{j_1} \pm h_{j_1}\right)$, $\left(\alpha_{j_2} \pm h_{j_2}\right)$ and $\left(\alpha_{j_3} \pm h_{j_3}\right)$. As has also been previously discussed, the values of h_{j_1}, h_{j_2} and h_{j_3}, respectively, must be chosen "judiciously" by "trial and error" for each of the parameters α_{j_1}, α_{j_2} and α_{j_3}. The finite difference formulas introduce their intrinsic "methodological errors" of order $\mathrm{O}\left(h_{j_1}^2, h_{j_2}^2, h_{j_3}^2\right)$, which are in addition to, and independent of, the errors that might be incurred in the computation of $\partial^3 R(\boldsymbol{\alpha})/\partial\alpha_{j_1}\partial\alpha_{j_2}\partial\alpha_{j_3}$.

Within the framework of the Forward Sensitivity Analysis Methodology, the third-order sensitivities $\partial^3 R(\boldsymbol{\alpha})/\partial\alpha_{j_1}\partial\alpha_{j_2}\partial\alpha_{j_3}, j_1, j_2, j_3 = 1, \ldots, TP$, of the response to the model parameters are obtained by computing the partial G-derivatives of the second-order sensitivities represented by (2.60), which yields the following result:

$$\begin{aligned}
\frac{\partial^3 R(\boldsymbol{\varphi};\boldsymbol{\psi};\boldsymbol{\alpha})}{\partial\alpha_{j_3}\partial\alpha_{j_2}\partial\alpha_{j_1}} &\triangleq \frac{1}{\delta\alpha_{j_3}}\left\{\frac{d}{d\varepsilon}R^{(2)}\left[j_2;j_1;\boldsymbol{\varphi}\left(\alpha_{j_3}^0+\varepsilon\delta\alpha_{j_3}\right);\right.\right. \\
&\boldsymbol{\psi}\left(\alpha_{j_3}^0+\varepsilon\delta\alpha_{j_3}\right);\alpha_{j_3}^0+\varepsilon\delta\alpha_{j_3};\frac{\partial\boldsymbol{\varphi}\left(\alpha_{j_3}^0+\varepsilon\delta\alpha_{j_3}\right)}{\partial\alpha_{j_2}};\frac{\partial\boldsymbol{\psi}\left(\alpha_{j_3}^0+\varepsilon\delta\alpha_{j_3}\right)}{\partial\alpha_{j_2}}; \\
&\left.\left.\frac{\partial^2\boldsymbol{\varphi}\left(\alpha_{j_3}^0+\varepsilon\delta\alpha_{j_3}\right)}{\partial\alpha_{j_1}\partial\alpha_{j_2}};\frac{\partial^2\boldsymbol{\psi}\left(\alpha_{j_3}^0+\varepsilon\delta\alpha_{j_3}\right)}{\partial\alpha_{j_1}\partial\alpha_{j_2}}\right]\right\}_{\varepsilon=0} = R^{(3)}\left[j_2;j_1;\boldsymbol{\varphi};\boldsymbol{\psi};\boldsymbol{\alpha};\right. \\
&\left.\frac{\partial\boldsymbol{\varphi}}{\partial\alpha_{j_3}};\frac{\partial\boldsymbol{\psi}}{\partial\alpha_{j_3}};\frac{\partial^2\boldsymbol{\varphi}}{\partial\alpha_{j_3}\partial\alpha_{j_2}};\frac{\partial^2\boldsymbol{\psi}}{\partial\alpha_{j_3}\partial\alpha_{j_2}};\frac{\partial^3\boldsymbol{\varphi}}{\partial\alpha_{j_1}\partial\alpha_{j_2}\delta\alpha_{j_3}};\frac{\partial^3\boldsymbol{\psi}}{\partial\alpha_{j_1}\partial\alpha_{j_2}\delta\alpha_{j_3}}\right].
\end{aligned} \tag{2.100}$$

As indicated by the functional dependence of the last term in (2.100), the evaluation of the third-order response sensitivities requires prior knowledge of the first- and second-order derivatives of the state functions with respect to the model parameters. The functions $\partial\boldsymbol{\varphi}(\mathbf{x})/\partial\alpha_{j_3}$ and $\partial\boldsymbol{\psi}(\mathbf{x})/\partial\alpha_{j_3}$ are obtained by solving the 1st-OFSS. The functions $\partial^2\boldsymbol{\varphi}/\partial\alpha_{j_3}\partial\alpha_{j_2}$ and $\partial^2\boldsymbol{\psi}/\partial\alpha_{j_3}\partial\alpha_{j_2}$ are the solutions of the 2nd-OFSS. The functions $\partial^3\boldsymbol{\varphi}/\partial\alpha_{j_1}\partial\alpha_{j_2}\partial\alpha_{j_3}$ and $\partial^3\boldsymbol{\psi}/\partial\alpha_{j_1}\partial\alpha_{j_2}\partial\alpha_{j_3}$ are the solutions of the following "*Third-Order Forward Sensitivity System*" (3rd-OFSS), which is obtained by taking the partial G-derivative of the 2nd-OFSS with respect to a generic model parameter α_{j_3}, to obtain the following system for $j_1, j_2, j_3 = 1, \ldots, TP$:

$$\mathbf{L}(\mathbf{x};\boldsymbol{\alpha}^0)\frac{\partial^3\boldsymbol{\varphi}}{\partial\alpha_{j_3}\partial\alpha_{j_2}\partial\alpha_{j_1}} = \left\{\mathbf{f}^{(3)}(j_3;j_2;j_1;\boldsymbol{\varphi};\boldsymbol{\alpha})\right\}_{\boldsymbol{\alpha}^0};\ \mathbf{x}\in\Omega_x; \tag{2.101}$$

$$\left\{\mathbf{B}_\varphi(\mathbf{x};\boldsymbol{\alpha})\right\}_{\boldsymbol{\alpha}^0}\frac{\partial^3\boldsymbol{\varphi}}{\partial\alpha_{j_3}\partial\alpha_{j_2}\partial\alpha_{j_1}} = \left\{\mathbf{c}_\varphi^{(3)}(j_3;j_2;j_1;\boldsymbol{\varphi};\boldsymbol{\alpha})\right\}_{\boldsymbol{\alpha}^0};\mathbf{x}\in\partial\Omega_x(\boldsymbol{\alpha}^0); \tag{2.102}$$

$$\mathbf{L}^*(\mathbf{x};\boldsymbol{\alpha}^0)\frac{\partial^3\boldsymbol{\psi}}{\partial\alpha_{j_3}\partial\alpha_{j_2}\partial\alpha_{j_1}} = \left\{\mathbf{g}^{(3)}(j_3;j_2;j_1;\boldsymbol{\psi};\boldsymbol{\alpha})\right\}_{\boldsymbol{\alpha}^0};\ \mathbf{x}\in\Omega_x; \tag{2.103}$$

$$\left\{\mathbf{B}_\psi(\mathbf{x};\boldsymbol{\alpha})\right\}_{\boldsymbol{\alpha}^0}\frac{\partial^3\boldsymbol{\psi}}{\partial\alpha_{j_3}\partial\alpha_{j_2}\partial\alpha_{j_1}} = \left\{\mathbf{c}_\psi^{(3)}(j_3;j_2;j_1;\boldsymbol{\psi};\boldsymbol{\alpha})\right\}_{\boldsymbol{\alpha}^0};\mathbf{x}\in\partial\Omega_x(\boldsymbol{\alpha}^0); \tag{2.104}$$

where

$$\mathbf{f}^{(3)}(j_3;j_2;j_1;\boldsymbol{\varphi};\boldsymbol{\alpha}) \triangleq \frac{\partial \mathbf{f}^{(2)}(j_2;j_1;\boldsymbol{\varphi};\boldsymbol{\alpha})}{\partial \alpha_{j_3}} - \frac{\partial \mathbf{L}(\mathbf{x};\boldsymbol{\alpha})}{\partial \alpha_{j_3}} \frac{\partial^2 \boldsymbol{\varphi}}{\partial \alpha_{j_2} \partial \alpha_{j_1}}; \tag{2.105}$$

$$\mathbf{g}^{(3)}(j_3;j_2;j_1;\boldsymbol{\psi};\boldsymbol{\alpha}) \triangleq \frac{\partial \mathbf{g}^{(2)}(j_2;j_1;\boldsymbol{\psi};\boldsymbol{\alpha})}{\partial \alpha_{j_3}} - \frac{\partial \mathbf{L}^*(\mathbf{x};\boldsymbol{\alpha})}{\partial \alpha_{j_3}} \frac{\partial^2 \boldsymbol{\psi}}{\partial \alpha_{j_2} \partial \alpha_{j_1}}; \tag{2.106}$$

$$\mathbf{c}_\varphi^{(3)}(j_3;j_2;j_1;\boldsymbol{\varphi};\boldsymbol{\alpha}) \triangleq \frac{\partial \mathbf{c}_\varphi^{(2)}(j_2;j_1;\boldsymbol{\varphi};\boldsymbol{\alpha})}{\partial \alpha_{j_3}} - \frac{\partial \mathbf{B}_\varphi(\mathbf{x};\boldsymbol{\alpha})}{\partial \alpha_{j_3}} \frac{\partial^2 \boldsymbol{\varphi}}{\partial \alpha_{j_2} \partial \alpha_{j_1}}; \tag{2.107}$$

$$\mathbf{c}_\psi^{(3)}(j_3;j_2;j_1;\boldsymbol{\psi};\boldsymbol{\alpha}) \triangleq \frac{\partial \mathbf{c}_\psi^{(2)}(j_2;j_1;\boldsymbol{\psi};\boldsymbol{\alpha})}{\partial \alpha_{j_3}} - \frac{\partial \mathbf{B}_\psi(\mathbf{x};\boldsymbol{\alpha})}{\partial \alpha_{j_3}} \frac{\partial^2 \boldsymbol{\psi}}{\partial \alpha_{j_2} \partial \alpha_{j_1}}. \tag{2.108}$$

Evidently, the computations of the third-order sensitivities $\partial^3 R(\boldsymbol{\alpha})/\partial \alpha_{j_3} \partial \alpha_{j_2} \partial \alpha_{j_1}$ require $2TP(TP + 1)(TP + 2)/3!$ large-scale computations to solve (2.101)–(2.104), followed by $2TP(TP + 1)(TP + 2)/3!$ small-scale computations for performing the integrations that define the various sensitivities. These computations would be in addition to the $2TP$ large-scale computations needed to determine the functions $\partial \boldsymbol{\varphi}(\mathbf{x})/\partial \alpha_{j_3}$ and $\partial \boldsymbol{\psi}(\mathbf{x})/\partial \alpha_{j_3}$ by solving the 1st-OFSS, and the $TP(TP + 1)$ large-scale computations needed to determine the functions $\partial^2 \boldsymbol{\varphi}/\partial \alpha_{j_3} \partial \alpha_{j_2}$ and $\partial^2 \boldsymbol{\psi}/\partial \alpha_{j_3} \partial \alpha_{j_2}$ by solving the 2nd-OFSS.

The third-order sensitivities of the response $R[\boldsymbol{\varphi}(\mathbf{x}), \boldsymbol{\psi}(\mathbf{x}); \boldsymbol{\alpha}]$ with respect to the model parameters are obtained by determining the first-order G-differential of $R^{(2)}\left[j_2;j_1;\mathbf{u}^{(2)}(\mathbf{x});\mathbf{a}^{(2)}(j_1;\mathbf{x});\boldsymbol{\alpha}\right] = \partial^2 R(\boldsymbol{\varphi};\boldsymbol{\psi};\boldsymbol{\alpha})/\partial \alpha_{j_1} \partial \alpha_{j_2}$ which was defined in (2.97). By definition, the total G-differential of $R^{(2)}[j_2;j_1;\mathbf{u}^{(2)}(\mathbf{x});\mathbf{a}^{(2)}(j_1;\mathbf{x});\boldsymbol{\alpha}]$ is obtained as follows:

$$\begin{aligned}
&\left\{\delta R^{(2)}\left[j_2;j_1;\mathbf{u}^{(2)};\mathbf{a}^{(2)}(j1);\boldsymbol{\alpha}\right]\right\}_{\boldsymbol{\alpha}^0} \triangleq \left\{\frac{d}{d\varepsilon}\left[\prod_{k=1}^{TI} \int\limits_{\lambda_k(\boldsymbol{\alpha}^0+\varepsilon\delta\boldsymbol{\alpha})}^{\omega_k(\boldsymbol{\alpha}^0+\varepsilon\delta\boldsymbol{\alpha})} dx_k S^{(2)}[j_2;j_1;\right.\right. \\
&\qquad \left.\left.\mathbf{u}^{(2),0} + \varepsilon\delta\mathbf{u}^{(2)};\mathbf{a}^{(2),0}(j_1) + \varepsilon\delta\mathbf{a}^{(2)}(j_1);\boldsymbol{\alpha}^0 + \varepsilon\delta\boldsymbol{\alpha}\right]\right\}_{\varepsilon=0} \\
&= \left\{\delta R^{(2)}\left[j_2;j_1;\mathbf{u}^{(2),0};\mathbf{a}^{(2),0}(j_1);\boldsymbol{\alpha}^0;\delta\boldsymbol{\alpha}\right]\right\}_{dir} \\
&+ \left\{\delta R^{(2)}\left[j_2;j_1;\mathbf{u}^{(2),0};\mathbf{a}^{(2),0}(j_1);\boldsymbol{\alpha}^0;\delta\mathbf{u}^{(2)};\delta\mathbf{a}^{(2)}(j_1)\right]\right\}_{ind},
\end{aligned} \tag{2.109}$$

where the direct-effect term $\{\delta R^{(2)}[j_2;j_1;\ldots]\}_{dir}$ depends directly on the parameter variations and is defined as follows:

$$\left\{\delta R^{(2)}\left[j_2;j_1;\mathbf{u}^{(2),0};\mathbf{a}^{(2),0}(j_1);\boldsymbol{\alpha}^0;\delta\boldsymbol{\alpha}\right]\right\}_{dir} \triangleq \left\{\frac{\partial}{\partial\boldsymbol{\alpha}}\prod_{k=1}^{TI}\int_{\lambda_k(\boldsymbol{\alpha})}^{\omega_k(\boldsymbol{\alpha})} dx_k S^{(2)}\left[j_2;j_1;\mathbf{u}^{(2)}(\mathbf{x});\mathbf{a}^{(2)}(j_1;\mathbf{x});\boldsymbol{\alpha}\right]\right\}_{\boldsymbol{\alpha}^0}\delta\boldsymbol{\alpha}, \tag{2.110}$$

while the "indirect-effect term" $\{\delta R^{(2)}[j_2;j_1;\ldots]\}_{ind}$ depends indirectly on the parameter variations through the variations in the forward and adjoint state functions and is defined as follows:

$$\begin{aligned}
&\left\{\delta R^{(2)}\left[j_2;j_1;\mathbf{u}^{(2),0};\mathbf{a}^{(2),0}(j_1);\boldsymbol{\alpha}^0;\delta\mathbf{u}^{(2)};\delta\mathbf{a}^{(2)}(j_1)\right]\right\}_{ind} \\
&\triangleq \left\{\prod_{k=1}^{TI}\int_{\lambda_k(\boldsymbol{\alpha})}^{\omega_k(\boldsymbol{\alpha})} dx_k \frac{\partial S^{(2)}\left[j_2;j_1;\mathbf{u}^{(2)}(\mathbf{x});\mathbf{a}^{(2)}(j_1;\mathbf{x});\boldsymbol{\alpha}\right]}{\partial\boldsymbol{\varphi}}\right\}_{\boldsymbol{\alpha}^0}\delta\boldsymbol{\varphi}(\mathbf{x}) \\
&+\left\{\prod_{k=1}^{TI}\int_{\lambda_k(\boldsymbol{\alpha})}^{\omega_k(\boldsymbol{\alpha})} dx_k \frac{\partial S^{(2)}\left[j_2;j_1;\mathbf{u}^{(2)}(\mathbf{x});\mathbf{a}^{(2)}(j_1;\mathbf{x});\boldsymbol{\alpha}\right]}{\partial\boldsymbol{\psi}}\right\}_{\boldsymbol{\alpha}^0}\delta\boldsymbol{\psi}(\mathbf{x}) \\
&+\sum_{m=1}^{2}\left\{\prod_{k=1}^{TI}\int_{\lambda_k(\boldsymbol{\alpha})}^{\omega_k(\boldsymbol{\alpha})} dx_k \frac{\partial S^{(2)}\left[j_2;j_1;\mathbf{u}^{(2)}(\mathbf{x});\mathbf{a}^{(2)}(j_1;\mathbf{x});\boldsymbol{\alpha}\right]}{\partial\mathbf{a}_m^{(1)}}\right\}_{\boldsymbol{\alpha}^0}\delta\mathbf{a}_m^{(1)}(\mathbf{x}) \\
&+\sum_{m=1}^{4}\left\{\prod_{k=1}^{TI}\int_{\lambda_k(\boldsymbol{\alpha})}^{\omega_k(\boldsymbol{\alpha})} dx_k \frac{\partial S^{(2)}\left[j_2;j_1;\mathbf{u}^{(2)}(\mathbf{x});\mathbf{a}^{(2)}(j_1;\mathbf{x});\boldsymbol{\alpha}\right]}{\partial\mathbf{a}_m^{(2)}(j_1)}\right\}_{\boldsymbol{\alpha}^0}\delta\mathbf{a}_m^{(2)}(j_1;\mathbf{x}).
\end{aligned} \tag{2.111}$$

The indirect-effect term defined in (2.111) can be computed only after having determined $\delta\mathbf{a}^{(2)}(j_1;\mathbf{x})\triangleq\left[\delta\mathbf{a}_1^{(2)}(j_1;\mathbf{x}),\delta\mathbf{a}_2^{(2)}(j_1;\mathbf{x}),\delta\mathbf{a}_3^{(2)}(j_1;\mathbf{x}),\delta\mathbf{a}_4^{(2)}(j_1;\mathbf{x})\right]^{\dagger}$ and $\delta\mathbf{u}^{(2)}(\mathbf{x})$. The vector of variations $\delta\mathbf{u}^{(2)}(\mathbf{x})$ is the solution of the 2nd-LVSS while the vector $\delta\mathbf{a}^{(2)}(\mathbf{x})$ is the solution of the system of equations obtained by G-differentiating the 2nd-LASS defined by (2.87) and (2.89).

Applying the definition of the total G-differential, cf. Eq. (2.26), to the 2nd-LASS yields the following set of equations for the vector of variations $\delta\mathbf{a}^{(2)}(\mathbf{x})$, all to be computed at the nominal values of the parameters and state functions (although this will not be indicated explicitly below, in order to keep the notation as simple as possible):

$$
\begin{aligned}
&\mathbf{L}^*(\boldsymbol{\alpha})\delta\mathbf{a}_1^{(2)}(j_1;\mathbf{x}) - \frac{\partial^2 S^{(1)}\left(j_1;\mathbf{u}^{(2)};\boldsymbol{\alpha}\right)}{\partial\boldsymbol{\varphi}\partial\boldsymbol{\varphi}}\delta\boldsymbol{\varphi} - \frac{\partial^2 S^{(1)}\left(j_1;\mathbf{u}^{(2)};\boldsymbol{\alpha}\right)}{\partial\boldsymbol{\varphi}\partial\boldsymbol{\psi}}\delta\boldsymbol{\psi}\\
&-\frac{\partial^2 S^{(1)}\left(j_1;\mathbf{u}^{(2)};\boldsymbol{\alpha}\right)}{\partial\boldsymbol{\varphi}\partial\mathbf{a}_1^{(1)}}\delta\mathbf{a}_1^{(1)} - \frac{\partial^2 S^{(1)}\left(j_1;\mathbf{u}^{(2)};\boldsymbol{\alpha}\right)}{\partial\boldsymbol{\varphi}\partial\mathbf{a}_2^{(1)}}\delta\mathbf{a}_2^{(1)}\\
&-\left[\frac{\partial^3 S\left(\mathbf{u}^{(1)};\boldsymbol{\alpha}\right)}{\partial\boldsymbol{\varphi}\partial\boldsymbol{\varphi}\partial\boldsymbol{\varphi}}\delta\boldsymbol{\varphi} + \frac{\partial^3 S\left(\mathbf{u}^{(1)};\boldsymbol{\alpha}\right)}{\partial\boldsymbol{\varphi}\partial\boldsymbol{\varphi}\partial\boldsymbol{\psi}}\delta\boldsymbol{\psi}\right]\mathbf{a}_3^{(2)}(j_1;\mathbf{x})\\
&-\frac{\partial^2 S\left(\mathbf{u}^{(1)};\boldsymbol{\alpha}\right)}{\partial\boldsymbol{\varphi}\partial\boldsymbol{\varphi}}\delta\mathbf{a}_3^{(2)}(j_1;\mathbf{x}) - \frac{\partial^2 S\left(\mathbf{u}^{(1)};\boldsymbol{\alpha}\right)}{\partial\boldsymbol{\psi}\partial\boldsymbol{\varphi}}\delta\mathbf{a}_4^{(2)}(j_1;\mathbf{x})\\
&-\left[\frac{\partial^3 S\left(\mathbf{u}^{(1)};\boldsymbol{\alpha}\right)}{\partial\boldsymbol{\psi}\partial\boldsymbol{\varphi}\partial\boldsymbol{\varphi}}\delta\boldsymbol{\varphi} + \frac{\partial^3 S\left(\mathbf{u}^{(1)};\boldsymbol{\alpha}\right)}{\partial\boldsymbol{\psi}\partial\boldsymbol{\psi}\partial\boldsymbol{\varphi}}\delta\boldsymbol{\psi}\right]\mathbf{a}_4^{(2)}(j_1;\mathbf{x})\\
&= \mathbf{p}_1^{(2)}\left[\mathbf{u}^{(2)};\mathbf{a}^{(2)}(j_1);\boldsymbol{\alpha};\delta\boldsymbol{\alpha}\right],
\end{aligned}
\tag{2.112}
$$

$$
\begin{aligned}
&\mathbf{L}(\boldsymbol{\alpha})\delta\mathbf{a}_2^{(2)}(j_1;\mathbf{x}) - \frac{\partial^2 S^{(1)}\left(j_1;\mathbf{u}^{(2)};\boldsymbol{\alpha}\right)}{\partial\boldsymbol{\varphi}\partial\boldsymbol{\psi}}\delta\boldsymbol{\varphi} - \frac{\partial^2 S^{(1)}\left(j_1;\mathbf{u}^{(2)};\boldsymbol{\alpha}\right)}{\partial\boldsymbol{\psi}\partial\boldsymbol{\psi}}\delta\boldsymbol{\psi}\\
&-\frac{\partial^2 S^{(1)}\left(j_1;\mathbf{u}^{(2)};\boldsymbol{\alpha}\right)}{\partial\mathbf{a}_1^{(1)}\partial\boldsymbol{\psi}}\delta\mathbf{a}_1^{(1)} - \frac{\partial^2 S^{(1)}\left(j_1;\mathbf{u}^{(2)};\boldsymbol{\alpha}\right)}{\partial\mathbf{a}_2^{(1)}\partial\boldsymbol{\psi}}\delta\mathbf{a}_2^{(1)}\\
&-\left[\frac{\partial^3 S\left(\mathbf{u}^{(1)};\boldsymbol{\alpha}\right)}{\partial\boldsymbol{\varphi}\partial\boldsymbol{\varphi}\partial\boldsymbol{\psi}}\delta\boldsymbol{\varphi} + \frac{\partial^3 S\left(\mathbf{u}^{(1)};\boldsymbol{\alpha}\right)}{\partial\boldsymbol{\varphi}\partial\boldsymbol{\psi}\partial\boldsymbol{\psi}}\delta\boldsymbol{\psi}\right]\mathbf{a}_3^{(2)}(j_1;\mathbf{x})\\
&-\frac{\partial^2 S\left(\mathbf{u}^{(1)};\boldsymbol{\alpha}\right)}{\partial\boldsymbol{\varphi}\partial\boldsymbol{\psi}}\delta\mathbf{a}_3^{(2)}(j_1;\mathbf{x}) - \frac{\partial^2 S\left(\mathbf{u}^{(1)};\boldsymbol{\alpha}\right)}{\partial\boldsymbol{\psi}\partial\boldsymbol{\psi}}\delta\mathbf{a}_4^{(2)}(j_1;\mathbf{x})\\
&-\left[\frac{\partial^3 S\left(\mathbf{u}^{(1)};\boldsymbol{\alpha}\right)}{\partial\boldsymbol{\varphi}\partial\boldsymbol{\psi}\partial\boldsymbol{\psi}}\delta\boldsymbol{\varphi} + \frac{\partial^3 S\left(\mathbf{u}^{(1)};\boldsymbol{\alpha}\right)}{\partial\boldsymbol{\psi}\partial\boldsymbol{\psi}\partial\boldsymbol{\psi}}\delta\boldsymbol{\psi}\right]\mathbf{a}_4^{(2)}(j_1;\mathbf{x})\\
&= \mathbf{p}_2^{(2)}\left[\mathbf{u}^{(2)};\mathbf{a}^{(2)}(j_1);\boldsymbol{\alpha};\delta\boldsymbol{\alpha}\right],
\end{aligned}
\tag{2.113}
$$

$$
\begin{aligned}
&\mathbf{L}(\boldsymbol{\alpha})\delta\mathbf{a}_3^{(2)}(j_1;\mathbf{x}) - \frac{\partial^2 S^{(1)}\left(j_1;\mathbf{u}^{(2)};\boldsymbol{\alpha}\right)}{\partial\boldsymbol{\varphi}\partial\mathbf{a}_1^{(1)}}\delta\boldsymbol{\varphi} - \frac{\partial^2 S^{(1)}\left(j_1;\mathbf{u}^{(2)};\boldsymbol{\alpha}\right)}{\partial\boldsymbol{\psi}\partial\mathbf{a}_1^{(1)}}\delta\boldsymbol{\psi}\\
&-\frac{\partial^2 S^{(1)}\left(j_1;\mathbf{u}^{(2)};\boldsymbol{\alpha}\right)}{\partial\mathbf{a}_1^{(1)}\partial\mathbf{a}_1^{(1)}}\delta\mathbf{a}_1^{(1)} - \frac{\partial^2 S^{(1)}\left(j_1;\mathbf{u}^{(2)};\boldsymbol{\alpha}\right)}{\partial\mathbf{a}_1^{(1)}\partial\mathbf{a}_2^{(1)}}\delta\mathbf{a}_2^{(1)}\\
&= \mathbf{p}_3^{(2)}\left[\mathbf{u}^{(2)};\mathbf{a}^{(2)}(j_1);\boldsymbol{\alpha};\delta\boldsymbol{\alpha}\right],
\end{aligned}
\tag{2.114}
$$

$$\mathbf{L}^*(\boldsymbol{\alpha})\delta\mathbf{a}_4^{(2)}(j_1;\mathbf{x}) - \frac{\partial^2 S^{(1)}(j_1;\mathbf{u}^{(2)};\boldsymbol{\alpha})}{\partial\boldsymbol{\varphi}\partial\mathbf{a}_2^{(1)}}\delta\boldsymbol{\varphi} - \frac{\partial^2 S^{(1)}(j_1;\mathbf{u}^{(2)};\boldsymbol{\alpha})}{\partial\boldsymbol{\psi}\partial\mathbf{a}_2^{(1)}}\delta\boldsymbol{\psi}$$
$$- \frac{\partial^2 S^{(1)}(j_1;\mathbf{u}^{(2)};\boldsymbol{\alpha})}{\partial\mathbf{a}_1^{(1)}\partial\mathbf{a}_2^{(1)}}\delta\mathbf{a}_1^{(1)} - \frac{\partial^2 S^{(1)}(j_1;\mathbf{u}^{(2)};\boldsymbol{\alpha})}{\partial\mathbf{a}_2^{(1)}\partial\mathbf{a}_2^{(1)}}\delta\mathbf{a}_2^{(1)} \tag{2.115}$$
$$= \mathbf{p}_4^{(2)}\left[\mathbf{u}^{(2)};\mathbf{a}^{(2)}(j_1);\boldsymbol{\alpha};\delta\boldsymbol{\alpha}\right],$$

where:

$$\mathbf{p}_1^{(2)}\left[\mathbf{u}^{(2)};\mathbf{a}^{(2)}(j_1);\boldsymbol{\alpha};\delta\boldsymbol{\alpha}\right] \triangleq -\frac{\partial\left[\mathbf{L}^*(\boldsymbol{\alpha})\mathbf{a}_1^{(2)}(j_1;\mathbf{x})\right]}{\partial\boldsymbol{\alpha}}\delta\boldsymbol{\alpha}$$
$$+\left[\frac{\partial^3 S(\mathbf{u}^{(1)};\boldsymbol{\alpha})}{\partial\boldsymbol{\alpha}\partial\boldsymbol{\varphi}\partial\boldsymbol{\varphi}}\delta\boldsymbol{\alpha}\right]\mathbf{a}_3^{(2)}(j_1;\mathbf{x}) + \left[\frac{\partial^3 S(\mathbf{u}^{(1)};\boldsymbol{\alpha})}{\partial\boldsymbol{\alpha}\partial\boldsymbol{\psi}\partial\boldsymbol{\varphi}}\delta\boldsymbol{\alpha}\right]\mathbf{a}_4^{(2)}(j_1;\mathbf{x}) \tag{2.116}$$
$$+\frac{\partial^2 S^{(1)}(j_1;\mathbf{u}^{(2)};\boldsymbol{\alpha})}{\partial\boldsymbol{\alpha}\partial\boldsymbol{\varphi}}\delta\boldsymbol{\alpha},$$

$$\mathbf{p}_2^{(2)}\left[\mathbf{u}^{(2)};\mathbf{a}^{(2)}(j_1);\boldsymbol{\alpha};\delta\boldsymbol{\alpha}\right] \triangleq -\frac{\partial\left[\mathbf{L}(\boldsymbol{\alpha})\mathbf{a}_2^{(2)}(j_1;\mathbf{x})\right]}{\partial\boldsymbol{\alpha}}\delta\boldsymbol{\alpha}$$
$$+\left[\frac{\partial^3 S(\mathbf{u}^{(1)};\boldsymbol{\alpha})}{\partial\boldsymbol{\alpha}\partial\boldsymbol{\varphi}\partial\boldsymbol{\psi}}\delta\boldsymbol{\alpha}\right]\mathbf{a}_3^{(2)}(j_1;\mathbf{x}) + \left[\frac{\partial^3 S(\mathbf{u}^{(1)};\boldsymbol{\alpha})}{\partial\boldsymbol{\alpha}\partial\boldsymbol{\psi}\partial\boldsymbol{\psi}}\delta\boldsymbol{\alpha}\right]\mathbf{a}_4^{(2)}(j_1;\mathbf{x}) \tag{2.117}$$
$$+\frac{\partial S^{(1)}(j_1;\mathbf{u}^{(2)};\boldsymbol{\alpha})}{\partial\boldsymbol{\alpha}\partial\boldsymbol{\psi}}\delta\boldsymbol{\alpha},$$

$$\mathbf{p}_3^{(2)}\left[\mathbf{u}^{(2)};\mathbf{a}^{(2)}(j_1);\boldsymbol{\alpha};\delta\boldsymbol{\alpha}\right]$$
$$\triangleq -\frac{\partial\left[\mathbf{L}(\boldsymbol{\alpha})\mathbf{a}_3^{(2)}(j_1;\mathbf{x})\right]}{\partial\boldsymbol{\alpha}}\delta\boldsymbol{\alpha} + \frac{\partial^2 S^{(1)}(j_1;\mathbf{u}^{(2)};\boldsymbol{\alpha})}{\partial\boldsymbol{\alpha}\partial\mathbf{a}_1^{(1)}}\delta\boldsymbol{\alpha}, \tag{2.118}$$

$$\mathbf{p}_4^{(2)}\left[\mathbf{u}^{(2)};\mathbf{a}^{(2)}(j_1);\boldsymbol{\alpha};\delta\boldsymbol{\alpha}\right]$$
$$\triangleq -\frac{\partial\left[\mathbf{L}^*(\boldsymbol{\alpha})\mathbf{a}_4^{(2)}(j_1;\mathbf{x})\right]}{\partial\boldsymbol{\alpha}}\delta\boldsymbol{\alpha} + \frac{\partial^2 S^{(1)}(j_1;\mathbf{u}^{(2)};\boldsymbol{\alpha})}{\partial\boldsymbol{\alpha}\partial\mathbf{a}_2^{(1)}}\delta\boldsymbol{\alpha}. \tag{2.119}$$

The vector of variations $\delta\mathbf{a}^{(2)}(\mathbf{x})$ is subject to the boundary conditions obtained by G-differentiating (2.89), which will be represented in operator form as follows, for $j_1 = 1, \ldots, TP$:

$$\left\{\delta\mathbf{b}_A^{(2)}\left[\mathbf{u}^{(2)};\mathbf{a}^{(2)}(j_1);\boldsymbol{\alpha};\delta\mathbf{u}^{(2)};\delta\mathbf{a}^{(2)}(j_1);\delta\boldsymbol{\alpha}\right]\right\}_{\boldsymbol{\alpha}^0}=\mathbf{0},\ \mathbf{x}\in\partial\Omega_x\left(\boldsymbol{\alpha}^0\right). \tag{2.120}$$

As (2.112)–(2.120) indicate, the vector of variations $\delta\mathbf{a}^{(2)}(\mathbf{x})$ and $\delta\mathbf{u}^{(2)}(\mathbf{x})\triangleq\left[\delta\mathbf{u}^{(1)}(\mathbf{x});\delta\mathbf{a}_1^{(1)}(\mathbf{x}),\delta\mathbf{a}_2^{(1)}(\mathbf{x})\right]^\dagger$ are related to each other, and must be determined by solving simultaneously the coupled system of equations obtained by concatenating (2.112)–(2.119) to the 2nd-LVSS, while being subject to the boundary conditions obtained by concatenating the boundary conditions which belong to the 2nd-LVSS with the boundary conditions represented by (2.120). This coupled system will comprise 8 (coupled) equations, which can be represented in the following matrix-form, for $j_1 = 1, \ldots, TP$:

$$\begin{aligned}&\mathbf{V}^{(3)}\left[j_1;\mathbf{u}^{(2)}(\mathbf{x})\right]\delta\mathbf{u}^{(3)}(j_1;\mathbf{x})\\&=\mathbf{q}^{(3)}\left[j_1;\mathbf{u}^{(2)}(\mathbf{x});\mathbf{a}^{(2)}(j_1;\mathbf{x});\boldsymbol{\alpha};\delta\boldsymbol{\alpha}\right],\mathbf{x}\in\Omega_x,\end{aligned} \tag{2.121}$$

$$\begin{aligned}&\mathbf{b}_V^{(3)}\left[j_1;\mathbf{u}^{(3)}(j_1;\mathbf{x});\boldsymbol{\alpha};\delta\mathbf{u}^{(3)}(j_1;\mathbf{x});\delta\boldsymbol{\alpha}\right]\\&\triangleq\begin{pmatrix}\mathbf{b}_V^{(2)}\left[\mathbf{u}^{(2)}(\mathbf{x});\boldsymbol{\alpha};\delta\mathbf{u}^{(2)}(\mathbf{x});\delta\boldsymbol{\alpha}\right]\\\delta\mathbf{b}_A^{(2)}\left[j_1;\mathbf{u}^{(3)}(j_1;\mathbf{x});\boldsymbol{\alpha};\delta\mathbf{u}^{(3)}(j_1;\mathbf{x});\delta\boldsymbol{\alpha}\right]\end{pmatrix}=\begin{pmatrix}\mathbf{0}\\\mathbf{0}\end{pmatrix},\mathbf{x}\in\partial\Omega_x\left(\boldsymbol{\alpha}^0\right),\end{aligned} \tag{2.122}$$

where all of the components of the matrices and vectors are to be computed at nominal parameter and state function values, although the corresponding indication (namely, the superscript "zero") has been omitted in order to simplify the notation. The 8x8 block-matrix $\mathbf{V}^{(3)}[j1;\mathbf{u}^{(2)}(\mathbf{x})]$ and the 8-component block-vectors $\delta\mathbf{u}^{(3)}(j1;\mathbf{x})$ and $\mathbf{q}^{(3)}[\mathbf{u}^{(3)}(j1;\mathbf{x});\boldsymbol{\alpha};\delta\boldsymbol{\alpha}]$ have the following structures:

$$\begin{aligned}\mathbf{u}^{(3)}(j_1;\mathbf{x})&=\left[\mathbf{u}^{(2)}(\mathbf{x});\mathbf{a}^{(2)}(j_1;\mathbf{x})\right]^\dagger;\\\delta\mathbf{u}^{(3)}(j_1;\mathbf{x})&=\left[\delta\mathbf{u}^{(2)}(\mathbf{x});\delta\mathbf{a}^{(2)}(j_1;\mathbf{x})\right]^\dagger;\end{aligned} \tag{2.123}$$

$$\begin{aligned}&\mathbf{q}^{(3)}\left[\mathbf{u}^{(3)}(j_1;\mathbf{x});\boldsymbol{\alpha};\delta\boldsymbol{\alpha}\right]\triangleq\begin{pmatrix}\mathbf{q}^{(2)}\left(\mathbf{u}^{(2)};\boldsymbol{\alpha};\delta\boldsymbol{\alpha}\right)\\\mathbf{p}^{(2)}\left[j_1;\mathbf{u}^{(2)}(\mathbf{x});\mathbf{a}^{(2)}(j_1;\mathbf{x});\boldsymbol{\alpha};\delta\boldsymbol{\alpha}\right]\end{pmatrix}\\&\triangleq\left[\mathbf{q}_1^{(3)},\mathbf{q}_2^{(3)},\mathbf{q}_3^{(3)},\mathbf{q}_4^{(3)},\mathbf{q}_5^{(3)}(j_1),\mathbf{q}_6^{(3)}(j_1),\mathbf{q}_7^{(3)}(j_1),\mathbf{q}_8^{(3)}(j_1)\right]^\dagger;\end{aligned} \tag{2.124}$$

$$\mathbf{V}^{(3)}\left[j_1;\mathbf{u}^{(2)}(\mathbf{x})\right] \triangleq \begin{pmatrix} \mathbf{V}^{(2)}\left(\mathbf{u}^{(1)}\right) & [\mathbf{0}]_{(4\times 4)} \\ \mathbf{V}_{21}^{(3)}\left(j_1;\mathbf{u}^{(2)}\right) & \mathbf{V}_{22}^{(3)}\left(\mathbf{u}^{(1)}\right) \end{pmatrix};$$

$$\mathbf{V}_{22}^{(3)}\left(\mathbf{u}^{(1)}\right) \triangleq \begin{pmatrix} \mathbf{L}^*(\boldsymbol{\alpha}) & \mathbf{0} & -\frac{\partial^2 S\left(\mathbf{u}^{(1)};\boldsymbol{\alpha}\right)}{\partial\boldsymbol{\varphi}\partial\boldsymbol{\varphi}} & -\frac{\partial^2 S\left(\mathbf{u}^{(1)};\boldsymbol{\alpha}\right)}{\partial\boldsymbol{\psi}\partial\boldsymbol{\varphi}} \\ \mathbf{0} & \mathbf{L}(\boldsymbol{\alpha}) & -\frac{\partial^2 S\left(\mathbf{u}^{(1)};\boldsymbol{\alpha}\right)}{\partial\boldsymbol{\varphi}\partial\boldsymbol{\psi}} & -\frac{\partial^2 S\left(\mathbf{u}^{(1)};\boldsymbol{\alpha}\right)}{\partial\boldsymbol{\psi}\partial\boldsymbol{\psi}} \\ \mathbf{0} & \mathbf{0} & \mathbf{L}(\boldsymbol{\alpha}) & \mathbf{0} \\ \mathbf{0} & \mathbf{0} & \mathbf{0} & \mathbf{L}^*(\boldsymbol{\alpha}) \end{pmatrix}; \tag{2.125}$$

and where the components $\mathbf{v}_{21}^{(3)}(i,j)$ of the matrix $\mathbf{V}_{21}^{(3)}(j_1) \triangleq \left[\mathbf{v}_{21}^{(3)}(i,j)\right]_{(4\times 4)}; i,j = 1,\ldots,4$ are defined as follows:

$$\begin{aligned} \mathbf{v}_{21}^{(3)}(1,1) = &-\frac{\partial^3 S\left(\mathbf{u}^{(1)};\boldsymbol{\alpha}\right)}{\partial\boldsymbol{\varphi}\partial\boldsymbol{\varphi}\partial\boldsymbol{\varphi}}\mathbf{a}_3^{(2)}(j_1;\mathbf{x}) \\ &-\frac{\partial^3 S\left(\mathbf{u}^{(1)};\boldsymbol{\alpha}\right)}{\partial\boldsymbol{\psi}\partial\boldsymbol{\varphi}\partial\boldsymbol{\varphi}}\mathbf{a}_4^{(2)}(j_1;\mathbf{x}) - \frac{\partial^2 S^{(1)}\left(j_1;\mathbf{u}^{(2)};\boldsymbol{\alpha}\right)}{\partial\boldsymbol{\varphi}\partial\boldsymbol{\varphi}}; \end{aligned} \tag{2.126}$$

$$\begin{aligned} \mathbf{v}_{21}^{(3)}(1,2) = &-\frac{\partial^3 S\left(\mathbf{u}^{(1)};\boldsymbol{\alpha}\right)}{\partial\boldsymbol{\varphi}\partial\boldsymbol{\varphi}\partial\boldsymbol{\psi}}\mathbf{a}_3^{(2)}(j_1;\mathbf{x}) \\ &-\frac{\partial^2 S\left(\mathbf{u}^{(1)};\boldsymbol{\alpha}\right)}{\partial\boldsymbol{\psi}\partial\boldsymbol{\psi}\partial\boldsymbol{\varphi}}\mathbf{a}_4^{(2)}(j_1;\mathbf{x}) - \frac{\partial^2 S^{(1)}\left(j_1;\mathbf{u}^{(2)};\boldsymbol{\alpha}\right)}{\partial\boldsymbol{\varphi}\partial\boldsymbol{\psi}}; \end{aligned} \tag{2.127}$$

$$\mathbf{v}_{21}^{(3)}(1,3) = -\frac{\partial^2 S^{(1)}\left(j_1;\mathbf{u}^{(2)};\boldsymbol{\alpha}\right)}{\partial\boldsymbol{\varphi}\partial\mathbf{a}_1^{(1)}};\quad \mathbf{v}_{21}^{(3)}(1,4) = -\frac{\partial^2 S^{(1)}\left(j_1;\mathbf{u}^{(2)};\boldsymbol{\alpha}\right)}{\partial\boldsymbol{\varphi}\partial\mathbf{a}_2^{(1)}}; \tag{2.128}$$

$$\begin{aligned} \mathbf{v}_{21}^{(3)}(2,1) = &-\frac{\partial^3 S\left(\mathbf{u}^{(2)};\boldsymbol{\alpha}\right)}{\partial\boldsymbol{\varphi}\partial\boldsymbol{\varphi}\partial\boldsymbol{\psi}}\mathbf{a}_3^{(2)}(j_1;\mathbf{x}) \\ &-\frac{\partial^3 S\left(\mathbf{u}^{(1)};\boldsymbol{\alpha}\right)}{\partial\boldsymbol{\varphi}\partial\boldsymbol{\psi}\partial\boldsymbol{\psi}}\mathbf{a}_4^{(2)}(j_1;\mathbf{x}) - \frac{\partial^2 S^{(1)}\left(j_1;\mathbf{u}^{(2)};\boldsymbol{\alpha}\right)}{\partial\boldsymbol{\varphi}\partial\boldsymbol{\psi}}; \end{aligned} \tag{2.129}$$

$$\begin{aligned} \mathbf{v}_{21}^{(3)}(2,2) = &-\frac{\partial^3 S\left(\mathbf{u}^{(1)};\boldsymbol{\alpha}\right)}{\partial\boldsymbol{\varphi}\partial\boldsymbol{\psi}\partial\boldsymbol{\psi}}\mathbf{a}_3^{(2)}(j_1;\mathbf{x}) \\ &-\frac{\partial^3 S\left(\mathbf{u}^{(1)};\boldsymbol{\alpha}\right)}{\partial\boldsymbol{\psi}\partial\boldsymbol{\psi}\partial\boldsymbol{\psi}}\mathbf{a}_4^{(2)}(j_1;\mathbf{x}) - \frac{\partial^2 S^{(1)}\left(j_1;\mathbf{u}^{(2)};\boldsymbol{\alpha}\right)}{\partial\boldsymbol{\psi}\partial\boldsymbol{\psi}}; \end{aligned} \tag{2.130}$$

$$\mathbf{v}_{21}^{(3)}(2,3) = -\frac{\partial^2 S^{(1)}\left(j_1;\mathbf{u}^{(2)};\boldsymbol{\alpha}\right)}{\partial \mathbf{a}_1^{(1)} \partial \boldsymbol{\psi}}; \quad \mathbf{v}_{21}^{(3)}(2,4) = -\frac{\partial^2 S^{(1)}\left(j_1;\mathbf{u}^{(2)};\boldsymbol{\alpha}\right)}{\partial \mathbf{a}_2^{(1)} \partial \boldsymbol{\psi}}; \tag{2.131}$$

$$\mathbf{v}_{21}^{(3)}(3,1) = -\frac{\partial^2 S^{(1)}\left(j_1;\mathbf{u}^{(2)};\boldsymbol{\alpha}\right)}{\partial \boldsymbol{\varphi} \partial \mathbf{a}_1^{(1)}}; \quad \mathbf{v}_{21}^{(3)}(3,2) = -\frac{\partial^2 S^{(1)}\left(j_1;\mathbf{u}^{(2)};\boldsymbol{\alpha}\right)}{\partial \boldsymbol{\psi} \partial \mathbf{a}_1^{(1)}}; \tag{2.132}$$

$$\mathbf{v}_{21}^{(3)}(3,3) = -\frac{\partial^2 S^{(1)}\left(j_1;\mathbf{u}^{(2)};\boldsymbol{\alpha}\right)}{\partial \mathbf{a}_1^{(1)} \partial \mathbf{a}_1^{(1)}}; \quad \mathbf{v}_{21}^{(3)}(3,4) = -\frac{\partial^2 S^{(1)}\left(j_1;\mathbf{u}^{(2)};\boldsymbol{\alpha}\right)}{\partial \mathbf{a}_2^{(1)} \partial \mathbf{a}_1^{(1)}}; \tag{2.133}$$

$$\mathbf{v}_{21}^{(3)}(4,1) = -\frac{\partial^2 S^{(1)}\left(j_1;\mathbf{u}^{(2)};\boldsymbol{\alpha}\right)}{\partial \boldsymbol{\varphi} \partial \mathbf{a}_2^{(1)}}; \quad \mathbf{v}_{21}^{(3)}(4,2) = -\frac{\partial^2 S^{(1)}\left(j_1;\mathbf{u}^{(2)};\boldsymbol{\alpha}\right)}{\partial \boldsymbol{\psi} \partial \mathbf{a}_2^{(1)}}; \tag{2.134}$$

$$\mathbf{v}_{21}^{(3)}(4,3) = -\frac{\partial^2 S^{(1)}\left(j_1;\mathbf{u}^{(2)};\boldsymbol{\alpha}\right)}{\partial \mathbf{a}_1^{(1)} \partial \mathbf{a}_2^{(1)}}; \quad \mathbf{v}_{21}^{(3)}(4,4) = -\frac{\partial^2 S^{(1)}\left(j_1;\mathbf{u}^{(2)};\boldsymbol{\alpha}\right)}{\partial \mathbf{a}_2^{(1)} \partial \mathbf{a}_2^{(1)}}. \tag{2.135}$$

The matrix-operator equations represented (2.121) together with the boundary conditions represented by (2.122) will be called the *Third-Level Variational Sensitivity System* (3rd-LVSS). The solution, $\delta\mathbf{u}^{(3)}(\mathbf{x})$, of the 3rd-LVSS will be called the "*third-level variational sensitivity function.*"

In the particular case when the response depends only on the forward function $\boldsymbol{\varphi}(\mathbf{x})$, i.e., if the response has the functional dependence $R[\boldsymbol{\varphi}(\mathbf{x});\boldsymbol{\alpha}]$, the quantities $\boldsymbol{\psi}$, $\mathbf{a}_2^{(1)}$, $\mathbf{a}_2^{(2)}(j_1;\mathbf{x})$, $\mathbf{a}_4^{(2)}(j_1;\mathbf{x})$ and related quantities would not exist. Consequently, the vector $\delta\mathbf{u}^{(3)}(j_1;\mathbf{x})$ would reduce to the following 4-component vector: $\delta\mathbf{u}^{(3)}(j_1;\mathbf{x}) = \left[\delta\boldsymbol{\varphi}(\mathbf{x}), \delta\mathbf{a}_1^{(1)}(\mathbf{x}), \delta\mathbf{a}_1^{(2)}, \delta\mathbf{a}_3^{(2)}\right]^\dagger$ and the 3rd-LVSS would reduce correspondingly to the following system of 4x4 matrix equations:

$$\begin{pmatrix} \mathbf{L}(\boldsymbol{\alpha}) & \mathbf{0} & \mathbf{0} & \mathbf{0} \\ -\dfrac{\partial^2 S(\boldsymbol{\varphi};\boldsymbol{\alpha}^0)}{\partial\boldsymbol{\varphi}\partial\boldsymbol{\varphi}} & \mathbf{L}^*(\boldsymbol{\alpha}) & \mathbf{0} & \mathbf{0} \\ \mathbf{F}_{31}^{(3)}[j_1] & \mathbf{F}_{32}^{(3)}[j_1] & \mathbf{L}^*(\boldsymbol{\alpha}) & -\dfrac{\partial^2 S(\boldsymbol{\varphi};\boldsymbol{\alpha})}{\partial\boldsymbol{\varphi}\partial\boldsymbol{\varphi}} \\ \mathbf{F}_{41}^{(3)}[j_1] & \mathbf{F}_{42}^{(3)}[j_1] & \mathbf{0} & \mathbf{L}(\boldsymbol{\alpha}) \end{pmatrix} \begin{pmatrix} \delta\boldsymbol{\varphi}(\mathbf{x}) \\ \delta\mathbf{a}_1^{(1)} \\ \delta\mathbf{a}_1^{(2)}(j_1;\mathbf{x}) \\ \delta\mathbf{a}_3^{(2)}(j_1;\mathbf{x}) \end{pmatrix} = \begin{pmatrix} \mathbf{q}_1^{(1)} \\ \mathbf{q}_3^{(2)} \\ \mathbf{q}_5^{(3)} \\ \mathbf{q}_7^{(3)} \end{pmatrix}, \tag{2.136}$$

where:

$$\mathbf{F}_{31}^{(3)}[j_1] = -\frac{\partial^3 S(\boldsymbol{\varphi};\boldsymbol{\psi};\boldsymbol{\alpha})}{\partial\boldsymbol{\varphi}\partial\boldsymbol{\varphi}\partial\boldsymbol{\varphi}}\mathbf{a}_3^{(2)}(j_1;\mathbf{x}) - \frac{\partial^2 S^{(1)}\left(j_1;\boldsymbol{\varphi};\boldsymbol{\psi};\mathbf{a}^{(1)};\boldsymbol{\alpha}\right)}{\partial\boldsymbol{\varphi}\partial\boldsymbol{\varphi}};$$

$$\mathbf{F}_{32}^{(3)}[j_1] = -\frac{\partial^2 S^{(1)}\left(j_1;\boldsymbol{\varphi};\mathbf{a}^{(1)};\boldsymbol{\alpha}\right)}{\partial\boldsymbol{\varphi}\partial\mathbf{a}_1^{(1)}};$$

$$\mathbf{F}_{41}^{(3)}[j_1] = -\frac{\partial^2 S^{(1)}\left(j_1;\boldsymbol{\varphi};\boldsymbol{\psi};\mathbf{a}^{(1)};\boldsymbol{\alpha}\right)}{\partial\boldsymbol{\varphi}\partial\mathbf{a}_1^{(1)}};$$

$$\mathbf{F}_{42}^{(3)}[j_1] = -\frac{\partial^2 S^{(1)}\left(j_1;\boldsymbol{\varphi};\boldsymbol{\psi};\mathbf{a}^{(1)};\boldsymbol{\alpha}\right)}{\partial\mathbf{a}_1^{(1)}\partial\mathbf{a}_1^{(1)}};$$

$$\mathbf{q}_1^{(1)} = \frac{\partial\left[\mathbf{q}_\varphi(\boldsymbol{\alpha}) - \mathbf{L}(\boldsymbol{\alpha})\boldsymbol{\varphi}\right]}{\partial\boldsymbol{\alpha}}\delta\boldsymbol{\alpha};$$

$$\mathbf{q}_3^{(2)} = \frac{\partial^2 S(\boldsymbol{\varphi};\boldsymbol{\psi};\boldsymbol{\alpha}^0)}{\partial\boldsymbol{\alpha}\partial\boldsymbol{\varphi}}\delta\boldsymbol{\alpha} - \frac{\partial\left[\mathbf{L}^*(\boldsymbol{\alpha}^0)\mathbf{a}_1^{(1)}\right]}{\partial\boldsymbol{\alpha}}\delta\boldsymbol{\alpha};$$

$$\mathbf{q}_5^{(3)} = -\frac{\partial\left[\mathbf{L}^*(\boldsymbol{\alpha})\mathbf{a}_1^{(2)}(j_1;\mathbf{x})\right]}{\partial\boldsymbol{\alpha}}\delta\boldsymbol{\alpha} + \frac{\partial^2 S^{(1)}(j_1)}{\partial\boldsymbol{\alpha}\partial\boldsymbol{\varphi}}\delta\boldsymbol{\alpha} + \left[\frac{\partial^3 S(\boldsymbol{\varphi};\boldsymbol{\alpha})}{\partial\boldsymbol{\alpha}\partial\boldsymbol{\varphi}\partial\boldsymbol{\varphi}}\delta\boldsymbol{\alpha}\right]\mathbf{a}_3^{(2)}(j_1;\mathbf{x});$$

$$\mathbf{q}_7^{(3)} = -\frac{\partial\left[\mathbf{L}(\boldsymbol{\alpha})\mathbf{a}_3^{(2)}(j_1;\mathbf{x})\right]}{\partial\boldsymbol{\alpha}}\delta\boldsymbol{\alpha} + \frac{\partial^2 S^{(1)}\left(j_1;\boldsymbol{\varphi};\mathbf{a}^{(1)};\boldsymbol{\alpha}\right)}{\partial\boldsymbol{\alpha}\partial\mathbf{a}_1^{(1)}}\delta\boldsymbol{\alpha}.$$

A reduction of the 3rd-LVSS to a 4×4 matrix-equation, which would be similar to the reduction shown in (2.136), would also occur for a response $R[\boldsymbol{\psi}(\mathbf{x});\boldsymbol{\alpha}]$ which depended only on the adjoint function $\boldsymbol{\psi}(\mathbf{x})$.

Since the 3rd-LVSS equations depend on the parameter variations $\delta\alpha_j$, solving them is prohibitively expensive computationally for large-scale systems involving many parameters. The need for solving the 3rd-LVSS can be avoided by expressing the indirect-effect term $\{\delta R^{(2)}(j_2;j_1;\ldots)\}_{ind}$ defined in (2.111) in an alternative way, namely in terms of the solutions of a *Third-Level Adjoint Sensitivity System* (3rd-LASS), which will not involve any variations in the state functions. This 3rd-LASS is constructed by implementing the same sequence of logical steps as used to construct the 1st-LASS and the 2nd-LASS, namely:

1. Define a Hilbert space, denoted as H_3, comprising vector-valued elements of the form $\boldsymbol{\eta}^{(3)}(\mathbf{x}) \triangleq \left[\boldsymbol{\eta}_1^{(3)}(\mathbf{x}), \ldots, \boldsymbol{\eta}_8^{(3)}(\mathbf{x})\right]^\dagger \in H_3$, with $\boldsymbol{\eta}_i^{(3)}(\mathbf{x}) \triangleq \left[\eta_{i,1}^{(3)}(\mathbf{x}), \ldots, \eta_{i,j}^{(3)}(\mathbf{x}), \ldots, \eta_{i,TD}^{(3)}(\mathbf{x})\right]^\dagger, i = 1, .., 8.$ The inner product between two elements, $\boldsymbol{\eta}^{(3)}(\mathbf{x}) \in H_3$ and $\boldsymbol{\xi}^{(3)}(\mathbf{x}) \in H_3$, of this Hilbert space, will be denoted as $\left\langle \boldsymbol{\eta}^{(3)}(\mathbf{x}), \boldsymbol{\xi}^{(3)}(\mathbf{x}) \right\rangle_3$ and is defined as follows:

$$\left\langle \boldsymbol{\eta}^{(3)}(\mathbf{x}), \boldsymbol{\xi}^{(3)}(\mathbf{x}) \right\rangle_3 \triangleq \sum_{i=1}^{8} \left\langle \boldsymbol{\eta}_i^{(3)}(\mathbf{x}), \boldsymbol{\xi}_i^{(3)}(\mathbf{x}) \right\rangle_0. \tag{2.137}$$

2. In the Hilbert space H_3, form the inner product of (2.121) with a set of vector-functions $\mathbf{a}^{(3)}(j_2;j_1;\mathbf{x}) \triangleq \left[\mathbf{a}_1^{(3)}(j_2;j_1;\mathbf{x}), \ldots, \mathbf{a}_8^{(3)}(j_2;j_1;\mathbf{x})\right]^\dagger \in H_3, j_1, j_2 = 1, \ldots,$ TP, which remain to be defined, to obtain the following relation:

$$\begin{gathered} \left\{ \left\langle \mathbf{a}^{(3)}(j_2;j_1;\mathbf{x}), \mathbf{V}^{(3)}\left[j_1;\mathbf{u}^{(2)}(\mathbf{x})\right] \delta\mathbf{u}^{(3)}(\mathbf{x}) \right\rangle_3 \right\}_{\boldsymbol{\alpha}^0} \\ = \left\{ \left\langle \mathbf{a}^{(3)}(j_2;j_1;\mathbf{x}), \mathbf{q}^{(3)}\left[j_1;\mathbf{u}^{(3)}(j_1;\mathbf{x});\boldsymbol{\alpha};\delta\boldsymbol{\alpha}\right] \right\rangle_3 \right\}_{\boldsymbol{\alpha}^0}. \end{gathered} \tag{2.138}$$

3. Using the definition of the adjoint operator in the Hilbert space H_3, recast the left side of (2.138) as follows:

$$\begin{gathered} \left\{ \left\langle \mathbf{a}^{(3)}(j_2;j_1;\mathbf{x}), \mathbf{V}^{(3)}\left[j_1;\mathbf{u}^{(2)}(\mathbf{x})\right] \delta\mathbf{u}^{(3)}(\mathbf{x}) \right\rangle_3 \right\}_{\boldsymbol{\alpha}^0} \\ = \left\{ \left\langle \delta\mathbf{u}^{(3)}(\mathbf{x}), \mathbf{A}^{(3)}(j_1)\mathbf{a}^{(3)}(j_2;j_1;\mathbf{x}) \right\rangle_3 \right\}_{\boldsymbol{\alpha}^0} \\ + \left\{ P^{(3)}\left[\delta\mathbf{u}^{(3)}(\mathbf{x}); \mathbf{a}^{(3)}(j_2;j_1;\mathbf{x}); \mathbf{u}^{(3)}(j_1;\mathbf{x}); \boldsymbol{\alpha}; \delta\boldsymbol{\alpha}\right] \right\}_{\boldsymbol{\alpha}^0}, \end{gathered} \tag{2.139}$$

where $\{P^{(3)}[\delta\mathbf{u}^{(3)}(\mathbf{x}); \mathbf{a}^{(3)}(j_2;j_1;\mathbf{x}); \mathbf{u}^{(3)}(j_1;\mathbf{x}); \boldsymbol{\alpha}; \delta\boldsymbol{\alpha}]\}_{\boldsymbol{\alpha}^0}$ denotes the bilinear concomitant defined on the phase-space boundary $\mathbf{x} \in \partial\Omega_x(\boldsymbol{\alpha}^0)$ and where $\mathbf{A}^{(3)}[j_1;\mathbf{u}^{(2)}(\mathbf{x})] \triangleq [\mathbf{V}^{(3)}(j_1;\mathbf{u}^{(2)})]^*$ is the operator formally adjoint to $\mathbf{V}^{(3)}[j_1;$ $\mathbf{u}^{(2)}(\mathbf{x})]$ and therefore has the following form:

$$\mathbf{A}^{(3)}\left[j_1;\mathbf{u}^{(2)}(\mathbf{x})\right] = \begin{pmatrix} \mathbf{A}^{(2)} & \left[\mathbf{V}_{21}^{(3)}(j_1;\mathbf{u}^{(2)})\right]^* \\ [\mathbf{0}]_{(4\times 4)} & \left[\mathbf{V}_{22}^{(3)}\right]^* \end{pmatrix}. \tag{2.140}$$

4. The first term on right side of (2.139) is now required to represent $\{\delta R^{(2)}[j_2;j_1;\mathbf{u}^{(2),\,0};\mathbf{a}^{(2),\,0}(j_1);\boldsymbol{\alpha}^0;\delta\mathbf{u}^{(2)};\delta\mathbf{a}^{(2)}(j_1)]\}_{ind}$, as defined in (2.111), which is accomplished by imposing the following relation on each function $\mathbf{a}^{(3)}(j_2;j_1;\mathbf{x})\triangleq\left[\mathbf{a}_1^{(3)}(j_2;j_1;\mathbf{x}),\ \ldots,\mathbf{a}_8^{(3)}(j_2;j_1;\mathbf{x})\right]^{\dagger}\in \mathrm{H}_3, j_1, j_2 = 1, \ldots, TP$:

$$\left\{\mathbf{A}^{(3)}\left[j_1;\mathbf{u}^{(2)}(\mathbf{x})\right]\mathbf{a}^{(3)}(j_2;j_1;\mathbf{x})\right\}_{\boldsymbol{\alpha}^0}=\left\{\mathbf{s}^{(3)}\left[j_2;j_1;\mathbf{u}^{(3)}(j_1;\mathbf{x});\boldsymbol{\alpha}\right]\right\}_{\boldsymbol{\alpha}^0}, \tag{2.141}$$

where the block-vector $\mathbf{s}^{(3)}[j_2;j_1;\mathbf{u}^{(3)}(j_1;\mathbf{x});\boldsymbol{\alpha}]$ comprises, for each $j_1, j_2 = 1, \ldots, TP$, eight components (which are themselves block-vectors) defined as follows:

$$\begin{aligned}&\mathbf{s}^{(3)}\left[j_2;j_1;\mathbf{u}^{(3)}(j_1;\mathbf{x});\boldsymbol{\alpha}\right]\\&\triangleq\left[\mathbf{s}_1^{(3)}\left(j_2;j_1;\mathbf{u}^{(3)}(j_1;\mathbf{x});\boldsymbol{\alpha}\right),\ \ldots,\mathbf{s}_8^{(3)}\left(j_2;j_1;\mathbf{u}^{(3)}(j_1;\mathbf{x});\boldsymbol{\alpha}\right)\right]^{\dagger},\end{aligned} \tag{2.142}$$

where

$$\mathbf{s}_1^{(3)}\left[j_2;j_1;\mathbf{u}^{(3)}(j1;\mathbf{x});\boldsymbol{\alpha}\right]\triangleq\frac{\partial S^{(2)}\left[j_2;j_1;\mathbf{u}^{(3)}(j_1;\mathbf{x});\boldsymbol{\alpha}\right]}{\partial\boldsymbol{\varphi}}, \tag{2.143}$$

$$\mathbf{s}_2^{(3)}\left[j_2;j_1;\mathbf{u}^{(3)}(j_1;\mathbf{x});\boldsymbol{\alpha}\right]\triangleq\frac{\partial S^{(2)}\left[j_2;j_1;\mathbf{u}^{(3)}(j_1;\mathbf{x});\boldsymbol{\alpha}\right]}{\partial\boldsymbol{\psi}}, \tag{2.144}$$

$$\mathbf{s}_3^{(3)}\left[j_2;j_1;\mathbf{u}^{(3)}(j_1;\mathbf{x});\boldsymbol{\alpha}\right]\triangleq\frac{\partial S^{(2)}\left[j_2;j_1;\mathbf{u}^{(3)}(j_1;\mathbf{x});\boldsymbol{\alpha}\right]}{\partial\mathbf{a}_1^{(1)}}, \tag{2.145}$$

$$\mathbf{s}_4^{(3)}\left[j_2;j_1;\mathbf{u}^{(3)}(j_1;\mathbf{x});\boldsymbol{\alpha}\right]\triangleq\frac{\partial S^{(2)}\left[j_2;j_1;\mathbf{u}^{(3)}(j_1;\mathbf{x});\boldsymbol{\alpha}\right]}{\partial\mathbf{a}_2^{(1)}}, \tag{2.146}$$

$$\mathbf{s}_5^{(3)}\left[j_2;j_1;\mathbf{u}^{(3)}(j_1;\mathbf{x});\boldsymbol{\alpha}\right]\triangleq\frac{\partial S^{(2)}\left[j_2;j_1;\mathbf{u}^{(3)}(j_1;\mathbf{x});\boldsymbol{\alpha}\right]}{\partial\mathbf{a}_1^{(2)}(j_1)}, \tag{2.147}$$

$$\mathbf{s}_6^{(3)}\left[j_2;j_1;\mathbf{u}^{(3)}(j_1;\mathbf{x});\boldsymbol{\alpha}\right]\triangleq\frac{\partial S^{(2)}\left[j_2;j_1;\mathbf{u}^{(3)}(j_1;\mathbf{x});\boldsymbol{\alpha}\right]}{\partial\mathbf{a}_2^{(2)}(j1)}, \tag{2.148}$$

$$\mathbf{s}_7^{(3)}\left[j_2;j_1;\mathbf{u}^{(3)}(j_1;\mathbf{x});\boldsymbol{\alpha}\right]\triangleq\frac{\partial S^{(2)}\left[j_2;j_1;\mathbf{u}^{(3)}(j_1;\mathbf{x});\boldsymbol{\alpha}\right]}{\partial\mathbf{a}_3^{(2)}(j_1)}, \tag{2.149}$$

$$\mathbf{s}_8^{(3)}\left[j_2;j_1;\mathbf{u}^{(3)}(j_1;\mathbf{x});\boldsymbol{\alpha}\right]\triangleq\frac{\partial S^{(2)}\left[j_2;j_1;\mathbf{u}^{(3)}(j_1;\mathbf{x});\boldsymbol{\alpha}\right]}{\partial\mathbf{a}_4^{(2)}(j_1)}. \tag{2.150}$$

5. The definition of the set of vector-functions $\mathbf{a}^{(3)}(j_2;j_1;\mathbf{x})\triangleq\left[\mathbf{a}_1^{(3)}(j_2;j_1;\mathbf{x}),\ \ldots,\mathbf{a}_8^{(3)}(j_2;j_1;\mathbf{x})\right]^{\dagger}\in \mathrm{H}_3, j_2, j_1 = 1, \ldots, TP$, will

now be completed by selecting boundary conditions to be satisfied by each of these vectors; the selected boundary conditions are represented in operator form as follows:

$$\left\{\mathbf{b}_A^{(3)}\left[\mathbf{a}^{(3)}(j_2;j_1;\mathbf{x});\mathbf{u}^{(3)}(j_1;\mathbf{x});\boldsymbol{\alpha}\right]\right\}_{\boldsymbol{\alpha}^0}=\mathbf{0},\mathbf{x}\in\partial\Omega_x(\boldsymbol{\alpha}^0). \quad (2.151)$$

The boundary conditions represented by (2.151) are selected so as to satisfy the following requirements:

1. The boundary conditions (2.151) together with the operator Eq. (2.141) constitute a well-posed problem for the functions $\mathbf{a}^{(3)}(j_2;j_1;\mathbf{x})$.
2. Implementation in (2.139) of the boundary conditions provided in (2.122) together with those provided in (2.151) eliminates all of the unknown values of the functions $\delta\mathbf{u}^{(3)}(\mathbf{x})$ and $\mathbf{a}^{(3)}(j_2;j_1;\mathbf{x})$ in the expression of the bilinear concomitant $\left\{P^{(3)}\left[\delta\mathbf{u}^{(3)}(\mathbf{x});\mathbf{a}^{(3)}(j_2;j_1;\mathbf{x});\mathbf{u}^{(3)}(j_1;\mathbf{x});\boldsymbol{\alpha};\delta\boldsymbol{\alpha}\right]\right\}_{(\boldsymbol{\alpha}^0)}$. This bilinear concomitant may vanish after implementing the boundary conditions represented by (2.151), but if it does not, it will be reduced to a residual quantity which will be denoted as $\widehat{P}^{(3)}\left[\mathbf{a}^{(3)}(j_2;j_1;\mathbf{x});\mathbf{u}^{(3)}(j_1;\mathbf{x});\boldsymbol{\alpha};\delta\boldsymbol{\alpha}\right]$ and which will comprise only known values of $\mathbf{a}^{(3)}(j_2;j_1;\mathbf{x})$, $\mathbf{u}^{(3)}(j_1;\mathbf{x})$, $\boldsymbol{\alpha}$, and $\delta\boldsymbol{\alpha}$.

The system of equations represented by (2.141) together with the boundary conditions represented by (2.151) constitute the *Third-Level Adjoint Sensitivity System* (3rd-LASS). The solution $\mathbf{a}^{(3)}(j_2;j_1;\mathbf{x})\triangleq\left[\mathbf{a}_1^{(3)}(j_2;j_1;\mathbf{x}),\ldots,\mathbf{a}_8^{(3)}(j_2;j_1;\mathbf{x})\right]^\dagger\in\mathrm{H}_3$, $j_1,j_2=1,\ldots,TP$, of the 3rd-LASS will be called the *third-level adjoint sensitivity function.*

The 3rd-LASS together with the results provided in (2.139) and (2.138) are employed in (2.111) to obtain the following expression for the indirect-effect term $\{\delta R^{(2)}[j_2;j_1;\mathbf{u}^{(3)}(j_1;\mathbf{x});\boldsymbol{\alpha};\delta\mathbf{u}^{(2)};\delta\mathbf{a}^{(2)}(j_1)]\}_{ind}$ in terms of the third-level adjoint sensitivity functions $\mathbf{a}^{(3)}(j_2;j_1;\mathbf{x})$, for $j_1,j_2=1,\ldots,TP$:

$$\begin{aligned}&\left\{\delta R^{(2)}\left[j_2;j_1;\mathbf{u}^{(3)}(j_1;\mathbf{x});\boldsymbol{\alpha};\delta\mathbf{u}^{(3)}(j_1;\mathbf{x})\right]\right\}_{ind}\\&=\left\{\left\langle\mathbf{a}^{(3)}(j_2;j_1;\mathbf{x}),\mathbf{q}^{(3)}\left[\mathbf{u}^{(3)}(j_1;\mathbf{x});\boldsymbol{\alpha};\delta\boldsymbol{\alpha}\right]\right\rangle_3\right\}_{\boldsymbol{\alpha}^0}\\&-\left\{\widehat{P}^{(3)}\left[\mathbf{u}^{(3)}(j_1;\mathbf{x});\mathbf{a}^{(3)}(j_2;j_1;\mathbf{x});\boldsymbol{\alpha};\delta\boldsymbol{\alpha}\right]\right\}_{\boldsymbol{\alpha}^0}\\&\equiv\left\{\delta R^{(2)}\left[j_2;j_1;\mathbf{u}^{(3)}(j_1;\mathbf{x});\mathbf{a}^{(3)}(j_2;j_1;\mathbf{x});\boldsymbol{\alpha};\delta\boldsymbol{\alpha}\right]\right\}_{ind}.\end{aligned} \quad (2.152)$$

As the identity in the last line of (2.152) indicates, the dependence of the indirect-effect term on the variations $\delta\mathbf{u}^{(2)}$ and $\delta\mathbf{a}^{(2)}(j_1)$ has been eliminated, by having replaced this dependence by the dependence on the third-level adjoint sensitivity function $\mathbf{a}^{(3)}(j_2;j_1;\mathbf{x})$.

Inserting the expressions of the components of the vector $\mathbf{q}^{(3)}[\mathbf{u}^{(3)}(j_1;\mathbf{x})\boldsymbol{\alpha};\delta\boldsymbol{\alpha}]$, cf. (2.124), into (2.152) and adding the resulting expression for the indirect-effect term $\{\delta R^{(2)}[j_2;j_1;\mathbf{u}^{(3)}(j_1;\mathbf{x});\boldsymbol{\alpha};\delta\mathbf{u}^{(2)};\delta\mathbf{a}^{(2)}(j_1)]\}_{ind}$ to the expression of the direct-effect term given in (2.110) yields the following expression for the total third-order G-differential of the response $R[\boldsymbol{\varphi}(\mathbf{x}), \boldsymbol{\psi}(\mathbf{x}); \boldsymbol{\alpha}]$, for $j_1, j_2 = 1, \ldots, TP$:

$$\begin{aligned}
&\left\{\delta R^{(2)}\left[j_2;j_1;\mathbf{u}^{(3)}(j_1;\mathbf{x});\mathbf{a}^{(3)}(j_2;j_1;\mathbf{x});\boldsymbol{\alpha};\delta\boldsymbol{\alpha}\right]\right\}_{\boldsymbol{\alpha}^0} \\
&= \left\{\frac{\partial}{\partial \boldsymbol{\alpha}} \prod_{k=1}^{TI} \int_{\lambda_k(\boldsymbol{\alpha})}^{\omega_k(\boldsymbol{\alpha})} dx_k S^{(2)}\left[j_2;j_1;\mathbf{u}^{(3)}(j_1;\mathbf{x});\boldsymbol{\alpha}\right]\right\}_{\boldsymbol{\alpha}^0} \delta\boldsymbol{\alpha} \\
&+ \left\{\sum_{m=1}^{8} \left\langle \mathbf{a}_m^{(3)}(j_2;j_1;\mathbf{x}), \mathbf{q}_m^{(3)}\left(\mathbf{u}^{(3)}(j_1;\mathbf{x});\boldsymbol{\alpha};\delta\boldsymbol{\alpha}\right)\right\rangle_0\right\}_{\boldsymbol{\alpha}^0} \\
&- \left\{\widehat{P}^{(3)}\left[\mathbf{u}^{(3)}(j_1;\mathbf{x});\mathbf{a}^{(3)}(j_2;j_1;\mathbf{x});\boldsymbol{\alpha};\delta\boldsymbol{\alpha}\right]\right\}_{\boldsymbol{\alpha}^0}.
\end{aligned} \tag{2.153}$$

Inserting the explicit expressions for the components of $\mathbf{q}^{(3)}[\mathbf{u}^{(3)}(j_1;\mathbf{x});\boldsymbol{\alpha};\delta\boldsymbol{\alpha}]$ into (2.153) leads to the following expression of the total third-order G-differential $\left\{\delta R^{(2)}\left[j_2;j_1;\mathbf{u}^{(3)}(j_1;\mathbf{x});\mathbf{a}^{(3)}(j_2;j_1;\mathbf{x});\boldsymbol{\alpha};\delta\boldsymbol{\alpha}\right]\right\}_{\boldsymbol{\alpha}^0}$ in terms of the third-order partial sensitivities (G-derivatives) of the response with respect to the model parameters:

$$\begin{aligned}
&\left\{\delta R^{(2)}\left[\mathbf{u}^{(3)}(j_1;\mathbf{x});\mathbf{a}^{(3)}(j_2;j_1;\mathbf{x});\boldsymbol{\alpha};\delta\boldsymbol{\alpha}\right]\right\}_{\boldsymbol{\alpha}^0} \\
&= \sum_{j_3=1}^{TP} \left\{R^{(3)}\left[j_3;j_2;j_1;\mathbf{u}^{(3)}(j_1;\mathbf{x});\mathbf{a}^{(3)}(j_2;j_1;\mathbf{x});\boldsymbol{\alpha}\right]\right\}_{\boldsymbol{\alpha}^0} \delta\alpha_{j_3},
\end{aligned} \tag{2.154}$$

where $R^{(3)}[j_3;j_2;j_1;\mathbf{u}^{(3)}(j_1;\mathbf{x});\mathbf{a}^{(3)}(j_2;j_1;\mathbf{x});\boldsymbol{\alpha}]$ denotes the *third-order partial sensitivity* of the response with respect to the model parameter α_{j3}, evaluated at the nominal parameter values $\boldsymbol{\alpha}^0$. The explicit expression of $R^{(3)}[j_3;j_2;j_1;\mathbf{u}^{(3)}(j_1;\mathbf{x});$ $\mathbf{a}^{(3)}(j_2;j_1;\mathbf{x});\boldsymbol{\alpha}]$ is obtained by using (2.153) in (2.154) and subsequently identifying the expression that multiplies the parameter $\delta\alpha_{j3}$. This process indicates that the *third-order partial sensitivity* $R^{(3)}[j_3;j_2;j_1;\mathbf{u}^{(3)}(j_1;\mathbf{x});\mathbf{a}^{(3)}(j_2;j_1;\mathbf{x});\boldsymbol{\alpha}]$ has the following explicit expression, for each $j_1, j_2, j_3 = 1, \ldots, TP$:

$$
\begin{aligned}
&R^{(3)}\left[j_3;j_2;j_1;\mathbf{u}^{(3)}(j_1;\mathbf{x});\mathbf{a}^{(3)}(j_2;j_1;\mathbf{x});\boldsymbol{\alpha}\right] \equiv \frac{\partial^3 R(\boldsymbol{\varphi};\boldsymbol{\psi};\boldsymbol{\alpha})}{\partial\alpha_{j_3}\partial\alpha_{j_2}\partial\alpha_{j_1}} \\
&= -\left\{\frac{\partial}{\partial\alpha_{j_3}}\widehat{P}^{(3)}\left[\mathbf{u}^{(3)}(j_1;\mathbf{x});\mathbf{a}^{(3)}(j_2;j_1;\mathbf{x});\boldsymbol{\alpha}\right]\right\}_{\boldsymbol{\alpha}^0} \\
&+\frac{\partial}{\partial\alpha_{j_3}}\left\{\prod_{k=1}^{TI}\int_{\lambda_k(\boldsymbol{\alpha})}^{u_k(\boldsymbol{\alpha})} dx_k S^{(2)}\left[j_2;j_1;\mathbf{u}^{(3)}(j_1;\mathbf{x});\boldsymbol{\alpha}\right]\right\}_{\boldsymbol{\alpha}^0} \\
&+\left\{\sum_{m=1}^{8}\left\langle \mathbf{a}_m^{(3)}(j_2;j_1;\mathbf{x}),\mathbf{q}_m^{(3)}\left(\mathbf{u}^{(3)}(j_1;\mathbf{x});\boldsymbol{\alpha};\delta\boldsymbol{\alpha}\right)\right\rangle_0\right\}_{\boldsymbol{\alpha}^0} \\
&\triangleq \prod_{k=1}^{TI}\int_{\lambda_k(\boldsymbol{\alpha})}^{\omega_k(\boldsymbol{\alpha})} dx_k S^{(3)}\left[j_3;j_2;j_1;\mathbf{u}^{(3)}(j_1;\mathbf{x});\mathbf{a}^{(3)}(j_2;j_1;\mathbf{x});\boldsymbol{\alpha}\right],
\end{aligned}
\tag{2.155}
$$

where:

$$
\begin{aligned}
&\left\langle \mathbf{a}_1^{(3)}(j_2;j_1;\mathbf{x}),\mathbf{q}_1^{(3)}\left(\mathbf{u}^{(3)}(j_1;\mathbf{x});\boldsymbol{\alpha};\delta\boldsymbol{\alpha}\right)\right\rangle_0 \\
&= \left\langle \mathbf{a}_1^{(3)}(j_2;j_1;\mathbf{x}),\frac{\partial\left[\mathbf{q}_\varphi(\boldsymbol{\alpha})-\mathbf{L}(\boldsymbol{\alpha})\boldsymbol{\varphi}(\mathbf{x})\right]}{\partial\alpha_{j_3}}\right\rangle_0,
\end{aligned}
\tag{2.156}
$$

$$
\begin{aligned}
&\left\langle \mathbf{a}_2^{(3)}(j_2;j_1;\mathbf{x}),\mathbf{q}_2^{(3)}\left(\mathbf{u}^{(3)}(j_1;\mathbf{x});\boldsymbol{\alpha};\delta\boldsymbol{\alpha}\right)\right\rangle_0 \\
&= \left\langle \mathbf{a}_2^{(3)}(j_2;j_1;\mathbf{x}),\frac{\partial\left[\mathbf{q}_\psi(\boldsymbol{\alpha})-\mathbf{L}^*(\boldsymbol{\alpha})\boldsymbol{\psi}(\mathbf{x})\right]}{\partial\alpha_{j_3}}\right\rangle_0,
\end{aligned}
\tag{2.157}
$$

$$
\begin{aligned}
&\left\langle \mathbf{a}_3^{(3)}(j_2;j_1;\mathbf{x}),\mathbf{q}_3^{(3)}\left(\mathbf{u}^{(3)}(j_1;\mathbf{x});\boldsymbol{\alpha};\delta\boldsymbol{\alpha}\right)\right\rangle_0 \\
&= \left\langle \mathbf{a}_3^{(3)}(j_2;j_1;\mathbf{x}),\frac{\partial^2 S(\mathbf{u}^{(1)};\boldsymbol{\alpha})}{\partial\alpha_{j_3}\partial\boldsymbol{\varphi}}-\frac{\partial\left[\mathbf{L}^*(\boldsymbol{\alpha})\mathbf{a}_1^{(1)}\right]}{\partial\alpha_{j_3}}\right\rangle_0,
\end{aligned}
\tag{2.158}
$$

$$
\begin{aligned}
&\left\langle \mathbf{a}_4^{(3)}(j_2;j_1;\mathbf{x}),\mathbf{q}_4^{(3)}\left(\mathbf{u}^{(3)}(j_1;\mathbf{x});\boldsymbol{\alpha};\delta\boldsymbol{\alpha}\right)\right\rangle_0 \\
&= \left\langle \mathbf{a}_4^{(3)}(j_2;j_1;\mathbf{x}),\frac{\partial^2 S(\mathbf{u}^{(1)};\boldsymbol{\alpha})}{\partial\alpha_{j_3}\partial\boldsymbol{\psi}}-\frac{\partial\left[\mathbf{L}(\boldsymbol{\alpha})\mathbf{a}_2^{(1)}\right]}{\partial\alpha_{j_3}}\right\rangle,
\end{aligned}
\tag{2.159}
$$

$$\left\langle \mathbf{a}_5^{(3)}(j_2;j_1;\mathbf{x}), \mathbf{q}_5^{(3)}\left(\mathbf{u}^{(3)}(j_1;\mathbf{x});\boldsymbol{\alpha};\delta\boldsymbol{\alpha}\right)\right\rangle_0 = \left\langle \mathbf{a}_5^{(3)}(j_2;j_1;\mathbf{x}), -\frac{\partial\left[\mathbf{L}^*(\boldsymbol{\alpha})\mathbf{a}_1^{(2)}(j_1;\mathbf{x})\right]}{\partial\alpha_{j_3}} + \frac{\partial^3 S\left(\mathbf{u}^{(1)};\boldsymbol{\alpha}\right)}{\partial\alpha_{j_3}\partial\boldsymbol{\varphi}\partial\boldsymbol{\varphi}}\mathbf{a}_3^{(2)}(j_1;\mathbf{x}) + \frac{\partial^3 S\left(\mathbf{u}^{(1)};\boldsymbol{\alpha}\right)}{\partial\alpha_{j_3}\partial\boldsymbol{\psi}\partial\boldsymbol{\varphi}}\mathbf{a}_4^{(2)}(j_1;\mathbf{x}) + \frac{\partial^2 S^{(1)}\left(j_1;\mathbf{u}^{(2)};\boldsymbol{\alpha}\right)}{\partial\alpha_{j_3}\partial\boldsymbol{\varphi}}\right\rangle_0, \tag{2.160}$$

$$\left\langle \mathbf{a}_6^{(3)}(j_2;j_1;\mathbf{x}), \mathbf{q}_6^{(3)}\left(\mathbf{u}^{(3)}(j_1;\mathbf{x});\boldsymbol{\alpha};\delta\boldsymbol{\alpha}\right)\right\rangle_0 = \left\langle \mathbf{a}_6^{(3)}(j_2;j_1;\mathbf{x}), -\frac{\partial\left[\mathbf{L}(\boldsymbol{\alpha})\mathbf{a}_2^{(2)}(j_1;\mathbf{x})\right]}{\partial\alpha_{j_3}} + \frac{\partial^3 S\left(\mathbf{u}^{(1)};\boldsymbol{\alpha}\right)}{\partial\alpha_{j_3}\partial\boldsymbol{\varphi}\partial\boldsymbol{\psi}}\mathbf{a}_3^{(2)}(j_1;\mathbf{x}) + \frac{\partial^3 S\left(\mathbf{u}^{(1)};\boldsymbol{\alpha}\right)}{\partial\alpha_{j_3}\partial\boldsymbol{\psi}\partial\boldsymbol{\psi}}\mathbf{a}_4^{(2)}(j_1;\mathbf{x}) + \frac{\partial^2 S^{(1)}\left(j_1;\mathbf{u}^{(2)};\boldsymbol{\alpha}\right)}{\partial\alpha_{j_3}\partial\boldsymbol{\psi}}\right\rangle_0, \tag{2.161}$$

$$\left\langle \mathbf{a}_7^{(3)}(j_2;j_1;\mathbf{x}), \mathbf{q}_7^{(3)}\left(\mathbf{u}^{(3)}(j_1;\mathbf{x});\boldsymbol{\alpha};\delta\boldsymbol{\alpha}\right)\right\rangle_0 = \left\langle \mathbf{a}_7^{(3)}(j_2;j_1;\mathbf{x}), -\frac{\partial\left[\mathbf{L}(\boldsymbol{\alpha})\mathbf{a}_3^{(2)}(j_1;\mathbf{x})\right]}{\partial\alpha_{j_3}} + \frac{\partial^2 S^{(1)}\left(j_1;\mathbf{u}^{(2)};\boldsymbol{\alpha}\right)}{\partial\alpha_{j_3}\partial\mathbf{a}_1^{(1)}}\right\rangle_0, \tag{2.162}$$

$$\left\langle \mathbf{a}_8^{(3)}(j_2;j_1;\mathbf{x}), \mathbf{q}_8^{(3)}\left(\mathbf{u}^{(3)}(j_1;\mathbf{x});\boldsymbol{\alpha};\delta\boldsymbol{\alpha}\right)\right\rangle_0 = \left\langle \mathbf{a}_8^{(3)}(j_2;j_1;\mathbf{x}), -\frac{\partial\left[\mathbf{L}^*(\boldsymbol{\alpha})\mathbf{a}_4^{(2)}(j_1;\mathbf{x})\right]}{\partial\alpha_{j_3}} + \frac{\partial^2 S^{(1)}\left(j_1;\mathbf{u}^{(2)};\boldsymbol{\alpha}\right)}{\partial\alpha_{j_3}\partial\mathbf{a}_2^{(1)}}\right\rangle_0. \tag{2.163}$$

The expression in (2.155) can be computed after having obtained the third-level adjoint sensitivity function $\mathbf{a}^{(3)}(j_2;j_1;\mathbf{x}) \triangleq \left[\mathbf{a}_1^{(3)}(j_2;j_1;\mathbf{x}), \ldots, \mathbf{a}_8^{(3)}(j_2;j_1;\mathbf{x})\right]^\dagger$. Since the 3rd-LASS is independent of parameter variations $\delta\boldsymbol{\alpha}$, the exact computation of all of the partial third-order sensitivities $R^{(3)}[j_3;j_2;j_1;\mathbf{u}^{(3)}(j_1;\mathbf{x});\mathbf{a}^{(3)}(j_2;j_1;\mathbf{x});\boldsymbol{\alpha}]$ requires at most $O(TP^2)$ large-scale (adjoint) computations using the 3rd-LASS, rather than $O(TP^3)$ large-scale computations as would be required by forward methods. In component form, the equations comprising the 3rd-LASS are solved at the nominal values for the model parameters and state functions (for each $j_1 = 1, \ldots, TP$; $j_2 = 1, \ldots, j_1$) in the following order:

$$\mathbf{L}(\boldsymbol{\alpha})\mathbf{a}_5^{(3)}(j_2;j_1;\mathbf{x}) = \mathbf{s}_5^{(3)}\left[j_2;j_1;\mathbf{u}^{(3)}(j_1;\mathbf{x});\boldsymbol{\alpha}\right], \tag{2.164}$$

$$\mathbf{L}^*(\boldsymbol{\alpha})\mathbf{a}_6^{(3)}(j_2;j_1;\mathbf{x}) = \mathbf{s}_6^{(3)}\left[j_2;j_1;\mathbf{u}^{(3)}(j_1;\mathbf{x});\boldsymbol{\alpha}\right], \tag{2.165}$$

$$\mathbf{L}^{*}(\boldsymbol{\alpha})\mathbf{a}_{7}^{(3)}(j_2;j_1;\mathbf{x}) = \mathbf{s}_{7}^{(3)}\left[j_2;j_1;\mathbf{u}^{(3)}(j_1;\mathbf{x});\boldsymbol{\alpha}\right] + \frac{\partial^2 S(\mathbf{u}^{(1)};\boldsymbol{\alpha})}{\partial\boldsymbol{\varphi}\partial\boldsymbol{\varphi}}\mathbf{a}_{5}^{(3)}(j_2;j_1;\mathbf{x}) + \frac{\partial^2 S(\mathbf{u}^{(1)};\boldsymbol{\alpha})}{\partial\boldsymbol{\varphi}\partial\boldsymbol{\psi}}\mathbf{a}_{6}^{(3)}(j_2;j_1;\mathbf{x}), \tag{2.166}$$

$$\mathbf{L}(\boldsymbol{\alpha})\mathbf{a}_{8}^{(3)}(j_2;j_1;\mathbf{x}) = \mathbf{s}_{8}^{(3)}\left[j_2;j_1;\mathbf{u}^{(3)}(j_1;\mathbf{x});\boldsymbol{\alpha}\right] + \frac{\partial^2 S(\mathbf{u}^{(1)};\boldsymbol{\alpha})}{\partial\boldsymbol{\psi}\partial\boldsymbol{\varphi}}\mathbf{a}_{5}^{(3)}(j_2;j_1;\mathbf{x}) + \frac{\partial^2 S(\mathbf{u}^{(1)};\boldsymbol{\alpha})}{\partial\boldsymbol{\psi}\partial\boldsymbol{\psi}}\mathbf{a}_{6}^{(3)}(j_2;j_1;\mathbf{x}), \tag{2.167}$$

$$\mathbf{L}(\boldsymbol{\alpha})\mathbf{a}_{3}^{(3)}(j_2;j_1;\mathbf{x}) = \mathbf{s}_{3}^{(3)}\left[j_2;j_1;\mathbf{u}^{(3)}(j_1;\mathbf{x});\boldsymbol{\alpha}\right] + \sum_{k=1}^{4}\mathbf{v}_{21}^{(3)}(k,3)\,\mathbf{a}_{k+4}^{(3)}(j_2;j_1;\mathbf{x}), \tag{2.168}$$

$$\mathbf{L}^{*}(\boldsymbol{\alpha})\mathbf{a}_{4}^{(3)}(j_2;j_1;\mathbf{x}) = \mathbf{s}_{4}^{(3)}\left[j_2;j_1;\mathbf{u}^{(3)}(j_1;\mathbf{x});\boldsymbol{\alpha}\right] + \sum_{k=1}^{4}\mathbf{v}_{21}^{(3)}(k,4)\,\mathbf{a}_{k+4}^{(3)}(j_2;j_1;\mathbf{x}), \tag{2.169}$$

$$\mathbf{L}^{*}(\boldsymbol{\alpha})\mathbf{a}_{1}^{(3)}(j_2;j_1;\mathbf{x}) = \mathbf{s}_{1}^{(3)}\left[j_2;j_1;\mathbf{u}^{(3)}(j_1;\mathbf{x});\boldsymbol{\alpha}\right] + \frac{\partial^2 S(\mathbf{u}^{(1)};\boldsymbol{\alpha})}{\partial\boldsymbol{\varphi}\partial\boldsymbol{\varphi}}\mathbf{a}_{3}^{(3)}(j_2;j_1;\mathbf{x}) + \frac{\partial^2 S(\mathbf{u}^{(1)};\boldsymbol{\alpha})}{\partial\boldsymbol{\psi}\partial\boldsymbol{\varphi}}\mathbf{a}_{4}^{(3)}(j_2;j_1;\mathbf{x}) + \sum_{k=1}^{4}\mathbf{v}_{21}^{(3)}(k,1)\,\mathbf{a}_{k+4}^{(3)}(j_2;j_1;\mathbf{x}), \tag{2.170}$$

$$\mathbf{L}(\boldsymbol{\alpha})\mathbf{a}_{2}^{(3)}(j_2;j_1;\mathbf{x}) = \mathbf{s}_{2}^{(3)}\left[j_2;j_1;\mathbf{u}^{(3)}(j_1;\mathbf{x});\boldsymbol{\alpha}\right] + \frac{\partial^2 S(\mathbf{u}^{(1)};\boldsymbol{\alpha})}{\partial\boldsymbol{\varphi}\partial\boldsymbol{\psi}}\mathbf{a}_{3}^{(3)}(j_2;j_1;\mathbf{x}) + \frac{\partial^2 S(\mathbf{u}^{(1)};\boldsymbol{\alpha})}{\partial\boldsymbol{\psi}\partial\boldsymbol{\psi}}\mathbf{a}_{4}^{(3)}(j_2;j_1;\mathbf{x}) + \sum_{k=1}^{4}\mathbf{v}_{21}^{(3)}(k,2)\,\mathbf{a}_{k+4}^{(3)}(j_2;j_1;\mathbf{x}). \tag{2.171}$$

The equations underlying the 3rd-LASS are solved successively, starting with (2.164) and (2.165) to compute the third-level adjoint sensitivity functions $\mathbf{a}_{5}^{(3)}(j_2;j_1;\mathbf{x})$ and $\mathbf{a}_{6}^{(3)}(j_2;j_1;\mathbf{x})$. Note that solving (2.164) and (2.165) would be performed by using the same forward and adjoint solvers (i.e., computer codes) as used for solving the original forward and adjoint systems. Only the corresponding source terms and boundary conditions would differ. After having obtained the third-level adjoint sensitivity functions $\mathbf{a}_{5}^{(3)}(j_2;j_1;\mathbf{x})$ and $\mathbf{a}_{6}^{(3)}(j_2;j_1;\mathbf{x})$, the next round of computations would be to solve (2.166) and (2.167) in order to determine the third-level adjoint sensitivity functions $\mathbf{a}_{7}^{(3)}(j_2;j_1;\mathbf{x})$ and $\mathbf{a}_{8}^{(3)}(j_2;j_1;\mathbf{x})$, respectively. Solving (2.166) and (2.167) would also be performed by using the same forward and adjoint solvers (i.e., computer codes) as used for solving the original forward and adjoint systems. The next round of computations would employ the forward and adjoint solvers to solve (2.168) and (2.169) in order to determine the third-level adjoint sensitivity functions $\mathbf{a}_3^{(3)}(j_2;j_1;\mathbf{x})$ and $\mathbf{a}_4^{(3)}(j_2;j_1;\mathbf{x})$, respectively. The final set of computations would require the use of the forward and adjoint solvers to solve (2.170) and (2.171) in order to determine the third-level adjoint sensitivity functions $\mathbf{a}_1^{(3)}(j_2;j_1;\mathbf{x})$ and $\mathbf{a}_2^{(3)}(j_2;j_1;\mathbf{x})$, respectively. Thus, solving the 3rd-LASS in order to determine the third-level adjoint sensitivity functions does not require any significant "code development," since the original forward and adjoint solvers (codes) do not need to be modified; only the right sides (i.e., "sources") and boundary conditions would need to be programmed accordingly. Using the 3rd-LASS enables the specific computation of the third-order sensitivities in the priority order set by the user, so that only the important third-order partial sensitivities $R^{(3)}[j_3;j_2;j_1;\mathbf{u}^{(3)}(j_1;\mathbf{x}); \mathbf{a}^{(3)}(j_2;j_1;\mathbf{x});\boldsymbol{\alpha}]$ would be computed. Information provided by the first- and second-order sensitivities is very likely to indicate which third-order sensitivities could be neglected.

2.6 The Fourth-Order Comprehensive Sensitivity Analysis Methodology for Response-Coupled Linear Forward and Adjoint Systems (4th-CASAM-L)

There are $TP(TP + 1)(TP + 2)(TP + 3)/4!$ distinct fourth-order sensitivities of the response with respect to the model and response parameters. The fourth-order unmixed sensitivities, $\partial^4 R(\boldsymbol{\alpha})/\partial \alpha_{j_1}^4$, of the leakage response, $R(\boldsymbol{\alpha})$, with respect to all the model parameters, can be approximately computed by re-computations using the well-known finite-difference formula presented below, for $j_1 = 1, \ldots, TP$:

$$\frac{\partial^4 R(\boldsymbol{\alpha})}{\partial \alpha_{j_1}^4} \approx \frac{1}{h_{j_1}{}^4}\left(R_{j_1+2} - 4R_{j_1+1} + 6R_{j_1} - 4R_{j_1-1} + R_{j_1-2}\right) + O\left(h_{j_1}^2\right), \tag{2.172}$$

where $R_{j_1+2} \triangleq R\left(\alpha_{j_1} + 2h_{j_1}\right)$, $R_{j_1+1} \triangleq R\left(\alpha_{j_1} + h_{j_1}\right)$, $R_{j_1-1} \triangleq R\left(\alpha_{j_1} - h_{j_1}\right)$, $R_{j_1-2} \triangleq R\left(\alpha_{j_1} - 2h_{j_1}\right)$, and h_{j_1} denote a "judiciously chosen" variation in the parameter α_{j_1} around its nominal value $\alpha_{j_1}^0$. The fourth-order mixed sensitivities, $\partial^4 R(\boldsymbol{\alpha})/\partial \alpha_{j_1} \partial \alpha_{j_2} \partial \alpha_{j_3} \partial \alpha_{j_4}$, can be calculated by using the following finite-difference formula:

$$\frac{\partial^4 R(\boldsymbol{\alpha})}{\partial \alpha_{j_1} \partial \alpha_{j_2} \partial \alpha_{j_3} \partial \alpha_{j_4}} \approx \frac{1}{16 h_{j_1} h_{j_2} h_{j_3} h_{j_4}} \Big(R_{j_1+1,j_2+1,j_3+1,j_4+1} - R_{j_1+1,j_2+1,j_3+1,j_4-1} \tag{2.173}$$
$$- R_{j_1+1,j_2+1,j_3-1,j_4+1} + R_{j_1+1,j_2+1,j_3-1,j_4-1}$$
$$- R_{j_1+1,j_2-1,j_3+1,j_4+1} + R_{j_1+1,j_2-1,j_3+1,j_4-1}$$
$$+ R_{j_1+1,j_2-1,j_3-1,j_4+1} - R_{j_1+1,j_2-1,j_3-1,j_4-1}$$
$$- R_{j_1-1,j_2+1,j_3+1,j_4+1} + R_{j_1-1,j_2+1,j_3+1,j_4-1}$$
$$+ R_{j_1-1,j_2+1,j_3-1,j_4+1} - R_{j_1-1,j_2+1,j_3 1,j_4-1}$$
$$+ R_{j_1-1,j_2-1,j_3+1,j_4+1} - R_{j_1-1,j_2-1,j_3+1,j_4-1}$$
$$- R_{j_1-1,j_2-1,j_3-1,j_4+1} + R_{j_1-1,j_2-1,j_3-1,j_4-1} \Big)$$
$$+ O\left(h_{j_1}^2, h_{j_2}^2, h_{j_3}^2, h_{j_4}^2\right),$$

where $R_{j_1+1,j_2+1,j_3+1,j_4+1} \triangleq R\left(\alpha_{j_1} + h_{j_1}, \alpha_{j_2} + h_{j_2}, \alpha_{j_3} + h_{j_3}, \alpha_{j_4} + h_{j_4}\right)$, etc. The finite difference formulas introduce their intrinsic "methodological errors" of order $O\left(h_{j_1}^2, h_{j_2}^2, h_{j_3}^2, h_{j_4}^2\right)$ which are in addition to, and independent of, the errors that might be incurred in the computation of $\partial^4 R(\boldsymbol{\alpha}) / \partial \alpha_{j_1} \partial \alpha_{j_2} \partial \alpha_{j_3} \partial \alpha_{j_4}$.

Within the framework of the Forward Sensitivity Analysis Methodology, the fourth-order sensitivities $\partial^4 R(\boldsymbol{\alpha}) / \partial \alpha_{j_1} \partial \alpha_{j_2} \partial \alpha_{j_3} \partial \alpha_{j_4}, j_1, j_2, j_3; j_4 = 1, \ldots, TP$, of the response to the model parameters are obtained by computing the partial G-derivatives of the third-order sensitivities represented by (2.100), which yields the following result:

$$\begin{aligned}
&\frac{\partial^4 R(\boldsymbol{\varphi}; \boldsymbol{\psi}; \boldsymbol{\alpha})}{\partial \alpha_{j_1} \partial \alpha_{j_2} \partial \alpha_{j_3} \partial \alpha_{j_4}} \triangleq \frac{1}{\delta \alpha_{j_4}} \left\{ \frac{d}{d\varepsilon} R^{(3)} \left[j_3; j_2; j_1; \alpha_{j_4}^0 + \varepsilon \delta \alpha_{j_4}; \right.\right. \\
&\boldsymbol{\varphi}\left(\alpha_{j_4}^0 + \varepsilon \delta \alpha_{j_4}\right); \boldsymbol{\psi}\left(\alpha_{j_4}^0 + \varepsilon \delta \alpha_{j_4}\right); \frac{\partial \boldsymbol{\varphi}\left(\alpha_{j_4}^0 + \varepsilon \delta \alpha_{j_4}\right)}{\partial \alpha_{j_3}}; \\
&\frac{\partial \boldsymbol{\psi}\left(\alpha_{j_4}^0 + \varepsilon \delta \alpha_{j_4}\right)}{\partial \alpha_{j_3}}; \frac{\partial^2 \boldsymbol{\varphi}\left(\alpha_{j_4}^0 + \varepsilon \delta \alpha_{j_4}\right)}{\partial \alpha_{j_3} \partial \alpha_{j_2}}; \frac{\partial^2 \boldsymbol{\psi}\left(\alpha_{j_4}^0 + \varepsilon \delta \alpha_{j_4}\right)}{\partial \alpha_{j_3} \partial \alpha_{j_2}}; \\
&\left.\left. \frac{\partial^3 \boldsymbol{\varphi}\left(\alpha_{j_4}^0 + \varepsilon \delta \alpha_{j_4}\right)}{\partial \alpha_{j_1} \partial \alpha_{j_2} \partial \alpha_{j_3}}; \frac{\partial^3 \boldsymbol{\psi}\left(\alpha_{j_4}^0 + \varepsilon \delta \alpha_{j_4}\right)}{\partial \alpha_{j_1} \partial \alpha_{j_2} \partial \alpha_{j_3}} \right] \right\}_{\varepsilon=0} = R^{(4)} \left(j_4; j_3; j_2; j_1; \right. \\
&\boldsymbol{\varphi}; \boldsymbol{\psi}; \boldsymbol{\alpha}; \frac{\partial \boldsymbol{\varphi}}{\partial \alpha_{j_4}}; \frac{\partial \boldsymbol{\psi}}{\partial \alpha_{j_4}}; \frac{\partial^2 \boldsymbol{\varphi}}{\partial \alpha_{j_4} \partial \alpha_{j_3}}; \frac{\partial^2 \boldsymbol{\psi}}{\partial \alpha_{j_4} \partial \alpha_{j_3}}; \frac{\partial^3 \boldsymbol{\varphi}}{\partial \alpha_{j_4} \partial \alpha_{j_2} \partial \alpha_{j_3}}; \\
&\left. \frac{\partial^3 \boldsymbol{\psi}}{\partial \alpha_{j_4} \partial \alpha_{j_2} \partial \alpha_{j_3}}; \frac{\partial^4 \boldsymbol{\varphi}}{\partial \alpha_{j_1} \partial \alpha_{j_2} \partial \alpha_{j_3} \partial \alpha_{j_4}}; \frac{\partial^4 \boldsymbol{\psi}}{\partial \alpha_{j_1} \partial \alpha_{j_2} \partial \alpha_{j_3} \partial \alpha_{j_4}} \right).
\end{aligned} \tag{2.174}$$

As indicated by the functional dependence of the last term in (2.174), the evaluation of the fourth-order sensitivities requires prior knowledge of the first-, second-, and third-order partial derivatives of the state functions with respect to the model parameters. The functions $\partial \boldsymbol{\varphi}(\mathbf{x})/\partial \alpha_{j_4}$ and $\partial \boldsymbol{\psi}(\mathbf{x})/\partial \alpha_{j_4}$ are obtained by solving the 1st-OFSS. The functions $\partial^2 \boldsymbol{\varphi}/\partial \alpha_{j_3}\partial \alpha_{j_4}$ and $\partial^2 \boldsymbol{\psi}/\partial \alpha_{j_3}\partial \alpha_{j_4}$ are the solutions of the 2nd-OFSS. The functions $\partial^3 \boldsymbol{\varphi}/\partial \alpha_{j_2}\partial \alpha_{j_3}\partial \alpha_{j_4}$ and $\partial^3 \boldsymbol{\psi}/\partial \alpha_{j_2}\partial \alpha_{j_3}\partial \alpha_{j_4}$ are the solutions of the 3rd-OFSS. The functions $\partial^4 \boldsymbol{\varphi}/\partial \alpha_{j_1}\partial \alpha_{j_2}\partial \alpha_{j_3}\partial \alpha_{j_4}$ and $\partial^4 \boldsymbol{\psi}/\partial \alpha_{j_1}\partial \alpha_{j_2}\partial \alpha_{j_3}\partial \alpha_{j_4}$ are the solutions of the following "*Fourth-Order Forward Sensitivity System*" (4th-OFSS), which is obtained by taking the partial G-derivative of the 3rd-OFSS with respect to a generic model parameter α_{j_4}; the resulting equations have the following expressions, for $j_1, \ldots, j_4 = 1, \ldots, TP$:

$$\mathbf{L}(\mathbf{x};\boldsymbol{\alpha}^0)\frac{\partial^4 \boldsymbol{\varphi}}{\partial \alpha_{j_4}\partial \alpha_{j_3}\partial \alpha_{j_2}\partial \alpha_{j_1}} = \left\{\mathbf{f}^{(4)}(j_4;j_3;j_2;j_1;\boldsymbol{\varphi};\boldsymbol{\alpha})\right\}_{\boldsymbol{\alpha}^0};\mathbf{x}\in\Omega_x; \tag{2.175}$$

$$\left\{\mathbf{B}_{\varphi}(\mathbf{x};\boldsymbol{\alpha})\right\}_{\boldsymbol{\alpha}^0}\frac{\partial^4 \boldsymbol{\varphi}}{\partial \alpha_{j_4}\partial \alpha_{j_3}\partial \alpha_{j_2}\partial \alpha_{j_1}} = \left\{\mathbf{c}_{\varphi}^{(4)}(j_4;j_3;j_2;j_1;\boldsymbol{\varphi};\boldsymbol{\alpha})\right\}_{\boldsymbol{\alpha}^0};\mathbf{x}\in\partial\Omega_x(\boldsymbol{\alpha}^0); \tag{2.176}$$

$$\mathbf{L}^*(\mathbf{x};\boldsymbol{\alpha}^0)\frac{\partial^4 \boldsymbol{\psi}}{\partial \alpha_{j_4}\partial \alpha_{j_3}\partial \alpha_{j_2}\partial \alpha_{j_1}} = \left\{\mathbf{g}^{(4)}(j_4;j_3;j_2;j_1;\boldsymbol{\psi};\boldsymbol{\alpha})\right\}_{\boldsymbol{\alpha}^0};\mathbf{x}\in\Omega_x; \tag{2.177}$$

$$\left\{\mathbf{B}_{\psi}(\mathbf{x};\boldsymbol{\alpha})\right\}_{\boldsymbol{\alpha}^0}\frac{\partial^4 \boldsymbol{\psi}}{\partial \alpha_{j_4}\partial \alpha_{j_3}\partial \alpha_{j_2}\partial \alpha_{j_1}} = \left\{\mathbf{c}_{\psi}^{(4)}(j_4;j_3;j_2;j_1;\boldsymbol{\psi};\boldsymbol{\alpha})\right\}_{\boldsymbol{\alpha}^0};\mathbf{x}\in\partial\Omega_x(\boldsymbol{\alpha}^0); \tag{2.178}$$

where

$$\mathbf{f}^{(4)}(j_4;j_3;j_2;j_1;\boldsymbol{\varphi};\boldsymbol{\alpha}) \triangleq \frac{\partial \mathbf{f}^{(3)}(j_3;j_2;j_1;\boldsymbol{\varphi};\boldsymbol{\alpha})}{\partial \alpha_{j_4}} - \frac{\partial \mathbf{L}(\mathbf{x};\boldsymbol{\alpha})}{\partial \alpha_{j_4}}\frac{\partial^3 \boldsymbol{\varphi}}{\partial \alpha_{j_3}\partial \alpha_{j_2}\partial \alpha_{j_1}}; \tag{2.179}$$

$$\mathbf{g}^{(4)}(j_4;j_3;j_2;j_1;\boldsymbol{\psi};\boldsymbol{\alpha}) \triangleq \frac{\partial \mathbf{g}^{(3)}(j_3;j_2;j_1;\boldsymbol{\psi};\boldsymbol{\alpha})}{\partial \alpha_{j_4}} - \frac{\partial \mathbf{L}^*(\mathbf{x};\boldsymbol{\alpha})}{\partial \alpha_{j_4}}\frac{\partial^3 \boldsymbol{\psi}}{\partial \alpha_{j_3}\partial \alpha_{j_2}\partial \alpha_{j_1}}; \tag{2.180}$$

$$\mathbf{c}_{\varphi}^{(4)}(j_4;j_3;j_2;j_1;\boldsymbol{\varphi};\boldsymbol{\alpha}) \triangleq \frac{\partial \mathbf{c}_{\varphi}^{(3)}(j_3;j_2;j_1;\boldsymbol{\varphi};\boldsymbol{\alpha})}{\partial \alpha_{j_4}} - \frac{\partial \mathbf{B}_{\varphi}(\mathbf{x};\boldsymbol{\alpha})}{\partial \alpha_{j_4}}\frac{\partial^3 \boldsymbol{\varphi}}{\partial \alpha_{j_3}\partial \alpha_{j_2}\partial \alpha_{j_1}};\mathbf{x}\in\partial\Omega_x(\boldsymbol{\alpha}^0); \tag{2.181}$$

$$\begin{aligned}&\mathbf{c}_{\psi}^{(4)}(j_4;j_3;j_2;j_1;\boldsymbol{\psi};\boldsymbol{\alpha})\\&\triangleq\frac{\partial\mathbf{c}_{\psi}^{(3)}(j_3;j_2;j_1;\boldsymbol{\psi};\boldsymbol{\alpha})}{\partial\alpha_{j_4}}-\frac{\partial\mathbf{B}_{\psi}(\mathbf{x};\boldsymbol{\alpha})}{\partial\alpha_{j_4}}\frac{\partial^3\boldsymbol{\psi}}{\partial\alpha_{j_3}\partial\alpha_{j_2}\partial\alpha_{j_1}};\mathbf{x}\in\partial\Omega_x(\boldsymbol{\alpha}^0).\end{aligned}\tag{2.182}$$

Evidently, the computations of the fourth-order sensitivities $\partial^4R(\boldsymbol{\alpha})/\partial\alpha_{j_4}\partial\alpha_{j_3}\partial\alpha_{j_2}\partial\alpha_{j_1}$ require $2TP(TP+1)(TP+2)(TP+3)/4!$ large-scale computations to solve (2.175)-(2.178), followed by just as many small-scale computations for performing the integrations that define the various sensitivities. These computations would be in addition to the $2TP$ large-scale computations needed to determine the functions $\partial\boldsymbol{\varphi}(\mathbf{x})/\partial\alpha_{j_4}$ and $\partial\boldsymbol{\psi}(\mathbf{x})/\partial\alpha_{j_4}$ by solving the 1st-OFSS, the $TP(TP+1)$ large-scale computations needed to determine the functions $\partial^2\boldsymbol{\varphi}/\partial\alpha_{j_3}\partial\alpha_{j_4}$ and $\partial^2\boldsymbol{\psi}/\partial\alpha_{j_3}\partial\alpha_{j_4}$ by solving the 2nd-OFSS, and the $TP(TP+1)(TP+2)/3!$ large-scale computations needed to determine the functions $\partial^3\boldsymbol{\varphi}/\partial\alpha_{j_2}\partial\alpha_{j_3}\partial\alpha_{j_4}$ and $\partial^3\boldsymbol{\psi}/\partial\alpha_{j_2}\partial\alpha_{j_3}\partial\alpha_{j_4}$ by solving the 3rd-OFSS. It is not practicable to compute these sensitivities using finite-difference formulas or the forward sensitivity analysis methodology.

Within the framework of the 4th-CASAM-L, the fourth-order sensitivities of the response $R[\boldsymbol{\varphi}(\mathbf{x}),\boldsymbol{\psi}(\mathbf{x});\boldsymbol{\alpha}]$ with respect to the model parameters are obtained by determining the first-order G-differential of the third-order sensitivities $R^{(3)}\left[j_3;j_2;j_1;\mathbf{u}^{(3)}(j_1;\mathbf{x});\mathbf{a}^{(3)}(j_2;j_1;\mathbf{x});\boldsymbol{\alpha}\right]\equiv\partial^3R(\boldsymbol{\varphi};\boldsymbol{\psi};\boldsymbol{\alpha})/\partial\alpha_{j_3}\partial\alpha_{j_2}\partial\alpha_{j_1}$, which were defined in (2.155). By definition, the total G-differential of $R^{(3)}[j_3;j_2;j_1;\mathbf{u}^{(3)}(j_1;\mathbf{x});\mathbf{a}^{(3)}(j_2;j_1;\mathbf{x});\boldsymbol{\alpha}]$ is obtained as follows:

$$\begin{aligned}&\left\{\delta R^{(3)}\left[j_3;j_2;j_1;\mathbf{u}^{(3)}(j_1;\mathbf{x});\mathbf{a}^{(3)}(j_2;j_1;\mathbf{x});\boldsymbol{\alpha};\delta\mathbf{u}^{(3)};\delta\mathbf{a}^{(3)};\delta\boldsymbol{\alpha}\right]\right\}_{\boldsymbol{\alpha}^0}\\&\triangleq\left\{\frac{d}{d\varepsilon}\prod_{k=1}^{TI}\int_{\lambda_k(\boldsymbol{\alpha})}^{\omega_k(\boldsymbol{\alpha})}dx_kS^{(3)}\left[j_3;j_2;j_1;\mathbf{u}^{(3),0}(j_1;\mathbf{x})+\varepsilon\delta\mathbf{u}^{(3)}(j_1;\mathbf{x});\right.\right.\\&\left.\left.\mathbf{a}^{(3),0}(j_2;j_1;\mathbf{x})+\varepsilon\delta\mathbf{a}^{(3)}(j_2;j_1);\boldsymbol{\alpha}^0+\varepsilon\delta\boldsymbol{\alpha}\right]\right\}_{\varepsilon=0}\\&=\left\{\delta R^{(3)}\left(j_3;j_2;j_1;\mathbf{u}^{(3)};\mathbf{a}^{(3)};\delta\boldsymbol{\alpha}\right)\right\}_{dir}\\&\quad+\left\{\delta R^{(3)}\left(j_3;j_2;j_1;\mathbf{u}^{(3)};\mathbf{a}^{(3)};\delta\mathbf{u}^{(3)};\delta\mathbf{a}^{(3)}\right)\right\}_{ind},\end{aligned}\tag{2.183}$$

where the direct-effect term $\{\delta R^{(3)}(j_3;j_2;j_1;\mathbf{u}^{(3)};\mathbf{a}^{(3)};\delta\boldsymbol{\alpha})\}_{dir}$ depends directly on the parameter variations and is defined as follows:

$$\left\{\delta R^{(3)}\left(j_3;j_2;j_1;\mathbf{u}^{(3)};\mathbf{a}^{(3)};\delta\boldsymbol{\alpha}\right)\right\}_{dir} \triangleq \left\{\frac{\partial}{\partial\boldsymbol{\alpha}}\prod_{k=1}^{TI}\int_{\lambda_k(\boldsymbol{\alpha})}^{\omega_k(\boldsymbol{\alpha})} dx_k S^{(3)}\left[j_3;j_2;j_1;\mathbf{u}^{(3)}(j_1;\mathbf{x});\mathbf{a}^{(3)}(j_2;j_1;\mathbf{x});\boldsymbol{\alpha}\right]\right\}_{\boldsymbol{\alpha}^0}\delta\boldsymbol{\alpha}, \tag{2.184}$$

while the "indirect-effect term" $\{\delta R^{(3)}(j_3;j_2;j_1;\mathbf{u}^{(3)};\mathbf{a}^{(3)};\delta\mathbf{u}^{(3)};\delta\mathbf{a}^{(3)})\}_{ind}$ depends indirectly on the parameter variations through the variations in the forward and adjoint state functions and is defined as follows:

$$\begin{aligned}&\left\{\delta R^{(3)}\left[j_3;j_2;j_1;\mathbf{u}^{(3)};\mathbf{a}^{(3)};\delta\mathbf{u}^{(3)}(j_1;\mathbf{x});\delta\mathbf{a}^{(3)}(j_2;j_1;\mathbf{x})\right]\right\}_{ind}\\ &\triangleq\left\{\prod_{k=1}^{TI}\int_{\lambda_k(\boldsymbol{\alpha})}^{\omega_k(\boldsymbol{\alpha})} dx_k\frac{\partial S^{(3)}\left[j_3;j_2;j_1;\mathbf{u}^{(3)}(j_1;\mathbf{x});\mathbf{a}^{(3)};\boldsymbol{\alpha}\right]}{\partial\mathbf{u}^{(3)}(j_1;\mathbf{x})}\right\}_{\boldsymbol{\alpha}^0}\delta\mathbf{u}^{(3)}(j_1;\mathbf{x})\\ &+\left\{\prod_{k=1}^{TI}\int_{\lambda_k(\boldsymbol{\alpha})}^{\omega_k(\boldsymbol{\alpha})} dx_k\frac{\partial S^{(3)}\left[j_3;j_2;j_1;\mathbf{u}^{(3)};\mathbf{a}^{(3)};\boldsymbol{\alpha}\right]}{\partial\mathbf{a}^{(3)}(j_2;j_1;\mathbf{x})}\right\}_{\boldsymbol{\alpha}^0}\delta\mathbf{a}^{(3)}(j_2;j_1;\mathbf{x}),\end{aligned} \tag{2.185}$$

where

$$\begin{aligned}\frac{\partial[]}{\partial\mathbf{u}^{(3)}(j_1;\mathbf{x})}\delta\mathbf{u}^{(3)}(j_1;\mathbf{x}) &\equiv \frac{\partial[]}{\partial\boldsymbol{\varphi}}\delta\boldsymbol{\varphi}(\mathbf{x})+\frac{\partial[]}{\partial\boldsymbol{\psi}}\delta\boldsymbol{\psi}(\mathbf{x})\\ &+\sum_{m=1}^{2}\frac{\partial[]}{\partial\mathbf{a}_m^{(1)}}\delta\mathbf{a}_m^{(1)}(\mathbf{x})+\sum_{m=1}^{4}\frac{\partial[]}{\partial\mathbf{a}_m^{(2)}(j_1)}\delta\mathbf{a}_m^{(2)}(j_1;\mathbf{x}),\end{aligned} \tag{2.186}$$

$$\frac{\partial[]}{\partial\mathbf{a}^{(3)}(j_2;j_1;\mathbf{x})}\delta\mathbf{a}^{(3)}(j_2;j_1;\mathbf{x})\equiv\sum_{m=1}^{8}\frac{\partial[]}{\partial\mathbf{a}_m^{(3)}(j_2;j_1;\mathbf{x})}\delta\mathbf{a}_m^{(3)}(j_2;j_1;\mathbf{x}). \tag{2.187}$$

The indirect-effect term $\{\delta R^{(3)}(j_3;j_2;j_1;\mathbf{u}^{(3)};\mathbf{a}^{(3)};\delta\mathbf{u}^{(3)};\delta\mathbf{a}^{(3)})\}_{ind}$ defined in (2.185) can be computed only after having computed the variations $\delta\mathbf{u}^{(3)}(j_1;\mathbf{x})$ and $\delta\mathbf{a}^{(3)}(j_2;j_1;\mathbf{x})$. Recall that the third-level variational sensitivity function $\delta\mathbf{u}^{(3)}(j_1;\mathbf{x})$ is the solution of the 3rd-LVSS. On the other hand, the vector $\delta\mathbf{a}^{(3)}(j_2;j_1;\mathbf{x})$ is the solution of the system of equations obtained by G-differentiating the 3rd-LASS. Applying the definition of the total G-differential, cf. (2.26), to the 3rd-LASS, cf. (2.141) and (2.151), yields the following set of equations for the vector of variations $\delta\mathbf{a}^{(3)}(j_2;j_1;\mathbf{x})$, all to be evaluated at the nominal values for the parameters and state functions, for $j_1, j_2 = 1, \ldots, TP$:

$$\frac{\partial\left\{\mathbf{A}^{(3)}\left[j_1;\mathbf{u}^{(2)}(\mathbf{x})\right]\mathbf{a}^{(3)}(j_2;j_1;\mathbf{x})-\mathbf{s}^{(3)}\left[j_2;j_1;\mathbf{u}^{(3)};\boldsymbol{\alpha}\right]\right\}}{\partial\mathbf{u}^{(3)}}\delta\mathbf{u}^{(3)} + \left\{\mathbf{A}^{(3)}\left(j_1;\mathbf{u}^{(2)}\right)\right\}_{\boldsymbol{\alpha}^0}\delta\mathbf{a}^{(3)} = \mathbf{p}^{(3)}\left[j_2;j_1;\mathbf{u}^{(3)};\mathbf{a}^{(3)};\boldsymbol{\alpha};\delta\boldsymbol{\alpha}\right], \tag{2.188}$$

$$\left\{\delta\mathbf{b}_A^{(3)}\left[\mathbf{a}^{(3)};\mathbf{u}^{(3)};\boldsymbol{\alpha};\delta\mathbf{a}^{(3)};\delta\mathbf{u}^{(3)};\delta\boldsymbol{\alpha}\right]\right\}_{\boldsymbol{\alpha}^0}=\mathbf{0},\ \ \mathbf{x}\in\partial\Omega_x\left(\boldsymbol{\alpha}^0\right), \tag{2.189}$$

where

$$\mathbf{p}^{(3)}\left[j_2;j_1;\mathbf{u}^{(3)}(j_1;\mathbf{x});\mathbf{a}^{(3)}(j_2;j_1;\mathbf{x});\boldsymbol{\alpha};\delta\boldsymbol{\alpha}\right] \triangleq \frac{\partial\left\{\mathbf{s}^{(3)}\left[j_2;j_1;\mathbf{u}^{(3)}(j_1;\mathbf{x});\boldsymbol{\alpha}\right]-\mathbf{A}^{(3)}\left(\mathbf{u}^{(2)}\right)\mathbf{a}^{(3)}(j_2;j_1;\mathbf{x})\right\}}{\partial\boldsymbol{\alpha}}\delta\boldsymbol{\alpha}. \tag{2.190}$$

As (2.188) and (2.189) indicate, the vectors of variations $\delta\mathbf{u}^{(3)}(j_1;\mathbf{x})$ and $\delta\mathbf{a}^{(3)}(j_2;j_1;\mathbf{x})$ are related to each other, and must be determined by solving simultaneously the coupled system of equations obtained by concatenating the 3rd-LVSS to (2.188) while being subject to the boundary conditions obtained by concatenating the boundary conditions which belong to the 3rd-LVSS with the boundary conditions represented by (2.189). Concatenating (2.121) and (2.188) yields the following 16x16 block-matrix operator equation:

$$\mathbf{V}^{(4)}\left(\mathbf{u}^{(3)}\right)\delta\mathbf{u}^{(4)}(j_2;j_1;\mathbf{x})=\mathbf{q}^{(4)}\left[j_2;j_1;\mathbf{u}^{(4)}(j_2;j_1;\mathbf{x});\boldsymbol{\alpha};\delta\boldsymbol{\alpha}\right], \tag{2.191}$$

where:

$$\mathbf{u}^{(4)}(j_2;j_1;\mathbf{x})\triangleq\begin{pmatrix}\mathbf{u}^{(3)}(j_1;\mathbf{x})\\ \mathbf{a}^{(3)}(j_2;j_1;\mathbf{x})\end{pmatrix};\quad \delta\mathbf{u}^{(4)}(j_2;j_1;\mathbf{x})\triangleq\begin{pmatrix}\delta\mathbf{u}^{(3)}(j_1;\mathbf{x})\\ \delta\mathbf{a}^{(3)}(j_2;j_1;\mathbf{x})\end{pmatrix}; \tag{2.192}$$

$$\mathbf{V}^{(4)}\left(\mathbf{u}^{(3)}\right)=\begin{pmatrix}\mathbf{V}^{(3)}\left(j_1;\mathbf{u}^{(2)}\right) & [\mathbf{0}]_{(8\times 8)}\\ \mathbf{V}_{21}^{(4)}(j_2;j_1) & \mathbf{V}_{22}^{(4)}(j_1)\end{pmatrix};\quad \mathbf{V}_{22}^{(4)}(j_1)\triangleq\mathbf{A}^{(3)}\left(j_1;\mathbf{u}^{(2)}\right);$$
$$\mathbf{V}_{21}^{(4)}\triangleq\frac{\partial\left\{\mathbf{A}^{(3)}\left[j_1;\mathbf{u}^{(2)}(\mathbf{x})\right]\mathbf{a}^{(3)}(j_2;j_1;\mathbf{x})\right\}}{\partial\mathbf{u}^{(3)}} - \frac{\partial\left\{\mathbf{s}^{(3)}\left[j_2;j_1;\mathbf{u}^{(3)}(j_1;\mathbf{x});\boldsymbol{\alpha}\right]\right\}}{\partial\mathbf{u}^{(3)}}; \tag{2.193}$$

$$
\begin{aligned}
&\mathbf{q}^{(4)}\left[j_2;j_1;\mathbf{u}^{(4)}(j_2;j_1;\mathbf{x});\boldsymbol{\alpha};\delta\boldsymbol{\alpha}\right] \\
&\triangleq \begin{pmatrix} \mathbf{q}^{(3)}\left[j_1;\mathbf{u}^{(3)}(j_1;\mathbf{x});\boldsymbol{\alpha};\delta\boldsymbol{\alpha}\right] \\ \mathbf{p}^{(3)}\left[j_2;j_1;\mathbf{u}^{(4)}(j_2;j_1;\mathbf{x});\boldsymbol{\alpha};\delta\boldsymbol{\alpha}\right] \end{pmatrix} \\
&\equiv \begin{pmatrix} \mathbf{q}_1^{(4)}\left[j_2;j_1;\mathbf{u}^{(4)}(j_2;j_1;\mathbf{x});\boldsymbol{\alpha};\delta\boldsymbol{\alpha}\right] \\ \cdot \\ \mathbf{q}_{16}^{(4)}\left[j_2;j_1;\mathbf{u}^{(4)}(j_2;j_1;\mathbf{x});\boldsymbol{\alpha};\delta\boldsymbol{\alpha}\right] \end{pmatrix}.
\end{aligned}
\tag{2.194}
$$

Concatenating the boundary conditions which belong to the 3rd-LVSS, Eq. (2.122), with the boundary conditions represented by (2.189) yields the following boundary conditions for the function $\delta\mathbf{u}^{(4)}(j_2;j_1;\mathbf{x})$:

$$
\begin{aligned}
&\mathbf{b}_V^{(4)}\left[j_2;j_1;\mathbf{u}^{(4)}(j_2;j_1;\mathbf{x});\boldsymbol{\alpha};\delta\mathbf{u}^{(4)}(j_2;j_1;\mathbf{x});\delta\boldsymbol{\alpha}\right] \\
&\triangleq \begin{pmatrix} \mathbf{b}_V^{(3)}\left[j_1;\mathbf{u}^{(3)}(j_1;\mathbf{x});\boldsymbol{\alpha};\delta\mathbf{u}^{(3)}(j_1;\mathbf{x});\delta\boldsymbol{\alpha}\right] \\ \delta\mathbf{b}_A^{(3)}\left[\mathbf{u}^{(4)}(j_2;j_1;\mathbf{x});\boldsymbol{\alpha};\delta\mathbf{u}^{(4)}(j_2;j_1;\mathbf{x});\delta\boldsymbol{\alpha}\right] \end{pmatrix} = \begin{pmatrix} \mathbf{0} \\ \mathbf{0} \end{pmatrix}, \mathbf{x} \in \partial\Omega_x(\boldsymbol{\alpha}^0).
\end{aligned}
\tag{2.195}
$$

The matrix-operator equations represented by (2.191) together with the boundary conditions represented by (2.195) will be called the *Fourth-Level Variational Sensitivity System* (4th-LVSS). The solution, $\delta\mathbf{u}^{(4)}(j_2;j_1;\mathbf{x})$, of the 4th-LVSS will be called the "*fourth-level variational sensitivity function.*"

Since the 4th-LVSS equations depend on the parameter variations $\delta\alpha_j$, solving them is prohibitively expensive computationally for large-scale systems involving many parameters. The need for solving the 4th-LVSS can be avoided by expressing the indirect-effect term $\{\delta R^{(3)}(j_3;j_2;j_1;\mathbf{u}^{(3)}(j_1;\mathbf{x});\mathbf{a}^{(3)}(j_2;j_1;\mathbf{x});\delta\mathbf{u}^{(3)};\delta\mathbf{a}^{(3)})\}_{ind}$ defined in (2.185) in an alternative way, which does not involve the function $\delta\mathbf{u}^{(4)}(j_2;j_1;\mathbf{x})$. This alternative expression will be derived by using the solution of a *Fourth-Level Adjoint Sensitivity Systems* (4th-LASS), which will be constructed by implementing the same sequence of logical steps as were followed for constructing the 1st, 2nd, and 3rd-LASS. The first step is to define a Hilbert space, denoted as H_4, comprising block-vector elements of the form $\boldsymbol{\eta}^{(4)}(\mathbf{x}) \triangleq \left[\boldsymbol{\eta}_1^{(4)}(\mathbf{x}),\ \ldots,\boldsymbol{\eta}_{16}^{(4)}(\mathbf{x})\right]^\dagger \in H_4$, with $\boldsymbol{\eta}_i^{(4)}(\mathbf{x}) \triangleq \left[\eta_{i,1}^{(4)}(\mathbf{x}),\ \ldots,\eta_{i,TD}^{(4)}(\mathbf{x})\right]^\dagger$, for $i = 1, \ldots, 16$. The inner product between two elements, $\boldsymbol{\eta}^{(4)}(\mathbf{x}) \in H_4$ and $\boldsymbol{\xi}^{(4)}(\mathbf{x}) \in H_4$, of the Hilbert space H_4, will be denoted as $\langle\boldsymbol{\eta}^{(4)}(\mathbf{x}),\boldsymbol{\xi}^{(4)}(\mathbf{x})\rangle_4$ and is defined as follows:

$$
\left\langle \boldsymbol{\eta}^{(4)}(\mathbf{x}),\boldsymbol{\xi}^{(4)}(\mathbf{x})\right\rangle_4 \triangleq \sum_{i=1}^{16}\left\langle \boldsymbol{\eta}_i^{(4)}(\mathbf{x}),\boldsymbol{\xi}_i^{(4)}(\mathbf{x})\right\rangle_0. \tag{2.196}
$$

Using the definition provided in (2.196), form the inner product in H_4 of (2.191) with a set of yet undefined vector-valued functions $\mathbf{a}^{(4)}(j_3;j_2;j_1\mathbf{x}) \triangleq \left[\mathbf{a}_1^{(4)}(j_3;j_2;j_1\mathbf{x}), \ldots, \mathbf{a}_{16}^{(4)}(j_3;j_2;j_1\mathbf{x})\right]^\dagger \in H_4,\ j_1, j_2, j_3 = 1, \ldots, TP$, to obtain the following relation:

$$\begin{aligned}
&\left\{\left\langle \mathbf{a}^{(4)}(j_3;j_2;j_1;\mathbf{x}), \mathbf{V}^{(4)}\left(\mathbf{u}^{(3)}\right)\delta\mathbf{u}^{(4)}(j_2;j_1;\mathbf{x})\right\rangle_4\right\}_{\boldsymbol{\alpha}^0} \\
&= \left\{\left\langle \mathbf{a}^{(4)}(j_3;j_2;j_1;\mathbf{x}), \mathbf{q}^{(4)}\left[j_2;j_1;\mathbf{u}^{(4)}(j_2;j_1;\mathbf{x});\boldsymbol{\alpha};\delta\boldsymbol{\alpha}\right]\right\rangle_4\right\}_{\boldsymbol{\alpha}^0}.
\end{aligned} \tag{2.197}$$

Using the definition of the adjoint operator in the Hilbert space H_4, recast the left side of (2.197) as follows:

$$\begin{aligned}
&\left\{\left\langle \mathbf{a}^{(4)}(j_3;j_2;j_1;\mathbf{x}), \mathbf{V}^{(4)}\left(\mathbf{u}^{(3)}\right)\delta\mathbf{u}^{(4)}(j_2;j_1;\mathbf{x})\right\rangle_4\right\}_{\boldsymbol{\alpha}^0} \\
&= \left\{\left\langle \delta\mathbf{u}^{(4)}(j_2;j_1;\mathbf{x}), \mathbf{A}^{(4)}\left(\mathbf{u}^{(3)}\right)\mathbf{a}^{(4)}(j_3;j_2;j_1;\mathbf{x})\right\rangle_4\right\}_{\boldsymbol{\alpha}^0} \\
&+\left\{P^{(4)}\left[\delta\mathbf{u}^{(4)}(j_2;j_1;\mathbf{x});\mathbf{a}^{(4)}(j_3;j_2;j_1;\mathbf{x});\boldsymbol{\alpha};\delta\boldsymbol{\alpha}\right]\right\}_{\boldsymbol{\alpha}^0},
\end{aligned} \tag{2.198}$$

where $\left\{P^{(4)}\left[\delta\mathbf{u}^{(4)}(j_2;j_1;\mathbf{x});\mathbf{a}^{(4)}(j_3;j_2;j_1;\mathbf{x});\boldsymbol{\alpha};\delta\boldsymbol{\alpha}\right]\right\}_{\boldsymbol{\alpha}^0}$ denotes the bilinear concomitant defined on the phase-space boundary $\mathbf{x} \in \partial\Omega_x(\boldsymbol{\alpha}^0)$ and where $\mathbf{A}^{(4)}(j_2;j_1;\mathbf{u}^{(3)}) = \{\mathbf{V}^{(4)}[j_2;j_1;\mathbf{u}^{(3)}(\mathbf{x})]\}^*$ is the operator formally adjoint to $\mathbf{V}^{(4)}[j_2;j_1;\mathbf{u}^{(3)}(\mathbf{x})]$ and has therefore the following form:

$$\begin{aligned}
\mathbf{A}^{(4)}\left(j_2;j_1;\mathbf{u}^{(3)}\right) &= \begin{pmatrix} \mathbf{V}^{(3)}\left(\mathbf{u}^{(1)}\right) & [\mathbf{0}]_{(4\times4)} \\ \mathbf{V}_{21}^{(4)}\left(j_1;\mathbf{u}^{(2)}\right) & \mathbf{V}_{22}^{(4)}\left(\mathbf{u}^{(1)}\right) \end{pmatrix}^* \\
&= \begin{pmatrix} \mathbf{A}^{(3)} & \left[\mathbf{V}_{21}^{(4)}\right]^* \\ [\mathbf{0}]_{(4\times4)} & \left[\mathbf{V}_{22}^{(4)}\right]^* \end{pmatrix}.
\end{aligned} \tag{2.199}$$

The first term on right side of (2.198) is now required to represent the indirect-effect term $\{\delta R^{(3)}(j_3;j_2;j_1;\mathbf{u}^{(3)};\mathbf{a}^{(3)};\delta\mathbf{u}^{(3)};\delta\mathbf{a}^{(3)})\}_{ind}$ defined in (2.185). This requirement is satisfied by imposing the following relation on each element $\mathbf{a}^{(4)}(j_3;j_2;j_1;\mathbf{x}) \triangleq \left[\mathbf{a}_1^{(4)}(j_3;j_2;j_1;\mathbf{x}), \ldots, \mathbf{a}_{16}^{(4)}(j_3;j_2;j_1;\mathbf{x})\right]^\dagger,\ j_1, j_2, j_3 = 1, \ldots, TP$:

$$\mathbf{A}^{(4)}\left(\mathbf{u}^{(3)}\right)\mathbf{a}^{(4)}(j_3;j_2;j_1;\mathbf{x}) = \mathbf{s}^{(4)}\left[j_3;j_2;j_1;\mathbf{u}^{(4)}(j_2;j_1;\mathbf{x});\boldsymbol{\alpha}\right], \tag{2.200}$$

where the 16-component block-vector

$$\mathbf{s}^{(4)}\left[j_3;j_2;j_1;\mathbf{u}^{(4)}(j_2;j_1;\mathbf{x});\boldsymbol{\alpha}\right] \triangleq \left[\mathbf{s}_1^{(4)}(j_3;j_2;j_1;\mathbf{x}),\ \ldots,\mathbf{s}_{16}^{(4)}(j_3;j_2;j_1;\mathbf{x})\right]^{\dagger}, \tag{2.201}$$

comprises, for each $j_1, j_2, j_3 = 1, \ldots, TP$, 16 components defined as follows:

$$\mathbf{s}_1^{(4)}\left[j_3;j_2;j_1;\mathbf{u}^{(4)}(j_2;j_1;\mathbf{x});\boldsymbol{\alpha}\right] \triangleq \frac{\partial S^{(3)}\left[j_3;j_2;j_1;\mathbf{u}^{(4)}(j_2;j_1;\mathbf{x});\boldsymbol{\alpha}\right]}{\partial\boldsymbol{\varphi}}, \tag{2.202}$$

$$\mathbf{s}_2^{(4)}\left[j_3;j_2;j_1;\mathbf{u}^{(4)}(j_2;j_1;\mathbf{x});\boldsymbol{\alpha}\right] \triangleq \frac{\partial S^{(3)}\left[j_3;j_2;j_1;\mathbf{u}^{(4)}(j_2;j_1;\mathbf{x});\boldsymbol{\alpha}\right]}{\partial\boldsymbol{\psi}}, \tag{2.203}$$

$$\mathbf{s}_{2+m}^{(4)}\left[j_3;j_2;j_1;\mathbf{u}^{(4)}(j_2;j_1;\mathbf{x});\boldsymbol{\alpha}\right] \triangleq \frac{\partial S^{(3)}\left[j_3;j_2;j_1;\mathbf{u}^{(4)}(j_2;j_1;\mathbf{x});\boldsymbol{\alpha}\right]}{\partial\mathbf{a}_m^{(1)}};\ m=1,2; \tag{2.204}$$

$$\mathbf{s}_{4+m}^{(4)}\left[j_3;j_2;j_1;\mathbf{u}^{(4)}(j_2;j_1;\mathbf{x});\boldsymbol{\alpha}\right] \triangleq \frac{\partial S^{(3)}\left[j_3;j_2;j_1;\mathbf{u}^{(4)}(j_2;j_1;\mathbf{x});\boldsymbol{\alpha}\right]}{\partial\mathbf{a}_m^{(2)}(j_1)};\ m=1,..,4; \tag{2.205}$$

$$\mathbf{s}_{8+m}^{(4)}\left[j_3;j_2;j_1;\mathbf{u}^{(4)}(j_2;j_1;\mathbf{x});\boldsymbol{\alpha}\right] \triangleq \frac{\partial S^{(3)}\left[j_3;j_2;j_1;\mathbf{u}^{(4)}(j_2;j_1;\mathbf{x});\boldsymbol{\alpha}\right]}{\partial\mathbf{a}_m^{(3)}(j_2;j_1)};\ m=1,..,8. \tag{2.206}$$

The definition of the set of vectors $\mathbf{a}^{(4)}(j_3;j_2;j_1;\mathbf{x}) \triangleq \left[\mathbf{a}_1^{(4)}(j_3;j_2;j_1;\mathbf{x}),\ \ldots,\mathbf{a}_{16}^{(4)}(j_3;j_2;j_1;\mathbf{x})\right]^{\dagger}$, for each j_1, j_2, $j_3 = 1$, ..., TP, will now be completed by selecting boundary conditions for this set of vectors, which will be represented in operator form as follows:

$$\left\{\mathbf{b}_A^{(4)}\left[\mathbf{a}^{(4)}(j_3;j_2;j_1;\mathbf{x});\mathbf{u}^{(4)}(j_2;j_1;\mathbf{x});\boldsymbol{\alpha}\right]\right\}_{\boldsymbol{\alpha}^0} = \mathbf{0}, \mathbf{x}\in\partial\Omega_x(\boldsymbol{\alpha}^0). \tag{2.207}$$

The boundary conditions represented by (2.207) are selected so as to satisfy the following requirements:

1. The boundary conditions (2.207) together with the operator Eq. (2.200) constitute a well-posed problem for the functions $\mathbf{a}^{(4)}(j_3;j_2;j_1;\mathbf{x})$.

2. Implementation in (2.198) of the boundary conditions (for the 3rd-LVSS) provided in (2.151) together with those provided in (2.207) eliminates all of the unknown values of the functions $\delta\mathbf{u}^{(4)}(j_2;j_1;\mathbf{x})$ and $\mathbf{a}^{(4)}(j_3;j_2;j_1;\mathbf{x})$ in the expression of the bilinear concomitant $\left\{P^{(4)}\left[\delta\mathbf{u}^{(4)}(j_2;j_1;\mathbf{x});\mathbf{a}^{(4)}(j_3;j_2;j_1;\mathbf{x});\boldsymbol{\alpha};\delta\boldsymbol{\alpha}\right]\right\}_{\boldsymbol{\alpha}^0}$. This bilinear concomitant may vanish after implementing the boundary conditions represented by (2.207); if it does not vanish, it will be reduced to a residual quantity which will comprise only known values of $\mathbf{a}^{(4)}(j_3;j_2;j_1;\mathbf{x})$, $\mathbf{u}^{(4)}(j_2;j_1;\mathbf{x})$, $\boldsymbol{\alpha}$, and $\delta\boldsymbol{\alpha}$, and will be denoted as $\left\{\widehat{P}^{(4)}\left[\mathbf{a}^{(4)}(j_3;j_2;j_1;\mathbf{x});\mathbf{u}^{(4)}(j_2;j_1;\mathbf{x});\boldsymbol{\alpha};\delta\boldsymbol{\alpha}\right]\right\}_{\boldsymbol{\alpha}^0}$.

The system of equations represented by (2.200) together with the boundary conditions represented by (2.207) constitute the *Fourth-Level Adjoint Sensitivity System* (4th-LASS).

The solution $\mathbf{a}^{(4)}(j_3;j_2;j_1;\mathbf{x})\triangleq\left[\mathbf{a}_1^{(4)}(j_3;j_2;j_1;\mathbf{x}),\ldots,\mathbf{a}_{16}^{(4)}(j_3;j_2;j_1;\mathbf{x})\right]^{\dagger}$, j_1, j_2, $j_3=1,\ldots,TP$, of the 4th-LASS will be called the *fourth-level adjoint sensitivity function.*

The 4th-LASS together with the results provided in (2.197) and (2.198) are employed in (2.185) to obtain the following expression for the indirect-effect term $\{\delta R^{(3)}[j_3;j_2;j_1;\mathbf{u}^{(3)}(j_1;\mathbf{x});\mathbf{a}^{(3)}(j_2;j_1;\mathbf{x});\delta\mathbf{u}^{(3)}(j_1;\mathbf{x});\delta\mathbf{a}^{(3)}(j_2;j_1;\mathbf{x})]\}_{ind}$ in terms of the fourth-level adjoint sensitivity functions $\mathbf{a}^{(4)}(j_3;j_2;j_1;\mathbf{x})$, for $j_1,j_2,j_3=1,\ldots,TP$:

$$\begin{aligned}
&\left\{\delta R^{(3)}\left[j_3;j_2;j_1;\mathbf{u}^{(3)}(j1;\mathbf{x});\mathbf{a}^{(3)}(j_2;j_1;\mathbf{x});\delta\mathbf{u}^{(3)}(j_1;\mathbf{x});\right.\right.\\
&\left.\left.\delta\mathbf{a}^{(3)}(j_2;j_1;\mathbf{x})\right]\right\}_{ind}=-\left\{\widehat{P}^{(4)}\left[\mathbf{a}^{(4)};\mathbf{u}^{(4)};\boldsymbol{\alpha};\delta\boldsymbol{\alpha}\right]\right\}_{\boldsymbol{\alpha}^0}\\
&+\left\{\left\langle\mathbf{a}^{(4)}(j_3;j_2;j_1;\mathbf{x}),\mathbf{q}^{(4)}\left[j_2;j_1;\mathbf{u}^{(4)}(j_2;j_1;\mathbf{x});\boldsymbol{\alpha};\delta\boldsymbol{\alpha}\right]\right\rangle_4\right\}_{\boldsymbol{\alpha}^0}\\
&\equiv\left\{\delta R^{(3)}\left[j_3;j_2;j_1;\mathbf{u}^{(4)}(j_2;j_1;\mathbf{x});\mathbf{a}^{(4)}(j_3;j_2;j_1;\mathbf{x});\boldsymbol{\alpha};\delta\boldsymbol{\alpha}\right]\right\}_{ind}.
\end{aligned}\tag{2.208}$$

As the identity in the last line of (2.208) indicates, the dependence of the indirect-effect term on the variations $\delta\mathbf{u}^{(3)}(j_1;\mathbf{x})$ and $\delta\mathbf{a}^{(3)}(j_2;j_1;\mathbf{x})$ has been eliminated by having replaced these functional-dependences by the dependence on the fourth-level adjoint sensitivity function $\mathbf{a}^{(4)}(j_3;j_2;j_1;\mathbf{x})$.

Representing the components of $\mathbf{q}^{(4)}[j_2;j_1;\mathbf{u}^{(4)}(j_2;j_1;\mathbf{x});\boldsymbol{\alpha};\delta\boldsymbol{\alpha}]$ in the following form:

$$\begin{aligned}
&\mathbf{q}_m^{(4)}\left(j_2;j_1;\mathbf{u}^{(4)};\boldsymbol{\alpha};\delta\boldsymbol{\alpha}\right)\\
&\triangleq\sum_{j_4=1}^{TP}\left\{\frac{\partial Q_m^{(4)}\left[j_4;j_3;j_2;j_1;\mathbf{u}^{(4)}(j_2;j_1;\mathbf{x});\boldsymbol{\alpha}\right]}{\partial\alpha_{j_4}}\right\}_{(\boldsymbol{\alpha}^0)}\delta\alpha_{j_4},\ m=1,\ldots,16,
\end{aligned}\tag{2.209}$$

and adding the expression obtained in (2.208) for the indirect-effect term with the expression of the direct-effect term given in (2.184) yields the following expression for the fourth-order total G-differential

$\left\{\delta R^{(3)}\left[j_3;j_2;j_1;\mathbf{u}^{(3)}(j_1;\mathbf{x});\mathbf{a}^{(3)}(j_2;j_1;\mathbf{x});\boldsymbol{\alpha};\delta\mathbf{u}^{(3)};\delta\mathbf{a}^{(3)};\delta\boldsymbol{\alpha}\right]\right\}_{\boldsymbol{\alpha}^0}$ of the response with respect to the model parameters, for $j_1,\, j_2,\, j_3 = 1, \ldots, TP$:

$$
\begin{aligned}
&\delta R^{(3)}\left[j_3;j_2;j_1;\mathbf{u}^{(4)}(j_2;j_1;\mathbf{x});\ \mathbf{a}^{(4)}(j_3;j_2;j_1;\mathbf{x});\boldsymbol{\alpha};\delta\boldsymbol{\alpha}\right] \\
&=\sum_{j_4=1}^{TP}\left\{R^{(4)}\left[j_4;j_3;j_2;j_1;\mathbf{u}^{(4)}(j_2;j_1;\mathbf{x});\mathbf{a}^{(4)}(j_3;j_2;j_1;\mathbf{x});\boldsymbol{\alpha}\right]\right\}_{(\boldsymbol{\alpha}^0)}\delta\alpha_{j_4},
\end{aligned}
\tag{2.210}
$$

where $R^{(4)}[j_4;j_3;j_2;j_1;\mathbf{u}^{(4)}(j_2;j_1;\mathbf{x});\mathbf{a}^{(4)}(j_3;\, j_2;j_1;\mathbf{x});\boldsymbol{\alpha}]$ denotes the *fourth-order partial sensitivity* of the response $R[\mathbf{u}^{(1)}(\mathbf{x});\boldsymbol{\alpha}]$ with respect to the model parameters, evaluated at the nominal parameter values $\boldsymbol{\alpha}^0$, and can be represented as follows:

$$
\begin{aligned}
&R^{(4)}\left[j_4;j_3;j_2;j_1;\mathbf{u}^{(4)}(j_2;j_1;\mathbf{x});\mathbf{a}^{(4)}(j_3;j_2;j_1;\mathbf{x});\boldsymbol{\alpha}\right]=\frac{\partial^4 R(\boldsymbol{\varphi};\boldsymbol{\psi};\boldsymbol{\alpha})}{\partial\alpha_{j_4}\partial\alpha_{j_3}\partial\alpha_{j_2}\partial\alpha_{j_1}} \\
&=-\left\{\frac{\partial}{\partial\alpha_{j_4}}\widehat{P}^{(4)}\left[j_3;j_2;j_1;\mathbf{u}^{(4)}(j_2;j_1;\mathbf{x});\mathbf{a}^{(4)}(j_3;j_2;j_1;\mathbf{x});\boldsymbol{\alpha}\right]\right\}_{(\boldsymbol{\alpha}^0)} \\
&+\frac{\partial}{\partial\alpha_{j_4}}\left\{\prod_{k=1}^{TI}\int_{\lambda_k(\boldsymbol{\alpha})}^{\omega_k(\boldsymbol{\alpha})}dx_k S^{(3)}\left[j_3;j_2;j_1;\mathbf{u}^{(3)}(j_1;\mathbf{x});\mathbf{a}^{(3)}(j_2;j_1;\mathbf{x});\boldsymbol{\alpha}\right]\right\}_{(\boldsymbol{\alpha}^0)} \\
&+\frac{\partial}{\partial\alpha_{j_4}}\left\{\sum_{m=1}^{16}\left\langle\mathbf{a}_m^{(4)}(j_3;j_2;j_1;\mathbf{x}),\right.\right. \\
&\left.\left.Q_m^{(4)}\left[j_4;j_3;j_2;j_1;\mathbf{u}^{(3)}(j1;\mathbf{x});\mathbf{a}^{(3)}(j_2;j_1;\mathbf{x});\boldsymbol{\alpha}\right]\right\rangle_0\right\}_{(\boldsymbol{\alpha}^0)}.
\end{aligned}
\tag{2.211}
$$

After obtaining the fourth-level adjoint sensitivity function $\mathbf{a}^{(4)}(j_3;j_2;j_1;\mathbf{x})$ by performing at most $O(TP^3)$ large-scale computations, the fourth-order response sensitivities to model parameters are obtained using the expression provided in (2.211).

2.7 The nth-Order Comprehensive Adjoint Sensitivity Analysis Methodology for Response-Coupled Forward/Adjoint Linear Systems (nth-CASAM-L)

The general framework of the "nth-Order Comprehensive Adjoint Sensitivity Analysis Methodology for Response-Coupled Forward/Adjoint Linear Systems (nth-CASAM-L)" will be established by using a "proof by induction." This proof will comprise the usual steps, as follows:

1. Establish the pattern underlying the nth-CASAM-L for $n = 1$.
2. Assume that the pattern is valid for an arbitrarily high-order, n.
3. Prove that the pattern is valid for $n + 1$.

2.7.1 *The Pattern Underlying the nth-CASAM-L for* n = *1, 2, 3, 4*

The mathematical frameworks of the 1st-, 2nd-, 3rd-, and 4th-CASAM-L are summarized in Tables 2.1, 2.2, 2.3, and 2.4. This information will be used in the next subsection to surmise the general mathematical framework underlying the *n*th-CASAM-L, for any order *n*, to enable the computations of response sensitivities of any order, *n*.

2.7.2 *The Pattern Underlying the nth-CASAM-L: Arbitrarily High-Order* n

Based on the pattern displayed in Tables 2.1, 2.2, 2.3, and 2.4, the pattern displayed in Table 2.5, below, is expected to be valid for any integer *n*.

Table 2.1 1st-CASAM-L: first-order sensitivities ($n = 1$)

Original system; Hilbert space; Inner product; Response	$\mathbf{F}^{(1)}(\boldsymbol{\alpha};\mathbf{x})\mathbf{u}^{(1)}(\mathbf{x}) = \mathbf{q}_F^{(1)}(\mathbf{x};\boldsymbol{\alpha})$; $\mathbf{b}_F^{(1)}(\mathbf{u}^{(1)};\boldsymbol{\alpha}) = \mathbf{0}$; $\mathbf{u}^{(1)}(\mathbf{x}) \triangleq [\boldsymbol{\varphi}(\mathbf{x}), \boldsymbol{\psi}(\mathbf{x})]^{\dagger}$; H_0; $\langle \boldsymbol{\varphi}(\mathbf{x}), \boldsymbol{\psi}(\mathbf{x})\rangle_0 \triangleq \prod_{i=1}^{TI} \int_{\lambda_i(\boldsymbol{\alpha})}^{\omega_i(\boldsymbol{\alpha})} \boldsymbol{\varphi}(\mathbf{x}) \cdot \boldsymbol{\psi}(\mathbf{x})\,dx$; $R[\boldsymbol{\varphi}(\mathbf{x}), \boldsymbol{\psi}(\mathbf{x}); \boldsymbol{\alpha}]$.
1st-LVSS	$\left[\mathbf{V}^{(1)}(\boldsymbol{\alpha};\mathbf{x})\right]_{2\times 2}\delta\mathbf{u}^{(1)}(\mathbf{x}) = \mathbf{q}_V^{(1)}\left(\mathbf{u}^{(1)};\boldsymbol{\alpha};\delta\boldsymbol{\alpha}\right)$; $\delta\mathbf{u}^{(1)}(\mathbf{x}) \triangleq [\delta\boldsymbol{\varphi}(\mathbf{x}), \delta\boldsymbol{\psi}(\mathbf{x})]^{\dagger}$; $[\mathbf{V}^{(1)}(\boldsymbol{\alpha};\mathbf{x})]_{2\times 2} \triangleq \mathbf{F}(\boldsymbol{\alpha};\mathbf{x})$; $\mathbf{b}_V^{(1)}\left(\mathbf{u}^{(1)};\boldsymbol{\alpha};\delta\mathbf{u}^{(1)};\delta\boldsymbol{\alpha}\right) = \delta\mathbf{b}_F\left(\mathbf{u}^{(1)};\boldsymbol{\alpha};\delta\mathbf{u}^{(1)};\delta\boldsymbol{\alpha}\right) = \mathbf{0}$.
First-level Hilbert space	H_1; $\left\langle \boldsymbol{\eta}^{(1)}(\mathbf{x}), \boldsymbol{\xi}^{(1)}(\mathbf{x})\right\rangle_1 \triangleq \sum_{i=1}^{2}\left\langle \boldsymbol{\eta}_i^{(1)}(\mathbf{x}), \boldsymbol{\xi}_i^{(1)}(\mathbf{x})\right\rangle_0$.
1st-LASS	$\left[\mathbf{A}^{(1)}(\boldsymbol{\alpha};\mathbf{x})\right]_{2\times 2}\mathbf{a}^{(1)}(\mathbf{x}) = \mathbf{s}_A^{(1)}\left[\mathbf{u}^{(1)}(\mathbf{x});\boldsymbol{\alpha}\right]$; $\left[\mathbf{A}^{(1)}(\boldsymbol{\alpha};\mathbf{x})\right]_{2\times 2} \triangleq \left[\mathbf{V}^{(1)}(\boldsymbol{\alpha};\mathbf{x})\right]_{2\times 2}^{*}$; $\mathbf{a}^{(1)}(\mathbf{x}) \triangleq \left[\mathbf{a}_1^{(1)}(\mathbf{x}), \mathbf{a}_2^{(1)}(\mathbf{x})\right]^{\dagger}$; $\mathbf{b}_A^{(1)}\left(\mathbf{u}^{(1)};\mathbf{a}^{(1)};\boldsymbol{\alpha}\right) = \mathbf{0}$.
First-order sensitivities	$R^{(1)}\left[j_1;\mathbf{u}^{(1)}(\mathbf{x});\mathbf{a}^{(1)}(\mathbf{x});\boldsymbol{\alpha}\right] \triangleq \partial R[\boldsymbol{\varphi}(\mathbf{x}), \boldsymbol{\psi}(\mathbf{x});\boldsymbol{\alpha}]/\partial \alpha_{j_1}$; $j_1 = 1, \ldots, TP$.

Table 2.2 2nd-CASAM-L: second-order sensitivities ($n = 2$)

2nd-LFSS	Original forward system + 1st-LASS: $\mathbf{F}^{(2)}(\boldsymbol{\alpha})\mathbf{u}^{(2)}(\mathbf{x}) = \mathbf{q}_F^{(2)}(\mathbf{u}^{(1)};\boldsymbol{\alpha})$; $\mathbf{u}^{(2)}(\mathbf{x}) \triangleq [\mathbf{u}^{(1)}(\mathbf{x}), \mathbf{a}^{(1)}(\mathbf{x})]^\dagger$; $\mathbf{F}^{(2)}(\boldsymbol{\alpha}) \triangleq diag\,(\mathbf{F}^{(1)}, \mathbf{A}^{(1)})$; $\mathbf{b}_F^{(2)}(\mathbf{u}^{(2)};\boldsymbol{\alpha}) \triangleq \left(\mathbf{b}_F^{(1)}, \mathbf{b}_A^{(1)}\right)^\dagger = [\mathbf{0},\mathbf{0}]^\dagger$.
2nd-LVSS	$\left[\mathbf{V}^{(2)}(\mathbf{u}^{(1)};\boldsymbol{\alpha})\right]_{2^2\times 2^2}\delta\mathbf{u}^{(2)}(\mathbf{x}) = \mathbf{q}_V^{(2)}(\mathbf{u}^{(2)};\boldsymbol{\alpha};\delta\boldsymbol{\alpha})$; $\delta\mathbf{u}^{(2)}(\mathbf{x}) \triangleq [\delta\mathbf{u}^{(1)}(\mathbf{x}), \delta\mathbf{a}^{(1)}(\mathbf{x})]^\dagger$; $\left[\mathbf{V}^{(2)}(\mathbf{u}^{(1)};\boldsymbol{\alpha})\right]_{2^2\times 2^2} \triangleq \begin{pmatrix} \mathbf{V}^{(1)} & [\mathbf{0}]_{2\times 2} \\ \mathbf{V}_{21}^{(2)} & \mathbf{V}_{22}^{(2)} \end{pmatrix}_{2^2\times 2^2}$; $\mathbf{b}_V^{(2)}\left[\mathbf{u}^{(2)};\boldsymbol{\alpha};\delta\mathbf{u}^{(2)}(\mathbf{x});\delta\boldsymbol{\alpha}\right] \triangleq \begin{pmatrix} \mathbf{b}_V^{(1)}(\mathbf{u}^{(1)};\boldsymbol{\alpha};\delta\mathbf{u}^{(1)};\delta\boldsymbol{\alpha}) \\ \delta\mathbf{b}_A^{(1)}(\mathbf{u}^{(2)};\boldsymbol{\alpha};\delta\mathbf{u}^{(2)};\delta\boldsymbol{\alpha}) \end{pmatrix} = \begin{pmatrix} \mathbf{0} \\ \mathbf{0} \end{pmatrix}$.
Second-level Hilbert space	H_2; $\left\langle \boldsymbol{\eta}^{(2)}(\mathbf{x}), \boldsymbol{\xi}^{(2)}(\mathbf{x})\right\rangle_2 = \sum_{i=1}^{4}\left\langle \boldsymbol{\eta}_i^{(2)}(\mathbf{x}), \boldsymbol{\xi}_i^{(2)}(\mathbf{x})\right\rangle_0$.
2nd-LASS	$\left[\mathbf{A}^{(2)}(\mathbf{u}^{(1)};\boldsymbol{\alpha})\right]_{2^2\times 2^2}\mathbf{a}^{(2)}(j_1;\mathbf{x}) = \mathbf{s}_A^{(2)}(j_1;\mathbf{u}^{(2)};\boldsymbol{\alpha})$; $j_1 = 1, \ldots, TP$; $\mathbf{a}^{(2)}(j_1;\mathbf{x}) \triangleq \left[\mathbf{a}_1^{(2)}(j_1;\mathbf{x}), \ldots, \mathbf{a}_k^{(2)}(j_1;\mathbf{x}), \ldots, \mathbf{a}_4^{(2)}(j_1;\mathbf{x})\right]^\dagger$; $k = 1, \ldots, 2^2$. $\left[\mathbf{A}^{(2)}(\mathbf{u}^{(1)};\boldsymbol{\alpha})\right]_{2^2\times 2^2} \triangleq \left[\mathbf{V}^{(2)}(\mathbf{u}^{(1)};\boldsymbol{\alpha})\right]^*_{2^2\times 2^2}$; $\mathbf{b}_A^{(2)}\left[\mathbf{u}^{(2)}(\mathbf{x});\mathbf{a}^{(2)}(j_1;\mathbf{x});\boldsymbol{\alpha}\right] = \mathbf{0}$.
Second-order sensitivities	$R^{(2)}\left[j_2;j_1;\mathbf{u}^{(2)};\mathbf{a}^{(2)};\boldsymbol{\alpha}\right] \triangleq \partial^2 R[\boldsymbol{\varphi}(\mathbf{x}), \boldsymbol{\psi}(\mathbf{x});\boldsymbol{\alpha}]/\partial\alpha_{j_1}\partial\alpha_{j_2}$; $j_2 = 1, \ldots, j_1$.

2.7.3 Proving That the Framework for the nth-CASAM-L also Holds for the (n + 1)th-CASAM-L

The pattern presented in Table 2.5 implies that the generic nth-order sensitivity of the response $R[\boldsymbol{\varphi}(\mathbf{x}), \boldsymbol{\psi}(\mathbf{x});\boldsymbol{\alpha}]$ to model parameters has the following form, for $j_1 = 1, \ldots, TP$; $j_2 = 1, \ldots, j_1$; ...; $j_n = 1, \ldots, j_{n-1}$:

$$\begin{aligned}
&\partial^n R[\boldsymbol{\varphi}(\mathbf{x}), \boldsymbol{\psi}(\mathbf{x});\boldsymbol{\alpha}]/\partial\alpha_{j_1}\ldots\partial\alpha_{j_n} \\
&\triangleq R^{(n)}\left[j_n, \ldots, j_1; \mathbf{u}^{(n)}(j_{n-2}, \ldots, j_1;\mathbf{x}); \mathbf{a}^{(n)}(j_{n-1}, \ldots, j_1;\mathbf{x}); \boldsymbol{\alpha}\right] \\
&\triangleq \prod_{i=1}^{TI}\int_{\lambda_i(\boldsymbol{\alpha})}^{\omega_i(\boldsymbol{\alpha})} dx_i S^{(n)}\left[j_n, \ldots, j_1; \mathbf{u}^{(n)}(j_{n-2}, \ldots, j_1;\mathbf{x}); \mathbf{a}^{(n)}(j_{n-1}, \ldots, j_1;\mathbf{x}); \boldsymbol{\alpha}\right],
\end{aligned} \tag{2.212}$$

where the function $\mathbf{u}^{(n)}(j_{n-2}, \ldots, j_1;\mathbf{x})$ is the solution of the following *nth-Level Forward Sensitivity System* (nth-LFSS):

Table 2.3 3rd-CASAM-L: third-order sensitivities ($n = 3$)

3rd-LFSS	2nd-LFSS + 2nd-LASS; $\mathbf{F}^{(3)}\mathbf{u}^{(3)}(\mathbf{x}) = \mathbf{q}_F^{(3)}$; $\mathbf{F}^{(3)} \triangleq diag\,(\mathbf{F}^{(2)}, \mathbf{A}^{(2)})$; $\mathbf{u}^{(3)}(j_1;\mathbf{x}) = [\mathbf{u}^{(2)}(\mathbf{x}), \mathbf{a}^{(2)}(j_1;\mathbf{x})]^\dagger$; $\mathbf{b}_F^{(3)}(\mathbf{u}^{(3)};\boldsymbol{\alpha}) \triangleq \left(\mathbf{b}_F^{(2)}, \mathbf{b}_A^{(2)}\right)^\dagger = [\mathbf{0},\mathbf{0}]^\dagger$.
3rd-LVSS	$\left[\mathbf{V}^{(3)}(j_1)\right]_{2^3\times 2^3}\delta\mathbf{u}^{(3)}(j_1;\mathbf{x}) = \mathbf{q}_V^{(3)}\left[j_1;\mathbf{u}^{(3)}(j_1;\mathbf{x});\boldsymbol{\alpha};\delta\boldsymbol{\alpha}\right]$; $\delta\mathbf{u}^{(3)}(j_1;\mathbf{x}) = [\delta\mathbf{u}^{(2)}(\mathbf{x}), \delta\mathbf{a}^{(2)}(j_1;\mathbf{x})]^\dagger$; $\left[\mathbf{V}^{(3)}\right]_{2^3\times 2^3} \triangleq \begin{pmatrix} \mathbf{V}^{(2)} & [\mathbf{0}]_{(4\times 4)} \\ \mathbf{V}_{21}^{(3)} & \mathbf{V}_{22}^{(3)} \end{pmatrix}$; $\mathbf{b}_V^{(3)} \triangleq \left[\mathbf{b}_V^{(2)}, \delta\mathbf{b}_A^{(2)}\right]^\dagger = [\mathbf{0},\mathbf{0}]^\dagger$.
Third-level Hilbert space	H_3; $\left\langle \boldsymbol{\eta}^{(3)}(\mathbf{x}), \boldsymbol{\xi}^{(3)}(\mathbf{x}) \right\rangle_3 \triangleq \sum_{i=1}^{2^3} \left\langle \boldsymbol{\eta}_i^{(3)}(\mathbf{x}), \boldsymbol{\xi}_i^{(3)}(\mathbf{x}) \right\rangle_0$.
3d-LASS	$\left[\mathbf{A}^{(3)}(j_1;\mathbf{u}^{(2)})\right]_{2^3\times 2^3}\mathbf{a}^{(3)}(j_2,j_1;\mathbf{x}) = \mathbf{s}_A^{(3)}(j_2,j_1;\mathbf{u}^{(3)};\boldsymbol{\alpha})$; $\left[\mathbf{A}^{(3)}(j_1;\mathbf{u}^{(2)})\right]_{2^3\times 2^3} \triangleq \left[\mathbf{V}^{(3)}(j_1;\mathbf{u}^{(2)})\right]^*_{2^3\times 2^3}$; $\mathbf{b}_A^{(3)}\left[\mathbf{a}^{(3)}(j_2;j_1;\mathbf{x});\mathbf{u}^{(3)}(j_1;\mathbf{x});\boldsymbol{\alpha}\right] = \mathbf{0}$; $\mathbf{a}^{(3)}(j_2,j_1;\mathbf{x}) \triangleq \left[\mathbf{a}_1^{(3)}(j_2,j_1;\mathbf{x}), \ldots, \mathbf{a}_8^{(3)}(j_2,j_1;\mathbf{x})\right]^\dagger$ $\triangleq \left[\ldots, \mathbf{a}_k^{(3)}(j_2,j_1;\mathbf{x}), \ldots\right]^\dagger$; $k = 1, \ldots, 2^3$; $j_1 = 1, \ldots, TP$; $j_2 = 1, \ldots, j_1$.
Third-order sensitivities	$R^{(3)}\left[j_3,j_2,j_1;\mathbf{u}^{(3)}(j_1;\mathbf{x});\mathbf{a}^{(3)}(j_2,j_1;\mathbf{x});\boldsymbol{\alpha}\right]$ $\triangleq \partial^3 R[\boldsymbol{\varphi}(\mathbf{x}), \boldsymbol{\psi}(\mathbf{x});\boldsymbol{\alpha}]/\partial\alpha_{j_1}\partial\alpha_{j_2}\partial\alpha_{j_3}; j_3 = 1, \ldots, j_2$.

Table 2.4 4th-CASAM-L: fourth-order sensitivities ($n = 4$)

4th-LFSS	3d-LFSS + 3rd-LASS: $\mathbf{F}^{(4)}\mathbf{u}^{(4)}(\mathbf{x}) = \mathbf{q}_F^{(4)}$; $\mathbf{u}^{(4)}(j_2,j_1;\mathbf{x}) = [\mathbf{u}^{(3)}(j_1;\mathbf{x}), \mathbf{a}^{(3)}(j_2,j_1;\mathbf{x})]^\dagger$; $\mathbf{F}^{(4)} \triangleq diag\,(\mathbf{F}^{(3)}, \mathbf{A}^{(3)})$; $\mathbf{b}_F^{(4)}(\mathbf{u}^{(4)};\boldsymbol{\alpha}) \triangleq \left(\mathbf{b}_F^{(3)}, \mathbf{b}_A^{(3)}\right)^\dagger = [\mathbf{0},\mathbf{0}]^\dagger$.
4th-LVSS	$\left[\mathbf{V}^{(4)}\right]_{2^4\times 2^4}\delta\mathbf{u}^{(4)}(j_2,j_1;\mathbf{x}) = \mathbf{q}_V^{(4)}\left[j_2,j_1;\mathbf{u}^{(4)}(j_2,j_1;\mathbf{x});\boldsymbol{\alpha};\delta\boldsymbol{\alpha}\right]$; $\mathbf{b}_V^{(4)} \triangleq \left[\mathbf{b}_V^{(3)}, \delta\mathbf{b}_A^{(3)}\right]^\dagger = [\mathbf{0},\mathbf{0}]^\dagger$; $\left[\mathbf{V}^{(4)}\right]_{2^4\times 2^4} \triangleq \begin{bmatrix} \mathbf{V}^{(3)} & [\mathbf{0}]_{2^3\times 2^3} \\ \mathbf{V}_{21}^{(4)} & \mathbf{V}_{22}^{(4)} \end{bmatrix}_{2^4\times 2^4}$; $\delta\mathbf{u}^{(4)}(j_2,j_1;\mathbf{x}) = [\delta\mathbf{u}^{(3)}(j_1;\mathbf{x}), \delta\mathbf{a}^{(3)}(j_2,j_1;\mathbf{x})]^\dagger$.
Fourth-level Hilbert space	H_4; $\left\langle \boldsymbol{\eta}^{(4)}(\mathbf{x}), \boldsymbol{\xi}^{(4)}(\mathbf{x}) \right\rangle_4 \triangleq \sum_{i=1}^{2^4} \left\langle \boldsymbol{\eta}_i^{(4)}(\mathbf{x}), \boldsymbol{\xi}_i^{(4)}(\mathbf{x}) \right\rangle_0$.
4th-LASS	$\left[\mathbf{A}^{(4)}\right]_{2^4\times 2^4}\mathbf{a}^{(4)}(j_3,j_2,j_1;\mathbf{x}) = \mathbf{s}_A^{(4)}\left[j_3,j_2,j_1;\mathbf{u}^{(4)}(j_2,j_1;\mathbf{x});\boldsymbol{\alpha}\right]$; $\left[\mathbf{A}^{(4)}\right]_{2^4\times 2^4} \triangleq \left[\mathbf{V}^{(4)}\right]^*_{2^4\times 2^4}$; $\mathbf{b}_A^{(4)}\left[\mathbf{a}^{(4)}(j_3,j_2,j_1;\mathbf{x});\mathbf{u}^{(4)}(j_2,j_1;\mathbf{x});\boldsymbol{\alpha}\right] = \mathbf{0}$; $\mathbf{a}^{(4)}(j_3,j_2,j_1;\mathbf{x}) \triangleq \left[\mathbf{a}_1^{(4)}(j_3,j_2,j_1;\mathbf{x}), \ldots, \mathbf{a}_{16}^{(4)}(j_3,j_2,j_1;\mathbf{x})\right]^\dagger$ $\triangleq \left[\ldots, \mathbf{a}_k^{(4)}(j_3,j_2,j_1;\mathbf{x}), \ldots\right]^\dagger$; $k = 1, \ldots, 2^4$.
Fourth-order sensitivities	$R^{(4)}\left[j_4,j_3,j_2,j_1;\mathbf{u}^{(4)}(j_2,j_1;\mathbf{x});\mathbf{a}^{(4)}(j_3,j_2,j_1;\mathbf{x});\boldsymbol{\alpha}\right]$ $\triangleq \partial^4 R[\boldsymbol{\varphi}(\mathbf{x}), \boldsymbol{\psi}(\mathbf{x});\boldsymbol{\alpha}]/\partial\alpha_{j_1}\partial\alpha_{j_2}\partial\alpha_{j_3}\partial\alpha_{j_4}; j_4 = 1, \ldots, j_3$.

Table 2.5 nth-CASAM-L: nth-order sensitivities

nth-LFSS	$(n-1)$th-LFSS + $(n-1)$th-LASS: $\mathbf{F}^{(n)}\mathbf{u}^{(n)}(\mathbf{x}) = \mathbf{q}_F^{(n)}$; $\mathbf{F}^{(n)} \triangleq diag\,(\mathbf{F}^{(n-1)}, \mathbf{A}^{(n-1)})$; $\mathbf{b}_F^{(n)}\left(\mathbf{u}^{(n)};\boldsymbol{\alpha}\right) \triangleq \left(\mathbf{b}_F^{(n-1)}, \mathbf{b}_A^{(n-1)}\right)^\dagger = [\mathbf{0},\mathbf{0}]^\dagger$; $\mathbf{u}^{(n)}(j_{n-2},\ldots,j_1;\mathbf{x}) = [\mathbf{u}^{(n-1)}(j_{n-3},\ldots,j_1;\mathbf{x}), \mathbf{a}^{(n-1)}(j_{n-2},\ldots,j_1;\mathbf{x})]^\dagger$.
nth-LVSS	$\left[\mathbf{V}^{(n)}\right]_{2^n\times 2^n}\delta\mathbf{u}^{(n)}(j_{n-2},\ldots,j_1;\mathbf{x}) = \mathbf{q}_V^{(n)}\left[j_{n-2},\ldots,j_1;\mathbf{u}^{(n)}(j_{n-2},\ldots,j_1;\mathbf{x});\boldsymbol{\alpha};\delta\boldsymbol{\alpha}\right]$; $\delta\mathbf{u}^{(n)}(j_{n-2},\ldots,j_1;\mathbf{x}) = \left[\delta\mathbf{u}^{(n-1)}(j_{n-3},\ldots,j_1;\mathbf{x}), \delta\mathbf{a}^{(n-1)}(j_{n-2},\ldots,j_1;\mathbf{x})\right]^\dagger$; $\mathbf{b}_V^{(n)} \triangleq \left[\mathbf{b}_V^{(n-1)}, \delta\mathbf{b}_A^{(n-1)}\right]^\dagger = [\mathbf{0},\mathbf{0}]^\dagger$; $\left[\mathbf{V}^{(n)}\right]_{2^n\times 2^n} \triangleq \begin{pmatrix} \mathbf{V}^{(n-1)} & [\mathbf{0}]_{(2^{n-1}\times 2^{n-1})} \\ \mathbf{V}_{21}^{(n)} & \mathbf{V}_{22}^{(n)} \end{pmatrix}_{2^n\times 2^n}$.
nth-level Hilbert space	H_n; $\left\langle \boldsymbol{\eta}^{(n)}(\mathbf{x}), \boldsymbol{\xi}^{(n)}(\mathbf{x}) \right\rangle_n \triangleq \sum_{i=1}^{2^n} \left\langle \boldsymbol{\eta}_i^{(n)}(\mathbf{x}), \boldsymbol{\xi}_i^{(n)}(\mathbf{x}) \right\rangle_0$.
nth-LASS	$\left[\mathbf{A}^{(n)}\right]_{2^n\times 2^n}\mathbf{a}^{(n)}(j_{n-1},\ldots,j_1;\mathbf{x}) = \mathbf{s}_A^{(n)}\left[j_{n-1},\ldots,j_1;\mathbf{u}^{(n)}(j_{n-2},\ldots,j_1;\mathbf{x});\boldsymbol{\alpha}\right]$; $\left[\mathbf{A}^{(n)}\right]_{2^n\times 2^n} \triangleq \left[\mathbf{V}^{(n)}\right]^*_{2^n\times 2^n}$; $\mathbf{b}_A^{(n)}\left[\mathbf{a}^{(n)}(j_{n-1},\ldots,j_1;\mathbf{x});\mathbf{u}^{(n)}(j_{n-2},\ldots,j_1;\mathbf{x});\boldsymbol{\alpha}\right] = \mathbf{0}$; $\mathbf{a}^{(n)}(j_{n-1},\ldots,j_1;\mathbf{x}) \triangleq \left[\ldots,\mathbf{a}_k^{(n)}(j_{n-1},\ldots,j_1;\mathbf{x}),\ldots\right]$; $k=1,\ldots,2^n$.
nth-order sensitivities	$R^{(n)}\left[j_n,\ldots,j_1;\mathbf{u}^{(n)}(j_{n-2},\ldots,j_1;\mathbf{x});\mathbf{a}^{(n)}(j_{n-1},\ldots,j_1;\mathbf{x});\boldsymbol{\alpha}\right] \triangleq \partial^n R[\boldsymbol{\varphi}(\mathbf{x}),\boldsymbol{\psi}(\mathbf{x});\boldsymbol{\alpha}]/\partial\alpha_{j_1}\ldots\partial\alpha_{j_n}; j_1 = 1,\ldots,TP$; $j_2 = 1,\ldots,j_1;\ldots;j_n = 1,\ldots,j_{n-1}$; Distinct Sensitivities: $TP(TP+1)(TP+2)\ldots(TP+n-1)/n!$

$$\mathbf{F}^{(n)}\mathbf{u}^{(n)}(\mathbf{x}) = \mathbf{q}_F^{(n)};\quad \mathbf{b}_F^{(n)}\left(\mathbf{u}^{(n)};\boldsymbol{\alpha}\right) = \mathbf{0}, \tag{2.213}$$

and where the *nth-level adjoint sensitivity function* $\mathbf{a}^{(n)}(j_{n-1};\ldots;j_1;\mathbf{x})$ is the solution of the following *nth-Level Adjoint Sensitivity System* (nth-LASS):

$$\left[\mathbf{A}^{(n)}\right]_{2^n\times 2^n}\mathbf{a}^{(n)}(j_{n-1},\ldots,j_1;\mathbf{x}) = \mathbf{s}_A^{(n)}\left[j_{n-1},\ldots,j_1;\mathbf{u}^{(n)}(j_{n-2},\ldots,j_1;\mathbf{x});\boldsymbol{\alpha}\right],$$
$$\mathbf{b}_A^{(n)}\left[\mathbf{a}^{(n)}(j_{n-1},\ldots,j_1;\mathbf{x});\mathbf{u}^{(n)}(j_{n-2},\ldots,j_1;\mathbf{x});\boldsymbol{\alpha}\right] = \mathbf{0}. \tag{2.214}$$

Concatenating (2.213) and (2.214) yields the following *(n + 1)th-Level Forward Sensitivity System* for $\mathbf{u}^{(n+1)}(j_{n-1},\ldots,j_1;\mathbf{x}) = [\mathbf{u}^{(n)}(j_{n-2},\ldots,j_1;\mathbf{x}); \mathbf{a}^{(n)}(j_{n-1},\ldots, j_1;\mathbf{x})]^\dagger$:

$$\begin{aligned}&\left\{\mathbf{F}^{(n+1)}(\boldsymbol{\alpha};\mathbf{x})\mathbf{u}^{(n+1)}(j_{n-1},\ldots,j_1;\mathbf{x})\right\}_{\boldsymbol{\alpha}^0}\\&=\left\{\mathbf{q}_F^{(n+1)}\left(j_{n-1},\ldots,j_1;\mathbf{u}^{(n)};\mathbf{x};\boldsymbol{\alpha}\right)\right\}_{\boldsymbol{\alpha}^0},\ \mathbf{x}\in\Omega_x\left(\boldsymbol{\alpha}^0\right),\end{aligned}\tag{2.215}$$

$$\left\{\mathbf{b}_F^{(n+1)}\left[\mathbf{u}^{(n+1)}(j_{n-1},\ldots,j_1;\mathbf{x});\boldsymbol{\alpha}\right]\right\}_{\boldsymbol{\alpha}^0}=\mathbf{0},\ \ \mathbf{x}\in\partial\Omega_x\left(\boldsymbol{\alpha}^0\right),\tag{2.216}$$

where:

$$\begin{aligned}&\mathbf{F}^{(n+1)}(\boldsymbol{\alpha};\mathbf{x})\triangleq\begin{pmatrix}\left[\mathbf{F}^{(n)}\right]_{2^n\times 2^n} & [\mathbf{0}]_{2^n\times 2^n}\\ [\mathbf{0}]_{2^n\times 2^n} & \left[\mathbf{A}^{(n)}\right]_{2^n\times 2^n}\end{pmatrix};\\&\mathbf{u}^{(n+1)}(j_{n-1},\ldots,j_1;\mathbf{x})\triangleq\begin{pmatrix}\mathbf{u}^{(n)}(j_{n-2},\ldots,j_1)\\ \mathbf{a}^{(n)}(j_{n-1},\ldots,j_1)\end{pmatrix};\\&\mathbf{q}_F^{(n+1)}(j_{n-1},\ldots,j_1;\mathbf{x};\boldsymbol{\alpha})\triangleq\begin{pmatrix}\mathbf{q}_F^{(n)}\\ \mathbf{s}_A^{(n)}\end{pmatrix};\\&\mathbf{b}_F^{(n+1)}\left[\mathbf{u}^{(n+1)}(j_{n-1},\ldots,j_1;\mathbf{x});\boldsymbol{\alpha}\right]\triangleq\begin{pmatrix}\mathbf{b}_F^{(n)}\\ \mathbf{b}_A^{(n)}\end{pmatrix}.\end{aligned}\tag{2.217}$$

In terms of $\mathbf{u}^{(n+1)}(j_{n-1},\ldots,j_1;\mathbf{x})\triangleq[\mathbf{u}^{(n)}(j_{n-2},\ldots,j_1;\mathbf{x});\mathbf{a}^{(n)}(j_{n-1},\ldots,j_1;\mathbf{x})]^\dagger$, each *nth*-order sensitivity can be written as follows:

$$\begin{aligned}\frac{\partial^n R(\boldsymbol{\varphi};\boldsymbol{\psi};\boldsymbol{\alpha})}{\partial\alpha_{j_1}\ldots\partial\alpha_{j_n}}&\equiv R^{(n)}\left(j_n,\ldots,j_1;\mathbf{u}^{(n+1)};\boldsymbol{\alpha}\right)\\&\equiv\prod_{i=1}^{TI}\int_{\lambda_i(\boldsymbol{\alpha})}^{\omega_i(\boldsymbol{\alpha})}dx_i S^{(n)}\left(j_n,\ldots,j_1;\mathbf{u}^{(n+1)};\boldsymbol{\alpha}\right).\end{aligned}\tag{2.218}$$

The total G-differential of $R^{(n)}(j_n,\ldots,j_1;\mathbf{u}^{(n+1)};\boldsymbol{\alpha})$ is obtained, by definition, as follows:

$$\begin{aligned}&\left\{\delta R^{(n)}\left(j_n,\ldots,j_1;\mathbf{u}^{(n+1)};\boldsymbol{\alpha};\delta\mathbf{u}^{(n+1)};\delta\boldsymbol{\alpha}\right)\right\}_{\boldsymbol{\alpha}^0}\\&\triangleq\left\{\frac{d}{d\varepsilon}\left[\prod_{i=1}^{TI}\int_{\lambda_i(\boldsymbol{\alpha})}^{\omega_i(\boldsymbol{\alpha})}dx_i S^{(n)}\left(j_n,\ldots,j_1;\mathbf{u}^{(n+1)}+\varepsilon\delta\mathbf{u}^{(n+1)};\boldsymbol{\alpha}+\varepsilon\delta\boldsymbol{\alpha}\right)\right]_{\boldsymbol{\alpha}^0}\right\}_{\varepsilon=0}\\&=\left\{\delta R^{(n)}\left(j_n,\ldots,j_1;\mathbf{u}^{(n+1)};\boldsymbol{\alpha};\delta\boldsymbol{\alpha}\right)_{\boldsymbol{\alpha}^0}\right\}_{dir}\\&+\left\{\delta R^{(n)}\left(j_n,\ldots,j_1;\mathbf{u}^{(n+1)};\boldsymbol{\alpha};\delta\mathbf{u}^{(n+1)}\right)_{\boldsymbol{\alpha}^0}\right\}_{ind},\end{aligned}\tag{2.219}$$

where the direct effect term $\left\{\delta R^{(n)}\left(j_n, \ldots, j_1; \mathbf{u}^{(n+1)}; \boldsymbol{\alpha}; \delta\boldsymbol{\alpha}\right)_{\boldsymbol{\alpha}^0}\right\}_{dir}$ depends directly on the vector of parameter variations $\delta\boldsymbol{\alpha}$, and is defined as follows, for $j_1 = 1, \ldots, TP; j_2 = 1, \ldots, j_1; \ldots; j_n = 1, \ldots, j_{n-1}$:

$$\left\{\delta R^{(n)}\left(j_n, \ldots, j_1; \mathbf{u}^{(n+1)}; \boldsymbol{\alpha}; \delta\boldsymbol{\alpha}\right)_{\boldsymbol{\alpha}^0}\right\}_{dir} \triangleq \left\{\frac{\partial}{\partial \boldsymbol{\alpha}} \prod_{i=1}^{TI} \int_{\lambda_i(\boldsymbol{\alpha})}^{\omega_i(\boldsymbol{\alpha})} dx_i S^{(n)}\left(j_n, \ldots, j_1; \mathbf{u}^{(n+1)}; \boldsymbol{\alpha}\right)\right\}_{\boldsymbol{\alpha}^0} \delta\boldsymbol{\alpha}. \tag{2.220}$$

The indirect-effect term $\left\{\delta R^{(n)}\left(j_n, \ldots, j_1; \mathbf{u}^{(n+1)}; \boldsymbol{\alpha}; \delta\mathbf{u}^{(n+1)}\right)_{\boldsymbol{\alpha}^0}\right\}_{ind}$ in (2.219) is defined as follows:

$$\left\{\delta R^{(n)}\left(j_n, \ldots, j_1; \mathbf{u}^{(n+1)}; \boldsymbol{\alpha}; \delta\mathbf{u}^{(n+1)}\right)_{\boldsymbol{\alpha}^0}\right\}_{ind} \triangleq \left\{\prod_{i=1}^{TI} \int_{\lambda_i(\boldsymbol{\alpha})}^{\omega_i(\boldsymbol{\alpha})} dx_i \frac{\partial S^{(n)}\left(j_n, \ldots, j_1; \mathbf{u}^{(n+1)}; \boldsymbol{\alpha}\right)}{\partial \mathbf{u}^{(n+1)}\left(j_{n-1}, \ldots, j_1; \mathbf{x}\right)}\right\}_{\boldsymbol{\alpha}^0} \delta\mathbf{u}^{(n+1)}, \tag{2.221}$$

where

$$\begin{aligned} &\frac{\partial []}{\partial \mathbf{u}^{(n+1)}\left(j_{n-1}, \ldots, j_1; \mathbf{x}\right)} \delta\mathbf{u}^{(n+1)}\left(j_{n-1}, \ldots, j_1; \mathbf{x}\right) \equiv \frac{\partial []}{\partial \boldsymbol{\varphi}} \delta\boldsymbol{\varphi}(\mathbf{x}) \\ &+ \frac{\partial []}{\partial \boldsymbol{\psi}} \delta\boldsymbol{\psi}(\mathbf{x}) + \sum_{k=1}^{2} \frac{\partial []}{\partial \mathbf{a}_k^{(1)}} \delta\mathbf{a}_k^{(1)}(\mathbf{x}) + \sum_{k=1}^{2^2} \frac{\partial []}{\partial \mathbf{a}_k^{(2)}(j_1)} \delta\mathbf{a}_k^{(2)}(j_1; \mathbf{x}) \\ &+ \cdots + \sum_{k=1}^{2^n} \frac{\partial []}{\partial \mathbf{a}_k^{(n)}\left(j_{n-1}, \ldots, j_1\right)} \delta\mathbf{a}_k^{(n)}\left(j_{n-1}, \ldots, j_1; \mathbf{x}\right). \end{aligned} \tag{2.222}$$

The vector $\delta\mathbf{u}^{(n+1)}(j_{n-1}, \ldots, j_1; \mathbf{x}) \triangleq [\delta\mathbf{u}^{(n)}(j_{n-2}, \ldots, j_1; \mathbf{x}); \delta\mathbf{a}^{(n)}(j_{n-1}, \ldots, j_1; \mathbf{x})]^\dagger$ is the solution of the G-differentiated $(n + 1)$th-LFSS comprising (2.215) and (2.216), which is equivalent to concatenating the nth-LVSS to the system of equations obtained by G-differentiating the nth-LASS while being subject to the corresponding boundary conditions. The nth-LVSS is assumed to be correct as given in Table 2.5, namely:

$$\left[\mathbf{V}^{(n)}\right]_{2^n \times 2^n} \delta\mathbf{u}^{(n)}\left(j_{n-2}, \ldots, j_1; \mathbf{x}\right) = \mathbf{q}_V^{(n)}\left[j_{n-2}, \ldots, j_1; \mathbf{u}^{(n)}\left(j_{n-2}, \ldots, j_1; \mathbf{x}\right); \boldsymbol{\alpha}; \delta\boldsymbol{\alpha}\right], \tag{2.223}$$

$$\mathbf{b}_V^{(n)} \triangleq \left[\mathbf{b}_V^{(n-1)}, \delta\mathbf{b}_A^{(n-1)}\right]^\dagger = [\mathbf{0}, \mathbf{0}]^\dagger. \tag{2.224}$$

The G-differential of the nth-LASS is obtained from (2.214) in the following form:

$$\frac{\partial\left\{\left[\mathbf{A}^{(n)}\right]_{2^n\times 2^n}\mathbf{a}^{(n)}(j_{n-1},\ldots,j_1;\mathbf{x})-\mathbf{s}_A^{(n)}\left[j_{n-1}\ldots j_1;\mathbf{u}^{(n)}(j_{n-2},\ldots,j_1;\mathbf{x});\boldsymbol{\alpha}\right]\right\}}{\partial\mathbf{u}^{(n)}}\delta\mathbf{u}^{(n)}$$
$$+\left[\mathbf{A}^{(n)}\right]_{2^n\times 2^n}\delta\mathbf{a}^{(n)}(j_{n-1},\ldots,j_1;\mathbf{x})=\mathbf{p}^{(n)}\left(j_{n-1},\ldots,j_1;\mathbf{u}^{(n+1)};\boldsymbol{\alpha};\delta\boldsymbol{\alpha}\right), \tag{2.225}$$

$$\delta\mathbf{b}_A^{(n)}\left[\mathbf{u}^{(n+1)}(j_{n-1},\ldots,j_1;\mathbf{x});\boldsymbol{\alpha};\delta\mathbf{u}^{(n+1)}(j_{n-1},\ldots,j_1;\mathbf{x});\delta\boldsymbol{\alpha}\right]=\mathbf{0}, \tag{2.226}$$

where

$$\mathbf{p}^{(n)}\left(j_{n-1},\ldots,j_1;\mathbf{u}^{(n+1)};\boldsymbol{\alpha};\delta\boldsymbol{\alpha}\right)$$
$$\triangleq\frac{\partial\left\{\mathbf{s}_A^{(n)}\left[j_{n-1},\ldots,j_1;\mathbf{u}^{(n)}(j_{n-2},\ldots,j_1;\mathbf{x});\boldsymbol{\alpha}\right]-\mathbf{A}^{(n)}\mathbf{a}^{(n)}\right\}}{\partial\boldsymbol{\alpha}}\delta\boldsymbol{\alpha}. \tag{2.227}$$

Concatenating (2.223)–(2.226) yields the following $(n + 1)$th-Level Variational Sensitivity System [$(n + 1)$th-LVSS] for the function $\delta\mathbf{u}^{(n+1)}(j_{n-1},\ldots,j_1;\mathbf{x})$:

$$\left[\mathbf{V}^{(n+1)}\right]_{2^{n+1}\times 2^{n+1}}\delta\mathbf{u}^{(n+1)}(j_{n-1},\ldots,j_1;\mathbf{x})$$
$$=\mathbf{q}_V^{(n+1)}\left[j_{n-1},\ldots,j_1;\mathbf{u}^{(n+1)}(j_{n-1},\ldots,j_1;\mathbf{x});\boldsymbol{\alpha};\delta\boldsymbol{\alpha}\right], \tag{2.228}$$

$$\mathbf{b}_V^{(n+1)}\left(\delta\mathbf{u}^{(n+1)};\delta\boldsymbol{\alpha}\right)\triangleq\begin{pmatrix}\mathbf{b}_V^{(n)}\left(\delta\mathbf{u}^{(n)};\delta\boldsymbol{\alpha}\right)\\ \delta\mathbf{b}_A^{(n)}\left(\delta\mathbf{a}^{(n)};\delta\mathbf{u}^{(n)};\delta\boldsymbol{\alpha}\right)\end{pmatrix}=\begin{pmatrix}[\mathbf{0}]\\ [\mathbf{0}]\end{pmatrix},\mathbf{x}\in\partial\Omega_x\left(\boldsymbol{\alpha}^0\right), \tag{2.229}$$

where:

$$\left[\mathbf{V}^{(n+1)}\right]_{2^{n+1}\times 2^{n+1}}\triangleq\begin{pmatrix}\left[\mathbf{V}^{(n)}\right]_{2^n\times 2^n} & [\mathbf{0}]_{2^n\times 2^n}\\ \left[\mathbf{V}_{21}^{(n+1)}\right]_{2^n\times 2^n} & \left[\mathbf{V}_{22}^{(n+1)}\right]_{2^n\times 2^n}\end{pmatrix}_{2^{n+1}\times 2^{n+1}};\mathbf{V}_{22}^{(n+1)}\triangleq\mathbf{A}^{(n)};$$
$$\mathbf{V}_{21}^{(n+1)}\triangleq\frac{\partial\left\{\left[\mathbf{A}^{(n)}\right]_{2^n\times 2^n}\mathbf{a}^{(n)}(j_{n-1},\ldots,j_1;\mathbf{x})\right\}}{\partial\mathbf{u}^{(n)}}$$
$$-\frac{\partial\left\{\mathbf{s}_A^{(n)}\left[j_{n-1},\ldots,j_1;\mathbf{u}^{(n)}(j_{n-2},\ldots,j_1;\mathbf{x});\boldsymbol{\alpha}\right]\right\}}{\partial\mathbf{u}^{(n)}}; \tag{2.230}$$

$$\mathbf{q}_V^{(n+1)}\left[j_{n-1},\ldots,j_1;\mathbf{u}^{(n+1)};\boldsymbol{\alpha};\delta\boldsymbol{\alpha}\right]\triangleq\begin{pmatrix}\mathbf{q}_V^{(n)}\left[j_{n-2},\ldots,j_1;\mathbf{u}^{(n)};\boldsymbol{\alpha};\delta\boldsymbol{\alpha}\right]\\ \mathbf{p}^{(n)}\left[j_{n-1},\ldots,j_1;\mathbf{u}^{(n+1)};\boldsymbol{\alpha};\delta\boldsymbol{\alpha}\right]\end{pmatrix} \tag{2.231}$$
$$\equiv\left[\cdots,\mathbf{q}_k^{(n+1)}\left(j_{n-1},\ldots,j_1;\mathbf{u}^{(n+1)};\boldsymbol{\alpha};\delta\boldsymbol{\alpha}\right),\cdots\right]^{\dagger},\ k=1,\ldots,2^{n+1}.$$

In principle, the $(n + 1)$th-LVSS, comprising (2.228) and (2.229), could be solved to determine the variational vector $\delta\mathbf{u}^{(n+1)}(j_{n-1},\ldots,j_1;\mathbf{x})$ which could be used, in turn, to determine the indirect-effect term $\left\{\delta R^{(n)}\left(j_n,\ldots,j_1;\mathbf{u}^{(n+1)};\boldsymbol{\alpha};\delta\mathbf{u}^{(n+1)}\right)_{\boldsymbol{\alpha}^0}\right\}_{ind}$. On the other hand, the need for solving the $(n + 1)$th-LVSS can be avoided by expressing the indirect-effect term $\left\{\delta R^{(n)}\left(j_n,\ldots,j_1;\mathbf{u}^{(n+1)};\boldsymbol{\alpha};\delta\mathbf{u}^{(n+1)}\right)_{\boldsymbol{\alpha}^0}\right\}_{ind}$ in an alternative way, which does not involve the function $\delta\mathbf{u}^{(n+1)}(j_{n-1},\ldots,j_1;\mathbf{x})$. This alternative expression will be obtained by using the solution of a $(n + 1)$th-Level Adjoint Sensitivity Systems [$(n + 1)$th-LASS], which will be constructed below by implementing the same sequence of logical steps as were followed for constructing the 1st-LASS (and all of the other lower-level adjoint sensitivity systems).

The $(n + 1)$th-LASS is constructed in a Hilbert space, denoted as H_{n+1}, comprising block-vector elements of the form $\boldsymbol{\eta}^{(n+1)}(\mathbf{x}) \in \mathrm{H}_{n+1}$, $\boldsymbol{\eta}^{(n+1)}(\mathbf{x}) \triangleq \left[\ldots,\boldsymbol{\eta}_k^{(n+1)}(\mathbf{x}),\ldots\right]^{\dagger}; k = 1,\ldots,2^{n+1}$ with elements having the following structure: $\boldsymbol{\eta}_k^{(n+1)}(\mathbf{x}) \triangleq \left[\eta_{k,1}^{(n+1)}(\mathbf{x}),\ldots,\eta_{k,TD}^{(n+1)}(\mathbf{x})\right]^{\dagger}$. The inner product between two elements, $\boldsymbol{\eta}^{(n+1)}(\mathbf{x}) \in \mathrm{H}_{n+1}$ and $\boldsymbol{\xi}^{(n+1)}(\mathbf{x}) \in \mathrm{H}_{n+1}$, of the Hilbert space H_{n+1}, will be denoted as $\langle\boldsymbol{\eta}^{(n+1)}(\mathbf{x}),\boldsymbol{\xi}^{(n+1)}(\mathbf{x})\rangle_{n+1}$ and is defined as follows:

$$\left\langle\boldsymbol{\eta}^{(n+1)}(\mathbf{x}),\boldsymbol{\xi}^{(n+1)}(\mathbf{x})\right\rangle_{n+1} \triangleq \sum_{k=1}^{2^{n+1}}\left\langle\boldsymbol{\eta}_k^{(n+1)}(\mathbf{x}),\boldsymbol{\xi}_k^{(n+1)}(\mathbf{x})\right\rangle_0. \tag{2.232}$$

Using the definition provided in (2.232), form the inner product in H_{n+1} of (2.223) with a yet undefined vector-valued function $\mathbf{a}^{(n+1)}(j_n,\ldots,j_1;\mathbf{x}) \triangleq \left[\ldots,\mathbf{a}_k^{(n+1)}(j_n,\ldots,j_1;\mathbf{x}),\ldots\right]^{\dagger} \in \mathrm{H}_{n+1}; k = 1,\ldots,2^{n+1}, j_1 = 1,\ldots,TP; j_2 = 1,\ldots,j_1; j_{n+1} = 1,\ldots,j_n$, to obtain the following relation:

$$\begin{aligned}
&\left\{\left\langle\mathbf{a}^{(n+1)}(j_n,\ldots,j_1;\mathbf{x}),\left[\mathbf{V}^{(n+1)}\right]\delta\mathbf{u}^{(n+1)}\right\rangle_{n+1}\right\}_{\boldsymbol{\alpha}^0} = \left\{\left\langle\mathbf{a}^{(n+1)}(j_n,\ldots,j_1;\mathbf{x}),\right.\right.\\
&\qquad\left.\left.\mathbf{q}_V^{(n+1)}\left[j_{n-1}\ldots j_1;\mathbf{u}^{(n+1)}(j_{n-1},\ldots,j_1;\mathbf{x});\boldsymbol{\alpha};\delta\boldsymbol{\alpha}\right]\right\rangle_{n+1}\right\}_{\boldsymbol{\alpha}^0}\\
&= \left\{\left\langle\delta\mathbf{u}^{(n+1)},\mathbf{A}^{(n+1)}\mathbf{a}^{(n+1)}(j_n,\ldots,j_1;\mathbf{x})\right\rangle_{n+1}\right\}_{\boldsymbol{\alpha}^0}\\
&+\left\{P^{(n+1)}\left[\delta\mathbf{u}^{(n+1)};\mathbf{a}^{(n+1)};\boldsymbol{\alpha};\delta\boldsymbol{\alpha}\right]\right\}_{\boldsymbol{\alpha}^0},
\end{aligned} \tag{2.233}$$

where $\left\{P^{(n+1)}\left[\delta\mathbf{u}^{(n+1)};\mathbf{a}^{(n+1)};\boldsymbol{\alpha};\delta\boldsymbol{\alpha}\right]\right\}_{\boldsymbol{\alpha}^0}$ denotes the bilinear concomitant defined on the phase-space boundary $\mathbf{x} \in \partial\Omega_x(\boldsymbol{\alpha}^0)$, evaluated at the nominal values of the model parameter and respective functions, and where:

$$\left[\mathbf{A}^{(n+1)}\right]_{2^{n+1}\times 2^{n+1}} \triangleq \left[\mathbf{V}^{(n+1)}\right]^*_{2^{n+1}\times 2^{n+1}} = \begin{pmatrix} \left[\mathbf{V}^{(n)}\right]_{2^n\times 2^n} & [\mathbf{0}]_{2^n\times 2^n} \\ \left[\mathbf{V}_{21}^{(n+1)}\right]_{2^n\times 2^n} & \left[\mathbf{V}_{22}^{(n+1)}\right]_{2^n\times 2^n} \end{pmatrix}^*$$
$$= \begin{pmatrix} \mathbf{A}^{(n)} & \left[\mathbf{V}_{21}^{(n+1)}\right]^* \\ [\mathbf{0}]_{2^n\times 2^n} & \left[\mathbf{V}_{22}^{(n+1)}\right]^* \end{pmatrix}. \tag{2.234}$$

The first term on right side of the second equality in (2.233) is now required to represent the indirect-effect term $\left\{\delta R^{(n)}\left(j_n, \ldots, j_1; \mathbf{u}^{(n+1)}; \boldsymbol{\alpha}; \delta\mathbf{u}^{(n+1)}\right)_{\boldsymbol{\alpha}^0}\right\}_{ind}$, which is achieved by requiring that the $(n + 1)$th-level adjoint sensitivity function $\mathbf{a}^{(n+1)}(j_n, \ldots, j_1; \mathbf{x}) \triangleq \left[\ldots, \mathbf{a}_k^{(n+1)}(j_n, \ldots, j_1; \mathbf{x}), \ldots\right]^\dagger \in \mathrm{H}_{n+1}; k = 1, \ldots, 2^{n+1}$, be the solution of the following $(n + 1)$th-LASS, for $j_1 = 1, \ldots, TP$; $j_2 = 1, \ldots, j_1; \ldots; j_n = 1, \ldots, j_{n-1}$:

$$\left[\mathbf{A}^{(n+1)}\right]_{2^{n+1}\times 2^{n+1}} \mathbf{a}^{(n+1)}(j_n, \ldots, j_1; \mathbf{x}) = \mathbf{s}_A^{(n+1)}(j_n, \ldots, j_1; \boldsymbol{\alpha}), \tag{2.235}$$

$$\left\{\mathbf{b}_A^{(n+1)}\left[\mathbf{a}^{(n+1)}(j_n, \ldots, j_1; \mathbf{x}); \mathbf{u}^{(n+1)}(j_{n-1}, \ldots, j_1; \mathbf{x}); \boldsymbol{\alpha}\right]\right\}_{\boldsymbol{\alpha}^0} = \mathbf{0}, \mathbf{x} \in \partial\Omega_x\left(\boldsymbol{\alpha}^0\right), \tag{2.236}$$

where the vector $\mathbf{s}_A^{(n+1)}(j_n, \ldots, j_1; \boldsymbol{\alpha}) \triangleq \left[\ldots, \mathbf{s}_k^{(n+1)}(j_n, \ldots, j_1; \boldsymbol{\alpha}), \ldots\right]^\dagger$, $k = 1, \ldots, 2^{n+1}$, comprises 2^{n+1} components defined as follows, for each $j_1 = 1, \ldots, TP$; $j_2 = 1, \ldots, j_1; \ldots; j_n = 1, \ldots, j_{n-1}$:

$$\mathbf{s}_1^{(n+1)}(j_n, \ldots, j_1; \boldsymbol{\alpha}) \triangleq \frac{\partial S^{(n)}\left(j_n, \ldots, j_1; \mathbf{u}^{(n+1)}; \boldsymbol{\alpha}\right)}{\partial \boldsymbol{\varphi}}; \tag{2.237}$$

$$\mathbf{s}_2^{(n+1)}(j_n, \ldots, j_1; \boldsymbol{\alpha}) \triangleq \frac{\partial S^{(n)}\left(j_n, \ldots, j_1; \mathbf{u}^{(n+1)}; \boldsymbol{\alpha}\right)}{\partial \boldsymbol{\psi}}; \tag{2.238}$$

$$\mathbf{s}_{2+k}^{(n+1)}(j_n, \ldots, j_1; \boldsymbol{\alpha}) \triangleq \frac{\partial S^{(n)}\left(j_n, \ldots, j_1; \mathbf{u}^{(n+1)}; \boldsymbol{\alpha}\right)}{\partial \mathbf{a}_k^{(1)}}; \quad k = 1, 2; \tag{2.239}$$

$$\mathbf{s}_{2^2+k}^{(n+1)}(j_n, \ldots, j_1; \boldsymbol{\alpha}) \triangleq \frac{\partial S^{(n)}\left(j_n, \ldots, j_1; \mathbf{u}^{(n+1)}; \boldsymbol{\alpha}\right)}{\partial \mathbf{a}_k^{(2)}(j_1)}; \quad k = 1, \ldots, 2^2; \tag{2.240}$$

$$\mathbf{s}_{2^3+k}^{(n+1)}(j_n,\ldots,j_1;\boldsymbol{\alpha})\triangleq\frac{\partial S^{(n)}\left(j_n,\ldots,j_1;\mathbf{u}^{(n+1)};\boldsymbol{\alpha}\right)}{\partial\mathbf{a}_k^{(3)}(j_2;j_1)};\ k=1,\ldots,2^3; \tag{2.241}$$

$$\mathbf{s}_{2^{n-1}+k}^{(n+1)}(j_n,\ldots,j_1;\boldsymbol{\alpha})\triangleq\frac{\partial S^{(n)}\left(j_n,\ldots,j_1;\mathbf{u}^{(n+1)};\boldsymbol{\alpha}\right)}{\partial\mathbf{a}_k^{(n-1)}(j_{n-2},\ldots,j_1)};\ k=1,\ldots,2^{n-1}; \tag{2.242}$$

$$\mathbf{s}_{2^n+k}^{(n+1)}(j_n,\ldots,j_1;\boldsymbol{\alpha})\triangleq\frac{\partial S^{(n)}\left(j_n,\ldots,j_1;\mathbf{u}^{(n+1)};\boldsymbol{\alpha}\right)}{\partial\mathbf{a}_k^{(n)}(j_{n-1},\ldots,j_1)};\ k=1,\ldots,2^n. \tag{2.243}$$

The $(n+1)$th-level adjoint boundary conditions represented by (2.236) are selected so as to eliminate, in conjunction with the boundary conditions represented by (2.229), all of the unknown values of the functions $\delta\mathbf{u}^{(n+1)}(j_{n-1},\ldots,j_1;\mathbf{x})$ in the expression of the bilinear concomitant $\left\{P^{(n+1)}\left[\delta\mathbf{u}^{(n+1)};\mathbf{a}^{(n+1)};\boldsymbol{\alpha};\delta\boldsymbol{\alpha}\right]\right\}_{\boldsymbol{\alpha}^0}$. This bilinear concomitant may vanish after implementing the boundary conditions represented by (2.229) and (2.236); if it does not vanish, this bilinear concomitant will be reduced to a residual quantity which will comprise only known values of $\mathbf{a}^{(n+1)}(j_n,\ldots,j_1;\mathbf{x})$, $\mathbf{u}^{(n+1)}(j_{n-1},\ldots,j_1;\mathbf{x})$, $\boldsymbol{\alpha}$, and $\delta\boldsymbol{\alpha}$, and which will be denoted as $\left\{\widehat{P}^{(n+1)}\left(\mathbf{a}^{(n+1)};\mathbf{u}^{(n+1)};\boldsymbol{\alpha};\delta\boldsymbol{\alpha}\right)\right\}_{\boldsymbol{\alpha}^0}$.

Using the equations underlying the nth-LASS together with the relation provided in (2.221) in (2.233) yields the following expression for the indirect-effect term $\left\{\delta R^{(n)}\left(j_n,\ldots,j_1;\mathbf{u}^{(n+1)};\boldsymbol{\alpha};\delta\mathbf{u}^{(n+1)}\right)_{\boldsymbol{\alpha}^0}\right\}_{ind}$ in terms of the *(n + 1)th-level adjoint sensitivity functions* $\mathbf{a}^{(n+1)}(j_n,\ldots,j_1;\mathbf{x})$, for each $j_1=1,\ldots,TP$; $j_2=1,\ldots,j_1$;...; $j_n=1,\ldots,j_{n-1}$:

$$\begin{aligned}&\left\{\delta R^{(n)}\left[j_n,\ldots,j_1;\mathbf{u}^{(n+1)};\boldsymbol{\alpha};\delta\mathbf{u}^{(n+1)}\right]_{\boldsymbol{\alpha}^0}\right\}_{ind}\\&=-\left\{\widehat{P}^{(n+1)}\left(\mathbf{a}^{(n+1)};\mathbf{u}^{(n+1)};\boldsymbol{\alpha};\delta\boldsymbol{\alpha}\right)\right\}_{\boldsymbol{\alpha}^0}+\left\{\langle\mathbf{a}^{(n+1)}(j_n,\ldots,j_1;\mathbf{x}),\right.\\&\quad\left.\mathbf{q}_V^{(n+1)}\left[j_{n-1},\ldots,j_1;\mathbf{u}^{(n+1)}(j_{n-1},\ldots,j_1;\mathbf{x});\boldsymbol{\alpha};\delta\boldsymbol{\alpha}\right]\rangle_{n+1}\right\}_{\boldsymbol{\alpha}^0}.\end{aligned} \tag{2.244}$$

Adding the result obtained in (2.244) for the indirect effect term to the result provided in (2.220) for the direct effect term yields the following expression for the total nth-order G-variation (of the response) defined in (2.219):

$$
\begin{aligned}
&\left\{\delta R^{(n)}\left(j_n,\ldots,j_1;\mathbf{u}^{(n+1)};\boldsymbol{\alpha};\delta\mathbf{u}^{(n+1)};\delta\boldsymbol{\alpha}\right)\right\}_{\boldsymbol{\alpha}^0} \\
&= -\left\{\widehat{P}^{(n+1)}\left(\mathbf{a}^{(n+1)};\mathbf{u}^{(n+1)};\boldsymbol{\alpha};\delta\boldsymbol{\alpha}\right)\right\}_{\boldsymbol{\alpha}^0} \\
&+\left\{\frac{\partial}{\partial\boldsymbol{\alpha}}\prod_{i=1}^{TI}\int\limits_{\lambda_i(\boldsymbol{\alpha})}^{\omega_i(\boldsymbol{\alpha})} dx_i S^{(n)}\left(j_n,\ldots,j_1;\mathbf{u}^{(n+1)};\boldsymbol{\alpha}\right)\right\}_{\boldsymbol{\alpha}^0}\delta\boldsymbol{\alpha} \\
&+\left\{\left\langle \mathbf{a}^{(n+1)}(j_n,\ldots,j_1;\mathbf{x}),\mathbf{q}_V^{(n+1)}\left[j_{n-1},\ldots,j_1;\mathbf{u}^{(n+1)}(j_{n-1},\ldots,j_1;\mathbf{x});\boldsymbol{\alpha};\delta\boldsymbol{\alpha}\right]\right\rangle_{n+1}\right\}_{\boldsymbol{\alpha}^0} \\
&\equiv \sum_{j_{n+1}=1}^{TP}\left\{R^{(n+1)}\left(j_{n+1},\ldots,j_1;\mathbf{u}^{(n+1)};\mathbf{a}^{(n+1)};\boldsymbol{\alpha}\right)\right\}_{(\boldsymbol{\alpha}^0)}\delta\alpha_{j_{n+1}},
\end{aligned}
\tag{2.245}
$$

where $R^{(n+1)}(j_{n+1},\ldots,j_1;\mathbf{u}^{(n+1)};\mathbf{a}^{(n+1)};\boldsymbol{\alpha})$ denotes the $(n+1)$th-order partial sensitivity of the response $R[\mathbf{u}^{(1)}(\mathbf{x});\boldsymbol{\alpha}]$ with respect to the model parameters, evaluated at the nominal parameter values $\boldsymbol{\alpha}^0$.

The results obtained in this subsection are summarized in Table 2.6. These results were obtained by applying the general principles underlying the 1st-CASAM-L to the nth-order sensitivities, considered as "model responses." The expressions of the $(n+1)$th-order sensitivities were obtained by determining the first-order differential of the nth-order sensitivities, which yields an expression that is identical to the expression that would be obtained by advancing the index, from n to $(n+1)$, of the expressions of the nth-order sensitivities. Thus, the proof by mathematical induction of the general mathematical framework underlying the nth-CASAM-L is thereby completed.

2.8 The Fifth-Order Comprehensive Adjoint Sensitivity Analysis Methodology for Coupled Forward/Adjoint Linear Systems (5th-CASAM-L)

As an additional application of the general framework of the nth-CASAM-L, this subsection will present the "5th-Order Comprehensive Adjoint Sensitivity Analysis Methodology for Coupled Forward/Adjoint Linear Systems" (5th-CASAM-L). The derivation of the 5th-CASAM-L will be performed explicitly, by starting with the results produced by the 4th-CASAM-L, which were summarized in Table 2.4. It will be shown that the results thus obtained for the 5th-CASAM-L coincide with the results that would be predicted by the nth-CASAM-L (Table 2.5) for the particular value $n = 5$.

The 5th-CASAM-L commences by recalling the end-products of the 4th-CASAM-L, namely the expressions of the fourth-order partial sensitivity

Table 2.6 $(n+1)$th-CASAM-L: $(n+1)$th-order sensitivities

$(n+1)$th-LFSS	nth-LFSS + nth-LASS: $\mathbf{F}^{(n+1)}\mathbf{u}^{(n+1)}(\mathbf{x}) = \mathbf{q}_F^{(n+1)}$; $\mathbf{F}^{(n+1)}(\boldsymbol{\alpha};\mathbf{x}) \triangleq diag\,(\mathbf{F}^{(n)},\mathbf{A}^{(n)})$; $\mathbf{b}_F^{(n+1)}\left(\mathbf{u}^{(n+1)};\boldsymbol{\alpha}\right) \triangleq \left(\mathbf{b}_F^{(n)},\mathbf{b}_A^{(n)}\right)^{\dagger} = [\mathbf{0},\mathbf{0}]^{\dagger}$; $\mathbf{u}^{(n+1)}(j_{n-1},\ldots,j_1;\mathbf{x}) = [\mathbf{u}^{(n)}(j_{n-2},\ldots,j_1;\mathbf{x}),\mathbf{a}^{(n)}(j_{n-1},\ldots,j_1;\mathbf{x})]^{\dagger}$.
$(n+1)$th-LVSS	$\left[\mathbf{V}^{(n+1)}\right]_{2^{n+1}\times 2^{n+1}}\delta\mathbf{u}^{(n+1)}(j_{n-1},\ldots,j_1;\mathbf{x})$ $= \mathbf{q}_V^{(n+1)}\left(j_{n-1},\ldots,j_1;\mathbf{u}^{(n+1)};\boldsymbol{\alpha};\delta\boldsymbol{\alpha}\right)$; $\delta\mathbf{u}^{(n+1)}(j_{n-1},\ldots,j_1;\mathbf{x}) = [\delta\mathbf{u}^{(n)}(j_{n-2},\ldots,j_1;\mathbf{x});$ $\delta\mathbf{a}^{(n)}(j_{n-1},\ldots,j_1;\mathbf{x})]^{\dagger}$; $\mathbf{b}_V^{(n+1)} \triangleq \left[\mathbf{b}_V^{(n)},\delta\mathbf{b}_A^{(n)}\right]^{\dagger} = [\mathbf{0},\mathbf{0}]^{\dagger}$; $\left[\mathbf{V}^{(n+1)}\right]_{2^{n+1}\times 2^{n+1}} \triangleq \begin{pmatrix} \mathbf{V}^{(n)} & [\mathbf{0}]_{2^n\times 2^n} \\ \mathbf{V}_{21}^{(n+1)} & \mathbf{V}_{22}^{(n+1)} \end{pmatrix}$.
$(n+1)$th-level Hilbert space	H_{n+1}; $\left\langle \boldsymbol{\eta}^{(n+1)}(\mathbf{x}),\boldsymbol{\xi}^{(n+1)}(\mathbf{x})\right\rangle_n \triangleq \sum_{i=1}^{2^{n+1}}\left\langle \boldsymbol{\eta}_i^{(n+1)}(\mathbf{x}),\boldsymbol{\xi}_i^{(n+1)}(\mathbf{x})\right\rangle_0$.
$(n+1)$th-LASS	$\left[\mathbf{A}^{(n+1)}\right]_{2^{n+1}\times 2^{n+1}}\mathbf{a}^{(n+1)}(j_n,\ldots,j_1;\mathbf{x})$ $= \mathbf{s}_A^{(n+1)}\left[j_n,\ldots,j_1;\mathbf{u}^{(n+1)}(j_{n-1},\ldots,j_1;\mathbf{x});\boldsymbol{\alpha}\right]$; $\left[\mathbf{A}^{(n+1)}\right]_{2^{n+1}\times 2^{n+1}} \triangleq \left[\mathbf{V}^{(n+1)}\right]^{*}_{2^{n+1}\times 2^{n+1}}$; $\mathbf{b}_A^{(n+1)}\left[\mathbf{a}^{(n+1)}(j_n,\ldots,j_1;\mathbf{x});\mathbf{u}^{(n+1)}(j_{n-1},\ldots,j_1;\mathbf{x});\boldsymbol{\alpha}\right] = \mathbf{0}$; $\mathbf{a}^{(n+1)}(j_n,\ldots,j_1;\mathbf{x}) \triangleq \left[\ldots,\mathbf{a}_k^{(n+1)}(j_n,\ldots,j_1;\mathbf{x}),\ldots\right]$; $k = 1,\ldots,2^{n+1}$.
$(n+1)$th-order sensitivities	$R^{(n+1)}\left[j_{n+1},\ldots,j_1;\mathbf{u}^{(n+1)}(j_{n-1},\ldots,j_1;\mathbf{x});\mathbf{a}^{(n+1)}(j_n,\ldots,j_1;\mathbf{x});\boldsymbol{\alpha}\right]$ $\triangleq \partial^{n+1}R[\boldsymbol{\varphi}(\mathbf{x}),\boldsymbol{\psi}(\mathbf{x});\boldsymbol{\alpha}]/\partial\alpha_{j_1}\ldots\partial\alpha_{j_{n+1}}$; $j_1 = 1,\ldots,TP; j_2 = 1,\ldots,j_1;\ldots;j_{n+1} = 1,\ldots,j_n$. Distinct Sensitivities: $TP(TP+1)(TP+2)\ldots(TP+n)/(n+1)!$

$\left\{R^{(4)}\left[j_4,j_3,j_2,j_1;\mathbf{u}^{(4)}(j_2,j_1;\mathbf{x});\mathbf{a}^{(4)}(j_3,j_2,j_1;\mathbf{x});\boldsymbol{\alpha}\right]\right\}_{(\boldsymbol{\alpha}^0)}$ of the response with respect to the model parameters, which can be written in the following form:

$$\begin{aligned}
&\left\{R^{(4)}\left[j_4,j_3,j_2,j_1;\mathbf{u}^{(5)}(j_3,j_2,j_1;\mathbf{x});\boldsymbol{\alpha}\right]\right\}_{(\boldsymbol{\alpha}^0)} \\
&\triangleq \partial^4 R[\boldsymbol{\varphi}(\mathbf{x}),\boldsymbol{\psi}(\mathbf{x});\boldsymbol{\alpha}]/\partial\alpha_{j_1}\ldots\partial\alpha_{j_4} \\
&\triangleq \prod_{i=1}^{TI}\int_{\lambda_i(\boldsymbol{\alpha})}^{\omega_i(\boldsymbol{\alpha})} dx_i S^{(4)}\left[j_4,\ldots,j_1;\mathbf{u}^{(5)}(j_3,j_2,j_1;\mathbf{x});\boldsymbol{\alpha}\right].
\end{aligned} \tag{2.246}$$

The function $\mathbf{u}^{(5)}(j_3,j_2,j_1;\mathbf{x}) \triangleq [\mathbf{u}^{(4)}(j_2,j_1;\mathbf{x}),\mathbf{a}^{(4)}(j_3,j_2,j_1;\mathbf{x})]^{\dagger}$ is the solution of the following Fifth-Level Forward Sensitivity System (5th-LFSS), which is obtained by concatenating the following equations provided in items (a) and (b), below:

(a) the Fourth-Level Forward Sensitivity System, which comprises the equations below:

$$\left\{\mathbf{F}^{(4)}(\boldsymbol{\alpha};\mathbf{x})\mathbf{u}^{(4)}(j_2,j_1;\mathbf{x})\right\}_{\boldsymbol{\alpha}^0} = \left\{\mathbf{q}_F^{(4)}\left(j_2,j_1;\mathbf{u}^{(3)};\mathbf{x};\boldsymbol{\alpha}\right)\right\}_{\boldsymbol{\alpha}^0}, \quad \mathbf{x}\in\Omega_x\left(\boldsymbol{\alpha}^0\right), \tag{2.247}$$

$$\left\{\mathbf{b}_F^{(4)}\left[\mathbf{u}^{(4)}(j_2,j_1;\mathbf{x});\boldsymbol{\alpha}\right]\right\}_{\boldsymbol{\alpha}^0} = \mathbf{0}, \quad \mathbf{x}\in\partial\Omega_x\left(\boldsymbol{\alpha}^0\right), \tag{2.248}$$

where:

$$\begin{aligned}
&\mathbf{F}^{(4)}(\boldsymbol{\alpha};\mathbf{x})\triangleq\begin{pmatrix}\left[\mathbf{F}^{(3)}(\boldsymbol{\alpha};\mathbf{x})\right]_{2^3\times 2^3} & \mathbf{0}\\ \mathbf{0} & \left[\mathbf{A}^{(3)}(j_1)\right]_{2^3\times 2^3}\end{pmatrix};\\
&\mathbf{u}^{(4)}(j_2,j_1;\mathbf{x})\triangleq\begin{pmatrix}\mathbf{u}^{(3)}(j_1;\mathbf{x})\\ \mathbf{a}^{(3)}(j_2,j_1;\mathbf{x})\end{pmatrix};\\
&\mathbf{q}_F^{(4)}\left(j_2,j_1;\mathbf{u}^{(3)};\mathbf{x};\boldsymbol{\alpha}\right)\triangleq\begin{pmatrix}\mathbf{q}_F^{(3)}\\ \mathbf{s}_A^{(3)}\end{pmatrix};\ \mathbf{b}_F^{(4)}\left[\mathbf{u}^{(4)}(j_2,j_1;\mathbf{x});\boldsymbol{\alpha}\right]\triangleq\begin{pmatrix}\mathbf{b}_F^{(3)}\\ \mathbf{b}_A^{(3)}\end{pmatrix}.
\end{aligned} \tag{2.249}$$

(b) the equations underlying the Fourth-Level Adjoint Sensitivity System, for $j_1 = 1, \ldots, TP; j_2 = 1, \ldots, j_1; j_3 = 1, \ldots, j_2$, which are reproduced below:

$$\left[\mathbf{A}^{(4)}\left(j_2,j_1;\mathbf{u}^{(3)}\right)\right]_{2^4\times 2^4}\mathbf{a}^{(4)}(j_3,j_2,j_1;\mathbf{x}) = \mathbf{s}_A^{(4)}\left[j_3,j_2,j_1;\mathbf{u}^{(4)}(j_2,j_1;\mathbf{x});\boldsymbol{\alpha}\right], \tag{2.250}$$

$$\left\{\mathbf{b}_A^{(4)}\left[\mathbf{a}^{(4)}(j_3,j_2,j_1;\mathbf{x});\mathbf{u}^{(4)}(j_2,j_1;\mathbf{x});\boldsymbol{\alpha}\right]\right\}_{\boldsymbol{\alpha}^0} = \mathbf{0}, \mathbf{x}\in\partial\Omega_x\left(\boldsymbol{\alpha}^0\right), \tag{2.251}$$

where $\mathbf{s}_A^{(4)}\left[j_3,j_2,j_1;\mathbf{u}^{(4)}(j_2,j_1;\mathbf{x});\boldsymbol{\alpha}\right]\triangleq\left[\mathbf{s}_1^{(4)}(j_3,j_2,j_1;\mathbf{x}), \ldots, \mathbf{s}_{16}^{(4)}(j_3,j_2,j_1;\mathbf{x})\right]^{\dagger}$ comprises, for each $j_1 = 1, \ldots, TP; j_2 = 1, \ldots, j_1; j_3 = 1, \ldots, j_2$, 16 components defined as follows:

$$\mathbf{s}_1^{(4)}\left[j_3,j_2,j_1;\mathbf{u}^{(4)}(j_2,j_1;\mathbf{x});\boldsymbol{\alpha}\right]\triangleq\frac{\partial S^{(3)}\left[j_3,j_2,j_1;\mathbf{u}^{(4)}(j_2,j_1;\mathbf{x});\boldsymbol{\alpha}\right]}{\partial\boldsymbol{\varphi}}; \tag{2.252}$$

$$\mathbf{s}_2^{(4)}\left[j_3,j_2,j_1;\mathbf{u}^{(4)}(j_2,j_1;\mathbf{x});\boldsymbol{\alpha}\right]\triangleq\frac{\partial S^{(3)}\left[j_3,j_2,j_1;\mathbf{u}^{(4)}(j_2,j_1;\mathbf{x});\boldsymbol{\alpha}\right]}{\partial\boldsymbol{\psi}}; \tag{2.253}$$

$$\mathbf{s}_{2+k}^{(4)}\left[j_3,j_2,j_1;\mathbf{u}^{(4)}(j_2,j_1;\mathbf{x});\boldsymbol{\alpha}\right]\triangleq\frac{\partial S^{(3)}\left[j_3,j_2,j_1;\mathbf{u}^{(4)}(j_2,j_1;\mathbf{x});\boldsymbol{\alpha}\right]}{\partial\mathbf{a}_k^{(1)}}; \tag{2.254}$$

$k = 1, 2;$

$$\mathbf{s}_{4+k}^{(4)}\left[j_3,j_2,j_1;\mathbf{u}^{(4)}(j_2,j_1;\mathbf{x});\boldsymbol{\alpha}\right]\triangleq\frac{\partial S^{(3)}\left[j_3,j_2,j_1;\mathbf{u}^{(4)}(j_2,j_1;\mathbf{x});\boldsymbol{\alpha}\right]}{\partial\mathbf{a}_k^{(2)}(j_1)};\tag{2.255}$$
$$k=1,..,2^2;$$

$$\mathbf{s}_{8+k}^{(4)}\left[j_3,j_2,j_1;\mathbf{u}^{(4)}(j_2,j_1;\mathbf{x});\boldsymbol{\alpha}\right]\triangleq\frac{\partial S^{(3)}\left[j_3,j_2,j_1;\mathbf{u}^{(4)}(j_2,j_1,\mathbf{x});\boldsymbol{\alpha}\right]}{\partial\mathbf{a}_k^{(3)}(j_2,j_1)};\tag{2.256}$$
$$k=1,..,2^3.$$

The Eqs. (2.247), (2.248), (2.250), and (2.251) are written in block-matrix form as follows:

$$\left\{\mathbf{F}^{(5)}(\boldsymbol{\alpha};\mathbf{x})\mathbf{u}^{(5)}(j_3,j_2,j_1;\mathbf{x})\right\}_{\boldsymbol{\alpha}^0}=\left\{\mathbf{q}_F^{(5)}\left(j_3,\ldots,j_1;\mathbf{u}^{(4)};\mathbf{x};\boldsymbol{\alpha}\right)\right\}_{\boldsymbol{\alpha}^0},\ \mathbf{x}\in\Omega_x\left(\boldsymbol{\alpha}^0\right),\tag{2.257}$$

$$\left\{\mathbf{b}_F^{(5)}\left[\mathbf{u}^{(5)}(j_3,\ldots,j_1;\mathbf{x});\boldsymbol{\alpha}\right]\right\}_{\boldsymbol{\alpha}^0}=\mathbf{0},\ \mathbf{x}\in\partial\Omega_x(\boldsymbol{\alpha}^0),\tag{2.258}$$

where:

$$\mathbf{F}^{(5)}(\boldsymbol{\alpha};\mathbf{x})\triangleq\begin{pmatrix}\left[\mathbf{F}^{(4)}\right]_{2^4\times2^4} & [\mathbf{0}]_{2^4\times2^4}\\ [\mathbf{0}]_{2^4\times2^4} & \left[\mathbf{A}^{(4)}\right]_{2^4\times2^4}\end{pmatrix};\ \mathbf{u}^{(5)}(j_3,j_2,j_1;\mathbf{x})\triangleq\begin{pmatrix}\mathbf{u}^{(4)}(j_2,j_1)\\ \mathbf{a}^{(4)}(j_3,j_2,j_1)\end{pmatrix};$$
$$\mathbf{q}_F^{(5)}(j_3,\ldots,j_1;\mathbf{x};\boldsymbol{\alpha})\triangleq\begin{pmatrix}\mathbf{q}_F^{(4)}\\ \mathbf{s}_A^{(4)}\end{pmatrix};\ \mathbf{b}_F^{(5)}\left[\mathbf{u}^{(5)}(j_3,j_2,j_1;\mathbf{x});\boldsymbol{\alpha}\right]\triangleq\begin{pmatrix}\mathbf{b}_F^{(4)}\\ \mathbf{b}_A^{(4)}\end{pmatrix}.\tag{2.259}$$

The fifth-order total differential of the response $R[\mathbf{u}^{(1)}(\mathbf{x});\boldsymbol{\alpha}]$ is obtained by G-differentiating the fourth-order total G-differential of the response, to obtain the following expression:

$$\begin{aligned}&\left\{\delta R^{(4)}\left(j_4,\ldots,j_1;\mathbf{u}^{(5)};\boldsymbol{\alpha};\delta\mathbf{u}^{(5)};\delta\boldsymbol{\alpha}\right)\right\}_{\boldsymbol{\alpha}^0}\\ &\triangleq\left\{\frac{d}{d\varepsilon}\left[\prod_{i=1}^{TI}\int_{\lambda_i(\boldsymbol{\alpha})}^{\omega_i(\boldsymbol{\alpha})}dx_i S^{(4)}\left(j_4,\ldots,j_1;\mathbf{u}^{(5)}+\varepsilon\delta\mathbf{u}^{(5)};\boldsymbol{\alpha}+\varepsilon\delta\boldsymbol{\alpha}\right)\right]_{\boldsymbol{\alpha}^0}\right\}_{\varepsilon=0}\\ &=\left\{\delta R^{(4)}\left(j_4,\ldots,j_1;\mathbf{u}^{(5)};\boldsymbol{\alpha};\delta\boldsymbol{\alpha}\right)_{\boldsymbol{\alpha}^0}\right\}_{dir}+\left\{\delta R^{(4)}\left(j_4,\ldots,j_1;\mathbf{u}^{(5)};\boldsymbol{\alpha};\delta\mathbf{u}^{(5)}\right)_{\boldsymbol{\alpha}^0}\right\}_{ind},\end{aligned}\tag{2.260}$$

where the direct effect term $\left\{\delta R^{(4)}\left(j_4,\ldots,j_1;\mathbf{u}^{(5)};\boldsymbol{\alpha};\delta\boldsymbol{\alpha}\right)_{\boldsymbol{\alpha}^0}\right\}_{dir}$ depends directly on the vector of parameter variations $\delta\boldsymbol{\alpha}$, and is defined as follows, for $j_1 = 1, \ldots, TP$; $j_2 = 1, \ldots, j_1; \ldots; j_4 = 1, \ldots, j_3$:

$$\left\{\delta R^{(4)}\left(j_4,\ldots,j_1;\mathbf{u}^{(5)};\boldsymbol{\alpha};\delta\boldsymbol{\alpha}\right)_{\boldsymbol{\alpha}^0}\right\}_{dir} \triangleq \left\{\frac{\partial}{\partial\boldsymbol{\alpha}}\prod_{i=1}^{TI}\int_{\lambda_i(\boldsymbol{\alpha})}^{\omega_i(\boldsymbol{\alpha})} dx_i S^{(4)}\left(j_4,\ldots,j_1;\mathbf{u}^{(5)};\boldsymbol{\alpha}\right)\right\}_{\boldsymbol{\alpha}^0}\delta\boldsymbol{\alpha}, \tag{2.261}$$

and where the indirect-effect term $\left\{\delta R^{(4)}\left(j_4,\ldots,j_1;\mathbf{u}^{(5)};\boldsymbol{\alpha};\delta\mathbf{u}^{(5)}\right)_{\boldsymbol{\alpha}^0}\right\}_{ind}$ is defined as follows:

$$\left\{\delta R^{(4)}\left(j_4,\ldots,j_1;\mathbf{u}^{(5)};\boldsymbol{\alpha};\delta\mathbf{u}^{(5)}\right)_{\boldsymbol{\alpha}^0}\right\}_{ind} \triangleq \left\{\prod_{i=1}^{TI}\int_{\lambda_i(\boldsymbol{\alpha})}^{\omega_i(\boldsymbol{\alpha})} dx_i \frac{\partial S^{(4)}\left(j_4,\ldots,j_1;\mathbf{u}^{(5)};\boldsymbol{\alpha}\right)}{\partial\mathbf{u}^{(5)}(j_3,j_2,j_1;\mathbf{x})}\right\}_{\boldsymbol{\alpha}^0}\delta\mathbf{u}^{(5)}, \tag{2.262}$$

with

$$\begin{aligned}
&\frac{\partial[]}{\partial\mathbf{u}^{(5)}(j_3,j_2,j_1;\mathbf{x})}\delta\mathbf{u}^{(5)}(j_3,j_2,j_1;\mathbf{x}) \equiv \frac{\partial[]}{\partial\boldsymbol{\varphi}}\delta\boldsymbol{\varphi}(\mathbf{x}) \\
&+\frac{\partial[]}{\partial\boldsymbol{\psi}}\delta\boldsymbol{\psi}(\mathbf{x}) + \sum_{k=1}^{2}\frac{\partial[]}{\partial\mathbf{a}_k^{(1)}}\delta\mathbf{a}_k^{(1)}(\mathbf{x}) + \sum_{k=1}^{2^2}\frac{\partial[]}{\partial\mathbf{a}_k^{(2)}(j_1)}\delta\mathbf{a}_k^{(2)}(j_1;\mathbf{x}) \\
&+\sum_{k=1}^{2^3}\frac{\partial[]}{\partial\mathbf{a}_k^{(3)}(j_2,j_1)}\delta\mathbf{a}_k^{(3)}(j_2,j_1;\mathbf{x}) \\
&+\sum_{k=1}^{2^4}\frac{\partial[]}{\partial\mathbf{a}_k^{(4)}(j_3,j_2,j_1)}\delta\mathbf{a}_k^{(4)}(j_3,j_2,j_1;\mathbf{x}).
\end{aligned} \tag{2.263}$$

The vector of variations $\delta\mathbf{u}^{(5)}(j_3,j_2,j_1;\mathbf{x}) \triangleq [\delta\mathbf{u}^{(4)}(j_2,j_1;\mathbf{x});\delta\mathbf{a}^{(4)}(j_3,j_2,j_1;\mathbf{x})]^{\dagger}$ is the solution of the G-differentiated 5th-LFSS represented by Eqs. (2.257) and (2.258), which is equivalent concatenating the 4th-LVSS with the system of equations obtained by G-differentiating the 4th-LASS while being subject to the corresponding boundary conditions. The G-differential of the 4th-LASS, cf. (2.250) and (2.251), has the following form:

$$\frac{\partial\left\{\left[\mathbf{A}^{(4)}\right]_{2^4\times 2^4}\mathbf{a}^{(4)}(j_3,j_2,j_1;\mathbf{x})-\mathbf{s}_A^{(4)}\left[j_3,j_2,j_1;\mathbf{u}^{(4)}(j_2,j_1;\mathbf{x});\boldsymbol{\alpha}\right]\right\}}{\partial\mathbf{u}^{(4)}}\delta\mathbf{u}^{(4)}$$
$$+\left[\mathbf{A}^{(4)}\right]_{2^4\times 2^4}\delta\mathbf{a}^{(4)}(j_3,j_2,j_1;\mathbf{x})=\mathbf{p}^{(4)}\left(j_3,j_2,j_1;\mathbf{u}^{(5)};\boldsymbol{\alpha};\delta\boldsymbol{\alpha}\right), \tag{2.264}$$

$$\delta\mathbf{b}_A^{(4)}\left[\mathbf{a}^{(4)}(j_3,j_2,j_1;\mathbf{x});\mathbf{u}^{(4)}(j_2,j_1;\mathbf{x});\boldsymbol{\alpha}\right]=\mathbf{0}, \tag{2.265}$$

where

$$\mathbf{p}^{(4)}\left(j_3,j_2,j_1;\mathbf{u}^{(5)};\boldsymbol{\alpha};\delta\boldsymbol{\alpha}\right)$$
$$\triangleq\frac{\partial\left\{\mathbf{s}_A^{(4)}\left[j_3,j_2,j_1;\mathbf{u}^{(4)}(j_2,j_1;\mathbf{x});\boldsymbol{\alpha}\right]-\mathbf{A}^{(4)}\mathbf{a}^{(4)}\right\}}{\partial\boldsymbol{\alpha}}\delta\boldsymbol{\alpha}. \tag{2.266}$$

Concatenating the 4th-LVSS with (2.264) and (2.265) yields the following Fifth-Level Variational Sensitivity System (5th-LVSS) for the vector-valued function $\delta\mathbf{u}^{(5)}(j_3,j_2,j_1;\mathbf{x})$:

$$\left[\mathbf{V}^{(5)}\right]_{2^5\times 2^5}\delta\mathbf{u}^{(5)}(j_3,j_2,j_1;\mathbf{x})=\mathbf{q}_V^{(5)}\left[j_3,j_2,j_1;\mathbf{u}^{(5)}(j_3,j_2,j_1;\mathbf{x});\boldsymbol{\alpha};\delta\boldsymbol{\alpha}\right], \tag{2.267}$$

$$\mathbf{b}_V^{(5)}\left(\delta\mathbf{u}^{(5)};\delta\boldsymbol{\alpha}\right)\triangleq\begin{pmatrix}\mathbf{b}_V^{(4)}\left(\delta\mathbf{u}^{(4)};\delta\boldsymbol{\alpha}\right)\\ \delta\mathbf{b}_A^{(4)}\left(\delta\mathbf{a}^{(4)};\delta\boldsymbol{\alpha}\right)\end{pmatrix}=\begin{pmatrix}[\mathbf{0}]\\ [\mathbf{0}]\end{pmatrix},\mathbf{x}\in\partial\Omega_x\left(\boldsymbol{\alpha}^0\right), \tag{2.268}$$

where

$$\left[\mathbf{V}^{(5)}\right]_{2^5\times 2^5}\triangleq\begin{pmatrix}\left[\mathbf{V}^{(4)}\right]_{2^4\times 2^4} & [\mathbf{0}]_{2^4\times 2^4}\\ \left[\mathbf{V}_{21}^{(5)}\right]_{2^4\times 2^4} & \left[\mathbf{V}_{22}^{(5)}\right]_{2^4\times 2^4}\end{pmatrix};$$
$$\mathbf{V}_{22}^{(5)}\triangleq\mathbf{A}^{(4)};\ \mathbf{V}_{21}^{(5)}\triangleq\frac{\partial\left\{\mathbf{A}^{(4)}\mathbf{a}^{(4)}-\mathbf{s}_A^{(4)}\right\}}{\partial\mathbf{u}^{(4)}}; \tag{2.269}$$

$$\mathbf{q}_V^{(5)}\left[j_3,j_3,j_1;\mathbf{u}^{(5)};\boldsymbol{\alpha};\delta\boldsymbol{\alpha}\right]\triangleq\begin{pmatrix}\mathbf{q}_V^{(4)}\\ \mathbf{p}^{(4)}\end{pmatrix}$$
$$\equiv\left[\ldots,q_k^{(5)}\left(j_3,j_2,j_1;\mathbf{u}^{(5)};\boldsymbol{\alpha};\delta\boldsymbol{\alpha}\right),\ldots\right]^{\dagger},\ k=1,\ldots,2^5. \tag{2.270}$$

In principle, the 5th-LVSS, comprising (2.267) and (2.268), could be solved to determine the vector $\delta\mathbf{u}^{(5)}(j_3,j_2,j_1;\mathbf{x})$ which could be used, in turn, to determine the indirect-effect term $\left\{\delta R^{(4)}\left(j_4,\ldots,j_1;\mathbf{u}^{(5)};\boldsymbol{\alpha};\delta\mathbf{u}^{(5)}\right)_{\boldsymbol{\alpha}^0}\right\}_{ind}$. However, the need for solving the 5th-LVSS can be avoided by expressing the indirect-effect term $\left\{\delta R^{(4)}\left(j_4,\ldots,j_1;\mathbf{u}^{(5)};\boldsymbol{\alpha};\delta\mathbf{u}^{(5)}\right)_{\boldsymbol{\alpha}^0}\right\}_{ind}$ in terms of the solution of a Fifth-Level Adjoint Sensitivity Systems (5th-LASS), which is constructed in a Hilbert space, denoted as H_5, comprising block-vector elements of the form $\boldsymbol{\eta}^{(5)}(\mathbf{x}) \in H_5$, $\boldsymbol{\eta}^{(5)}(\mathbf{x}) \triangleq \left[\ldots, \boldsymbol{\eta}_k^{(5)}(\mathbf{x}), \ldots\right]^\dagger; k = 1, \ldots, 2^5$, with elements having the following structure: $\boldsymbol{\eta}_k^{(5)}(\mathbf{x}) \triangleq \left[\eta_{k,1}^{(5)}(\mathbf{x}), \ldots, \eta_{k,TD}^{(5)}(\mathbf{x})\right]^\dagger$.

The inner product between two elements, $\boldsymbol{\eta}^{(5)}(\mathbf{x}) \in H_5$ and $\boldsymbol{\xi}^{(5)}(\mathbf{x}) \in H_5$ will be denoted as $\langle \boldsymbol{\eta}^{(5)}(\mathbf{x}), \boldsymbol{\xi}^{(5)}(\mathbf{x})\rangle_5$ and is defined as follows:

$$\left\langle \boldsymbol{\eta}^{(5)}(\mathbf{x}), \boldsymbol{\xi}^{(5)}(\mathbf{x}) \right\rangle_5 \triangleq \sum_{k=1}^{2^5} \left\langle \boldsymbol{\eta}_k^{(5)}(\mathbf{x}), \boldsymbol{\xi}_k^{(5)}(\mathbf{x}) \right\rangle_0. \tag{2.271}$$

Using the definition provided in (2.271), form the inner product in H_5 of (2.267) with a yet undefined vector-valued function $\mathbf{a}^{(5)}(j_4, \ldots, j_1; \mathbf{x}) \triangleq \left[\ldots, \mathbf{a}_k^{(5)}(j_4, \ldots, j_1; \mathbf{x}), \ldots\right]^\dagger \in H_5; k = 1, \ldots, 2^5$, $j_1 = 1, \ldots, TP$; $j_2 = 1, \ldots, j_1; \ldots; j_4 = 1, \ldots, j_3$, to obtain the following relation:

$$\begin{aligned}
&\left\{\left\langle \mathbf{a}^{(5)}(j_4, \ldots, j_1; \mathbf{x}), \mathbf{V}^{(5)} \delta\mathbf{u}^{(5)} \right\rangle_5\right\}_{\boldsymbol{\alpha}^0} \\
&= \left\{\left\langle \mathbf{a}^{(5)}(j_4, \ldots, j_1; \mathbf{x}), \mathbf{q}_V^{(5)}\left[j_3, j_2, j_1; \mathbf{u}^{(5)}(j_3, j_2, j_1; \mathbf{x}); \boldsymbol{\alpha}; \delta\boldsymbol{\alpha}\right] \right\rangle_5\right\}_{\boldsymbol{\alpha}^0} \\
&= \left\{\left\langle \delta\mathbf{u}^{(5)}, \mathbf{A}^{(5)} \mathbf{a}^{(5)}(j_4, \ldots, j_1; \mathbf{x}) \right\rangle_5\right\}_{\boldsymbol{\alpha}^0} + \left\{P^{(5)}\left[\delta\mathbf{u}^{(5)}; \mathbf{a}^{(5)}; \boldsymbol{\alpha}; \delta\boldsymbol{\alpha}\right]\right\}_{\boldsymbol{\alpha}^0},
\end{aligned} \tag{2.272}$$

where $\left\{P^{(5)}\left[\delta\mathbf{u}^{(5)}; \mathbf{a}^{(5)}; \boldsymbol{\alpha}; \delta\boldsymbol{\alpha}\right]\right\}_{\boldsymbol{\alpha}^0}$ denotes the bilinear concomitant defined on the phase-space boundary $\mathbf{x} \in \partial\Omega_x(\boldsymbol{\alpha}^0)$, evaluated at the nominal values of the model parameter and respective functions, and where:

$$\begin{aligned}
\left[\mathbf{A}^{(5)}\right]_{2^5 \times 2^5} \triangleq \left[\mathbf{V}^{(5)}\right]^*_{2^5 \times 2^5} &= \begin{pmatrix} \left[\mathbf{V}^{(4)}\right]_{2^4 \times 2^4} & [\mathbf{0}]_{2^4 \times 2^4} \\ \left[\mathbf{V}_{21}^{(5)}\right]_{2^4 \times 2^4} & \left[\mathbf{V}_{22}^{(5)}\right]_{2^4 \times 2^4} \end{pmatrix}^* \\
&= \begin{pmatrix} \mathbf{A}^{(4)} & \left[\mathbf{V}_{21}^{(5)}\right]^* \\ [\mathbf{0}]_{2^4 \times 2^4} & \left[\mathbf{V}_{22}^{(5)}\right]^* \end{pmatrix}.
\end{aligned} \tag{2.273}$$

The first term on right side of the second equality in (2.272) is now required to represent the indirect-effect term $\left\{\delta R^{(4)}\left(j_4, \ldots, j_1; \mathbf{u}^{(5)}; \boldsymbol{\alpha}; \delta\mathbf{u}^{(5)}\right)_{\boldsymbol{\alpha}^0}\right\}_{ind}$, which is achieved by requiring that the fifth-level adjoint sensitivity function $\mathbf{a}^{(5)}(j_4, \ldots, j_1; \mathbf{x}) \triangleq \left[\ldots, \mathbf{a}_k^{(5)}(j_4, \ldots, j_1; \mathbf{x}), \ldots\right]^\dagger \in H_5; k = 1, \ldots, 2^5$, be the solution of the following Fifth-Level Adjoint Sensitivity System (5th-LASS), for $j_1 = 1, \ldots, TP; j_2 = 1, \ldots, j_1; \ldots; j_4 = 1, \ldots, j_3$:

$$\left[\mathbf{A}^{(5)}\right]_{2^5\times 2^5}\mathbf{a}^{(5)}(j_4,\ldots,j_1;\mathbf{x}) = \mathbf{s}_A^{(5)}(j_4,\ldots,j_1;\boldsymbol{\alpha}), \tag{2.274}$$

$$\left\{\mathbf{b}_A^{(5)}\left[\mathbf{a}^{(5)}(j_4,\ldots,j_1;\mathbf{x});\mathbf{u}^{(5)}(j_3,j_2,j_1;\mathbf{x});\boldsymbol{\alpha}\right]\right\}_{\boldsymbol{\alpha}^0} = \mathbf{0}, \mathbf{x}\in\partial\Omega_x(\boldsymbol{\alpha}^0), \tag{2.275}$$

where the vector $\mathbf{s}_A^{(5)}(j_4,\ldots,j_1;\boldsymbol{\alpha}) \triangleq \left[\ldots,\mathbf{s}_k^{(5)}(j_4,\ldots,j_1;\boldsymbol{\alpha}),\ldots\right]^\dagger$, $k = 1, \ldots, 2^5$, comprises 2^5 components defined as follows, for each $j_1 = 1, \ldots, TP$; $j_2 = 1, \ldots, j_1$;. . .; $j_4 = 1, \ldots, j_3$:

$$\mathbf{s}_1^{(5)}(j_4,\ldots,j_1;\boldsymbol{\alpha}) \triangleq \frac{\partial S^{(4)}\left(j_4,\ldots,j_1;\mathbf{u}^{(5)};\boldsymbol{\alpha}\right)}{\partial\boldsymbol{\varphi}}; \tag{2.276}$$

$$\mathbf{s}_2^{(5)}(j_4,\ldots,j_1;\boldsymbol{\alpha}) \triangleq \frac{\partial S^{(4)}\left(j_4,\ldots,j_1;\mathbf{u}^{(5)};\boldsymbol{\alpha}\right)}{\partial\boldsymbol{\psi}}; \tag{2.277}$$

$$\mathbf{s}_{2+k}^{(5)}(j_4,\ldots,j_1;\boldsymbol{\alpha}) \triangleq \frac{\partial S^{(4)}\left(j_4,\ldots,j_1;\mathbf{u}^{(5)};\boldsymbol{\alpha}\right)}{\partial\mathbf{a}_k^{(1)}};\ k = 1,2; \tag{2.278}$$

$$\mathbf{s}_{2^2+k}^{(5)}(j_4,\ldots,j_1;\boldsymbol{\alpha}) \triangleq \frac{\partial S^{(4)}\left(j_4,\ldots,j_1;\mathbf{u}^{(5)};\boldsymbol{\alpha}\right)}{\partial\mathbf{a}_k^{(2)}(j_1)};\ k = 1,\ldots,2^2; \tag{2.279}$$

$$\mathbf{s}_{2^3+k}^{(5)}(j_4,\ldots,j_1;\boldsymbol{\alpha}) \triangleq \frac{\partial S^{(4)}\left(j_4,\ldots,j_1;\mathbf{u}^{(5)};\boldsymbol{\alpha}\right)}{\partial\mathbf{a}_k^{(3)}(j_2;j_1)};\ k = 1,\ldots,2^3; \tag{2.280}$$

$$\mathbf{s}_{2^4+k}^{(5)}(j_4,\ldots,j_1;\boldsymbol{\alpha}) \triangleq \frac{\partial S^{(4)}\left(j_4,\ldots,j_1;\mathbf{u}^{(5)};\boldsymbol{\alpha}\right)}{\partial\mathbf{a}_k^{(4)}(j_3,j_2,j_1)};\ k = 1,\ldots,2^4. \tag{2.281}$$

The fifth-level adjoint boundary conditions represented by (2.275) are selected so as to eliminate, in conjunction with the boundary conditions represented by (2.268), all of the unknown values of the functions $\delta\mathbf{u}^{(5)}(j_3,j_2,j_1;\mathbf{x})$ in the expression of $\left\{P^{(5)}\left[\delta\mathbf{u}^{(5)};\mathbf{a}^{(5)};\boldsymbol{\alpha};\delta\boldsymbol{\alpha}\right]\right\}_{\boldsymbol{\alpha}^0}$. This bilinear concomitant may vanish after implementing the boundary conditions represented by (2.268) and (2.275); if it does not vanish, this bilinear concomitant will be reduced to a residual quantity which will comprise only known values of $\mathbf{a}^{(5)}(j_4,\ldots,j_1;\mathbf{x})$, $\mathbf{u}^{(5)}(j_3,j_2,j_1;\mathbf{x})$, $\boldsymbol{\alpha}$, and $\delta\boldsymbol{\alpha}$, and will be denoted as $\left\{\widehat{P}^{(5)}\left(\mathbf{a}^{(5)};\mathbf{u}^{(5)};\boldsymbol{\alpha};\delta\boldsymbol{\alpha}\right)\right\}_{\boldsymbol{\alpha}^0}$.

Using the equations underlying the 5th-LASS together with the relation provided in (2.272) in (2.262) yields the following expression for the indirect-effect term $\left\{\delta R^{(4)}\left(j_4,\ldots,j_1;\mathbf{u}^{(5)};\boldsymbol{\alpha};\delta\mathbf{u}^{(5)}\right)_{\boldsymbol{\alpha}^0}\right\}_{ind}$ in terms of the fifth-level adjoint sensitivity functions $\mathbf{a}^{(5)}(j_4,\ldots,j_1;\mathbf{x})$, for each $j_1 = 1,\ldots,TP$; $j_2 = 1,\ldots,j_1$; . . .; $j_4 = 1,\ldots,j_3$:

$$\begin{aligned}&\left\{\delta R^{(4)}\left(j_4,\ldots,j_1;\mathbf{u}^{(5)};\boldsymbol{\alpha};\delta\mathbf{u}^{(5)}\right)_{\boldsymbol{\alpha}^0}\right\}_{ind} = -\left\{\widehat{P}^{(5)}\left(\mathbf{a}^{(5)};\mathbf{u}^{(5)};\boldsymbol{\alpha};\delta\boldsymbol{\alpha}\right)\right\}_{\boldsymbol{\alpha}^0}\\&+\left\{\left\langle\mathbf{a}^{(5)}(j_4,\ldots,j_1;\mathbf{x}),\mathbf{q}_V^{(5)}\left[j_3,j_2,j_1;\mathbf{u}^{(5)}(j_3,j_2,j_1;\mathbf{x});\boldsymbol{\alpha};\delta\boldsymbol{\alpha}\right]\right\rangle_5\right\}_{\boldsymbol{\alpha}^0}.\end{aligned} \tag{2.282}$$

Adding the result obtained in (2.282) for the indirect effect term to the result provided in (2.261) for the direct effect term yields the following expression for the total nth-order G-variation (of the response) defined in (2.260):

$$\begin{aligned}
&\left\{\delta R^{(4)}\left(j_4,\ldots,j_1;\mathbf{u}^{(5)};\boldsymbol{\alpha};\delta\mathbf{u}^{(5)};\delta\boldsymbol{\alpha}\right)\right\}_{\boldsymbol{\alpha}^0} \\
&=\left\{\frac{\partial}{\partial\boldsymbol{\alpha}}\prod_{i=1}^{TI}\int_{\lambda_i(\boldsymbol{\alpha})}^{\omega_i(\boldsymbol{\alpha})} dx_i S^{(4)}\left(j_4,\ldots,j_1;\mathbf{u}^{(5)};\boldsymbol{\alpha}\right)\right\}_{\boldsymbol{\alpha}^0}\delta\boldsymbol{\alpha}-\left\{\widehat{P}^{(5)}\left(\mathbf{a}^{(5)};\mathbf{u}^{(5)};\boldsymbol{\alpha};\delta\boldsymbol{\alpha}\right)\right\}_{\boldsymbol{\alpha}^0} \\
&+\left\{\left\langle\mathbf{a}^{(5)}(j_4,\ldots,j_1;\mathbf{x}),\mathbf{q}_V^{(5)}\left[j_3,j_2,j_1;\mathbf{u}^{(5)}(j_3,j_2,j_1;\mathbf{x});\boldsymbol{\alpha};\delta\boldsymbol{\alpha}\right]\right\rangle_5\right\}_{\boldsymbol{\alpha}^0} \\
&\equiv\sum_{j_5=1}^{TP}\left\{R^{(5)}\left(j_5,\ldots,j_1;\mathbf{u}^{(5)};\mathbf{a}^{(5)};\boldsymbol{\alpha}\right)\right\}_{\boldsymbol{\alpha}^0}\delta\alpha_{j_5},
\end{aligned}
\tag{2.283}$$

where $R^{(5)}(j_5,\ldots,j_1;\mathbf{u}^{(5)};\mathbf{a}^{(5)};\boldsymbol{\alpha})$ denotes the fifth-order partial sensitivity of the response $R[\mathbf{u}^{(1)}(\mathbf{x});\boldsymbol{\alpha}]$ with respect to the model parameters, evaluated at the nominal parameter values $\boldsymbol{\alpha}^0$.

The results obtained in this subsection are summarized in Table 2.7.

2.9 Chapter Summary

The nth-CASAM-L (where "n" is a finite number that indicates any desired, arbitrarily high, order) enables the efficient computation of the exact expressions of the nth-order functional derivatives ("sensitivities") of a general system response, which depends on both the forward and adjoint state functions, with respect to all of the parameters (including boundary and initial conditions) underlying the respective forward and adjoint systems. Since nonlinear operators do not admit adjoint operators (only linearized versions of nonlinear operators can admit a bona fide adjoint operator), responses that simultaneously depend on forward and adjoint functions can arise only in conjunction with linear systems. Therefore, the sensitivity analysis of responses that simultaneously involve both forward and adjoint state functions makes it necessary to treat linear models/systems in their own right, rather than treating them as particular cases of nonlinear systems (in which the responses can depend only on the forward functions). Particularly important model responses that involve both the forward and adjoint functions are various forms of Lagrangian functionals, the Roussopoulos functional for computing reaction rates, the Raleigh quotient for computing eigenvalues and/or separation constants when solving partial differential equations, the Schwinger functional for "normalization-free" solutions, and many others. Since these functionals play a fundamental role in various

Table 2.7 5th-CASAM-L: fifth-order sensitivities ($n = 5$)

5th-LFSS	4th-LFSS + 4th-LASS: $\mathbf{F}^{(5)}\mathbf{u}^{(5)}(\mathbf{x}) = \mathbf{q}_F^{(5)}$; $\mathbf{u}^{(5)}(j_3, j_2, j_1; \mathbf{x}) = [\mathbf{u}^{(4)}(j_2, j_1; \mathbf{x}); \mathbf{a}^{(4)}(j_3, j_2, j_1; \mathbf{x})]^{\dagger}$; $\mathbf{F}^{(5)} \triangleq diag\,(\mathbf{F}^{(4)}, \mathbf{A}^{(4)})$; $\mathbf{b}_F^{(5)}\left(\mathbf{u}^{(5)}; \boldsymbol{\alpha}\right) \triangleq \left(\mathbf{b}_F^{(4)}, \mathbf{b}_A^{(4)}\right)^{\dagger} = [\mathbf{0}, \mathbf{0}]^{\dagger}$.
5th-LVSS	$\left[\mathbf{V}^{(5)}\right]_{2^5 \times 2^5} \delta\mathbf{u}^{(5)}(j_3, j_2, j_1; \mathbf{x})$ $= \mathbf{q}_V^{(5)}\left[j_3, j_2, j_1; \mathbf{u}^{(5)}(j_3; j_2, j_1; \mathbf{x}); \boldsymbol{\alpha}; \delta\boldsymbol{\alpha}\right]$; $\delta\mathbf{u}^{(5)}(j_3, j_2, j_1; \mathbf{x}) = (\delta\mathbf{u}^{(4)}, \delta\mathbf{a}^{(4)})^{\dagger}$; $\mathbf{b}_V^{(5)} \triangleq \left(\mathbf{b}_V^{(4)}, \delta\mathbf{b}_A^{(4)}\right)^{\dagger} = [\mathbf{0}, \mathbf{0}]^{\dagger};\ \left[\mathbf{V}^{(5)}\right]_{2^5 \times 2^5} \triangleq \begin{pmatrix} \mathbf{V}^{(4)} & [\mathbf{0}]_{(2^4 \times 2^4)} \\ \mathbf{V}_{21}^{(5)} & \mathbf{V}_{22}^{(5)} \end{pmatrix}_{2^5 \times 2^5}$.
Fifth-level Hilbert space	$\mathsf{H}_5;\ \left\langle \boldsymbol{\eta}^{(5)}(\mathbf{x}), \boldsymbol{\xi}^{(5)}(\mathbf{x}) \right\rangle_5 \triangleq \sum_{i=1}^{2^5} \left\langle \boldsymbol{\eta}_i^{(5)}(\mathbf{x}), \boldsymbol{\xi}_i^{(5)}(\mathbf{x}) \right\rangle_0$.
5th-LASS	$\left[\mathbf{A}^{(5)}\right]_{2^5 \times 2^5} \mathbf{a}^{(5)}(j_4, \ldots, j_1; \mathbf{x}) = \mathbf{s}_A^{(5)}\left[j_4, \ldots, j_1; \mathbf{u}^{(5)}(j_3, j_2, j_1; \mathbf{x}); \boldsymbol{\alpha}\right]$; $\mathbf{a}^{(5)}(j_4, \ldots, j_1; \mathbf{x}) \triangleq \left[\ldots, \mathbf{a}_k^{(5)}(j_4, \ldots, j_1; \mathbf{x}), \ldots\right];\ k = 1, \ldots, 2^5$; $\left[\mathbf{A}^{(5)}\right]_{2^5 \times 2^5} \triangleq \left[\mathbf{V}^{(5)}\right]_{2^5 \times 2^5}^{*}$; $\mathbf{b}_A^{(5)}\left[\mathbf{a}^{(5)}(j_4, \ldots, j_1; \mathbf{x}); \mathbf{u}^{(5)}(j_3, j_2, j_1; \mathbf{x}); \boldsymbol{\alpha}\right] = \mathbf{0}$.
Fifth-order sensitivities	$R^{(5)}\left[j_5, \ldots, j_1; \mathbf{u}^{(5)}(j_3, j_2, j_1; \mathbf{x}); \mathbf{a}^{(5)}(j_4, \ldots, j_1; \mathbf{x}); \boldsymbol{\alpha}\right]$ $\triangleq \partial^5 R[\boldsymbol{\varphi}(\mathbf{x}), \boldsymbol{\psi}(\mathbf{x}); \boldsymbol{\alpha}] / \partial\alpha_{j_1} \ldots \partial\alpha_{j_5}$; $j_1 = 1, \ldots, TP; j_2 = 1, \ldots, j_1; \ldots; j_5 = 1, \ldots, j_4$. Distinct sensitivities: $TP(TP + 1)(TP + 2)(TP + 3)(TP + 4)/5!$

optimization and control procedures, derivation of numerical methods for solving (differential, integral, integro-differential) equations, etc., they will be analyzed in detail in Chap. 4 in this book.

It has been shown in this chapter that the mathematical framework underlying the nth-CASAM-L is set in linearly increasing higher-dimensional Hilbert spaces, as opposed to exponentially increasing parameter-dimensional spaces. In particular, for a scalar-valued response associated with a linear model comprising *TP* parameters, the 1st-CASAM-L requires 1 additional large-scale adjoint computation (as opposed to *TP* large-scale computations, as required by other methods) for computing exactly all of the first-order response sensitivities. All of the (mixed) second-order sensitivities are computed exactly by the 2nd-CASAM-L in at most *TP* computations, as opposed to TP^2 computations required by other methods, and so on. For every lower-order sensitivity of interest, the nth-CASAM-L computes the *TP* "next-higher-order" sensitivities in one adjoint computation performed in a linearly increasing higher-dimensional Hilbert space. Very importantly, the nth-CASAM-L computes the higher-level adjoint sensitivity functions using the same forward and adjoint solvers (i.e., computer codes) as used for solving the original forward and adjoint systems, thus requiring relatively minor additional software development for computing the various-order sensitivities.

The nth-CASAM-L is the only practically implementable methodology for obtaining the exact expressions (i.e., free of methodologically introduced approximations) of arbitrarily high-order sensitivities (functional derivatives) of model responses to model parameters, for coupled forward/adjoint linear systems. By enabling the practical computation of arbitrarily high-order response sensitivities to model parameters, the nth-CASAM-L makes it possible to compare the relative values of the sensitivities of various order, in order to assess which sensitivities are important and which may actually be neglected. Such quantitative comparisons enable investigations of the convergence of the (multivariate) Taylor-series expansion of the response in terms of parameter variations, as well as investigating the actual validity of expressions that are derived from Taylor-expansion of the response (e.g., response variances/covariance, skewness, kurtosis) as functions of the model's parameters. The larger the number of model parameters, the more efficient the nth-CASAM-L becomes for computing arbitrarily high-order response sensitivities. The nth-CASAM-L also enables the direct and exact derivation (first on paper, then programing the exact expression thus obtained for numerical evaluation) of the expression of a specific high-order sensitivity of interest, which can subsequently be computed directly; such a direct computation is not possible with any statistical method.

The nth-CASAM-L presented in this chapter provides a fundamentally important step in the quest to overcome the "curse of dimensionality" in sensitivity analysis, uncertainty quantification and predictive modeling. Volume 3 of this book will present the "nth-Order Comprehensive Adjoint Sensitivity Analysis Methodology for Nonlinear Systems" (abbreviation: nth-CASAM-N) for the practical, efficient, and exact computation of arbitrarily high-order sensitivities of responses to model parameters for nonlinear systems. The nth-CASAM-L, together with the nth-CASAM-N, are expected to revolutionize all of the fields of activities which require response sensitivities, including the fields of uncertainty quantification, model validation, optimization, data assimilation, model calibration, sensor fusion, reduced-order modeling, inverse problems, and predictive modeling.

Chapter 3
Illustrative Applications of the nth-CASAM-L to Paradigm Physical Systems with Imprecisely Known Properties, Internal and External Boundaries

3.1 Introduction

The application of the *nth-Order Comprehensive Adjoint Sensitivity Analysis Methodology for Response-Coupled Forward/Adjoint Linear Systems* (abbreviated as "nth-CASAM-L") is illustrated in this chapter by considering two paradigm models, as follows: (1) one of the models describes the spatial transmission of particles (e.g., neutrons and/or photons) through homogenized materials, as encountered in radiation protection and shielding; and (2) the other model simulates a typical initial-condition problem, describing the evolution in time of a quantity that obeys a linear conservation equation in a homogeneous mixture of materials, starting from an imprecisely known initial quantity, measured at an imprecisely known initial-time value and evolving toward an imprecisely known final-time value. Both of these models comprise many parameters, including internal and external boundaries and initial/final-time conditions, all of which are considered to be imprecisely known, subject to uncertainties.

The particle transmission model is described in Sect. 3.2. The responses considered in this model are as follows: (1) a point-detector response, which simulates a measurement of the particle flux within or on the surface of the medium/material in which the particles propagate; (2) a particle leakage response, which accounts for the particles leaving trough the outer boundary of the medium; (3) a reaction rate response, which simulates the average interaction rate of particles with a detector within or on the surface of the medium; (4) a "contributon-response flux," which is defined as the product of the forward and adjoint particle fluxes within the medium, and which is of importance to radiation shielding and protection. The first-order sensitivities of these responses to the model's imprecisely known internal parameters, interfaces and outer boundaries, are determined by applying the 1st-CASAM-L and are presented in Sect. 3.3. The corresponding second-order sensitivities are determined by applying the 2nd-CASAM-L and are presented in Sect. 3.4. Illustrative computations of third-order sensitivities to imprecisely known boundaries and

D. G. Cacuci, *The nth-Order Comprehensive Adjoint Sensitivity Analysis Methodology, Volume I*, https://doi.org/10.1007/978-3-030-96364-4_3

sources are determined by applying the 3rd-CASAM-L and are presented in Sect. 3.5. The first-, second-, and third-order sensitivities of responses that depend on both the forward and/or adjoint particle fluxes are obtained exactly, in closed form, underscoring the principles and methodology underlying the nth-CASAM-L.

The principles underlying the nth-CASAM-L are also illustrated in Sect. 3.6 by its application to a paradigm time-evolution model which simulates the typical evolution in time of a quantity that evolves according to the following linear conservation equation law from an imprecisely known initial quantity measured at an imprecisely known initial-time toward an imprecisely known final-time. The complete set of first- and second-order sensitivities of a response that simulates the measurement at an instance in time of the model's state function are determined in Sects. 3.6.1 and 3.6.2.1–3.6.2.6. Subsequently, Sects. 3.6.2.7 and 3.6.2.8 present the important concept of aggregating fundamental model parameters in order to reduce drastically the number of large-scale computation required for determining sensitivities of order higher than first. Since a response that simulates a measurement at a certain instance in time may not depend on the model's final-time, Sect. 3.6.3 illustrates the determination of first- and second-order sensitivities for an average reaction rate response which depends both on the initial- and final-time values.

As highlighted in the concluding discussion presented in Sect. 3.7, the models analyzed in this chapter were selected because they involve many uncertain parameters, yet the expressions of the responses and their sensitivities of various orders can be obtained exactly, in closed form, in terms of elementary functions. This feature conveniently enables the verification of all results for the various adjoint functions and response sensitivities, thus serving the didactic aims of the material presented in this chapter.

3.2 Transmission of Particles Through Media

Consider a source of monoenergetic particles (e.g., neutrons or gamma ray photons), denoted as q, which is distributed within a one-dimensional homogenized inner slab of thickness $2b_1$ [cm]. The various physical quantities that characterize this inner slab will be denoted using the subscript "s," to indicate "source" material properties. Thus, the inner slab is considered to comprise M_s distinct nuclides, each nuclide being characterized by its total microscopic interaction cross section with the source particles, denoted as $\sigma_{s,i}$, and by its atomic number density, denoted as $N_{s,i}$, for $i = 1, \ldots, M_s$. The inner slab is shielded on each of its outer surfaces by an adjacent outer slab of thickness b_2 [cm], which does not contain sources. Each of the outer slabs is made of homogenized material comprising M distinct nuclides, each nuclide being characterized by its total microscopic cross section, denoted as σ_i, and by its atomic number density, denoted as N_i, for $i = 1, \ldots, M$. All of the number densities and cross sections are considered to be imprecisely known. The thicknesses of the slabs are also considered to be subject to manufacturing tolerances; hence, the internal boundaries (i.e., internal interfaces between the slabs) as well as the outer boundaries are imprecisely known. Since this arrangement of slabs is symmetrical with respect to the midplane of the inner slab, it is convenient to choose the origin of the one-dimensional coordinate system, which will de denoted as the "z-direction," to be at the inner slab's midplane. Consequently, the transport equation governing the uncollided angular flux of monoenergetic particles, denoted as $u_s(z,\omega)$, through the inner slab in the positive z-direction has the following form:

$$\omega \frac{du_s(z,\omega)}{dz} + \mu_s u_s(z,\omega) = q, \quad 0 < z < b_1, \ 0 \le \omega \le 1, \tag{3.1}$$

where ω denotes the cosine of the angle between the particle's direction and the z-axis, while μ_s denotes the interaction coefficient of the particles (photons or neutrons) with the inner slab's homogenized material and is defined as follows:

$$\mu_s \triangleq \sum_{i=1}^{M_s} N_{s,i} \sigma_{s,i}. \tag{3.2}$$

Due to the model's symmetry (by choosing the origin of the coordinate system at the inner slab's midplane), the appropriate boundary condition for the inner slab is:

$$\frac{du_s(z,\omega)}{dz} = 0, \quad \text{at } z = 0, \ 0 \le \omega \le 1. \tag{3.3}$$

The solution of (3.1) and (3.3) is the following constant function:

$$u_s(z,\omega) = \frac{q}{\mu_s}, \quad 0 < z < b_1, \ 0 \le \omega \le 1. \tag{3.4}$$

Due to the model's symmetry, the expression obtained in (3.4) remains valid for all particle directions, i.e., for $-1 \le \omega \le 1$.

The particle transport equation governing the uncollided angular flux of monoenergetic particles, denoted as $\varphi(z,\omega)$, through the outer slab in particle directions $0 \le \omega \le 1$ has the following form:

$$\omega \frac{d\varphi(z,\omega)}{dz} + \mu\varphi(z,\omega) = 0, \quad b_1 < z < b_2, \ 0 \le \omega \le 1, \tag{3.5}$$

where the interaction coefficient μ is defined as follows:

$$\mu \triangleq \sum_{i=1}^{M} N_i \sigma_i. \tag{3.6}$$

The uncollided flux is continuous across the interface at $z = b_1$, i.e.,

$$\varphi(b_1,\omega)=u_s(b_1,\omega)=\frac{q}{\mu_s}. \tag{3.7}$$

For subsequent verification of the expressions to be obtained for the response sensitivities to parameters, the closed-form expression of the uncollided flux $\varphi(z,\omega)$, which is the solution of (3.5) and (3.7), is provided below:

$$\varphi(z,\omega)=\frac{q}{\mu_s}\exp\left[\frac{\mu}{\omega}(b_1-z)\right],\ b_1\le z\le b_2,\ 0\le\omega\le 1. \tag{3.8}$$

The linear particle transport model described by (3.5) and (3.7) is naturally set in a Hilbert space endowed with an inner-product, denoted as $\langle\varphi(z,\omega),\psi(z,\omega)\rangle_0$, between two square integrable functions $\varphi(z,\omega)$ and $\psi(z,\omega)$, defined as follows:

$$\langle\varphi(z,\omega),\psi(z,\omega)\rangle_0\triangleq\int_0^1 d\omega\int_{b_1}^{b_2}\varphi(z,\omega)\psi(z,\omega)dz. \tag{3.9}$$

The Hilbert space endowed with the inner product defined in (3.9) will be denoted as H_0, where the subscript "zero" denotes "zeroth-level" or "original." Higher-level Hilbert spaces, which will be denoted as H_1, H_2, etc., will also be introduced and used in this and subsequent sections in this chapter.

In the Hilbert space H_0, the adjoint model corresponding to the original forward model represented by (3.5) and (3.7) is readily constructed by using the inner product defined in (3.9) to obtain the following relation for the adjoint function $\psi(z,\omega)$:

$$\begin{aligned}&\int_0^1 d\omega\int_{b_1}^{b_2}\psi(z,\omega)dz\left[\omega\frac{d\varphi(z,\omega)}{dz}+\mu\varphi(z,\omega)\right]\\&=\int_0^1 d\omega\int_{b_1}^{b_2}\varphi(z,\omega)dz\left[-\omega\frac{d\psi(z,\omega)}{dz}+\mu\psi(z,\omega)\right]\\&+\int_0^1 \omega d\omega[\psi(b_2,\omega)\,\varphi(b_2,\omega)-\psi(b_1,\omega)\,\varphi(b_1,\omega)].\end{aligned} \tag{3.10}$$

It follows from the relation provided in (3.10) that the adjoint model for the adjoint function $\psi(z,\omega)$ is as follows:

$$-\omega\frac{d\psi(z,\omega)}{dz}+\mu\psi(z,\omega)=q^*(z,\omega),\quad b_1<z<b_2,\quad 0\le\omega\le 1, \tag{3.11}$$

$$\psi(b_2,\omega)=0, \tag{3.12}$$

where $q^*(z, \omega)$ is a source-term which is usually determined by the model response under consideration, while the boundary condition provided in (3.12) has been chosen in order to eliminate the appearance of the unknown function $\varphi(b_2, \omega)$ from the bilinear concomitant represented by the second term on the right side of (3.10).

3.2.1 Point-Detector Response

Doses of uncollided particles, including from point-doses, particle leakage and reaction rates, are of interest in many shielding applications (Shultis and Faw 2000; Stacey 1974; Williams and Engle 1977; Williams 1986). In particular, the result of a measurement of the particle flux at any phase-space point (z_d, ω_d) within or on the surface of the medium, $b_1 \le z_d \le b_2$, $0 < \omega_d < 1$, which will be denoted as $R(\varphi; \boldsymbol{\alpha})$, is represented by the following mathematical expression:

$$R(\varphi;\boldsymbol{\alpha})=\mu_d\int_0^1 d\omega\,\delta(\omega-\omega_d)\int_{b_1}^{b_2}\varphi(z,\omega)\,\delta(z-z_d)dz, \tag{3.13}$$
$$0<\omega_d<1, b_1<z_d<b_2.$$

The phase-space coordinates (z_d, ω_d) of the detector's location will also be considered to be imprecisely known. The detector's effective macroscopic cross section (also called "interaction coefficient") is considered to be an imprecisely known parameter denoted as μ_d and defined as follows:

$$\mu_d \triangleq \sum_{i=1}^{M_d} N_{d,i}\sigma_{d,i}. \tag{3.14}$$

In (3.14), the quantity M_d denotes the number of distinct nuclides comprising the detector's response function, $\sigma_{d,i}$ denotes the total microscopic interaction cross section of the respective nuclide with the source particles, and $N_{d,i}$ denotes the respective nuclide's atomic number density, for $i = 1, \ldots, M_d$.

The uncertain parameters that describe the properties (microscopic cross sections, atomic number densities, source, slab boundaries) characterizing the two materials and the detector will be generically denoted as α_i, $i = 1, \ldots, TP$, where TP denotes the "total number of model parameters." These parameters will be considered to be the components of the (column) "vector of imprecisely known model parameters" defined as follows:

$$\boldsymbol{\alpha} \triangleq (\alpha_1, \ldots, \alpha_{TP})^\dagger \triangleq [N_{s,1}, \ldots, N_{s,M_s}; \sigma_{s,1}, \ldots, \sigma_{s,M_s}; \\ N_1, \ldots, N_M; \sigma_1, \ldots, \sigma_M; N_{d,1}, \ldots, N_{d,M_d}; \sigma_{d,1}, \ldots, \sigma_{d,M_d}; q; z_d, \omega_d; b_1, b_2]^\dagger. \tag{3.15}$$

The dagger "(†)" denotes "transposition" and all vectors are considered to be column vectors except if the contrary is explicitly stated.

All of the components of the vector of parameters $\boldsymbol{\alpha}$ are considered to be imprecisely known (i.e., uncertain). It is assumed that only the parameters' nominal values, which will be denoted using the superscript "0" (i.e., α_i^0, q^0, z_d^0, a^0), are known/available. The nominal values of the model parameters are considered to be components of the column vector of "nominal parameter values," denoted as $\boldsymbol{\alpha}^0$ and defined as follows:

$$\boldsymbol{\alpha}^0 \triangleq \left(\alpha_1^0, \ldots, \alpha_{TP}^0\right)^\dagger \triangleq \left[N_{s,1}^0, \ldots, N_{s,M_s}^0; \sigma_{s,1}^0, \ldots, \sigma_{s,M_s}^0; \\ N_1^0, \ldots, N_M^0; \sigma_1^0, \ldots, \sigma_{M_s}^0; N_{d,1}^0, \ldots, N_{d,M_d}^0; \sigma_{d,1}^0, \ldots, \sigma_{d,M_d}^0; q^0; z_d^0, \omega_d^0; b_1^0, b_2^0\right]^\dagger. \tag{3.16}$$

The parameters which appear as components of the vector defined in (3.15) are the model's *fundamental parameters*. A variation in any of these *fundamental* parameters is independent of any variation in any of the other fundamental parameters. Not all of the fundamental parameters necessarily appear explicitly in the model's equations or in the definition of the model's response. A model and/or response may depend explicitly on derived parameters or correlations, which are functions of the fundamental parameters. For example, in the particle transport model under consideration, the interaction coefficients appear explicitly in the definition of the model's response and in the particle conservation equations that underly the paradigm model. On the other hand, the nuclide number densities and/or microscopic cross sections (which are independent/fundamental parameters) do not appear explicitly either in the definition of the response or in the equations that underly the model. Parameters such as the interaction coefficients could be called *derived parameters* and they are important because they may act like independent parameters in the definition of the model's underlying equations and/or response, in which case it suffices to determine the sensitivities of the response to these derived parameters using the model. Subsequently, the sensitivities of the response to the underlying fundamental parameters can be computed separately, by simple differentiations. For example, in the paradigm model under consideration, after having determined the sensitivities of the form $\partial R/\partial \mu$ of the detector response with respect to the interaction coefficient using the model's equation, the sensitivities to the underlying fundamental parameters (i.e., corresponding nuclide number densities and microscopic cross sections) can be determined by using the chain rule, without involving the model's equations, since $\partial R/\partial \sigma_i = [\partial R/\partial \mu][\partial \mu/\partial \sigma_i] = N_i[\partial R/\partial \mu]$, and so on. This observation may enable a significant reduction in the number of computations needed to determine the sensitivities of the response to the fundamental parameters. Such simplifications often occur in *linear systems*, since the parameters that characterize such systems are not functions of the dependent variable (e.g., the interaction coefficients are independent of the particle flux). In linear heat conduction models, for example, the conductivity is independent of the medium's temperature (dependent variable), so the conductivity can be considered either a primary parameter (if it does not stem from some correlation) or a derived parameters (if it stems from a temperature-independent correlation).

In view of the above discussion, it is convenient to refer to the model parameters which appear explicitly in the model's equations, boundary/initial conditions, and in the model's response as *primary parameters*, since the response sensitivities will be computed initially with respect to these parameters. The primary parameters could be either fundamental parameters or derived parameters, as is also the case for the illustrative paradigm particle transport model considered in this work. After obtaining the response sensitivities to the model's primary parameters, if any of these parameters are "derived parameters," the sensitivities of the response to the fundamental parameters which are comprised in the respective "derived parameter" can be computed efficiently and exactly without needing to use the model's complete set of equations, thus gaining considerable accuracy and savings of computational resources. These issues will be illustrated by the derivations which will be performed in the remainder of this chapter.

In view of the above discussion, the interaction coefficients will be denoted as $\mu(\boldsymbol{\alpha})$, $\mu_s(\boldsymbol{\alpha})$, and $\mu_d(\boldsymbol{\alpha})$, even though they only depend on some, but not all, of the components of $\boldsymbol{\alpha}$. This short-hand notation conveniently simplifies the list of the actual arguments (i.e., the corresponding nuclide number densities and microscopic cross sections) of the respective interaction coefficients.

Inserting the expression obtained in (3.8) into (3.13) yields the following exact, closed-form, expression for the model's response:

$$R(z_d, \omega_d; \boldsymbol{\alpha}) = \frac{q\mu_d(\boldsymbol{\alpha})}{\mu_s(\boldsymbol{\alpha})} \exp\left[\frac{\mu(\boldsymbol{\alpha})}{\omega_d}(b_1 - z_d)\right], \; b_1 \le z_d \le b_2, \; 0 \le \omega_d \le 1. \quad (3.17)$$

In practice, the closed-form solution provided in (3.8) is unavailable. Instead, the nominal value $\varphi^0(z,\omega)$ is obtained by numerically solving (3.5) and (3.7) using the nominal parameter values represented by (3.16). The exact, closed-form expression of the model response provided in (3.17) is also unavailable in practice, so the value of the response must be computed by evaluating numerically the expression provided in (3.13) using the nominal value $\varphi^0(z,\omega)$ of the model's state function together with the nominal parameter values $\boldsymbol{\alpha}^0$.

Notably, the response $R(z_d, \omega_d; \boldsymbol{\alpha})$ can also be represented in terms of an adjoint function $\psi(z,\omega)$ by using the relation provided in (3.10); the resulting expression is:

$$R(z_d, \omega_d; \boldsymbol{\alpha}) = \frac{q}{\mu_s} \int_0^1 \omega d\omega \int_{b_1}^{b_2} \psi(z, \omega) \delta(z - b_1) dz, \tag{3.18}$$

where the adjoint function $\psi(z, \omega)$ is the solution of the following adjoint system:

$$-\omega \frac{d\psi(z, \omega)}{dz} + \mu\psi(z, \omega) = \mu_d \delta(\omega - \omega_d) \delta(z - z_d);$$
$$b_1 < z < b_2; \; 0 \leq \omega \leq 1; \tag{3.19}$$

$$\psi(b_2, \omega) = 0. \tag{3.20}$$

The adjoint system comprising (3.19) and (3.20) indicates that the adjoint function $\psi(z, \omega)$ plays the role of a Green's function for the point-detector response $R(z_d, \omega_d; \boldsymbol{\alpha})$. As highlighted by the expression on the right side of (3.18), the adjoint function $\psi(z, \omega)$ also plays the role of an "importance function" in weighing the particle transmitted from the source q to the response $R(z_d, \omega_d; \boldsymbol{\alpha})$. Notably, the model response $R(z_d, \omega_d; \boldsymbol{\alpha})$ itself plays the role of a Green's function in slab geometry, in that it can be integrated over the phase-space coordinates (z_d, ω_d) to obtain other responses of interest involving uncollided doses from distributed sources. This property will be illustrated in subsequent sections in this chapter.

3.2.2 *Particle Leakage Response*

Another response of significant interest is the leakage of particles through the outer slab's surface at $z = b_2$. Such a response will be denoted as $R_L(\varphi; \boldsymbol{\alpha})$ and is represented mathematically by the following expression:

$$R_L[\varphi(z, \omega); \boldsymbol{\alpha}] \triangleq \mu_d \int_0^1 d\omega \int_{b_1}^{b_2} \varphi(z, \omega) \delta(z - b_2) dz. \tag{3.21}$$

Replacing the expression of $\varphi(z, \omega)$ obtained in (3.8) into (3.21) yields the following expression for the particle leakage response $R_L(\varphi; \boldsymbol{\alpha})$:

$$R_L(\varphi; \boldsymbol{\alpha}) = \frac{q\mu_d(\boldsymbol{\alpha})}{\mu_s(\boldsymbol{\alpha})} E_2[(b_2 - b_1)\mu(\boldsymbol{\alpha})], \tag{3.22}$$

where the exponential-integral function is defined as follows:

$$E_n(z) \triangleq \int_0^1 y^{n-2} e^{-z/y} dy; \;\; E_0(z) = \frac{e^{-z}}{z};$$
$$E_1(z) \triangleq \int_0^1 y^{-1} e^{-z/y} dy = -Ei(-z); \;\; E_2(0) = 1. \tag{3.23}$$

The expression of the leakage response shown in (3.22) can evidently be obtained by setting $z_d = b_2$ in the expression of the point-detector response $R(z_d, \omega_d; \boldsymbol{\alpha})$ and by considering the point ω_d in (3.17) to be a phase-space variable (rather than a fixed-point). Thus, integrating (3.17) over ω_d yields the expression in (3.22), i.e.,

$$\int_0^1 d\omega_d \, R(b_2, \omega_d; \boldsymbol{\alpha}) = \frac{q\mu_d(\boldsymbol{\alpha})}{\mu_s(\boldsymbol{\alpha})} E_2[(b_2 - b_1)\mu(\boldsymbol{\alpha})] = R_L[\varphi(z, \omega); \boldsymbol{\alpha}]. \tag{3.24}$$

3.2.3 Reaction Rate Response

Interaction/reaction rates of particles within the medium they travel through are also of significant interest. In particular, the interaction/reaction rate of uncollided particles within the entire slab will be denoted as $R_r(\varphi; \boldsymbol{\alpha})$ and is represented mathematically by the following expression:

$$R_r[\varphi(z, \omega); \boldsymbol{\alpha}] \triangleq \mu_d \int_0^1 d\omega \int_{b_1}^{b_2} \varphi(z, \omega) \, dz. \tag{3.25}$$

Replacing the expression of $\varphi(z, \omega)$ obtained in (3.8) into (3.25) yields the following closed-form expression for the reaction rate response $R_r(\varphi; \boldsymbol{\alpha})$:

$$R_r(\varphi; \boldsymbol{\alpha}) = \frac{q\mu_d(\boldsymbol{\alpha})}{\mu_s(\boldsymbol{\alpha})\mu(\boldsymbol{\alpha})} \left\{ \frac{1}{2} - E_3[(b_2 - b_1)\mu(\boldsymbol{\alpha})] \right\}. \tag{3.26}$$

Just as in the case of the leakage response, cf. (3.24), the reaction-rate response $R_r(\varphi; \boldsymbol{\alpha})$ can also be obtained by considering the phase-space points (z_d, ω_d) in the expression of the point-detector response $R(z_d, \omega_d; \boldsymbol{\alpha})$ as being variables that can be integrated over the domain; thus, integrating over (z_d, ω_d) yields the following result:

$$\int_0^1 d\omega_d \int_{b_1}^{b_2} R(z_d, \omega_d; \boldsymbol{\alpha})\, dz_d$$
$$= \frac{q\mu_d(\boldsymbol{\alpha})}{\mu_s(\boldsymbol{\alpha})\mu(\boldsymbol{\alpha})}\left\{\frac{1}{2} - E_3[(b_2 - b_1)\mu(\boldsymbol{\alpha})]\right\} = R_r(\varphi; \boldsymbol{\alpha}). \tag{3.27}$$

The results obtained in (3.24) and (3.27) underscore the role as "Green's Function" (or "kernel") played by the point-detector response $R(z_d, \omega_d; \boldsymbol{\alpha})$.

3.2.4 Contributon-Response Flux

Bilinear functionals of the forward and adjoint particle fluxes of the form $\int O(\mathbf{x})\varphi(\mathbf{x})\psi(\mathbf{x})\, d\mathbf{x}$, where $\mathbf{x}$ denotes the vector of independent variable, $\varphi(\mathbf{x})$ and $\psi(\mathbf{x})$ denote the forward and adjoint particle fluxes, respectively, and $O(\mathbf{x})$ denotes a function that does not depend on the fluxes, occur in many Lagrangian, Raleigh, Roussopoulos, and Schwinger functionals (to mention just the most prominent such functionals); these functionals are used in many practical applications (see, e.g., Stacey 1974). Perhaps the simplest example of a response that depends simultaneously on both the forward and the adjoint particle flux is the so-called "contributon-response flux," which has the form $\int \delta(\mathbf{x} - \mathbf{x}_0)\varphi(\mathbf{x})\psi(\mathbf{x})d\mathbf{x}$ and which is encountered in "channel theory" or "contributon theory," with applications in particle shielding analysis (Williams and Engle 1977; Williams 1986). The contributon-response flux retains the essential features which characterize the sensitivity analysis of responses that depend simultaneously on both the forward and the adjoint particle fluxes.

For the illustrative paradigm model of uncollided particle transmission through the slab $b_1 \le z \le b_2$ governed by (3.5) and (3.7), the "contributon-response flux" at a location z is defined as follows:

$$\rho(z) = \int_0^1 \varphi(z, \omega)\psi(z, \omega)d\omega, \;\; b_1 \le z \le b_2, \tag{3.28}$$

where the function $\psi(z, \omega)$ is the solution of the following adjoint particle transport model:

$$-\omega \frac{d\psi(z, \omega)}{dz} + \mu\psi(z, \omega) = q^*, \;\; b_1 \le z \le b_2, \;\; 0 \le \omega \le 1, \tag{3.29}$$

$$\psi(b_2, \omega) = 0, \;\; 0 \le \omega \le 1. \tag{3.30}$$

The adjoint source parameter q^* characterizes particles that reach a detector within or on the outside surface of the slab. The adjoint particle transport model comprising (3.29) and (3.30) has the following closed-form solution:

$$\psi(z,\omega) = \frac{q^*}{\mu(\boldsymbol{\alpha})}\left\{1 - \exp\left[(z-b_2)\frac{\mu(\boldsymbol{\alpha})}{\omega}\right]\right\}, \; b_1 \le z \le b_2, \;\; 0 \le \omega \le 1. \quad (3.31)$$

The model response which will be subsequently considered in this chapter is the "contributon-response flux" at a location z_0 within the slab $b_1 \le z_0 \le b_2$, which is defined as follows:

$$\rho(z_0) = \int_0^1 d\omega \int_{b_1}^{b_1} \varphi(z,\omega)\psi(z,\omega)\delta(z-z_0)dz, \;\; b_1 \le z_0 \le b_2. \quad (3.32)$$

The closed-form expression of $\rho(z_0)$ is readily obtained by inserting into (3.32) the expressions of $\varphi(z,\omega)$ and $\psi(z,\omega)$ from (3.8) and (3.31), respectively, to obtain the following result:

$$\rho(z_0) = \frac{qq^*}{\mu_s\mu}\{E_2[(z_0-b_1)\mu] - E_2[(b_2-b_1)\mu]\}. \quad (3.33)$$

3.3 Application of the 1st-CASAM-L to Compute First-Order Response Sensitivities to Imprecisely Known Parameters

Consider arbitrary parameter variations $\delta\boldsymbol{\alpha}$ around the parameters' nominal values, $\boldsymbol{\alpha}^0$, where the vector $\delta\boldsymbol{\alpha}$ is defined as follows:

$$\delta\boldsymbol{\alpha} \triangleq (\delta\alpha_1, \ldots, \delta\alpha_{TP})^\dagger \triangleq [\delta N_{s,1}, \ldots, \delta N_{s,M_s}; \delta\sigma_{s,1}, \ldots, \delta\sigma_{s,M_s}; \delta N_1, \ldots, \delta N_M; \delta\sigma_1, \ldots, \delta\sigma_M; \delta N_{d,1}, \ldots, \delta N_{d,M_d}; \delta\sigma_{d,1}, \ldots, \delta\sigma_{d,M_d}; \delta q; \delta z_d, \delta\omega_d; \delta b_1, \delta b_2]^\dagger. \quad (3.34)$$

Such parameter variations will cause variations in the model's response, both directly and indirectly, through variations they cause—via the model's underlying equations—in the model's state functions (i.e., dependent variables).

3.3.1 Point-Detector Response

The total first-order sensitivity of the response defined in (3.13) to the model's primary parameters (including interface and boundary locations) is provided by the Gateaux- (G-) differential $\{\delta R(\varphi;\boldsymbol{\alpha};\delta\varphi;\delta\boldsymbol{\alpha})\}_{\boldsymbol{\alpha}^0}$ of the response $R(\varphi;\boldsymbol{\alpha})$ for arbitrary variations $(\delta\varphi,\delta\boldsymbol{\alpha})$ around the nominal parameter and state functions values, which is defined as follows:

$$\begin{aligned}\delta R(\varphi;\boldsymbol{\alpha};\delta\varphi;\delta\boldsymbol{\alpha})_{\boldsymbol{\alpha}^0}&\triangleq\frac{d}{d\varepsilon}\{[(\mu_d+\varepsilon\delta\mu_d)\int_0^1 d\omega\int_{b_1+\varepsilon\delta b_1}^{b_2+\varepsilon\delta b_2}(\varphi+\varepsilon\delta\varphi)\\&\times\delta(z-z_d-\varepsilon\delta z_d)\delta(\omega-\omega_d-\varepsilon\delta\omega_d)]_{\boldsymbol{\alpha}^0}dz\}_{\varepsilon=0}\\&=\{\delta R(\varphi;\boldsymbol{\alpha};\delta\boldsymbol{\alpha})_{\boldsymbol{\alpha}^0}\}_{dir}+\{\delta R(\varphi;\boldsymbol{\alpha};\delta\varphi)_{\boldsymbol{\alpha}^0}\}_{ind},\end{aligned}\tag{3.35}$$

where the "direct-effect" term depends only on variations $\delta\boldsymbol{\alpha}$ in the parameters and is defined as follows:

$$\begin{aligned}\{\delta R(\varphi;\boldsymbol{\alpha};\delta\boldsymbol{\alpha})_{\boldsymbol{\alpha}^0}\}_{dir}&\triangleq\left\{\delta\mu_d\int_0^1 d\omega\,\delta(\omega-\omega_d)\int_{b_1}^{b_2}dz\,\varphi(z,\omega)\delta(z-z_d)\right\}_{\boldsymbol{\alpha}^0}\\&-(\delta z_d)\left\{\mu_d\int_0^1 d\omega\,\delta(\omega-\omega_d)\int_{b_1}^{b_2}\varphi(z,\omega)\,\delta'(z-z_d)dz\right\}_{\boldsymbol{\alpha}^0}\\&-(\delta\omega_d)\left\{\mu_d\int_0^1 d\omega\,\delta'(\omega-\omega_d)\int_{b_1}^{b_2}\varphi(z,\omega)\,\delta(z-z_d)dz\right\}_{\boldsymbol{\alpha}^0},\end{aligned}\tag{3.36}$$

and where the "indirect-effect" term depends only on variations $\delta\varphi(z,\omega)$ in the dependent variable (state function) and is defined as follows:

$$\begin{aligned}&\{\delta R(\varphi;\boldsymbol{\alpha};\delta\varphi)_{\boldsymbol{\alpha}^0}\}_{ind}\triangleq\left\{\mu_d\int_0^1 d\omega\,\delta(\omega-\omega_d)\int_{b_1}^{b_2}\delta\varphi(z,\omega)\,\delta(z-z_d)dz\right\}_{\boldsymbol{\alpha}^0},\\&b_1\le z_d\le b_2,\;\;0\le\omega_d\le 1.\end{aligned}\tag{3.37}$$

The direct-effect term in (3.36) can be computed immediately at this stage, since the parameter variations are known, i.e.,

$$\delta\mu_s = \sum_{i=1}^{M_s} \left[(\delta N_{s,i})\sigma_{s,i} + N_{s,i}(\delta\sigma_{s,i})\right]; \delta\mu = \sum_{i=1}^{M} \left[(\delta N_i)\sigma_i + N_i(\delta\sigma_i)\right];$$
$$\delta\mu_d = \sum_{i=1}^{M_d} \left[(\delta N_{d,i})\sigma_{d,i} + N_{d,i}(\delta\sigma_{d,i})\right]. \tag{3.38}$$

On the contrary, the "indirect-effect" terms in (3.37) cannot be computed at this stage since the first-order variation, $\delta\varphi(z,\omega)$, is not available at this stage, but could be determined by solving the First-Level Variational Sensitivity System (1st-LVSS) obtained by G-differentiating (3.5) and (3.7) at the nominal parameter values, to obtain:

$$\left\{\frac{d}{d\varepsilon}\left[\omega\frac{d(\varphi+\varepsilon\delta\varphi)}{dz} + (\mu+\varepsilon\delta\mu)(\varphi+\varepsilon\delta\varphi)\right]_{\boldsymbol{\alpha}^0}\right\}_{\varepsilon=0} = 0,$$
$$b_1^0 < z < b_2^0,\ 0 \le \omega \le 1, \tag{3.39}$$

$$\left\{\frac{d}{d\varepsilon}\left[\varphi\left(z=b_1^0+\varepsilon\delta b_1;\omega\right) + \varepsilon\delta\varphi\left(z=b_1^0+\varepsilon\delta b_1;\omega\right)\right]\right\}_{\varepsilon=0}$$
$$= \left\{\frac{d}{d\varepsilon}\left[\frac{q+\varepsilon\delta q}{\mu_s(\boldsymbol{\alpha})+\varepsilon\delta\mu_s}\right]_{\boldsymbol{\alpha}^0}\right\}_{\varepsilon=0}. \tag{3.40}$$

Carrying out the differentiations with respect to ε in (3.39) and (3.40) yields the following 1st-LVSS to be satisfied by the first-order variation $\delta\varphi(z,\omega)$:

$$\left\{\omega\frac{d}{dz}[\delta\varphi(z,\omega)] + \mu[\delta\varphi(z,\omega)]\right\}_{\boldsymbol{\alpha}^0} = -(\delta\mu)\{\varphi(z,\omega)\}_{\boldsymbol{\alpha}^0};$$
$$b_1^0 < z < b_2^0,\ 0 \le \omega \le 1, \tag{3.41}$$

$$\{\delta\varphi(z,\omega)\}_{z=b_1^0} = \left\{\frac{(\delta q)}{\mu_s(\boldsymbol{\alpha})} - \frac{q\delta\mu_s(\boldsymbol{\alpha})}{\mu_s^2(\boldsymbol{\alpha})}\right\}_{\boldsymbol{\alpha}^0} - \delta b_1\left\{\frac{d\varphi(z,\omega)}{dz}\right\}_{z=b_1^0};$$
$$z = b_1^0, 0 \le \omega \le 1. \tag{3.42}$$

The quantity $\{d\varphi/dz\}_{z=b_1^0}$, which appears in (3.42), can be evaluated by using either (3.5) or (3.8), to obtain the following boundary condition for the variation $\delta\varphi(z,\omega)$:

$$\{\delta\varphi(z,\omega)\}_{z=b_1^0} = \left\{\frac{(\delta q)}{\mu_s(\boldsymbol{\alpha})} - \frac{q\delta\mu_s(\boldsymbol{\alpha})}{\mu_s^2(\boldsymbol{\alpha})}\right\}_{\boldsymbol{\alpha}^0} + \delta b_1\left\{\frac{q\mu(\boldsymbol{\alpha})}{\omega\mu_s(\boldsymbol{\alpha})}\right\}_{\boldsymbol{\alpha}^0};$$
$$z = b_1^0, 0 \le \omega \le 1. \tag{3.43}$$

In principle, it is possible to solve the 1st-LVSS for each parameter variation, $\delta\alpha_i$, $i = 1, \ldots, TP$, as follows:

1. For each parameter variation $\delta\alpha_i$, $i = 1, \ldots, TP$, the 1st-LVSS is equivalent to the following system of equations:

$$\left\{\omega\frac{d}{dz}[\delta\varphi_i(z,\omega)] + \mu[\delta\varphi_i(z,\omega)]\right\}_{\boldsymbol{\alpha}^0} = -(\delta\alpha_i)\left\{\frac{\partial\mu(\boldsymbol{\alpha})}{\partial\alpha_i}\varphi(z,\omega)\right\}_{\boldsymbol{\alpha}^0}; \quad b_1^0 < z < b_2^0,\ 0 \le \omega \le 1, \tag{3.44}$$

$$\{\delta\varphi_i(z,\omega)\}_{z=b_1^0} = \left\{\frac{(\delta q)}{\mu_s(\boldsymbol{\alpha})} - \frac{q}{\mu_s^2(\boldsymbol{\alpha})}\frac{\partial\mu_s(\boldsymbol{\alpha})}{\partial\alpha_i}\delta\alpha_i\right\}_{\boldsymbol{\alpha}^0} - \delta b_1\left\{\frac{d\varphi(z,\omega)}{dz}\right\}_{z=b_1^0}; z = b_1^0, 0 \le \omega \le 1. \tag{3.45}$$

2. If the parameter variations $\delta\alpha_i$, $i = 1, \ldots, TP$, are independent of each other, (3.44) and (3.45) can be solved by setting $\delta\alpha_i = \delta_{ij}$, $i, j = 1, \ldots, TP$, where δ_{ij} represents the Kronecker delta-functional, which is defined as follows: $\delta_{ij} = 1$, $i = j$; $\delta_{ij} = 0$, $i \neq j$.

It is evident that solving (3.41) and (3.42) would require at least TP large-scale computations (i.e., at least one large-scale computation for each parameter variation) to obtain the first-order variation $\delta\varphi(z,\omega)$ for every parameter variation. Performing so many large-scale computations is impractical for large-scale systems involving many parameters. The alternative to solving repeatedly the 1st-LVSS for every possible variation in the system's imprecisely known parameters is to use the First-Level Adjoint Sensitivity System (1st-LASS), which is constructed by applying the following sequence of steps:

1. Consider that the function $\delta\varphi(z,\omega)$ belongs to a Hilbert space, which will be denoted as $H_1(\Omega_z)$, $\Omega_z \triangleq \{\omega \in [0,1] \otimes z \in [b_1, b_2]\}$. This Hilbert space is endowed with an inner product of two functions $\varphi(z,\omega) \in H_1$ and $\psi(z,\omega) \in H_1$, which is denoted as $\langle\varphi(z,\omega), \psi(z,\omega)\rangle_1$ and is defined as follows:

$$\langle\varphi(z,\omega), \psi(z,\omega)\rangle_1 \triangleq \left\{\int_0^1 d\omega \int_{b_1}^{b_2} \varphi(z,\omega)\psi(z,\omega)\, dz\right\}_{\boldsymbol{\alpha}^0}. \tag{3.46}$$

2. Using the definition provided in (3.46), construct the inner product of a function $\psi_1(z,\omega) \in H_1$ with (3.41) to obtain

$$\left\{ \int_0^1 d\omega \int_{b_1}^{b_2} \psi_1(z,\omega) \left[\omega \frac{d}{dz} \delta\varphi(z,\omega) + \mu \delta\varphi(z,\omega) \right] dz \right\}_{\boldsymbol{\alpha}^0} = - \left\{ (\delta\mu) \int_0^1 d\omega \int_{b_1}^{b_2} \psi_1(z,\omega)\varphi(z,\omega) dz \right\}_{\boldsymbol{\alpha}^0}. \tag{3.47}$$

3. Integrate by parts, over the independent variable z, the left side of (3.47) to obtain

$$\begin{aligned} &\left\{ \int_0^1 d\omega \int_{b_1}^{b_2} \psi_1(z,\omega) \left[\omega \frac{d}{dz} \delta\varphi(z,\omega) + \mu \delta\varphi(z,\omega) \right] dz \right\}_{\boldsymbol{\alpha}^0} \\ &= \int_0^1 \omega\, d\omega \{ \psi_1(b_2,\omega)\delta\varphi(b_2,\omega) - \psi_1(b_1,\omega)\delta\varphi(b_1,\omega) \}_{\boldsymbol{\alpha}^0} \\ &+ \left\{ \int_0^1 d\omega \int_{b_1}^{b_2} \delta\varphi(z,\omega) \left[-\omega \frac{d}{dz} \psi_1(z,\omega) + \mu \psi_1(z,\omega) \right] dz \right\}_{\boldsymbol{\alpha}^0}. \end{aligned} \tag{3.48}$$

4. Require the last term on the right side of (3.48) to represent the indirect-effect term defined in (3.37) by imposing the following relationship:

$$\left\{ \int_0^1 d\omega \int_{b_1}^{b_2} \delta\varphi(z,\omega) \left[-\omega \frac{d}{dz} \psi_1(z,\omega) + \mu \psi_1(z,\omega) \right] dz \right\}_{\boldsymbol{\alpha}^0} = \left\{ \mu_d \int_0^1 d\omega\, \delta(\omega - \omega_d) \int_{b_1}^{b_2} \delta\varphi(z,\omega)\, \delta(z - z_d) dz \right\}_{\boldsymbol{\alpha}^0}. \tag{3.49}$$

The relation in (3.49) implies that the following equation holds in the weak sense at the nominal parameter values:

$$\left\{ -\omega \frac{d}{dz} \psi_1(z,\omega) + \mu \psi_1(z,\omega) \right\}_{\boldsymbol{\alpha}^0} = \{ \mu_d \delta(\omega - \omega_d) \delta(z - z_d) \}_{\boldsymbol{\alpha}^0}, \quad b_1^0 \le z \le b_2^0,\ 0 \le \omega \le 1. \tag{3.50}$$

5. Complete the definition of the function $\psi_1(z,\omega)$ by requiring that the unknown value of the function $\delta\varphi(z,\omega)$ be eliminated from appearing on the right side of (3.48). This requirement is met by imposing the following condition to be satisfied by the function $\psi_1(z,\omega)$ on the model's outer boundary:

$$\psi_1\left(b_2^0,\omega\right)=0,\ \ 0\le\omega\le 1. \tag{3.51}$$

Altogether, (3.50) and (3.51) constitute a well-posed system, called the "First-Level Adjoint Sensitivity System" (1st-LASS), for determining the function $\psi_1(z,\omega)\in \mathrm{H}_1$. The function $\psi_1(z,\omega)\in \mathrm{H}_1$ is called the "first-level adjoint sensitivity function." Since the 1st-LASS does not depend on the parameter variations, it needs to be solved only once in order to obtain the first-level adjoint sensitivity function $\psi_1(z,\omega)$. For the paradigm model under consideration, the 1st-LASS given by (3.50) and (3.51) can be solved exactly to obtain the following closed-form expression for the first-level adjoint function $\psi_1(z,\omega)$:

$$\psi_1\left(z,\omega;z_d^0,\omega_d^0\right)=\left\{\frac{\mu_d\delta(\omega-\omega_d)}{\omega}[1-H(z-z_d)]\exp\left[\frac{\mu(z-z_d)}{\omega}\right]\right\}_{\boldsymbol{\alpha}^0}, \tag{3.52}$$

where $H(z-z_d)$ denotes the Heaviside functional defined as

$$H(z-z_d)=\begin{cases}1, & z\ge z_d;\\ 0, & z<z_d.\end{cases} \tag{3.53}$$

Collecting the results obtained in (3.47) through (3.51) leads to the following expression for the indirect-effect term:

$$\begin{aligned}&\left\{\delta R(\varphi;\boldsymbol{\alpha};\delta\varphi)_{\boldsymbol{\alpha}^0}\right\}_{ind}\\&=\left\{\int_0^1\psi_1(b_1,\omega)\left[\frac{(\delta q)}{\mu_s(\boldsymbol{\alpha})}-\frac{q\delta\mu_s(\boldsymbol{\alpha})}{\mu_s^2(\boldsymbol{\alpha})}+\frac{\delta b_1}{\omega}\mu\varphi(b_1,\omega)\right]\omega d\omega\right\}_{\boldsymbol{\alpha}^0}\\&-\left\{(\delta\mu)\int_0^1 d\omega\int_{b_1}^{b_2}\psi_1(z,\omega)\varphi(z,\omega)dz\right\}_{\boldsymbol{\alpha}^0}\triangleq\{\delta R(\varphi;\psi_1;\boldsymbol{\alpha};\delta\boldsymbol{\alpha})\}_{ind}.\end{aligned} \tag{3.54}$$

The appearance of the first-level adjoint function $\psi_1(z,\omega)$ in the list of arguments of the indirect-effect term $\{\delta R(\varphi;\psi_1;\boldsymbol{\alpha};\delta\boldsymbol{\alpha})\}_{ind}$ in (3.54) emphasizes the fact the function $\delta\varphi(z,\omega)$, which depends implicitly on parameter variations, has been eliminated. Instead, the indirect-effect term is now expressed in terms of the first-level adjoint function $\psi_1(z,\omega)$, which does not depend on the model parameter variations. Hence, solving the 1st-LASS to obtain the function $\psi_1(z,\omega)$, which requires a single large-scale computation comparable to solving the original equations underlying the model, suffices to determine subsequently all of the partial response sensitivities included in the indirect-effect term.

Adding the results obtained in (3.54) and (3.36) yields the complete expression for the total first-order sensitivity $\{\delta R(\varphi;\boldsymbol{\alpha};\delta\varphi;\delta\boldsymbol{\alpha})\}_{\boldsymbol{\alpha}^0}$. The specific expression of each first-order partial sensitivity of the response $R(\varphi;\boldsymbol{\alpha})$ to each uncertain parameter is obtained by identifying the expression that multiplies the respective parameter variation, which yields the following results:

$$\left\{\frac{\partial R(\varphi;\boldsymbol{\alpha})}{\partial \mu(\boldsymbol{\alpha})}\right\}_{\boldsymbol{\alpha}^0} = -\left\{\int_0^1 d\omega \int_{b_1}^{b_2} \psi_1(z,\omega)\varphi(z,\omega)dz\right\}_{\boldsymbol{\alpha}^0}, \tag{3.55}$$

$$\left\{\frac{\partial R(\varphi;\boldsymbol{\alpha})}{\partial q}\right\}_{\boldsymbol{\alpha}^0} = \left\{\frac{1}{\mu_s(\boldsymbol{\alpha})}\int_0^1 \psi_1(b_1,\omega)\,\omega\,d\omega\right\}_{\boldsymbol{\alpha}^0}, \tag{3.56}$$

$$\left\{\frac{\partial R(\varphi;\boldsymbol{\alpha})}{\partial \mu_s(\boldsymbol{\alpha})}\right\}_{\boldsymbol{\alpha}^0} = -\left\{\frac{q}{\mu_s^2(\boldsymbol{\alpha})}\int_0^1 \psi_1(b_1,\omega)\,\omega\,d\omega\right\}_{\boldsymbol{\alpha}^0}, \tag{3.57}$$

$$\left\{\frac{\partial R(\varphi;\boldsymbol{\alpha})}{\partial \mu_d(\boldsymbol{\alpha})}\right\}_{\boldsymbol{\alpha}^0} = \left\{\int_0^1 d\omega\,\delta(\omega-\omega_d)\int_{b_1}^{b_2} dz\,\varphi(z,\omega)\delta(z-z_d)\right\}_{\boldsymbol{\alpha}^0}, \tag{3.58}$$

$$\left\{\frac{\partial R(\varphi;\boldsymbol{\alpha})}{\partial \omega_d}\right\}_{\boldsymbol{\alpha}^0} = -\left\{\mu_d\int_0^1 d\omega\,\delta'(\omega-\omega_d)\int_{b_1}^{b_2} \varphi(z,\omega)\,\delta(z-z_d)dz\right\}_{\boldsymbol{\alpha}^0}, \tag{3.59}$$

$$\left\{\frac{\partial R(\varphi;\boldsymbol{\alpha})}{\partial z_d}\right\}_{\boldsymbol{\alpha}^0} = -\left\{\mu_d\int_0^1 d\omega\,\delta(\omega-\omega_d)\int_{b_1}^{b_2} \varphi(z,\omega)\,\delta'(z-z_d)dz\right\}_{\boldsymbol{\alpha}^0}, \tag{3.60}$$

$$\left\{\frac{\partial R(\varphi;\boldsymbol{\alpha})}{\partial b_1}\right\}_{\boldsymbol{\alpha}^0} = \left\{\mu\int_0^1 \psi_1(b_1,\omega)\,\varphi(b_1,\omega)\,d\omega\right\}_{\boldsymbol{\alpha}^0}. \tag{3.61}$$

If the detector is placed inside the medium, the sensitivity of the response $R(\varphi;\boldsymbol{\alpha})$ to the boundary parameter b_2 is zero, as would be expected, since the response would not depend on b_2 in such a situation. On the other hand, if the detector lies on the outer surface, at $z_d = b_2$, then the response sensitivity to b_2 is provided by the expression in (3.60), which would be evaluated at $z_d = b_2$, providing the following results:

$$\left\{\frac{\partial R(\varphi;\boldsymbol{\alpha})}{\partial b_2}\right\}_{\boldsymbol{\alpha}^0} = 0, z_d \neq b_2;$$

$$\left\{\frac{\partial R(\varphi;\boldsymbol{\alpha})}{\partial b_2}\right\}_{\boldsymbol{\alpha}^0} = -\left\{\mu_d \int_0^1 d\omega\, \delta(\omega-\omega_d) \int_{b_1}^{b_2} \varphi(z,\omega)\, \delta'(z-b_2) dz\right\}_{\boldsymbol{\alpha}^0}, z_d = b_2. \tag{3.62}$$

The following conclusions can be drawn based on the results obtained in (3.55)–(3.62):

1. The indirect-effect term provides the complete first-order partial sensitivities of the response $R(\varphi;\boldsymbol{\alpha})$ with respect to the parameters $\mu(\boldsymbol{\alpha})$, q, and $\mu_s(\boldsymbol{\alpha})$, which are parameters that occur solely in the equations underlying the paradigm model.
2. The direct-effect term provides the complete first-order partial sensitivities of the response $R(\varphi;\boldsymbol{\alpha})$ with respect to the parameters $\mu_d(\boldsymbol{\alpha})$, z_d, and ω_d, which are parameters that occur solely in the definition of the model's response.
3. Both the direct- and the indirect-effect terms may, in general, contribute to the partial sensitivities of the response $R(\varphi;\boldsymbol{\alpha})$ with respect to the interface/boundary parameters b_1 and b_2. In this particular case, the indirect-effect provides the entire contribution for the response sensitivity with respect to the boundary parameter b_1, while the direct-effect term provides the entire contribution for the response sensitivity with respect to the boundary parameter b_2, when the detector is placed on the outer boundary.

Notably, obtaining all of the first-order partial response sensitivities to the primary model parameters has necessitated a single "large-scale" computation for determining the first-level adjoint function $\psi_1(z,\omega)$. This is in contradistinction with the use of the 1st-LVSS, which would require at least *TP* large-scale computations to obtain the corresponding *TP* first-order sensitivities. Furthermore, the expressions obtained for these sensitivities were exact, as opposed to approximate, as would have been the case if these sensitivities would have been computed by, e.g., finite-difference or statistical procedures. Also, many more large-scale computations would have been necessary to compute these first-order partial response sensitivities by any other (finite-difference or statistical) procedure. Thus, the 1st-CASAM-L is the most efficient computational method for obtaining the exact expressions of the first-order partial response sensitivities to the primary model parameters.

The primary model parameters q, z_d, ω_d, b_1, b_2 are fundamental parameters. On the other hand, the primary model parameters $\mu_s(\boldsymbol{\alpha})$, $\mu(\boldsymbol{\alpha})$, and $\mu_d(\boldsymbol{\alpha})$ are derived parameters, being themselves functions of the respective nuclide number densities and cross sections (which are fundamental parameters). Hence, the response sensitivities to the corresponding fundamental parameters can now be determined by taking into account the respective functional dependencies of the primary model parameters $\mu_s(\boldsymbol{\alpha})$, $\mu(\boldsymbol{\alpha})$, and $\mu_d(\boldsymbol{\alpha})$ on the corresponding nuclide number densities and cross sections. Thus, the response sensitivities to the nuclide number densities and cross sections included in $\mu_s(\boldsymbol{\alpha})$ are obtained as follows:

$$\left\{\frac{\partial R(\varphi;\boldsymbol{\alpha})}{\partial N_{s,i}}\right\}_{\boldsymbol{\alpha}^0}=\left\{\frac{\partial R(\varphi;\boldsymbol{\alpha})}{\partial \mu_s(\boldsymbol{\alpha})}\frac{\partial \mu_s(\boldsymbol{\alpha})}{\partial N_{s,i}}\right\}_{\boldsymbol{\alpha}^0}$$
$$=-\left\{\frac{\sigma_{s,i}q}{\mu_s^2(\boldsymbol{\alpha})}\int_0^1 \psi_1(b_1,\omega)\,\omega\,d\omega\right\}_{\boldsymbol{\alpha}^0}\quad ;i=1,\ldots,M_s; \tag{3.63}$$

$$\left\{\frac{\partial R(\varphi;\boldsymbol{\alpha})}{\partial \sigma_{s,i}}\right\}_{\boldsymbol{\alpha}^0}=\left\{\frac{\partial R(\varphi;\boldsymbol{\alpha})}{\partial \mu_s(\boldsymbol{\alpha})}\frac{\partial \mu_s(\boldsymbol{\alpha})}{\partial \sigma_{s,i}}\right\}_{\boldsymbol{\alpha}^0}$$
$$=-\left\{\frac{N_{s,i}q}{\mu_s^2(\boldsymbol{\alpha})}\int_0^1 \psi_1(b_1,\omega)\,\omega\,d\omega\right\}_{\boldsymbol{\alpha}^0}\quad ;i=1,\ldots,M_s. \tag{3.64}$$

The response sensitivities to the remaining nuclide number densities and cross sections are determined similarly, to obtain the following expressions:

$$\left\{\frac{\partial R(\varphi;\boldsymbol{\alpha})}{\partial N_i}\right\}_{\boldsymbol{\alpha}^0}=\left\{\frac{\partial R(\varphi;\boldsymbol{\alpha})}{\partial \mu(\boldsymbol{\alpha})}\frac{\partial \mu(\boldsymbol{\alpha})}{\partial N_i}\right\}_{\boldsymbol{\alpha}^0}$$
$$=-\left\{\sigma_i\int_0^1 d\omega\int_{b_1}^{b_2}\psi_1(z,\omega)\varphi(z,\omega)dz\right\}_{\boldsymbol{\alpha}^0}\quad ;i=1,\ldots,M; \tag{3.65}$$

$$\left\{\frac{\partial R(\varphi;\boldsymbol{\alpha})}{\partial \sigma_i}\right\}_{\boldsymbol{\alpha}^0}=\left\{\frac{\partial R(\varphi;\boldsymbol{\alpha})}{\partial \mu(\boldsymbol{\alpha})}\frac{\partial \mu(\boldsymbol{\alpha})}{\partial \sigma_i}\right\}_{\boldsymbol{\alpha}^0}$$
$$=-\left\{N_i\int_0^1 d\omega\int_{b_1}^{b_2}\psi_1(z,\omega)\varphi(z,\omega)dz\right\}_{\boldsymbol{\alpha}^0}\quad ;i=1,\ldots,M; \tag{3.66}$$

$$\left\{\frac{\partial R(\varphi;\boldsymbol{\alpha})}{\partial N_{d,i}}\right\}_{\boldsymbol{\alpha}^0}=\left\{\frac{\partial R(\varphi;\boldsymbol{\alpha})}{\partial \mu_d(\boldsymbol{\alpha})}\frac{\partial \mu_d(\boldsymbol{\alpha})}{\partial N_{d,i}}\right\}_{\boldsymbol{\alpha}^0}$$
$$=\left\{\sigma_{d,i}\int_0^1 d\omega\,\delta(\omega-\omega_d)\int_{b_1}^{b_2} dz\,\varphi(z,\omega)\delta(z-z_d)\right\}_{\boldsymbol{\alpha}^0}\quad ;i=1,\ldots,M_d; \tag{3.67}$$

$$\left\{\frac{\partial R(\varphi;\boldsymbol{\alpha})}{\partial\sigma_{d,i}}\right\}_{\boldsymbol{\alpha}^0}=\left\{\frac{\partial R(\varphi;\boldsymbol{\alpha})}{\partial\mu_d(\boldsymbol{\alpha})}\frac{\partial\mu_d(\boldsymbol{\alpha})}{\partial\sigma_{d,i}}\right\}_{\boldsymbol{\alpha}^0}$$
$$=\left\{N_{d,i}\int_0^1 d\omega\,\delta(\omega-\omega_d)\int_{b_1}^{b_2}dz\,\varphi(z,\omega)\delta(z-z_d)\right\}_{\boldsymbol{\alpha}^0};i=1,\ldots,M_d. \tag{3.68}$$

In practice, the integrals involving the first-level adjoint function $\psi_1(z,\omega)$ in (3.55)–(3.68) would be performed numerically, after having obtained $\psi_1(z,\omega)$ by solving numerically the 1st-LASS (which would be the only large-scale computation required to obtain all of the *TP* first-order sensitivities).

It is noteworthy that the relative sensitivities of the response with respect to the nuclide densities have the same values as the relative sensitivities of the response with respect to the cross sections, i.e.,

$$\left\{\frac{\partial R(\varphi;\boldsymbol{\alpha})}{\partial\mu_s(\boldsymbol{\alpha})}\frac{\partial\mu_s(\boldsymbol{\alpha})}{\partial N_{s,i}}\frac{N_{s,i}}{R(\varphi;\boldsymbol{\alpha})}\right\}_{\boldsymbol{\alpha}^0}=-\left\{\frac{N_{s,i}\sigma_{s,i}}{R(\varphi;\boldsymbol{\alpha})}\frac{q}{\mu_s^2(\boldsymbol{\alpha})}\int_0^1\psi_1(b_1,\omega)\,\omega\,d\omega\right\}_{\boldsymbol{\alpha}^0}$$
$$=\left\{\frac{\partial R(\varphi;\boldsymbol{\alpha})}{\partial\mu_s(\boldsymbol{\alpha})}\frac{\partial\mu_s(\boldsymbol{\alpha})}{\partial\sigma_{s,i}}\frac{\sigma_{s,i}}{R(\varphi;\boldsymbol{\alpha})}\right\}_{\boldsymbol{\alpha}^0};i=1,\ldots,M_s; \tag{3.69}$$

$$\left\{\frac{\partial R(\varphi;\boldsymbol{\alpha})}{\partial\mu(\boldsymbol{\alpha})}\frac{\partial\mu(\boldsymbol{\alpha})}{\partial N_i}\frac{N_i}{R(\varphi;\boldsymbol{\alpha})}\right\}_{\boldsymbol{\alpha}^0}=-\left\{\frac{N_i\sigma_i}{R(\varphi;\boldsymbol{\alpha})}\int_0^1 d\omega\int_{b_1}^{b_2}\psi_1(z,\omega)\varphi(z,\omega)dz\right\}_{\boldsymbol{\alpha}^0}$$
$$=\left\{\frac{\partial R(\varphi;\boldsymbol{\alpha})}{\partial\mu(\boldsymbol{\alpha})}\frac{\partial\mu(\boldsymbol{\alpha})}{\partial\sigma_i}\frac{\sigma_i}{R(\varphi;\boldsymbol{\alpha})}\right\}_{\boldsymbol{\alpha}^0};i=1,\ldots,M; \tag{3.70}$$

$$\left\{\frac{\partial R(\varphi;\boldsymbol{\alpha})}{\partial\mu_d(\boldsymbol{\alpha})}\frac{\partial\mu_d(\boldsymbol{\alpha})}{\partial N_{d,i}}\frac{N_{d,i}}{R(\varphi;\boldsymbol{\alpha})}\right\}_{\boldsymbol{\alpha}^0}$$
$$=\left\{\frac{\sigma_{d,i}N_{d,i}}{R(\varphi;\boldsymbol{\alpha})}\int_0^1 d\omega\,\delta(\omega-\omega_d)\int_{b_1}^{b_2}dz\,\varphi(z,\omega)\delta(z-z_d)\right\}_{\boldsymbol{\alpha}^0} \tag{3.71}$$
$$=\left\{\frac{\partial R(\varphi;\boldsymbol{\alpha})}{\partial\mu_d(\boldsymbol{\alpha})}\frac{\partial\mu_d(\boldsymbol{\alpha})}{\partial\sigma_{d,i}}\frac{\sigma_{d,i}}{R(\varphi;\boldsymbol{\alpha})}\right\}_{\boldsymbol{\alpha}^0};i=1,\ldots,M_d.$$

The fact that so many relative sensitivities have the same values, as indicated in (3.69), (3.70), and (3.71), respectively, render quasi-useless the statistical methods that attempt to rank the sensitivities by their relative values, because such statistical methods break down in situations when many relative sensitivities have equal values. The only reliable method for computing response sensitivities accurately (and also most efficiently, from a computational standpoint) in such situations is the 1st-CASAM-L.

For the paradigm model considered here, the closed-form expressions of $\varphi(z,\omega)$ and $\psi_1(z,\omega)$ are available (by design, of course), so the closed-form expressions of the first-order sensitivities are obtained by inserting the expression of $\varphi(z,\omega)$ from (3.8) and the expression of the first-level adjoint function $\psi_1(z,\omega)$ from (3.52) into (3.55)–(3.62), and performing the respective integrations. The results of these operations are the following closed-form expressions for the partial first-order sensitivities of the response $R(\varphi;\boldsymbol{\alpha})$ with respect to the primary parameters:

$$\left\{\frac{\partial R(\varphi;\boldsymbol{\alpha})}{\partial \mu(\boldsymbol{\alpha})}\right\}_{\boldsymbol{\alpha}^0} = \left\{\frac{(b_1 - z_d) q \mu_d(\boldsymbol{\alpha})}{\omega_d \mu_s(\boldsymbol{\alpha})} \exp\left[\frac{(b_1 - z_d)\mu(\boldsymbol{\alpha})}{\omega_d}\right]\right\}_{\boldsymbol{\alpha}^0}, \tag{3.72}$$

$$\left\{\frac{\partial R(\varphi;\boldsymbol{\alpha})}{\partial q}\right\}_{\boldsymbol{\alpha}^0} = \left\{\frac{\mu_d(\boldsymbol{\alpha})}{\mu_s(\boldsymbol{\alpha})} \exp\left[\frac{(b_1 - z_d)\mu(\boldsymbol{\alpha})}{\omega_d}\right]\right\}_{\boldsymbol{\alpha}^0}, \tag{3.73}$$

$$\left\{\frac{\partial R(\varphi;\boldsymbol{\alpha})}{\partial \mu_s(\boldsymbol{\alpha})}\right\}_{\boldsymbol{\alpha}^0} = -\left\{\frac{q \mu_d(\boldsymbol{\alpha})}{\mu_s^2(\boldsymbol{\alpha})} \exp\left[\frac{(b_1 - z_d)\mu(\boldsymbol{\alpha})}{\omega_d}\right]\right\}_{\boldsymbol{\alpha}^0}, \tag{3.74}$$

$$\left\{\frac{\partial R(\varphi;\boldsymbol{\alpha})}{\partial \mu_d(\boldsymbol{\alpha})}\right\}_{\boldsymbol{\alpha}^0} = \left\{\frac{q}{\mu_s(\boldsymbol{\alpha})} \exp\left[\frac{(b_1 - z_d)\mu(\boldsymbol{\alpha})}{\omega_d}\right]\right\}_{\boldsymbol{\alpha}^0}, \tag{3.75}$$

$$\left\{\frac{\partial R(\varphi;\boldsymbol{\alpha})}{\partial \omega_d}\right\}_{\boldsymbol{\alpha}^0} = -\left\{\frac{q \mu_d(\boldsymbol{\alpha})}{\mu_s(\boldsymbol{\alpha})} \frac{(b_1 - z_d)\mu(\boldsymbol{\alpha})}{(\omega_d)^2} \exp\left[\frac{(b_1 - z_d)\mu(\boldsymbol{\alpha})}{\omega_d}\right]\right\}_{\boldsymbol{\alpha}^0}, \tag{3.76}$$

$$\left\{\frac{\partial R(\varphi;\boldsymbol{\alpha})}{\partial z_d}\right\}_{\boldsymbol{\alpha}^0} = -\left\{\frac{q \mu_d(\boldsymbol{\alpha})}{\mu_s(\boldsymbol{\alpha})} \frac{\mu(\boldsymbol{\alpha})}{\omega_d} \exp\left[\frac{(b_1 - z_d)\mu(\boldsymbol{\alpha})}{\omega_d}\right]\right\}_{\boldsymbol{\alpha}^0}, \tag{3.77}$$

$$\left\{\frac{\partial R(\varphi;\boldsymbol{\alpha})}{\partial b_1}\right\}_{\boldsymbol{\alpha}^0} = \left\{\frac{q \mu_d(\boldsymbol{\alpha})}{\mu_s(\boldsymbol{\alpha})} \frac{\mu(\boldsymbol{\alpha})}{\omega_d} \exp\left[\frac{(b_1 - z_d)\mu(\boldsymbol{\alpha})}{\omega_d}\right]\right\}_{\boldsymbol{\alpha}^0}, \tag{3.78}$$

$$\begin{aligned} &\left\{\frac{\partial R(\varphi;\boldsymbol{\alpha})}{\partial b_2}\right\}_{\boldsymbol{\alpha}^0} = 0, \text{ if } z_d \neq b_2; \\ &\left\{\frac{\partial R(\varphi;\boldsymbol{\alpha})}{\partial b_2}\right\}_{\boldsymbol{\alpha}^0} = -\left\{\frac{q \mu_d(\boldsymbol{\alpha})}{\mu_s(\boldsymbol{\alpha})} \frac{\mu(\boldsymbol{\alpha})}{\omega_d} \exp\left[\frac{(b_1 - b_2)\mu(\boldsymbol{\alpha})}{\omega_d}\right]\right\}_{\boldsymbol{\alpha}^0}, \text{ if } z_d = b_2. \end{aligned} \tag{3.79}$$

The response sensitivities with respect to the nuclide number densities and cross sections included in $\mu_s(\boldsymbol{\alpha})$, $\mu(\boldsymbol{\alpha})$, and $\mu_d(\boldsymbol{\alpha})$, respectively, have the following closed-form expressions:

$$\left\{\frac{\partial R(\varphi;\boldsymbol{\alpha})}{\partial N_{s,i}}\right\}_{\boldsymbol{\alpha}^0}=\left\{\frac{\partial R(\varphi;\boldsymbol{\alpha})}{\partial \mu_s(\boldsymbol{\alpha})}\frac{\partial \mu_s(\boldsymbol{\alpha})}{\partial N_{s,i}}\right\}_{\boldsymbol{\alpha}^0}$$
$$=-\sigma_{s,i}\left\{\frac{q\mu_d(\boldsymbol{\alpha})}{\mu_s^2(\boldsymbol{\alpha})}\exp\left[\frac{(b_1-z_d)\mu(\boldsymbol{\alpha})}{\omega_d}\right]\right\}_{\boldsymbol{\alpha}^0};i=1,\ldots,M_s; \tag{3.80}$$

$$\left\{\frac{\partial R(\varphi;\boldsymbol{\alpha})}{\partial \sigma_{s,i}}\right\}_{\boldsymbol{\alpha}^0}=\left\{\frac{\partial R(\varphi;\boldsymbol{\alpha})}{\partial \mu_s(\boldsymbol{\alpha})}\frac{\partial \mu_s(\boldsymbol{\alpha})}{\partial \sigma_{s,i}}\right\}_{\boldsymbol{\alpha}^0}$$
$$=-N_{s,i}\left\{\frac{q\mu_d(\boldsymbol{\alpha})}{\mu_s^2(\boldsymbol{\alpha})}\exp\left[\frac{(b_1-z_d)\mu(\boldsymbol{\alpha})}{\omega_d}\right]\right\}_{\boldsymbol{\alpha}^0};i=1,\ldots,M_s; \tag{3.81}$$

$$\left\{\frac{\partial R(\varphi;\boldsymbol{\alpha})}{\partial N_i}\right\}_{\boldsymbol{\alpha}^0}=\left\{\frac{\partial R(\varphi;\boldsymbol{\alpha})}{\partial \mu(\boldsymbol{\alpha})}\frac{\partial \mu(\boldsymbol{\alpha})}{\partial N_i}\right\}_{\boldsymbol{\alpha}^0}$$
$$=\sigma_i\left\{\frac{(b_1-z_d)q\mu_d\mu_d(\boldsymbol{\alpha})}{\omega_d\mu_s(\boldsymbol{\alpha})}\exp\left[\frac{(b_1-z_d)\mu(\boldsymbol{\alpha})}{\omega_d}\right]\right\}_{\boldsymbol{\alpha}^0};i=1,\ldots,M; \tag{3.82}$$

$$\left\{\frac{\partial R(\varphi;\boldsymbol{\alpha})}{\partial \sigma_i}\right\}_{\boldsymbol{\alpha}^0}=\left\{\frac{\partial R(\varphi;\boldsymbol{\alpha})}{\partial \mu(\boldsymbol{\alpha})}\frac{\partial \mu(\boldsymbol{\alpha})}{\partial \sigma_i}\right\}_{\boldsymbol{\alpha}^0}$$
$$=N_i\left\{\frac{(b_1-z_d)q\mu_d\mu_d(\boldsymbol{\alpha})}{\omega_d\mu_s(\boldsymbol{\alpha})}\exp\left[\frac{(b_1-z_d)\mu(\boldsymbol{\alpha})}{\omega_d}\right]\right\}_{\boldsymbol{\alpha}^0};i=1,\ldots,M; \tag{3.83}$$

$$\left\{\frac{\partial R(\varphi;\boldsymbol{\alpha})}{\partial N_{d,i}}\right\}_{\boldsymbol{\alpha}^0}=\left\{\frac{\partial R(\varphi;\boldsymbol{\alpha})}{\partial \mu_d(\boldsymbol{\alpha})}\frac{\partial \mu_d(\boldsymbol{\alpha})}{\partial N_{d,i}}\right\}_{\boldsymbol{\alpha}^0}$$
$$=\sigma_{d,i}\left\{\frac{q}{\mu_s(\boldsymbol{\alpha})}\exp\left[\frac{(b_1-z_d)\mu(\boldsymbol{\alpha})}{\omega_d}\right]\right\}_{\boldsymbol{\alpha}^0};i=1,\ldots,M_d; \tag{3.84}$$

$$\left\{\frac{\partial R(\varphi;\boldsymbol{\alpha})}{\partial \sigma_{d,i}}\right\}_{\boldsymbol{\alpha}^0}=\left\{\frac{\partial R(\varphi;\boldsymbol{\alpha})}{\partial \mu_d(\boldsymbol{\alpha})}\frac{\partial \mu_d(\boldsymbol{\alpha})}{\partial \sigma_{d,i}}\right\}_{\boldsymbol{\alpha}^0}$$
$$=N_{d,i}\left\{\frac{q}{\mu_s(\boldsymbol{\alpha})}\exp\left[\frac{(b_1-z_d)\mu(\boldsymbol{\alpha})}{\omega_d}\right]\right\}_{\boldsymbol{\alpha}^0};i=1,\ldots,M_d. \tag{3.85}$$

The closed-form expressions obtained in (3.72)–(3.85) can be verified directly by differentiating the closed-form expression of the response provided in (3.17) with respect to each of the model parameters.

3.3.2 *Particle Leakage Response*

The first-order total sensitivity of $R_L(\varphi;\boldsymbol{\alpha})$ is provided by the first-order G-differential $\delta R_L(\varphi;\boldsymbol{\alpha};\delta\varphi;\delta\boldsymbol{\alpha})_{\boldsymbol{\alpha}^0}$ of $R_L(\varphi;\boldsymbol{\alpha})$ at the nominal parameter values, which is determined by applying the definition of the G-differential to (3.21). This operation yields the following expression:

$$\delta R_L(\varphi;\boldsymbol{\alpha};\delta\varphi;\delta\boldsymbol{\alpha})_{\boldsymbol{\alpha}^0}$$
$$\triangleq \left\{ \frac{d}{d\varepsilon} \left[(\mu_d + \varepsilon\delta\mu_d) \int_0^1 d\omega \int_{b_1+\varepsilon\delta b_1}^{b_2+\varepsilon\delta b_2} (\varphi + \varepsilon\delta\varphi)\delta(z - b_2 - \varepsilon\delta b_2)dz \right]_{\boldsymbol{\alpha}^0} \right\}_{\varepsilon=0}$$
$$= \left\{ \delta R_L(\varphi;\boldsymbol{\alpha};\delta\boldsymbol{\alpha})_{\boldsymbol{\alpha}^0} \right\}_{dir} + \left\{ \delta R_L(\varphi;\boldsymbol{\alpha};\delta\varphi)_{\boldsymbol{\alpha}^0} \right\}_{ind}, \tag{3.86}$$

where the direct-effect term $\left\{ \delta R_L(\varphi;\boldsymbol{\alpha};\delta\boldsymbol{\alpha})_{\boldsymbol{\alpha}^0} \right\}_{dir}$ depends only on the parameter variations $\delta\boldsymbol{\alpha}$ and is defined as follows:

$$\left\{ \delta R_L(\varphi;\boldsymbol{\alpha};\delta\boldsymbol{\alpha})_{\boldsymbol{\alpha}^0} \right\}_{dir} \triangleq (\delta\mu_d) \left\{ \int_0^1 d\omega \int_{b_1}^{b_2} \varphi(z,\omega)\delta(z - b_2)dz \right\}_{\boldsymbol{\alpha}^0}$$
$$-(\delta b_2) \left\{ \mu_d \int_0^1 d\omega \int_{b_1}^{b_2} \varphi(z,\omega)\delta'(z - b_2)dz \right\}_{\boldsymbol{\alpha}^0}, \tag{3.87}$$

and where the indirect-effect term $\left\{ \delta R_L(\varphi;\boldsymbol{\alpha};\delta\varphi)_{\boldsymbol{\alpha}^0} \right\}_{ind}$ depends only on the variations $\delta\varphi$ and is defined as follows:

$$\left\{ \delta R_L(\varphi;\boldsymbol{\alpha};\delta\varphi)_{\boldsymbol{\alpha}^0} \right\}_{ind} \triangleq \left\{ \mu_d \int_0^1 d\omega \int_{b_1}^{b_2} \delta\varphi(z,\omega)\delta(z - b_2)dz \right\}_{\boldsymbol{\alpha}^0}. \tag{3.88}$$

The need for computing the function $\delta\varphi(z,\omega)$ is circumvented by expressing the indirect-effect term $\left\{ \delta R_L(\varphi;\boldsymbol{\alpha};\delta\varphi)_{\boldsymbol{\alpha}^0} \right\}_{ind}$ defined in (3.88) in terms of a first-level adjoint function, which will be denoted as $\psi_L(z,\omega)$, and which is the solution of a First-Level Adjoint Sensitivity System (1st-LASS) constructed by performing the same sequence of operations as indicated in (3.46)–(3.48) but replacing the right side of (3.49) by the right side of (3.88). Performing this sequence of operations leads to the following 1st-LASS for the first-level adjoint function $\psi_L(z,\omega)$:

$$\left\{ -\omega \frac{d}{dz} \psi_L(z,\omega) + \mu \psi_L(z,\omega) \right\}_{\boldsymbol{\alpha}^0} = \{\mu_d \delta(z - b_2)\}_{\boldsymbol{\alpha}^0},$$
$$b_1^0 \le z \le b_2^0, \ 0 \le \omega \le 1, \tag{3.89}$$

$$\psi_L(b_2^0,\omega) = 0, \ 0 \le \omega \le 1. \tag{3.90}$$

In terms of the first-level adjoint function $\psi_L(z,\omega)$, the indirect-effect term $\left\{ \delta R_L(\varphi;\boldsymbol{\alpha};\delta\varphi)_{\boldsymbol{\alpha}^0} \right\}_{ind}$ will have the same formal expression as in (3.54) except that the adjoint function $\psi_1(z,\omega)$ is replaced by the adjoint function $\psi_L(z,\omega)$, i.e.,

$$\left\{\delta R_L(\varphi;\boldsymbol{\alpha};\delta\varphi)_{\boldsymbol{\alpha}^0}\right\}_{ind}$$
$$=\left\{\int_0^1 \psi_L(b_1,\omega)\left[\frac{(\delta q)}{\mu_s(\boldsymbol{\alpha})}-\frac{q\delta\mu_s(\boldsymbol{\alpha})}{\mu_s^2(\boldsymbol{\alpha})}+\frac{\delta b_1}{\omega}\mu\varphi(b_1,\omega)\right]\omega d\omega\right\}_{\boldsymbol{\alpha}^0}$$
$$-\left\{(\delta\mu)\int_0^1 d\omega\int_{b_1}^{b_2}\psi_L(z,\omega)\varphi(z,\omega)dz\right\}_{\boldsymbol{\alpha}^0}\triangleq\left\{\delta R_L(\varphi;\psi_L;\boldsymbol{\alpha};\delta\boldsymbol{\alpha})\right\}_{ind}. \tag{3.91}$$

The total first-order G-differential $\{\delta R_L(\varphi;\psi_L;\boldsymbol{\alpha};\delta\boldsymbol{\alpha})\}$ is obtained by adding the expression for the indirect-effect term obtained in (3.91) with the expression of the direct-effect term obtained in (3.87). Identifying in the resulting expression for $\{\delta R_L(\varphi;\psi_L;\boldsymbol{\alpha};\delta\boldsymbol{\alpha})\}$ the quantities that multiply the individual parameter variations yields the following expressions for the partial sensitivities of the response $R_L(\varphi;\boldsymbol{\alpha})$ with respect to the various parameters:

$$\left\{\frac{\partial R_L(\varphi;\boldsymbol{\alpha})}{\partial\mu(\boldsymbol{\alpha})}\right\}_{\boldsymbol{\alpha}^0}=-\left\{\int_0^1 d\omega\int_{b_1}^{b_2}\psi_L(z,\omega)\varphi(z,\omega)dz\right\}_{\boldsymbol{\alpha}^0}, \tag{3.92}$$

$$\left\{\frac{\partial R_L(\varphi;\boldsymbol{\alpha})}{\partial q}\right\}_{\boldsymbol{\alpha}^0}=\left\{\frac{1}{\mu_s(\boldsymbol{\alpha})}\int_0^1\psi_L(b_1,\omega)\,\omega\,d\omega\right\}_{\boldsymbol{\alpha}^0}, \tag{3.93}$$

$$\left\{\frac{\partial R_L(\varphi;\boldsymbol{\alpha})}{\partial\mu_s(\boldsymbol{\alpha})}\right\}_{\boldsymbol{\alpha}^0}=-\left\{\frac{q}{\mu_s^2(\boldsymbol{\alpha})}\int_0^1\psi_L(b_1,\omega)\,\omega\,d\omega\right\}_{\boldsymbol{\alpha}^0}, \tag{3.94}$$

$$\left\{\frac{\partial R_L(\varphi;\boldsymbol{\alpha})}{\partial b_1}\right\}_{\boldsymbol{\alpha}^0}=\left\{\mu(\boldsymbol{\alpha})\int_0^1\psi_L(b_1,\omega)\,\varphi(b_1,\omega)\,d\omega\right\}_{\boldsymbol{\alpha}^0}. \tag{3.95}$$

The sensitivities provided in (3.92)–(3.95) stem solely from the indirect-effect term given by (3.91). On the other hand, the direct-effect term defined in (3.87) gives rise to the following sensitivities:

$$\left\{\frac{\partial R_L(\varphi;\boldsymbol{\alpha})}{\partial\mu_d(\boldsymbol{\alpha})}\right\}_{\boldsymbol{\alpha}^0}=\left\{\int_0^1 d\omega\int_{b_1}^{b_2}dz\,\varphi(z,\omega)\delta(z-b_2)\right\}_{\boldsymbol{\alpha}^0}, \tag{3.96}$$

$$\left\{\frac{\partial R_L(\varphi;\boldsymbol{\alpha})}{\partial b_2}\right\}_{\boldsymbol{\alpha}^0} = -\left\{\mu_d \int_0^1 d\omega \int_{b_1}^{b_2} \varphi(z,\omega)\,\delta'(z-b_2)dz\right\}_{\boldsymbol{\alpha}^0}. \tag{3.97}$$

Solving (3.89) and (3.90) yields the following expression for the first-level adjoint function $\psi_L(z,\omega)$:

$$\psi_L\left(z,\omega;b_2^0\right) = \left\{\frac{\mu_d}{\omega}[1-H(z-b_2)]\exp\left[\frac{\mu(z-b_2)}{\omega}\right]\right\}_{\boldsymbol{\alpha}^0}, \tag{3.98}$$

where $H(z - b_2)$ denotes the Heaviside functional. Replacing the expressions obtained in (3.98) and (3.8) into (3.92)–(3.97) and performing the respective integrations yields the following closed-form expressions for the first-order partial sensitivities of the response $R_L(\varphi;\boldsymbol{\alpha})$ with respect to the model parameters:

$$\left\{\frac{\partial R_L(\varphi;\boldsymbol{\alpha})}{\partial \mu(\boldsymbol{\alpha})}\right\}_{\boldsymbol{\alpha}^0} = \left\{\frac{q\mu_d(\boldsymbol{\alpha})}{\mu_s(\boldsymbol{\alpha})}(b_1-b_2)E_1[(b_2-b_1)\mu(\boldsymbol{\alpha})]\right\}_{\boldsymbol{\alpha}^0}, \tag{3.99}$$

$$\left\{\frac{\partial R_L(\varphi;\boldsymbol{\alpha})}{\partial q}\right\}_{\boldsymbol{\alpha}^0} = \left\{\frac{\mu_d(\boldsymbol{\alpha})}{\mu_s(\boldsymbol{\alpha})}E_2[(b_2-b_1)\mu(\boldsymbol{\alpha})]\right\}_{\boldsymbol{\alpha}^0}, \tag{3.100}$$

$$\left\{\frac{\partial R_L(\varphi;\boldsymbol{\alpha})}{\partial \mu_s(\boldsymbol{\alpha})}\right\}_{\boldsymbol{\alpha}^0} = -\left\{\frac{q\mu_d(\boldsymbol{\alpha})}{\mu_s^2(\boldsymbol{\alpha})}E_2[(b_2-b_1)\mu(\boldsymbol{\alpha})]\right\}_{\boldsymbol{\alpha}^0}, \tag{3.101}$$

$$\left\{\frac{\partial R_L(\varphi;\boldsymbol{\alpha})}{\partial b_1}\right\}_{\boldsymbol{\alpha}^0} = \left\{\frac{q\mu_d(\boldsymbol{\alpha})\mu(\boldsymbol{\alpha})}{\mu_s(\boldsymbol{\alpha})}E_1[(b_2-b_1)\mu(\boldsymbol{\alpha})]\right\}_{\boldsymbol{\alpha}^0}, \tag{3.102}$$

$$\left\{\frac{\partial R_L(\varphi;\boldsymbol{\alpha})}{\partial \mu_d(\boldsymbol{\alpha})}\right\}_{\boldsymbol{\alpha}^0} = \left\{\frac{q}{\mu_s(\boldsymbol{\alpha})}E_2[(b_2-b_1)\mu(\boldsymbol{\alpha})]\right\}_{\boldsymbol{\alpha}^0}, \tag{3.103}$$

$$\left\{\frac{\partial R_L(\varphi;\boldsymbol{\alpha})}{\partial b_2}\right\}_{\boldsymbol{\alpha}^0} = -\left\{\frac{q\mu_d(\boldsymbol{\alpha})\mu(\boldsymbol{\alpha})}{\mu_s(\boldsymbol{\alpha})}E_1[(b_2-b_1)\mu(\boldsymbol{\alpha})]\right\}_{\boldsymbol{\alpha}^0}. \tag{3.104}$$

The expressions obtained in (3.99)–(3.104) may be verified by differentiating directly (with respect to the various parameters) the expression of $R_L(\varphi;\boldsymbol{\alpha})$ provided in (3.22), while also considering the following relations:

$$\begin{aligned} &E_n(z) = \frac{1}{n-1}\left[e^{-z} - zE_{n-1}(z)\right], for\ n>1; \\ &\frac{dE_n(z)}{dz} = -E_{n-1}(z); E_n(z) = \int_z^\infty E_{n-1}(y)dy. \end{aligned} \tag{3.105}$$

As expected, the expressions of the sensitivities of $R_L(\varphi;\boldsymbol{\alpha})$ with respect to the model parameters can be obtained directly from the corresponding sensitivities of the point-detector response $R(\varphi;\boldsymbol{\alpha})$ provided in (3.72)–(3.79), as follows: (1) set

$z_d = b_2$ in (3.72)–(3.79); (2) consider that the parameter ω_d plays the role of a "phase-space variable" and integrate over ω_d the resulting expressions. The specific results of these operations are as follows:

$$\int_0^1 \left\{ \frac{\partial R(\varphi;\boldsymbol{\alpha})}{\partial \mu(\boldsymbol{\alpha})} \right\}_{z_d=b_2} d\omega_d = \frac{(b_1 - b_2) q \mu_d(\boldsymbol{\alpha})}{\mu_s(\boldsymbol{\alpha})} E_1[(b_2 - b_1)\mu(\boldsymbol{\alpha})] = \frac{\partial R_L(\varphi;\boldsymbol{\alpha})}{\partial \mu(\boldsymbol{\alpha})}, \tag{3.106}$$

$$\int_0^1 \left\{ \frac{\partial R(\varphi;\boldsymbol{\alpha})}{\partial q} \right\}_{z_d=b_2} d\omega_d = \frac{\mu_d(\boldsymbol{\alpha})}{\mu_s(\boldsymbol{\alpha})} E_2[(b_2 - b_1)\mu(\boldsymbol{\alpha})] = \frac{\partial R_L(\varphi;\boldsymbol{\alpha})}{\partial q}, \tag{3.107}$$

$$\int_0^1 \left\{ \frac{\partial R(\varphi;\boldsymbol{\alpha})}{\partial \mu_s(\boldsymbol{\alpha})} \right\}_{z_d=b_2} d\omega_d = -\frac{q \mu_d(\boldsymbol{\alpha})}{\mu_s^2(\boldsymbol{\alpha})} E_2[(b_2 - b_1)\mu(\boldsymbol{\alpha})] = \frac{\partial R_L(\varphi;\boldsymbol{\alpha})}{\partial \mu_s(\boldsymbol{\alpha})}, \tag{3.108}$$

$$\int_0^1 \left\{ \frac{\partial R(\varphi;\boldsymbol{\alpha})}{\partial b_1} \right\}_{z_d=b_2} d\omega_d = \frac{q \mu_d(\boldsymbol{\alpha}) \mu(\boldsymbol{\alpha})}{\mu_s(\boldsymbol{\alpha})} E_1[(b_2 - b_1)\mu(\boldsymbol{\alpha})] = \frac{\partial R_L(\varphi;\boldsymbol{\alpha})}{\partial b_1}, \tag{3.109}$$

$$\int_0^1 \left\{ \frac{\partial R(\varphi;\boldsymbol{\alpha})}{\partial \mu_d(\boldsymbol{\alpha})} \right\}_{z_d=b_2} d\omega_d = \frac{q}{\mu_s(\boldsymbol{\alpha})} E_2[(b_2 - b_1)\mu(\boldsymbol{\alpha})] = \frac{\partial R_L(\varphi;\boldsymbol{\alpha})}{\partial \mu_d(\boldsymbol{\alpha})}, \tag{3.110}$$

$$\int_0^1 \left\{ \frac{\partial R(\varphi;\boldsymbol{\alpha})}{\partial z_d} \right\}_{z_d=b_2} d\omega_d = \int_0^1 \frac{\partial R(\varphi;\boldsymbol{\alpha})}{\partial b_2} d\omega_d = \frac{\partial R_L(\varphi;\boldsymbol{\alpha})}{\partial b_2}. \tag{3.111}$$

The explicit indication that the expressions in (3.106)–(3.111) are to be evaluated at the nominal parameter and state function values has been omitted, in order to simplify the respective notation.

3.3.3 Reaction Rate Response

The first-order total sensitivity of $R_r(\varphi;\boldsymbol{\alpha})$ is obtained by determining the first-order G-differential $\delta R_r(\varphi;\boldsymbol{\alpha};\delta\varphi;\delta\boldsymbol{\alpha})_{\boldsymbol{\alpha}^0}$ of $R_r(\varphi;\boldsymbol{\alpha})$ at the nominal parameter values. By definition, the G-differential of (3.25) is obtained as follows:

$$\delta R_r(\varphi;\boldsymbol{\alpha};\delta\varphi;\delta\boldsymbol{\alpha})_{\boldsymbol{\alpha}^0} \triangleq \left\{ \frac{d}{d\varepsilon} \left[(\mu_d + \varepsilon\delta\mu_d) \int_0^1 d\omega \int_{b_1+\varepsilon\delta b_1}^{b_2+\varepsilon\delta b_2} (\varphi + \varepsilon\delta\varphi) dz \right]_{\boldsymbol{\alpha}^0} \right\}_{\varepsilon=0}$$
$$= \left\{ \delta R_r(\varphi;\boldsymbol{\alpha};\delta\boldsymbol{\alpha})_{\boldsymbol{\alpha}^0} \right\}_{dir} + \left\{ \delta R_r(\varphi;\boldsymbol{\alpha};\delta\varphi)_{\boldsymbol{\alpha}^0} \right\}_{ind}, \tag{3.112}$$

where

$$\left\{ \delta R_r(\varphi;\boldsymbol{\alpha};\delta\boldsymbol{\alpha})_{\boldsymbol{\alpha}^0} \right\}_{dir} \triangleq (\delta\mu_d) \left\{ \int_0^1 d\omega \int_{b_1}^{b_2} \varphi(z,\omega) dz \right\}_{\boldsymbol{\alpha}^0}$$
$$+ (\delta b_2) \left\{ \mu_d \int_0^1 d\omega \, \varphi(b_2,\omega) \right\}_{\boldsymbol{\alpha}^0} - (\delta b_1) \left\{ \mu_d \int_0^1 d\omega \, \varphi(b_1,\omega) \right\}_{\boldsymbol{\alpha}^0}, \tag{3.113}$$

and

$$\left\{ \delta R_r(\varphi;\boldsymbol{\alpha};\delta\varphi)_{\boldsymbol{\alpha}^0} \right\}_{ind} \triangleq \left\{ \mu_d \int_0^1 d\omega \int_{b_1}^{b_2} \delta\varphi(z,\omega) dz \right\}_{\boldsymbol{\alpha}^0}. \tag{3.114}$$

The need for computing the function $\delta\varphi(z, \omega)$ is circumvented by expressing the indirect-effect term $\left\{ \delta R_r(\varphi;\boldsymbol{\alpha};\delta\varphi)_{\boldsymbol{\alpha}^0} \right\}_{ind}$ defined in (3.114) in terms of a first-level adjoint function, which will be denoted as $\psi_r(z, \omega)$, and which is the solution of the following 1st-LASS:

$$\left\{ -\omega \frac{d}{dz} \psi_r(z,\omega) + \mu \psi_r(z,\omega) \right\}_{\boldsymbol{\alpha}^0} = \{\mu_d(\boldsymbol{\alpha})\}_{\boldsymbol{\alpha}^0}, \; b_1^0 \le z \le b_2^0, \; 0 \le \omega \le 1; \tag{3.115}$$

$$\psi_r\left(b_2^0, \omega\right) = 0, \;\; 0 \le \omega \le 1. \tag{3.116}$$

In terms of the first-level adjoint function $\psi_r(z, \omega)$, the indirect-effect term $\left\{ \delta R_r(\varphi;\boldsymbol{\alpha};\delta\varphi)_{\boldsymbol{\alpha}^0} \right\}_{ind}$ will have the same formal expression as in (3.54) except that the adjoint function $\psi_1(z, \omega)$ is replaced by the adjoint function $\psi_r(z, \omega)$, i.e.,

$$\{\delta R_r(\varphi;\boldsymbol{\alpha};\delta\varphi)_{\boldsymbol{\alpha}^0}\}_{ind}$$
$$=\left\{\int_0^1 \psi_r(b_1,\omega)\left[\frac{(\delta q)}{\mu_s(\boldsymbol{\alpha})}-\frac{q\delta\mu_s(\boldsymbol{\alpha})}{\mu_s^2(\boldsymbol{\alpha})}+\frac{\delta b_1}{\omega}\mu\varphi(b_1,\omega)\right]\omega d\omega\right\}_{\boldsymbol{\alpha}^0}$$
$$-\left\{(\delta\mu)\int_0^1 d\omega\int_{b_1}^{b_2}\psi_r(z,\omega)\varphi(z,\omega)dz\right\}_{\boldsymbol{\alpha}^0}\triangleq\{\delta R_r(\varphi;\psi_r;\boldsymbol{\alpha};\delta\boldsymbol{\alpha})\}_{ind}. \tag{3.117}$$

The total first-order G-differential $\{\delta R_r(\varphi;\psi_r;\boldsymbol{\alpha};\delta\boldsymbol{\alpha})\}$ is obtained by adding the expression for the indirect-effect term obtained in (3.117) with the expression of the direct-effect term obtained in (3.113). Identifying in the resulting expression for $\{\delta R_r(\varphi;\psi_r;\boldsymbol{\alpha};\delta\boldsymbol{\alpha})\}$ the quantities that multiply the individual parameter variations yields the following expressions for the partial sensitivities of the response $R_r(\varphi;\boldsymbol{\alpha})$ with respect to the various parameters:

$$\left\{\frac{\partial R_r(\varphi;\boldsymbol{\alpha})}{\partial\mu(\boldsymbol{\alpha})}\right\}_{\boldsymbol{\alpha}^0}=-\left\{\int_0^1 d\omega\int_{b_1}^{b_2}\psi_r(z,\omega)\varphi(z,\omega)dz\right\}_{\boldsymbol{\alpha}^0}, \tag{3.118}$$

$$\left\{\frac{\partial R_r(\varphi;\boldsymbol{\alpha})}{\partial q}\right\}_{\boldsymbol{\alpha}^0}=\left\{\frac{1}{\mu_s(\boldsymbol{\alpha})}\int_0^1 \psi_r(b_1,\omega)\,\omega\,d\omega\right\}_{\boldsymbol{\alpha}^0}, \tag{3.119}$$

$$\left\{\frac{\partial R_r(\varphi;\boldsymbol{\alpha})}{\partial\mu_s(\boldsymbol{\alpha})}\right\}_{\boldsymbol{\alpha}^0}=-\left\{\frac{q}{\mu_s^2(\boldsymbol{\alpha})}\int_0^1 \psi_r(b_1,\omega)\,\omega\,d\omega\right\}_{\boldsymbol{\alpha}^0}. \tag{3.120}$$

The sensitivities provided in (3.118)–(3.123) stem solely from the indirect-effect term given by (3.117) and have expressions that are formally identical to the corresponding expressions of the respective sensitivities of the point-detector response $R(\varphi;\boldsymbol{\alpha})$ and leakage response $R_L(\varphi;\boldsymbol{\alpha})$, except that the corresponding first-level adjoint functions differ from one another.

On the other hand, the sensitivities $\partial R_r(\varphi;\boldsymbol{\alpha})/\partial\mu_d(\boldsymbol{\alpha})$ and $\partial R_r(\varphi;\boldsymbol{\alpha})/\partial b_2$ stem solely from the direct-effect term defined in (3.113), namely:

$$\left\{\frac{\partial R_r(\varphi;\boldsymbol{\alpha})}{\partial\mu_d(\boldsymbol{\alpha})}\right\}_{\boldsymbol{\alpha}^0}=\left\{\int_0^1 d\omega\int_{b_1}^{b_2}\varphi(z,\omega)dz\right\}_{\boldsymbol{\alpha}^0}, \tag{3.121}$$

$$\left\{\frac{\partial R_r(\varphi;\boldsymbol{\alpha})}{\partial b_2}\right\}_{\boldsymbol{\alpha}^0} = \left\{\mu_d \int_0^1 d\omega\, \varphi(b_2,\omega)\right\}_{\boldsymbol{\alpha}^0}. \tag{3.122}$$

Both the direct-effect and indirect-effect terms contribute to the sensitivity $\partial R_r(\varphi;\boldsymbol{\alpha})/\partial b_1$, which has the following expression:

$$\left\{\frac{\partial R_r(\varphi;\boldsymbol{\alpha})}{\partial b_1}\right\}_{\boldsymbol{\alpha}^0} = \left\{\mu(\boldsymbol{\alpha}) \int_0^1 \psi_r(b_1,\omega)\,\varphi(b_1,\omega)\,d\omega\right\}_{\boldsymbol{\alpha}^0} - \left\{\mu_d \int_0^1 d\omega\, \varphi(b_1,\omega)\right\}_{\boldsymbol{\alpha}^0}. \tag{3.123}$$

Solving (3.115) and (3.116) yields the following expression for the first-level adjoint function $\psi_r(z,\omega)$:

$$\psi_r(z,\omega) = \left\{\frac{\mu_d(\boldsymbol{\alpha})}{\mu(\boldsymbol{\alpha})}\left\{1 - \exp\left[-\frac{(b_2 - z)\mu(\boldsymbol{\alpha})}{\omega}\right]\right\}\right\}_{\boldsymbol{\alpha}^0}. \tag{3.124}$$

Replacing the expressions obtained in (3.124) and (3.8) into (3.118)–(3.123) and performing the respective integrations yields the following closed-form expressions for the first-order partial sensitivities of the response $R_r(\varphi;\boldsymbol{\alpha})$ with respect to the model parameters:

$$\frac{\partial R_r(\varphi;\boldsymbol{\alpha})}{\partial \mu(\boldsymbol{\alpha})} = \frac{q}{\mu_s(\boldsymbol{\alpha})}\frac{\mu_d(\boldsymbol{\alpha})}{\mu(\boldsymbol{\alpha})}\{(b_2 - b_1)E_2[(b_2 - b_1)\mu(\boldsymbol{\alpha})] + \frac{1}{\mu(\boldsymbol{\alpha})}E_3[(b_2 - b_1)\mu(\boldsymbol{\alpha})] - \frac{1}{2\mu(\boldsymbol{\alpha})}\}, \tag{3.125}$$

$$\frac{\partial R_r(\varphi;\boldsymbol{\alpha})}{\partial q} = \frac{\mu_d(\boldsymbol{\alpha})}{\mu_s(\boldsymbol{\alpha})\mu(\boldsymbol{\alpha})}\left\{\frac{1}{2} - E_3[(b_2 - b_1)\mu(\boldsymbol{\alpha})]\right\}, \tag{3.126}$$

$$\frac{\partial R_r(\varphi;\boldsymbol{\alpha})}{\partial \mu_s(\boldsymbol{\alpha})} = -\frac{q\mu_d(\boldsymbol{\alpha})}{\mu_s^2(\boldsymbol{\alpha})\mu(\boldsymbol{\alpha})}\left\{\frac{1}{2} - E_3[(b_2 - b_1)\mu(\boldsymbol{\alpha})]\right\}, \tag{3.127}$$

$$\frac{\partial R_r(\varphi;\boldsymbol{\alpha})}{\partial \mu_d(\boldsymbol{\alpha})} = \frac{q}{\mu_s(\boldsymbol{\alpha})\mu(\boldsymbol{\alpha})}\left\{\frac{1}{2} - E_3[(b_2 - b_1)\mu(\boldsymbol{\alpha})]\right\}, \tag{3.128}$$

$$\frac{\partial R_r(\varphi;\boldsymbol{\alpha})}{\partial b_2} = \frac{q\mu_d}{\mu_s(\boldsymbol{\alpha})}E_2[(b_2 - b_1)\mu(\boldsymbol{\alpha})], \tag{3.129}$$

$$\frac{\partial R_r(\varphi;\boldsymbol{\alpha})}{\partial b_1} = -\frac{q\mu_d(\boldsymbol{\alpha})}{\mu_s(\boldsymbol{\alpha})}E_2[(b_2 - b_1)\mu(\boldsymbol{\alpha})]. \tag{3.130}$$

The explicit indication that the expressions in (3.125)–(3.130) are to be evaluated at the nominal parameter and state function values has been omitted, in order to simplify the respective notation.

As expected, the expressions of the sensitivities of $R_r(\varphi;\boldsymbol{\alpha})$ with respect to the model parameters can be obtained directly by integrating the corresponding sensitivities of the point-detector response $R(\varphi;\boldsymbol{\alpha})$ provided in (3.72)–(3.79) over z_d and ω_d, which are considered for this purpose to be independent phase-space variables (as done when using Green's functions).

3.3.4 Contributon-Response

The model response considered in this subsection is the "contributon-response flux" at a location z_0 within the slab $b_1 \le z_0 \le b_2$, which was defined in (3.28). The contributon-response flux, $\rho(z_0)$, depends on a "vector of fundamental model parameters" $\boldsymbol{\alpha}$, which, in this case, has the following components to be considered for determining the sensitivities of $\rho(z_0)$:

$$\begin{aligned}&\boldsymbol{\alpha} \triangleq (\alpha_1, \ldots, \alpha_{TP})^\dagger \triangleq [N_{s,1}, \ldots, N_{s,M_s}; \sigma_{s,1}, \ldots, \sigma_{s,M_s};\\ &N_1, \ldots, N_M; \sigma_1, \ldots, \sigma_M; q; q^*; z_0; b_1, b_2]^\dagger.\end{aligned} \tag{3.131}$$

The contributon-response flux $\rho(z_0)$ depends on the following primary model parameters: q, q^*, z_0, b_1, b_2, $\mu_s(\boldsymbol{\alpha})$, $\mu(\boldsymbol{\alpha})$. The interaction coefficients $\mu_s(\boldsymbol{\alpha})$ and $\mu(\boldsymbol{\alpha})$ are "derived primary parameters" that depend, as before, on the respective number densities and microscopic cross sections, which remain "fundamental parameters."

The first-order total sensitivity of $\rho(\varphi,\psi;\boldsymbol{\alpha})$ is obtained by determining the first-order G-differential $\delta\rho(\varphi,\psi;\boldsymbol{\alpha};\delta\varphi;\delta\psi;\delta\boldsymbol{\alpha})_{\boldsymbol{\alpha}^0}$ of $\rho(\varphi,\psi;\boldsymbol{\alpha})$ at the nominal parameter values. By definition, the G-differential of (3.32) is obtained as follows:

$$\begin{aligned}&\delta\rho(\varphi,\psi;\boldsymbol{\alpha};\delta\varphi;\delta\psi;\delta\boldsymbol{\alpha})_{\boldsymbol{\alpha}^0}\\ &\triangleq \left\{\frac{d}{d\varepsilon}\left[\int_0^1 d\omega \int_{b_1+\varepsilon\delta b_1}^{b_2+\varepsilon\delta b_2} (\varphi+\varepsilon\delta\varphi)(\psi+\varepsilon\delta\psi)\delta(z-z_0-\varepsilon\delta z_0)dz\right]_{\boldsymbol{\alpha}^0}\right\}_{\varepsilon=0}\\ &= \left\{\delta\rho(\varphi,\psi;\boldsymbol{\alpha};\delta\boldsymbol{\alpha})_{\boldsymbol{\alpha}^0}\right\}_{dir} + \left\{\delta\rho(\varphi,\psi;\boldsymbol{\alpha};\delta\varphi;\delta\psi)_{\boldsymbol{\alpha}^0}\right\}_{ind},\end{aligned} \tag{3.132}$$

where

$$\begin{aligned}&\left\{\delta\rho(\varphi,\psi;\boldsymbol{\alpha};\delta\boldsymbol{\alpha})_{\boldsymbol{\alpha}^0}\right\}_{dir} \triangleq\\ &-(\delta z_0)\left\{\int_0^1 d\omega \int_{b_1}^{b_2} \varphi(z,\omega)\psi(z,\omega)\delta'(z-z_0)dz\right\}_{\boldsymbol{\alpha}^0},\end{aligned} \tag{3.133}$$

and

$$\left\{\delta\rho(\varphi,\psi;\boldsymbol{\alpha};\delta\varphi;\delta\psi)_{\boldsymbol{\alpha}^0}\right\}_{ind}\triangleq\left\{\int_0^1 d\omega\int_{b_1}^{b_2}\delta\varphi(z,\omega)\psi(z,\omega)\delta(z-z_0)dz\right\}_{\boldsymbol{\alpha}^0}$$
$$+\left\{\int_0^1 d\omega\int_{b_1}^{b_2}\varphi(z,\omega)\delta\psi(z,\omega)\delta(z-z_0)dz\right\}_{\boldsymbol{\alpha}^0}. \tag{3.134}$$

The function $\delta\varphi(z,\omega)$, which appears in the first term on the right side of (3.134), is the solution of (3.41) and (3.42). Moreover, the function $\delta\psi(z,\omega)$, which appears in the second term on the right side of (3.134), is the solution of the equations obtained by G-differentiating (3.29) and (3.30). Applying the definition of the G-differential to (3.29) and (3.30) yields the following system of equations:

$$\left\{-\omega\frac{d}{dz}[\delta\psi(z,\omega)]+\mu(\boldsymbol{\alpha})[\delta\psi(z,\omega)]\right\}_{\boldsymbol{\alpha}^0}$$
$$=\delta q^*-(\delta\mu)\{\psi(z,\omega)\}_{\boldsymbol{\alpha}^0};\quad b_1^0<z<b_2^0,\ 0\le\omega\le 1, \tag{3.135}$$

$$\{\delta\psi(b_2,\omega)\}_{\boldsymbol{\alpha}^0}=-(\delta b_2)\left\{\frac{d\psi(z,\omega)}{dz}\right\}_{z=b_2^0}=(\delta b_2)\frac{q^*}{\omega},\quad 0\le\omega\le 1. \tag{3.136}$$

The First-Level Variational Sensitivity System (1st-LVSS) to be solved for obtaining the functions $\delta\varphi(z,\omega)$ and $\delta\psi(z,\omega)$ comprises the boundary conditions provided in (3.43) and (3.136), together with (3.41) and (3.135), which are written in the following matrix form:

$$\left\{\mathbf{V}^{(1)}\delta\mathbf{u}^{(1)}(z,\omega)\right\}_{\boldsymbol{\alpha}^0}=\left\{\mathbf{q}_V^{(1)}\left(\mathbf{u}^{(1)};\boldsymbol{\alpha};\delta\boldsymbol{\alpha}\right)\right\}_{\boldsymbol{\alpha}^0}, \tag{3.137}$$

where

$$\mathbf{V}^{(1)}\triangleq\begin{pmatrix}\omega\frac{d}{dz}+\mu(\boldsymbol{\alpha}) & 0\\ 0 & -\omega\frac{d}{dz}+\mu(\boldsymbol{\alpha})\end{pmatrix};$$
$$\mathbf{u}^{(1)}(z,\omega)\triangleq\begin{pmatrix}\varphi(z,\omega)\\ \psi(z,\omega)\end{pmatrix};\quad \delta\mathbf{u}^{(1)}(z,\omega)\triangleq\begin{pmatrix}\delta\varphi(z,\omega)\\ \delta\psi(z,\omega)\end{pmatrix}; \tag{3.138}$$

$$\mathbf{q}_V^{(1)}\left(\mathbf{u}^{(1)};\boldsymbol{\alpha};\delta\boldsymbol{\alpha}\right)=\begin{pmatrix}-(\delta\mu)\varphi(z,\omega)\\ \delta q^*-(\delta\mu)\psi(z,\omega)\end{pmatrix}. \tag{3.139}$$

The need for solving the 1st-LVSS can be circumvented by replacing the appearance of the function $\delta\mathbf{u}^{(1)}(z,\omega)$ in the indirect-effect term $\left\{\delta\rho\left(\varphi,\psi;\boldsymbol{\alpha};\delta\mathbf{u}^{(1)}\right)_{\boldsymbol{\alpha}^0}\right\}_{ind}$

with a first-level adjoint function which is independent of parameter variations. This first-level adjoint function is the solution of a First-Level Adjoint Sensitivity System (1st-LASS) which is constructing by introducing a Hilbert space, denoted as H_1, which comprises square-integrable vector-valued elements of the form $\boldsymbol{\eta}^{(1)}(z,\omega)\triangleq\left[\eta_1^{(1)}(z,\omega),\eta_2^{(1)}(z,\omega)\right]^{\dagger}$, and which is endowed with an inner product between two elements, $\boldsymbol{\eta}^{(1)}(z,\omega)\in\mathrm{H}_1$, $\boldsymbol{\xi}^{(1)}(z,\omega)\in\mathrm{H}_1$, denoted as $\langle\boldsymbol{\eta}^{(1)}(\mathbf{x}),\boldsymbol{\xi}^{(1)}(\mathbf{x})\rangle_1$ and defined as follows:

$$
\begin{aligned}
&\left\langle\boldsymbol{\eta}^{(1)}(z,\omega),\boldsymbol{\xi}^{(1)}(z,\omega)\right\rangle_1\triangleq\sum_{i=1}^{2}\left\langle\eta_i^{(1)}(z,\omega),\xi_i^{(1)}(z,\omega)\right\rangle_0\\
&=\sum_{i=1}^{2}\left\{\int_0^1 d\omega\int_{b_1}^{b_2}\eta_i^{(1)}(z,\omega)\xi_i^{(1)}(z,\omega)dz\right\}_{\boldsymbol{\alpha}^0}.
\end{aligned}
\tag{3.140}
$$

Forming the inner product of (3.137) with a yet-undefined vector-valued function $\mathbf{a}^{(1)}(z,\omega)\triangleq\left[a_1^{(1)}(z,\omega),a_2^{(1)}(z,\omega)\right]^{\dagger}\in\mathrm{H}_1$ in the Hilbert H_1 yields the following relation:

$$
\begin{aligned}
&\left\{\left\langle\mathbf{a}^{(1)}(z,\omega),\mathbf{V}^{(1)}\delta\mathbf{u}^{(1)}(z,\omega)\right\rangle_1\right\}_{\boldsymbol{\alpha}^0}\\
&=\left\{\int_0^1 d\omega\int_{b_1}^{b_2}a_1^{(1)}(z,\omega)\left[\omega\frac{d}{dz}\delta\varphi(z,\omega)+\mu(\boldsymbol{\alpha})\delta\varphi(z,\omega)\right]dz\right\}_{\boldsymbol{\alpha}^0}\\
&+\left\{\int_0^1 d\omega\int_{b_1}^{b_2}a_2^{(1)}(z,\omega)\left[-\omega\frac{d}{dz}\delta\psi(z,\omega)+\mu(\boldsymbol{\alpha})\delta\psi(z,\omega)\right]dz\right\}_{\boldsymbol{\alpha}^0}\\
&=\left\{\int_0^1 d\omega\int_{b_1}^{b_2}a_1^{(1)}(z,\omega)[-(\delta\mu)\varphi(z,\omega)]dz\right\}_{\boldsymbol{\alpha}^0}\\
&+\left\{\int_0^1 d\omega\int_{b_1}^{b_2}a_2^{(1)}(z,\omega)[\delta q^*-(\delta\mu)\psi(z,\omega)]dz\right\}_{\boldsymbol{\alpha}^0}.
\end{aligned}
\tag{3.141}
$$

Integrating by parts the terms containing derivatives with respect to z in (3.141) and using the boundary conditions provided in (3.43) and (3.136) yields the following relation:

$$
\begin{aligned}
&\left\{\int_0^1 d\omega \int_{b_1}^{b_2} \delta\varphi(z,\omega)\left[-\omega\frac{da_1^{(1)}(z,\omega)}{dz}+\mu(\boldsymbol{\alpha})a_1^{(1)}(z,\omega)\right]dz\right\}_{\boldsymbol{\alpha}^0}\\
&+\left\{\int_0^1 d\omega \int_{b_1}^{b_2} \delta\psi(z,\omega)\left[\omega\frac{da_2^{(1)}(z,\omega)}{dz}+\mu(\boldsymbol{\alpha})a_2^{(1)}(z,\omega)\right]dz\right\}_{\boldsymbol{\alpha}^0}\\
&=\left\{\int_0^1 d\omega \int_{b_1}^{b_2} a_1^{(1)}(z,\omega)[-(\delta\mu)\varphi(z,\omega)]dz\right\}_{\boldsymbol{\alpha}^0}\\
&+\left\{\int_0^1 d\omega \int_{b_1}^{b_2} a_2^{(1)}(z,\omega)[\delta q^* - (\delta\mu)\psi(z,\omega)]dz\right\}_{\boldsymbol{\alpha}^0}\\
&-\left\{\int_0^1 \omega d\omega\left[\delta\varphi(b_2,\omega)a_1^{(1)}(b_2,\omega)-\delta\varphi(b_1,\omega)a_1^{(1)}(b_1,\omega)\right]\right\}_{\boldsymbol{\alpha}^0}\\
&-\left\{\int_0^1 \omega d\omega\left[-\delta\psi(b_2,\omega)a_2^{(1)}(b_2,\omega)+\delta\psi(b_1,\omega)a_2^{(1)}(b_1,\omega)\right]\right\}_{\boldsymbol{\alpha}^0}.
\end{aligned}
\tag{3.142}
$$

The first two terms on the left side of (3.142) are required to represent the indirect-effect term $\left\{\delta\rho(\varphi,\psi;\boldsymbol{\alpha};\delta\varphi;\delta\psi)_{\boldsymbol{\alpha}^0}\right\}_{ind}$ defined in (3.134) by imposing the following relations:

$$
\left\{-\omega\frac{da_1^{(1)}(z,\omega)}{dz}+\mu(\boldsymbol{\alpha})a_1^{(1)}(z,\omega)\right\}_{\boldsymbol{\alpha}^0}=\{\psi(z,\omega)\delta(z-z_0)\}_{\boldsymbol{\alpha}^0},\quad b_1^0<z<b_2^0,\ 0\le\omega\le 1,
\tag{3.143}
$$

$$
\left\{\omega\frac{da_2^{(1)}(z,\omega)}{dz}+\mu(\boldsymbol{\alpha})a_2^{(1)}(z,\omega)\right\}_{\boldsymbol{\alpha}^0}=\{\varphi(z,\omega)\delta(z-z_0)\}_{\boldsymbol{\alpha}^0};\quad b_1^0<z<b_2^0,\ 0\le\omega\le 1.
\tag{3.144}
$$

The unknown quantities $\delta\varphi(b_2,\omega)$ and $\delta\psi(b_1,\omega)$, which appear in the last two terms on the right side of (3.142), are eliminated by choosing the following boundary conditions for the first-level adjoint functions $a_1^{(1)}(z,\omega)$ and $a_2^{(1)}(z,\omega)$:

$$
\left\{a_1^{(1)}(b_2,\omega)\right\}_{\boldsymbol{\alpha}^0}=0,\ z=b_2^0;\ 0\le\omega\le 1;
\tag{3.145}
$$

$$\left\{a_2^{(1)}(b_1,\omega)\right\}_{\boldsymbol{\alpha}^0}=0,\ \ z=b_1^0;\ \ 0\le\omega\le 1. \tag{3.146}$$

The system of equations comprising (3.143)–(3.146) constitutes the First-Level Adjoint Sensitivity System (1st-LASS). The solution $\mathbf{a}^{(1)}(z,\omega)\triangleq\left[a_1^{(1)}(z,\omega),a_2^{(1)}(z,\omega)\right]^\dagger$ of the 1st-LASS is called the first-level adjoint function. Recall that the 1st-LASS is called "first-level" (as opposed to "first-order") because it does not contain any differential or functional-derivatives, but its solution, $\mathbf{a}^{(1)}(z,\omega)$, will be used below to compute the first-order sensitivities of the response with respect to the model parameters.

Using the relations underlying the 1st-LASS together with boundary conditions provided in (3.43) and (3.136) in (3.142) yields the following expression for the indirect-effect term $\{\delta\rho(\varphi,\psi;\boldsymbol{\alpha};\delta\varphi;\delta\psi)_{\boldsymbol{\alpha}^0}\}_{ind}$ in terms of the first-level adjoint function $\mathbf{a}^{(1)}(z,\omega)\triangleq\left[a_1^{(1)}(z,\omega),a_2^{(1)}(z,\omega)\right]^\dagger$:

$$\begin{aligned}
\{\delta\rho(\varphi,\psi;\boldsymbol{\alpha};\delta\varphi;\delta\psi)_{\boldsymbol{\alpha}^0}\}_{ind} &= -(\delta\mu)\left\{\int_0^1 d\omega\int_{b_1}^{b_2} a_1^{(1)}(z,\omega)\varphi(z,\omega)dz\right\}_{\boldsymbol{\alpha}^0}\\
&+\left\{(\delta q^*)\int_0^1 d\omega\int_{b_1}^{b_2} a_2^{(1)}(z,\omega)dz\right\}_{\boldsymbol{\alpha}^0}-(\delta\mu)\left\{\int_0^1 d\omega\int_{b_1}^{b_2} a_2^{(1)}(z,\omega)\psi(z,\omega)dz\right\}_{\boldsymbol{\alpha}^0}\\
&+\left\{\frac{(\delta q)}{\mu_s(\boldsymbol{\alpha})}-\frac{q\delta\mu_s(\boldsymbol{\alpha})}{\mu_s^2(\boldsymbol{\alpha})}\right\}_{\boldsymbol{\alpha}^0}\left\{\int_0^1 a_1^{(1)}(b_1,\omega)\omega d\omega\right\}_{\boldsymbol{\alpha}^0}\\
&+(\delta b_1)\left\{\frac{q\mu(\boldsymbol{\alpha})}{\mu_s(\boldsymbol{\alpha})}\right\}_{\boldsymbol{\alpha}^0}\left\{\int_0^1 a_1^{(1)}(b_1,\omega)d\omega\right\}_{\boldsymbol{\alpha}^0}\\
&+\left\{(\delta b_2)q^*\int_0^1 a_2^{(1)}(b_2,\omega)d\omega\right\}_{\boldsymbol{\alpha}^0}=\left\{\delta\rho\left(\varphi,\psi;\boldsymbol{\alpha};\mathbf{a}^{(1)}\right)_{\boldsymbol{\alpha}^0}\right\}_{ind}.
\end{aligned} \tag{3.147}$$

The last equality in (3.147) highlights the fact that the indirect-effect term no longer depends on the function $\delta\mathbf{u}^{(1)}(z,\omega)$, but depends on the first-level adjoint function $\mathbf{a}^{(1)}(z,\omega)\triangleq\left[a_1^{(1)}(z,\omega),a_2^{(1)}(z,\omega)\right]^\dagger$, which is independent of parameter variations.

By identifying the expressions multiplying the individual parameter variations in (3.147), it follows that the indirect-effect term $\left\{\delta\rho\left(\varphi,\psi;\boldsymbol{\alpha};\mathbf{a}^{(1)}\right)_{\boldsymbol{\alpha}^0}\right\}_{ind}$ contributes the expressions of the following first-order sensitivities:

$$\frac{\partial \rho(\varphi,\psi;\boldsymbol{\alpha})}{\partial \mu(\boldsymbol{\alpha})} = -\left\{ \int_0^1 d\omega \int_{b_1}^{b_2} a_1^{(1)}(z,\omega)\varphi(z,\omega)dz \right\}_{\boldsymbol{\alpha}^0} - \left\{ \int_0^1 d\omega \int_{b_1}^{b_2} a_2^{(1)}(z,\omega)\psi(z,\omega)dz \right\}_{\boldsymbol{\alpha}^0}, \tag{3.148}$$

$$\frac{\partial \rho(\varphi,\psi;\boldsymbol{\alpha})}{\partial \mu_s(\boldsymbol{\alpha})} = -\left\{ \frac{q}{\mu_s^2(\boldsymbol{\alpha})} \int_0^1 a_1^{(1)}(b_1,\omega)\omega d\omega \right\}_{\boldsymbol{\alpha}^0}, \tag{3.149}$$

$$\frac{\partial \rho(\varphi,\psi;\boldsymbol{\alpha})}{\partial q} = \left\{ \frac{1}{\mu_s(\boldsymbol{\alpha})} \int_0^1 a_1^{(1)}(b_1,\omega)\omega d\omega \right\}_{\boldsymbol{\alpha}^0}, \tag{3.150}$$

$$\frac{\partial \rho(\varphi,\psi;\boldsymbol{\alpha})}{\partial q^*} = \left\{ \int_0^1 d\omega \int_{b_1}^{b_2} a_2^{(1)}(z,\omega)dz \right\}_{\boldsymbol{\alpha}^0}, \tag{3.151}$$

$$\frac{\partial \rho(\varphi,\psi;\boldsymbol{\alpha})}{\partial b_1} = \left\{ \frac{q\mu(\boldsymbol{\alpha})}{\mu_s(\boldsymbol{\alpha})} \right\}_{\boldsymbol{\alpha}^0} \left\{ \int_0^1 a_1^{(1)}(b_1,\omega)d\omega \right\}_{\boldsymbol{\alpha}^0}, \tag{3.152}$$

$$\frac{\partial \rho(\varphi,\psi;\boldsymbol{\alpha})}{\partial b_2} = q^* \int_0^1 a_2^{(1)}(b_2,\omega)d\omega. \tag{3.153}$$

It follows from (3.133) that the direct-effect term contributes the expressions of the following first-order sensitivity:

$$\frac{\partial \rho(\varphi,\psi;\boldsymbol{\alpha})}{\partial z_0} = -\left\{ \int_0^1 d\omega \int_{b_1}^{b_2} \varphi(z,\omega)\psi(z,\omega)\delta'(z-z_0)dz \right\}_{\boldsymbol{\alpha}^0}. \tag{3.154}$$

The 1st-LASS needs to be solved only once since it is independent of parameter variations (as opposed to the 1st-LVSS, which would need to be solved anew for each parameter variation). Subsequently, the sensitivities of the contributon-response flux $\rho(\varphi,\psi;\boldsymbol{\alpha})$ to the primary model parameters can be computed efficiently and exactly by simply performing the integrations over the adjoint function $\mathbf{a}^{(1)}(z,\omega) \triangleq \left[a_1^{(1)}(z,\omega), a_2^{(1)}(z,\omega)\right]^{\dagger}$, as indicated by the expressions provided in (3.148)–(3.153). The sensitivities of $\rho(\varphi,\psi;\boldsymbol{\alpha})$ to the respective nuclide number densities and microscopic cross sections can be obtained directly from the sensitivities $\partial\rho(\varphi,\psi;\boldsymbol{\alpha})/\partial\mu(\boldsymbol{\alpha})$ and $\partial\rho(\varphi,\psi;\boldsymbol{\alpha})/\partial\mu_s(\boldsymbol{\alpha})$, respectively, and have the following expressions:

$$\left\{\frac{\partial\rho(\varphi,\psi;\boldsymbol{\alpha})}{\partial N_i}\right\}_{\boldsymbol{\alpha}^0}=\left\{\frac{\partial\rho(\varphi,\psi;\boldsymbol{\alpha})}{\partial\mu(\boldsymbol{\alpha})}\frac{\partial\mu(\boldsymbol{\alpha})}{\partial N_i}\right\}_{\boldsymbol{\alpha}^0}$$
$$=\left\{\sigma_i\frac{\partial\rho(\varphi,\psi;\boldsymbol{\alpha})}{\partial\mu(\boldsymbol{\alpha})}\right\}_{\boldsymbol{\alpha}^0};i=1,\ldots,M; \tag{3.155}$$

$$\left\{\frac{\partial\rho(\varphi,\psi;\boldsymbol{\alpha})}{\partial \sigma_i}\right\}_{\boldsymbol{\alpha}^0}=\left\{\frac{\partial\rho(\varphi,\psi;\boldsymbol{\alpha})}{\partial\mu(\boldsymbol{\alpha})}\frac{\partial\mu(\boldsymbol{\alpha})}{\partial \sigma_i}\right\}_{\boldsymbol{\alpha}^0}$$
$$=\left\{N_i\frac{\partial\rho(\varphi,\psi;\boldsymbol{\alpha})}{\partial\mu(\boldsymbol{\alpha})}\right\}_{\boldsymbol{\alpha}^0};i=1,\ldots,M; \tag{3.156}$$

$$\left\{\frac{\partial\rho(\varphi,\psi;\boldsymbol{\alpha})}{\partial N_{s,i}}\right\}_{\boldsymbol{\alpha}^0}=\left\{\frac{\partial\rho(\varphi,\psi;\boldsymbol{\alpha})}{\partial\mu_s(\boldsymbol{\alpha})}\frac{\partial\mu_s(\boldsymbol{\alpha})}{\partial N_{s,i}}\right\}_{\boldsymbol{\alpha}^0}$$
$$=\left\{\sigma_{s,i}\frac{\partial\rho(\varphi,\psi;\boldsymbol{\alpha})}{\partial\mu_s(\boldsymbol{\alpha})}\right\}_{\boldsymbol{\alpha}^0};i=1,\ldots,M_s; \tag{3.157}$$

$$\left\{\frac{\partial\rho(\varphi,\psi;\boldsymbol{\alpha})}{\partial \sigma_{s,i}}\right\}_{\boldsymbol{\alpha}^0}=\left\{\frac{\partial\rho(\varphi,\psi;\boldsymbol{\alpha})}{\partial\mu_s(\boldsymbol{\alpha})}\frac{\partial\mu_s(\boldsymbol{\alpha})}{\partial \sigma_{s,i}}\right\}_{\boldsymbol{\alpha}^0}$$
$$=\left\{N_{s,i}\frac{\partial\rho(\varphi,\psi;\boldsymbol{\alpha})}{\partial\mu_s(\boldsymbol{\alpha})}\right\}_{\boldsymbol{\alpha}^0};i=1,\ldots,M_s. \tag{3.158}$$

The closed-form expressions of the solutions $a_1^{(1)}(z,\omega)$ and $a_2^{(1)}(z,\omega)$ of the 1st-LASS are obtained as follows:

$$\left\{a_1^{(1)}(z,\omega)\right\}_{\boldsymbol{\alpha}^0}=\left\{\frac{\psi(z_0,\omega)}{\omega}[1-H(z-z_0)]\exp\left[\frac{\mu(z-z_0)}{\omega}\right]\right\}_{\boldsymbol{\alpha}^0}, \tag{3.159}$$

$$\left\{a_2^{(1)}(z,\omega)\right\}_{\boldsymbol{\alpha}^0}=\left\{\frac{\varphi(z_0,\omega)}{\omega}H(z-z_0)\exp\left[-\frac{\mu(z-z_0)}{\omega}\right]\right\}_{\boldsymbol{\alpha}^0}$$
$$=\left\{\frac{H(z-z_0)q}{\omega\mu_s}\exp\left[\frac{(b_1-z)\mu}{\omega}\right]\right\}_{\boldsymbol{\alpha}^0}. \tag{3.160}$$

The sensitivities expressed by (3.148)–(3.154) can now be obtained by inserting the expressions of $\varphi(z,\omega)$ $\psi(z,\omega)$, $a_1^{(1)}(z,\omega)$, and $a_2^{(1)}(z,\omega)$ into these equations. Using the expressions provided in (3.8), (3.31), (3.159), and (3.160), respectively, into (3.148)–(3.154) and performing the respective integrations over the independent variables leads to the following expressions for the first-order sensitivities of the "contributon-response flux" to the model's parameters:

$$\left\{\frac{\partial \rho(\varphi,\psi;\boldsymbol{\alpha})}{\partial \mu(\boldsymbol{\alpha})}\right\}_{\boldsymbol{\alpha}^0} = \left\{\frac{qq^*}{\mu_s(\boldsymbol{\alpha})\mu^2(\boldsymbol{\alpha})}\right\}_{\boldsymbol{\alpha}^0}\{E_2[(b_2-b_1)\mu(\boldsymbol{\alpha})]$$
$$-E_2[(z_0-b_1)\mu(\boldsymbol{\alpha})]\}_{\boldsymbol{\alpha}^0} + \left\{\frac{qq^*}{\mu_s(\boldsymbol{\alpha})\mu(\boldsymbol{\alpha})}\right\}_{\boldsymbol{\alpha}^0} \tag{3.161}$$
$$\times\{(b_1-z_0)E_1[(z_0-b_1)\mu(\boldsymbol{\alpha})] + (b_2-b_1)E_1[(b_2-b_1)\mu(\boldsymbol{\alpha})]\}_{\boldsymbol{\alpha}^0},$$

$$\left\{\frac{\partial \rho(\varphi,\psi;\boldsymbol{\alpha})}{\partial \mu_s(\boldsymbol{\alpha})}\right\}_{\boldsymbol{\alpha}^0} = -\left\{\frac{qq^*}{\mu_s^2(\boldsymbol{\alpha})\mu(\boldsymbol{\alpha})}\right\}_{\boldsymbol{\alpha}^0} \times\{E_2[(z_0-b_1)\mu(\boldsymbol{\alpha})] - E_2[(b_2-b_1)\mu(\boldsymbol{\alpha})]\}_{\boldsymbol{\alpha}^0}, \tag{3.162}$$

$$\left\{\frac{\partial \rho(\varphi,\psi;\boldsymbol{\alpha})}{\partial q}\right\}_{\boldsymbol{\alpha}^0} = \left\{\frac{q^*}{\mu_s(\boldsymbol{\alpha})\mu(\boldsymbol{\alpha})}\right\}_{\boldsymbol{\alpha}^0} \times\{E_2[(z_0-b_1)\mu(\boldsymbol{\alpha})] - E_2[(b_2-b_1)\mu(\boldsymbol{\alpha})]\}_{\boldsymbol{\alpha}^0}, \tag{3.163}$$

$$\left\{\frac{\partial \rho(\varphi,\psi;\boldsymbol{\alpha})}{\partial q^*}\right\}_{\boldsymbol{\alpha}^0} = \left\{\frac{q}{\mu_s(\boldsymbol{\alpha})\mu(\boldsymbol{\alpha})}\right\}_{\boldsymbol{\alpha}^0} \times\{E_2[(z_0-b_1)\mu(\boldsymbol{\alpha})] - E_2[(b_2-b_1)\mu(\boldsymbol{\alpha})]\}_{\boldsymbol{\alpha}^0}, \tag{3.164}$$

$$\left\{\frac{\partial \rho(\varphi,\psi;\boldsymbol{\alpha})}{\partial b_1}\right\}_{\boldsymbol{\alpha}^0} = \left\{\frac{qq^*}{\mu_s(\boldsymbol{\alpha})}\right\}_{\boldsymbol{\alpha}^0} \times\{E_1[(z_0-b_1)\mu(\boldsymbol{\alpha})] - E_1[(b_2-b_1)\mu(\boldsymbol{\alpha})]\}_{\boldsymbol{\alpha}^0}, \tag{3.165}$$

$$\left\{\frac{\partial \rho(\varphi,\psi;\boldsymbol{\alpha})}{\partial b_2}\right\}_{\boldsymbol{\alpha}^0} = \left\{\frac{qq^*}{\mu_s(\boldsymbol{\alpha})}E_1[(b_2-b_1)\mu(\boldsymbol{\alpha})]\right\}_{\boldsymbol{\alpha}^0}, \tag{3.166}$$

$$\left\{\frac{\partial \rho(\varphi,\psi;\boldsymbol{\alpha})}{\partial z_0}\right\}_{\boldsymbol{\alpha}^0} = -\left\{\frac{qq^*}{\mu_s(\boldsymbol{\alpha})}E_1[(z_0-b_1)\mu(\boldsymbol{\alpha})]\right\}_{\boldsymbol{\alpha}^0}. \tag{3.167}$$

Remarkably, even though the contributon-response flux $\rho(z_0)$ vanishes at the external boundary $z_0 = b_2$, i.e., $\rho(z_0 = b_2) = 0$, not all of the first-order sensitivities of $\rho(z_0)$ with respect to the model parameters and boundaries vanish. The following results are obtained by setting $z_0 = b_2$ in (3.161)–(3.167):

$$\left\{\frac{\partial \rho(\varphi,\psi;\boldsymbol{\alpha})}{\partial \mu(\boldsymbol{\alpha})}\right\}_{z_0=b_2} = \left\{\frac{\partial \rho(\varphi,\psi;\boldsymbol{\alpha})}{\partial \mu_s(\boldsymbol{\alpha})}\right\}_{z_0=b_2} = \left\{\frac{\partial \rho(\varphi,\psi;\boldsymbol{\alpha})}{\partial q}\right\}_{z_0=b_2}$$
$$= \left\{\frac{\partial \rho(\varphi,\psi;\boldsymbol{\alpha})}{\partial q^*}\right\}_{z_0=b_2} = \left\{\frac{\partial \rho(\varphi,\psi;\boldsymbol{\alpha})}{\partial b_1}\right\}_{z_0=b_2} = 0, \tag{3.168}$$

$$\left\{\frac{\partial \rho(\varphi,\psi;\boldsymbol{\alpha})}{\partial b_2}\right\}_{z_0=b_2} = \left\{\frac{qq^*}{\mu_s(\boldsymbol{\alpha})}E_1[(b_2-b_1)\mu(\boldsymbol{\alpha})]\right\}_{\boldsymbol{\alpha}^0}$$
$$= -\left\{\frac{\partial \rho(\varphi,\psi;\boldsymbol{\alpha})}{\partial z_0}\right\}_{z_0=b_2}. \tag{3.169}$$

All of the first-order sensitivities of $\rho(\varphi, \psi; \boldsymbol{\alpha})$ have finite values for finite non-zero parameter values, except for the sensitivity $\partial \rho(\varphi, \psi; \boldsymbol{\alpha})/\partial z_0$, which becomes unbounded at the inner interface, as $z_0 \to b_1$. The sensitivity $\partial \rho(\varphi, \psi; \boldsymbol{\alpha})/\partial b_2$ is non-zero for all finite non-zero parameter values.

3.4 Application of the 2nd-CASAM-L to Compute Second-Order Response Sensitivities to Imprecisely Known Parameters

The second-order response sensitivities to the model parameters will be obtained by applying the general principles presented in Chap. 2. In essence, each of the first-order response sensitivities will be considered a new "model response" and the principles underlying the 1st-CASAM-L will be applied to determine the "first-order sensitivities of the first-order sensitivities," which by definition are the sought-after second-order sensitivities. The expressions obtained in (3.148)–(3.154) indicate the following characteristics:

1. The sensitivity $\partial \rho(\varphi, \psi; \boldsymbol{\alpha})/\partial z_0$ depends on the product $\varphi(z, \omega)\psi(z, \omega)$ of the original forward and adjoint state functions, just as the contributon-response flux $\rho(\varphi, \psi; \boldsymbol{\alpha})$ does. Hence, the determination of the second-order sensitivities derived from $\partial \rho(\varphi, \psi; \boldsymbol{\alpha})/\partial z_0$ will involve the exact same steps as those that were involved in determining the first-order sensitivities of $\rho(\varphi, \psi; \boldsymbol{\alpha})$.
2. The sensitivity $\partial \rho(\varphi, \psi; \boldsymbol{\alpha})/\partial \mu(\boldsymbol{\alpha})$ depends on all state functions, i.e., on both components $a_1^{(1)}(z, \omega)$ and $a_2^{(1)}(z, \omega)$ of the first-level adjoint function and also on the original forward and adjoint functions $\varphi(z, \omega)$ and $\psi(z, \omega)$. Hence, the determination of the second-order sensitivities stemming from $\partial \rho(\varphi, \psi; \boldsymbol{\alpha})/\partial \mu(\boldsymbol{\alpha})$ will require the complete sequence of steps underlying the 2nd-CASAM-L, requiring the determination of a four-component vector-valued second-level adjoint function.
3. The sensitivities $\partial \rho(\varphi, \psi; \boldsymbol{\alpha})/\partial \mu_s(\boldsymbol{\alpha})$, $\partial \rho(\varphi, \psi; \boldsymbol{\alpha})/\partial q$, $\partial \rho(\varphi, \psi; \boldsymbol{\alpha})/\partial q^*$, $\partial \rho(\varphi, \psi; \boldsymbol{\alpha})/\partial b_1$, and $\partial \rho(\varphi, \psi; \boldsymbol{\alpha})/\partial b_2$ involve only one of the components of the first-level adjoint function, i.e., these sensitivities involve either $a_1^{(1)}(z, \omega)$ or $a_2^{(1)}(z, \omega)$. Hence, the determination of the second-order sensitivities stemming from these first-order sensitivities will involve a simplified application of the 2nd-CASAM-L. The actual application of the 2nd-CASAM-L for these simpler situations will be illustrated by considering the sensitivity $\partial \rho(\varphi, \psi; \boldsymbol{\alpha})/\partial b_2$ as the "model response," since this sensitivity remains non-zero for all finite non-zero parameter values.

3.4.1 Determination of the Second-Order Sensitivities of the Form $\partial^2 \rho(\varphi, \psi; \alpha)/\partial \alpha_i \partial \mu(\alpha)$, i = 1, ..., TP

The first-order G-differential of the first-order sensitivity $\partial\rho(\varphi,\psi;\boldsymbol{\alpha})/\partial\mu(\boldsymbol{\alpha})$ will comprise all of the second-order sensitivities of the form $\partial^2\rho(\varphi,\psi;\boldsymbol{\alpha})/\partial\alpha_i\partial\mu(\boldsymbol{\alpha})$, $i = 1, \ldots, TP$, and will be obtained by considering that the expression provided in (3.148) plays the role of a "model response." For this purpose, it is convenient to introduce the following notation:

$$\rho^{(1)}\left[\mathbf{u}^{(2)}(z,\omega);\boldsymbol{\alpha}\right] \triangleq \frac{\partial\rho(\varphi,\psi;\boldsymbol{\alpha})}{\partial\mu(\boldsymbol{\alpha})}$$

$$= -\int_0^1 d\omega \int_{b_1}^{b_2} a_1^{(1)}(z,\omega)\varphi(z,\omega)dz - \int_0^1 d\omega \int_{b_1}^{b_2} a_2^{(1)}(z,\omega)\psi(z,\omega)dz, \tag{3.170}$$

$$\mathbf{u}^{(2)}(z,\omega) \triangleq \begin{pmatrix} \mathbf{u}^{(1)}(z,\omega) \\ \mathbf{a}^{(1)}(z,\omega) \end{pmatrix} \triangleq \left[\varphi(z,\omega),\psi(z,\omega),a_1^{(1)}(z,\omega),a_2^{(1)}(z,\omega)\right]^\dagger.$$

Applying the definition of the first-order total G-differential to (3.170) yields the following relation:

$$\delta\rho^{(1)}\left[\mathbf{u}^{(2)};\boldsymbol{\alpha};\delta\mathbf{u}^{(2)};\delta\boldsymbol{\alpha}\right]_{\boldsymbol{\alpha}^0}$$

$$\triangleq -\left\{\frac{d}{d\varepsilon}\left[\int_0^1 d\omega \int_{b_1+\varepsilon\delta b_1}^{b_2+\varepsilon\delta b_2} \left(a_1^{(1)} + \varepsilon\delta a_1^{(1)}\right)(\varphi + \varepsilon\delta\varphi)dz\right]_{\boldsymbol{\alpha}^0}\right\}_{\varepsilon=0}$$

$$-\left\{\frac{d}{d\varepsilon}\left[\int_0^1 d\omega \int_{b_1+\varepsilon\delta b_1}^{b_2+\varepsilon\delta b_2} \left(a_2^{(1)} + \varepsilon\delta a_2^{(1)}\right)(\psi + \varepsilon\delta\psi)dz\right]_{\boldsymbol{\alpha}^0}\right\}_{\varepsilon=0} \tag{3.171}$$

$$= \left\{\delta\rho^{(1)}\left[\mathbf{u}^{(2)};\boldsymbol{\alpha};\delta\boldsymbol{\alpha}\right]_{\boldsymbol{\alpha}^0}\right\}_{dir} + \left\{\delta\rho^{(1)}\left[\mathbf{u}^{(2)};\boldsymbol{\alpha};\delta\mathbf{u}^{(2)}\right]_{\boldsymbol{\alpha}^0}\right\}_{ind},$$

where

$$\left\{\delta\rho^{(1)}\left[\mathbf{u}^{(2)};\boldsymbol{\alpha};\delta\boldsymbol{\alpha}\right]_{\boldsymbol{\alpha}^0}\right\}_{dir}$$

$$\triangleq -(\delta b_2)\left\{\int_0^1 d\omega\left[a_1^{(1)}(b_2,\omega)\varphi(b_2,\omega) + a_2^{(1)}(b_2,\omega)\psi(b_2,\omega)\right]\right\}_{\boldsymbol{\alpha}^0} \tag{3.172}$$

$$+(\delta b_1)\left\{\int_0^1 d\omega\left[a_1^{(1)}(b_1,\omega)\varphi(b_1,\omega) + a_2^{(1)}(b_1,\omega)\psi(b_1,\omega)\right]\right\}_{\boldsymbol{\alpha}^0},$$

and

$$\left\{\delta\rho^{(1)}\left[\mathbf{u}^{(2)};\boldsymbol{\alpha};\delta\mathbf{u}^{(2)}\right]_{\boldsymbol{\alpha}^0}\right\}_{ind} \triangleq -\left\{\int_0^1 d\omega \int_{b_1}^{b_2}\left[a_1^{(1)}\delta\varphi + a_2^{(1)}\delta\psi + \delta a_1^{(1)}\varphi + \delta a_2^{(1)}\psi\right]dz\right\}_{\boldsymbol{\alpha}^0}. \tag{3.173}$$

The direct-effect term $\left\{\delta\rho^{(1)}\left[\mathbf{u}^{(2)};\boldsymbol{\alpha};\delta\boldsymbol{\alpha}\right]_{\boldsymbol{\alpha}^0}\right\}_{dir}$ can be computed immediately, since all of the quantities appearing in its definition on the right side of (3.172) are known. However, the computation of the indirect-effect term $\left\{\delta\rho^{(1)}\left[\mathbf{u}^{(2)};\boldsymbol{\alpha};\delta\mathbf{u}^{(2)}\right]_{\boldsymbol{\alpha}^0}\right\}_{ind}$ requires the prior determination of the vector of variations

$$\delta\mathbf{u}^{(2)}(z,\omega)\triangleq\left[\delta\varphi(z,\omega),\delta\psi(z,\omega),\delta a_1^{(1)}(z,\omega),\delta a_2^{(1)}(z,\omega)\right]^{\dagger}. \tag{3.174}$$

The vector of variations $\delta\mathbf{u}^{(1)}(z,\omega) \triangleq [\delta\varphi(z,\omega),\delta\psi(z,\omega)]^{\dagger}$ is the solution of the 1st-LVSS defined in (3.137) together with the boundary conditions provided in (3.43) and (3.136). Furthermore, the vector of variations $\delta\mathbf{a}^{(1)}(z,\omega)\triangleq\left[\delta a_1^{(1)}(z,\omega),\delta a_2^{(1)}(z,\omega)\right]^{\dagger}$ is the solution of the system of equations obtained by G-differentiating the 1st-LASS, which comprises (3.143)–(3.146). Taking the G-differentials of (3.143)–(3.146) at the nominal parameter values yields the following system of equations and boundary conditions:

$$\begin{aligned}&\left\{-\omega\frac{d}{dz}\delta a_1^{(1)}(z,\omega)+\mu\delta a_1^{(1)}(z,\omega)-\delta(z-z_0)\delta\psi(z,\omega)\right\}_{\boldsymbol{\alpha}^0}\\&=\left\{-(\delta\mu)a_1^{(1)}(z,\omega)-(\delta z_0)\psi(z,\omega)\delta'(z-z_0)\right\}_{\boldsymbol{\alpha}^0},\end{aligned} \tag{3.175}$$

$$\begin{aligned}&\left\{\omega\frac{d}{dz}\delta a_2^{(1)}(z,\omega)+\mu\delta a_2^{(1)}(z,\omega)-\delta(z-z_0)\delta\varphi(z,\omega)\right\}_{\boldsymbol{\alpha}^0}\\&=\left\{-(\delta\mu)a_2^{(1)}(z,\omega)-(\delta z_0)\varphi(z,\omega)\delta'(z-z_0)\right\}_{\boldsymbol{\alpha}^0},\end{aligned} \tag{3.176}$$

$$\left\{\delta a_1^{(1)}(b_2,\omega)\right\}_{\boldsymbol{\alpha}^0}=-(\delta b_2)\left\{\frac{da_1^{(1)}(z,\omega)}{dz}\right\}_{z=b_2^0}=0,\ z=b_2^0;0\le\omega\le1, \tag{3.177}$$

$$\left\{\delta a_2^{(1)}(b_1,\omega)\right\}_{\boldsymbol{\alpha}^0}=-(\delta b_1)\left\{\frac{da_2^{(1)}(z,\omega)}{dz}\right\}_{z=b_1^0}=0,\ z=b_1^0;0\le\omega\le1. \tag{3.178}$$

Concatenating the 1st-LVSS together with (3.175) and (3.176) yields the following system of equations written in matrix form:

$$\left\{\mathbf{V}^{(2)}\delta\mathbf{u}^{(2)}(z,\omega)\right\}_{\boldsymbol{\alpha}^0} = \left\{\mathbf{q}_V^{(2)}(z,\omega)\right\}_{\boldsymbol{\alpha}^0}, \tag{3.179}$$

where:

$$\mathbf{V}^{(2)} \triangleq \begin{pmatrix} \omega\frac{d}{dz}+\mu & 0 & 0 & 0 \\ 0 & -\omega\frac{d}{dz}+\mu & 0 & 0 \\ 0 & -\delta(z-z_0) & -\omega\frac{d}{dz}+\mu & 0 \\ -\delta(z-z_0) & 0 & 0 & \omega\frac{d}{dz}+\mu \end{pmatrix}; \tag{3.180}$$

$$\delta\mathbf{u}^{(2)}(z,\omega) = \begin{pmatrix} \delta\varphi(z,\omega) \\ \delta\psi(z,\omega) \\ \delta a_1^{(1)}(z,\omega) \\ \delta a_2^{(1)}(z,\omega) \end{pmatrix};$$

$$\mathbf{q}_V^{(2)}(z,\omega) \triangleq \begin{pmatrix} -(\delta\mu)\varphi(z,\omega) \\ \delta q^* - (\delta\mu)\psi(z,\omega) \\ -(\delta\mu)a_1^{(1)}(z,\omega) - (\delta z_0)\psi(z,\omega)\delta'(z-z_0) \\ -(\delta\mu)a_2^{(1)}(z,\omega) - (\delta z_0)\varphi(z,\omega)\delta'(z-z_0) \end{pmatrix}. \tag{3.181}$$

The system comprising (3.179) together with the corresponding boundary conditions given in (3.43), (3.136), (3.177), and (3.178) is the "Second-Level Variational Sensitivity System" (2nd-LVSS), and its solution, $\delta\mathbf{u}^{(2)}(z,\omega)$, is the second-level variational function.

The application of the 2nd-CASAM-L avoids the need for solving the 2nd-LVSS (repeatedly, for every parameter variation) by expressing the indirect-effect term $\left\{\delta\rho^{(1)}\left[\mathbf{u}^{(2)};\boldsymbol{\alpha};\delta\mathbf{u}^{(2)}\right]_{\boldsymbol{\alpha}^0}\right\}_{ind}$ defined in (3.173) in an alternative manner, in terms of the solution of a Second-Level Adjoint Sensitivity System (2nd-LASS), which is constructed so as to eliminate the appearance of the second-level variational vector $\delta\mathbf{u}^{(2)}(z,\omega)$ in the alternative expression of the indirect-effect term $\left\{\delta\rho^{(1)}\left[\mathbf{u}^{(2)};\boldsymbol{\alpha};\delta\mathbf{u}^{(2)}\right]_{\boldsymbol{\alpha}^0}\right\}_{ind}$. The construction of the requisite 2nd-LASS commences by introducing a Hilbert space, denoted as H_2, comprising square-integrable vector-valued elements of the form $\boldsymbol{\eta}^{(2)}(z,\omega)\triangleq\left[\eta_1^{(2)}(z,\omega),\eta_2^{(2)}(z,\omega),\eta_3^{(2)}(z,\omega),\eta_4^{(2)}(z,\omega)\right]^{\dagger}\in \mathrm{H}_2$. The inner product between two elements $\boldsymbol{\eta}^{(2)}(z,\omega)\in \mathrm{H}_2$ and $\boldsymbol{\xi}^{(2)}(z,\omega)\in \mathrm{H}_2$ will be denoted as $\langle\boldsymbol{\eta}^{(2)}(z,\omega),\boldsymbol{\xi}^{(2)}(z,\omega)\rangle_2$ and is defined as follows:

$$\left\langle\boldsymbol{\eta}^{(2)}(z,\omega),\boldsymbol{\xi}^{(2)}(z,\omega)\right\rangle_2\triangleq\sum_{k=1}^{4}\left\{\int_0^1 d\omega\int_{b_1}^{b_2}\eta_k^{(2)}(z,\omega),\xi_k^{(2)}(z,\omega)dz\right\}_{\boldsymbol{\alpha}^0}. \tag{3.182}$$

Forming the inner product of (3.179) with a yet-undefined function $\mathbf{a}^{(2)}(z,\omega)\triangleq\left[a_1^{(2)}(z,\omega),a_2^{(2)}(z,\omega),a_3^{(2)}(z,\omega),a_4^{(2)}(z,\omega)\right]^{\dagger}\in \mathrm{H}_2$ in the Hilbert H_2 yields the following relation:

$$\left\{\left\langle\mathbf{a}^{(2)}(z,\omega),\mathbf{V}^{(2)}\delta\mathbf{u}^{(2)}(z,\omega)\right\rangle_2\right\}_{\boldsymbol{\alpha}^0}=\left\{\left\langle\mathbf{a}^{(2)}(z,\omega),\mathbf{q}_V^{(2)}(z,\omega)\right\rangle_2\right\}_{\boldsymbol{\alpha}^0}. \tag{3.183}$$

The left side and right side of (3.183) have the following detailed forms, respectively:

$$\begin{aligned}
&\left\{\left\langle\mathbf{a}^{(2)}(z,\omega),\mathbf{V}^{(2)}\delta\mathbf{u}^{(2)}(z,\omega)\right\rangle_2\right\}_{\boldsymbol{\alpha}^0}\\
&=\left\{\int_0^1 d\omega\int_{b_1}^{b_2}a_1^{(2)}(z,\omega)\left[\omega\frac{d}{dz}\delta\varphi(z,\omega)+\mu(\boldsymbol{\alpha})\delta\varphi(z,\omega)\right]dz\right\}_{\boldsymbol{\alpha}^0}\\
&+\left\{\int_0^1 d\omega\int_{b_1}^{b_2}a_2^{(2)}(z,\omega)\left[-\omega\frac{d}{dz}\delta\psi(z,\omega)+\mu(\boldsymbol{\alpha})\delta\psi(z,\omega)\right]dz\right\}_{\boldsymbol{\alpha}^0}\\
&+\left\{\int_0^1 d\omega\int_{b_1}^{b_2}a_3^{(2)}(z,\omega)\left[-\omega\frac{d}{dz}\delta a_1^{(1)}(z,\omega)\right.\right.\\
&\qquad\left.\left.+\mu\delta a_1^{(1)}(z,\omega)-\delta(z-z_0)\delta\psi(z,\omega)\right]dz\right\}_{\boldsymbol{\alpha}^0}\\
&+\left\{\int_0^1 d\omega\int_{b_1}^{b_2}a_4^{(2)}(z,\omega)dz\left[\omega\frac{d}{dz}\delta a_2^{(1)}(z,\omega)\right.\right.\\
&\qquad\left.\left.+\mu\delta a_2^{(1)}(z,\omega)-\delta(z-z_0)\delta\varphi(z,\omega)\right]dz\right\}_{\boldsymbol{\alpha}^0},
\end{aligned} \tag{3.184}$$

and

$$
\left\{\left\langle \mathbf{a}^{(2)}(z,\omega),\mathbf{q}_V^{(2)}(z,\omega)\right\rangle_2\right\}_{\boldsymbol{\alpha}^0} = \left\{\int_0^1 d\omega \int_{b_1}^{b_2} a_1^{(2)}(z,\omega)[-(\delta\mu)\varphi(z,\omega)]dz\right\}_{\boldsymbol{\alpha}^0}
$$
$$
+\left\{\int_0^1 d\omega \int_{b_1}^{b_2} a_2^{(2)}(z,\omega)[\delta q^* - (\delta\mu)\psi(z,\omega)]dz\right\}_{\boldsymbol{\alpha}^0}
$$
$$
+\left\{\int_0^1 d\omega \int_{b_1}^{b_2} a_3^{(2)}(z,\omega)\left[-(\delta\mu)a_1^{(1)}(z,\omega) - (\delta z_0)\psi(z,\omega)\delta'(z-z_0)\right]dz\right\}_{\boldsymbol{\alpha}^0}
$$
$$
+\left\{\int_0^1 d\omega \int_{b_1}^{b_2} a_4^{(2)}(z,\omega)\left[-(\delta\mu)a_2^{(1)}(z,\omega) - (\delta z_0)\varphi(z,\omega)\delta'(z-z_0)\right]dz\right\}_{\boldsymbol{\alpha}^0}. \tag{3.185}
$$

Integrating by parts the terms containing derivatives with respect to z in (3.184) and re-arranging terms yields the following relation:

$$
\begin{aligned}
&\left\{\left\langle \mathbf{a}^{(2)}(z,\omega),\mathbf{V}^{(2)}\delta\mathbf{u}^{(2)}(z,\omega)\right\rangle_2\right\}_{\boldsymbol{\alpha}^0} = \left\{P^{(2)}\left[\delta\mathbf{u}^{(2)};\mathbf{u}^{(1)};\mathbf{a}^{(2)};\boldsymbol{\alpha};\delta\boldsymbol{\alpha}\right]\right\}_{\boldsymbol{\alpha}^0} \\
&+\left\{\int_0^1 d\omega \int_{b_1}^{b_2} \delta\varphi(z,\omega)\left[-\omega\frac{da_1^{(2)}(z,\omega)}{dz}\right.\right. \\
&\left.\left.+\mu(\boldsymbol{\alpha})a_1^{(2)}(z,\omega) - \delta(z-z_0)a_4^{(2)}(z,\omega)\right]dz\right\}_{\boldsymbol{\alpha}^0} \\
&+\left\{\int_0^1 d\omega \int_{b_1}^{b_2} \delta\psi(z,\omega)\left[\omega\frac{da_2^{(2)}(z,\omega)}{dz}\right.\right. \\
&\left.\left.+\mu(\boldsymbol{\alpha})a_2^{(2)}(z,\omega) - \delta(z-z_0)a_3^{(2)}(z,\omega)\right]dz\right\}_{\boldsymbol{\alpha}^0} \\
&+\left\{\int_0^1 d\omega \int_{b_1}^{b_2} \delta a_1^{(1)}(z,\omega)\left[\omega\frac{da_3^{(2)}(z,\omega)}{dz} + \mu a_3^{(2)}(z,\omega)\right]dz\right\}_{\boldsymbol{\alpha}^0} \\
&+\left\{\int_0^1 d\omega \int_{b_1}^{b_2} \delta a_2^{(1)}(z,\omega)\left[-\omega\frac{da_4^{(2)}(z,\omega)}{dz} + \mu a_4^{(2)}(z,\omega)\right]dz\right\}_{\boldsymbol{\alpha}^0}.
\end{aligned} \tag{3.186}
$$

where the bilinear concomitant $\left\{P^{(2)}\left[\delta\mathbf{u}^{(2)};\mathbf{u}^{(1)};\mathbf{a}^{(2)};\boldsymbol{\alpha};\delta\boldsymbol{\alpha}\right]\right\}_{\boldsymbol{\alpha}^0}$ has the following expression:

$$\begin{aligned}
&\left\{P^{(2)}\left[\delta\mathbf{u}^{(2)};\mathbf{u}^{(1)};\mathbf{a}^{(2)};\boldsymbol{\alpha};\delta\boldsymbol{\alpha}\right]\right\}_{\boldsymbol{\alpha}^0} \\
&= \int_0^1 \omega d\omega\left[\delta\varphi(b_2,\omega)a_1^{(2)}(b_2,\omega) - \delta\varphi(b_1,\omega)a_1^{(2)}(b_1,\omega)\right] \\
&- \int_0^1 \omega d\omega\left[\delta\psi(b_2,\omega)a_2^{(2)}(b_2,\omega) - \delta\psi(b_1,\omega)a_2^{(2)}(b_1,\omega)\right] \\
&- \int_0^1 \omega d\omega\left[\delta a_1^{(1)}(b_2,\omega)a_3^{(2)}(b_2,\omega) - \delta a_1^{(1)}(b_1,\omega)a_3^{(2)}(b_1,\omega)\right] \\
&+ \int_0^1 \omega d\omega\left[\delta a_2^{(1)}(b_2,\omega)a_4^{(2)}(b_2,\omega) - \delta a_2^{(1)}(b_1,\omega)a_4^{(2)}(b_1,\omega)\right].
\end{aligned} \tag{3.187}$$

The right side of (3.186) is required to represent the indirect-effect term $\left\{\delta\rho^{(1)}\left[\mathbf{u}^{(2)};\boldsymbol{\alpha};\delta\mathbf{u}^{(2)}\right]_{\boldsymbol{\alpha}^0}\right\}_{ind}$ defined in (3.173) by imposing the following relations for $b_1^0 < z < b_2^0,\ 0 \le \omega \le 1$:

$$\begin{aligned}
&\left\{-\omega\frac{da_1^{(2)}(z,\omega)}{dz} + \mu(\boldsymbol{\alpha})a_1^{(2)}(z,\omega)\right\}_{\boldsymbol{\alpha}^0} \\
&\quad = \left\{\delta(z-z_0)a_4^{(2)}(z,\omega) - a_1^{(1)}(z,\omega)\right\}_{\boldsymbol{\alpha}^0},
\end{aligned} \tag{3.188}$$

$$\left\{\omega\frac{da_2^{(2)}(z,\omega)}{dz} + \mu(\boldsymbol{\alpha})a_2^{(2)}(z,\omega)\right\}_{\boldsymbol{\alpha}^0} = \left\{\delta(z-z_0)a_3^{(2)}(z,\omega) - a_2^{(1)}(z,\omega)\right\}_{\boldsymbol{\alpha}^0}, \tag{3.189}$$

$$\left\{\omega\frac{da_3^{(2)}(z,\omega)}{dz} + \mu(\boldsymbol{\alpha})a_3^{(2)}(z,\omega)\right\}_{\boldsymbol{\alpha}^0} = -\{\varphi(z,\omega)\}_{\boldsymbol{\alpha}^0}, \tag{3.190}$$

$$\left\{-\omega\frac{da_4^{(2)}(z,\omega)}{dz} + \mu(\boldsymbol{\alpha})a_4^{(2)}(z,\omega)\right\}_{\boldsymbol{\alpha}^0} = -\{\psi(z,\omega)\}_{\boldsymbol{\alpha}^0}. \tag{3.191}$$

The unknown quantities $\delta\varphi(b_2,\omega)$, $\delta\psi(b_1,\omega)$, $\delta a_1^{(1)}(b_1,\omega)$, and $\delta a_2^{(1)}(b_2,\omega)$, which appear in bilinear concomitant expressed by (3.187), are eliminated from further consideration by choosing the following boundary conditions for the second-level adjoint function $\mathbf{a}^{(2)}(z,\omega)\triangleq\left[a_1^{(2)}(z,\omega),a_2^{(2)}(z,\omega),a_3^{(2)}(z,\omega),a_4^{(2)}(z,\omega)\right]^{\dagger}$:

$$\left\{a_1^{(2)}(b_2,\omega)\right\}_{\boldsymbol{\alpha}^0}=0,\ z=b_2^0;\ 0\le\omega\le 1; \tag{3.192}$$

$$\left\{a_2^{(2)}(b_1,\omega)\right\}_{\boldsymbol{\alpha}^0}=0,\ z=b_1^0;\ 0\le\omega\le 1; \tag{3.193}$$

$$\left\{a_3^{(2)}(b_1,\omega)\right\}_{\boldsymbol{\alpha}^0}=0,\ z=b_1^0;\ 0\le\omega\le 1; \tag{3.194}$$

$$\left\{a_4^{(2)}(b_2,\omega)\right\}_{\boldsymbol{\alpha}^0}=0,\ z=b_2^0;\ 0\le\omega\le 1. \tag{3.195}$$

The system comprising (3.188)–(3.195) constitute the Second-Level Adjoint Sensitivity System for the second-level adjoint function $\mathbf{a}^{(2)}(z,\omega)\triangleq\left[a_1^{(2)}(z,\omega),a_2^{(2)}(z,\omega),a_3^{(2)}(z,\omega),a_4^{(2)}(z,\omega)\right]^{\dagger}$. Very importantly, the equations underlying this 2nd-LASS need not be solved simultaneously. Instead, they are solved successively, decoupled from one another, by first solving separately (3.190) and (3.191) to obtain the functions $a_3^{(2)}(z,\omega)$ and $a_4^{(2)}(z,\omega)$, respectively. The solution $a_3^{(2)}(z,\omega)$ is used in the source term of (3.189), which is subsequently solved to obtain the function $a_2^{(2)}(z,\omega)$. Separately, the solution $a_4^{(2)}(z,\omega)$ of (3.191) is used in the source term of (3.188), which is then solved to obtain the function $a_1^{(2)}(z,\omega)$.

Using in (3.186) the equations underlying the 2nd-LASS together with the boundary conditions provided in (3.43), (3.136), (3.177), and (3.178) and recalling that the right side of (3.186) represents the indirect-effect term $\left\{\delta\rho^{(1)}\left[\mathbf{u}^{(2)};\boldsymbol{\alpha};\delta\mathbf{u}^{(2)}\right]_{\boldsymbol{\alpha}^0}\right\}_{ind}$ eliminates the appearance of $\delta\mathbf{u}^{(2)}$, yielding the following expression for $\left\{\delta\rho^{(1)}\left[\mathbf{u}^{(2)};\boldsymbol{\alpha};\delta\mathbf{u}^{(2)}\right]_{\boldsymbol{\alpha}^0}\right\}_{ind}$ in terms of the second-level adjoint function $\mathbf{a}^{(2)}(z,\omega)$:

$$
\begin{aligned}
&\left\{\delta\rho^{(1)}\left[\mathbf{u}^{(2)};\boldsymbol{\alpha};\delta\mathbf{u}^{(2)}\right]_{\boldsymbol{\alpha}^0}\right\}_{ind}=\left\{\left\langle\mathbf{a}^{(2)}(z,\omega),\mathbf{q}^{(2)}\left(\mathbf{u}^{(2)};\boldsymbol{\alpha};\delta\boldsymbol{\alpha}\right)\right\rangle_2\right\}_{\boldsymbol{\alpha}^0}\\
&-\left\{P^{(2)}\left[\delta\mathbf{u}^{(2)};\mathbf{u}^{(1)};\mathbf{a}^{(2)};\boldsymbol{\alpha};\delta\boldsymbol{\alpha}\right]\right\}_{\boldsymbol{\alpha}^0}=\left\{\int_0^1 d\omega\int_{b_1}^{b_2}a_1^{(2)}(z,\omega)[-(\delta\mu)\varphi(z,\omega)]dz\right\}_{\boldsymbol{\alpha}^0}\\
&+\left\{\int_0^1 d\omega\int_{b_1}^{b_2}a_2^{(2)}(z,\omega)[\delta q^*-(\delta\mu)\psi(z,\omega)]dz\right\}_{\boldsymbol{\alpha}^0}\\
&+\left\{\int_0^1 d\omega\int_{b_1}^{b_2}a_3^{(2)}(z,\omega)\left[-(\delta\mu)a_1^{(1)}(z,\omega)-(\delta z_0)\psi(z,\omega)\delta'(z-z_0)\right]dz\right\}_{\boldsymbol{\alpha}^0}\\
&+\left\{\int_0^1 d\omega\int_{b_1}^{b_2}a_4^{(2)}(z,\omega)\left[-(\delta\mu)a_2^{(1)}(z,\omega)-(\delta z_0)\varphi(z,\omega)\delta'(z-z_0)\right]dz\right\}_{\boldsymbol{\alpha}^0}\\
&+\left\{\int_0^1\left[\frac{(\delta q)}{\mu_s(\boldsymbol{\alpha})}-\frac{q\delta\mu_s(\boldsymbol{\alpha})}{\mu_s^2(\boldsymbol{\alpha})}+\frac{(\delta b_1)q\mu(\boldsymbol{\alpha})}{\omega\mu_s(\boldsymbol{\alpha})}\right]a_1^{(2)}(b_1,\omega)\omega d\omega\right\}_{\boldsymbol{\alpha}^0}\\
&+\left\{(\delta b_2)q^*\int_0^1 a_2^{(2)}(b_2,\omega)d\omega\right\}_{\boldsymbol{\alpha}^0}.
\end{aligned}
\tag{3.196}
$$

Adding the expression of the indirect-effect term obtained in (3.196) with the expression of the direct-effect term provided in (3.172) and subsequently identifying the expressions that multiply the various parameter variations provides the expressions of the second-order sensitivities of the form $\partial^2\rho(\varphi,\psi;\boldsymbol{\alpha})/\partial\alpha_i\partial\mu(\boldsymbol{\alpha})$, $i=1,\ldots,$ TP. These expressions are as follows:

$$
\begin{aligned}
&\left\{\frac{\partial^2\rho(\varphi,\psi;\boldsymbol{\alpha})}{\partial\mu(\boldsymbol{\alpha})\partial\mu(\boldsymbol{\alpha})}\right\}_{\boldsymbol{\alpha}^0}=-\left\{\int_0^1 d\omega\int_{b_1}^{b_2}a_1^{(2)}(z,\omega)\varphi(z,\omega)dz\right\}_{\boldsymbol{\alpha}^0}\\
&-\left\{\int_0^1 d\omega\int_{b_1}^{b_2}a_2^{(2)}(z,\omega)\psi(z,\omega)dz\right\}_{\boldsymbol{\alpha}^0}-\left\{\int_0^1 d\omega\int_{b_1}^{b_2}a_3^{(2)}(z,\omega)a_1^{(1)}(z,\omega)dz\right\}_{\boldsymbol{\alpha}^0}\\
&-\left\{\int_0^1 d\omega\int_{b_1}^{b_2}a_4^{(2)}(z,\omega)a_2^{(1)}(z,\omega)dz\right\}_{\boldsymbol{\alpha}^0},
\end{aligned}
\tag{3.197}
$$

$$\left\{\frac{\partial^2 \rho(\varphi,\psi;\boldsymbol{\alpha})}{\partial \mu_s(\boldsymbol{\alpha})\partial \mu(\boldsymbol{\alpha})}\right\}_{\boldsymbol{\alpha}^0} = -\left\{\frac{q}{\mu_s^2(\boldsymbol{\alpha})}\int_0^1 a_1^{(2)}(b_1,\omega)\omega d\omega\right\}_{\boldsymbol{\alpha}^0}, \tag{3.198}$$

$$\left\{\frac{\partial^2 \rho(\varphi,\psi;\boldsymbol{\alpha})}{\partial q\partial \mu(\boldsymbol{\alpha})}\right\}_{\boldsymbol{\alpha}^0} = \left\{\frac{1}{\mu_s(\boldsymbol{\alpha})}\int_0^1 a_1^{(2)}(b_1,\omega)\omega d\omega\right\}_{\boldsymbol{\alpha}^0}, \tag{3.199}$$

$$\left\{\frac{\partial^2 \rho(\varphi,\psi;\boldsymbol{\alpha})}{\partial q^*\partial \mu(\boldsymbol{\alpha})}\right\}_{\boldsymbol{\alpha}^0} = \left\{\int_0^1 d\omega\int_{b_1}^{b_2} a_2^{(2)}(z,\omega)dz\right\}_{\boldsymbol{\alpha}^0}, \tag{3.200}$$

$$\begin{aligned}\left\{\frac{\partial^2 \rho(\varphi,\psi;\boldsymbol{\alpha})}{\partial b_1\partial \mu(\boldsymbol{\alpha})}\right\}_{\boldsymbol{\alpha}^0} &= \left\{\frac{q\mu(\boldsymbol{\alpha})}{\mu_s(\boldsymbol{\alpha})}\int_0^1 a_1^{(2)}(b_1,\omega)d\omega\right\}_{\boldsymbol{\alpha}^0}\\ &+\left\{\int_0^1 d\omega\left[a_1^{(1)}(b_1,\omega)\varphi(b_1,\omega)+a_2^{(1)}(b_1,\omega)\psi(b_1,\omega)\right]\right\}_{\boldsymbol{\alpha}^0},\end{aligned} \tag{3.201}$$

$$\begin{aligned}\left\{\frac{\partial^2 \rho(\varphi,\psi;\boldsymbol{\alpha})}{\partial b_2\partial \mu(\boldsymbol{\alpha})}\right\}_{\boldsymbol{\alpha}^0} &= \left\{q^*\int_0^1 a_2^{(2)}(b_2,\omega)d\omega\right\}_{\boldsymbol{\alpha}^0}\\ &-\left\{\int_0^1 d\omega\left[a_1^{(1)}(b_2,\omega)\varphi(b_2,\omega)+a_2^{(1)}(b_2,\omega)\psi(b_2,\omega)\right]\right\}_{\boldsymbol{\alpha}^0},\end{aligned} \tag{3.202}$$

$$\left\{\frac{\partial^2 \rho(\varphi,\psi;\boldsymbol{\alpha})}{\partial z_0\partial \mu(\boldsymbol{\alpha})}\right\}_{\boldsymbol{\alpha}^0} = \left\{\int_0^1 d\omega\left[\frac{d}{dz}\left(a_3^{(2)}\psi+a_4^{(2)}\varphi\right)\right]_{z=z_0}\right\}_{\boldsymbol{\alpha}^0}. \tag{3.203}$$

The various second-order sensitivities with respect to nuclide number densities and microscopic cross sections, respectively, can be obtained by using in (3.197)–(3.203) the relations provided in (3.155)–(3.158). For example, using (3.155)–(3.158) and (3.197) yields the following relations:

$$\begin{aligned}\left\{\frac{\partial^2 \rho(\varphi,\psi;\boldsymbol{\alpha})}{\partial N_i\partial N_i}\right\}_{\boldsymbol{\alpha}^0} &= \left\{\frac{\partial^2 \rho(\varphi,\psi;\boldsymbol{\alpha})}{\partial \mu(\boldsymbol{\alpha})\partial \mu(\boldsymbol{\alpha})}\left[\frac{\partial \mu(\boldsymbol{\alpha})}{\partial N_i}\right]^2\right\}_{\boldsymbol{\alpha}^0}\\ &= \left\{(\sigma_i)^2\frac{\partial^2 \rho(\varphi,\psi;\boldsymbol{\alpha})}{\partial \mu(\boldsymbol{\alpha})\partial \mu(\boldsymbol{\alpha})}\right\}_{\boldsymbol{\alpha}^0};i=1,\ \ldots,M;\end{aligned} \tag{3.204}$$

$$\left\{\frac{\partial^2\rho(\varphi,\psi;\boldsymbol{\alpha})}{\partial N_i\partial\sigma_i}\right\}_{\boldsymbol{\alpha}^0}=\left\{\frac{\partial^2\rho(\varphi,\psi;\boldsymbol{\alpha})}{\partial\mu(\boldsymbol{\alpha})\partial\mu(\boldsymbol{\alpha})}\frac{\partial\mu(\boldsymbol{\alpha})}{\partial N_i}\frac{\partial\mu(\boldsymbol{\alpha})}{\partial\sigma_i}\right\}_{\boldsymbol{\alpha}^0}$$
$$=\left\{\sigma_i N_i\frac{\partial^2\rho(\varphi,\psi;\boldsymbol{\alpha})}{\partial\mu(\boldsymbol{\alpha})\partial\mu(\boldsymbol{\alpha})}\right\}_{\boldsymbol{\alpha}^0};i=1,\ldots,M; \tag{3.205}$$

$$\left\{\frac{\partial^2\rho(\varphi,\psi;\boldsymbol{\alpha})}{\partial\sigma_i\partial\sigma_i}\right\}_{\boldsymbol{\alpha}^0}=\left\{\frac{\partial^2\rho(\varphi,\psi;\boldsymbol{\alpha})}{\partial\mu(\boldsymbol{\alpha})\partial\mu(\boldsymbol{\alpha})}\left[\frac{\partial\mu(\boldsymbol{\alpha})}{\partial\sigma_i}\right]^2\right\}_{\boldsymbol{\alpha}^0}$$
$$=\left\{(N_i)^2\frac{\partial^2\rho(\varphi,\psi;\boldsymbol{\alpha})}{\partial\mu(\boldsymbol{\alpha})\partial\mu(\boldsymbol{\alpha})}\right\}_{\boldsymbol{\alpha}^0};i=1,\ldots,M. \tag{3.206}$$

It is important to note that all of the relative second-order sensitivities of $\rho(\varphi,\psi;\boldsymbol{\alpha})$ with respect to N_i and σ_i have the same values, i.e.,

$$\left\{\frac{(N_i)^2}{\rho(\varphi,\psi;\boldsymbol{\alpha})}\frac{\partial^2\rho(\varphi,\psi;\boldsymbol{\alpha})}{\partial N_i\partial N_i}\right\}_{\boldsymbol{\alpha}^0}=\left\{\frac{N_i\sigma_i}{\rho(\varphi,\psi;\boldsymbol{\alpha})}\frac{\partial^2\rho(\varphi,\psi;\boldsymbol{\alpha})}{\partial N_i\partial\sigma_i}\right\}_{\boldsymbol{\alpha}^0}$$
$$=\left\{\frac{(\sigma_i)^2}{\rho(\varphi,\psi;\boldsymbol{\alpha})}\frac{\partial^2\rho(\varphi,\psi;\boldsymbol{\alpha})}{\partial\sigma_i\partial\sigma_i}\right\}_{\boldsymbol{\alpha}^0}=\left\{\frac{(N_i\sigma_i)^2}{\rho(\varphi,\psi;\boldsymbol{\alpha})}\frac{\partial^2\rho(\varphi,\psi;\boldsymbol{\alpha})}{\partial\mu(\boldsymbol{\alpha})\partial\mu(\boldsymbol{\alpha})}\right\}_{\boldsymbol{\alpha}^0};i=1,\ldots,M. \tag{3.207}$$

Distinct relative sensitivities which have the same values are notoriously difficult to compute using statistical methods, even when they are just of first order. Situation such as presented in (3.207) by the respective second-order sensitivities pose extreme, even insurmountable, difficulties when attempting to estimate them using statistical methods.

3.4.2 *Determination of the Second-Order Sensitivities of the Form* $\partial^2\rho(\varphi,\psi;\boldsymbol{\alpha})/\partial\alpha_i\partial b_2$, i = *1*, ..., TP

The explicit determination of the second-order sensitivities of the form $\partial^2\rho(\varphi,\psi;\boldsymbol{\alpha})/\partial\alpha_i\partial b_2$, which involve the imprecisely known domain boundary at $z=b_2$, will be presented in this subsection. These second-order sensitivities will be determined by considering that the expression provided in (3.153) for the first-order sensitivity $\partial\rho(\varphi,\psi;\boldsymbol{\alpha})/\partial b_2$ plays the role of a "model response." The expressions of the second-order sensitivities $\partial^2\rho(\varphi,\psi;\boldsymbol{\alpha})/\partial\alpha_i\partial b_2$, $i=1,\ldots,TP$, will thus be determined by taking the first-order G-differential of the first-order sensitivity $\partial\rho(\varphi,\psi;\boldsymbol{\alpha})/\partial b_2$. Applying the definition of the first-order total G-differential to (3.153) yields the following relation:

$$\left\{\delta\left[\frac{\partial\rho(\varphi,\psi;\boldsymbol{\alpha})}{\partial b_2}\right]\right\}_{\boldsymbol{\alpha}^0}=\left\{\delta\left(\frac{\partial\rho}{\partial b_2}\right)_{\boldsymbol{\alpha}^0}\right\}_{dir}+\left\{\delta\left(\frac{\partial\rho}{\partial b_2}\right)_{\boldsymbol{\alpha}^0}\right\}_{ind}$$
$$\triangleq\left\{\frac{d}{d\varepsilon}\left[(q^*+\varepsilon\delta q^*)\int_0^1 d\omega\int_{b_1+\varepsilon\delta b_1}^{b_2+\varepsilon\delta b_2}\left(a_2^{(1)}+\varepsilon\delta a_2^{(1)}\right)\delta(z-b_2-\varepsilon\delta b_2)dz\right]_{\boldsymbol{\alpha}^0}\right\}_{\varepsilon=0}, \tag{3.208}$$

where

$$\left\{\delta\left(\frac{\partial\rho}{\partial b_2}\right)_{\boldsymbol{\alpha}^0}\right\}_{dir}\triangleq(\delta q^*)\left\{\int_0^1 a_2^{(1)}(b_2,\omega)d\omega\right\}_{\boldsymbol{\alpha}^0}$$
$$+(\delta b_2)\left\{q^*\int_0^1\left[\frac{da_2^{(1)}(z,\omega)}{dz}\right]_{z=b_2}d\omega\right\}_{\boldsymbol{\alpha}^0}, \tag{3.209}$$

and

$$\left\{\delta\left(\frac{\partial\rho}{\partial b_2}\right)_{\boldsymbol{\alpha}^0}\right\}_{ind}\triangleq\left\{q^*\int_0^1 d\omega\int_{b_1}^{b_2}\delta a_2^{(1)}(z,\omega)\delta(z-b_2)dz\right\}_{\boldsymbol{\alpha}^0}. \tag{3.210}$$

The direct-effect term defined in (3.209) can be computed immediately, but the computation of the indirect-effect term defined in (3.210) requires the prior determination of the function $\delta a_2^{(1)}(z,\omega)$. It has already been shown that $\delta a_2^{(1)}(z,\omega)$ is the solution of (3.176) subject to the boundary condition provided in (3.178). But (3.176) involves the function $\delta\varphi(z,\omega)$, which is the solution of (3.41) subject to the boundary condition provided in (3.43). Concatenating (3.176) and (3.41) yields the following 2nd-LVSS:

$$\left\{\begin{pmatrix}\omega\frac{d}{dz}+\mu & 0\\ -\delta(z-z_0) & \omega\frac{d}{dz}+\mu\end{pmatrix}\right\}_{\boldsymbol{\alpha}^0}\begin{pmatrix}\delta\varphi(z,\omega)\\ \delta a_2^{(1)}(z,\omega)\end{pmatrix}$$
$$=\left\{\begin{pmatrix}-(\delta\mu)\varphi(z,\omega)\\ -(\delta\mu)a_2^{(1)}(z,\omega)-(\delta z_0)\varphi(z,\omega)\delta'(z-z_0)\end{pmatrix}\right\}_{\boldsymbol{\alpha}^0}. \tag{3.211}$$

The 2nd-LVSS represented by (3.211), together with the corresponding boundary conditions provided in (3.178) and (3.43), depends on the parameter variations and would need to be solved as many times as there are parameter variations. The need for solving the 2nd-LVSS can be circumvented by applying the principles of the 2nd-CASAM-L to construct a 2nd-LASS. In view of the structure of (3.211), the Hilbert space appropriate for constructing the corresponding 2nd-LASS will be endowed with the inner product defined in (3.140). Hence, the 2nd-LASS corresponding to (3.211) is constructed as follows:

1. Form the inner product of a 2-component vector function $\mathbf{c}^{(2)}(z,\omega) \triangleq \left[c_1^{(2)}(z,\omega), c_2^{(2)}(z,\omega)\right]^{\dagger}$ with (3.211) to obtain the following relation:

$$
\begin{aligned}
&\left\{ \int_0^1 d\omega \int_{b_1}^{b_2} c_1^{(2)}(z,\omega) \left[\omega \frac{d}{dz} \delta\varphi(z,\omega) + \mu \delta\varphi(z,\omega) \right] dz \right\}_{\boldsymbol{\alpha}^0} \\
&+ \left\{ \int_0^1 d\omega \int_{b_1}^{b_2} c_2^{(2)}(z,\omega) [-\delta(z - z_0) \delta\varphi(z,\omega) \right. \\
&\qquad \left. + \omega \frac{d}{dz} \delta a_2^{(1)}(z,\omega) + \mu \delta a_2^{(1)}(z,\omega) \Big] dz \right\}_{\boldsymbol{\alpha}^0} \\
&= -(\delta\mu) \left\{ \int_0^1 d\omega \int_{b_1}^{b_2} \varphi(z,\omega) c_1^{(2)}(z,\omega) dz \right\}_{\boldsymbol{\alpha}^0} \\
&\quad -(\delta\mu) \left\{ \int_0^1 d\omega \int_{b_1}^{b_2} c_2^{(2)}(z,\omega) a_2^{(1)}(z,\omega)\, dz \right\}_{\boldsymbol{\alpha}^0} \\
&\quad -(\delta z_0) \left\{ \int_0^1 d\omega \int_{b_1}^{b_2} c_2^{(2)}(z,\omega) \varphi(z,\omega) \delta'(z - z_0) dz \right\}_{\boldsymbol{\alpha}^0} .
\end{aligned}
\tag{3.212}
$$

2. Integrate by parts the left side of (3.212) to obtain the following relation:

$$
\begin{aligned}
&\left\{\int_0^1 d\omega \int_{b_1}^{b_2} c_1^{(2)}(z,\omega)\left[\omega\frac{d}{dz}\delta\varphi(z,\omega)+\mu\delta\varphi(z,\omega)\right]dz\right\}_{\boldsymbol{\alpha}^0} \\
&+\left\{\int_0^1 d\omega \int_{b_1}^{b_2} c_2^{(2)}(z,\omega)[-\delta(z-z_0)\delta\varphi(z,\omega) \right. \\
&\left. +\omega\frac{d}{dz}\delta a_2^{(1)}(z,\omega)+\mu\delta a_2^{(1)}(z,\omega)\right]dz\Bigg\}_{\boldsymbol{\alpha}^0} \\
&=\left\{\int_0^1 d\omega \int_{b_1}^{b_2} \delta\varphi(z,\omega)\left[-\omega\frac{dc_1^{(2)}(z,\omega)}{dz}\right.\right. \\
&\left.\left. +\mu c_1^{(2)}(z,\omega)-\delta(z-z_0)c_2^{(2)}(z,\omega)\right]dz\right\}_{\boldsymbol{\alpha}^0} \\
&+\left\{\int_0^1 d\omega \int_{b_1}^{b_2} \delta a_2^{(1)}(z,\omega)\left[-\omega\frac{dc_2^{(2)}(z,\omega)}{dz}+\mu c_2^{(2)}(z,\omega)\right]dz\right\}_{\boldsymbol{\alpha}^0} \\
&+\left\{\int_0^1 \omega d\omega\left[\delta\varphi(b_2,\omega)c_1^{(2)}(b_2,\omega)-\delta\varphi(b_1,\omega)c_1^{(2)}(b_1,\omega)\right]\right\}_{\boldsymbol{\alpha}^0} \\
&+\left\{\int_0^1 \omega d\omega\left[\delta a_2^{(1)}(b_2,\omega)c_2^{(2)}(b_2,\omega)-\delta a_2^{(1)}(b_1,\omega)c_2^{(2)}(b_1,\omega)\right]\right\}_{\boldsymbol{\alpha}^0}.
\end{aligned}
\tag{3.213}
$$

3. Require the first two terms on the right side of (3.213) to represent the indirect-effect term defined in (3.210), which is accomplished by imposing the following relations:

$$
\left\{-\omega\frac{dc_1^{(2)}(z,\omega)}{dz}+\mu c_1^{(2)}(z,\omega)-\delta(z-z_0)c_2^{(2)}(z,\omega)\right\}_{\boldsymbol{\alpha}^0}=0, \tag{3.214}
$$

$$
\left\{-\omega\frac{dc_2^{(2)}(z,\omega)}{dz}+\mu c_2^{(2)}(z,\omega)\right\}_{\boldsymbol{\alpha}^0}=\{q^*\delta(z-b_2)\}_{\boldsymbol{\alpha}^0}. \tag{3.215}
$$

4. Determine the boundary conditions for the functions $c_1^{(2)}(z,\omega)$ and $c_2^{(2)}(z,\omega)$ by eliminating the unknown quantities $\delta\varphi(b_2,\omega)$ and $\delta a_2^{(1)}(b_2,\omega)$ from (3.213). This is accomplished by imposing the following boundary conditions:

$$\left\{c_1^{(2)}(b_2,\omega)\right\}_{\boldsymbol{\alpha}^0}=0,\ z=b_2^0;\ 0\le\omega\le 1, \tag{3.216}$$

$$\left\{c_2^{(2)}(b_2,\omega)\right\}_{\boldsymbol{\alpha}^0}=0,\ z=b_2^0;\ 0\le\omega\le 1. \tag{3.217}$$

The relations provided in (3.214)–(3.217) constitute the 2nd-LASS for the second-level adjoint function $\mathbf{c}^{(2)}(z,\omega)\triangleq\left[c_1^{(2)}(z,\omega),c_2^{(2)}(z,\omega)\right]^{\dagger}$. Notably, both components $c_1^{(2)}(z,\omega)$ and $c_2^{(2)}(z,\omega)$ satisfy adjoint-like equations.

5. Use in (3.210) the relations comprising the 2nd-LASS defined by (3.214)–(3.217) together with the relations provided in (3.212) and (3.213) and the boundary conditions provided in (3.178) and (3.43) for the functions $\delta\varphi(z,\omega)$ and $\delta a_2^{(1)}(z,\omega)$ to obtain the following expression for the indirect-effect term:

$$\begin{aligned}
\left\{\delta\left(\frac{\partial\rho}{\partial b_2}\right)_{\boldsymbol{\alpha}^0}\right\}_{ind}&=-(\delta\mu)\left\{\int_0^1 d\omega\int_{b_1}^{b_2}\varphi(z,\omega)c_1^{(2)}(z,\omega)dz\right\}_{\boldsymbol{\alpha}^0}\\
&-(\delta\mu)\left\{\int_0^1 d\omega\int_{b_1}^{b_2}a_2^{(1)}(z,\omega)c_2^{(2)}(z,\omega)\,dz\right\}_{\boldsymbol{\alpha}^0}\\
&-(\delta z_0)\left\{\int_0^1 d\omega\int_{b_1}^{b_2}\varphi(z,\omega)c_2^{(2)}(z,\omega)\delta'(z-z_0)dz\right\}_{\boldsymbol{\alpha}^0}\\
&+(\delta b_1)\left\{\frac{q\mu(\boldsymbol{\alpha})}{\mu_s(\boldsymbol{\alpha})}\int_0^1 c_1^{(2)}(b_1,\omega)d\omega\right\}_{\boldsymbol{\alpha}^0}\\
&+\left\{\left[\frac{(\delta q)}{\mu_s(\boldsymbol{\alpha})}-\frac{q\delta\mu_s(\boldsymbol{\alpha})}{\mu_s^2(\boldsymbol{\alpha})}\right]\int_0^1 c_1^{(2)}(b_1,\omega)\omega d\omega\right\}_{\boldsymbol{\alpha}^0}.
\end{aligned} \tag{3.218}$$

Inserting the expression obtained in (3.218) together with the expression of the direct-effect term obtained in (3.209) into (3.208) and identifying in the resulting expression the quantities that multiply the various parameter variations yields the following expressions for the corresponding second-order sensitivities:

$$\left\{\frac{\partial^2\rho}{\partial\mu(\boldsymbol{\alpha})\partial b_2}\right\}_{\boldsymbol{\alpha}^0} = -\left\{\int_0^1 d\omega \int_{b_1}^{b_2} \left[\varphi(z,\omega)c_1^{(2)}(z,\omega) + a_2^{(1)}(z,\omega)c_2^{(2)}(z,\omega)\right]dz\right\}_{\boldsymbol{\alpha}^0}, \tag{3.219}$$

$$\left\{\frac{\partial^2\rho}{\partial\mu_s(\boldsymbol{\alpha})\partial b_2}\right\}_{\boldsymbol{\alpha}^0} = \left\{-\frac{q}{\mu_s^2(\boldsymbol{\alpha})}\int_0^1 c_1^{(2)}(b_1,\omega)\omega d\omega\right\}_{\boldsymbol{\alpha}^0}, \tag{3.220}$$

$$\left\{\frac{\partial^2\rho}{\partial q\partial b_2}\right\}_{\boldsymbol{\alpha}^0} = \left\{\frac{1}{\mu_s(\boldsymbol{\alpha})}\int_0^1 c_1^{(2)}(b_1,\omega)\omega d\omega\right\}_{\boldsymbol{\alpha}^0}, \tag{3.221}$$

$$\left\{\frac{\partial^2\rho}{\partial q^*\partial b_2}\right\}_{\boldsymbol{\alpha}^0} = \left\{\int_0^1 a_2^{(1)}(b_2,\omega)d\omega\right\}_{\boldsymbol{\alpha}^0}, \tag{3.222}$$

$$\left\{\frac{\partial^2\rho}{\partial b_1\partial b_2}\right\}_{\boldsymbol{\alpha}^0} = \left\{\frac{q\mu(\boldsymbol{\alpha})}{\mu_s(\boldsymbol{\alpha})}\int_0^1 c_1^{(2)}(b_1,\omega)d\omega\right\}_{\boldsymbol{\alpha}^0}, \tag{3.223}$$

$$\left\{\frac{\partial^2\rho}{\partial z_0\partial b_2}\right\}_{\boldsymbol{\alpha}^0} = -\left\{\int_0^1 d\omega \int_{b_1}^{b_2} c_2^{(2)}(z,\omega)\varphi(z,\omega)\delta'(z-z_0)dz\right\}_{\boldsymbol{\alpha}^0}, \tag{3.224}$$

$$\left\{\frac{\partial^2\rho}{\partial b_2\partial b_2}\right\}_{\boldsymbol{\alpha}^0} = \left\{q^*\int_0^1 \left[\frac{da_2^{(1)}(z,\omega)}{dz}\right]_{z=b_2} d\omega\right\}. \tag{3.225}$$

Solving the 2nd-LASS yields the following expressions for the components of the second-level adjoint function $\mathbf{c}^{(2)}(z,\omega)\triangleq\left[c_1^{(2)}(z,\omega), c_2^{(2)}(z,\omega)\right]^\dagger$:

$$\left\{c_1^{(2)}(z,\omega)\right\}_{\boldsymbol{\alpha}^0} = \left\{\frac{q^*}{\omega^2}[1-H(z-z_0)]\exp\left[\frac{\mu(z-b_2)}{\omega}\right]\right\}_{\boldsymbol{\alpha}^0}, \tag{3.226}$$

$$\left\{c_2^{(2)}(z,\omega)\right\}_{\boldsymbol{\alpha}^0} = \left\{\frac{q^*}{\omega}[1-H(z-b_2)]\exp\left[\frac{\mu(z-b_2)}{\omega}\right]\right\}_{\boldsymbol{\alpha}^0}. \tag{3.227}$$

Replacing the expressions obtained in (3.226) and (3.227) together with the expressions provided for $\varphi(z,\omega)$ and $a_2^{(1)}(z,\omega)$ in (3.8) and (3.160), respectively, into (3.219) yields the following expression:

$$\left\{ \frac{\partial^2 \rho}{\partial \mu(\boldsymbol{\alpha}) \partial b_2} \right\}_{\boldsymbol{\alpha}^0} = \left\{ \frac{q q^* (b_1 - b_2)}{\mu_s(\boldsymbol{\alpha})} E_0[(b_2 - b_1)\mu(\boldsymbol{\alpha})] \right\}_{\boldsymbol{\alpha}^0} \\ = -\left\{ \frac{q q^* \exp\left[(b_1 - b_2)\mu(\boldsymbol{\alpha})\right]}{\mu(\boldsymbol{\alpha})\mu_s(\boldsymbol{\alpha})} \right\}_{\boldsymbol{\alpha}^0}. \tag{3.228}$$

The expression obtained in (3.228) is identical to the expression that would be obtained by taking the derivative $\partial/\partial b_2$ of (3.161), which provides a mutual verification of the correctness of the derivations (in particular: the derivation/computation of the adjoint functions involved) underlying the determination of the mixed second-order sensitivity $\partial^2 \rho/\partial \mu(\boldsymbol{\alpha}) \partial b_2$.

Replacing the expression obtained in (3.226) into (3.220) and performing the respective integration yields the following expression:

$$\left\{ \frac{\partial^2 \rho}{\partial \mu_s(\boldsymbol{\alpha}) \partial b_2} \right\}_{\boldsymbol{\alpha}^0} = \left\{ -\frac{q q^*}{\mu_s^2(\boldsymbol{\alpha})} E_1[(b_2 - b_1)\mu(\boldsymbol{\alpha})] \right\}_{\boldsymbol{\alpha}^0}. \tag{3.229}$$

The expression obtained in (3.229) is identical to the expression that would be obtained by taking the derivative $\partial/\partial b_2$ of (3.162), which provides a mutual verification of the correctness of the derivations (in particular: the derivation/computation of the adjoint functions involved) underlying the determination of the mixed second-order sensitivity $\partial^2 \rho/\partial \mu_s(\boldsymbol{\alpha}) \partial b_2$.

Replacing the expression obtained in (3.226) into (3.221) and performing the respective integration yields the following expression:

$$\left\{ \frac{\partial^2 \rho}{\partial q \partial b_2} \right\}_{\boldsymbol{\alpha}^0} = \left\{ \frac{q^*}{\mu_s(\boldsymbol{\alpha})} E_1[(b_2 - b_1)\mu(\boldsymbol{\alpha})] \right\}_{\boldsymbol{\alpha}^0}. \tag{3.230}$$

The expression obtained in (3.230) is identical to the expression that would be obtained by taking the derivative $\partial/\partial b_2$ of (3.163), which provides a mutual verification of the correctness of the derivations (in particular: the derivation/computation of the adjoint functions involved) underlying the determination of the mixed second-order sensitivity $\partial^2 \rho/\partial q \partial b_2$.

Replacing the expression obtained in (3.160) into (3.222) and performing the respective integration yields the following expression:

$$\left\{ \frac{\partial^2 \rho}{\partial q^* \partial b_2} \right\}_{\boldsymbol{\alpha}^0} = \left\{ \frac{q}{\mu_s(\boldsymbol{\alpha})} E_1[(b_2 - b_1)\mu(\boldsymbol{\alpha})] \right\}_{\boldsymbol{\alpha}^0}. \tag{3.231}$$

The expression obtained in (3.231) is identical to the expression that would be obtained by taking the derivative $\partial/\partial b_2$ of (3.164), which provides a mutual verification of the correctness of the derivations (in particular: the derivation/computation of the adjoint functions involved) underlying the determination of the mixed second-order sensitivity $\partial^2 \rho/\partial q^* \partial b_2$.

Replacing the expression obtained in (3.226) into (3.223) and performing the respective integration yields the following expression:

$$\begin{aligned} \left\{ \frac{\partial^2 \rho}{\partial b_1 \partial b_2} \right\}_{\boldsymbol{\alpha}^0} &= \left\{ \frac{q^* q \mu(\boldsymbol{\alpha})}{\mu_s(\boldsymbol{\alpha})} E_0[(b_2 - b_1)\mu(\boldsymbol{\alpha})] \right\}_{\boldsymbol{\alpha}^0} \\ &= \left\{ \frac{q^* q}{\mu_s(\boldsymbol{\alpha})} \frac{\exp\left[(b_1 - b_2)\mu(\boldsymbol{\alpha})\right]}{(b_2 - b_1)} \right\}_{\boldsymbol{\alpha}^0}. \end{aligned} \tag{3.232}$$

The expression obtained in (3.232) is identical to the expression that would be obtained by taking the derivative $\partial/\partial b_2$ of (3.165), which provides a mutual verification of the correctness of the derivations (in particular: the derivation/computation of the adjoint functions involved) underlying the determination of the mixed second-order sensitivity $\partial^2 \rho / \partial b_1 \partial b_2$.

Using the expressions obtained in (3.8) and (3.227) in (3.224) and performing the respective integrations yields the following expression:

$$\begin{aligned} \left\{ \frac{\partial^2 \rho}{\partial z_0 \partial b_2} \right\}_{\boldsymbol{\alpha}^0} &= -\left\{ \int_0^1 d\omega \int_{b_1}^{b_2} c_2^{(2)}(z,\omega)\varphi(z,\omega)\delta'(z - z_0)dz \right\}_{\boldsymbol{\alpha}^0} \\ &= \frac{q}{\mu_s} \int_0^1 \frac{q^*}{\omega} \exp\left[\frac{\mu}{\omega}(b_1 - b_2)\right] d\omega \int_{b_1}^{b_2} \delta(z - b_2)\delta(z - z_0)dz = 0. \end{aligned} \tag{3.233}$$

The expression obtained in (3.233) is identical to the expression that would be obtained by taking the derivative $\partial/\partial b_2$ of (3.167), which provides a mutual verification of the correctness of the derivations (in particular: the derivation/computation of the adjoint functions involved) underlying the determination of the mixed second-order sensitivity $\partial^2 \rho / \partial z_0 \partial b_2$.

Using the expression obtained in (3.160) in (3.225) and performing the respective integration yields the following expression:

$$\begin{aligned} \left\{ \frac{\partial^2 \rho}{\partial b_2 \partial b_2} \right\}_{\boldsymbol{\alpha}^0} &= -\left\{ \frac{q q^* \mu(\boldsymbol{\alpha})}{\mu_s(\boldsymbol{\alpha})} E_0[(b_2 - b_1)\mu(\boldsymbol{\alpha})] \right\} \\ &= -\left\{ \frac{q^* q}{\mu_s(\boldsymbol{\alpha})} \frac{\exp\left[(b_1 - b_2)\mu(\boldsymbol{\alpha})\right]}{(b_2 - b_1)} \right\}_{\boldsymbol{\alpha}^0}. \end{aligned} \tag{3.234}$$

Since $\partial^2 \rho / \partial b_2 \partial b_2$ is an unmixed second-order sensitivity, its expression cannot be verified independently in terms of an equivalent expression using different adjoint functions, as is the case for the mixed second-order sensitivities, the expressions of which can always be independently verified in terms of alternative expressions involving different adjoint and/or forward functions.

3.4.3 Summary of Main Features Underlying the Computation of the Second-Order Sensitivities $\partial^2 \rho(\varphi, \psi; \alpha)/\partial\alpha_i \partial\alpha_j$, i, j = 1, ..., TP

The second-order sensitivities corresponding to the remaining first-order sensitivities provided in (3.149)–(3.154) are obtained by following the same conceptual steps as those that led to the second-order sensitivities of the form $\partial^2 \rho(\varphi, \psi; \boldsymbol{\alpha})/\partial \alpha_i \partial \mu(\boldsymbol{\alpha})$, which are presented in (3.197)–(3.203) and/or the steps that led to the sensitivities of the form $\partial^2 \rho(\varphi, \psi; \boldsymbol{\alpha})/\partial \alpha_i \partial b_2$. The salient characteristics underlying the determination of the remaining second-order sensitivities of the contributon-response flux $\rho(\varphi, \psi; \boldsymbol{\alpha})$ are summarized below:

1. As shown in (3.151), the first-order sensitivity $\partial \rho(\varphi, \psi; \boldsymbol{\alpha})/\partial q^*$ depends on the first-level adjoint function $a_2^{(1)}(z, \omega)$, just like the sensitivity $\partial \rho(\varphi, \psi; \boldsymbol{\alpha})/\partial b_2$. Consequently, the second-order sensitivities of the form $\partial^2 \rho(\varphi, \psi; \boldsymbol{\alpha})/\partial \alpha_i \partial q^*$ can be determined by following procedure as used for determining the second-order sensitivities of the form $\partial^2 \rho(\varphi, \psi; \boldsymbol{\alpha})/\partial \alpha_i \partial b_2$. Consequently, the second-order sensitivities of the form $\partial^2 \rho(\varphi, \psi; \boldsymbol{\alpha})/\partial \alpha_i \partial q^*$ will be expressed in terms of a two-component second-level adjoint function denoted, e.g., as $\mathbf{d}^{(2)}(z, \omega) \triangleq \left[d_1^{(2)}(z, \omega), d_2^{(2)}(z, \omega) \right]^\dagger$, which will be the solution of a 2nd-LASS having the same form as in (3.214) and (3.215), but with different source terms on the right sides of these equations.
2. As shown in (3.149), (3.150), and (3.152), the first-order sensitivities $\partial \rho(\varphi, \psi; \boldsymbol{\alpha})/\partial \mu_s(\boldsymbol{\alpha})$, $\partial \rho(\varphi, \psi; \boldsymbol{\alpha})/\partial q$, and $\partial \rho(\varphi, \psi; \boldsymbol{\alpha})/\partial b_1$ depend on the first-level adjoint function $a_1^{(1)}(z, \omega)$. In turn, the first-level adjoint function $a_1^{(1)}(z, \omega)$ depends on the original adjoint function $\psi(z, \omega)$, as indicated in (3.143). The second-order sensitivities of the form $\partial^2 \rho(\varphi, \psi; \boldsymbol{\alpha})/\partial \alpha_i \partial \mu_s(\boldsymbol{\alpha})$ will ultimately be expressed in terms of a two-component second-level adjoint function, denoted, e.g., as $\mathbf{e}^{(2)}(z, \omega) \triangleq \left[e_1^{(2)}(z, \omega), e_2^{(2)}(z, \omega) \right]^\dagger$, one component of which [e.g., $e_1^{(2)}(z, \omega)$] will correspond to the variational function $\delta\psi(z, \omega)$, while the other component [e.g., $e_2^{(2)}(z, \omega)$] will correspond to the variational function $\delta a_1^{(1)}(z, \omega)$. Similarly, the second-order sensitivities of the form $\partial^2 \rho(\varphi, \psi; \boldsymbol{\alpha})/\partial \alpha_i \partial q$ will ultimately be expressed in terms of a two-component second-level adjoint function designated as, e.g., $\mathbf{f}^{(2)}(z, \omega) \triangleq \left[f_1^{(2)}(z, \omega), f_2^{(2)}(z, \omega) \right]^\dagger$, while the second-order sensitivities of the form $\partial^2 \rho(\varphi, \psi; \boldsymbol{\alpha})/\partial \alpha_i \partial b_1$ will ultimately be expressed in terms of a two-component second-level adjoint function designated as, e.g., $\mathbf{g}^{(2)}(z, \omega) \triangleq \left[g_1^{(2)}(z, \omega), g_2^{(2)}(z, \omega) \right]^\dagger$. The second-level adjoint functions $\mathbf{e}^{(2)}(z, \omega)$, $\mathbf{f}^{(2)}(z, \omega)$, and $\mathbf{g}^{(2)}(z, \omega)$ will be solutions of Second-Level Adjoint Sensitivity Systems similar to the 2nd-LASS satisfied by the second-level adjoint function $\mathbf{c}^{(2)}(z, \omega)$.

3. As shown in (3.154), the first-order sensitivity $\partial\rho(\varphi,\psi;\boldsymbol{\alpha})/\partial z_0$ depends on the original functions $\varphi(z,\omega)$ and $\psi(z,\omega)$. This dependence is similar to the dependence of $\rho(\varphi,\psi;\boldsymbol{\alpha})$. Consequently, the procedure for determining the second-order sensitivities of the form $\partial^2\rho(\varphi,\psi;\boldsymbol{\alpha})/\partial\alpha_i\partial z_0$ is similar to the procedure followed for determining the first-order sensitivities of $\rho(\varphi,\psi;\boldsymbol{\alpha})$. Therefore, the second-order sensitivities of the form $\partial^2\rho(\varphi,\psi;\boldsymbol{\alpha})/\partial\alpha_i\partial z_0$ will ultimately be expressed in terms of a two-component second-level adjoint function, denoted, e.g., as $\mathbf{h}^{(2)}(z,\omega)\triangleq\left[h_1^{(2)}(z,\omega),h_2^{(2)}(z,\omega)\right]^\dagger$. One of these components, e.g., $h_1^{(2)}(z,\omega)$, will correspond to the function $\delta\varphi(z,\omega)$, while the second component, e.g., $h_2^{(2)}(z,\omega)$, will correspond to the function $\delta\psi(z,\omega)$. The second-level adjoint function $\mathbf{h}^{(2)}(z,\omega)\triangleq\left[h_1^{(2)}(z,\omega),h_2^{(2)}(z,\omega)\right]^\dagger$ will be the solution of a 2nd-LASS having a form similar to the 1st-LASS, but with sources that would correspond to the expression of $\partial\rho(\varphi,\psi;\boldsymbol{\alpha})/\partial z_0$ (and would therefore differ from the source terms of the 1st-LASS).
4. The symmetry property $\partial^2\rho(\varphi,\psi;\boldsymbol{\alpha})/\partial\alpha_i\partial\alpha_j;\ j\neq i$ of the mixed second-order sensitivities enables independent verification of the computational accuracy of the solvers used to compute the various first-level and second-level adjoint functions, since the adjoint functions involved in the expression of $\partial^2\rho(\varphi,\psi;\boldsymbol{\alpha})/\partial\alpha_i\partial\alpha_j$ differ from the adjoint functions involved in the expression of $\partial^2\rho(\varphi,\psi;\boldsymbol{\alpha})/\partial\alpha_j\partial\alpha_i$.

In summary, the exact computation of all of the partial second-order sensitivities requires a total of 7 large-scale (adjoint) computations to determine the 7 second-level adjoint functions $\mathbf{a}^{(2)}(z,\omega)\triangleq\left[a_1^{(2)}(z,\omega),a_2^{(2)}(z,\omega),a_3^{(2)}(z,\omega),a_4^{(2)}(z,\omega)\right]^\dagger$, $\mathbf{c}^{(2)}(z,\omega)\triangleq\left[c_1^{(2)}(z,\omega),c_2^{(2)}(z,\omega)\right]^\dagger$, $\mathbf{d}^{(2)}(z,\omega)\triangleq\left[d_1^{(2)}(z,\omega),d_2^{(2)}(z,\omega)\right]^\dagger$, $\mathbf{e}^{(2)}(z,\omega)\triangleq\left[e_1^{(2)}(z,\omega),e_2^{(2)}(z,\omega)\right]^\dagger$, $\mathbf{f}^{(2)}(z,\omega)\triangleq\left[f_1^{(2)}(z,\omega),f_2^{(2)}(z,\omega)\right]^\dagger$, $\mathbf{g}^{(2)}(z,\omega)\triangleq\left[g_1^{(2)}(z,\omega),g_2^{(2)}(z,\omega)\right]^\dagger$ and $\mathbf{h}^{(2)}(z,\omega)\triangleq\left[h_1^{(2)}(z,\omega),h_2^{(2)}(z,\omega)\right]^\dagger$, by solving the corresponding Second-Level Adjoint Sensitivity Systems. As has been mentioned under item (4), above, the second-order mixed sensitivities will be computed twice, in two different ways (i.e., using distinct second-level adjoint functions), thereby providing an independent intrinsic (numerical) verification that the first- and second-order response sensitivities are computed accurately. The information provided by the first-order sensitivities usually indicates which second-order sensitivities are important and which could be neglected. Therefore, it is useful to prioritize the computation of the second-order sensitivities by using the rankings of the relative magnitudes of the first-order sensitivities as a "priority indicator": the larger the magnitude of the relative first-order sensitivity, the higher the priority for computing the corresponding second-order sensitivities. Also, since vanishing first-order sensitivities may indicate critical points of the response in the phase-space of model parameters, it is also of interest to compute the second-order sensitivities that correspond to vanishing first-order sensitivities. In practice, only those second-order partial sensitivities which are deemed important would need to be computed.

3.5 Application of the 3rd-CASAM-L to Compute Third-Order Response Sensitivities to Imprecisely Known Parameters

The third-order response sensitivities to the model parameters will be obtained by applying the general principles presented in Chap. 2. In essence, each of the second-order response sensitivities will be considered a new "model response" and the principles underlying the 1st-CASAM-L will be applied to determine the "first-order sensitivities of the second-order sensitivities," which by definition are the sought-after third-order sensitivities. The application of these principles will be illustrated in Sect. 3.5.1 by considering as "model response" the second-order unmixed sensitivity $\partial^2 \rho(\varphi, \psi; \boldsymbol{\alpha})/\partial \mu(\boldsymbol{\alpha}) \partial \mu(\boldsymbol{\alpha})$ of the "contributon-response flux" with respect to the interaction coefficient $\mu(\boldsymbol{\alpha})$. This sensitivity is chosen to demonstrate the principles underlying the nth-CASAM-L for computing high-order sensitivities because its expression, as provided in (3.197), involves all of the original (forward and adjoint) and first-level and second-level adjoint state functions. Consequently, the computation of the third-order sensitivities $\partial^3 \rho(\varphi, \psi; \boldsymbol{\alpha})/\partial \alpha_i \partial \mu(\boldsymbol{\alpha}) \partial \mu(\boldsymbol{\alpha})$, $i = 1, \ldots, TP$, embodies all of the conceptual aspects underlying the computation of third-order sensitivities. The determination of the other third-order sensitivities involves only partially the mathematical complexities that are needed to determine the third-order sensitivities which will arise from $\partial^2 \rho(\varphi, \psi; \boldsymbol{\alpha})/\partial \mu(\boldsymbol{\alpha}) \partial \mu(\boldsymbol{\alpha})$.

Section 3.5.2 will present the explicit determination of the third-order sensitivities of the form $\partial^3 \rho(\varphi, \psi; \boldsymbol{\alpha})/\partial \alpha_i \partial b_1 \partial b_2$, $i = 1, \ldots, TP$. These sensitivities involve both the inner (interface) and the outer boundary parameters, b_1 and b_2, respectively. The determination of the sensitivities $\partial^3 \rho(\varphi, \psi; \boldsymbol{\alpha})/\partial \alpha_i \partial b_1 \partial b_2$ illustrates the general applicability of the 3rd-CASAM-L to determine the third-order sensitivities that involve boundary and interface parameters.

3.5.1 *Determination of the Third-Order Sensitivities of the Form* $\partial^3 \rho(\boldsymbol{\varphi}, \boldsymbol{\psi}; \boldsymbol{\alpha})/\partial \alpha_i \partial \mu(\boldsymbol{\alpha}) \partial \mu(\boldsymbol{\alpha})$, i = *1*, ..., TP

To determine the third-order sensitivities which will arise from $\partial^2 \rho(\varphi, \psi; \boldsymbol{\alpha})/\partial \mu(\boldsymbol{\alpha}) \partial \mu(\boldsymbol{\alpha})$, it is convenient to introduce the following notation:

$$\rho^{(2)}\left[\mathbf{u}^{(3)}(z,\omega);\boldsymbol{\alpha}\right]\triangleq\left\{\frac{\partial^2\rho(\varphi,\psi;\boldsymbol{\alpha})}{\partial\mu(\boldsymbol{\alpha})\partial\mu(\boldsymbol{\alpha})}\right\}_{\boldsymbol{\alpha}^0}$$
$$=-\left\{\int_0^1 d\omega\int_{b_1}^{b_2}a_1^{(2)}(z,\omega)\varphi(z,\omega)dz\right\}_{\boldsymbol{\alpha}^0}-\left\{\int_0^1 d\omega\int_{b_1}^{b_2}a_2^{(2)}(z,\omega)\psi(z,\omega)dz\right\}_{\boldsymbol{\alpha}^0}$$
$$-\left\{\int_0^1 d\omega\int_{b_1}^{b_2}a_3^{(2)}a_1^{(1)}dz\right\}_{\boldsymbol{\alpha}^0}-\left\{\int_0^1 d\omega\int_{b_1}^{b_2}a_4^{(2)}a_2^{(1)}dz\right\}_{\boldsymbol{\alpha}^0},\tag{3.235}$$

where:

$$\mathbf{u}^{(3)}(z,\omega)\triangleq\begin{pmatrix}\mathbf{u}^{(2)}(z,\omega)\\ \mathbf{a}^{(2)}(z,\omega)\end{pmatrix}\triangleq\left[\varphi,\psi;a_1^{(1)},a_2^{(1)};a_1^{(2)},a_2^{(2)},a_3^{(2)},a_4^{(4)}\right]^\dagger.\tag{3.236}$$

The first-order G-differential of $\partial^2\rho(\varphi,\psi;\boldsymbol{\alpha})/\partial\mu(\boldsymbol{\alpha})\partial\mu(\boldsymbol{\alpha})$ will comprise all of the third-order sensitivities of the form $\partial^3\rho(\varphi,\psi;\boldsymbol{\alpha})/\partial\alpha_i\partial\mu(\boldsymbol{\alpha})\partial\mu(\boldsymbol{\alpha})$, $i=1,\ldots,TP$, and will be obtained by considering that the expression provided in (3.235) plays the role of a "model response."

Applying the definition of the first-order total G-differential to (3.235) yields the following relation:

$$\delta\rho^{(2)}\left[\mathbf{u}^{(3)}(z,\omega);\boldsymbol{\alpha};\delta\mathbf{u}^{(3)};\delta\boldsymbol{\alpha}\right]_{\boldsymbol{\alpha}^0}$$
$$\triangleq-\left\{\frac{d}{d\varepsilon}\left[\int_0^1 d\omega\int_{b_1+\varepsilon\delta b_1}^{b_2+\varepsilon\delta b_2}\left(a_1^{(2)}+\varepsilon\delta a_1^{(2)}\right)(\varphi+\varepsilon\delta\varphi)dz\right]_{\boldsymbol{\alpha}^0}\right\}_{\varepsilon=0}$$
$$-\left\{\frac{d}{d\varepsilon}\left[\int_0^1 d\omega\int_{b_1+\varepsilon\delta b_1}^{b_2+\varepsilon\delta b_2}\left(a_2^{(2)}+\varepsilon\delta a_2^{(2)}\right)(\psi+\varepsilon\delta\psi)dz\right]_{\boldsymbol{\alpha}^0}\right\}_{\varepsilon=0}$$
$$-\left\{\frac{d}{d\varepsilon}\left[\int_0^1 d\omega\int_{b_1+\varepsilon\delta b_1}^{b_2+\varepsilon\delta b_2}\left(a_3^{(2)}+\varepsilon\delta a_3^{(2)}\right)\left(a_1^{(1)}+\varepsilon\delta a_1^{(1)}\right)dz\right]_{\boldsymbol{\alpha}^0}\right\}_{\varepsilon=0}\tag{3.237}$$
$$-\left\{\frac{d}{d\varepsilon}\left[\int_0^1 d\omega\int_{b_1+\varepsilon\delta b_1}^{b_2+\varepsilon\delta b_2}\left(a_4^{(2)}+\varepsilon\delta a_4^{(2)}\right)\left(a_2^{(1)}+\varepsilon\delta a_2^{(1)}\right)dz\right]_{\boldsymbol{\alpha}^0}\right\}_{\varepsilon=0}$$
$$=\left\{\delta\rho^{(2)}\left[\mathbf{u}^{(3)}(z,\omega);\boldsymbol{\alpha};\delta\boldsymbol{\alpha}\right]_{\boldsymbol{\alpha}^0}\right\}_{dir}+\left\{\delta\rho^{(2)}\left[\mathbf{u}^{(3)}(z,\omega);\boldsymbol{\alpha};\delta\mathbf{u}^{(3)}\right]_{\boldsymbol{\alpha}^0}\right\}_{ind},$$

where

$$
\begin{aligned}
&\left\{\delta\rho^{(2)}\left[\mathbf{u}^{(3)}(z,\omega);\boldsymbol{\alpha};\delta\boldsymbol{\alpha}\right]_{\boldsymbol{\alpha}^0}\right\}_{dir} \\
&\triangleq -(\delta b_2)\left\{\int_0^1 d\omega\left[a_1^{(2)}(b_2,\omega)\varphi(b_2,\omega)+a_2^{(2)}(b_2,\omega)\psi(b_2,\omega)\right]\right\}_{\boldsymbol{\alpha}^0} \\
&-(\delta b_2)\left\{\int_0^1 d\omega\left[a_3^{(2)}(b_2,\omega)a_1^{(1)}(b_2,\omega)+a_4^{(2)}(b_2,\omega)a_2^{(1)}(b_2,\omega)\right]\right\}_{\boldsymbol{\alpha}^0} \\
&+(\delta b_1)\left\{\int_0^1 d\omega\left[a_1^{(2)}(b_1,\omega)\varphi(b_1,\omega)+a_2^{(2)}(b_1,\omega)\psi(b_1,\omega)\right]\right\}_{\boldsymbol{\alpha}^0} \\
&+(\delta b_1)\left\{\int_0^1 d\omega\left[a_3^{(2)}(b_1,\omega)a_1^{(1)}(b_1,\omega)+a_4^{(2)}(b_1,\omega)a_2^{(1)}(b_1,\omega)\right]\right\}_{\boldsymbol{\alpha}^0},
\end{aligned}
\tag{3.238}
$$

and

$$
\begin{aligned}
&\left\{\delta\rho^{(2)}\left[\mathbf{u}^{(3)}(z,\omega);\boldsymbol{\alpha};\delta\mathbf{u}^{(3)}\right]_{\boldsymbol{\alpha}^0}\right\}_{ind} \\
&\triangleq -\left\{\int_0^1 d\omega\int_{b_1}^{b_2}\left[a_1^{(2)}\delta\varphi+a_2^{(2)}\delta\psi+a_3^{(2)}\delta a_1^{(1)}+a_4^{(2)}\delta a_2^{(1)}\right]dz\right\}_{\boldsymbol{\alpha}^0} \\
&-\left\{\int_0^1 d\omega\int_{b_1}^{b_2}\left[\varphi\delta a_1^{(2)}+\psi\delta a_2^{(2)}+a_1^{(1)}\delta a_3^{(2)}+a_2^{(1)}\delta a_4^{(2)}\right]dz\right\}_{\boldsymbol{\alpha}^0}.
\end{aligned}
\tag{3.239}
$$

The direct-effect term $\left\{\delta\rho^{(2)}\left[\mathbf{u}^{(3)}(z,\omega);\boldsymbol{\alpha};\delta\boldsymbol{\alpha}\right]_{\boldsymbol{\alpha}^0}\right\}_{dir}$ can be computed immediately, since all of the quantities appearing in its definition on the right side of (3.238) are known. However, the computation of the indirect-effect term $\left\{\delta\rho^{(2)}\left[\mathbf{u}^{(3)}(z,\omega);\boldsymbol{\alpha};\delta\mathbf{u}^{(3)}\right]_{\boldsymbol{\alpha}^0}\right\}_{ind}$ requires the prior determination of the following vector of variations:

$$
\delta\mathbf{u}^{(3)}(z,\omega)\triangleq\begin{pmatrix}\delta\mathbf{u}^{(2)}(z,\omega)\\ \delta\mathbf{a}^{(2)}(z,\omega)\end{pmatrix}\triangleq\left[\delta\varphi,\delta\psi;\delta a_1^{(1)},\delta a_2^{(1)};\delta a_1^{(2)},\delta a_2^{(2)},\delta a_3^{(2)},\delta a_4^{(2)}\right]^{\dagger}.
\tag{3.240}
$$

The vector of variations $\delta\mathbf{u}^{(2)}(z,\omega)$ is the solution of the 2nd-LVSS defined in (3.179) together with the boundary conditions provided in (3.43), (3.136), (3.177), and (3.178). Furthermore, the vector of variations $\delta\mathbf{a}^{(2)}(z,\omega)\triangleq\left[\delta a_1^{(2)},\delta a_2^{(2)},\delta a_3^{(2)},\delta a_4^{(2)}\right]^{\dagger}$ is the solution of the system of equations obtained by G-differentiating the 2nd-LASS, which comprises (3.188)–(3.195).

Taking the G-differentials of (3.188)–(3.195) at the nominal parameter values yields the following system of equations and boundary conditions:

$$\left\{-\omega\frac{d}{dz}\delta a_1^{(2)}(z,\omega)+\mu\delta a_1^{(2)}(z,\omega)-\delta(z-z_0)\delta a_4^{(2)}(z,\omega)+\delta a_1^{(1)}(z,\omega)\right\}_{\boldsymbol{\alpha}^0}$$
$$=\left\{-(\delta\mu)a_1^{(2)}(z,\omega)-(\delta z_0)\delta'(z-z_0)a_4^{(2)}(z,\omega)\right\}_{\boldsymbol{\alpha}^0}, \tag{3.241}$$

$$\left\{\omega\frac{d}{dz}\delta a_2^{(2)}(z,\omega)+\mu\delta a_2^{(2)}(z,\omega)-\delta(z-z_0)\delta a_3^{(2)}(z,\omega)+\delta a_2^{(1)}(z,\omega)\right\}_{\boldsymbol{\alpha}^0} \tag{3.242}$$
$$=\left\{-(\delta\mu)a_2^{(2)}(z,\omega)-(\delta z_0)\delta'(z-z_0)a_3^{(2)}(z,\omega)\right\}_{\boldsymbol{\alpha}^0},$$

$$\left\{\omega\frac{d}{dz}\delta a_3^{(2)}(z,\omega)+\mu\delta a_3^{(2)}(z,\omega)+\delta\varphi(z,\omega)\right\}_{\boldsymbol{\alpha}^0}=-(\delta\mu)\left\{a_3^{(2)}(z,\omega)\right\}_{\boldsymbol{\alpha}^0}, \tag{3.243}$$

$$\left\{-\omega\frac{d}{dz}\delta a_4^{(2)}(z,\omega)+\mu\delta a_4^{(2)}(z,\omega)+\delta\psi(z,\omega)\right\}_{\boldsymbol{\alpha}^0}$$
$$=-(\delta\mu)\left\{a_4^{(2)}(z,\omega)\right\}_{\boldsymbol{\alpha}^0}, \tag{3.244}$$

$$\left\{\delta a_1^{(2)}(b_2,\omega)\right\}_{\boldsymbol{\alpha}^0}=0,\ z=b_2^0;\ 0\le\omega\le 1, \tag{3.245}$$

$$\left\{\delta a_2^{(2)}(b_1,\omega)\right\}_{\boldsymbol{\alpha}^0}=0,\ z=b_1^0;\ 0\le\omega\le 1, \tag{3.246}$$

$$\left\{\delta a_3^{(2)}(b_1,\omega)\right\}_{\boldsymbol{\alpha}^0}=0,\ z=b_1^0;\ 0\le\omega\le 1, \tag{3.247}$$

$$\left\{\delta a_4^{(2)}(b_2,\omega)\right\}_{\boldsymbol{\alpha}^0}=0,\ z=b_2^0;\ 0\le\omega\le 1. \tag{3.248}$$

Concatenating (3.241)–(3.244) together with the 2nd-LVSS yields the following system of equations written in matrix form:

$$\left\{\mathbf{V}^{(3)}\delta\mathbf{u}^{(3)}(z,\omega)\right\}_{\boldsymbol{\alpha}^0}=\left\{\mathbf{q}_V^{(3)}(z,\omega)\right\}_{\boldsymbol{\alpha}^0}, \tag{3.249}$$

where:

$$\mathbf{V}^{(3)}\triangleq\begin{pmatrix}\mathbf{V}^{(2)} & [\mathbf{0}]_{4\times4}\\ \mathbf{V}_{21}^{(3)} & \mathbf{V}_{22}^{(3)}\end{pmatrix};\quad \mathbf{q}_V^{(3)}(z,\omega)\triangleq\begin{pmatrix}\mathbf{q}_V^{(2)}(z,\omega)\\ \mathbf{p}^{(2)}(z,\omega)\end{pmatrix};$$
$$\mathbf{V}_{21}^{(3)}\triangleq\begin{pmatrix}0&0&1&0\\0&0&0&1\\1&0&0&0\\0&1&0&0\end{pmatrix}; \tag{3.250}$$

$$\mathbf{V}_{22}^{(3)} \triangleq \begin{pmatrix} -\omega\frac{d}{dz}+\mu & 0 & 0 & -\delta(z-z_0) \\ 0 & \omega\frac{d}{dz}+\mu & -\delta(z-z_0) & 0 \\ 0 & 0 & \omega\frac{d}{dz}+\mu & 0 \\ 0 & 0 & 0 & -\omega\frac{d}{dz}+\mu \end{pmatrix}; \tag{3.251}$$

$$\mathbf{p}^{(2)}(z,\omega) \triangleq \begin{pmatrix} -(\delta\mu)a_1^{(2)}(z,\omega)-(\delta z_0)\delta'(z-z_0)a_4^{(2)}(z,\omega) \\ -(\delta\mu)a_2^{(2)}(z,\omega)-(\delta z_0)\delta'(z-z_0)a_3^{(2)}(z,\omega) \\ -(\delta\mu)a_3^{(2)}(z,\omega) \\ -(\delta\mu)a_4^{(2)}(z,\omega) \end{pmatrix}. \tag{3.252}$$

The system represented by (3.249) together with the corresponding boundary conditions given in (3.43), (3.136), (3.177), (3.178), and (3.245)–(3.248) is the Third-Level Variational Sensitivity System (3rd-LVSS) and its solution, $\delta\mathbf{u}^{(3)}(z,\omega)$, is the third-level variational function. Evidently, the solution $\delta\mathbf{u}^{(3)}(z,\omega)$ depends on each parameter variation; hence, its computation is prohibitively expensive. By applying the general principles underlying the 3rd-CASAM-L, the need for computing $\delta\mathbf{u}^{(3)}(z,\omega)$ can be circumvented by eliminating the appearance of $\delta\mathbf{u}^{(3)}(z,\omega)$ from the expression of the indirect-effect term $\left\{\delta\rho^{(2)}\left[\mathbf{u}^{(3)}(z,\omega);\boldsymbol{\alpha};\delta\mathbf{u}^{(3)}\right]_{\boldsymbol{\alpha}^0}\right\}_{ind}$ defined in (3.239). The elimination of $\delta\mathbf{u}^{(3)}(z,\omega)$ from the expression of the indirect-effect term is accomplished by recasting the expression of $\left\{\delta\rho^{(2)}\left[\mathbf{u}^{(3)}(z,\omega);\boldsymbol{\alpha};\delta\mathbf{u}^{(3)}\right]_{\boldsymbol{\alpha}^0}\right\}_{ind}$ in terms of the solution of a Third-Level Adjoint Sensitivity System (3rd-LASS) which is constructed by implementing the same sequence of steps as used to construct the 1st-LASS and the 2nd-LASS. The 3rd-LASS is constructed in a Hilbert space, denoted as H_3, which comprises square-integrable vector-valued elements of the form $\boldsymbol{\eta}^{(3)}(z,\omega)\triangleq\left[\eta_1^{(3)}(z,\omega),\ \ldots,\eta_8^{(3)}(z,\omega)\right]^\dagger \in H_3$. The inner product between two elements $\boldsymbol{\eta}^{(3)}(z,\omega)\in H_3$ and $\boldsymbol{\xi}^{(3)}(z,\omega)\in H_3$ will be denoted as $\langle\boldsymbol{\eta}^{(3)}(z,\omega),\boldsymbol{\xi}^{(3)}(z,\omega)\rangle_3$ and is defined as follows:

$$\left\langle\boldsymbol{\eta}^{(3)}(z,\omega),\boldsymbol{\xi}^{(3)}(z,\omega)\right\rangle_3 \triangleq \sum_{k=1}^{8}\left\{\int_0^1 d\omega\int_{b_1}^{b_2}\eta_k^{(3)}(z,\omega)\xi_k^{(3)}(z,\omega)dz\right\}_{\boldsymbol{\alpha}^0}. \tag{3.253}$$

Forming the inner product of (3.249) with a yet-undefined function $\mathbf{a}^{(3)}(z,\omega)\triangleq\left[a_1^{(3)}(z,\omega),\ \ldots,a_8^{(3)}(z,\omega)\right]^\dagger \in H_3$ in the Hilbert H_3 yields the following relation:

$$\left\{\left\langle \mathbf{a}^{(3)}(z,\omega), \mathbf{V}^{(3)}\delta\mathbf{u}^{(3)}(z,\omega)\right\rangle_3\right\}_{\boldsymbol{\alpha}^0} = \left\{\left\langle \mathbf{a}^{(3)}(z,\omega), \mathbf{q}_V^{(3)}(z,\omega)\right\rangle_3\right\}_{\boldsymbol{\alpha}^0}. \tag{3.254}$$

The left side and the right side of (3.254) have the following detailed forms, respectively:

$$\begin{aligned}
&\left\{\left\langle \mathbf{a}^{(3)}(z,\omega), \mathbf{V}^{(3)}\delta\mathbf{u}^{(3)}(z,\omega)\right\rangle_3\right\}_{\boldsymbol{\alpha}^0} \\
&= \left\{\int_0^1 d\omega \int_{b_1}^{b_2} a_1^{(3)}(z,\omega)\left[\omega\frac{d}{dz}\delta\varphi(z,\omega) + \mu\delta\varphi(z,\omega)\right]dz\right\}_{\boldsymbol{\alpha}^0} \\
&+ \left\{\int_0^1 d\omega \int_{b_1}^{b_2} \left[a_2^{(3)}\left(-\omega\frac{d}{dz}\delta\psi + \mu\delta\psi\right)\right]dz\right\}_{\boldsymbol{\alpha}^0} \\
&+ \left\{\int_0^1 d\omega \int_{b_1}^{b_2} \left[a_3^{(3)}\left(-\omega\frac{d}{dz}\delta a_1^{(1)} + \mu\delta a_1^{(1)} - \delta(z-z_0)\delta\psi\right)\right]dz\right\}_{\boldsymbol{\alpha}^0} \\
&+ \left\{\int_0^1 d\omega \int_{b_1}^{b_2} a_4^{(3)} dz\left[\omega\frac{d}{dz}\delta a_2^{(1)} + \mu\delta a_2^{(1)} - \delta(z-z_0)\delta\varphi\right]\right\}_{\boldsymbol{\alpha}^0} \\
&+ \left\{\int_0^1 d\omega \int_{b_1}^{b_2} a_5^{(3)}\left[-\omega\frac{d}{dz}\delta a_1^{(2)} + \mu\delta a_1^{(2)} - \delta(z-z_0)\delta a_4^{(2)} + \delta a_1^{(1)}\right]dz\right\}_{\boldsymbol{\alpha}^0} \\
&+ \left\{\int_0^1 d\omega \int_{b_1}^{b_2} a_6^{(3)}\left[\omega\frac{d}{dz}\delta a_2^{(2)} + \mu\delta a_2^{(2)} - \delta(z-z_0)\delta a_3^{(2)} + \delta a_2^{(1)}\right]dz\right\}_{\boldsymbol{\alpha}^0} \\
&+ \left\{\int_0^1 d\omega \int_{b_1}^{b_2} \left[a_7^{(3)}\left(\omega\frac{d}{dz}\delta a_3^{(2)} + \mu\delta a_3^{(2)} + \delta\varphi\right)\right]dz\right\}_{\boldsymbol{\alpha}^0} \\
&+ \left\{\int_0^1 d\omega \int_{b_1}^{b_2} \left[a_8^{(3)}\left(-\omega\frac{d}{dz}\delta a_4^{(2)} + \mu\delta a_4^{(2)} + \delta\psi\right)\right]dz\right\}_{\boldsymbol{\alpha}^0},
\end{aligned} \tag{3.255}$$

and

$$
\begin{aligned}
&\left\{\left\langle \mathbf{a}^{(3)}(z,\omega),\mathbf{q}_V^{(3)}(z,\omega)\right\rangle_3\right\}_{\boldsymbol{\alpha}^0}\\
&=\left\{\int_0^1 d\omega\int_{b_1}^{b_2}\left[-(\delta\mu)a_1^{(3)}\varphi+(\delta q^*)a_2^{(3)}-(\delta\mu)a_2^{(3)}\psi\right]dz\right\}_{\boldsymbol{\alpha}^0}\\
&-\left\{\int_0^1 d\omega\int_{b_1}^{b_2}a_3^{(3)}\left[(\delta\mu)a_1^{(1)}+(\delta z_0)\psi\delta'(z-z_0)\right]dz\right\}_{\boldsymbol{\alpha}^0}\\
&-\left\{\int_0^1 d\omega\int_{b_1}^{b_2}a_4^{(3)}\left[(\delta\mu)a_2^{(1)}+(\delta z_0)\varphi\delta'(z-z_0)\right]dz\right\}_{\boldsymbol{\alpha}^0}\\
&-\left\{\int_0^1 d\omega\int_{b_1}^{b_2}a_5^{(3)}\left[(\delta\mu)a_1^{(2)}+(\delta z_0)\delta'(z-z_0)a_4^{(2)}\right]dz\right\}_{\boldsymbol{\alpha}^0}\\
&-\left\{\int_0^1 d\omega\int_{b_1}^{b_2}a_6^{(3)}\left[(\delta\mu)a_2^{(2)}+(\delta z_0)\delta'(z-z_0)a_3^{(2)}\right]dz\right\}_{\boldsymbol{\alpha}^0}\\
&-(\delta\mu)\left\{\int_0^1 d\omega\int_{b_1}^{b_2}\left[a_7^{(3)}a_3^{(2)}+a_8^{(3)}a_4^{(2)}\right]dz\right\}_{\boldsymbol{\alpha}^0}.
\end{aligned}
\tag{3.256}
$$

Integrating by parts the terms containing derivatives with respect to z in (3.255) and re-arranging terms yields the following relation:

$$
\begin{aligned}
&\left\{\left\langle \mathbf{a}^{(3)}(z,\omega),\mathbf{V}^{(3)}\delta\mathbf{u}^{(3)}(z,\omega)\right\rangle_3\right\}_{\boldsymbol{\alpha}^0}\\
&=\left\{\left\langle \delta\mathbf{u}^{(3)}(z,\omega),\mathbf{A}^{(3)}\mathbf{a}^{(3)}(z,\omega)\right\rangle_3\right\}_{\boldsymbol{\alpha}^0}+\left\{P^{(3)}\left[\delta\mathbf{u}^{(3)};\mathbf{a}^{(3)};\boldsymbol{\alpha};\delta\boldsymbol{\alpha}\right]\right\}_{\boldsymbol{\alpha}^0},
\end{aligned}
\tag{3.257}
$$

where the operator $\mathbf{A}^{(3)}$ is the formal adjoint of the operator $\mathbf{V}^{(3)}$ and is defined as follows:

$$
\mathbf{A}^{(3)}\triangleq\left[\mathbf{V}^{(3)}\right]^*=\begin{pmatrix}\left[\mathbf{V}^{(2)}\right]^* & \left[\mathbf{V}_{21}^{(3)}\right]^*\\ [\mathbf{0}]_{4\times4} & \left[\mathbf{V}_{22}^{(3)}\right]^*\end{pmatrix},
\tag{3.258}
$$

and where the bilinear concomitant $\left\{P^{(3)}\left[\delta\mathbf{u}^{(3)};\mathbf{a}^{(3)};\boldsymbol{\alpha};\delta\boldsymbol{\alpha}\right]\right\}_{\boldsymbol{\alpha}^0}$ has the following expression:

$$
\begin{aligned}
&\left\{P^{(3)}\left[\delta\mathbf{u}^{(3)};\mathbf{a}^{(3)};\boldsymbol{\alpha};\delta\boldsymbol{\alpha}\right]\right\}_{\boldsymbol{\alpha}^0}\\
&=\int_0^1 \omega d\omega\left[\delta\varphi(b_2,\omega)a_1^{(3)}(b_2,\omega)-\delta\varphi(b_1,\omega)a_1^{(3)}(b_1,\omega)\right]\\
&-\int_0^1 \omega d\omega\left[\delta\psi(b_2,\omega)a_2^{(3)}(b_2,\omega)-\delta\psi(b_1,\omega)a_2^{(3)}(b_1,\omega)\right]\\
&-\int_0^1 \omega d\omega\left[\delta a_1^{(1)}(b_2,\omega)a_3^{(3)}(b_2,\omega)-\delta a_1^{(1)}(b_1,\omega)a_3^{(3)}(b_1,\omega)\right]\\
&+\int_0^1 \omega d\omega\left[\delta a_2^{(1)}(b_2,\omega)a_4^{(3)}(b_2,\omega)-\delta a_2^{(1)}(b_1,\omega)a_4^{(3)}(b_1,\omega)\right]\\
&-\int_0^1 \omega d\omega\left[\delta a_1^{(2)}(b_2,\omega)a_5^{(3)}(b_2,\omega)-\delta a_1^{(2)}(b_1,\omega)a_5^{(3)}(b_1,\omega)\right]\\
&+\int_0^1 \omega d\omega\left[\delta a_2^{(2)}(b_2,\omega)a_6^{(3)}(b_2,\omega)-\delta a_2^{(2)}(b_1,\omega)a_6^{(3)}(b_1,\omega)\right]\\
&+\int_0^1 \omega d\omega\left[\delta a_3^{(2)}(b_2,\omega)a_7^{(3)}(b_2,\omega)-\delta a_3^{(2)}(b_1,\omega)a_7^{(3)}(b_1,\omega)\right]\\
&-\int_0^1 \omega d\omega\left[\delta a_4^{(2)}(b_2,\omega)a_8^{(3)}(b_2,\omega)-\delta a_4^{(2)}(b_1,\omega)a_8^{(3)}(b_1,\omega)\right].
\end{aligned}
\tag{3.259}
$$

The quantity $\left\{\left\langle\delta\mathbf{u}^{(3)}(z,\omega),\mathbf{A}^{(3)}\mathbf{a}^{(3)}(z,\omega)\right\rangle_3\right\}_{\boldsymbol{\alpha}^0}$ on the right side of (3.257) is required to represent the indirect-effect term $\left\{\delta\rho^{(2)}\left[\mathbf{u}^{(3)}(z,\omega);\boldsymbol{\alpha};\delta\mathbf{u}^{(3)}\right]_{\boldsymbol{\alpha}^0}\right\}_{ind}$ defined in (3.239) by imposing the following relation:

$$\left\{\mathbf{A}^{(3)}\mathbf{a}^{(3)}(z,\omega)\right\}_{\boldsymbol{\alpha}^0} = -\left[a_1^{(2)},a_2^{(2)},a_3^{(2)},a_4^{(2)},\varphi,\psi,a_1^{(1)},a_2^{(1)}\right]^\dagger. \qquad (3.260)$$

The unknown quantities $\delta\varphi(b_2,\omega)$, $\delta\psi(b_1,\omega)$, $\delta a_1^{(1)}(b_1,\omega)$, $\delta a_2^{(1)}(b_2,\omega)$, $\delta a_1^{(2)}(b_1,\omega)$, $\delta a_2^{(2)}(b_2,\omega)$, $\delta a_3^{(2)}(b_2,\omega)$, and $\delta a_4^{(2)}(b_1,\omega)$, which appear in the bilinear concomitant expressed by (3.259), are eliminated from further consideration by choosing the following boundary conditions for the third-level adjoint function $\mathbf{a}^{(3)}(z,\omega)\triangleq\left[a_1^{(3)}(z,\omega),\ \ldots,a_8^{(3)}(z,\omega)\right]^\dagger$:

$$\begin{aligned}&\left\{a_1^{(3)}(b_2,\omega)\right\}_{\boldsymbol{\alpha}^0} = \left\{a_2^{(3)}(b_1,\omega)\right\}_{\boldsymbol{\alpha}^0} = \left\{a_3^{(3)}(b_1,\omega)\right\}_{\boldsymbol{\alpha}^0} \\ &= \left\{a_4^{(3)}(b_2,\omega)\right\}_{\boldsymbol{\alpha}^0} = \left\{a_5^{(3)}(b_1,\omega)\right\}_{\boldsymbol{\alpha}^0} = \left\{a_6^{(3)}(b_2,\omega)\right\}_{\boldsymbol{\alpha}^0} \\ &= \left\{a_7^{(3)}(b_2,\omega)\right\}_{\boldsymbol{\alpha}^0} = \left\{a_8^{(3)}(b_1,\omega)\right\}_{\boldsymbol{\alpha}^0} = 0,\ 0\le\omega\le 1.\end{aligned} \qquad (3.261)$$

The system comprising (3.260) and (3.261) constitutes the 3rd-LASS for the third-level adjoint function $\mathbf{a}^{(3)}(z,\omega)\triangleq\left[a_1^{(3)}(z,\omega),\ \ldots,a_8^{(3)}(z,\omega)\right]^\dagger$. Very importantly, the equations underlying the 3rd-LASS need not be solved simultaneously since they can be readily decoupled when written in component form, and can therefore be solved successively, as follows:

1. Solve the following equations to determine the third-level adjoint function $a_5^{(3)}(b_1,\omega)$:

$$\left\{\omega\frac{da_5^{(3)}(z,\omega)}{dz} + \mu a_5^{(3)}(z,\omega)\right\}_{\boldsymbol{\alpha}^0} = -\{\varphi(z,\omega)\}_{\boldsymbol{\alpha}^0}, \qquad (3.262)$$

$$\left\{a_5^{(3)}(b_1,\omega)\right\}_{\boldsymbol{\alpha}^0} = 0,\ \ z=b_1^0;\ \ 0\le\omega\le 1. \qquad (3.263)$$

2. Solve the following equations to determine the third-level adjoint function $a_6^{(3)}(z,\omega)$:

$$\left\{-\omega\frac{da_6^{(3)}(z,\omega)}{dz} + \mu a_6^{(3)}(z,\omega)\right\}_{\boldsymbol{\alpha}^0} = -\{\psi(z,\omega)\}_{\boldsymbol{\alpha}^0}, \qquad (3.264)$$

$$\left\{a_6^{(3)}(b_2,\omega)\right\}_{\boldsymbol{\alpha}^0} = 0,\ \ z=b_2^0;\ \ 0\le\omega\le 1. \qquad (3.265)$$

3. Use the function $a_6^{(3)}(z,\omega)$ and determine the third-level adjoint function $a_7^{(3)}(z,\omega)$ by solving the following equations:

$$\left\{-\omega\frac{da_7^{(3)}(z,\omega)}{dz}+\mu a_7^{(3)}(z,\omega)\right\}_{\boldsymbol{\alpha}^0}=\left\{-a_1^{(1)}(z,\omega)+\delta(z-z_0)a_6^{(3)}(z,\omega)\right\}_{\boldsymbol{\alpha}^0}, \tag{3.266}$$

$$\left\{a_7^{(3)}(b_2,\omega)\right\}_{\boldsymbol{\alpha}^0}=0,\ z=b_2^0;\ 0\le\omega\le 1. \tag{3.267}$$

4. Use the function $a_5^{(3)}(z,\omega)$ and determine the third-level adjoint function $a_8^{(3)}(z,\omega)$ by solving the following equations:

$$\begin{aligned}&\left\{\omega\frac{da_8^{(3)}(z,\omega)}{dz}+\mu a_8^{(3)}(z,\omega)\right\}_{\boldsymbol{\alpha}^0}\\&=\left\{-a_2^{(1)}(z,\omega)+\delta(z-z_0)a_5^{(3)}(z,\omega)\right\}_{\boldsymbol{\alpha}^0},\end{aligned} \tag{3.268}$$

$$\left\{a_8^{(3)}(b_1,\omega)\right\}_{\boldsymbol{\alpha}^0}=0,\ z=b_1^0;\ 0\le\omega\le 1. \tag{3.269}$$

5. Use the function $a_5^{(3)}(z,\omega)$ and determine the third-level adjoint function $a_3^{(3)}(z,\omega)$ by solving the following equations:

$$\left\{\omega\frac{da_3^{(3)}(z,\omega)}{dz}+\mu a_3^{(3)}(z,\omega)\right\}_{\boldsymbol{\alpha}^0}=-\left\{a_3^{(2)}(z,\omega)+a_5^{(3)}(z,\omega)\right\}_{\boldsymbol{\alpha}^0}, \tag{3.270}$$

$$\left\{a_3^{(3)}(b_1,\omega)\right\}_{\boldsymbol{\alpha}^0}=0,\ z=b_1^0;\ 0\le\omega\le 1. \tag{3.271}$$

6. Use the function $a_6^{(3)}(z,\omega)$ and determine the third-level adjoint function $a_4^{(3)}(z,\omega)$ by solving the following equations:

$$\left\{-\omega\frac{da_4^{(3)}(z,\omega)}{dz}+\mu a_4^{(3)}(z,\omega)\right\}_{\boldsymbol{\alpha}^0}=-\left\{a_4^{(2)}(z,\omega)+a_6^{(3)}(z,\omega)\right\}_{\boldsymbol{\alpha}^0}, \tag{3.272}$$

$$\left\{a_4^{(3)}(b_2,\omega)\right\}_{\boldsymbol{\alpha}^0}=0,\ z=b_2^0;\ 0\le\omega\le 1. \tag{3.273}$$

7. Use the functions $a_4^{(3)}(z,\omega)$ and $a_7^{(3)}(z,\omega)$ to determine the third-level adjoint function $a_1^{(3)}(z,\omega)$ by solving the following equations:

$$\left\{-\omega\frac{da_1^{(3)}(z,\omega)}{dz}+\mu a_1^{(3)}(z,\omega)\right\}_{\boldsymbol{\alpha}^0}=\left\{-a_1^{(2)}(z,\omega)+\delta(z-z_0)a_4^{(3)}(z,\omega)-a_7^{(3)}(z,\omega)\right\}_{\boldsymbol{\alpha}^0},\tag{3.274}$$

$$\left\{a_1^{(3)}(b_2,\omega)\right\}_{\boldsymbol{\alpha}^0}=0,\ z=b_2^0;\ 0\le\omega\le 1.\tag{3.275}$$

8. Use the functions $a_3^{(3)}(z,\omega)$ and $a_8^{(3)}(z,\omega)$ to determine the third-level adjoint function $a_2^{(3)}(z,\omega)$ by solving the following equations:

$$\left\{\omega\frac{da_2^{(3)}(z,\omega)}{dz}+\mu a_2^{(3)}(z,\omega)\right\}_{\boldsymbol{\alpha}^0}=\left\{-a_2^{(2)}(z,\omega)+\delta(z-z_0)a_3^{(3)}(z,\omega)-a_8^{(3)}(z,\omega)\right\}_{\boldsymbol{\alpha}^0},\tag{3.276}$$

$$\left\{a_2^{(3)}(b_1,\omega)\right\}_{\boldsymbol{\alpha}^0}=0,\ z=b_1^0;\ 0\le\omega\le 1.\tag{3.277}$$

Inserting the boundary conditions provided in (3.43), (3.136), (3.177), (3.178), (3.245)–(3.248), and (3.261) into (3.259) reduces the bilinear concomitant to the following expression:

$$\begin{aligned}\left\{P^{(3)}\left[\delta\mathbf{u}^{(3)};\mathbf{a}^{(3)};\boldsymbol{\alpha};\delta\boldsymbol{\alpha}\right]\right\}_{\boldsymbol{\alpha}^0}&=-(\delta b_2)\left\{q^*\int_0^1 d\omega\, a_2^{(3)}(b_2,\omega)\right\}_{\boldsymbol{\alpha}^0}\\&-\left\{\int_0^1\omega d\omega\left[\frac{(\delta q)}{\mu_s(\boldsymbol{\alpha})}-\frac{q\delta\mu_s(\boldsymbol{\alpha})}{\mu_s^2(\boldsymbol{\alpha})}+(\delta b_1)\frac{q\mu(\boldsymbol{\alpha})}{\omega\mu_s(\boldsymbol{\alpha})}\right]a_1^{(3)}(b_1,\omega)\right\}_{\boldsymbol{\alpha}^0}.\end{aligned}\tag{3.278}$$

Replacing the results obtained in (3.278), (3.260), and (3.254) into (3.257) yields the following expression for the indirect-effect term defined in (3.239)

$$
\begin{aligned}
&\left\{\delta\rho^{(2)}\left[\mathbf{u}^{(3)}(z,\omega);\boldsymbol{\alpha};\delta\mathbf{u}^{(3)}\right]_{\boldsymbol{\alpha}^0}\right\}_{ind}\\
&=\left\{\int_0^1 \omega d\omega\left[\frac{(\delta q)}{\mu_s(\boldsymbol{\alpha})}-\frac{q\delta\mu_s(\boldsymbol{\alpha})}{\mu_s^2(\boldsymbol{\alpha})}+(\delta b_1)\frac{q\mu(\boldsymbol{\alpha})}{\omega\mu_s(\boldsymbol{\alpha})}\right]a_1^{(3)}(b_1,\omega)\right\}_{\boldsymbol{\alpha}^0}\\
&+(\delta b_2)\left\{q^*\int_0^1 d\omega\, a_2^{(3)}(b_2,\omega)\right\}_{\boldsymbol{\alpha}^0}\\
&+\left\{\int_0^1 d\omega\int_{b_1}^{b_2}\left[-(\delta\mu)a_1^{(3)}\varphi+(\delta q^*)a_2^{(3)}-(\delta\mu)a_2^{(3)}\psi\right]dz\right\}_{\boldsymbol{\alpha}^0}\\
&-\left\{\int_0^1 d\omega\int_{b_1}^{b_2} a_3^{(3)}\left[(\delta\mu)a_1^{(1)}+(\delta z_0)\psi\delta'(z-z_0)\right]dz\right\}_{\boldsymbol{\alpha}^0}\\
&-\left\{\int_0^1 d\omega\int_{b_1}^{b_2} a_4^{(3)}\left[(\delta\mu)a_2^{(1)}+(\delta z_0)\varphi\delta'(z-z_0)\right]dz\right\}_{\boldsymbol{\alpha}^0}\\
&-\left\{\int_0^1 d\omega\int_{b_1}^{b_2} a_5^{(3)}\left[(\delta\mu)a_1^{(2)}+(\delta z_0)\delta'(z-z_0)a_4^{(2)}\right]dz\right\}_{\boldsymbol{\alpha}^0}\\
&-\left\{\int_0^1 d\omega\int_{b_1}^{b_2} a_6^{(3)}\left[(\delta\mu)a_2^{(2)}+(\delta z_0)\delta'(z-z_0)a_3^{(2)}\right]dz\right\}_{\boldsymbol{\alpha}^0}\\
&-(\delta\mu)\left\{\int_0^1 d\omega\int_{b_1}^{b_2}\left[a_7^{(3)}a_3^{(2)}+a_8^{(3)}a_4^{(2)}\right]dz\right\}_{\boldsymbol{\alpha}^0}.
\end{aligned}
\tag{3.279}
$$

Replacing the expressions obtained in (3.279) and (3.238) into (3.237) and subsequently identifying the expressions that multiply the parameter variations yields the following expressions for the third-order sensitivities of the form $\partial^3\rho(\varphi,\psi;\boldsymbol{\alpha})/\partial\alpha_i\partial\mu(\boldsymbol{\alpha})\partial\mu(\boldsymbol{\alpha})$, $i=1,\ldots,TP$:

$$\left\{\frac{\partial^3 \rho(\varphi,\psi;\boldsymbol{\alpha})}{\partial\mu(\boldsymbol{\alpha})\partial\mu(\boldsymbol{\alpha})\partial\mu(\boldsymbol{\alpha})}\right\}_{\boldsymbol{\alpha}^0} = -\left\{\int_0^1 d\omega \int_{b_1}^{b_2}\left[a_1^{(3)}\varphi + a_2^{(3)}\psi + a_3^{(3)}a_1^{(1)} + a_4^{(3)}a_2^{(1)}\right]dz\right\}_{\boldsymbol{\alpha}^0} - \left\{\int_0^1 d\omega \int_{b_1}^{b_2}\left[a_5^{(3)}a_1^{(2)} + a_6^{(3)}a_2^{(2)} + a_7^{(3)}a_3^{(2)} + a_8^{(3)}a_4^{(2)}\right]dz\right\}_{\boldsymbol{\alpha}^0}; \tag{3.280}$$

$$\left\{\frac{\partial^3 \rho(\varphi,\psi;\boldsymbol{\alpha})}{\partial\mu_s(\boldsymbol{\alpha})\partial\mu(\boldsymbol{\alpha})\partial\mu(\boldsymbol{\alpha})}\right\}_{\boldsymbol{\alpha}^0} = -\left\{\frac{q}{\mu_s^2(\boldsymbol{\alpha})}\int_0^1 \omega d\omega\, a_1^{(3)}(b_1,\omega)\right\}_{\boldsymbol{\alpha}^0}; \tag{3.281}$$

$$\left\{\frac{\partial^3 \rho(\varphi,\psi;\boldsymbol{\alpha})}{\partial q\partial\mu(\boldsymbol{\alpha})\partial\mu(\boldsymbol{\alpha})}\right\}_{\boldsymbol{\alpha}^0} = \left\{\frac{1}{\mu_s(\boldsymbol{\alpha})}\int_0^1 \omega d\omega\, a_1^{(3)}(b_1,\omega)\right\}_{\boldsymbol{\alpha}^0}; \tag{3.282}$$

$$\left\{\frac{\partial^3 \rho(\varphi,\psi;\boldsymbol{\alpha})}{\partial q^*\partial\mu(\boldsymbol{\alpha})\partial\mu(\boldsymbol{\alpha})}\right\}_{\boldsymbol{\alpha}^0} = \left\{\int_0^1 d\omega\int_{b_1}^{b_2} a_2^{(3)}(z,\omega)dz\right\}_{\boldsymbol{\alpha}^0}; \tag{3.283}$$

$$\left\{\frac{\partial^3 \rho(\varphi,\psi;\boldsymbol{\alpha})}{\partial b_1\partial\mu(\boldsymbol{\alpha})\partial\mu(\boldsymbol{\alpha})}\right\}_{\boldsymbol{\alpha}^0} = \left\{\frac{q\mu(\boldsymbol{\alpha})}{\mu_s(\boldsymbol{\alpha})}\int_0^1 d\omega a_1^{(3)}(b_1,\omega)\right\}_{\boldsymbol{\alpha}^0} + \left\{\int_0^1 d\omega\left[a_1^{(2)}(b_1,\omega)\varphi(b_1,\omega) + a_2^{(2)}(b_1,\omega)\psi(b_1,\omega)\right]\right\}_{\boldsymbol{\alpha}^0} + \left\{\int_0^1 d\omega\left[a_3^{(2)}(b_1,\omega)a_1^{(1)}(b_1,\omega) + a_4^{(2)}(b_1,\omega)a_2^{(1)}(b_1,\omega)\right]\right\}_{\boldsymbol{\alpha}^0}; \tag{3.284}$$

$$\left\{\frac{\partial^3 \rho(\varphi,\psi;\boldsymbol{\alpha})}{\partial b_2\partial\mu(\boldsymbol{\alpha})\partial\mu(\boldsymbol{\alpha})}\right\}_{\boldsymbol{\alpha}^0} = \left\{q^*\int_0^1 d\omega a_2^{(3)}(b_2,\omega)\right\}_{\boldsymbol{\alpha}^0} - \left\{\int_0^1 d\omega\left[a_1^{(2)}(b_2,\omega)\varphi(b_2,\omega) + a_2^{(2)}(b_2,\omega)\psi(b_2,\omega)\right]\right\}_{\boldsymbol{\alpha}^0} - \left\{\int_0^1 d\omega\left[a_3^{(2)}(b_2,\omega)a_1^{(1)}(b_2,\omega) + a_4^{(2)}(b_2,\omega)a_2^{(1)}(b_2,\omega)\right]\right\}_{\boldsymbol{\alpha}^0}; \tag{3.285}$$

$$\left\{\frac{\partial^3 \rho(\varphi, \psi; \boldsymbol{\alpha})}{\partial z_0 \partial \mu(\boldsymbol{\alpha}) \partial \mu(\boldsymbol{\alpha})}\right\}_{\boldsymbol{\alpha}^0} = \left\{\int_0^1 d\omega \, \frac{d}{dz}\left[a_3^{(3)}(z, \omega)\psi(z, \omega)\right.\right.$$
$$\left.\left.+a_4^{(3)}(z, \omega)\varphi(z, \omega) + a_5^{(3)}(z, \omega)a_4^{(2)}(z, \omega) + a_6^{(3)}(z, \omega)a_3^{(2)}(z, \omega)\right]_{z=z_0}\right\}_{\boldsymbol{\alpha}^0}. \tag{3.286}$$

The computation of the unmixed third-order sensitivity obtained in (3.280) is the most demanding since it involves all 8 components of the third-level adjoint function $\mathbf{a}^{(3)}(z, \omega) \triangleq \left[a_1^{(3)}(z, \omega), \ldots, a_8^{(3)}(z, \omega)\right]^\dagger$. The computations of the mixed third-order sensitivities determined in (3.281)–(3.286) are less demanding.

The various third-order sensitivities with respect to the nuclide number densities and microscopic cross sections, respectively, can be obtained by using in (3.280)–(3.286) the relations provided in (3.155)–(3.158) together with the relations provided in (3.204)–(3.206). The relative third-order sensitivities corresponding to these parameters have equal values.

3.5.2 Determination of the Third-Order Sensitivities of the Form $\partial^3 \rho(\varphi, \psi; \alpha)/\partial \alpha_i \partial b_1 \partial b_2$, $i = 1, \ldots, TP$

This subsection will present the explicit determination of third-order sensitivities which involve both the inner (interface) and the outer boundary parameters, b_1 and b_2, in order to illustrate the general applicability of the 3rd-CASAM-L to determine the third-order sensitivities that involve boundary and interface parameters. These third-order sensitivities are obtained by considering the sensitivity $\partial^2 \rho / \partial b_1 \partial b_2$ as the "model response" and by taking its first G-differential. The expression of $\partial^2 \rho / \partial b_1 \partial b_2$ is given in (3.223). By definition, the G-differential of $\partial^2 \rho / \partial b_1 \partial b_2$ is obtained as follows:

$$\left\{\delta\left[\frac{\partial^2 \rho}{\partial b_1 \partial b_2}\right]\right\}_{\boldsymbol{\alpha}^0} \triangleq \left\{\frac{d}{d\varepsilon}\left[\frac{(q + \varepsilon \delta q)(\mu + \varepsilon \delta \mu)}{\mu_s + \varepsilon \delta \mu_s}\right.\right.$$
$$\left.\left.\times \int_0^1 d\omega \int_{b_1 + \varepsilon \delta b_1}^{b_2 + \varepsilon \delta b_2} \left(c_1^{(2)} + \varepsilon \delta c_1^{(2)}\right) \delta(z - b_1 - \varepsilon \delta b_1) dz\right]\right\}_{\varepsilon = 0} \tag{3.287}$$
$$= \left\{\delta\left(\frac{\partial^2 \rho}{\partial b_1 \partial b_2}\right)_{\boldsymbol{\alpha}^0}\right\}_{dir} + \left\{\delta\left(\frac{\partial^2 \rho}{\partial b_1 \partial b_2}\right)_{\boldsymbol{\alpha}^0}\right\}_{ind},$$

where

$$\left\{ \delta \left(\frac{\partial^2 \rho}{\partial b_1 \partial b_2} \right)_{\boldsymbol{\alpha}^0} \right\}_{dir} \triangleq -(\delta \mu_s) \left\{ \frac{q\mu}{\mu_s^2(\boldsymbol{\alpha})} \int_0^1 c_1^{(2)}(b_1, \omega) d\omega \right\}_{\boldsymbol{\alpha}^0}$$
$$+ (\delta q) \left\{ \frac{\mu(\boldsymbol{\alpha})}{\mu_s(\boldsymbol{\alpha})} \int_0^1 c_1^{(2)}(b_1, \omega) d\omega \right\}_{\boldsymbol{\alpha}^0} + (\delta \mu) \left\{ \frac{q}{\mu_s(\boldsymbol{\alpha})} \int_0^1 c_1^{(2)}(b_1, \omega) d\omega \right\}_{\boldsymbol{\alpha}^0}$$
$$+ (\delta b_1) \left\{ \frac{q\mu(\boldsymbol{\alpha})}{\mu_s(\boldsymbol{\alpha})} \int_0^1 \left[\frac{dc_1^{(2)}(z, \omega)}{dz} \right]_{z=b_1} d\omega \right\}, \tag{3.288}$$

and

$$\left\{ \delta \left(\frac{\partial^2 \rho}{\partial b_1 \partial b_2} \right)_{\boldsymbol{\alpha}^0} \right\}_{ind} \triangleq \left\{ \frac{q\mu(\boldsymbol{\alpha})}{\mu_s(\boldsymbol{\alpha})} \int_0^1 d\omega \int_{b_1}^{b_2} \delta c_1^{(2)}(z, \omega) \delta(z - b_1) dz \right\}_{\boldsymbol{\alpha}^0}. \tag{3.289}$$

The direct-effect term defined in (3.288) can be computed immediately, but the computation of the indirect-effect term defined in (3.289) requires the prior determination of the function $\delta c_1^{(2)}(z, \omega)$. The function $\delta c_1^{(2)}(z, \omega)$ is the solution of the system obtained by taking the first-order G-differential of (3.214)–(3.217). Applying the definition of the G-differential to (3.214)–(3.217) yields the following system of equations:

$$\left\{ -\omega \frac{d}{dz} \delta c_1^{(2)}(z, \omega) + \mu \delta c_1^{(2)}(z, \omega) - \delta(z - z_0) \delta c_2^{(2)}(z, \omega) \right\}_{\boldsymbol{\alpha}^0}$$
$$= -(\delta \mu) \left\{ c_1^{(2)}(z, \omega) \right\}_{\boldsymbol{\alpha}^0} - (\delta z_0) \left\{ \delta'(z - z_0) c_2^{(2)}(z, \omega) \right\}_{\boldsymbol{\alpha}^0}, \tag{3.290}$$

$$\left\{ -\omega \frac{d}{dz} \delta c_2^{(2)}(z, \omega) + \mu \delta c_2^{(2)}(z, \omega) \right\}_{\boldsymbol{\alpha}^0} = -(\delta \mu) \left\{ c_2^{(2)}(z, \omega) \right\}_{\boldsymbol{\alpha}^0}$$
$$+ (\delta q^*) \{ \delta(z - b_2) \}_{\boldsymbol{\alpha}^0} - (\delta b_2) \{ q^* \delta'(z - b_2) \}_{\boldsymbol{\alpha}^0}, \tag{3.291}$$

$$\left\{ \delta c_1^{(2)}(b_2, \omega) \right\}_{\boldsymbol{\alpha}^0} = 0, \ z = b_2^0; \ 0 \le \omega \le 1, \tag{3.292}$$

$$\left\{ \delta c_2^{(2)}(b_2, \omega) \right\}_{\boldsymbol{\alpha}^0} = 0, \ z = b_2^0; \ 0 \le \omega \le 1. \tag{3.293}$$

The 3rd-LVSS represented by (3.290)–(3.293) depends on the parameter variations and would need to be solved as many times as there are parameter variations. The need for solving this 3rd-LVSS can be circumvented by applying the principles of the 3rd-CASAM-L to construct a corresponding 3rd-LASS. In view of the structure of (3.290)–(3.293), the Hilbert space appropriate for constructing the corresponding adjoint system is the Hilbert space H_1, which comprises two-component vector-valued elements and which is endowed with the inner product defined in (3.140). The 3rd-LASS corresponding to (3.290)–(3.293) is constructed as follows:

1. Form the inner product of a 2-component vector function $\mathbf{c}^{(3)}(z,\omega) \triangleq \left[c_1^{(3)}(z,\omega), c_2^{(3)}(z,\omega)\right]^\dagger$ with (3.290) and (3.291) to obtain the following relation:

$$
\begin{aligned}
&\left\{\int_0^1 d\omega \int_{b_1}^{b_2} c_1^{(3)}(z,\omega)\left[-\omega\frac{d}{dz}\delta c_1^{(2)}(z,\omega)\right.\right.\\
&\left.\left.+\mu\delta c_1^{(2)}(z,\omega) - \delta(z-z_0)\delta c_2^{(2)}(z,\omega)\right]dz\right\}_{\boldsymbol{\alpha}^0}\\
&+\left\{\int_0^1 d\omega \int_{b_1}^{b_2} c_2^{(3)}(z,\omega)\left[-\omega\frac{d}{dz}\delta c_2^{(2)}(z,\omega) + \mu\delta c_2^{(2)}(z,\omega)\right]dz\right\}_{\boldsymbol{\alpha}^0}\\
&= -(\delta\mu)\left\{\int_0^1 d\omega \int_{b_1}^{b_2} c_1^{(3)}(z,\omega)\, c_1^{(2)}(z,\omega)dz\right\}_{\boldsymbol{\alpha}^0}\\
&-(\delta z_0)\left\{\int_0^1 d\omega \int_{b_1}^{b_2} c_1^{(3)}(z,\omega)\, \delta'(z-z_0)c_2^{(2)}(z,\omega)dz\right\}_{\boldsymbol{\alpha}^0}\\
&-(\delta\mu)\left\{\int_0^1 d\omega \int_{b_1}^{b_2} c_2^{(3)}(z,\omega)c_2^{(2)}(z,\omega)dz\right\}_{\boldsymbol{\alpha}^0}\\
&+(\delta q^*)\left\{\int_0^1 d\omega \int_{b_1}^{b_2} c_2^{(3)}(z,\omega)\delta(z-b_2)dz\right\}_{\boldsymbol{\alpha}^0}\\
&-(\delta b_2)\left\{q^*\int_0^1 d\omega \int_{b_1}^{b_2} c_2^{(3)}(z,\omega)\delta'(z-b_2)dz\right\}_{\boldsymbol{\alpha}^0}.
\end{aligned}
\tag{3.294}
$$

2. Integrate by parts the left side of (3.294) to obtain the following relation:

$$
\begin{aligned}
&\left\{\int_0^1 d\omega \int_{b_1}^{b_2} c_1^{(3)}(z,\omega)\left[-\omega\frac{d}{dz}\delta c_1^{(2)}(z,\omega)\right.\right.\\
&\left.\left.+\mu\delta c_1^{(2)}(z,\omega)-\delta(z-z_0)\delta c_2^{(2)}(z,\omega)\right]dz\right\}_{\boldsymbol{\alpha}^0}\\
&+\left\{\int_0^1 d\omega \int_{b_1}^{b_2} c_2^{(3)}(z,\omega)\left[-\omega\frac{d}{dz}\delta c_2^{(2)}(z,\omega)+\mu\delta c_2^{(2)}(z,\omega)\right]dz\right\}_{\boldsymbol{\alpha}^0}\\
&=\left\{\int_0^1 d\omega \int_{b_1}^{b_2} \delta c_1^{(2)}(z,\omega)\left[\omega\frac{dc_1^{(3)}(z,\omega)}{dz}+\mu c_1^{(3)}(z,\omega)\right]dz\right\}_{\boldsymbol{\alpha}^0}\\
&+\left\{\int_0^1 d\omega \int_{b_1}^{b_2} \delta c_2^{(2)}(z,\omega)\left[\omega\frac{dc_2^{(3)}(z,\omega)}{dz}\right.\right.\\
&\left.\left.+\mu c_2^{(3)}(z,\omega)-\delta(z-z_0)c_1^{(3)}(z,\omega)\right]dz\right\}_{\boldsymbol{\alpha}^0}\\
&-\left\{\int_0^1 \omega d\omega\left[\delta c_1^{(2)}(b_2,\omega)c_1^{(3)}(b_2,\omega)-\delta c_1^{(2)}(b_1,\omega)c_1^{(3)}(b_1,\omega)\right]\right\}_{\boldsymbol{\alpha}^0}\\
&-\left\{\int_0^1 \omega d\omega\left[\delta c_2^{(2)}(b_2,\omega)c_2^{(3)}(b_2,\omega)-\delta c_2^{(2)}(b_1,\omega)c_2^{(3)}(b_1,\omega)\right]\right\}_{\boldsymbol{\alpha}^0}.
\end{aligned}
\tag{3.295}
$$

3. Require the first two terms on the right side of (3.295) to represent the indirect-effect term defined in (3.289) which is accomplished by imposing the following relations:

$$
\left\{\omega\frac{dc_1^{(3)}(z,\omega)}{dz}+\mu c_1^{(3)}(z,\omega)\right\}_{\boldsymbol{\alpha}^0}=\left\{\frac{q\mu(\boldsymbol{\alpha})}{\mu_s(\boldsymbol{\alpha})}\delta(z-b_1)\right\}_{\boldsymbol{\alpha}^0}, \tag{3.296}
$$

$$
\left\{\omega\frac{dc_2^{(3)}(z,\omega)}{dz}+\mu c_2^{(3)}(z,\omega)\right\}_{\boldsymbol{\alpha}^0}=\left\{\delta(z-z_0)c_1^{(3)}(z,\omega)\right\}_{\boldsymbol{\alpha}^0}. \tag{3.297}
$$

4. Determine the boundary conditions for the functions $c_1^{(3)}(z,\omega)$ and $c_2^{(3)}(z,\omega)$ by eliminating the unknown quantities $\delta c_1^{(2)}(b_1,\omega)$ and $\delta c_2^{(2)}(b_1,\omega)$ from (3.295). This is accomplished by imposing the following boundary conditions:

$$
\left\{c_1^{(3)}(b_1,\omega)\right\}_{\boldsymbol{\alpha}^0}=0,\ z=b_1^0;\ 0\le\omega\le 1, \tag{3.298}
$$

$$\left\{c_2^{(3)}(b_1,\omega)\right\}_{\boldsymbol{\alpha}^0}=0,\ z=b_1^0;\ 0\le\omega\le 1. \tag{3.299}$$

The relations provided in (3.296)–(3.299) constitute the 3rd-LASS for the third-level adjoint function $\mathbf{c}^{(3)}(z,\omega)\triangleq\left[c_1^{(3)}(z,\omega),c_2^{(3)}(z,\omega)\right]^{\dagger}$. Notably, both components $c_1^{(3)}(z,\omega)$ and $c_2^{(3)}(z,\omega)$ satisfy forward-like equations.

5. Use in (3.295) the relation provided in (3.294) together with the equations underlying the 3rd-LASS defined by (3.296)–(3.299) and the equations underlying the 3rd-LVSS provided in (3.290)-(3.293) to obtain the following expression for the indirect-effect term defined by (3.289):

$$\begin{aligned}
\left\{\delta\left(\frac{\partial^2\rho}{\partial b_1\partial b_2}\right)_{\boldsymbol{\alpha}^0}\right\}_{ind}&=-(\delta b_2)\left\{q^*\int_0^1 d\omega\int_{b_1}^{b_2}c_2^{(3)}(z,\omega)\delta'(z-b_2)dz\right\}_{\boldsymbol{\alpha}^0}\\
&-(\delta\mu)\left\{\int_0^1 d\omega\int_{b_1}^{b_2}c_1^{(3)}(z,\omega)\,c_1^{(2)}(z,\omega)dz\right\}_{\boldsymbol{\alpha}^0}\\
&-(\delta z_0)\left\{\int_0^1 d\omega\int_{b_1}^{b_2}c_1^{(3)}(z,\omega)\,\delta'(z-z_0)c_2^{(2)}(z,\omega)dz\right\}_{\boldsymbol{\alpha}^0}\\
&-(\delta\mu)\left\{\int_0^1 d\omega\int_{b_1}^{b_2}c_2^{(3)}(z,\omega)c_2^{(2)}(z,\omega)dz\right\}_{\boldsymbol{\alpha}^0}\\
&+(\delta q^*)\left\{\int_0^1 d\omega\int_{b_1}^{b_2}c_2^{(3)}(z,\omega)\delta(z-b_2)dz\right\}_{\boldsymbol{\alpha}^0}.
\end{aligned} \tag{3.300}$$

Inserting the expression obtained in (3.300) together with the expression of the direct-effect term obtained in (3.288) into (3.287) and identifying in the resulting expression the quantities that multiply the various parameter variations yields the following expressions for the corresponding third-order sensitivities:

$$\begin{aligned}
\left(\frac{\partial^3\rho}{\partial\mu\partial b_1\partial b_2}\right)_{\boldsymbol{\alpha}^0}&=\left\{\frac{q}{\mu_s(\boldsymbol{\alpha})}\int_0^1 c_1^{(2)}(b_1,\omega)d\omega\right\}_{\boldsymbol{\alpha}^0}\\
&-\left\{\int_0^1 d\omega\int_{b_1}^{b_2}\left[c_1^{(3)}(z,\omega)c_1^{(2)}(z,\omega)+c_2^{(3)}(z,\omega)c_2^{(2)}(z,\omega)\right]dz\right\}_{\boldsymbol{\alpha}^0},
\end{aligned} \tag{3.301}$$

$$\left(\frac{\partial^3 \rho}{\partial \mu_s \partial b_1 \partial b_2}\right)_{\boldsymbol{\alpha}^0} = -\left\{\frac{q\mu}{\mu_s^2(\boldsymbol{\alpha})}\int_0^1 c_1^{(2)}(b_1,\omega)d\omega\right\}_{\boldsymbol{\alpha}^0}, \tag{3.302}$$

$$\left(\frac{\partial^3 \rho}{\partial q \partial b_1 \partial b_2}\right)_{\boldsymbol{\alpha}^0} = \left\{\frac{\mu(\boldsymbol{\alpha})}{\mu_s(\boldsymbol{\alpha})}\int_0^1 c_1^{(2)}(b_1,\omega)d\omega\right\}_{\boldsymbol{\alpha}^0}, \tag{3.303}$$

$$\left(\frac{\partial^3 \rho}{\partial q^* \partial b_2 \partial b_1}\right)_{\boldsymbol{\alpha}^0} = \left\{\int_0^1 d\omega \int_{b_1}^{b_2} c_2^{(3)}(z,\omega)\delta(z-b_2)dz\right\}_{\boldsymbol{\alpha}^0}, \tag{3.304}$$

$$\left(\frac{\partial^3 \rho}{\partial b_1 \partial b_1 \partial b_2}\right)_{\boldsymbol{\alpha}^0} = \left\{\frac{q\mu(\boldsymbol{\alpha})}{\mu_s(\boldsymbol{\alpha})}\int_0^1 \left[\frac{dc_1^{(2)}(z,\omega)}{dz}\right]_{z=b_1} d\omega\right\}, \tag{3.305}$$

$$\left(\frac{\partial^3 \rho}{\partial b_2 \partial b_1 \partial b_2}\right)_{\boldsymbol{\alpha}^0} = -\left\{q^* \int_0^1 d\omega \int_{b_1}^{b_2} c_2^{(3)}(z,\omega)\delta'(z-b_2)dz\right\}_{\boldsymbol{\alpha}^0}, \tag{3.306}$$

$$\left(\frac{\partial^3 \rho}{\partial z_0 \partial b_1 \partial b_2}\right)_{\boldsymbol{\alpha}^0} = -\left\{\int_0^1 d\omega \int_{b_1}^{b_2} c_1^{(3)}(z,\omega)\, c_2^{(2)}(z,\omega)\, \delta'(z-z_0)dz\right\}_{\boldsymbol{\alpha}^0}. \tag{3.307}$$

The explicit expressions of the third-order sensitivities provided in (3.301)–(3.307) are obtained after solving the 3rd-LASS to obtain the expressions of the components of the third-level adjoint function $\mathbf{c}^{(3)}(z,\omega) \triangleq \left[c_1^{(3)}(z,\omega), c_2^{(3)}(z,\omega)\right]^\dagger$. Thus, solving (3.296)–(3.299) yields the following expressions:

$$\left\{c_1^{(3)}(z,\omega)\right\}_{\boldsymbol{\alpha}^0} = \left\{\frac{q\mu(\boldsymbol{\alpha})}{\mu_s(\boldsymbol{\alpha})}\frac{1}{\omega}H(z-b_1)\exp\left[-\frac{\mu(z-b_1)}{\omega}\right]\right\}_{\boldsymbol{\alpha}^0}, \tag{3.308}$$

$$\left\{c_2^{(3)}(z,\omega)\right\}_{\boldsymbol{\alpha}^0} = \left\{\frac{q\mu(\boldsymbol{\alpha})}{\mu_s(\boldsymbol{\alpha})}\frac{1}{\omega^2}H(z-z_0)\exp\left[-\frac{\mu(z-b_1)}{\omega}\right]\right\}_{\boldsymbol{\alpha}^0}. \tag{3.309}$$

Inserting the expressions obtained in (3.308) and (3.309), along with the expressions of the other needed functions, into the right sides of (3.301)–(3.307) yields the following closed-form expressions:

$$\left(\frac{\partial^3 \rho}{\partial \mu \partial b_1 \partial b_2}\right)_{\boldsymbol{\alpha}^0} = -\left\{\frac{q^* q}{\mu_s(\boldsymbol{\alpha})}\exp\left[(b_1-b_2)\mu(\boldsymbol{\alpha})\right]\right\}_{\boldsymbol{\alpha}^0}, \tag{3.310}$$

$$\left(\frac{\partial^3 \rho}{\partial \mu_s \partial b_1 \partial b_2}\right)_{\boldsymbol{\alpha}^0} = -\left\{\frac{q^* q\mu(\boldsymbol{\alpha})}{\mu_s^2(\boldsymbol{\alpha})} E_0[(b_2 - b_1)\mu(\boldsymbol{\alpha})]\right\}_{\boldsymbol{\alpha}^0}, \tag{3.311}$$

$$\left(\frac{\partial^3 \rho}{\partial q \partial b_1 \partial b_2}\right)_{\boldsymbol{\alpha}^0} = \left\{\frac{q^* \mu(\boldsymbol{\alpha})}{\mu_s(\boldsymbol{\alpha})} E_0[(b_2 - b_1)\mu(\boldsymbol{\alpha})]\right\}_{\boldsymbol{\alpha}^0}, \tag{3.312}$$

$$\left(\frac{\partial^3 \rho}{\partial q^* \partial b_1 \partial b_2}\right)_{\boldsymbol{\alpha}^0} = \left\{\frac{q\mu(\boldsymbol{\alpha})}{\mu_s(\boldsymbol{\alpha})} E_0[\mu(b_2 - b_1)]\right\}_{\boldsymbol{\alpha}^0}, \tag{3.313}$$

$$\left(\frac{\partial^3 \rho}{\partial b_1 \partial b_1 \partial b_2}\right)_{\boldsymbol{\alpha}^0} = \left\{\frac{qq^*}{\mu_s} \frac{e^{(b_1 - b_2)\mu(\boldsymbol{\alpha})}}{b_2 - b_1}\left[\mu(\boldsymbol{\alpha}) + \frac{1}{b_2 - b_1}\right]\right\}_{\boldsymbol{\alpha}^0}, \tag{3.314}$$

$$\left(\frac{\partial^3 \rho}{\partial b_2 \partial b_1 \partial b_2}\right)_{\boldsymbol{\alpha}^0} = -\left\{\frac{qq^*}{\mu_s} \frac{e^{(b_1 - b_2)\mu(\boldsymbol{\alpha})}}{b_2 - b_1}\left[\mu(\boldsymbol{\alpha}) + \frac{1}{b_2 - b_1}\right]\right\}_{\boldsymbol{\alpha}^0}, \tag{3.315}$$

$$\left(\frac{\partial^3 \rho}{\partial z_0 \partial b_1 \partial b_2}\right)_{\boldsymbol{\alpha}^0} = 0. \tag{3.316}$$

The closed-form expressions obtained in (3.310)–(3.316) can be verified by appropriate differentiations of the second-order sensitivities obtained in (3.228)–(3.234).

3.6 Illustrative Application of the 4th-CASAM-L to a Paradigm Time-Evolution Model

The paradigm model considered in this section simulates the typical evolution in time of a quantity that will be denoted as $\rho(t)$, which evolves according to the following linear conservation equation:

$$\frac{d\rho(t)}{dt} + \rho(t)\sum_{i=1}^{N} n_i\sigma_i = 0, \quad 0 \le \beta_\ell \le t \le \beta_u < \infty, \tag{3.317}$$

$$\rho(\beta_\ell) = \rho_{in}, \quad \text{at } t = \beta_\ell. \tag{3.318}$$

The simple evolution system represented by (3.317) and (3.318) occurs in the mathematical modeling of many physical systems. For example, the dependent variable $\rho(t)$ could represent (Cacuci et al. 2016) the schematic evolution of the concentration of a substance in a homogeneous mixture of N materials, from an imprecisely known initial quantity, denoted as ρ_{in}, measured at an initial-time value $t = \beta_\ell$ toward an imprecisely known final-time value $t = \beta_u$. The quantities n_i and σ_i would represent various imprecisely known material (e.g., chemical) properties of the ith-material ($i = 1, \ldots, N$) in the homogeneous mixture. This illustrative problem admits readily obtainable closed-form analytic solutions for convenient verification of all results for the various adjoint functions and response sensitivities. In order to keep the various analytical expressions as simple as possible while illustrating all of the steps needed when applying the 2nd-CASAM-L, no sources are considered on the right side of (3.317) since their inclusion would just complicate the mathematical derivations without bringing any new mathematical or physical insights.

A typical response of interest for the physical problem modeled by (3.317) and (3.318) would be a measurement, denoted as $\rho(t_d)$, of $\rho(t)$ at some time instance $t = t_d$, $0 \le \beta_\ell \le t_d \le \beta_u < \infty$. The following functional, denoted as $R_1(\rho; \boldsymbol{\alpha})$, can represent mathematically such a measurement:

$$R_1(\rho; \boldsymbol{\alpha}) \triangleq \int_{\beta_\ell}^{\beta_u} \rho(t)\, \delta(t - t_d) dt, \tag{3.319}$$

where $\delta(t - t_d)$ denotes the well-known Dirac-delta (impulse) functional. In (3.319), the vector $\boldsymbol{\alpha}$ denotes the "vector of model parameters," which for this paradigm model is defined as follows:

$$\boldsymbol{\alpha} \triangleq (\alpha_1, \ldots, \alpha_{TP})^\dagger \triangleq (n_1, \ldots, n_N, \sigma_1, \ldots, \sigma_N, \rho_{in}, t_d, \beta_\ell, \beta_u)^\dagger. \tag{3.320}$$

As has already been mentioned, the symbol "$\triangleq$" denotes "is defined as" or "is by definition equal to" while the "dagger" (†) superscript is used to denote "transposition."

Although the model parameters ρ_{in}, n_i, σ_i, t_d, β_ℓ, and β_u are considered to be imperfectly known and subject to uncertainties, the actual probability distributions of these parameters are not known in practice. Usually, only the "nominal" (or "mean") values and the respective variations from the nominal values (e.g., standard deviations) of the respective components are known. The nominal values will be denoted using the superscript "zero" so that the vector comprising the nominal values of the model parameters, denoted as $\boldsymbol{\alpha}^0$, will be defined for the system under consideration as follows:

$$\boldsymbol{\alpha}^0 \triangleq \left(\alpha_1^0, \ldots, \alpha_{TP}^0\right)^\dagger \triangleq \left(n_1^0, \ldots, n_N^0, \sigma_1^0, \ldots, \sigma_N^0, \rho_{in}^0, t_d^0, \beta_\ell^0, \beta_u^0\right)^\dagger. \tag{3.321}$$

Altogether, the physical system modeled by (3.317)–(3.319) comprises 2 boundary parameters and $2N$+2 model parameters, for a total number (denoted as "TP") of $TP = 2N + 4$ uncertain parameters, which can be a large number for realistic problems. For example, the total number of uncertain parameters involved in the spent fuel dissolver model analyzed by Cacuci et al. (2016), which involves equations similar to (3.317)–(3.319), is $TP = 1292$ parameters.

For subsequent verification of the expressions that will be obtained for various response sensitivities, the closed-form solution of (3.317) and (3.318) is provided below, in (3.322):

$$\rho(t) = \rho_{in} \exp\left[(\beta_\ell - t) \sum_{i=1}^{N} n_i \sigma_i \right]. \quad (3.322)$$

In practice, the nominal solution, denoted as $\rho^0(t)$, is computed by solving numerically (3.317) and (3.318) using the nominal values for the model parameters. For this illustrative example, the nominal solution of (3.317) and (3.318) has the following expression:

$$\rho^0(t) = \rho_{in}^0 \exp\left[(\beta_\ell^0 - t) \sum_{i=1}^{N} n_i^0 \sigma_i^0 \right] \triangleq \left\{ \rho_{in} \exp\left[(\beta_\ell - t) \sum_{i=1}^{N} n_i \sigma_i \right] \right\}_{\boldsymbol{\alpha}^0}. \quad (3.323)$$

As before, the notation $\{\}_{\boldsymbol{\alpha}^0}$ indicates that the quantity in brackets is to be evaluated at the nominal values of the parameters and state functions.

Using (3.323) in (3.319) yields the following expression for the response $R_1(\rho; \boldsymbol{\alpha})$, which is to be evaluated at the nominal values of the parameters and state function:

$$\{R_1(\rho; \boldsymbol{\alpha})\}_{\boldsymbol{\alpha}^0} \triangleq R_1(\rho^0; \boldsymbol{\alpha}^0) = \left\{ \rho_{in} \exp\left[(\beta_\ell - t_d) \sum_{i=1}^{N} n_i \sigma_i \right] \right\}_{\boldsymbol{\alpha}^0}. \quad (3.324)$$

Of course, the closed-form analytical expression for the problem's dependent variable(s), as provided in (3.322), and the closed-form expression for the response $R_1(\rho; \boldsymbol{\alpha})$, as given in (3.324), will not be available for the large-scale systems encountered in practice.

3.6.1 Applying the 1st-CASAM-L to Compute the first-Order Sensitivities to Model Parameters, Including Imprecisely Known Initial- and Final-Time Conditions

The variations between the true and the nominal values of the model and boundary parameters will be considered to constitute the components of the vectors $\delta\boldsymbol{\alpha}$, respectively, defined as follows:

$$\delta\boldsymbol{\alpha} \triangleq (\delta\alpha_1, \ldots, \delta\alpha_{TP}), \delta\alpha_i \triangleq \alpha_i - \alpha_i^0. \quad (3.325)$$

Since the state function is related to the model and boundary parameters $\boldsymbol{\alpha}$ through (3.317) and (3.318), it follows that the variations $\delta\boldsymbol{\alpha}$ in the model parameters will cause a corresponding variation in the state function $\rho(t)$ around the nominal solution $\rho^0(t)$. In turn, these variations will cause variations $\delta R_1(\rho^0; \boldsymbol{\alpha}^0; \delta\rho; \delta\boldsymbol{\alpha})$ in the responses $R_1(\rho; \boldsymbol{\alpha})$ around the response's nominal value.

The total first-order sensitivity of the response $R_1(\rho;\boldsymbol{\alpha})$ defined in (3.319) is provided by the first-order total G-differential $\delta R_1(\rho^0;\boldsymbol{\alpha}^0;\delta\rho;\delta\boldsymbol{\alpha})$ evaluated at the nominal values of the parameters and state function, which is computed by applying the definition of the first-order G-differential to (3.319), to obtain the following expression:

$$\delta R_1\left(\rho^0;\boldsymbol{\alpha}^0;\delta\rho;\delta\boldsymbol{\alpha}\right)\triangleq\{\delta R_1(\rho;\boldsymbol{\alpha};\delta\rho;\delta\boldsymbol{\alpha})\}_{\boldsymbol{\alpha}^0}=(\delta R_1)_{ind}+(\delta R_1)_{dir}$$
$$\triangleq\frac{d}{d\varepsilon}\left\{\int_{\beta_\ell^0+\varepsilon\delta\beta_\ell}^{\beta_u^0+\varepsilon\delta\beta_u}\left[\rho^0(t)+\varepsilon\delta\rho(t)\right]\delta\left(t-t_d^0-\varepsilon\delta t_d\right)dt\right\}_{\boldsymbol{\alpha}^0}, \tag{3.326}$$

where the indirect-effect term $(\delta R_1)_{ind}$ and, respectively, the direct-effect term $(\delta R_1)_{dir}$ are defined as

$$(\delta R_1)_{ind}\triangleq\int_{\beta_\ell^0}^{\beta_u^0}\delta\rho(t)\,\delta\left(t-t_d^0\right)dt, \tag{3.327}$$

$$(\delta R_1)_{dir}\triangleq-(\delta t_d)\int_{\beta_\ell^0}^{\beta_u^0}\rho^0(t)\,\delta'\left(t-t_d^0\right)dt=(\delta t_d)\left\{\frac{\partial\rho(t)}{\partial t}\right\}_{t=t_d^0}. \tag{3.328}$$

The variation $\delta\rho(t)$, of the state function $\rho(t)$, which appears in (3.327) is the solution of the following First-Level Variational Sensitivity System (1st-LVSS) obtained by G-differentiating (3.317) and (3.318) around the nominal parameter values:

$$\frac{d[\delta\rho(t)]}{dt}+\delta\rho(t)\sum_{i=1}^{N}n_i^0\sigma_i^0,$$
$$=-\rho^0(t)\sum_{i=1}^{N}\left(n_i^0\delta\sigma_i+\sigma_i^0\delta n_i\right),\ 0\le\beta_\ell^0\le t\le\beta_u^0<\infty, \tag{3.329}$$

$$\delta\rho\left(\beta_\ell^0\right)+\delta\beta_\ell\left\{\frac{d\rho(t)}{dt}\right\}_{t=\beta_\ell^0}=\delta\rho_{in},\ \ at\ t=\beta_\ell^0. \tag{3.330}$$

Since the closed-form solution represented by (3.323) is not available in practice, the direct-effect term, $(\delta R_1)_{dir}$, defined by (3.328) can be computed by differentiating (numerically, in practice) the solution of (3.317) and (3.318). Also, in practice, the sensitivities included in the indirect effect, $(\delta R_1)_{ind}$, defined by (3.327) could be computed only by successively setting all but one of the parameter variations to zero in the 1st-LVSS and solving numerically the corresponding forms of the resulting

1st-LVSS. Thus, using the 1st-LVSS to compute the sensitivities of the response $R_1(\rho; \boldsymbol{\alpha})$ would require $2N + 4$ large-scale computations.

The need for performing these $2N + 4$ large-scale computations can be avoided by applying the 1st-CASAM-L presented in Chap. 2. In order to apply the 1st-CASAM-L, the function $\delta\rho(t)$ is considered to be an element of a Hilbert space $\mathrm{H}_1(\Omega_t)$, $\Omega_t \triangleq (\beta_\ell^0, \beta_u^0)$, endowed with the following inner product, denoted as $\langle \rho_1(t), \rho_2(t) \rangle$, between two (square-integrable) functions $\rho_1(t) \in \mathrm{H}_1(\Omega_t)$ and $\rho_2(t) \in \mathrm{H}_1(\Omega_t)$:

$$\langle \rho_1(t), \rho_2(t) \rangle \triangleq \int_{\beta_\ell^0}^{\beta_u^0} \rho_1(t)\rho_2(t)\, dt. \tag{3.331}$$

The construction of the requisite First-Level Adjoint Sensitivity System (1st-LASS) commences by multiplying (3.329) by a square-integrable function $\psi^{(1)}(t) \in \mathrm{H}_1(\Omega_t)$ and integrating the left side of the resulting equation by parts once, so as to transfer the differential operation from $\delta\rho(t)$ onto $\psi^{(1)}(t)$. This sequence of steps yields the following relation:

$$\begin{aligned} &\int_{\beta_\ell^0}^{\beta_u^0} \psi^{(1)}(t) \left[\frac{d[\delta\rho(t)]}{dt} + \delta\rho(t) \sum_{i=1}^{N} n_i^0 \sigma_i^0 \right] dt = \int_{\beta_\ell^0}^{\beta_u^0} dt\, \delta\rho(t) \left[-\frac{d\psi^{(1)}(t)}{dt} \right. \\ &\left. + \psi^{(1)}(t) \sum_{i=1}^{N} n_i^0 \sigma_i^0 \right] + \psi^{(1)}\left(\beta_u^0\right) \delta\rho\left(\beta_u^0\right) - \psi^{(1)}\left(\beta_\ell^0\right) \delta\rho\left(\beta_\ell^0\right). \end{aligned} \tag{3.332}$$

The following sequence of operations is performed next using (3.332):

1. Require that the first term on the right side of (3.332) be identical with the indirect effect $(\delta R_1)_{ind}$ defined in (3.327).
2. Use the right side of (3.329) to replace the term multiplying $\psi^{(1)}(t)$ on the left side of (3.332).
3. Eliminate the unknown quantity $\delta\rho\left(\beta_u^0\right)$ on the right side of (3.332) by imposing the condition $\psi^{(1)}\left(\beta_u^0\right) = 0$.
4. Insert the boundary condition provided in (3.330) into (3.332).

The result of the above sequence of operations is the following expression for $(\delta R_1)_{ind}$:

$$(\delta R_1)_{ind} = -\sum_{i=1}^{N} \left(n_i^0 \delta\sigma_i + \sigma_i^0 \delta n_i\right) \int_{\beta_\ell^0}^{\beta_u^0} \psi^{(1)}(t)\rho^0(t)\,dt$$
$$+\psi^{(1)}\left(\beta_\ell^0\right)\left[\delta\rho_{in} - \delta\beta_\ell \left\{\frac{d\rho(t)}{dt}\right\}_{t=\beta_\ell^0}\right], \quad (3.333)$$

where the first-level adjoint function $\psi^{(1)}(t)$ appearing in (3.333) is the solution of the following First-Level Adjoint Sensitivity System (1st-LASS):

$$-\frac{d\psi^{(1)}(t)}{dt} + \psi^{(1)}(t)\sum_{i=1}^{N} n_i^0 \sigma_i^0 = \delta\left(t - t_d^0\right), \quad 0 \le \beta_\ell^0 \le t \le \beta_u^0 < \infty, \quad (3.334)$$

$$\psi^{(1)}\left(\beta_u^0\right) = 0. \quad (3.335)$$

In terms of the first-level adjoint function $\psi^{(1)}(t)$, the partial sensitivities of $R_1(\rho;\boldsymbol{\alpha})$ with respect to the variations in the model parameters are the quantities in (3.333) that multiply the respective parameter variations, namely:

$$\left(\frac{\partial R_1}{\partial \sigma_i}\right)_{\boldsymbol{\alpha}^0} = -n_i^0 \int_{\beta_\ell^0}^{\beta_u^0} \psi^{(1)}(t)\rho^0(t)dt, \quad i = 1, \ldots, N; \quad (3.336)$$

$$\left(\frac{\partial R_1}{\partial n_i}\right)_{\boldsymbol{\alpha}^0} = -\sigma_i^0 \int_{\beta_\ell^0}^{\beta_u^0} \psi^{(1)}(t)\rho^0(t)\,dt, \quad i = 1, \ldots, N; \quad (3.337)$$

$$\left(\frac{\partial R_1}{\partial \rho_{in}}\right)_{\boldsymbol{\alpha}^0} = \psi^{(1)}\left(\beta_\ell^0\right); \quad (3.338)$$

$$\left(\frac{\partial R_1}{\partial \beta_\ell}\right)_{\boldsymbol{\alpha}^0} = -\psi^{(1)}\left(\beta_\ell^0\right)\left\{\frac{d\rho(t)}{dt}\right\}_{t=\beta_\ell^0}. \quad (3.339)$$

Recalling the expression of the direct-effect term, $(\delta R_1)_{dir}$, defined in (3.328), yields the following additional first-order sensitivity:

$$\frac{\partial R_1}{\partial t_d} = -\int_{\beta_\ell^0}^{\beta_u^0} \rho^0(t)\,\delta'\left(t - t_d^0\right)dt = \left\{\frac{d\rho(t)}{dt}\right\}_{t=t_d^0}. \quad (3.340)$$

Since neither the direct-effect nor the indirect-effect terms depend on the variation $\delta\beta_u$, it follows that

$$\frac{\partial R_1}{\partial \beta_u} \equiv 0. \tag{3.341}$$

It is evident from (3.336)–(3.340) that the sensitivities of the response $R_1(\rho;\boldsymbol{\alpha})$ can be computed by computationally inexpensive quadrature methods applied to the integrals appearing in these expressions, after the first-level adjoint function $\psi^{(1)}(t)$ has been obtained by solving once the 1st-LASS, which comprises (3.334) and (3.335). Notably, the 1st-LASS needs to be solved once only since it does not depend on variations in the model parameters or state functions. Particularly important is the response sensitivity to the "initial condition" ρ_{in} since, as (3.338) indicates, the value of the first-level adjoint function $\psi^{(1)}(t)$ at the "initial time-value" $t = \beta_\ell^0$ is proportional to the response sensitivity to the "initial condition." Since the value of the first-level adjoint function $\psi^{(1)}(t)$ at $t = \beta_\ell^0$ can be obtained only after computing the entire evolution of $\psi^{(1)}(t)$ "backwards in time," from the "final-time" $t = \beta_u^0$ to the "initial-time" $t = \beta_\ell^0$, it becomes apparent that response sensitivities to initial conditions provide a stringent verification procedure for assessing the accuracy of the solution of the 1st-LASS.

Solving the 1st-LASS, comprising (3.334) and (3.335), yields the following expression for the first-level adjoint function $\psi^{(1)}(t)$:

$$\psi^{(1)}(t) = \left[1 - H\left(t - t_d^0\right)\right] \exp\left[\left(t - t_d^0\right)\sum_{i=1}^{N} n_i^0 \sigma_i^0\right], \tag{3.342}$$

where $H(t - t_d)$ is the customary Heaviside unit-step functional, defined as

$$H(t - t_d) \triangleq \begin{cases} 1, & t \geq t_d; \\ 0, & t < t_d. \end{cases} \tag{3.343}$$

Inserting the result from (3.342) into (3.336)–(3.339), respectively, yields the following expressions:

$$\left(\frac{\partial R_1}{\partial \sigma_i}\right)_{\boldsymbol{\alpha}^0} = n_i^0 \rho_{in}^0 \left(\beta_\ell^0 - t_d^0\right) \exp\left[\left(\beta_\ell^0 - t_d^0\right)\sum_{i=1}^{N} n_i^0 \sigma_i^0\right], \quad i = 1, \ldots, N; \tag{3.344}$$

$$\left(\frac{\partial R_1}{\partial n_i}\right)_{\boldsymbol{\alpha}^0} = \sigma_i^0 \rho_{in}^0 \left(\beta_\ell^0 - t_d^0\right) \exp\left[\left(\beta_\ell^0 - t_d^0\right)\sum_{i=1}^{N} n_i^0 \sigma_i^0\right], \quad i = 1, \ldots, N; \tag{3.345}$$

$$\left(\frac{\partial R_1}{\partial \rho_{in}}\right)_{\boldsymbol{\alpha}^0} = \exp\left[\left(\beta_\ell^0 - t_d^0\right)\sum_{i=1}^{N} n_i^0 \sigma_i^0\right]; \tag{3.346}$$

$$\left(\frac{\partial R_1}{\partial \beta_\ell}\right)_{\boldsymbol{\alpha}^0} = \rho_{in}^0\left(\sum_{i=1}^{N} n_i^0\sigma_i^0\right)\exp\left[\left(\beta_\ell^0 - t_d^0\right)\sum_{i=1}^{N} n_i^0\sigma_i^0\right]; \tag{3.347}$$

$$\left(\frac{\partial R_1}{\partial t_d}\right)_{\boldsymbol{\alpha}^0} = -\rho_{in}^0\left(\sum_{i=1}^{N} n_i^0\sigma_i^0\right)\exp\left[\left(\beta_\ell^0 - t_d^0\right)\sum_{i=1}^{N} n_i^0\sigma_i^0\right]. \tag{3.348}$$

The magnitudes of the first-order relative sensitivities provide a quantitative measure for ranking the importance of the respective parameters in affecting the response (e.g., the importance of the respective parameter's uncertainty in contributing to the overall uncertainty in the response). For the paradigm evolution system considered in this section, (3.336) and (3.337) indicate the important fact that the relative sensitivities of the response to the parameters σ_i, $(\partial R_1/\partial \sigma_i)(\sigma_i/R_1)$, and the relative sensitivities of the response to the parameters n_i, $(\partial R_1/\partial n_i)(n_i/R_1)$, respectively, happen to be identical, for all of these $2N$ model parameters, since

$$\frac{\partial R_1}{\partial \sigma_i}\frac{\sigma_i}{R_1} = \frac{\sigma_i n_i}{R_1}\int_{\beta_\ell^0}^{\beta_u^0}\left[-\psi^{(1)}(t)\rho(t)\right]dt \equiv \frac{\partial R_1}{\partial n_i}\frac{n_i}{R_1}, \quad i = 1, \ldots, N. \tag{3.349}$$

Therefore, statistical methods that use a priori screening techniques to reduce the number of model parameters that are actually considered in the respective statistical uncertainty/sensitivity analysis will very likely fail to achieve their goal for problems that have many parameters with identical relative sensitivities, as is the case shown in (3.349). Hence, this illustrative paradigm evolution system, which has many model parameters that have identical relative sensitivities, would be a prime candidate for testing the various statistical methods for sensitivity and uncertainty analysis. In contrast, a single large-scale computation for obtaining the adjoint function $\psi^{(1)}(t)$ suffices for computing exactly and efficiently, using just quadrature methods, the $2N + 4$ sensitivities of the response $R_1(\rho; \boldsymbol{\alpha})$ with respect to all model and boundary parameters.

In the particular case when the response $R_1(\rho; \boldsymbol{\alpha})$ is located at $t_d = \beta_u$, the expressions of the response sensitivities provided in (3.344)–(3.348) remain valid, with the stipulation that $t_d = \beta_u$.

The results for the first-order response sensitivities obtained in this section can also be verified by noting that the solution of the 1st-LVSS, comprising (3.329) and (3.330), has the following expression:

$$\delta\rho(t) = \left\{ \rho_{in}^{0}(\delta\beta_{\ell}) \sum_{i=1}^{N} n_i^0 \sigma_i^0 + \delta\rho_{in} + \rho_{in}^{0}(\beta_{\ell}^{0} - t) \sum_{i=1}^{N} \left(n_i^0 \delta\sigma_i + \sigma_i^0 \delta n_i \right) \right\}$$
$$\times \exp\left[(\beta_{\ell}^{0} - t) \sum_{i=1}^{N} n_i^0 \sigma_i^0 \right], \quad 0 \le \beta_{\ell}^{0} \le t \le \beta_{u}^{0} < \infty. \tag{3.350}$$

3.6.2 Applying the 2nd-CASAM-L to Compute the Second-Order Response Sensitivities to Model Parameters, Including Imprecisely Known Initial- and Final-Time Conditions

The starting point for obtaining expressions of the second-order response sensitivities is provided by the G-differentials of the expressions shown in (3.336)–(3.340). To keep the notation as simple as possible, the superscript "zero" will henceforth be omitted (except where stringently needed) when denoting "nominal values," since it will be clear from the derivations to follow that all first- and second-order sensitivities are to be evaluated at the nominal values of parameters.

3.6.2.1 Second-Order Sensitivities Corresponding to $\partial R_1(\rho;\alpha)/\partial\sigma_i$, $i = 1, \ldots, N$

The first-order G-differential of (3.336) yields:

$$\left\{ \delta\left(\frac{\partial R_1}{\partial \sigma_i}\right) \right\}_{\boldsymbol{\alpha}^0} \triangleq -\frac{d}{d\varepsilon} \{ (n_i + \varepsilon\delta n_i) \int_{\beta_{\ell}^{0}+\varepsilon\delta\beta_{\ell}}^{\beta_{u}^{0}+\varepsilon\delta\beta_{u}} dt \left[\psi^{(1)}(t) + \varepsilon\delta\psi^{(1)}(t) \right]$$
$$\times [\rho(t) + \varepsilon\delta\rho(t)] \}_{\boldsymbol{\alpha}^0} = \left\{ \left[\delta\left(\frac{\partial R_1}{\partial \sigma_i}\right) \right]_{dir} \right\}_{\boldsymbol{\alpha}^0} + \left\{ \left[\delta\left(\frac{\partial R_1}{\partial \sigma_i}\right) \right]_{ind} \right\}_{\boldsymbol{\alpha}^0}, \tag{3.351}$$

where the following definitions were used for $i = 1, \ldots, N$:

$$\left\{ \left[\delta\left(\frac{\partial R_1}{\partial \sigma_i}\right) \right]_{dir} \right\}_{\boldsymbol{\alpha}^0} \triangleq -(\delta n_i) \int_{\beta_{\ell}^{0}}^{\beta_{u}^{0}} \psi^{(1)}(t)\rho(t)dt + (\delta\beta_{\ell}) n_i \psi^{(1)}(\beta_{\ell}^{0})\rho(\beta_{\ell}^{0}), \tag{3.352}$$

$$\left\{\left[\delta\left(\frac{\partial R_1}{\partial \sigma_i}\right)\right]_{ind}\right\}_{\boldsymbol{\alpha}^0} \triangleq \left\{ -n_i \int_{\beta_\ell^0}^{\beta_u^0} \delta\psi^{(1)}(t)\rho(t)dt - n_i \int_{\beta_\ell^0}^{\beta_u^0} \psi^{(1)}(t)\delta\rho(t)dt \right\}_{\boldsymbol{\alpha}^0}. \quad (3.353)$$

The direct-effect term defined by (3.352) can be computed immediately, since the adjoint function $\psi^{(1)}(t)$ and the forward function $\rho(t)$ are known. However, the indirect-effect term defined by (3.353) contains the variation $\delta\psi^{(1)}(t)$ in the adjoint function and, respectively, the variation $\delta\rho(t)$ in the forward function, both of which depend on parameter variations and neither of which is immediately available. The variation $\delta\psi^{(1)}(t)$ of the first-level adjoint function $\psi^{(1)}(t)$ is related to the parameter variations through the G-differential of the 1st-LASS, which is derived by applying the definition of the G-differential to (3.334) and (3.335). Thus, taking the G-differential of the 1st-LASS yields the following equations evaluated at the nominal parameter values:

$$\begin{aligned} &-\frac{d\left[\delta\psi^{(1)}(t)\right]}{dt} + \delta\psi^{(1)}(t)\sum_{k=1}^{N} n_k\sigma_k = -(\delta t_d)\delta'(t-t_d) \\ &-\psi^{(1)}(t)\sum_{k=1}^{N}[(\delta n_k)\sigma_k + n_k(\delta\sigma_k)], \quad 0 \le \beta_\ell^0 \le t \le \beta_u^0 < \infty, \end{aligned} \quad (3.354)$$

$$\left\{\delta\psi^{(1)}(t) + \left(\delta\beta_u\right)\frac{d\psi^{(1)}(t)}{dt}\right\}_{t=\beta_u^0} = \delta\psi^{(1)}\left(\beta_u^0\right) = 0. \quad (3.355)$$

Taken together, (3.329), (3.330), (3.354), and (3.355) constitute a well-posed system of equations which could, in principle, be solved to obtain the variations $\delta\psi^{(1)}(t)$ and $\delta\rho(t)$ in terms of the parameter variations. However, such a procedure would be just as impractical computationally as solving the 1st-LVSS. Therefore, the need for solving these equations (which depend on parameter variations) will be circumvented by expressing the indirect-effect term defined in (3.353) in an alternative way, so as to eliminate the appearance of $\delta\psi^{(1)}(t)$ and $\delta\rho(t)$, by using the solution of an appropriately constructed Second-Level Adjoint Sensitivity System (2nd-LASS). This 2nd-LASS will be constructed in a Hilbert space denoted as $\mathrm{H}_2(\Omega_t)$, $\Omega_t \triangleq \left(\beta_\ell^0, \beta_u^0\right)$, which comprises, as elements, two-component vectors of the form $\boldsymbol{\psi}_1^{(2)}(i;t) \triangleq \left[\psi_{11}^{(2)}(i;t), \psi_{12}^{(2)}(i;t)\right]$, with square-integrable functions $\psi_{1j}^{(2)}(i;t), j = 1,2$. The index "$i$" which is displayed in the argument of the functions $\psi_{1j}^{(2)}(i;t), j = 1,2$, indicates that there will be more than one such function needed for computing the second-order sensitivities. The inner product between two elements $\mathbf{u}_1^{(2)}(t) \triangleq \left[u_{11}^{(2)}(t), u_{12}^{(2)}(t)\right]$ and $\mathbf{v}_1^{(2)}(t) \triangleq \left[v_{11}^{(2)}(t), v_{12}^{(2)}(t)\right]$ of the Hilbert space $\mathrm{H}_2(\Omega_t)$ will be denoted as $\left\langle \mathbf{u}_1^{(2)}(t), \mathbf{v}_1^{(2)}(t) \right\rangle_2$ and is defined as follows:

$$\left\langle \mathbf{u}_1^{(2)}(t), \mathbf{v}_1^{(2)}(t) \right\rangle_2 \triangleq \sum_{j=1}^{2} \int_{\beta_\ell^0}^{\beta_u^0} u_{1j}^{(2)}(t) v_{1j}^{(2)}(t) dt. \tag{3.356}$$

Equations (3.329) and (3.354) are written in matrix form, as follows:

$$\begin{pmatrix} -\dfrac{d}{dt} + \sum_{k=1}^{N} n_k \sigma_k & 0 \\ 0 & \dfrac{d}{dt} + \sum_{k=1}^{N} n_k \sigma_k \end{pmatrix} \begin{pmatrix} \delta\psi^{(1)}(t) \\ \delta\rho(t) \end{pmatrix} = \begin{pmatrix} -(\delta t_d)\delta'(t - t_d) - \psi^{(1)}(t) \sum_{k=1}^{N} [(\delta n_k)\sigma_k + n_k(\delta\sigma_k)] \\ -\rho(t) \sum_{k=1}^{N} [(\delta n_k)\sigma_k + n_k(\delta\sigma_k)] \end{pmatrix}. \tag{3.357}$$

Using the definition given in (3.356), construct the inner product of (3.357) with a square integrable two-component function $\boldsymbol{\psi}_1^{(2)}(i;t) \triangleq \left[\psi_{11}^{(2)}(i;t), \psi_{12}^{(2)}(i;t)\right] \in \mathrm{H}_2(\Omega_t)$ to obtain the following relation:

$$\begin{aligned} &\int_{\beta_\ell^0}^{\beta_u^0} \left[\psi_{11}^{(2)}(i;t), \psi_{12}^{(2)}(i;t)\right] \begin{pmatrix} -\dfrac{d}{dt} + \sum_{k=1}^{N} n_k \sigma_k & 0 \\ 0 & \dfrac{d}{dt} + \sum_{i=1}^{N} n_k \sigma_k \end{pmatrix} \begin{pmatrix} \delta\psi^{(1)}(t) \\ \delta\rho(t) \end{pmatrix} dt \\ &= \int_{\beta_\ell^0}^{\beta_u^0} \left[\psi_{11}^{(2)}(i;t), \psi_{12}^{(2)}(i;t)\right] \\ &\times \begin{pmatrix} -(\delta t_d)\delta'(t - t_d) - \psi^{(1)}(t) \sum_{k=1}^{N} [(\delta n_k)\sigma_k + n_k(\delta\sigma_k)] \\ -\rho(t) \sum_{k=1}^{N} [(\delta n_k)\sigma_k + n_k(\delta\sigma_k)] \end{pmatrix} dt. \end{aligned} \tag{3.358}$$

Integrating by parts the left side of (3.358) so as to transfer the differential operations on $\delta\psi^{(1)}(t)$ and $\delta\rho(t)$ to differential operations on $\psi_{11}^{(2)}(i;t)$ and $\psi_{12}^{(2)}(i;t)$ yields the following relation:

$$
\begin{aligned}
&\int_{\beta_\ell^0}^{\beta_u^0} \psi_{11}^{(2)}(i;t)\left[-\frac{d}{dt}\delta\psi^{(1)}(t)+\delta\psi^{(1)}(t)\sum_{k=1}^{N}n_k\sigma_k\right]dt+\int_{\beta_\ell^0}^{\beta_u^0}\psi_{12}^{(2)}(i;t)\left[\frac{d}{dt}\delta\rho(t)\right.\\
&\left.+\delta\rho(t)\sum_{k=1}^{N}n_k\sigma_k\right]dt=-\psi_{11}^{(2)}\left(i;\beta_u^0\right)\delta\psi^{(1)}\left(\beta_u^0\right)+\psi_{11}^{(2)}\left(i;\beta_\ell^0\right)\delta\psi^{(1)}\left(\beta_\ell^0\right)\\
&+\psi_{12}^{(2)}\left(i;\beta_u^0\right)\delta\rho\left(\beta_u^0\right)-\psi_{12}^{(2)}\left(i;\beta_\ell^0\right)\delta\rho\left(\beta_\ell^0\right)+\int_{\beta_\ell^0}^{\beta_u^0}\delta\psi^{(1)}(t)\left[\frac{d\psi_{11}^{(2)}(i;t)}{dt}\right.\\
&\left.+\psi_{11}^{(2)}(i;t)\sum_{k=1}^{N}n_k\sigma_k\right]dt+\int_{\beta_\ell^0}^{\beta_u^0}\delta\rho(t)\left[-\frac{d\psi_{12}^{(2)}(i;t)}{dt}+\psi_{12}^{(2)}(i;t)\sum_{k=1}^{N}n_k\sigma_k\right]dt.
\end{aligned}
\tag{3.359}
$$

The last two terms on the right side of (3.359) will represent the indirect-effect term defined in (3.353) by requiring that

$$
\frac{d\psi_{11}^{(2)}(i;t)}{dt}+\psi_{11}^{(2)}(i;t)\sum_{k=1}^{N}n_k\sigma_k=-n_i\rho(t),\ i=1,\ldots,N;
\tag{3.360}
$$

$$
-\frac{d\psi_{12}^{(2)}(i;t)}{dt}+\psi_{12}^{(2)}(i;t)\sum_{k=1}^{N}n_k\sigma_k=-n_i\psi^{(1)}(t),\quad i=1,\ldots,N.
\tag{3.361}
$$

The boundary conditions for (3.360) and (3.361) are established by requiring that the contributions involving the unknown quantities $\delta\psi^{(1)}\left(\beta_\ell^0\right)$ and $\delta\rho\left(\beta_u^0\right)$ in (3.359) vanish, which can be accomplished by imposing the following conditions:

$$
\psi_{11}^{(2)}\left(i;\beta_\ell^0\right)=0,\quad \psi_{12}^{(2)}\left(i;\beta_u^0\right)=0,\ i=1,\ldots,N.
\tag{3.362}
$$

The system of equations comprising (3.360)–(3.362) constitutes the Second-Level Adjoint Sensitivity System (2nd-LASS) for the two-component vector-valued function $\boldsymbol{\psi}_1^{(2)}(i;t)\triangleq\left[\psi_{11}^{(2)}(i;t),\psi_{12}^{(2)}(i;t)\right]\in \mathrm{H}_2(\Omega_t)$, which is called the second-level adjoint function. It is important to note that the 2nd-LASS is independent of parameter variations.

Replacing the left side of (3.359) by the right side of (3.358) and taking into account (3.330), (3.355), and (3.360)–(3.362) yields the following expression for the indirect-effect term defined in (3.353), for $i = 1, \ldots, N$:

$$\left[\delta\left(\frac{\partial R_1}{\partial \sigma_i}\right)\right]_{ind} = \psi_{12}^{(2)}\left(i;\beta_\ell^0\right)\left\{\delta\rho_{in} - \delta\beta_\ell\left[\frac{d\rho(t)}{dt}\right]_{t=\beta_\ell^0}\right\}$$
$$-\int_{\beta_\ell^0}^{\beta_u^0} \psi_{11}^{(2)}(i;t)\left\{(\delta t_d)\delta'(t-t_d) + \psi^{(1)}(t)\sum_{k=1}^{N}\left[(\delta n_k)\sigma_k + n_k(\delta\sigma_k)\right]\right\}dt \tag{3.363}$$
$$-\int_{\beta_\ell^0}^{\beta_u^0} \psi_{12}^{(2)}(i;t)\left\{\rho(t)\sum_{k=1}^{N}\left[(\delta n_k)\sigma_k + n_k(\delta\sigma_k)\right]\right\}dt.$$

Adding the direct-effect term defined in (3.352) to (3.363) and identifying in the resulting expression the coefficients multiplying the variations $\delta\sigma_i$, δn_i, $\delta\rho_{in}$, δt_d, $\delta\beta_\ell$, and $\delta\beta_u$ yields the following expression for the respective second-order sensitivities of the response $R_1(\rho;\boldsymbol{\alpha})$:

$$\frac{\partial^2 R_1}{\partial \sigma_j \partial \sigma_i} = -n_j \int_{\beta_\ell^0}^{\beta_u^0} \left[\psi_{11}^{(2)}(i;t)\psi^{(1)}(t) + \psi_{12}^{(2)}(i;t)\rho(t)\right]dt;\ i,j = 1,\ldots,N; \tag{3.364}$$

$$\frac{\partial^2 R_1}{\partial n_j \partial \sigma_i} = -\sigma_j \int_{\beta_\ell^0}^{\beta_u^0} \left[\psi_{11}^{(2)}(i;t)\psi^{(1)}(t) + \psi_{12}^{(2)}(i;t)\rho(t)\right]dt$$
$$-\delta_{ij}\int_{\beta_\ell^0}^{\beta_u^0} \psi^{(1)}(t)\rho(t)\,dt;\ i,j = 1,\ldots,N;\ \delta_{ij} = \begin{cases} 1, & i=j; \\ 0, & i \neq j; \end{cases} \tag{3.365}$$

$$\frac{\partial^2 R_1}{\partial \rho_{in} \partial \sigma_i} = \psi_{12}^{(2)}(i;\beta_\ell);\quad i = 1,\ldots,N; \tag{3.366}$$

$$\frac{\partial^2 R_1}{\partial t_d \partial \sigma_i} = -\int_{\beta_\ell^0}^{\beta_u^0} \psi_{11}^{(2)}(i;t)\,\delta'(t-t_d) = \left[\frac{d\psi_{11}^{(2)}(i;t)}{dt}\right]_{t=t_d};\quad i = 1,\ldots,N; \tag{3.367}$$

$$\frac{\partial^2 R_1}{\partial \beta_u \partial \sigma_i} = 0;\ i = 1,\ldots,N; \tag{3.368}$$

$$\frac{\partial^2 R_1}{\partial \beta_\ell \partial \sigma_i} = n_i\psi^{(1)}(\beta_\ell)\rho(\beta_\ell) - \psi_{12}^{(2)}(i;\beta_\ell)\left[\frac{d\rho(t)}{dt}\right]_{t=\beta_\ell^0};\ i = 1,\ldots,N. \tag{3.369}$$

The second-order sensitivities shown in (3.364)–(3.369) can be computed after having determined the second-level adjoint function $\boldsymbol{\psi}_1^{(2)}(i;t) \triangleq \left[\psi_{11}^{(2)}(i;t), \psi_{12}^{(2)}(i;t)\right]$

by solving the 2nd-LASS comprising (3.360)–(3.362) using the nominal parameter values (the superscript "zero," which indicates "nominal values," has been omitted, for simplicity). Evidently, the second-level adjoint sensitivity function $\boldsymbol{\psi}_1^{(2)}(i;t) \triangleq \left[\psi_{11}^{(2)}(i;t), \psi_{12}^{(2)}(i;t)\right]$ depends on the index $i = 1, \ldots, N$ because of the model parameters n_i, which depend on the index $i = 1, \ldots, N$. Hence, in the most unfavorable situation, the 2nd-LASS, comprising (3.360)–(3.362) would need to be solved numerically for each distinct value n_i, for a total of N-times. Even in such a "worse-case scenario," however, only the right sides (i.e., "sources") of (3.360) and (3.361) would need to be modified, which is relatively easy to implement computationally. The left sides of these equations remain unchanged since they are independent of the index $i = 1, \ldots, N$.

In many practical situations, however, it is possible to reduce drastically the number of computations involving the 2nd-LASS by changing the dependent and/or the independent variables. For example, in the case of the 2nd-LASS comprising (3.360)–(3.362), implementing the following change of the dependent variables $\psi_{11}^{(2)}(i;t)$ and $\psi_{12}^{(2)}(i;t)$:

$$\psi_{11}^{(2)}(i;t) = n_i \varphi_{11}^{(2)}(t), \quad \psi_{12}^{(2)}(i;t) = n_i \varphi_{12}^{(2)}(t), \tag{3.370}$$

would transform (3.360)–(3.362) into the following form:

$$\frac{d\varphi_{11}^{(2)}(t)}{dt} + \varphi_{11}^{(2)}(t) \sum_{k=1}^{N} n_k \sigma_k = -\rho(t), \tag{3.371}$$

$$-\frac{d\varphi_{12}^{(2)}(t)}{dt} + \varphi_{12}^{(2)}(t) \sum_{k=1}^{N} n_k \sigma_k = -\psi^{(1)}(t), \tag{3.372}$$

$$\varphi_{11}^{(2)}(\beta_\ell) = 0, \quad \varphi_{12}^{(2)}(\beta_u) = 0. \tag{3.373}$$

The above (alternative) 2nd-LASS, comprising (3.371)–(3.373), is independent of the index $i = 1, \ldots, N$, and would need to be solved (numerically, in practice) only once, to obtain the following expressions for the functions $\varphi_{11}^{(2)}(t)$ and $\varphi_{12}^{(2)}(t)$:

$$\varphi_{11}^{(2)}(t) = \rho_{in}(\beta_\ell - t) \exp\left[(\beta_\ell - t) \sum_{k=1}^{N} n_k \sigma_k\right], \tag{3.374}$$

$$\varphi_{12}^{(2)}(t) = (t - t_d)[1 - H(t - t_d)] \exp\left[(t - t_d) \sum_{k=1}^{N} n_k \sigma_k\right]. \tag{3.375}$$

The components of the second-level adjoint function $\boldsymbol{\psi}_1^{(2)}(i;t)$ can now be obtained by multiplying the functions $\varphi_{11}^{(2)}(t)$ and $\varphi_{12}^{(2)}(t)$ by the respective model parameters n_i, as indicated in (3.370), to obtain the following expressions for the components of the second-level adjoint function $\boldsymbol{\psi}_1^{(2)}(i;t) \triangleq \left[\psi_{11}^{(2)}(i;t), \psi_{12}^{(2)}(i;t)\right]$:

$$\psi_{11}^{(2)}(i;t) = n_i \rho_{in}(\beta_\ell - t) \exp\left[(\beta_\ell - t)\sum_{k=1}^{N} n_k \sigma_k\right], \tag{3.376}$$

$$\psi_{12}^{(2)}(i;t) = n_i(t - t_d)[1 - H(t - t_d)] \exp\left[(t - t_d)\sum_{k=1}^{N} n_k \sigma_k\right]. \tag{3.377}$$

Using (3.376) and (3.377) in (3.364)–(3.369) and performing the respective operations yields the following results for the respective partial second-order sensitivities:

$$\frac{\partial^2 R_1}{\partial \sigma_j \partial \sigma_i} = \rho_{in} n_j n_i (\beta_\ell - t_d)^2 \exp\left[(\beta_\ell - t_d)\sum_{k=1}^{N} n_k \sigma_k\right]; \; i,j = 1, \ldots, N; \tag{3.378}$$

$$\begin{aligned} \frac{\partial^2 R_1}{\partial n_j \partial \sigma_i} &= \left[\sigma_j n_i (\beta_\ell - t_d) + \delta_{ij}\right] \rho_{in}(\beta_\ell - t_d) \\ &\times \exp\left[(\beta_\ell - t_d)\sum_{k=1}^{N} n_k \sigma_k\right]; \; i,j = 1, \ldots, N; \end{aligned} \tag{3.379}$$

$$\frac{\partial^2 R_1}{\partial \rho_{in} \partial \sigma_i} = n_i(\beta_\ell - t_d) \exp\left[(\beta_\ell - t_d)\sum_{i=1}^{N} n_i \sigma_i\right]; \; i = 1, \ldots, N; \tag{3.380}$$

$$\begin{aligned} \frac{\partial^2 R_1}{\partial t_d \partial \sigma_i} &= -\rho_{in} n_i \left[1 + (\beta_\ell - t_d)\sum_{k=1}^{N} n_k \sigma_k\right] \\ &\times \exp\left[(\beta_\ell - t_d)\sum_{k=1}^{N} n_k \sigma_k\right]; \; i = 1, \ldots, N; \end{aligned} \tag{3.381}$$

$$\frac{\partial^2 R_1}{\partial \beta_u \partial \sigma_i} = 0; \; i = 1, \ldots, N; \tag{3.382}$$

$$\begin{aligned} \frac{\partial^2 R_1}{\partial \beta_\ell \partial \sigma_i} &= \rho_{in} n_i \left[1 + (\beta_\ell - t_d)\sum_{k=1}^{N} n_k \sigma_k\right] \\ &\times \exp\left[(\beta_\ell - t_d)\sum_{k=1}^{N} n_k \sigma_k\right]; \; i = 1, \ldots, N. \end{aligned} \tag{3.383}$$

As before, the right sides of expressions shown in (3.378)–(3.383) are to be evaluated at the nominal values for the parameters, but the superscript "zero," which indicates "nominal values," has been omitted, for notational simplicity.

3.6.2.2 Second-Order Sensitivities Corresponding to $\partial R_1(\rho;\alpha)/\partial n_i$, $i = 1, \ldots, N$

Computing the first-order G-differential of (3.337), at the nominal parameter values, yields:

$$\left\{\delta\left(\frac{\partial R_1}{\partial n_i}\right)\right\}_{\boldsymbol{\alpha}^0} \triangleq -\frac{d}{d\varepsilon}\{(\sigma_i+\varepsilon\delta\sigma_i)\int_{\beta_\ell^0+\varepsilon\delta\beta_\ell}^{\beta_u^0+\varepsilon\delta\beta_u}\left[\psi^{(1)}(t)+\varepsilon\delta\psi^{(1)}(t)\right]$$
$$\times[\rho(t)+\varepsilon\delta\rho(t)]dt\}_{\boldsymbol{\alpha}^0} = \left\{\left[\delta\left(\frac{\partial R_1}{\partial n_i}\right)\right]_{dir}\right\}_{\boldsymbol{\alpha}^0} + \left\{\left[\delta\left(\frac{\partial R_1}{\partial n_i}\right)\right]_{ind}\right\}_{\boldsymbol{\alpha}^0}, \tag{3.384}$$

where the direct-effect and indirect-effect terms, respectively, are evaluated at the nominal parameter values and are defined as follows, $i = 1, \ldots, N$:

$$\left\{\left[\delta\left(\frac{\partial R_1}{\partial n_i}\right)\right]_{dir}\right\}_{\boldsymbol{\alpha}^0} \triangleq -(\delta\sigma_i)\int_{\beta_\ell^0}^{\beta_u^0}\psi^{(1)}(t)\rho(t)dt + (\delta\beta_\ell)\sigma_i\psi^{(1)}(\beta_\ell^0)\rho(\beta_\ell^0), \tag{3.385}$$

$$\left\{\left[\delta\left(\frac{\partial R_1}{\partial n_i}\right)\right]_{ind}\right\}_{\boldsymbol{\alpha}^0} \triangleq -\left\{\sigma_i\int_{\beta_\ell^0}^{\beta_u^0}\left[\delta\psi^{(1)}(t)\rho(t)+\psi^{(1)}(t)\delta\rho(t)\right]dt\right\}_{\boldsymbol{\alpha}^0}. \tag{3.386}$$

The direct-effect term defined in (3.385) can be computed immediately, since the adjoint function $\psi^{(1)}(t)$ and the forward function $\rho(t)$ are known. On the other hand, the indirect-effect term defined by (3.386) contains the variation $\delta\psi^{(1)}(t)$ in the adjoint function and, respectively, the variation $\delta\rho(t)$ in the forward function, both of which depend on parameter variations and neither of which is immediately available. Comparing (3.386) to (3.353) readily indicates that the right sides of these equations differ only in that the model parameter σ_i plays in (3.386) the same role as the model parameter n_i plays in (3.353). Thus, the same procedure that has been previously used to obtain an alternative expression for the indirect-effect term defined in (3.353), by means of a second-level adjoint function which will be denoted as $\boldsymbol{\psi}_2^{(2)}(i;t) \triangleq \left[\psi_{21}^{(2)}(i;t), \psi_{22}^{(2)}(i;t)\right]$, is applied to (3.386), which yields the following result, for $i = 1, \ldots, N$:

$$\left[\delta\left(\frac{\partial R_1}{\partial n_i}\right)\right]_{ind} = \psi_{22}^{(2)}\left(i;\beta_\ell^0\right)\left\{\delta\rho_{in} - \delta\beta_\ell\left[\frac{d\rho(t)}{dt}\right]_{t=\beta_\ell^0}\right\}_{\boldsymbol{\alpha}^0} + \int_{\beta_\ell^0}^{\beta_u^0}\left[\psi_{21}^{(2)}(i;t), \psi_{22}^{(2)}(i;t)\right]$$
$$\times\begin{pmatrix} -(\delta t_d)\delta'(t-t_d) - \psi^{(1)}(t)\sum_{k=1}^{N}\left[(\delta n_k)\sigma_k + n_k(\delta\sigma_k)\right] \\ -\rho(t)\sum_{k=1}^{N}\left[(\delta n_k)\sigma_k + n_k(\delta\sigma_k)\right]\end{pmatrix}_{\boldsymbol{\alpha}^0} dt, \tag{3.387}$$

where the second-level adjoint function $\boldsymbol{\psi}_2^{(2)}(i;t) \triangleq \left[\psi_{21}^{(2)}(i;t), \psi_{22}^{(2)}(i;t)\right]$ satisfies the following 2nd-LASS:

$$\frac{d\psi_{21}^{(2)}(i;t)}{dt} + \psi_{21}^{(2)}(i;t)\sum_{k=1}^{N} n_k\sigma_k = -\sigma_i\rho(t), \; i = 1, \ldots, N; \tag{3.388}$$

$$-\frac{d\psi_{22}^{(2)}(i;t)}{dt} + \psi_{22}^{(2)}(i;t)\sum_{k=1}^{N} n_k\sigma_k = -\sigma_i\psi^{(1)}(t), \; i = 1, \ldots, N; \tag{3.389}$$

$$\psi_{21}^{(2)}\left(i;\beta_\ell\right) = 0, \quad \psi_{22}^{(2)}\left(i;\beta_u\right) = 0. \tag{3.390}$$

The sources on the right sides of the 2nd-LASS defined by (3.388)–(3.390) are to be evaluated at the nominal values for the parameters, but the superscript "zero," which indicates "nominal values," has been omitted, for notational simplicity.

Comparing (3.388)–(3.390) to (3.360)–(3.362) and recalling (3.371)–(3.373) indicates that the components of the second-level adjoint function $\boldsymbol{\psi}_2^{(2)}(i;t) \triangleq \left[\psi_{21}^{(2)}(i;t), \psi_{22}^{(2)}(i;t)\right]$ have the following expressions:

$$\psi_{21}^{(2)}(i;t) = \sigma_i\varphi_{11}^{(2)}(t) = \sigma_i\rho_{in}(\beta_\ell - t)\exp\left[(\beta_\ell - t)\sum_{k=1}^{N} n_k\sigma_k\right], \tag{3.391}$$

$$\psi_{22}^{(2)}(i;t) = \sigma_i\varphi_{12}^{(2)}(t) = \sigma_i(t - t_d)[1 - H(t - t_d)]\exp\left[(t - t_d)\sum_{k=1}^{N} n_k\sigma_k\right]. \tag{3.392}$$

Adding the direct-effect term defined in (3.385) to the expression for the indirect-effect term shown in (3.387) and identifying in the resulting expression the coefficients multiplying the variations $\delta\sigma_i$, δn_i, $\delta\rho_{in}$, δt_d, $\delta\beta_\ell$, and $\delta\beta_u$ yields the following expression for the corresponding second-order sensitivities of the response $R_1(\rho;\boldsymbol{\alpha})$:

$$\frac{\partial^2 R_1}{\partial \sigma_j \partial n_i} = -n_j \int_{\beta_\ell^0}^{\beta_u^0} \left[\psi_{21}^{(2)}(i;t)\psi^{(1)}(t) + \psi_{22}^{(2)}(i;t)\rho(t) \right] dt$$
$$-\delta_{ij} \int_{\beta_\ell^0}^{\beta_u^0} \psi^{(1)}(t)\rho(t)\,dt; \; i,j = 1, \ldots, N; \tag{3.393}$$

$$\frac{\partial^2 R_1}{\partial n_j \partial n_i} = -\sigma_j \int_{\beta_\ell^0}^{\beta_u^0} \left[\psi_{21}^{(2)}(i;t)\psi^{(1)}(t) + \psi_{22}^{(2)}(i;t)\rho(t) \right] dt; \; i,j = 1, \ldots, N; \tag{3.394}$$

$$\frac{\partial^2 R_1}{\partial \rho_{in} \partial n_i} = \psi_{22}^{(2)}(i;\beta_\ell); \quad i = 1, \ldots, N; \tag{3.395}$$

$$\frac{\partial^2 R_1}{\partial t_d \partial n_i} = -\int_{\beta_\ell^0}^{\beta_u^0} \psi_{21}^{(2)}(i;t)\delta'(t - t_d) = \left[\frac{d\psi_{21}^{(2)}(i;t)}{dt} \right]_{t=t_d}; \quad i = 1, \ldots, N; \tag{3.396}$$

$$\frac{\partial^2 R_1}{\partial \beta_u \partial n_i} = 0; \; i = 1, \ldots, N; \tag{3.397}$$

$$\frac{\partial^2 R_1}{\partial \beta_\ell \partial n_i} = \sigma_i \psi^{(1)}(\beta_\ell)\rho(\beta_\ell) - \psi_{22}^{(2)}(i;\beta_\ell)\left[\frac{d\rho(t)}{dt} \right]_{t=\beta_\ell^0}; \quad i = 1, \ldots, N. \tag{3.398}$$

Inserting the expressions obtained in (3.391) and (3.392) for the components of the second-level adjoint function $\boldsymbol{\psi}_2^{(2)}(i;t) \triangleq \left[\psi_{21}^{(2)}(i;t), \psi_{22}^{(2)}(i;t) \right]$ into (3.393)–(3.398) yields the following expressions, where all quantities are to be evaluated at the nominal parameter values:

$$\frac{\partial^2 R_1}{\partial \sigma_j \partial n_i} = \left[n_j \sigma_i (\beta_\ell - t_d) + \delta_{ij} \right] \rho_{in}(\beta_\ell - t_d)$$
$$\times \exp\left[(\beta_\ell - t_d) \sum_{k=1}^{N} n_k \sigma_k \right]; \quad i,j = 1, \ldots, N; \tag{3.399}$$

$$\frac{\partial^2 R_1}{\partial n_j \partial n_i} = \rho_{in} \sigma_j \sigma_i (\beta_\ell - t_d)^2 \exp\left[(\beta_\ell - t_d) \sum_{k=1}^{N} n_k \sigma_k \right]; \; i,j = 1, \ldots, N; \tag{3.400}$$

$$\frac{\partial^2 R_1}{\partial \rho_{in}\,\partial n_i} = \sigma_i(\beta_\ell - t_d)\exp\left[(\beta_\ell - t_d)\sum_{k=1}^{N} n_k\sigma_k\right];\quad i = 1,\ldots,N; \tag{3.401}$$

$$\begin{aligned}\frac{\partial^2 R_1}{\partial t_d\,\partial n_i} &= -\rho_{in}\sigma_i\left[1 + (\beta_\ell - t_d)\sum_{k=1}^{N} n_k\sigma_k\right]\\ &\times\exp\left[(\beta_\ell - t_d)\sum_{k=1}^{N} n_k\sigma_k\right];\ i = 1,\ldots,N;\end{aligned} \tag{3.402}$$

$$\frac{\partial^2 R_1}{\partial \beta_u\,\partial n_i} = 0;\ i = 1,\ldots,N; \tag{3.403}$$

$$\frac{\partial^2 R_1}{\partial \beta_\ell\,\partial n_i} = \rho_{in}\sigma_i\left[1 + (\beta_\ell - t_d)\sum_{k=1}^{N} n_k\sigma_k\right]\exp\left[(\beta_\ell - t_d)\sum_{k=1}^{N} n_k\sigma_k\right];\ i = 1,\ldots,N. \tag{3.404}$$

3.6.2.3 Second-Order Sensitivities Corresponding to $\partial R_1(\rho;\alpha)/\partial\rho_{in}$

The second-order response sensitivities corresponding to $\partial R_1(\rho;\boldsymbol{\alpha})/\partial\rho_{in}$ will be calculated by taking the G-differential of (3.338). Since the model responses need to be written in the form of an inner product in order to apply the adjoint sensitivity analysis methodology, (3.338) is re-written in the following form:

$$\left(\frac{\partial R_1}{\partial \rho_{in}}\right) = \psi^{(1)}(\beta_\ell) = \int_{\beta_\ell}^{\beta_u} \psi^{(1)}(t)\delta(t - \beta_\ell)dt. \tag{3.405}$$

Taking the G-differential of (3.405) yields

$$\begin{aligned}\delta\left(\frac{\partial R_1}{\partial \rho_{in}}\right) &= \left\{\frac{d}{d\varepsilon}\int_{\beta_\ell^0+\varepsilon\,\delta\beta_\ell}^{\beta_u^0+\varepsilon\,\delta\beta_u}\left[\psi^{(1)}(t) + \varepsilon\delta\psi^{(1)}(t)\right]\delta\left(t - \beta_\ell^0 - \varepsilon\delta\beta_\ell\right)dt\right\}_{\varepsilon=0}\\ &= \left\{\delta\left(\frac{\partial R_1}{\partial \rho_{in}}\right)\right\}_{dir} + \left\{\delta\left(\frac{\partial R_1}{\partial \rho_{in}}\right)\right\}_{ind},\end{aligned} \tag{3.406}$$

where

$$\left\{\delta\left(\frac{\partial R_1}{\partial \rho_{in}}\right)\right\}_{dir} \triangleq -(\delta\beta_\ell)\int_{\beta_\ell^0}^{\beta_u^0} \psi^{(1)}(t)\delta'\left(t-\beta_\ell^0\right)dt, \tag{3.407}$$

and

$$\left\{\delta\left(\frac{\partial R_1}{\partial \rho_{in}}\right)\right\}_{ind} \triangleq \int_{\beta_\ell^0}^{\beta_u^0} \delta\psi^{(1)}(t)\delta\left(t-\beta_\ell^0\right)dt. \tag{3.408}$$

The direct-effect term defined in (3.407) can be computed immediately, since the adjoint function $\psi^{(1)}(t)$ is known. Noteworthy, the indirect-effect term defined in (3.408) only contains the variation $\delta\psi^{(1)}(t)$ in the first-level adjoint function, but does not contain the variation $\delta\rho(t)$ in the forward function. Therefore, the second-level adjoint function that would be needed to recast the indirect-effect term defined in (3.408) would be a one-component (as opposed to a "two-component" vector) function. Thus, the 2nd-LASS needed to recast the indirect-effect term defined in (3.408) is constructed by using the definition provided in (3.331) to construct the inner product of a square-integrable function $\psi_{31}^{(2)}(t) \in \mathrm{H}_1(\Omega_t)$ with (3.354) and integrating the left side of the resulting equation by parts once, so as to transfer the differential operation from $\delta\psi^{(1)}(t)$ onto $\psi_{31}^{(2)}(t)$. This sequence of steps yields the following relation:

$$\begin{aligned}\int_{\beta_\ell^0}^{\beta_u^0} \psi_{31}^{(2)}(t)\left[-\frac{d}{dt}\delta\psi^{(1)}(t)+\delta\psi^{(1)}(t)\sum_{k=1}^{N} n_k\sigma_k\right]dt &= -\psi_{31}^{(2)}\left(\beta_u^0\right)\delta\psi^{(1)}\left(\beta_u^0\right)\\ +\psi_{31}^{(2)}\left(\beta_\ell^0\right)\delta\psi^{(1)}\left(\beta_\ell^0\right)+\int_{\beta_\ell^0}^{\beta_u^0} \delta\psi^{(1)}(t)&\left[\frac{d\psi_{31}^{(2)}(t)}{dt}+\psi_{31}^{(2)}(t)\sum_{k=1}^{N} n_k\sigma_k\right]dt.\end{aligned} \tag{3.409}$$

The last term on the right side of (3.409) is now required to represent the indirect-effect term defined in (3.408). This is accomplished by requiring that

$$\frac{d\psi_{31}^{(2)}(t)}{dt}+\psi_{31}^{(2)}(t)\sum_{k=1}^{N} n_k\sigma_k = \delta\left(t-\beta_\ell^0\right). \tag{3.410}$$

The boundary condition for (3.410) is established by requiring that the contribution involving the unknown quantity $\delta\psi^{(1)}\left(\beta_\ell^0\right)$ in (3.409) vanish, which can be accomplished by imposing the following condition:

$$\psi_{31}^{(2)}\left(\beta_{\ell}^{0}\right) = 0. \tag{3.411}$$

As before, (3.410) and (3.411), which comprise the 2nd-LASS for the second-level adjoint function $\psi_{31}^{(2)}(t)$, are to be solved at the nominal parameter values. Note that the 2nd-LASS comprising (3.410) and (3.411) does not depend on any index and therefore needs to be solved only once to determine the second-level adjoint function $\psi_{31}^{(2)}(t)$.

Replacing the right side of (3.354) in the left side of (3.409) and taking into account (3.342), (3.355), and (3.411) yields the following expression for the indirect-effect term defined in (3.408):

$$\left\{\delta\left(\frac{\partial R_1}{\partial \rho_{in}}\right)\right\}_{ind} = -\int_{\beta_{\ell}^{0}}^{\beta_{u}^{0}} \psi_{31}^{(2)}(t)\left[(\delta t_d)\delta'(t-t_d) + \psi^{(1)}(t)\sum_{k=1}^{N}\left[(\delta n_k)\sigma_k + n_k(\delta\sigma_k)\right]\right] dt. \tag{3.412}$$

Adding the direct-effect term defined in (3.407) to (3.412) and identifying in the resulting expression the coefficients multiplying the variations $\delta\sigma_i$, δn_i, $\delta\rho_{in}$, δt_d, $\delta\beta_{\ell}$, and $\delta\beta_u$ yields the following expressions for the respective second-order sensitivities of the response $R_1(\rho;\boldsymbol{\alpha})$:

$$\frac{\partial^2 R_1}{\partial \sigma_j \partial \rho_{in}} = -n_j \int_{\beta_{\ell}^{0}}^{\beta_{u}^{0}} \psi_{31}^{(2)}(t)\psi^{(1)}(t)\,dt;\; j = 1,\ldots,N; \tag{3.413}$$

$$\frac{\partial^2 R_1}{\partial n_j \partial \rho_{in}} = -\sigma_j \int_{\beta_{\ell}^{0}}^{\beta_{u}^{0}} \psi_{31}^{(2)}(t)\psi^{(1)}(t)\,dt;\; j = 1,\ldots,N; \tag{3.414}$$

$$\frac{\partial^2 R_1}{\partial t_d\, \partial \rho_{in}} = -\int_{\beta_{\ell}^{0}}^{\beta_{u}^{0}} \psi_{31}^{(2)}(t)\delta'(t-t_d)dt; \tag{3.415}$$

$$\frac{\partial^2 R_1}{\partial \rho_{in}\, \partial \rho_{in}} = 0; \quad \frac{\partial^2 R_1}{\partial \beta_u\, \partial \rho_{in}} = 0; \tag{3.416}$$

$$\frac{\partial^2 R_1}{\partial \beta_\ell \, \partial \rho_{in}} = - \int_{\beta_\ell^0}^{\beta_u^0} \psi^{(1)}(t) \delta'\left(t - \beta_\ell^0\right) dt = \left\{ \frac{d\psi^{(1)}(t)}{dt} \right\}_{t=\beta_\ell^0}. \tag{3.417}$$

The closed-form expression of the solution of the 2nd-LASS provided in (3.410) and (3.411) is:

$$\psi_{31}^{(2)}(t) = H\left(t - \beta_\ell^0\right) \exp\left[\left(\beta_\ell^0 - t\right) \sum_{k=1}^{N} n_k \sigma_k\right]. \tag{3.418}$$

Replacing the result for the second-level adjoint function obtained in (3.418) into (3.413)–(3.415) and carrying out the respective operations yields the following expressions, which are to be evaluated at the nominal parameter values:

$$\frac{\partial^2 R_1}{\partial \sigma_j \partial \rho_{in}} = n_j \left(\beta_\ell - t_d\right) \exp\left[\left(\beta_\ell - t_d\right) \sum_{k=1}^{N} n_k \sigma_k\right]; \; j = 1, \ldots, N; \tag{3.419}$$

$$\frac{\partial^2 R_1}{\partial n_j \partial \rho_{in}} = \sigma_j \left(\beta_\ell - t_d\right) \exp\left[\left(\beta_\ell - t_d\right) \sum_{k=1}^{N} n_k \sigma_k\right]; \; j = 1, \ldots, N; \tag{3.420}$$

$$\frac{\partial^2 R_1}{\partial t_d \, \partial \rho_{in}} = -\left(\sum_{k=1}^{N} n_k \sigma_k\right) \exp\left[\left(\beta_\ell - t_d\right) \sum_{k=1}^{N} n_k \sigma_k\right]; \tag{3.421}$$

$$\frac{\partial^2 R_1}{\partial \beta_\ell \, \partial \rho_{in}} = \left(\sum_{k=1}^{N} n_k \sigma_k\right) \exp\left[\left(\beta_\ell - t_d\right) \sum_{k=1}^{N} n_k \sigma_k\right]. \tag{3.422}$$

3.6.2.4 Second-Order Sensitivities Corresponding to $\partial R_1(\rho; \alpha)/\partial t_d$

The second-order response sensitivities corresponding to $\partial R_1(\rho; \boldsymbol{\alpha})/\partial t_d$ will be calculated by taking the G-differential of (3.340), which yields the following expression:

$$\begin{aligned} \delta\left(\frac{\partial R_1}{\partial t_d}\right) &= -\left\{ \frac{d}{d\varepsilon} \int_{\beta_\ell^0 + \varepsilon\,\delta\beta_\ell}^{\beta_u^0 + \varepsilon\,\delta\beta_u} \left\{\left[\rho^0 + \varepsilon\delta\rho\right]\right\} \delta'\left(t - t_d^0 - \varepsilon\delta t_d\right) dt \right\}_{\varepsilon=0} \\ &= \left\{\delta\left(\frac{\partial R_1}{\partial t_d}\right)\right\}_{dir} + \left\{\delta\left(\frac{\partial R_1}{\partial t_d}\right)\right\}_{ind}, \end{aligned} \tag{3.423}$$

where

$$\left\{\delta\left(\frac{\partial R_1}{\partial t_d}\right)\right\}_{dir} = -(\delta t_d)\int_{\beta_\ell^0}^{\beta_u^0} \rho(t)\delta''\left(t - t_d^0\right) dt$$
$$= (\delta t_d)\rho_{in}\left(\sum_{k=1}^{N} n_k\sigma_k\right)^2 \exp\left[(\beta_\ell - t_d)\sum_{k=1}^{N} n_k\sigma_k\right], \tag{3.424}$$

and

$$\left\{\delta\left(\frac{\partial R_1}{\partial t_d}\right)\right\}_{ind} = -\int_{\beta_\ell^0}^{\beta_u^0} \delta\rho(t)\delta'\left(t - t_d^0\right)dt. \tag{3.425}$$

Noteworthy, the indirect-effect term defined in (3.425) only contains the variation $\delta\rho(t)$ in the forward function but does not contain the variation $\delta\psi^{(1)}(t)$ in the first-level adjoint function. Therefore, the second-level adjoint function that would be needed to recast the indirect-effect term defined in (3.425) would be a one-component (as opposed to a two-component vector) function. Thus, the 2nd-LASS needed to recast the indirect-effect term defined in (3.425) is constructed by applying the definition provided in (3.331) to construct the inner product of a square-integrable function $\psi_{41}^{(2)}(t) \in \mathrm{H}_1(\Omega_t)$ with (3.329) and integrating the left side of the resulting equation by parts once, so as to transfer the differential operation from $\delta\rho(t)$ onto the function $\psi_{41}^{(2)}(t)$. This sequence of steps yields the following relation [which is analogous to (3.332)]:

$$\int_{\beta_\ell^0}^{\beta_u^0} \psi_{41}^{(2)}(t)\left[\frac{d[\delta\rho(t)]}{dt} + \delta\rho(t)\sum_{k=1}^{N} n_k\sigma_k\right] dt = \psi_{41}^{(2)}\left(\beta_u^0\right)\delta\rho\left(\beta_u^0\right)$$
$$-\psi_{41}^{(2)}\left(\beta_\ell^0\right)\delta\rho\left(\beta_\ell^0\right) + \int_{\beta_\ell^0}^{\beta_u^0} \delta\rho(t)\left[-\frac{d\psi_{41}^{(2)}(t)}{dt} + \psi_{41}^{(2)}(t)\sum_{k=1}^{N} n_k\sigma_k\right] dt. \tag{3.426}$$

The following sequence of operations is now performed using (3.426):

1. Require that the last term on the right side of (3.426) be identical with the indirect effect $(\delta R_1)_{ind}$ defined in (3.425).
2. Use the right side of (3.329) to replace the term multiplying $\psi_{41}^{(2)}(t)$ on the left side of (3.426).

3. Eliminate the unknown quantity $\delta\rho\left(\beta_u^0\right)$ on the right side of (3.426) by imposing the condition $\psi_{41}^{(2)}\left(\beta_u^0\right) = 0$.
4. Insert the boundary condition provided in (3.330) into (3.426).

The result of the above sequence of operations is the following expression for the indirect-effect term defined in (3.425):

$$\left\{\delta\left(\frac{\partial R_1}{\partial t_d}\right)\right\}_{ind} = -\sum_{k=1}^{N}\left[(\delta n_k)\sigma_k + n_k(\delta\sigma_k)\right]\int_{\beta_\ell^0}^{\beta_u^0}\psi_{41}^{(2)}(t)\rho(t)\,dt$$
$$+\psi_{41}^{(2)}\left(\beta_\ell^0\right)\left[\delta\rho_{in} - \delta\beta_\ell\left\{\frac{d\rho(t)}{dt}\right\}_{t=\beta_\ell^0}\right], \tag{3.427}$$

where the first-level adjoint function $\psi_{41}^{(2)}(t)$ appearing in (3.427) is the solution of the following 2nd-LASS, evaluated at the nominal parameter values:

$$-\frac{d\psi_{41}^{(2)}(t)}{dt} + \psi_{41}^{(2)}(t)\sum_{k=1}^{N} n_k\sigma_k = -\delta'\left(t - t_d^0\right), \quad 0 \le \beta_\ell^0 \le t \le \beta_u^0 < \infty, \tag{3.428}$$

$$\psi_{41}^{(2)}\left(\beta_u^0\right) = 0. \tag{3.429}$$

The solution of (3.428) and (3.429) is:

$$\psi_{41}^{(2)}(t) = \delta(t - t_d) + \left(\sum_{i=1}^{N} n_i\sigma_i\right)\left[H(t - t_d) - 1\right]\exp\left[(t - t_d)\sum_{i=1}^{N} n_i\sigma_i\right]. \tag{3.430}$$

In terms of the second-level adjoint function $\psi_{41}^{(2)}(t)$, the partial second-order sensitivities corresponding to $\partial R_1(\rho;\boldsymbol{\alpha})/\partial t_d$ are obtained by adding (3.424) and (3.427), and subsequently identifying in the resulting expression the coefficients multiplying the variations $\delta\sigma_i$, δn_i, $\delta\rho_{in}$, δt_d, $\delta\beta_\ell$, and $\delta\beta_u$. This sequence of operations yields the following expressions:

$$\frac{\partial^2 R_1}{\partial\sigma_j\,\partial t_d} = -n_j\int_{\beta_\ell^0}^{\beta_u^0}\psi_{41}^{(2)}(t)\rho(t)\,dt; \; j = 1,\ldots,N; \tag{3.431}$$

$$\frac{\partial^2 R_1}{\partial n_j \, \partial t_d} = -\sigma_j \int_{\beta_\ell^0}^{\beta_u^0} \psi_{41}^{(2)}(t)\rho(t)\, dt; \; j = 1, \ldots, N; \tag{3.432}$$

$$\frac{\partial^2 R_1}{\partial \rho_{in} \, \partial t_d} = \psi_{41}^{(2)}\left(\beta_\ell^0\right); \tag{3.433}$$

$$\frac{\partial^2 R_1}{\partial t_d^2} = \rho_{in}\left(\sum_{k=1}^{N} n_k \sigma_k\right)^2 \exp\left[(\beta_\ell - t_d)\sum_{k=1}^{N} n_k \sigma_k\right]; \tag{3.434}$$

$$\frac{\partial^2 R_1}{\partial \beta_u \, \partial t_d} = 0; \tag{3.435}$$

$$\frac{\partial^2 R_1}{\partial \beta_\ell \, \partial t_d} = -\psi_{41}^{(2)}\left(\beta_\ell^0\right)\left\{\frac{d\rho(t)}{dt}\right\}_{t=\beta_\ell^0}. \tag{3.436}$$

Replacing the result for the second-level adjoint function obtained in (3.430) into (3.431)–(3.436) and carrying out the respective operations yields the following expressions, which are to be evaluated at the nominal parameter values:

$$\frac{\partial^2 R_1}{\partial \sigma_j \, \partial t_d} = -n_j \rho_{in}\left[1 + (\beta_\ell - t_d)\sum_{k=1}^{N} n_k \sigma_k\right] \exp\left[(\beta_\ell - t_d)\sum_{k=1}^{N} n_k \sigma_k\right]; j = 1, \ldots, N; \tag{3.437}$$

$$\begin{aligned} \frac{\partial^2 R_1}{\partial n_j \, \partial t_d} &= -\sigma_j \rho_{in}\left[1 + (\beta_\ell - t_d)\left(\sum_{k=1}^{N} n_k \sigma_k\right)\right] \\ &\times \exp\left[(\beta_\ell - t_d)\sum_{k=1}^{N} n_k \sigma_k\right]; j = 1, \ldots, N; \end{aligned} \tag{3.438}$$

$$\frac{\partial^2 R_1}{\partial \rho_{in} \, \partial t_d} = -\left(\sum_{k=1}^{N} n_k \sigma_k\right) \exp\left[(\beta_\ell - t_d)\sum_{k=1}^{N} n_k \sigma_k\right]; \tag{3.439}$$

$$\frac{\partial^2 R_1}{\partial t_d^2} = \rho_{in}\left(\sum_{k=1}^{N} n_k \sigma_k\right)^2 \exp\left[(\beta_\ell - t_d)\sum_{k=1}^{N} n_k \sigma_k\right]; \tag{3.440}$$

$$\frac{\partial^2 R_1}{\partial \beta_u \, \partial t_d} \equiv 0; \tag{3.441}$$

$$\frac{\partial^2 R_1}{\partial \beta_\ell \partial t_d} = -\rho_{in} \left(\sum_{k=1}^{N} n_k \sigma_k \right)^2 \exp\left[(\beta_\ell - t_d) \sum_{k=1}^{N} n_k \sigma_k \right]. \tag{3.442}$$

3.6.2.5 Second-Order Sensitivities Corresponding to $\partial R_1(\rho;\alpha)/\partial \beta_\ell$

The second-order response sensitivities corresponding to $\partial R_1(\rho;\boldsymbol{\alpha})/\partial \beta_\ell$ will be calculated by taking the G-differential of (3.339), which first needs to be written in the form of an inner product in order to apply the 2nd-CASAM-L, as follows:

$$\frac{\partial R_1}{\partial \beta_\ell} = -\left\{ \psi^{(1)}(\beta_\ell) \frac{d\rho(t)}{dt} \right\}_{t=\beta_\ell^0} = -\int_{\beta_\ell}^{\beta_u} \left[\psi^{(1)}(t) \frac{d\rho(t)}{dt} \right] \delta(t-\beta_\ell) dt. \tag{3.443}$$

Taking the G-differential of (3.443) at the nominal parameter values (the superscript "zero," denoting nominal values, is again omitted) yields:

$$\begin{aligned} \delta\left(\frac{\partial R_1}{\partial \beta_\ell}\right) &= -\left\{ \frac{d}{d\varepsilon} \int_{\beta_\ell^0 + \varepsilon\,\delta\beta_\ell}^{\beta_u^0 + \varepsilon\,\delta\beta_u} dt \left[\psi^{(1)} + \varepsilon \delta \psi^{(1)} \right] \frac{d}{dt} [\rho + \varepsilon \delta \rho] \right. \\ &\left. \times \delta\left(t - \beta_\ell^0 - \varepsilon \delta\beta_\ell\right) \right\}_{\varepsilon=0} = (\delta\beta_\ell) \int_{\beta_\ell^0}^{\beta_u^0} \left[\psi^{(1)}(t) \frac{d\rho(t)}{dt} \right] \delta'\left(t - \beta_\ell^0\right) dt \\ &+ (\delta\beta_\ell) \left[\psi^{(1)}(t) \frac{d\rho(t)}{dt} \right]_{\beta_\ell^0} - \int_{\beta_\ell^0}^{\beta_u^0} \left[\delta\psi^{(1)} \frac{d\rho}{dt} + \psi^{(1)} \frac{d(\delta\rho)}{dt} \right] \delta\left(t - \beta_\ell^0\right) dt. \end{aligned} \tag{3.444}$$

Using (3.342) and (3.322) in the first term on the right side of (3.444) yields the following result:

$$\int_{\beta_\ell^0}^{\beta_u^0} \left[\psi^{(1)}(t) \frac{d\rho(t)}{dt} \right] \delta'\left(t - \beta_\ell^0\right) dt = 0. \tag{3.445}$$

It is convenient to replace the quantity $d(\delta\rho)/dt$, which appears in the last term on the right side of (3.444), by using (3.329) which, together with the result obtained in (3.445), makes it possible to express the relation in (3.444) in the following form:

$$\delta\left(\frac{\partial R_1}{\partial \beta_\ell}\right) = \left\{\delta\left(\frac{\partial R_1}{\partial \beta_\ell}\right)\right\}_{dir} + \left\{\delta\left(\frac{\partial R_1}{\partial \beta_\ell}\right)\right\}_{ind}, \tag{3.446}$$

where

$$\left\{\delta\left(\frac{\partial R_1}{\partial \beta_\ell}\right)\right\}_{dir} = \rho\left(\beta_\ell^0\right)\psi^{(1)}\left(\beta_\ell^0\right)\sum_{k=1}^{N}\left[(\delta n_k)\sigma_k + n_k(\delta\sigma_k)\right], \tag{3.447}$$

and

$$\begin{aligned}\left\{\delta\left(\frac{\partial R_1}{\partial \beta_\ell}\right)\right\}_{ind} = \int_{\beta_\ell^0}^{\beta_u^0}\Bigg[&-\delta\psi^{(1)}(t)\frac{d\rho(t)}{dt} \\ &+\left(\sum_{k=1}^{N} n_k\sigma_k\right)\psi^{(1)}(t)\delta\rho(t)\Bigg]\delta\left(t-\beta_\ell^0\right)dt.\end{aligned} \tag{3.448}$$

The indirect-effect term defined in (3.448) will be expressed in terms of a square integrable two-component adjoint function $\boldsymbol{\psi}_5^{(2)}(t) \triangleq \left[\psi_{51}^{(2)}(t), \psi_{52}^{(2)}(t)\right] \in \mathrm{H}_2(\Omega_t)$ by first constructing the inner product of (3.357) with $\boldsymbol{\psi}_5^{(2)}(t)$ to obtain the following relation:

$$\begin{aligned}&\int_{\beta_\ell^0}^{\beta_u^0}\left[\psi_{51}^{(2)}(t), \psi_{52}^{(2)}(t)\right]\begin{pmatrix} -\dfrac{d}{dt} + \sum\limits_{k=1}^{N} n_k\sigma_k & 0 \\ 0 & \dfrac{d}{dt} + \sum\limits_{k=1}^{N} n_k\sigma_k \end{pmatrix}\begin{pmatrix}\delta\psi^{(1)}(t) \\ \delta\rho(t)\end{pmatrix}dt \\ &= \int_{\beta_\ell^0}^{\beta_u^0}\left[\psi_{51}^{(2)}(t), \psi_{52}^{(2)}(t)\right]\begin{pmatrix} -(\delta t_d)\delta'(t-t_d) - \psi^{(1)}(t)\sum\limits_{k=1}^{N}\left[(\delta n_k)\sigma_k + n_k(\delta\sigma_k)\right] \\ -\rho(t)\sum\limits_{k=1}^{N}\left[(\delta n_k)\sigma_k + n_k(\delta\sigma_k)\right]\end{pmatrix}dt.\end{aligned} \tag{3.449}$$

Integrating by parts the left side of (3.449) so as to transfer the differential operations on $\delta\psi^{(1)}(t)$ and $\delta\rho(t)$ to differential operations on $\psi_{51}^{(2)}(t)$ and $\psi_{52}^{(2)}(t)$ yields the following result:

$$\int_{\beta_\ell^0}^{\beta_u^0} \psi_{51}^{(2)}(t)\left[-\frac{d}{dt}\delta\psi^{(1)}(t)+\delta\psi^{(1)}(t)\sum_{k=1}^{N}n_k\sigma_k\right]dt+\int_{\beta_\ell^0}^{\beta_u^0}\psi_{52}^{(2)}(t)\left[\frac{d}{dt}\delta\rho(t)\right.$$

$$\left.+\delta\rho(t)\sum_{k=1}^{N}n_k\sigma_k\right]dt=-\psi_{51}^{(2)}\left(\beta_u^0\right)\delta\psi^{(1)}\left(\beta_u^0\right)+\psi_{51}^{(2)}\left(\beta_\ell^0\right)\delta\psi^{(1)}\left(\beta_\ell^0\right)$$

$$+\psi_{52}^{(2)}\left(\beta_u^0\right)\delta\rho\left(\beta_u^0\right)-\psi_{52}^{(2)}\left(\beta_\ell^0\right)\delta\rho\left(\beta_\ell^0\right) \tag{3.450}$$

$$+\int_{\beta_\ell^0}^{\beta_u^0}\delta\psi^{(1)}(t)\left[\frac{d\psi_{51}^{(2)}(t)}{dt}+\psi_{51}^{(2)}(t)\sum_{k=1}^{N}n_k\sigma_k\right]dt$$

$$+\int_{\beta_\ell^0}^{\beta_u^0}\delta\rho(t)\left[-\frac{d\psi_{52}^{(2)}(t)}{dt}+\psi_{52}^{(2)}(t)\sum_{k=1}^{N}n_k\sigma_k\right]dt.$$

The last two terms on the right side of (3.450) will represent the indirect-effect term defined in (3.448) by imposing the following relations:

$$\frac{d\psi_{51}^{(2)}(t)}{dt}+\psi_{51}^{(2)}(t)\sum_{k=1}^{N}n_k\sigma_k=-\left(\frac{d\rho}{dt}\right)\delta\left(t-\beta_\ell^0\right), \tag{3.451}$$

$$-\frac{d\psi_{52}^{(2)}(t)}{dt}+\psi_{52}^{(2)}(t)\sum_{k=1}^{N}n_k\sigma_k=\left(\sum_{k=1}^{N}n_k\sigma_k\right)\psi^{(1)}(t)\delta\left(t-\beta_\ell^0\right). \tag{3.452}$$

The boundary conditions for (3.451) and (3.452) are established by requiring that the contributions involving the unknown quantities $\delta\psi^{(1)}\left(\beta_\ell^0\right)$ and $\delta\rho\left(\beta_u^0\right)$ in (3.450) vanish, which can be accomplished by imposing the following conditions:

$$\psi_{51}^{(2)}\left(\beta_\ell^0\right)=0,\quad \psi_{52}^{(2)}\left(\beta_u^0\right)=0. \tag{3.453}$$

Solving (3.451)–(3.453) yields the following expressions for the components of the second-level adjoint function $\boldsymbol{\psi}_5^{(2)}(t)\triangleq\left[\psi_{51}^{(2)}(t),\psi_{52}^{(2)}(t)\right]$, to be evaluated at the nominal parameter values:

$$\psi_{51}^{(2)}(t)=\rho_{in}H(t-\beta_\ell)\sum_{k=1}^{N}n_k\sigma_k\exp\left[(\beta_\ell-t)\sum_{k=1}^{N}n_k\sigma_k\right], \tag{3.454}$$

$$\psi_{52}^{(2)}(t) = \left(\sum_{k=1}^{N} n_k \sigma_k\right)\left[1 - H\left(t - \beta_\ell^0\right)\right] \exp\left[\left(t - t_d^0\right)\sum_{k=1}^{N} n_k \sigma_k\right]. \tag{3.455}$$

Using (3.449)–(3.453) and (3.330) in (3.448) yields the following expression for the indirect-effect term defined in (3.448), for $i = 1, \ldots, N$:

$$\left\{\delta\left(\frac{\partial R_1}{\partial \beta_\ell}\right)\right\}_{ind} = \psi_{52}^{(2)}(\beta_\ell)\left\{\delta\rho_{in} - \delta\beta_\ell\left[\frac{d\rho(t)}{dt}\right]_{t=\beta_\ell^0}\right\}$$
$$+ \int_{\beta_\ell^0}^{\beta_u^0} \left[\psi_{51}^{(2)}(t), \psi_{52}^{(2)}(t)\right] \begin{pmatrix} -(\delta t_d)\delta'(t - t_d) - \psi^{(1)}(t)\sum_{k=1}^{N}\left[(\delta n_k)\sigma_k + n_k(\delta\sigma_k)\right] \\ -\rho(t)\sum_{k=1}^{N}\left[(\delta n_k)\sigma_k + n_k(\delta\sigma_k)\right] \end{pmatrix} dt. \tag{3.456}$$

Adding the direct-effect term defined in (3.447) to (3.456) and identifying in the resulting expression the coefficients multiplying the variations $\delta\sigma_i$, δn_i, $\delta\rho_{in}$, δt_d, $\delta\beta_\ell$, and $\delta\beta_u$ yields the following expressions for the respective second-order sensitivities of the response $R_1(\rho; \boldsymbol{\alpha})$:

$$\frac{\partial^2 R_1}{\partial \sigma_j \partial \beta_\ell} = -n_j \int_{\beta_\ell^0}^{\beta_u^0} \psi_{51}^{(2)}(t)\psi^{(1)}(t)dt$$
$$-n_j \int_{\beta_\ell^0}^{\beta_u^0} \psi_{52}^{(2)}(t)\rho(t)\,dt \; + n_j\rho\left(\beta_\ell^0\right)\psi^{(1)}\left(\beta_\ell^0\right); j = 1, \ldots, N; \tag{3.457}$$

$$\frac{\partial^2 R_1}{\partial n_j \partial \beta_\ell} = -\sigma_j \int_{\beta_\ell^0}^{\beta_u^0} \psi_{51}^{(2)}(t)\psi^{(1)}(t)dt$$
$$-\sigma_j \int_{\beta_\ell^0}^{\beta_u^0} \psi_{52}^{(2)}(t)\rho(t)\,dt + \sigma_j\rho\left(\beta_\ell^0\right)\psi^{(1)}\left(\beta_\ell^0\right); j = 1, \ldots, N; \tag{3.458}$$

$$\frac{\partial^2 R_1}{\partial \rho_{in} \partial \beta_\ell} = \psi_{52}^{(2)}(\beta_\ell); \tag{3.459}$$

$$\frac{\partial^2 R_1}{\partial t_d \, \partial \beta_\ell} = -\int_{\beta_\ell^0}^{\beta_u^0} \psi_{51}^{(2)}(t)\,\delta'(t - t_d) = \left[\frac{d\psi_{51}^{(2)}(t)}{dt}\right]_{t=t_d}; \tag{3.460}$$

$$\frac{\partial^2 R_1}{\partial \beta_u \, \partial \beta_\ell} = 0; \tag{3.461}$$

$$\frac{\partial^2 R_1}{\partial \beta_\ell \, \partial \beta_\ell} = \left\{\left[-\psi_{52}^{(2)}(t)\right]\frac{d\rho(t)}{dt}\right\}_{t=\beta_\ell^0}. \tag{3.462}$$

Inserting the expressions obtained in (3.454) and (3.455) for the components of the second-level adjoint function $\boldsymbol{\psi}_5^{(2)}(t) \triangleq \left[\psi_{51}^{(2)}(t), \psi_{52}^{(2)}(t)\right]$ into (3.457)–(3.462) yields the following expressions, where all quantities are to be evaluated at the nominal parameter values:

$$\frac{\partial^2 R_1}{\partial \sigma_j \, \partial \beta_\ell} = n_j \rho_{in}\left[1 + (\beta_\ell - t_d)\sum_{k=1}^{N} n_k \sigma_k\right] \exp\left[(\beta_\ell - t_d)\sum_{k=1}^{N} n_k \sigma_k\right]; j = 1, \ldots, N; \tag{3.463}$$

$$\frac{\partial^2 R_1}{\partial n_j \, \partial \beta_\ell} = \sigma_j \rho_{in}\left[1 + (\beta_\ell - t_d)\sum_{k=1}^{N} n_k \sigma_k\right] \exp\left[(\beta_\ell - t_d)\sum_{k=1}^{N} n_k \sigma_k\right]; j = 1, \ldots, N; \tag{3.464}$$

$$\frac{\partial^2 R_1}{\partial \rho_{in} \, \partial \beta_\ell} = \left(\sum_{k=1}^{N} n_k \sigma_k\right) \exp\left[(\beta_\ell - t_d)\sum_{k=1}^{N} n_k \sigma_k\right]; \tag{3.465}$$

$$\frac{\partial^2 R_1}{\partial t_d \, \partial \beta_\ell} = -\rho_{in}\left(\sum_{k=1}^{N} n_k \sigma_k\right)^2 \exp\left[(\beta_\ell - t_d)\sum_{k=1}^{N} n_k \sigma_k\right]; \tag{3.466}$$

$$\frac{\partial^2 R_1}{\partial \beta_u \, \partial \beta_\ell} = 0; \tag{3.467}$$

$$\frac{\partial^2 R_1}{\partial \beta_\ell \, \partial \beta_\ell} = \rho_{in}\left(\sum_{k=1}^{N} n_k \sigma_k\right)^2 \exp\left[(\beta_\ell - t_d)\sum_{k=1}^{N} n_k \sigma_k\right]. \tag{3.468}$$

3.6.2.6 Independent Mutual Verification of Adjoint Sensitivity Functions

Due to the symmetry of the mixed second-order sensitivities, the following identities hold:

1. The adjoint computations are considered "large-scale computations" because they require, in practice, the inversion of the same type of operators (differential, integral, matrices, etc.) as occurring in the original forward model. The computation of the first-order sensitivities requires a single adjoint computation, regardless of the number of model parameters. The total number of adjoint computations is at most $TP = 2N + 4$ (equal to the total number of model parameters) or less, depending on the model response under consideration. For the response $R_1(\rho; \boldsymbol{\alpha})$, for example, to total number of adjoint computations needed for computing the second-order sensitivities is $2N + 3$ (since the response does not depend on the parameter β_u), comprising the sum of the following computations: (1) N adjoint computations to determine $\boldsymbol{\psi}_1^{(2)}(i;t)$; (2) N adjoint computations to determine $\boldsymbol{\psi}_2^{(2)}(i;t)$; (3) one adjoint computation to determine $\boldsymbol{\psi}_3^{(2)}(t)$; (4) one adjoint computation to determine $\boldsymbol{\psi}_4^{(2)}(t)$; and (5) one adjoint computation to determine $\boldsymbol{\psi}_5^{(2)}(t)$.
2. The unmixed second-order sensitivities will be computed once. The mixed second-order sensitivities will be computed twice, with two distinct adjoint functions, as detailed in items (3)–(12), below.
3. The expression provided in (3.393) must be identical to the expression provided in (3.365). This identity provides an independent mutual verification of the accuracy of the computations of the second-level adjoint functions $\boldsymbol{\psi}_2^{(2)}(i;t) \triangleq \left[\psi_{21}^{(2)}(i;t), \psi_{22}^{(2)}(i;t)\right]$ and $\boldsymbol{\psi}_1^{(2)}(i;t) \triangleq \left[\psi_{11}^{(2)}(i;t), \psi_{12}^{(2)}(i;t)\right]$.
4. The expression provided in (3.413) must be identical to the expression provided in (3.366). This identity provides an independent mutual verification of the accuracy of the computations of the second-level adjoint functions $\boldsymbol{\psi}_3^{(2)}(t) \triangleq \left[\psi_{31}^{(2)}(t), \psi_{32}^{(2)}(t)\right]$ and $\boldsymbol{\psi}_1^{(2)}(i;t) \triangleq \left[\psi_{11}^{(2)}(i;t), \psi_{12}^{(2)}(i;t)\right]$.
5. The expression provided in (3.414) must be identical to the expression provided in (3.395). This identity provides an independent mutual verification of the accuracy of the computations of the second-level adjoint functions $\boldsymbol{\psi}_3^{(2)}(t) \triangleq \left[\psi_{31}^{(2)}(t), \psi_{32}^{(2)}(t)\right]$ and $\boldsymbol{\psi}_2^{(2)}(i;t) \triangleq \left[\psi_{21}^{(2)}(i;t), \psi_{22}^{(2)}(i;t)\right]$.
6. The expression provided in (3.431) must be identical to the expression provided in (3.367). This identity provides an independent mutual verification of the accuracy of the computations of the second-level adjoint functions $\boldsymbol{\psi}_4^{(2)}(t) \triangleq \left[\psi_{41}^{(2)}(t), \psi_{42}^{(2)}(t)\right]$ and $\boldsymbol{\psi}_1^{(2)}(i;t) \triangleq \left[\psi_{11}^{(2)}(i;t), \psi_{12}^{(2)}(i;t)\right]$.
7. The expression provided in (3.432) must be identical to the expression provided in (3.396). This identity provides an independent mutual verification of the

accuracy of the computations of the second-level adjoint functions $\boldsymbol{\psi}_4^{(2)}(t) \triangleq \left[\psi_{41}^{(2)}(t), \psi_{42}^{(2)}(t)\right]$ and $\boldsymbol{\psi}_2^{(2)}(i;t) \triangleq \left[\psi_{21}^{(2)}(i;t), \psi_{22}^{(2)}(i;t)\right]$.

8. The expression provided in (3.433) must be identical to the expression provided in (3.415). This identity provides an independent mutual verification of the accuracy of the computations of the second-level adjoint functions $\boldsymbol{\psi}_4^{(2)}(t) \triangleq \left[\psi_{41}^{(2)}(t), \psi_{42}^{(2)}(t)\right]$ and $\boldsymbol{\psi}_3^{(2)}(t) \triangleq \left[\psi_{31}^{(2)}(t), \psi_{32}^{(2)}(t)\right]$.
9. The expression provided in (3.457) must be identical to the expression provided in (3.369). This identity provides an independent mutual verification of the accuracy of the computations of the second-level adjoint functions $\boldsymbol{\psi}_5^{(2)}(t) \triangleq \left[\psi_{51}^{(2)}(t), \psi_{52}^{(2)}(t)\right]$ and $\boldsymbol{\psi}_1^{(2)}(i;t) \triangleq \left[\psi_{11}^{(2)}(i;t), \psi_{12}^{(2)}(i;t)\right]$.
10. The expression provided in (3.458) must be identical to the expression provided in (3.398). This identity provides an independent mutual verification of the accuracy of the computations of the second-level adjoint functions $\boldsymbol{\psi}_5^{(2)}(t) \triangleq \left[\psi_{51}^{(2)}(t), \psi_{52}^{(2)}(t)\right]$ and $\boldsymbol{\psi}_2^{(2)}(i;t) \triangleq \left[\psi_{21}^{(2)}(i;t), \psi_{22}^{(2)}(i;t)\right]$.
11. The expression provided in (3.459) must be identical to the expression provided in (3.417). This identity provides an independent mutual verification of the accuracy of the computations of the adjoint functions $\boldsymbol{\psi}_5^{(2)}(t) \triangleq \left[\psi_{51}^{(2)}(t), \psi_{52}^{(2)}(t)\right]$ and $\psi^{(1)}(t)$.
12. The expression provided in (3.460) must be identical to the expression provided in (3.436). This identity provides an independent mutual verification of the accuracy of the computations of the second-level adjoint functions $\boldsymbol{\psi}_5^{(2)}(t) \triangleq \left[\psi_{51}^{(2)}(t), \psi_{52}^{(2)}(t)\right]$ and $\boldsymbol{\psi}_4^{(2)}(t) \triangleq \left[\psi_{41}^{(2)}(t), \psi_{42}^{(2)}(t)\right]$.

The point-detector response $R_1(\rho; \boldsymbol{\alpha})$, which simulates a measurement at a given instance in time, turned out to be independent of the imprecisely known final-time point β_u, except when the measurement response is taken at the uncertain final-time (i.e., when the nominal values of the quantities t_d and β_u coincide). In this case, the expressions of the first- and second-order response sensitivities derived in this work remain valid, but with the stipulation that t_d is replaced by β_u.

3.6.2.7 Aggregating Model Parameters to Reduce the Number of Large-Scale Adjoint Computations for Determining the Second-Order Sensitivities

As has been discussed in Sect. 3.2.1, the structure of the original model often makes it possible to aggregate the "fundamental" model parameters into "derived" parameters, which enables significant reductions in the number of adjoint computations needed for determining the second-level (and, subsequently, the higher-level) adjoint sensitivity functions. Such an aggregation is also possible for the evolution model considered in (this) Sect. 3.6, by considering that the sum $\sum_{i=1}^{N} n_i \sigma_i$ of the products of atomic number densities n_i and cross sections σ_i (which are "fundamental" parameters) is a derived "primary" model parameter, namely an "interaction coefficient" denoted as $\mu(\mathbf{n}, \boldsymbol{\sigma})$ and defined as follows:

$$\mu(\mathbf{n}, \boldsymbol{\sigma}) \triangleq \sum_{i=1}^{N} n_i \sigma_i; \quad \mathbf{n} \triangleq [n_1, \ldots, n_N]^\dagger; \quad \boldsymbol{\sigma} \triangleq [\sigma_1, \ldots, \sigma_N]^\dagger. \tag{3.469}$$

In view of (3.469), the evolution model described by (3.317) and (3.318) takes on the following form:

$$\frac{d\rho(t)}{dt} + \rho(t)\mu(\mathbf{n}, \boldsymbol{\sigma}) = 0, \quad 0 \le \beta_\ell \le t \le \beta_u < \infty, \tag{3.470}$$

$$\rho(\beta_\ell) = \rho_{in}, \quad at\ t = \beta_\ell. \tag{3.471}$$

The model comprising (3.470) and (3.471) will be called the "aggregated parameter evolution model" in order to distinguish it from the original model comprising (3.317) and (3.318).

Using the "aggregated parameter evolution model," the total first-order sensitivity of the response $R_1(\rho; \boldsymbol{\alpha})$ defined in (3.319) is determined by implementing the same steps as were implemented in Sect. 3.6.1. The indirect-effect and direct-effect terms remain as provided in (3.327) and (3.328), respectively. However, the function $\delta\rho(t)$, which appears in (3.327) will be the solution of the following First-Level Variational Sensitivity System (1st-LVSS):

$$\left\{ \frac{d[\delta\rho(t)]}{dt} + \delta\rho(t)\mu(\mathbf{n}, \boldsymbol{\sigma}) \right\}_{\boldsymbol{\alpha}^0} = -\delta\mu(\mathbf{n}, \boldsymbol{\sigma}) \{\rho(t)\}_{\boldsymbol{\alpha}^0}, 0 \le \beta_\ell^0 \le t \le \beta_u^0 < \infty, \tag{3.472}$$

$$\delta\rho\left(\beta_\ell^0\right) + \delta\beta_\ell \left\{ \frac{d\rho(t)}{dt} \right\}_{t=\beta_\ell^0} = \delta\rho_{in}, \quad at\ t = \beta_\ell^0. \tag{3.473}$$

Of course, the 1st-LVSS comprising (3.472) and (3.473) is equivalent to the 1st-LVSS given by (3.329) and (3.330).

Using the inner product provided in (3.331) and following the same steps as implemented in Sect. 3.6.1 leads to the following expression of the indirect-effect term $(\delta R_1)_{ind}$, which is equivalent to (3.333):

$$(\delta R_1)_{ind} = -\delta\mu(\mathbf{n},\boldsymbol{\sigma}) \int_{\beta_\ell^0}^{\beta_u^0} \psi_a^{(1)}(t)\rho^0(t)\,dt + \psi_a^{(1)}\left(\beta_\ell^0\right)\left[\delta\rho_{in} - \delta\beta_\ell\left\{\frac{d\rho(t)}{dt}\right\}_{t=\beta_\ell^0}\right], \quad (3.474)$$

where the "alternative" first-level adjoint function $\psi_a^{(1)}(t)$ is the solution of the following First-Level Adjoint Sensitivity System (1st-LASS):

$$\left\{-\frac{d\psi_a^{(1)}(t)}{dt} + \mu(\mathbf{n},\boldsymbol{\sigma})\psi_a^{(1)}(t)\right\}_{\boldsymbol{\alpha}^0} = \delta\left(t - t_d^0\right),\ 0 \le \beta_\ell^0 \le t \le \beta_u^0 < \infty, \quad (3.475)$$

$$\psi_a^{(1)}\left(\beta_u^0\right) = 0. \quad (3.476)$$

As would be expected, the 1st-LASS comprising (3.475) and (3.476) is equivalent to the 1st-LASS comprising (3.334) and (3.335), so the "alternative" first-level adjoint function $\psi_a^{(1)}(t)$ is expected to have the same expression as the function $\psi^{(1)}(t)$, but written in terms of the "aggregated" parameter $\mu(\mathbf{n},\boldsymbol{\sigma})$ rather than in terms of the fundamental parameters n_i and σ_i.

Adding the expression of the direct-effect term, $(\delta R_1)_{dir}$, defined in (3.328), to the expression of the indirect-effect term, $(\delta R_1)_{ind}$, provided in (3.474), and identifying in the resulting expression the quantities that multiply the respective parameter variations yields the following expressions for the partial sensitivities of $R_1(\rho;\boldsymbol{\alpha})$ in terms of the alternative first-level adjoint function $\psi_a^{(1)}(t)$:

$$\left(\frac{\partial R_1}{\partial \mu(\mathbf{n},\boldsymbol{\sigma})}\right)_{\boldsymbol{\alpha}^0} = -\int_{\beta_\ell^0}^{\beta_u^0} \psi_a^{(1)}(t)\rho^0(t)dt; \quad (3.477)$$

$$\left(\frac{\partial R_1}{\partial \rho_{in}}\right)_{\boldsymbol{\alpha}^0} = \psi_a^{(1)}\left(\beta_\ell^0\right); \quad (3.478)$$

$$\left(\frac{\partial R_1}{\partial \beta_\ell}\right)_{\boldsymbol{\alpha}^0} = -\psi_a^{(1)}\left(\beta_\ell^0\right)\left\{\frac{d\rho(t)}{dt}\right\}_{t=\beta_\ell^0}; \quad (3.479)$$

$$\frac{\partial R_1}{\partial t_d} = -\int_{\beta_\ell^0}^{\beta_u^0} \rho^0(t)\,\delta'\left(t - t_d^0\right)dt = \left\{\frac{d\rho(t)}{dt}\right\}_{t=t_d^0}; \quad (3.480)$$

$$\frac{\partial R_1}{\partial \beta_u} \equiv 0. \quad (3.481)$$

The sensitivities of $R_1(\rho;\boldsymbol{\alpha})$ with respect to the "fundamental" parameters n_i and σ_i, respectively, are obtained by using the expression given in (3.469) together with the "chain rule" on the expression obtained in (3.477), which yields the following expressions:

$$\left\{\frac{\partial R_1}{\partial \sigma_i}\right\}_{\boldsymbol{\alpha}^0} = \left\{\frac{\partial R_1}{\partial \mu(\mathbf{n},\boldsymbol{\sigma})}\frac{\partial \mu(\mathbf{n},\boldsymbol{\sigma})}{\partial \sigma_i}\right\}_{\boldsymbol{\alpha}^0} = -\left\{n_i\int_{\beta_\ell^0}^{\beta_u^0}\psi_a^{(1)}(t)\rho(t)dt\right\}_{\boldsymbol{\alpha}^0};i=1,\ldots,N; \tag{3.482}$$

$$\left\{\frac{\partial R_1}{\partial n_i}\right\}_{\boldsymbol{\alpha}^0} = \left\{\frac{\partial R_1}{\partial \mu(\mathbf{n},\boldsymbol{\sigma})}\frac{\partial \mu(\mathbf{n},\boldsymbol{\sigma})}{\partial n_i}\right\}_{\boldsymbol{\alpha}^0} = -\left\{\sigma_i\int_{\beta_\ell^0}^{\beta_u^0}\psi_a^{(1)}(t)\rho(t)dt\right\}_{\boldsymbol{\alpha}^0};\ i=1,\ldots,N. \tag{3.483}$$

Solving (3.475) and (3.476) yields the following expression for the "alternative" first-level adjoint function $\psi_a^{(1)}(t)$:

$$\psi_a^{(1)}(t) = \{[1-H(t-t_d)]\exp[(t-t_d)\mu(\mathbf{n},\boldsymbol{\sigma})]\}_{\boldsymbol{\alpha}^0} = \psi^{(1)}(t), \tag{3.484}$$

which coincides, as expected, with the expression obtained in (3.342). Evidently, the expressions obtained in (3.478)–(3.483) coincide with the corresponding expressions obtained in (3.336)–(3.341).

It is evident that aggregating model parameters does not affect the number of large-scale adjoint computations: *only a single adjoint computation is needed to determine the first-level adjoint function* (which may be vector-valued if the original forward state function is vector-valued), *regardless of the number of model parameters.*

For the computation of the second-order response sensitivities, however, it is necessary to compute as many second-level adjoint functions as there are first-order sensitivities, so the number of large-scale adjoint computations is evidently influenced by the number of first-order sensitivities under consideration. Consequently, it is expected that aggregating model parameters in order to reduce the number of *primary parameters* will correspondingly reduce the number of second-level adjoint functions to be computed. This expectation is justified, as can be seen by considering the computation of the second-order sensitivities corresponding to $\partial R_1/\partial\mu(\mathbf{n},\boldsymbol{\sigma})$ using the expression given in (3.477). These second-order sensitivities are determined by taking the G-differential of (3.477), to obtain the following expression, which contains only an "indirect-effect" term:

$$\delta\left(\frac{\partial R_1}{\partial \mu(\boldsymbol{\alpha})}\right) = -\int_{\beta_\ell^0}^{\beta_u^0} \left\{\delta\psi_a^{(1)}(t)\rho(t) + \psi_a^{(1)}(t)\delta\rho(t)\right\}_{\boldsymbol{\alpha}^0} dt. \tag{3.485}$$

The function $\delta\rho(t)$ is the solution of (3.472) and (3.473) while the function $\delta\psi_a^{(1)}(t)$ is the solution of the system obtained by G-differentiating (3.475) and (3.476), which yields:

$$\begin{aligned}&\left\{-\frac{d\left[\delta\psi_a^{(1)}(t)\right]}{dt} + \mu(\mathbf{n},\boldsymbol{\sigma})\delta\psi_a^{(1)}(t)\right\}_{\boldsymbol{\alpha}^0} \\ &= -\left\{(\delta t_d)\delta'(t-t_d) + \delta\mu(\mathbf{n},\boldsymbol{\sigma})\psi_a^{(1)}(t)\right\}_{\boldsymbol{\alpha}^0},\end{aligned} \tag{3.486}$$

$$\delta\psi_a^{(1)}\left(\beta_u^0\right) = 0. \tag{3.487}$$

Concatenating (3.472) and (3.486) yields the following 2nd-LVSS:

$$\begin{aligned}&\begin{pmatrix} -\frac{d}{dt} + \mu(\mathbf{n},\boldsymbol{\sigma}) & 0 \\ 0 & \frac{d}{dt} + \mu(\mathbf{n},\boldsymbol{\sigma}) \end{pmatrix}\begin{pmatrix}\delta\psi_a^{(1)}(t) \\ \delta\rho(t)\end{pmatrix} \\ &= \begin{pmatrix} -(\delta t_d)\delta'(t-t_d) - \delta\mu(\mathbf{n},\boldsymbol{\sigma})\psi_a^{(1)}(t) \\ -\delta\mu(\mathbf{n},\boldsymbol{\sigma})\rho(t)\end{pmatrix}.\end{aligned} \tag{3.488}$$

Using the definition given in (3.356), construct the inner product of (3.357) with a square integrable two-component function $\boldsymbol{\psi}_a^{(2)}(t) \triangleq \left[\psi_{a1}^{(2)}(t), \psi_{a2}^{(2)}(t)\right] \in \mathrm{H}_2(\Omega_t)$ to obtain the following relation:

$$\begin{aligned}&\int_{\beta_\ell^0}^{\beta_u^0} \left[\psi_{a1}^{(2)}(t), \psi_{a2}^{(2)}(t)\right]\begin{pmatrix} -\frac{d}{dt} + \mu(\mathbf{n},\boldsymbol{\sigma}) & 0 \\ 0 & \frac{d}{dt} + \mu(\mathbf{n},\boldsymbol{\sigma}) \end{pmatrix}\begin{pmatrix}\delta\psi_a^{(1)}(t) \\ \delta\rho(t)\end{pmatrix} dt \\ &= \int_{\beta_\ell^0}^{\beta_u^0} \left[\psi_{a1}^{(2)}(t), \psi_{a2}^{(2)}(t)\right]\begin{pmatrix} -(\delta t_d)\delta'(t-t_d) - \delta\mu(\mathbf{n},\boldsymbol{\sigma})\psi_a^{(1)}(t) \\ -\delta\mu(\mathbf{n},\boldsymbol{\sigma})\rho(t)\end{pmatrix} dt.\end{aligned} \tag{3.489}$$

Integrating by parts the left side of (3.489) so as to transfer the differential operations on $\delta\psi_a^{(1)}(t)$ and $\delta\rho(t)$ to differential operations on $\psi_{a1}^{(2)}(t)$ and $\psi_{a2}^{(2)}(t)$ yields the following relation:

$$
\begin{aligned}
&\int_{\beta_\ell^0}^{\beta_u^0} \psi_{a1}^{(2)}(t)\left[-\frac{d}{dt}\delta\psi_a^{(1)}(t)+\delta\psi_a^{(1)}(t)\mu(\mathbf{n},\boldsymbol{\sigma})\right]dt+\int_{\beta_\ell^0}^{\beta_u^0} \psi_{a2}^{(2)}(t)\left[\frac{d}{dt}\delta\rho(t)\right.\\
&\left.+\delta\rho(t)\mu(\mathbf{n},\boldsymbol{\sigma})\right]dt=-\psi_{a1}^{(2)}\left(\beta_u^0\right)\delta\psi_a^{(1)}\left(\beta_u^0\right)+\psi_{a1}^{(2)}\left(\beta_\ell^0\right)\delta\psi_a^{(1)}\left(\beta_\ell^0\right)\\
&+\psi_{a2}^{(2)}\left(\beta_u^0\right)\delta\rho\left(\beta_u^0\right)-\psi_{a2}^{(2)}\left(\beta_\ell^0\right)\delta\rho\left(\beta_\ell^0\right)+\int_{\beta_\ell^0}^{\beta_u^0}\delta\psi_a^{(1)}(t)\left[\frac{d\psi_{a1}^{(2)}(t)}{dt}\right.\\
&\left.+\psi_{a1}^{(2)}(t)\mu(\mathbf{n},\boldsymbol{\sigma})\right]dt+\int_{\beta_\ell^0}^{\beta_u^0}\delta\rho(t)\left[-\frac{d\psi_{a2}^{(2)}(t)}{dt}+\psi_{a2}^{(2)}(t)\mu(\mathbf{n},\boldsymbol{\sigma})\right]dt.
\end{aligned}
\tag{3.490}
$$

The last two terms on the right side of (3.490) will represent the G-differential defined in (3.485) by requiring that

$$
\frac{d\psi_{a1}^{(2)}(t)}{dt}+\psi_{a1}^{(2)}(t)\mu(\mathbf{n},\boldsymbol{\sigma})=-\rho(t), \tag{3.491}
$$

$$
-\frac{d\psi_{a2}^{(2)}(t)}{dt}+\psi_{a2}^{(2)}(t)\mu(\mathbf{n},\boldsymbol{\sigma})=-\psi_a^{(1)}(t). \tag{3.492}
$$

The boundary conditions for (3.491) and (3.492) are established by requiring that the contributions involving the unknown quantities $\delta\psi_a^{(1)}\left(\beta_\ell^0\right)$ and $\delta\rho\left(\beta_u^0\right)$ in (3.490) vanish, which is accomplished by imposing the following conditions:

$$
\psi_{a1}^{(2)}\left(\beta_\ell^0\right)=0, \quad \psi_{a2}^{(2)}\left(\beta_u^0\right)=0. \tag{3.493}
$$

The system of equations comprising (3.491)–(3.493) constitutes the Second-Level Adjoint Sensitivity System (2nd-LASS) for the two-component second-level adjoint function $\boldsymbol{\psi}_a^{(2)}(t)\triangleq\left[\psi_{a1}^{(2)}(t),\psi_{a2}^{(2)}(t)\right]\in \mathrm{H}_2(\Omega_t)$. It is important to note that the 2nd-LASS is independent not only of parameter variations but also of any additional indices, in contradistinction with the "un-aggregated" second-level adjoint function $\boldsymbol{\psi}_1^{(2)}(i;t)\triangleq\left[\psi_{11}^{(2)}(i;t),\psi_{12}^{(2)}(i;t)\right]$, which depends on the index $i = 1, \ldots, N$.

Replacing the left side of (3.490) by the right side of (3.489) and taking into account (3.330), (3.487), and (3.491)–(3.493) yields the following expression for the G-differential defined in (3.485):

$$\delta\left(\frac{\partial R_1}{\partial \mu(\mathbf{n},\boldsymbol{\sigma})}\right) = -\int_{\beta_\ell^0}^{\beta_u^0} \psi_{a1}^{(2)}(t)\left[(\delta t_d)\delta'(t-t_d) + \delta\mu(\mathbf{n},\boldsymbol{\sigma})\psi_a^{(1)}(t)\right]dt$$
$$-\delta\mu(\mathbf{n},\boldsymbol{\sigma})\int_{\beta_\ell^0}^{\beta_u^0} \psi_{a2}^{(2)}(t)\rho(t)dt + \psi_{a2}^{(2)}\left(\beta_\ell^0\right)\left\{\delta\rho_{in} - \delta\beta_\ell\left[\frac{d\rho(t)}{dt}\right]_{t=\beta_\ell^0}\right\}. \tag{3.494}$$

Identifying in the resulting expression the coefficients multiplying the variations $\delta\mu(\boldsymbol{\alpha})$, $\delta\rho_{in}$, δt_d, $\delta\beta_\ell$, and $\delta\beta_u$ yields the following expression for the respective second-order sensitivities of the response $R_1(\rho;\boldsymbol{\alpha})$:

$$\frac{\partial^2 R_1}{\partial\mu(\mathbf{n},\boldsymbol{\sigma})\partial\mu(\mathbf{n},\boldsymbol{\sigma})} = -\int_{\beta_\ell^0}^{\beta_u^0}\left[\psi_{a1}^{(2)}(t)\psi_a^{(1)}(t) + \psi_{a2}^{(2)}(t)\rho(t)\right]dt; \tag{3.495}$$

$$\frac{\partial^2 R_1}{\partial\rho_{in}\,\partial\mu(\mathbf{n},\boldsymbol{\sigma})} = \psi_{a2}^{(2)}(\beta_\ell); \tag{3.496}$$

$$\frac{\partial^2 R_1}{\partial t_d\,\partial\mu(\mathbf{n},\boldsymbol{\sigma})} = -\int_{\beta_\ell^0}^{\beta_u^0}\psi_{a1}^{(2)}(t)\,\delta'(t-t_d) = \left[\frac{d\psi_{a1}^{(2)}(t)}{dt}\right]_{t=t_d}; \tag{3.497}$$

$$\frac{\partial^2 R_1}{\partial\beta_\ell\,\partial\mu(\mathbf{n},\boldsymbol{\sigma})} = -\psi_{a2}^{(2)}(\beta_\ell)\left[\frac{d\rho(t)}{dt}\right]_{t=\beta_\ell^0}; \tag{3.498}$$

$$\frac{\partial^2 R_1}{\partial\beta_u\,\partial\mu(\mathbf{n},\boldsymbol{\sigma})} = 0. \tag{3.499}$$

The second-order sensitivities shown in (3.495)–(3.498) can be computed after having determined the second-level adjoint function $\boldsymbol{\psi}_a^{(2)}(t)\triangleq\left[\psi_{a1}^{(2)}(t),\psi_{a2}^{(2)}(t)\right]$ by solving the 2nd-LASS, which comprises (3.491)–(3.493), using the nominal parameter values (the superscript "zero," which indicates "nominal values," has been omitted, for simplicity). Evidently, the determination of the "aggregate" second-level adjoint function $\boldsymbol{\psi}_a^{(2)}(t)\triangleq\left[\psi_{a1}^{(2)}(t),\psi_{a2}^{(2)}(t)\right]$ is obtained after a single "large-scale" computations required for solving (3.495)–(3.498).

The second-order sensitivities of the response $R_1(\rho;\boldsymbol{\alpha})$ with respect to the fundamental model parameters n_i and σ_i can be obtained by using the definition of $\mu(\boldsymbol{\alpha})$ provided in (3.469) to obtain relations similar to the relations (3.204)–(3.207) in Sect. 3.4.1, namely:

$$\left\{ \frac{\partial^2 R_1(\rho; \boldsymbol{\alpha})}{\partial n_i \partial n_i} \right\}_{\boldsymbol{\alpha}^0} = \left\{ (\sigma_i)^2 \frac{\partial^2 R_1(\rho; \boldsymbol{\alpha})}{\partial \mu(\mathbf{n}, \boldsymbol{\sigma}) \partial \mu(\mathbf{n}, \boldsymbol{\sigma})} \right\}_{\boldsymbol{\alpha}^0} ; i = 1, \ldots, N; \tag{3.500}$$

$$\left\{ \frac{\partial^2 R_1(\rho; \boldsymbol{\alpha})}{\partial n_i \partial \sigma_i} \right\}_{\boldsymbol{\alpha}^0} = \left\{ \sigma_i N_i \frac{\partial^2 R_1(\rho; \boldsymbol{\alpha})}{\partial \mu(\mathbf{n}, \boldsymbol{\sigma}) \partial \mu(\mathbf{n}, \boldsymbol{\sigma})} \right\}_{\boldsymbol{\alpha}^0} ; i = 1, \ldots, N; \tag{3.501}$$

$$\left\{ \frac{\partial^2 R_1(\rho; \boldsymbol{\alpha})}{\partial \sigma_i \partial \sigma_i} \right\}_{\boldsymbol{\alpha}^0} = \left\{ (n_i)^2 \frac{\partial^2 R_1(\rho; \boldsymbol{\alpha})}{\partial \mu(\mathbf{n}, \boldsymbol{\sigma}) \partial \mu(\mathbf{n}, \boldsymbol{\sigma})} \right\}_{\boldsymbol{\alpha}^0} ; i = 1, \ldots, N. \tag{3.502}$$

It is evident from (3.500)–(3.502) that, after having obtained the second-order sensitivities $\partial^2 R_1(\rho; \boldsymbol{\alpha})/\partial \mu(\mathbf{n}, \boldsymbol{\sigma}) \partial \mu(\mathbf{n}, \boldsymbol{\sigma})$ using (3.495), the second-order sensitivities with respect to the fundamental model parameters n_i and σ_i are obtained practically free of additional computational costs.

In summary, using the "aggregated parameter" $\mu(\mathbf{n}, \boldsymbol{\sigma})$ has reduced the number of large-scale (adjoint) computations from $2N + 3$ (when using the primary un-aggregated parameters) to 4 large-scale adjoint computations, as follows:

1. One adjoint computation for determining the second-order sensitivities corresponding to $\partial R_1/\partial \mu(\mathbf{n}, \boldsymbol{\sigma})$.
2. One adjoint computation for determining the second-order sensitivities corresponding to $\partial R_1/\partial \rho_{in}$, which is expressed by (3.478).
3. One adjoint computation for determining the second-order sensitivities corresponding to $\partial R_1/\partial \beta_\ell$, which is expressed by (3.479).
4. One adjoint computation for determining the second-order sensitivities corresponding to $\partial R_1/\partial t_d$, which is expressed by (3.480).

The computational advantages of using "aggregated" model parameters increase almost exponentially when computing higher-order sensitivities: the higher-order the sensitivities, the larger the computational savings. For example, the computational resources needed for computing the third-order sensitivities of $R_1(\rho; \boldsymbol{\alpha})$ would be as follows:

A. There are $TN \triangleq TP - 1 = 2N + 3$ non-zero first-order sensitivities when using the non-aggregated primary parameters. Correspondingly, there will be $TN(TN + 1)/2 = (2N + 3)(N + 2)$ distinct non-zero second-order response sensitivities. Each of these second-order sensitivities would be considered as a response when computing the third-order sensitivities. For each of these responses, there would correspond one third-level adjoint sensitivity function. Thus, $(2N + 3)(N + 2)$ adjoint computations would be necessary for computing all of the third-order sensitivities.

B. On the other hand, there are 4 non-zero first-order sensitivities when using the aggregated parameter $\mu(\mathbf{n}, \boldsymbol{\sigma})$. Correspondingly, there will be 6 distinct non-zero second-order response sensitivities. Each of these second-order sensitivities would play the role of a response when computing the third-order sensitivities. For each of these responses, there would correspond one third-level adjoint sensitivity function. Thus, 6 adjoint computations would be necessary for computing all of the third-order sensitivities. The third-order sensitivities corresponding to the fundamental model parameters n_i and σ_i would be obtained by using formulas similar to those shown in (3.500)–(3.502), but the additional computational costs incurred due the extra multiplications needed for computing these third-order sensitivities are negligible by comparison to the computational costs associated with the numerical solution of the various forward or adjoint models.

3.6.2.8 Illustrative Computation of Third- and Fourth-Order Sensitivities Using Aggregated Model Parameters

The most complex of the expressions for the second-order sensitivities obtained is that for $\partial^2 R_1/\partial^2 \mu$, since it involves all (four) state functions of the original forward system, 1st-LASS and 2nd-LASS. However, since the steps involved in computation of the third-order sensitivities that correspond to $\partial^2 R_1/\partial^2 \mu$ are conceptually the same as outlined in Sect. 3.5.1, the computation of the third- and fourth-order sensitivities of $R_1(\rho;\boldsymbol{\alpha})$ will be illustrated in this section by considering the mixed second-order sensitivity $\partial^2 R_1/\partial \rho_{in}\,\partial \mu(\mathbf{n},\boldsymbol{\sigma})$ of $R_1(\rho;\boldsymbol{\alpha})$ with respect to the initial condition ρ_{in} and the aggregated parameter $\mu(\mathbf{n},\boldsymbol{\sigma})$. Thus, the corresponding third-order sensitivities are determined by the total first-order G-differential of (3.496), which is obtained as follows:

$$\delta\left(\frac{\partial^2 R_1}{\partial \rho_{in}\,\partial \mu}\right) = \left\{\frac{d}{d\varepsilon}\int_{\beta_\ell^0+\varepsilon\,\delta\beta_\ell}^{\beta_u^0+\varepsilon\,\delta\beta_u} \left\{\left[\psi_{a2}^{(2)}(t)+\varepsilon\delta\psi_{a2}^{(2)}(t)\right]\delta\left(t-\beta_\ell^0-\varepsilon\delta\beta_\ell\right)\right\}_{\boldsymbol{\alpha}^0} dt\right\}_{\varepsilon=0}$$
$$= \left\{\delta\left(\frac{\partial^2 R_1}{\partial \rho_{in}\,\partial \mu}\right)\right\}_{dir} + \left\{\delta\left(\frac{\partial^2 R_1}{\partial \rho_{in}\,\partial \mu}\right)\right\}_{ind}, \tag{3.503}$$

where

$$\left\{\delta\left(\frac{\partial^2 R_1}{\partial \rho_{in}\,\partial \mu}\right)\right\}_{dir} \triangleq -\left(\delta\beta_\ell\right)\int_{\beta_\ell^0}^{\beta_u^0} \psi_{a2}^{(2)}(t)\,\delta'\left(t-\beta_\ell^0\right)dt, \tag{3.504}$$

and

$$\left\{\delta\left(\frac{\partial^2 R_1}{\partial \rho_{in}\,\partial \mu}\right)\right\}_{ind} \triangleq \int_{\beta_\ell^0}^{\beta_u^0} \delta\psi_{a2}^{(2)}(t)\delta\left(t-\beta_\ell^0\right)dt. \tag{3.505}$$

The direct-effect defined in (3.504) can be computed immediately, since the adjoint function $\psi_{a2}^{(2)}(t)$ is known. On the other hand, the computation of the indirect-effect term defined in (3.505) involved the function $\delta\psi_{a2}^{(2)}(t)$, which is the solution of the system obtained by the G-differentiation of (3.492) and (3.493), which is provided below:

$$-\frac{d\left[\delta\psi_{a2}^{(2)}(t)\right]}{dt} + \mu\delta\psi_{a2}^{(2)}(t) + \delta\psi_{a}^{(1)}(t) = -(\delta\mu)\,\psi_{a2}^{(2)}(t), \tag{3.506}$$

$$\delta\psi_{a2}^{(2)}\left(\beta_u^0\right) + \delta\beta_u^0\left\{\frac{d\psi_{a2}^{(2)}(t)}{dt}\right\}_{t=\beta_u^0} = \delta\psi_{a2}^{(2)}\left(\beta_u^0\right) = 0. \tag{3.507}$$

It follows from (3.506), (3.507), (3.486), and (3.487) that the 3rd-LVSS for the function $\delta\mathbf{u}^{(3)}(t)\triangleq\left[\delta\psi_{a}^{(1)}(t),\delta\psi_{a2}^{(2)}(t)\right]$ is obtained by concatenating these equations, which have the following matrix-operator form:

$$\begin{pmatrix} -\frac{d}{dt}+\mu & 0 \\ 1 & -\frac{d}{dt}+\mu \end{pmatrix}\begin{pmatrix} \delta\psi_{a}^{(1)}(t) \\ \delta\psi_{a2}^{(2)}(t) \end{pmatrix} = \begin{pmatrix} -(\delta t_d)\delta'(t-t_d) - (\delta\mu)\psi_{a}^{(1)}(t) \\ -(\delta\mu)\,\psi_{a2}^{(2)}(t) \end{pmatrix}. \tag{3.508}$$

It is noteworthy that the 3rd-LVSS comprising (3.508) and the final-time conditions given in (3.507) and (3.487) is a "final-time value problem" (as opposed to an "initial-value problem").

Using the definition given in (3.356), construct the inner product of (3.508) with a square integrable two-component function $\boldsymbol{\psi}_{a}^{(3)}(t)\triangleq\left[\psi_{a1}^{(3)}(t),\psi_{a2}^{(3)}(t)\right]\in \mathrm{H}_2(\Omega_t)$ to obtain the following relation:

$$\int_{\beta_\ell^0}^{\beta_u^0}\left[\psi_{a1}^{(3)}(t),\psi_{a2}^{(3)}(t)\right]\begin{pmatrix}-\dfrac{d}{dt}+\mu & 0\\ 1 & -\dfrac{d}{dt}+\mu\end{pmatrix}\begin{pmatrix}\delta\psi_a^{(1)}(t)\\ \delta\psi_{a2}^{(2)}(t)\end{pmatrix}dt$$
$$=\int_{\beta_\ell^0}^{\beta_u^0}\left[\psi_{a1}^{(3)}(t),\psi_{a2}^{(3)}(t)\right]\begin{pmatrix}-(\delta t_d)\delta'(t-t_d)-(\delta\mu)\psi_a^{(1)}(t)\\ -(\delta\mu)\psi_{a2}^{(2)}(t)\end{pmatrix}dt. \tag{3.509}$$

Integrating by parts the left side of (3.509) so as to transfer the differential operations on $\delta\psi_a^{(1)}(t)$ and $\delta\psi_{a2}^{(2)}(t)$ to differential operations on $\psi_{a1}^{(3)}(t)$ and $\psi_{a2}^{(3)}(t)$ yields the following relation:

$$\int_{\beta_\ell^0}^{\beta_u^0}\psi_{a1}^{(3)}(t)\left[-\frac{d}{dt}\delta\psi_a^{(1)}(t)+\mu\delta\psi_a^{(1)}(t)\right]dt+\int_{\beta_\ell^0}^{\beta_u^0}\psi_{a2}^{(3)}(t)\left[-\frac{d}{dt}\delta\psi_{a2}^{(2)}(t)\right.$$
$$\left.+\mu\delta\psi_{a2}^{(2)}(t)+\delta\psi_a^{(1)}(t)\right]dt=-\psi_{a1}^{(3)}\left(\beta_u^0\right)\delta\psi_a^{(1)}\left(\beta_u^0\right)+\psi_{a1}^{(3)}\left(\beta_\ell^0\right)\delta\psi_a^{(1)}\left(\beta_\ell^0\right)$$
$$-\psi_{a2}^{(3)}\left(\beta_u^0\right)\delta\psi_{a2}^{(2)}\left(\beta_u^0\right)+\psi_{a2}^{(3)}\left(\beta_\ell^0\right)\delta\psi_{a2}^{(2)}\left(\beta_\ell^0\right)+\int_{\beta_\ell^0}^{\beta_u^0}\delta\psi_a^{(1)}(t)\left[\frac{d\psi_{a1}^{(3)}(t)}{dt}\right. \tag{3.510}$$
$$\left.+\mu\psi_{a1}^{(3)}(t)+\psi_{a2}^{(3)}(t)\right]dt+\int_{\beta_\ell^0}^{\beta_u^0}\delta\psi_{a2}^{(2)}(t)\left[\frac{d\psi_{a2}^{(3)}(t)}{dt}+\psi_{a2}^{(3)}(t)\mu\right]dt.$$

The last two terms on the right side of (3.510) will represent the G-differential defined in (3.505) by requiring that:

$$\frac{d\psi_{a1}^{(3)}(t)}{dt}+\mu(\mathbf{n},\boldsymbol{\sigma})\psi_{a1}^{(3)}(t)+\psi_{a2}^{(3)}(t)=0, \tag{3.511}$$

$$\frac{d\psi_{a2}^{(3)}(t)}{dt}+\mu(\mathbf{n},\boldsymbol{\sigma})\psi_{a2}^{(3)}(t)=\delta\left(t-\beta_\ell^0\right). \tag{3.512}$$

The boundary conditions for (3.511) and (3.512) are established by requiring that the contributions involving the unknown quantities $\delta\psi_a^{(1)}\left(\beta_\ell^0\right)$ and $\delta\psi_{a2}^{(2)}\left(\beta_\ell^0\right)$ in (3.510) vanish, which is accomplished by imposing the following conditions:

$$\psi_{a1}^{(3)}\left(\beta_\ell^0\right)=0,\quad \psi_{a2}^{(3)}\left(\beta_\ell^0\right)=0. \tag{3.513}$$

The system of equations comprising (3.511)–(3.513) constitutes the Third-Level Adjoint Sensitivity System (3rd-LASS) for the two-component third-level adjoint function $\boldsymbol{\psi}_a^{(3)}(t) \triangleq \left[\psi_{a1}^{(3)}(t), \psi_{a2}^{(3)}(t)\right]$.

Replacing the left side of (3.510) by the right side of (3.509) and taking into account the equations underlying the 3rd-LASS yields the following expression for the indirect-effect term defined in (3.505):

$$\left\{\delta\left(\frac{\partial^2 R_1}{\partial \rho_{in}\,\partial \mu}\right)\right\}_{ind} = -\int_{\beta_\ell^0}^{\beta_u^0} \psi_{a1}^{(3)}(t)\left[(\delta t_d)\delta'(t-t_d) + (\delta\mu)\psi_a^{(1)}(t)\right]dt$$
$$-(\delta\mu)\int_{\beta_\ell^0}^{\beta_u^0} \psi_{a2}^{(3)}(t)\psi_{a2}^{(2)}(t)\,dt. \tag{3.514}$$

Identifying in the resulting expression the coefficients multiplying the variations $\delta\mu$, δt_d, and $\delta\beta_\ell$ in the direct-effect term defined in (3.504) and the indirect-effect term given by (3.514) yields the following expressions for the respective third-order sensitivities of the response $R_1(\rho; \boldsymbol{\alpha})$:

$$\frac{\partial^3 R_1}{\partial \rho_{in}\,\partial \mu \partial \mu} = -\int_{\beta_\ell^0}^{\beta_u^0} \left[\psi_{a1}^{(3)}(t)\psi_a^{(1)}(t) + \psi_{a2}^{(3)}(t)\psi_{a2}^{(2)}(t)\right]dt, \tag{3.515}$$

$$\frac{\partial^2 R_1}{\partial \rho_{in}\,\partial \mu \partial \beta_\ell} = -\int_{\beta_\ell^0}^{\beta_u^0} \psi_{a2}^{(2)}(t)\delta'(t-\beta_\ell^0)\,dt = \left\{\frac{d\psi_{a2}^{(2)}(t)}{dt}\right\}_{t=\beta_\ell^0}, \tag{3.516}$$

$$\frac{\partial^2 R_1}{\partial \rho_{in}\,\partial \mu \partial t_d} = -\int_{\beta_\ell^0}^{\beta_u^0} \psi_{a1}^{(3)}(t)\delta'(t-t_d)\,dt = \left\{\frac{d\psi_{a1}^{(3)}(t)}{dt}\right\}_{t=\beta_\ell^0}. \tag{3.517}$$

Of course, all of the quantities entering in the expressions of the above third-order sensitivities are to be evaluated at the nominal values of the parameters and state functions.

The computation of the fourth-order sensitivities is performed by considering each of the third-order sensitivities as a "response" and determining the total first-order G-differential of each of the respective third-order sensitivity. As an illustrative example, consider the determination of the fourth-order sensitivities that correspond to the third-order sensitivity $\partial^3 R_1/\partial \rho_{in}\,\partial \mu^2$ obtained in (3.515). The total G-differential of the expression obtained in (3.515) is determined, by definition, as follows:

$$\delta\left(\frac{\partial^3 R_1}{\partial \rho_{in}\,\partial \mu^2}\right) = -(\delta\beta_u)\left\{\left[\psi_{a1}^{(3)}(\beta_u)\psi_a^{(1)}(\beta_u) + \psi_{a2}^{(3)}(\beta_u)\psi_{a2}^{(2)}(\beta_u)\right]\right\}_{\boldsymbol{\alpha}^0}$$
$$+(\delta\beta_\ell)\left\{\left[\psi_{a1}^{(3)}(\beta_\ell)\psi_a^{(1)}(\beta_\ell) + \psi_{a2}^{(3)}(\beta_\ell)\psi_{a2}^{(2)}(\beta_\ell)\right]\right\}_{\boldsymbol{\alpha}^0}$$
$$-\left\{\frac{d}{d\varepsilon}\int_{\beta_\ell^0+\varepsilon\,\delta\beta_\ell}^{\beta_u^0+\varepsilon\,\delta\beta_u}\left\{\left[\psi_{a1}^{(3)}(t) + \varepsilon\delta\psi_{a1}^{(3)}(t)\right]\left[\psi_a^{(1)}(t) + \varepsilon\delta\psi_a^{(1)}(t)\right]\right\}_{\boldsymbol{\alpha}^0} dt\right\}_{\varepsilon=0} \tag{3.518}$$
$$-\left\{\frac{d}{d\varepsilon}\int_{\beta_\ell^0+\varepsilon\,\delta\beta_\ell}^{\beta_u^0+\varepsilon\,\delta\beta_u}\left\{\left[\psi_{a2}^{(3)}(t) + \varepsilon\delta\psi_{a2}^{(3)}(t)\right]\left[\psi_{a2}^{(2)}(t) + \varepsilon\delta\psi_{a2}^{(2)}(t)\right]\right\}_{\boldsymbol{\alpha}^0} dt\right\}_{\varepsilon=0}.$$

After performing the indicated differentiations and evaluations in the above relation, including the use of (3.513), (3.476), and (3.493), the expression (3.518) becomes:

$$\delta\left(\frac{\partial^3 R_1}{\partial \rho_{in}\,\partial \mu^2}\right) = -\int_{\beta_\ell^0}^{\beta_u^0}\left\{\psi_{a1}^{(3)}(t)\delta\psi_a^{(1)}(t) + \psi_a^{(1)}(t)\delta\psi_{a1}^{(3)}(t)\right\}_{\boldsymbol{\alpha}^0} dt$$
$$-\int_{\beta_\ell^0}^{\beta_u^0}\left\{\psi_{a2}^{(3)}(t)\delta\psi_{a2}^{(2)}(t) + \psi_{a2}^{(2)}(t)\delta\psi_{a2}^{(3)}(t)\right\}_{\boldsymbol{\alpha}^0} dt. \tag{3.519}$$

The functions $\delta\psi_{a1}^{(3)}(t)$ and $\delta\psi_{a2}^{(3)}(t)$, which appear in (3.519), are the solutions of the system obtained by G-differentiating the 3rd-LASS defined by (3.511)–(3.513), which yields the following system of equations:

$$\frac{d\left[\delta\psi_{a1}^{(3)}(t)\right]}{dt} + \mu\delta\psi_{a1}^{(3)}(t) + \delta\psi_{a2}^{(3)}(t) = -(\delta\mu)\psi_{a1}^{(3)}(t), \tag{3.520}$$

$$\frac{d\left[\delta\psi_{a2}^{(3)}(t)\right]}{dt} + \mu\delta\psi_{a2}^{(3)}(t) = -(\delta\mu)\psi_{a2}^{(3)}(t) - (\delta\beta_\ell)\delta'\left(t-\beta_\ell^0\right), \tag{3.521}$$

$$\delta\psi_{a1}^{(3)}\left(\beta_\ell^0\right) = 0, \quad \delta\psi_{a2}^{(3)}\left(\beta_\ell^0\right) = 0. \tag{3.522}$$

Altogether, the functions $\delta\psi_a^{(1)}(t)$, $\delta\psi_{a2}^{(2)}(t)$, $\delta\psi_{a1}^{(3)}(t)$, and $\delta\psi_{a2}^{(3)}(t)$, which appear in (3.519), are considered to be the components of a fourth-level variational vector denoted as $\delta\mathbf{u}^{(4)}(t)$ and defined as follows:

$$\delta \mathbf{u}^{(4)}(t) \triangleq \left[\delta \psi_a^{(1)}(t), \delta \psi_{a2}^{(2)}(t), \delta \psi_{a1}^{(3)}(t), \delta \psi_{a2}^{(3)}(t)\right]^{\dagger}. \tag{3.523}$$

The vector $\delta \mathbf{u}^{(4)}(t)$ is the solutions of 4th-Level Variational Sensitivity System (4th-LVSS) of equations obtained by concatenating (3.520) and (3.521) with the 3rd-LVSS for the function $\delta \mathbf{u}^{(3)}(t) \triangleq \left[\delta \psi_a^{(1)}(t), \delta \psi_{a2}^{(2)}(t)\right]$ given in (3.508). This 4th-LVSS can be written in following matrix form:

$$\mathbf{V}^{(4)} \delta \mathbf{u}^{(4)}(t) = \mathbf{q}^{(4)}(t), \tag{3.524}$$

where

$$\mathbf{V}^{(4)} \triangleq \begin{pmatrix} -\frac{d}{dt} + \mu & 0 & 0 & 0 \\ 1 & -\frac{d}{dt} + \mu & 0 & 0 \\ 0 & 0 & \frac{d}{dt} + \mu & 1 \\ 0 & 0 & 0 & \frac{d}{dt} + \mu \end{pmatrix}, \tag{3.525}$$

$$\mathbf{q}^{(4)}(t) \triangleq \begin{pmatrix} -(\delta t_d)\delta'(t - t_d) - (\delta \mu)\psi_a^{(1)}(t) \\ -(\delta \mu)\psi_{a2}^{(2)}(t) \\ -(\delta \mu)\psi_{a1}^{(3)}(t) \\ -(\delta \mu)\psi_{a2}^{(3)}(t) - (\delta \beta_\ell)\delta'\left(t - \beta_\ell^0\right) \end{pmatrix}. \tag{3.526}$$

The initial/final-time conditions for the 4th-LVSS given in (3.524) are as given in (3.487), (3.507), and (3.522), which are reproduced below, for convenience:

$$\delta \psi_a^{(1)}\left(\beta_u^0\right) = 0; \;\; \delta \psi_{a2}^{(2)}\left(\beta_u^0\right) = 0; \delta \psi_{a1}^{(3)}\left(\beta_\ell^0\right) = 0; \;\; \delta \psi_{a2}^{(3)}\left(\beta_\ell^0\right) = 0. \tag{3.527}$$

Solving the 4th-LVSS given in (3.524), which depends on the parameter variations, can be avoided by constructing an appropriate 4th-Level Adjoint Sensitivity System (4th-LASS) for a fourth-level adjoint sensitivity function denoted as $\boldsymbol{\psi}_a^{(4)}(t) \triangleq \left[\psi_{a1}^{(4)}(t), \psi_{a2}^{(4)}(t), \psi_{a3}^{(4)}(t), \psi_{a4}^{(4)}(t)\right]$, by applying the principles of the 4th-CASAM-L. The steps involved in construction this 4th-LASS are as follows:

1. Construct the inner product of (3.524) with a four-component function $\boldsymbol{\psi}_a^{(4)}(t) \triangleq \left[\psi_{a1}^{(4)}(t), \psi_{a2}^{(4)}(t), \psi_{a3}^{(4)}(t), \psi_{a4}^{(4)}(t)\right]^\dagger$, having square integrable components, to obtain the following relation:

$$\int_{\beta_\ell^0}^{\beta_u^0} \left[\boldsymbol{\psi}_a^{(4)}(t)\right]^\dagger \cdot \mathbf{V}^{(4)} \delta\mathbf{u}^{(4)}(t)\, dt = \int_{\beta_\ell^0}^{\beta_u^0} \left[\boldsymbol{\psi}_a^{(4)}(t)\right]^\dagger \cdot \mathbf{q}^{(4)}(t) dt. \tag{3.528}$$

2. Integrate by parts the left side of (3.528) so as to transfer the differential operations on $\delta\mathbf{u}^{(4)}(t)$ to differential operations on $\boldsymbol{\psi}_a^{(4)}(t)$, thereby obtaining the following relation:

$$\int_{\beta_\ell^0}^{\beta_u^0} \left[\boldsymbol{\psi}_a^{(4)}\right]^\dagger \cdot \mathbf{V}^{(4)} \delta\mathbf{u}^{(4)} dt = \int_{\beta_\ell^0}^{\beta_u^0} \left[\delta\mathbf{u}^{(4)}\right]^\dagger \cdot \mathbf{A}^{(4)} \boldsymbol{\psi}_a^{(4)} dt + P^{(4)}\left[\boldsymbol{\psi}_a^{(4)}, \delta\mathbf{u}^{(4)}\right], \tag{3.529}$$

where the matrix $\mathbf{A}^{(4)}$ is the adjoint of the matrix $\mathbf{V}^{(4)}$, while $P^{(4)}\left[\boldsymbol{\psi}_a^{(4)}, \delta\mathbf{u}^{(4)}\right]$ represents the bilinear concomitant of the terms evaluated at the respective initial/final times, having the following expressions:

$$\mathbf{A}^{(4)} \triangleq \left[\mathbf{V}^{(4)}\right]^* = \begin{pmatrix} \frac{d}{dt} + \mu & 1 & 0 & 0 \\ 0 & \frac{d}{dt} + \mu & 0 & 0 \\ 0 & 0 & -\frac{d}{dt} + \mu & 0 \\ 0 & 0 & 1 & -\frac{d}{dt} + \mu \end{pmatrix}, \tag{3.530}$$

$$\begin{aligned} P^{(4)}\left[\boldsymbol{\psi}_a^{(4)}, \delta\mathbf{u}^{(4)}\right] \triangleq & -\psi_{a1}^{(4)}(\beta_u^0)\delta\psi_a^{(1)}(\beta_u^0) + \psi_{a1}^{(4)}(\beta_\ell^0)\delta\psi_a^{(1)}(\beta_\ell^0) \\ & -\psi_{a2}^{(4)}(\beta_u^0)\delta\psi_{a2}^{(2)}(\beta_u^0) + \psi_{a2}^{(4)}(\beta_\ell^0)\delta\psi_{a2}^{(2)}(\beta_\ell^0) + \psi_{a3}^{(4)}(\beta_u^0)\delta\psi_{a1}^{(3)}(\beta_u^0) \\ & -\psi_{a3}^{(4)}(\beta_\ell^0)\delta\psi_{a1}^{(3)}(\beta_\ell^0) + \psi_{a4}^{(4)}(\beta_u^0)\delta\psi_{a2}^{(3)}(\beta_u^0) - \psi_{a4}^{(4)}(\beta_\ell^0)\delta\psi_{a2}^{(3)}(\beta_\ell^0). \end{aligned} \tag{3.531}$$

3. Require the first term on the right side of (3.529) to represent the G-differential defined in (3.519) by imposing the following relation:

$$\mathbf{A}^{(4)} \boldsymbol{\psi}_a^{(4)} = -\left[\psi_{a1}^{(3)}(t), \psi_{a2}^{(3)}(t), \psi_a^{(1)}(t), \psi_{a2}^{(2)}(t)\right]^\dagger. \tag{3.532}$$

4. The boundary conditions for (3.511) and (3.512) are established by requiring that the contributions involving the unknown quantities $\delta\psi_a^{(1)}(\beta_\ell^0)$, $\delta\psi_{a2}^{(2)}(\beta_\ell^0)$,

$\delta\psi_{a1}^{(3)}(\beta_u^0)$, and $\delta\psi_{a2}^{(3)}(\beta_u^0)$ in the bilinear concomitant expressed by (3.531) to vanish. This is accomplished by imposing the following initial/final-time conditions:

$$\psi_{a1}^{(4)}(\beta_\ell^0) = 0; \; \psi_{a2}^{(4)}(\beta_\ell^0) = 0; \; \psi_{a3}^{(4)}(\beta_u^0) = 0; \; \psi_{a4}^{(4)}(\beta_u^0) = 0. \tag{3.533}$$

The system of equations comprising (3.511)–(3.513) constitutes the 4th-Level Adjoint Sensitivity System (4th-LASS) for the four-component fourth-level adjoint function $\boldsymbol{\psi}_a^{(4)}(t) \triangleq \left[\psi_{a1}^{(4)}(t), \psi_{a2}^{(4)}(t), \psi_{a3}^{(4)}(t), \psi_{a4}^{(4)}(t)\right]$. Notably, the bilinear concomitant vanishes after imposing on $\boldsymbol{\psi}_a^{(4)}(t)$ the initial/final-time conditions required in (3.533).

It follows from (3.519), (3.528), and (3.529) that the G-differential $\delta(\partial^3 R_1/\partial\rho_{in}\,\partial\mu^2)$ is given by the following expression in terms of the fourth-level adjoint function $\boldsymbol{\psi}_a^{(4)}(t) \triangleq \left[\psi_{a1}^{(4)}(t), \psi_{a2}^{(4)}(t), \psi_{a3}^{(4)}(t), \psi_{a4}^{(4)}(t)\right]$:

$$\begin{aligned}
\delta\left(\frac{\partial^3 R_1}{\partial\rho_{in}\,\partial\mu^2}\right) &= \int_{\beta_\ell^0}^{\beta_u^0} \left[\boldsymbol{\psi}_a^{(4)}(t)\right]^\dagger \cdot \mathbf{q}^{(4)}(t)dt \\
&= -\int_{\beta_\ell^0}^{\beta_u^0} \left\{\psi_{a1}^{(4)}(t)\left[(\delta t_d)\delta'(t-t_d) + (\delta\mu)\psi_a^{(1)}(t)\right]\right\}_{\boldsymbol{\alpha}^0} dt \\
&\quad -(\delta\mu)\int_{\beta_\ell^0}^{\beta_u^0} \left\{\psi_{a2}^{(4)}(t)\psi_{a2}^{(2)}(t)\right\}_{\boldsymbol{\alpha}^0} dt - (\delta\mu)\int_{\beta_\ell^0}^{\beta_u^0} \left\{\psi_{a3}^{(4)}(t)\psi_{a1}^{(3)}(t)\right\}_{\boldsymbol{\alpha}^0} dt \\
&\quad -\int_{\beta_\ell^0}^{\beta_u^0} \left\{\psi_{a4}^{(4)}(t)\left[(\delta\mu)\psi_{a2}^{(3)}(t) + (\delta\beta_\ell)\delta'\left(t-\beta_\ell^0\right)\right]\right\}_{\boldsymbol{\alpha}^0} dt.
\end{aligned} \tag{3.534}$$

Identifying in (3.534) the coefficients multiplying the variations $\delta\mu$, δt_d, and $\delta\beta_\ell$ yields the following expressions for the respective fourth-order sensitivities of the response $R_1(\rho;\boldsymbol{\alpha})$:

$$\frac{\partial^4 R_1}{\partial\rho_{in}\,\partial\mu^3} = -\int_{\beta_\ell^0}^{\beta_u^0} \left\{\psi_{a1}^{(4)}(t)\psi_a^{(1)}(t) + \psi_{a2}^{(4)}(t)\psi_{a2}^{(2)}(t) + \psi_{a3}^{(4)}(t)\psi_{a1}^{(3)}(t) + \psi_{a4}^{(4)}(t)\psi_{a2}^{(3)}(t)\right\}_{\boldsymbol{\alpha}^0} dt; \tag{3.535}$$

$$\frac{\partial^4 R_1}{\partial \rho_{in}\, \partial \mu^2 \partial t_d} = -\int_{\beta_\ell^0}^{\beta_u^0} \left\{ \psi_{a1}^{(4)}(t)\delta'(t - t_d) \right\}_{\boldsymbol{\alpha}^0} dt = \left\{ \frac{d\psi_{a1}^{(4)}(t)}{dt} \right\}_{t=t_d}; \tag{3.536}$$

$$\frac{\partial^4 R_1}{\partial \rho_{in}\, \partial \mu^2 \partial \beta_\ell} = -\int_{\beta_\ell^0}^{\beta_u^0} \left\{ \psi_{a4}^{(4)}(t)\delta'\left(t - \beta_\ell^0\right) \right\}_{\boldsymbol{\alpha}^0} dt = \left\{ \frac{d\psi_{a4}^{(4)}(t)}{dt} \right\}_{t=\beta_\ell^0}. \tag{3.537}$$

3.6.3 *Applying the nth-CASAM-L to Compute Sensitivities of the Average Concentration Response to Model Parameters, Including Imprecisely Known Initial- and Final-Time Conditions*

An important response of interest for the physical problem modeled by (3.317) and (3.318) is a time-average reaction rate response, which will be denoted as $R_2(\rho; \boldsymbol{\alpha})$, and which is represented mathematically by the following functional:

$$R_2(\rho; \boldsymbol{\alpha}) = \Sigma_d \int_{\beta_\ell}^{\beta_u} \rho(t)\, dt, \tag{3.538}$$

where Σ_d represents a reaction/interaction cross section. For the evolution system and the response defined in (3.538), the "vector of model parameters" $\boldsymbol{\alpha}$ is defined as follows:

$$\boldsymbol{\alpha} \triangleq (\alpha_1, \ldots, \alpha_{TP})^\dagger \triangleq (n_1, \ldots, n_N, \sigma_1, \ldots, \sigma_N, \rho_{in}, \Sigma_d, \beta_\ell, \beta_u)^\dagger. \tag{3.539}$$

The closed-form expression of $R_2(\rho; \boldsymbol{\alpha})$ is readily obtained by inserting the expression of $\rho(t)$ provided in (3.322) into (3.538) and performing the respective integration, which yields the following expression:

$$R_2(\rho; \boldsymbol{\alpha}) = \frac{\rho_{in}\Sigma_d}{\sum_{i=1}^{N} n_i \sigma_i} \left\{ 1 - \exp\left[(\beta_\ell - \beta_u) \sum_{i=1}^{N} n_i \sigma_i \right] \right\}. \tag{3.540}$$

As indicated by the expression in (3.540), even though the forward function $\rho(t)$ is independent of the boundary point β_u, the response $R_2(\rho; \boldsymbol{\alpha})$ evidently depends on β_u, so its sensitivities with respect to β_u will not vanish.

3.6.3.1 First-Order Sensitivities

The first-order total sensitivity of the response $R_2(\rho; \boldsymbol{\alpha})$ defined in (3.538) is provided by the first-order G-differential of this response, which is obtained by G-differentiating (3.538) to obtain the following expression:

$$\delta R_2(\rho; \boldsymbol{\alpha}; \delta\rho; \delta\boldsymbol{\alpha}) = (\delta R_2)_{ind} + (\delta R_2)_{dir}, \tag{3.541}$$

where the indirect-effect term $(\delta R_2)_{ind}$ and the direct-effect term $(\delta R_2)_{dir}$, respectively, are defined as follows:

$$(\delta R_2)_{ind} \triangleq \Sigma_d^0 \int_{\beta_\ell^0}^{\beta_u^0} \delta\rho(t)\, dt, \tag{3.542}$$

$$(\delta R_2)_{dir} \triangleq (\delta\Sigma_d) \int_{\beta_\ell^0}^{\beta_u^0} \rho(t)\, dt + \Sigma_d^0 \{\rho(t)\}_{t=\beta_u^0} \delta\beta_u - \Sigma_d^0 \{\rho(t)\}_{t=\beta_\ell^0} \delta\beta_\ell. \tag{3.543}$$

The direct-effect term, $(\delta R_2)_{dir}$, can be computed directly by using in (3.543) the nominal solution provided in (3.323) to obtain:

$$\begin{aligned}(\delta R_2)_{dir} \triangleq & (\delta\Sigma_d) \frac{\rho_{in}^0}{\sum_{i=1}^{N} n_i^0 \sigma_i^0} \left\{ 1 - \exp\left[\left(\beta_\ell^0 - \beta_u^0\right) \sum_{i=1}^{N} n_i^0 \sigma_i^0 \right] \right\} \\ & + (\delta\beta_u) \Sigma_d^0 \rho_{in}^0 \exp\left[\left(\beta_\ell^0 - \beta_u^0\right) \sum_{i=1}^{N} n_i^0 \sigma_i^0 \right] - (\delta\beta_\ell) \Sigma_d^0 \rho_{in}^0.\end{aligned} \tag{3.544}$$

The indirect-effect term, $(\delta R_2)_{ind}$, depends on the unknown variation $\delta\rho(t)$, of the state function $\rho(t)$, which is the solution of the 1st-LVSS obtained in (3.329) and (3.330). Using the inner product defined in (3.331), form the inner product of (3.329) with a square-integrable function $\theta^{(1)}(t) \in \mathrm{H}_1(\Omega_t)$ and integrate the left side of the resulting equation by parts once, so as to transfer the differential operation from $\delta\rho(t)$ onto $\theta^{(1)}(t)$, to obtain the following relation:

$$\begin{aligned}\int_{\beta_\ell^0}^{\beta_u^0} \theta^{(1)}(t) \left[\frac{d[\delta\rho(t)]}{dt} + \delta\rho(t) \sum_{i=1}^{N} n_i^0 \sigma_i^0 \right] dt = & \theta^{(1)}\left(\beta_u^0\right) \delta\rho\left(\beta_u^0\right) \\ - \theta^{(1)}\left(\beta_\ell^0\right) \delta\rho\left(\beta_\ell^0\right) + & \int_{\beta_\ell^0}^{\beta_u^0} \delta\rho(t) \left[-\frac{d\theta^{(1)}(t)}{dt} + \theta^{(1)}(t) \sum_{i=1}^{N} n_i^0 \sigma_i^0 \right] dt.\end{aligned} \tag{3.545}$$

The following sequence of operations is performed next using (3.545):

1. Require that the last term on the right side of (3.545) be identical with the indirect effect $(\delta R_2)_{ind}$ defined in (3.542).
2. Use the right side of (3.329) to replace the term multiplying $\theta^{(1)}(t)$ on the left side of (3.545).
3. Eliminate the unknown quantity $\delta\rho\left(\beta_u^0\right)$ on the right side of (3.545) by imposing the condition $\theta^{(1)}\left(\beta_u^0\right) = 0$.
4. Insert the boundary condition provided in (3.330) into the resulting expression.

The result of the above sequence of operations is the following expression for the indirect-effect term $(\delta R_2)_{ind}$:

$$(\delta R_2)_{ind} = -\sum_{i=1}^{N}\left(n_i^0\delta\sigma_i + \sigma_i^0\delta n_i\right)\int_{\beta_\ell^0}^{\beta_u^0}\theta^{(1)}(t)\rho^0(t)\,dt + \theta^{(1)}\left(\beta_\ell^0\right)\left[\delta\rho_{in} - \delta\beta_\ell\left\{\frac{\partial\rho(t)}{\partial t}\right\}_{t=\beta_\ell^0}\right], \tag{3.546}$$

where the first-level adjoint function $\theta^{(1)}(t)$ appearing in (3.546) is the solution of the following 1st-LASS:

$$-\frac{d\theta^{(1)}(t)}{dt} + \theta^{(1)}(t)\sum_{i=1}^{N}n_i^0\sigma_i^0 = \Sigma_d^0,\quad 0 \le \beta_\ell^0 \le t \le \beta_u^0 < \infty, \tag{3.547}$$

$$\theta^{(1)}\left(\beta_u^0\right) = 0. \tag{3.548}$$

Solving the 1st-LASS defined by (3.547) and (3.548) yields the following expression for the first-level adjoint function $\theta^{(1)}(t)$, which can be subsequently used to verify the expressions of various response sensitivities:

$$\theta^{(1)}(t) = \frac{\Sigma_d^0}{\sum_{i=1}^{N}n_i^0\sigma_i^0}\left\{1 - \exp\left[\left(t-\beta_u^0\right)\sum_{i=1}^{N}n_i^0\beta_i^0\right]\right\}. \tag{3.549}$$

Adding (3.546) and (3.543) and identifying the quantities that multiply the respective parameter variations yields the following expressions for the first-order sensitivities of $R_2(\rho;\boldsymbol{\alpha})$ in terms of the first-level adjoint function $\theta^{(1)}(t)$:

$$\left(\frac{\partial R_2}{\partial \sigma_i}\right)_{\boldsymbol{\alpha}^0} = -n_i^0 \int_{\beta_\ell^0}^{\beta_u^0} \theta^{(1)}(t)\rho^0(t)dt, \;\; i = 1, \ldots, N, \tag{3.550}$$

$$\left(\frac{\partial R_2}{\partial n_i}\right)_{\boldsymbol{\alpha}^0} = -\sigma_i^0 \int_{\beta_\ell^0}^{\beta_u^0} \theta^{(1)}(t)\rho^0(t)\,dt, \;\; i = 1, \ldots, N, \tag{3.551}$$

$$\left(\frac{\partial R_2}{\partial \rho_{in}}\right)_{\boldsymbol{\alpha}^0} = \theta^{(1)}\left(\beta_\ell^0\right), \tag{3.552}$$

$$\left(\frac{\partial R_2}{\partial \beta_u}\right)_{\boldsymbol{\alpha}^0} = \Sigma_d^0\{\rho(t)\}_{t=\beta_u^0}, \tag{3.553}$$

$$\left(\frac{\partial R_2}{\partial \beta_\ell}\right)_{\boldsymbol{\alpha}^0} = -\Sigma_d^0\{\rho(t)\}_{t=\beta_\ell^0} - \theta^{(1)}\left(\beta_\ell^0\right)\left\{\frac{\partial \rho(t)}{\partial t}\right\}_{t=\beta_\ell^0}, \tag{3.554}$$

$$\left(\frac{\partial R_2}{\partial \Sigma_d}\right)_{\boldsymbol{\alpha}^0} = \int_{\beta_\ell^0}^{\beta_u^0} \rho(t)\,dt. \tag{3.555}$$

The results in (3.550) and (3.551) indicate that the relative sensitivities of the response to the parameters σ_i, $[\partial R_2/\partial \sigma_i][\sigma_i/R_2]$, are identical to the relative sensitivities $[\partial R_2/\partial n_i][n_i/R_2]$ of the response to the parameters n_i, for each $i = 1, \ldots, N$, for all of these $2N$ model parameters, since

$$\frac{\partial R_2}{\partial \sigma_i}\frac{\sigma_i}{R_2} = \frac{\sigma_i n_i}{R_2}\int_{\beta_\ell^0}^{\beta_u^0}\left[-\theta^{(1)}(t)\rho(t)\right]dt \equiv \frac{\partial R_2}{\partial n_i}\frac{n_i}{R_2}, \;\; i = 1, \ldots, N. \tag{3.556}$$

As indicated in (3.555) and (3.553), the first-order sensitivities of $R_2(\rho;\boldsymbol{\alpha})$ with respect to the model/detector parameter Σ_d and the boundary parameter β_u stem exclusively from the direct-effect term, $(\delta R_2)_{dir}$, and can therefore be computed directly using the solution $\rho(t)$, provided in (3.322), of the forward model [cf., (3.317) and (3.318)]. Thus, inserting the expression of $\rho(t)$ and performing the operations indicated in (3.555) and (3.553), respectively, yields the following expressions:

$$\left(\frac{\partial R_2}{\partial \Sigma_d}\right)_{\boldsymbol{\alpha}^0} = \frac{\rho_{in}^0}{\sum_{i=1}^{N} n_i^0\sigma_i^0}\left\{1 - \exp\left[\left(\beta_\ell^0 - \beta_u^0\right)\sum_{i=1}^{N} n_i^0\sigma_i^0\right]\right\}, \tag{3.557}$$

$$\left(\frac{\partial R_2}{\partial \beta_u}\right)_{\boldsymbol{\alpha}^0} = \Sigma_d^0 \rho_{in}^0 \exp\left[\left(\beta_\ell^0 - \beta_u^0\right)\sum_{i=1}^{N} n_i^0 \sigma_i^0\right]. \tag{3.558}$$

On the other hand, as indicated in (3.550)–(3.552), the first-order sensitivities of $R_2(\rho; \boldsymbol{\alpha})$ with respect to the model parameters ρ_{in}, σ_i, and n_i stem exclusively from the indirect-effect term $(\delta R_2)_{ind}$ and can therefore be computed after having obtained the first-level adjoint function $\theta^{(1)}(t)$ by solving the 1st-LASS, namely (3.547) and (3.548). Finally, the sensitivity of $R_2(\rho; \boldsymbol{\alpha})$ with respect to the boundary parameter β_ℓ stems from contributions arising from both the direct- and the indirect-effect terms, as indicated by (3.554).

Inserting the result for the first-level adjoint function $\theta^{(1)}(t)$ obtained in (3.549) into (3.550)–(3.552) and performing the respective operations yields the following results:

$$\left(\frac{\partial R_2}{\partial \rho_{in}}\right)_{\boldsymbol{\alpha}^0} = \frac{\Sigma_d^0}{\sum_{i=1}^{N} n_i^0 \sigma_i^0}\left\{1 - \exp\left[\left(\beta_\ell^0 - \beta_u^0\right)\sum_{i=1}^{N} n_i^0 \sigma_i^0\right]\right\}, \tag{3.559}$$

$$\begin{aligned}\left(\frac{\partial R_2}{\partial \sigma_i}\right)_{\boldsymbol{\alpha}^0} &= -\rho_{in}^0 \Sigma_d^0 n_i^0 \left(\sum_{i=1}^{N} n_i^0 \sigma_i^0\right)^{-2} \\ &\times \left\{1 + \left[\left(\beta_\ell^0 - \beta_u^0\right)\sum_{i=1}^{N} n_i^0 \sigma_i^0 - 1\right]\exp\left[\left(\beta_\ell^0 - \beta_u^0\right)\sum_{i=1}^{N} n_i^0 \sigma_i^0\right]\right\},\end{aligned} \tag{3.560}$$

$$\begin{aligned}\left(\frac{\partial R_2}{\partial n_i}\right)_{\boldsymbol{\alpha}^0} &= -\rho_{in}^0 \Sigma_d^0 \sigma_i^0 \left(\sum_{i=1}^{N} n_i^0 \sigma_i^0\right)^{-2} \\ &\times \left\{1 + \left[\left(\beta_\ell^0 - \beta_u^0\right)\sum_{i=1}^{N} n_i^0 \sigma_i^0 - 1\right]\exp\left[\left(\beta_\ell^0 - \beta_u^0\right)\sum_{i=1}^{N} n_i^0 \sigma_i^0\right]\right\}.\end{aligned} \tag{3.561}$$

Finally, the sensitivity of $R_2(\rho; \boldsymbol{\alpha})$ with respect to the boundary parameter β_ℓ is computed by using in (3.554) the expression of first-level adjoint function $\theta^{(1)}(t)$ obtained in (3.549) together with the expression of $\rho^0(t)$ provided in (3.323), to obtain the following result:

$$\left(\frac{\partial R_2}{\partial \beta_\ell}\right)_{\boldsymbol{\alpha}^0} = -\Sigma_d^0 \rho_{in}^0 \exp\left[\left(\beta_\ell^0 - \beta_u^0\right)\sum_{i=1}^{N} n_i^0 \sigma_i^0\right]. \tag{3.562}$$

Notably for the result obtained in (3.562), the contribution from the indirect-effect term involving $\theta^{(1)}(t)$ partially offsets the contribution from the direct-effect term involving the expression of $\rho\left(t = \beta_\ell^0\right)$.

In general, the detector's interaction cross section, Σ_d, may itself be a function of the fundamental parameters that characterize the materials in the detector under consideration, i.e., $\Sigma_d = \Sigma_d\left(\sigma_i^d, n_i^d, w_i^d;\ i = 1, \ldots, N_d\right)$, where σ_i^d, n_i^d, and w_i^d are the microscopic cross sections, atomic number densities, and weighting functions of the N_d materials that constitute the detector's active materials. The sensitivities of the response $R_2(\rho;\boldsymbol{\alpha})$ with respect to such (fundamental) parameters would be obtained from the expression of $\partial R_2/\partial\Sigma_d$ provided in (3.555).

After having determined the first-order response sensitivities to all of the model's fundamental parameters, the *relative* sensitivities (which are unitless scalars) should be ranked in the order of their respective magnitudes. Such a ranking enables decisions regarding the subsequent computation of second-order sensitivities. As a matter of good practice, the second-order *relative* sensitivities corresponding to the largest first-order sensitivities should be computed in any case, to ensure that they are either important or unimportant in contributing to the uncertainty in the response, etc. In preparation for computing second-order sensitivities, it is paramount to aggregate fundamental parameters, to the largest possible extent, in order to reduce, as much as possible, the number of second-level adjoint functions that would need to be computed. This concept has been illustrated in Sect. 3.6.2.7 for the response $R_1(\rho;\boldsymbol{\alpha})$, where it has been shown that there are $TN \triangleq TP - 1 = 2N + 3$ non-zero first-order sensitivities when using the non-aggregated primary parameters but only 4 non-zero first-order sensitivities when using the aggregated parameter $\mu(\mathbf{n},\boldsymbol{\sigma}) \triangleq \sum_{i=1}^{N} n_i\sigma_i$. Hence, there were $2N + 3$ second-level adjoint functions (each corresponding to one first-order sensitivity) to be computed when using the fundamental parameters but only *four* second-level adjoint functions (each corresponding to one first-order sensitivity) to be computed when using the aggregated parameter $\mu(\mathbf{n},\boldsymbol{\sigma})$. A similar reduction in the number of needed second-level adjoint functions can be achieved for the response $R_2(\rho;\boldsymbol{\alpha})$ by using the aggregated parameter $\mu(\mathbf{n},\boldsymbol{\sigma})$, as will be shown below.

In terms of the aggregated parameter $\mu(\mathbf{n},\boldsymbol{\sigma})$, the 1st-LVSS remains as written in (3.472) and (3.473) while the 1st-LASS corresponding to the indirect-effect term (3.542) retains the form shown in (3.547) and (3.548), but written in terms of $\mu(\mathbf{n},\boldsymbol{\sigma})$, namely:

$$\left\{-\frac{d\theta_a^{(1)}(t)}{dt} + \theta_a^{(1)}(t)\mu(\mathbf{n},\boldsymbol{\sigma})\right\}_{\boldsymbol{\alpha}^0} = \Sigma_d^0,\ \ 0 \le \beta_\ell^0 \le t \le \beta_u^0 < \infty, \tag{3.563}$$

$$\theta^{(1)}\left(\beta_u^0\right) = 0, \tag{3.564}$$

where the "aggregated" first-level adjoint function $\theta_a^{(1)}(t)$ has been introduced in preparation for subsequently deriving the 2nd-LASS, even though $\theta_a^{(1)}(t)$ is actually identical with the "un-aggregated" first-level adjoint function $\theta^{(1)}(t)$.

In terms of the aggregated parameter $\mu(\mathbf{n},\boldsymbol{\sigma})$ and the "aggregated" first-level adjoint function $\theta_a^{(1)}(t)$, the total first-order G-differential $\delta R_2(\rho;\boldsymbol{\alpha};\delta\rho;\delta\boldsymbol{\alpha})$ of

$R_2(\rho; \boldsymbol{\alpha})$, which is given by the sum of (3.546) and (3.543), takes on the following form

$$\delta R_2(\rho; \boldsymbol{\alpha}; \delta\rho; \delta\boldsymbol{\alpha}) = (\delta\Sigma_d) \int_{\beta_\ell^0}^{\beta_u^0} \rho(t)\, dt + \Sigma_d^0 \{\rho(t)\}_{t=\beta_u^0} \delta\beta_u - \Sigma_d^0 \{\rho(t)\}_{t=\beta_\ell^0} \delta\beta_\ell$$
$$-\delta\mu(\mathbf{n}, \boldsymbol{\sigma}) \int_{\beta_\ell^0}^{\beta_u^0} \theta_a^{(1)}(t)\rho^0(t)\, dt + \theta_a^{(1)}\left(\beta_\ell^0\right) \left[\delta\rho_{in} - \delta\beta_\ell \left\{\frac{\partial\rho(t)}{\partial t}\right\}_{t=\beta_\ell^0}\right]. \tag{3.565}$$

Identifying the expressions that multiply the various parameter variations shows that the expressions of the first-order sensitivities shown in (3.552)–(3.555) remain unchanged [except that $\theta^{(1)}(t)$ is formally replaced by $\theta_a^{(1)}(t)$] while the first-order sensitivities shown in (3.550) and (3.551) are replaced by the first-order sensitivity $\partial R_2/\partial\mu$, which has the following expression provided by (3.565):

$$\left(\frac{\partial R_2}{\partial\mu(\mathbf{n}, \boldsymbol{\sigma})}\right)_{\boldsymbol{\alpha}^0} = -\int_{\beta_\ell^0}^{\beta_u^0} \theta_a^{(1)}(t)\rho^0(t) dt. \tag{3.566}$$

3.6.3.2 Second-Order Sensitivities

The second-order sensitivities of $R_2(\rho; \boldsymbol{\alpha})$ are obtained by computing successively the first-order G-differentials of the expressions obtained for the first-order sensitivities. It has been shown in Sect. 3.6.2.7 for the response $R_1(\rho; \boldsymbol{\alpha})$, it is very advantageous to aggregate "fundamental" parameters into "derived" parameters, whenever possible, in order to reduce the number of first-order sensitivities that would need to be computed directly in terms of the forward and/or first-level adjoint functions. This fact also remains valid for the computation of the second-order sensitivities of $R_2(\rho; \boldsymbol{\alpha})$, as will be illustrated below.

In terms of the fundamental parameters listed in (3.539), the second-order sensitivities of $R_2(\rho; \boldsymbol{\alpha})$ are obtained by computing successively the first-order G-differentials of the expressions obtained for the first-order sensitivities obtained in (3.550)–(3.555). The detailed steps will be illustrated by computing the second-order sensitivities corresponding to $\partial R_2/\partial\sigma_i$, obtained in (3.550), which is representative of the steps that would be repeated for the computations of the other second-order sensitivities of $R_2(\rho; \boldsymbol{\alpha})$, which would be obtained from the first-order G-differentials of the expressions presented in (3.551)–(3.555). The first-order G-differential of the expression provided in (3.550) is obtained as follows:

$$\left\{\delta\left(\frac{\partial R_2}{\partial \sigma_i}\right)\right\}_{\boldsymbol{\alpha}^0} \triangleq -\left\{\frac{d}{d\varepsilon}\left[(n_i+\varepsilon\delta n_i)_{\boldsymbol{\alpha}^0}\int\limits_{\beta_\ell^0+\varepsilon\delta\beta_\ell}^{\beta_u^0+\varepsilon\delta\beta_u}\left(\theta^{(1)}+\varepsilon\delta\theta^{(1)}\right)_{\boldsymbol{\alpha}^0}\right.\right.$$
$$\left.\left.\times(\rho+\varepsilon\delta\rho)_{\boldsymbol{\alpha}^0}dt\right]\right\}_{\varepsilon=0} = \left[\delta\left(\frac{\partial R_2}{\partial \sigma_i}\right)\right]_{dir} + \left[\delta\left(\frac{\partial R_2}{\partial \sigma_i}\right)\right]_{ind}, \tag{3.567}$$

where, for each $i = 1, \ldots, N$, the direct-effect and, respectively, indirect-effect terms are defined as follows:

$$\left[\delta\left(\frac{\partial R_2}{\partial \sigma_i}\right)\right]_{dir} \triangleq -(\delta n_i)\int\limits_{\beta_\ell^0}^{\beta_u^0}\theta^{(1)}(t)\rho(t)dt + (\delta\beta_\ell)n_i\theta^{(1)}\left(\beta_\ell^0\right)\rho\left(\beta_\ell^0\right), \tag{3.568}$$

$$\left[\delta\left(\frac{\partial R_2}{\partial \sigma_i}\right)\right]_{ind} \triangleq -\left\{n_i\int\limits_{\beta_\ell^0}^{\beta_u^0}\delta\theta^{(1)}(t)\rho(t)dt + n_i\int\limits_{\beta_\ell^0}^{\beta_u^0}\theta^{(1)}(t)\delta\rho(t)dt\right\}_{\boldsymbol{\alpha}^0}. \tag{3.569}$$

The direct-effect term defined by (3.568) can be computed immediately, since the first-level adjoint function $\theta^{(1)}(t)$ and the forward function $\rho(t)$ are known. However, the indirect-effect term defined by (3.569) contains the variation $\delta\theta^{(1)}(t)$ in the first-level adjoint function and, respectively, the variation $\delta\rho(t)$ in the forward function, both of which depend on parameter variations and neither of which is immediately available. The variation $\delta\theta^{(1)}(t)$ of the first-level adjoint function $\theta^{(1)}(t)$ is related to the parameter variations through the G-differential of the 1st-LASS, which is derived by applying the definition of the G-differential to (3.547) and (3.548) to obtain the following equations (which are evaluated at the nominal parameter values, but this is not explicitly indicated below):

$$-\frac{d\left[\delta\theta^{(1)}(t)\right]}{dt} + \delta\theta^{(1)}(t)\sum_{k=1}^{N}n_k\sigma_k = (\delta\Sigma_d)$$
$$-\theta^{(1)}(t)\sum_{k=1}^{N}\left[n_k(\delta\sigma_k)+\sigma_k(\delta n_k)\right],\quad 0\le\beta_\ell^0\le t\le\beta_u^0<\infty, \tag{3.570}$$

$$\left\{\delta\theta^{(1)}(t) + \left(\delta\beta_u\right)\frac{d\theta^{(1)}(t)}{dt}\right\}_{t=\beta_u^0} = 0. \tag{3.571}$$

Since (3.570) and (3.571) depend on the parameter variations, solving them would be just as impractical computationally as solving the 1st-LVSS. Therefore, the need for solving these equations will be circumvented by expressing the indirect-effect term defined in (3.569) in an alternative way so as to eliminate the appearance of $\delta\theta^{(1)}(t)$ and $\delta\rho(t)$. This elimination will be accomplished by using a Hilbert space, denoted as $H_2(\Omega_t)$, $\Omega_t \triangleq (\beta_\ell^0, \beta_u^0)$, endowed with the inner product defined in (3.356). Using the definition given in (3.356), construct the inner product of (3.570) and (3.329) with a square integrable two-component function $\boldsymbol{\theta}_1^{(2)}(i;t) \triangleq \left[\theta_{11}^{(2)}(i;t), \theta_{12}^{(2)}(i;t)\right] \in H_2(\Omega_t)$ to obtain the following relation:

$$\int_{\beta_\ell^0}^{\beta_u^0} \left[\theta_{11}^{(2)}(i;t), \theta_{12}^{(2)}(i;t)\right] \begin{pmatrix} -\frac{d}{dt} + \sum_{k=1}^{N} n_k\sigma_k & 0 \\ 0 & \frac{d}{dt} + \sum_{k=1}^{N} n_k\sigma_k \end{pmatrix} \begin{pmatrix} \delta\theta^{(1)}(t) \\ \delta\rho(t) \end{pmatrix} dt$$
$$= \int_{\beta_\ell^0}^{\beta_u^0} \left[\theta_{11}^{(2)}(i;t), \theta_{12}^{(2)}(i;t)\right] \begin{pmatrix} (\delta\Sigma_d) - \theta^{(1)}(t) \sum_{k=1}^{N} \left[n_k(\delta\sigma_k) + \sigma_k(\delta n_k)\right] \\ -\rho(t) \sum_{k=1}^{N} \left[n_k(\delta\sigma_k) + \sigma_k(\delta n_k)\right] \end{pmatrix} dt. \tag{3.572}$$

Integrating by parts the left side of (3.572) so as to transfer the differential operations on $\delta\theta^{(1)}(t)$ and $\delta\rho(t)$ to differential operations on $\theta_{11}^{(2)}(i;t)$ and $\theta_{12}^{(2)}(i;t)$ yields the following result:

$$\int_{\beta_\ell^0}^{\beta_u^0} \theta_{11}^{(2)}(i;t) \left[-\frac{d}{dt}\delta\theta^{(1)}(t) + \delta\theta^{(1)}(t) \sum_{k=1}^{N} n_k\sigma_k \right] dt + \int_{\beta_\ell^0}^{\beta_u^0} \theta_{12}^{(2)}(i;t) \left[\frac{d}{dt}\delta\rho(t) \right.$$
$$\left. + \delta\rho(t) \sum_{k=1}^{N} n_k\sigma_k \right] dt = -\theta_{11}^{(2)}(i;\beta_u^0)\delta\theta^{(1)}(\beta_u^0) + \theta_{11}^{(2)}(i;\beta_\ell^0)\delta\theta^{(1)}(\beta_\ell^0)$$
$$+ \theta_{12}^{(2)}(i;\beta_u^0)\delta\rho(\beta_u^0) - \theta_{12}^{(2)}(i;\beta_\ell^0)\delta\rho(\beta_\ell^0) + \int_{\beta_\ell^0}^{\beta_u^0} \delta\theta^{(1)}(t) \left[\frac{d\theta_{11}^{(2)}(i;t)}{dt} \right. \tag{3.573}$$
$$\left. + \theta_{11}^{(2)}(i;t) \sum_{k=1}^{N} n_k\sigma_k \right] dt + \int_{\beta_\ell^0}^{\beta_u^0} \delta\rho(t) \left[-\frac{d\theta_{12}^{(2)}(i;t)}{dt} + \theta_{12}^{(2)}(i;t) \sum_{k=1}^{N} n_k\sigma_k \right] dt.$$

The last two terms on the right side of (3.573) are now required to represent the indirect-effect term defined in (3.569) by imposing the following relations:

$$\frac{d\theta_{11}^{(2)}(i;t)}{dt} + \theta_{11}^{(2)}(i;t)\sum_{k=1}^{N} n_k\sigma_k = -n_i\rho(t), i = 1, \ldots, N; \tag{3.574}$$

$$-\frac{d\theta_{12}^{(2)}(i;t)}{dt} + \theta_{12}^{(2)}(i;t)\sum_{k=1}^{N} n_k\sigma_k = -n_i\theta^{(1)}(t), i = 1, \ldots, N. \tag{3.575}$$

The boundary conditions for (3.574) and (3.575) are established by requiring the contributions involving the unknown quantities $\delta\theta^{(1)}(\beta_\ell)$ and $\delta\rho(\beta_u)$ in (3.573) to vanish, which can be accomplished by imposing the following conditions:

$$\theta_{11}^{(2)}(i;\beta_\ell) = 0, \quad \theta_{12}^{(2)}(i;\beta_u) = 0; \; i = 1, \ldots, N. \tag{3.576}$$

The system of equations comprising (3.574)–(3.576), which is independent of parameter variations, constitutes the 2nd-LASS for the two-component second-level adjoint function $\boldsymbol{\theta}_1^{(2)}(i;t) \triangleq \left[\theta_{11}^{(2)}(i;t), \theta_{12}^{(2)}(i;t)\right]$.

Replacing the left side of (3.573) by the right side of (3.572) and taking into account (3.574)–(3.576) yields the following expression for the indirect-effect term defined in (3.569), for $i = 1, \ldots, N$:

$$\left[\delta\left(\frac{\partial R_2}{\partial \sigma_i}\right)\right]_{ind} = \theta_{11}^{(2)}(i;\beta_u^0)\delta\theta^{(1)}(\beta_u^0) + \theta_{12}^{(2)}(i;\beta_\ell^0)\delta\rho(\beta_\ell^0)$$
$$+\int_{\beta_\ell^0}^{\beta_u^0} dt\left[\theta_{11}^{(2)}(i;t), \theta_{12}^{(2)}(i;t)\right]\begin{pmatrix} (\delta\Sigma_d) - \theta^{(1)}(t)\sum_{k=1}^{N}[n_k(\delta\sigma_k) + \sigma_k(\delta n_k)] \\ -\rho(t)\sum_{k=1}^{N}[n_k(\delta\sigma_k) + \sigma_k(\delta n_k)] \end{pmatrix}. \tag{3.577}$$

Using the conditions given in (3.330) and (3.571) in the first two terms on the right side of (3.577) yields the following expression for the indirect-effect term, $i = 1, \ldots, N$:

$$\left[\delta\left(\frac{\partial R_2}{\partial \sigma_i}\right)\right]_{ind} = \theta_{12}^{(2)}(i;\beta_\ell)\left\{\delta\rho_{in} - \delta\beta_\ell\left[\frac{d\rho(t)}{dt}\right]_{t=\beta_\ell^0}\right\} + \theta_{11}^{(2)}(i;\beta_u)\left\{-(\delta\beta_u)\frac{d\theta^{(1)}(t)}{dt}\right\}_{t=\beta_u^0}$$
$$+\int_{\beta_\ell^0}^{\beta_u^0} dt\left[\theta_{11}^{(2)}(i;t), \theta_{12}^{(2)}(i;t)\right]\begin{pmatrix} (\delta\Sigma_d) - \theta^{(1)}(t)\sum_{k=1}^{N}[n_k(\delta\sigma_k) + \sigma_k(\delta n_k)] \\ -\rho(t)\sum_{k=1}^{N}[n_k(\delta\sigma_k) + \sigma_k(\delta n_k)] \end{pmatrix}. \tag{3.578}$$

Adding the direct-effect term defined in (3.568) to (3.578) and identifying in the resulting expression the coefficients multiplying the variations $\delta\sigma_k$, δn_k, $\delta\rho_{in}$, δt_d, $\delta\beta_\ell$, and $\delta\beta_u$ yields the following expression for the respective second-order sensitivities of the response $R_2(\rho;\boldsymbol{\alpha})$:

$$\frac{\partial^2 R_2}{\partial\sigma_j\partial\sigma_i} = -n_j\int_{\beta_\ell^0}^{\beta_u^0}\left[\theta_{11}^{(2)}(i;t)\theta^{(1)}(t)+\theta_{12}^{(2)}(i;t)\rho(t)\right]dt;\ i,j=1,\ldots,N; \tag{3.579}$$

$$\begin{aligned}\frac{\partial^2 R_2}{\partial n_j\,\partial\sigma_i} &= -\sigma_j\int_{\beta_\ell^0}^{\beta_u^0}\left[\theta_{11}^{(2)}(i;t)\theta^{(1)}(t)+\theta_{12}^{(2)}(i;t)\rho(t)\right]dt\\ &-\delta_{ij}\int_{\beta_\ell^0}^{\beta_u^0}\theta^{(1)}(t)\rho(t)\,dt;\ i,j=1,\ldots,N;\ \delta_{ij}=\begin{cases}1, & i=j;\\ 0, & i\neq j;\end{cases}\end{aligned} \tag{3.580}$$

$$\frac{\partial^2 R_2}{\partial\rho_{in}\,\partial\sigma_i} = \theta_{12}^{(2)}(i;\beta_\ell);\quad i=1,\ldots,N; \tag{3.581}$$

$$\frac{\partial^2 R_2}{\partial\Sigma_d\,\partial\sigma_i} = \int_{\beta_\ell^0}^{\beta_u^0}\theta_{11}^{(2)}(i;t)\,dt;\quad i=1,\ldots,N; \tag{3.582}$$

$$\frac{\partial^2 R_2}{\partial\beta_u\,\partial\sigma_i} = -\theta_{11}^{(2)}(i;\beta_u)\left[\frac{d\theta^{(1)}(t)}{dt}\right]_{t=\beta_u^0};\quad i=1,\ldots,N; \tag{3.583}$$

$$\frac{\partial^2 R_2}{\partial\beta_\ell\,\partial\sigma_i} = n_i\theta^{(1)}(\beta_\ell)\rho(\beta_\ell)-\theta_{12}^{(2)}(i;\beta_\ell)\left[\frac{d\rho(t)}{dt}\right]_{t=\beta_\ell^0};\quad i=1,\ldots,N. \tag{3.584}$$

The second-order sensitivities obtained in (3.579)–(3.584) can be computed after having determined the second-level adjoint function $\boldsymbol{\theta}_1^{(2)}(t)\triangleq\left[\theta_{11}^{(2)}(i;t),\theta_{12}^{(2)}(i;t)\right]$ by solving the 2nd-LASS comprising (3.574)–(3.576) using the nominal parameter values (the superscript "zero," which indicates "nominal values," has been omitted, for simplicity). In the most unfavorable situation, the 2nd-LASS, comprising (3.574)–(3.576), would need to be solved numerically for each distinct value n_i, for a total of N-times. Even in such a "worse-case scenario," however, only the right sides (i.e., "sources") of (3.574) and (3.575) would need to be modified, which is relatively easy to implement computationally. The left sides of these equations remain unchanged since they are independent of the index $i = 1, \ldots, N$. Notably, the following change of the dependent variables:

$$\theta_{11}^{(2)}(i;t) = n_i u_{11}^{(2)}(t),\quad \theta_{12}^{(2)}(i;t) = n_i u_{12}^{(2)}(t), \tag{3.585}$$

would transform (3.574)–(3.576) into the following form:

$$\frac{du_{11}^{(2)}(t)}{dt}+u_{11}^{(2)}(t)\sum_{k=1}^{N}n_k\sigma_k=-\rho(t), \tag{3.586}$$

$$-\frac{du_{12}^{(2)}(t)}{dt}+u_{12}^{(2)}(t)\sum_{k=1}^{N}n_k\sigma_k=-\theta^{(1)}(t), \tag{3.587}$$

$$u_{11}^{(2)}(\beta_\ell)=0,\quad u_{12}^{(2)}(\beta_u)=0. \tag{3.588}$$

The alternative 2nd-LASS, which comprises (3.586)–(3.588), is independent of the index $i = 1, \ldots, N$, and would need to be solved (numerically or analytically) only once, to determine the functions $u_{11}^{(2)}(t)$ and $u_{12}^{(2)}(t)$, which have the following closed-form expressions:

$$u_{11}^{(2)}(t)=\rho_{in}(\beta_\ell-t)\exp\left[(\beta_\ell-t)\sum_{k=1}^{N}n_k\sigma_k\right], \tag{3.589}$$

$$u_{12}^{(2)}(t)=\Sigma_d\left(\sum_{k=1}^{N}n_k\sigma_k\right)^{-1} \times\left\{\left[\beta_u-t+\left(\sum_{k=1}^{N}n_k\sigma_k\right)^{-1}\right]\exp\left[(t-\beta_u)\sum_{k=1}^{N}n_k\sigma_k\right]-\left(\sum_{k=1}^{N}n_k\sigma_k\right)^{-1}\right\}. \tag{3.590}$$

The components of the second-level adjoint function $\boldsymbol{\theta}_1^{(2)}(i;t)$ are obtained by multiplying the functions $u_{11}^{(2)}(t)$ and $u_{12}^{(2)}(t)$ by the respective model parameters n_i, to obtain the following expressions for the components of the second-level adjoint function $\boldsymbol{\theta}_1^{(2)}(i;t)\triangleq\left[\theta_{11}^{(2)}(i;t),\theta_{12}^{(2)}(i;t)\right]$, for $i = 1, \ldots, N$:

$$\theta_{11}^{(2)}(i;t)=n_i\rho_{in}(\beta_\ell-t)\exp\left[(\beta_\ell-t)\sum_{k=1}^{N}n_k\sigma_k\right], \tag{3.591}$$

$$\theta_{12}^{(2)}(i;t)=-\frac{n_i\Sigma_d}{\left(\sum_{k=1}^{N}n_k\sigma_k\right)^2} \times\left\{1+\left[(t-\beta_u)\sum_{k=1}^{N}n_k\sigma_k-1\right]\exp\left[(t-\beta_u)\sum_{k=1}^{N}n_k\sigma_k\right]\right\}. \tag{3.592}$$

Using (3.591) and (3.592) in (3.579)–(3.584) and performing the respective operations yields the following results for the corresponding partial second-order sensitivities, for $i = 1, \ldots, N$:

$$\frac{\partial^2 R_2}{\partial \sigma_j \partial \sigma_i} = \frac{2n_i n_j \rho_{in} \Sigma_d}{\left(\sum_{k=1}^{N} n_k \sigma_k\right)^3} \left\{1 + \left[(\beta_\ell - \beta_u) \sum_{k=1}^{N} n_k \sigma_k - 1\right] \exp\left[(\beta_\ell - \beta_u) \sum_{k=1}^{N} n_k \sigma_k\right]\right\}$$
$$- \frac{n_i n_j (\beta_\ell - \beta_u) \rho_{in} \Sigma_d}{\left(\sum_{k=1}^{N} n_k \sigma_k\right)^2} \left[(\beta_\ell - \beta_u) \sum_{k=1}^{N} n_k \sigma_k - 1\right] \exp\left[(\beta_\ell - \beta_u) \sum_{k=1}^{N} n_k \sigma_k\right]$$
$$- \frac{n_i n_j (\beta_\ell - \beta_u) \rho_{in} \Sigma_d}{\left(\sum_{k=1}^{N} n_k \sigma_k\right)^2} \exp\left[(\beta_\ell - \beta_u) \sum_{k=1}^{N} n_k \sigma_k\right], \quad i,j = 1, \ldots, N; \tag{3.593}$$

$$\frac{\partial^2 R_2}{\partial n_j \partial \sigma_i} = \rho_{in} \Sigma_d \left(\sum_{k=1}^{N} n_k \sigma_k\right)^{-3} \left[2 n_i \sigma_j - \delta_{ij} \sum_{k=1}^{N} n_k \sigma_k\right]$$
$$\times \left\{1 + \left[(\beta_\ell - \beta_u) \sum_{k=1}^{N} n_k \sigma_k - 1\right] \exp\left[(\beta_\ell - \beta_u) \sum_{k=1}^{N} n_k \sigma_k\right]\right\}$$
$$- \frac{n_i \sigma_j (\beta_\ell - \beta_u) \rho_{in} \Sigma_d}{\left(\sum_{k=1}^{N} n_k \sigma_k\right)^2} \left[(\beta_\ell - \beta_u) \sum_{k=1}^{N} n_k \sigma_k - 1\right] \exp\left[(\beta_\ell - \beta_u) \sum_{k=1}^{N} n_k \sigma_k\right]$$
$$- \frac{n_i \sigma_j (\beta_\ell - \beta_u) \rho_{in} \Sigma_d}{\left(\sum_{k=1}^{N} n_k \sigma_k\right)^2} \exp\left[(\beta_\ell - \beta_u) \sum_{k=1}^{N} n_k \sigma_k\right], \quad i,j = 1, \ldots, N; \tag{3.594}$$

$$\frac{\partial^2 R_2}{\partial \rho_{in} \partial \sigma_i} = - \frac{n_i \Sigma_d \left\{1 + \left[(\beta_\ell - \beta_u) \sum_{k=1}^{N} n_k \sigma_k - 1\right] \exp\left[(\beta_\ell - \beta_u) \sum_{k=1}^{N} n_k \sigma_k\right]\right\}}{\left(\sum_{k=1}^{N} n_k \sigma_k\right)^2}; \tag{3.595}$$

$$\frac{\partial^2 R_2}{\partial \Sigma_d \partial \sigma_i} = - \frac{n_i \rho_{in} \left\{1 + \left[(\beta_\ell - \beta_u) \sum_{k=1}^{N} n_k \sigma_k - 1\right] \exp\left[(\beta_\ell - \beta_u) \sum_{k=1}^{N} n_k \sigma_k\right]\right\}}{\left(\sum_{k=1}^{N} n_k \sigma_k\right)^2}; \tag{3.596}$$

$$\frac{\partial^2 R_2}{\partial \beta_u \, \partial \sigma_i} = n_i \rho_{in} \Sigma_d (\beta_\ell - \beta_u) \exp\left[(\beta_\ell - \beta_u) \sum_{k=1}^{N} n_k \sigma_k \right]; \tag{3.597}$$

$$\frac{\partial^2 R_2}{\partial \beta_\ell \, \partial \sigma_i} = -n_i \rho_{in} \Sigma_d (\beta_\ell - \beta_u) \exp\left[(\beta_\ell - \beta_u) \sum_{k=1}^{N} n_k \sigma_k \right]. \tag{3.598}$$

As before, the right sides of expressions shown in (3.593)–(3.598) are to be evaluated at the nominal values for the parameters, but the superscript "zero," which indicates "nominal values," has been omitted, for notational simplicity.

The expressions of the remaining second-order sensitivities of $R_2(\rho; \boldsymbol{\alpha})$ can be derived similarly, by applying the 2nd-CASAM-L. All in all, using the fundamental $2N + 4$ parameters, a total number of $2N + 4$ second-level adjoint functions would need to be computed to determine the second-order sensitivities, comprising the following computations:

1. N computations for solving the 2nd-LASS to obtain the second-level adjoint function $\boldsymbol{\theta}_1^{(2)}(i; t)$ which corresponds to the "response" $\partial R_2/\partial \sigma_i$, as shown so far in this subsection.
2. N computations for solving the 2nd-LASS to obtain the second-level adjoint function which corresponds to the "response" $\partial R_2/\partial n_i$.
3. One computation for solving the 2nd-LASS to obtain the second-level adjoint function which corresponds to the "response" $\partial R_2/\partial \rho_{in}$.
4. One computation for solving the 2nd-LASS to obtain the second-level adjoint function which corresponds to the "response" $\partial R_2/\partial \beta_u$.
5. One computation for solving the 2nd-LASS to obtain the second-level adjoint function which corresponds to the "response" $\partial R_2/\partial \beta_\ell$.
6. One computation for solving the 2nd-LASS to obtain the second-level adjoint function which corresponds to the "response" $\partial R_2/\partial \Sigma_d$.

On the other hand, using the "aggregated" parameter $\mu(\mathbf{n}, \boldsymbol{\sigma})$ and the "aggregated" first-level adjoint function $\theta_a^{(1)}(t)$, the $2N$ computations mentioned above in items (1) and (2) would be replaced by a single adjoint computations that would correspond to the "response" $\partial R_2/\partial \mu$. Hence, the total number of "large-scale" adjoint computations would be reduced from $2N + 4$ to 5.

The steps involved in the computation of the second-order sensitivities corresponding to $\partial R_2/\partial \mu$ are provided below:

$$\begin{aligned} \delta R_2(\rho; \boldsymbol{\alpha}; \delta\rho; \delta\boldsymbol{\alpha}) = (\delta\Sigma_d) \int_{\beta_\ell^0}^{\beta_u^0} \rho(t)\, dt + \Sigma_d^0 \{\rho(t)\}_{t=\beta_u^0} \delta\beta_u - \Sigma_d^0 \{\rho(t)\}_{t=\beta_\ell^0} \delta\beta_\ell \\ - \delta\mu(\mathbf{n}, \boldsymbol{\sigma}) \int_{\beta_\ell^0}^{\beta_u^0} \theta_a^{(1)}(t) \rho^0(t)\, dt + \theta_a^{(1)}(\beta_\ell^0) \left[\delta\rho_{in} - \delta\beta_\ell \left\{ \frac{\partial \rho(t)}{\partial t} \right\}_{t=\beta_\ell^0} \right]. \end{aligned} \tag{3.599}$$

Identifying the expressions that multiply the various parameter variations shows that the expressions of the first-order sensitivities given in (3.552)–(3.555) remain unchanged [except that $\theta^{(1)}(t)$ is formally replaced by $\theta_a^{(1)}(t)$] while the first-order sensitivities shown in (3.550) and (3.551) are replaced by the first-order sensitivity $\partial R_2/\partial \mu$, which has the following expression provided by (3.565):

$$\left(\frac{\partial R_2}{\partial \mu(\mathbf{n},\boldsymbol{\sigma})}\right)_{\boldsymbol{\alpha}^0} = -\int_{\beta_\ell^0}^{\beta_u^0} \theta_a^{(1)}(t)\rho^0(t)dt. \tag{3.600}$$

The computation of the second-order sensitivities that correspond to the first-order sensitivity $\partial R_2/\partial \mu$ would follow the same conceptual steps as outlined in Sect. 3.6.2.7.

The third-order sensitivities of $R_2(\rho;\boldsymbol{\alpha})$ are obtained by computing successively the first-order G-differentials of the expressions for the second-order sensitivities, which play the roles of "responses." Thus, if the expressions of the second-order sensitivities obtained in (3.579)–(3.584) for the fundamental parameters were used, there would be $2N + 4$ "responses," so the corresponding $2N + 4$ third-level adjoint sensitivity functions would need to be determined by solving $2N + 4$ Third-Level Adjoint Sensitivity Systems. On the other hand, if the "aggregate" parameter $\mu(\mathbf{n},\boldsymbol{\sigma})$ is used, there would be only *five* second-order sensitivities as "responses," so there would be only *five* Third-Level Adjoint Sensitivity Systems to be solved to determine the corresponding third-level adjoint sensitivity functions. Clearly, using "aggregated parameter" such as $\mu(\mathbf{n},\boldsymbol{\sigma})$, whenever possible, greatly reduces the number of large-scale adjoint computations needed for determining the second- and higher-order sensitivities.

3.7 Chapter Summary

The application of the nth-CASAM-L has been illustrated in this chapter by considering two paradigm models characterized by uncertain model parameters. One model involved imprecisely known parameters, internal and external boundaries in a spatial domain. The other model involved imprecisely known parameters, initial- and final-times and conditions in a temporal domain.

It has been shown that the first-order response sensitivities are computed efficiently and exactly in terms of the first-level adjoint sensitivity function by simply performing inexpensive integrations (using quadrature formulas) rather than by needing to solve operator (differential, integral, matrix) equations, which represent "large-scale computations." Notably, the first-level adjoint function is obtained by a single large-scale computation, which is needed for solving the 1st-LASS once, regardless of the number of model parameters. Contributions to the response sensitivities with respect to the uncertain boundary parameters can arise from either the model's boundary conditions, from the definition of the model's response or from both. In some cases, the contributions from the model's boundary conditions may cancel partially the "direct-effect" contributions stemming from the response's definition. The models admit many relative sensitivities that have identical values. For the particle transmission model, for example, the relative first-order sensitivities of the response with respect to the nuclide number densities have the same values as the first-order sensitivities of the response with respect to the microscopic cross sections. The precise and efficient computation of many relative sensitivities that have identical values does not pose any difficulties to the nth-CASAM-L methodology but is known to pose extreme challenges to conventional statistical methods which rely on sensitivities being readily "sortable" by having relative values that clearly differ from each other.

All of the second-order sensitivities are obtained by considering each of the first-order sensitivity to be a "model response" and by determining the total G-differential of this response. This consideration stems from the basic definition of the "second-order differential" which is defined as the "first-order differential of the first-order differential." Consequently, to each first-order sensitivity, there corresponds a Second-Level Adjoint Sensitivity System (2nd-LASS) and a second-level adjoint sensitivity function, which enables the determination of all of the second-order partial sensitivities that stem from the response (partial first-order sensitivity) under consideration. The mixed second-order sensitivities are obtained twice, stemming from two distinct Second-Level Adjoint Sensitivity Systems. Hence, the 2nd-CASAM-L provides an inherent independent verification of the correctness and accuracy of the second-level adjoint sensitivity functions that are used to compute the respective mixed second-order sensitivities. In all cases, though, the equations underlying the 2nd-LASS need not be solved simultaneously. Instead, they are solved successively, partially decoupled from one another.

The information provided by the first-order sensitivities usually indicates which second-order sensitivities are important and which could be neglected. Therefore, it is useful to prioritize the computation of the second-order sensitivities by using the rankings of the relative magnitudes of the first-order sensitivities as a "priority indicator": the larger the magnitude of the relative first-order sensitivity, the higher the priority for computing the corresponding second-order sensitivities. Also, since vanishing first-order sensitivities may indicate critical points of the response in the phase-space of model parameters, it is also of interest to compute the second-order sensitivities that correspond to vanishing first-order sensitivities. In practice, only those second-order partial sensitivities which are deemed important would need to be computed.

Each successively higher-order sensitivities are computed using the same principles as used for computing the second-order sensitivities. The computation of the third-order sensitivities can be prioritized by using the magnitudes/importance of the first-order and second-order sensitivities as guiding indicators, and so on. It has also been shown that aggregating several parameters into a single one significantly reduces the number of adjoint computations needed for the computation of second- and higher-order sensitivities. Aggregating parameters is not necessary for the computation of the first-order sensitivities since the computation of all of the first-order sensitivities requires just a single adjoint-computation for solving the 1st-LASS.

The illustrative sensitivity analysis of the paradigm models considered in this chapter has confirmed the fundamental advantage of applying the nth-CASAM-L (as opposed to statistical and/or finite-difference methods) for sensitivity analysis of large-scale models with many parameters. It has been shown that the exact expressions are obtained for all sensitivities, of any order, and each sensitivity is computed in a minimum number of large-scale computations, as follows: for a model having TP-parameters (where "TP" denotes "total number of parameters"), for every response and/or sensitivity of interest, the nth-CASAM-L computes the TP "next-higher-order" sensitivities needing just one adjoint ("large-scale") computation performed in a linearly increasing higher-dimensional Hilbert space. Very importantly, the nth-CASAM-L computes the higher-level adjoint functions using the same forward and adjoint solvers (i.e., computer codes) as used for solving the original forward and adjoint systems, thus requiring relatively minor additional software development for computing the various-order sensitivities.

Since the models analyzed in the chapter involve numerous parameters yet are sufficiently simple to admit closed-form exact solutions for all quantities (forward and adjoint state functions, responses, sensitivities, etc.) of interest, these paradigm models can be used as a benchmarks for comparing the nth-CASAM-L to conventional statistical and/or finite-difference methods.

The nth-CASAM-L is the only practically implementable methodology for obtaining the exact expressions (i.e., free of methodologically introduced approximations) of arbitrarily high-order sensitivities (functional derivatives) of model responses with respect to model parameters, for coupled forward/adjoint linear systems. By enabling the practical computation of any, arbitrarily high-order, response sensitivities to model parameters, the nth-CASAM-L makes it possible to compare the relative values of the sensitivities of various order, in order to assess which sensitivities are important and which may actually be neglected. Such quantitative comparisons enable investigations of the convergence of the (multivariate) Taylor series expansion of the response in terms of parameter variations, as well as investigating the actual validity of expressions that are derived from Taylor expansion of the response (e.g., response variances/covariance, skewness, kurtosis) as a function of the model's parameters. The larger the number of model parameters, the more efficient the nth-CASAM-L becomes for computing arbitrarily high-order response sensitivities. The nth-CASAM-L enables the direct theoretical derivation ("on paper") of the expression of a specific high-order sensitivity of interest, which can subsequently be computed directly; such a selective direct computation of a sensitivity of particular interest is not possible using statistical methods.

Chapter 4
Sensitivity Analysis of Neutron Transport Modeled by the Forward and Adjoint Linear Boltzmann Equations

4.1 Introduction

This chapter illustrates the application of the 2nd-CASAM-L to perform sensitivity analysis of ratios of generic functionals of the forward and adjoint neutron flux distributions in a multiplying subcritical system which comprises a non-fission neutron source. The forward and adjoint flux distributions in such a system are modeled by the forward and, respectively, adjoint time-independent integro-differential neutron transport Boltzmann equations (see, e.g., Davison 1957; Stacey 1974, 2001). This physical system has been chosen as a paradigm application of the 2nd-CASAM-L because the Boltzmann equation comprises both differential and integral operators, in six independent variables, as well as a large number of model parameters. Using the integro-differential Boltzmann equation makes it possible to illustrate simultaneously the construction of adjoint operators both for differential and for integral operators. Solving the integro-differential Boltzmann equation epitomizes "large-scale computations." The system response considered for the application of the 2nd-CASAM-L is a ratio of two generic nonlinear functionals of the forward and adjoint particle fluxes, while the model parameters include isotopic number densities, microscopic cross sections, fission spectrum parameters, neutron sources, and detector response functions. Such responses represent the most general functionals of fundamental importance in reactor physics. Notably, the modeling of multiplying subcritical system using the Boltzmann equation considered in this chapter greatly generalizes the modeling of the spatial transmission of particles through homogenized materials considered in Chap. 3. Of course, closed-form analytical expressions for the response sensitivities of various order, such as obtained in Chap. 3, are no longer possible to obtain for the complex subcritical multiplying physical system which will be analyzed in this chapter.

The material presented in this chapter is organized as follows: the mathematical modeling of neutron transport in a multiplying medium with internal and external sources, together with the definition of a generic model response—namely a ratio of

D. G. Cacuci, *The nth-Order Comprehensive Adjoint Sensitivity Analysis Methodology, Volume I*, https://doi.org/10.1007/978-3-030-96364-4_4

nonlinear functionals of the forward and adjoint neutron fluxes—are presented in Sect. 4.2. The application of the 2nd-CASAM-L to determine the first-order and, respectively, second-order sensitivities of this generic response to the model parameters is presented in Sects. 4.3 and 4.4, respectively. The material presented in Sects. 4.2–4.4 is based on the work by Cacuci (2019c, 2020). Section 4.5 presents the application of the 2nd-CASAM-L to determine the first-order and, respectively, second-order response sensitivities of the Schwinger functional (Levine and Schwinger 1949) and the Roussopoulos-functional (Roussopolos 1953) with respect to imprecisely known model parameters. The material presented in Sect. 4.5 is based on the work by Cacuci (2019a, b). The summary presented in Sect. 4.6 highlights the significance of the methodology and results presented in this chapter.

4.2 Paradigm Physical System: Neutron Transport in a Multiplying Medium with Source

The physical system chosen in this chapter to illustrate the application of the general theoretical developments presented in Chap. 2 is the transport of neutrons in a multiplying medium of convex volume V which contains fission and non-fission sources of neutrons. Such a physical system is modeled by the linear integro-differential Boltzmann transport equation. Using the integro-differential Boltzmann equation makes it possible to illustrate simultaneously the construction of adjoint operators both for differential and for integral operators. The medium is considered to be placed in vacuum and the medium's outer boundary, denoted as ∂V, is considered to be perfectly well known in order to simplify the mathematical treatment by disregarding possible effects of boundary perturbations. The stationary distribution of neutrons in such a system is modeled by using the standard form of the time-independent integro-differential forward Boltzmann transport equation, namely:

$$\begin{aligned} L(\boldsymbol{\alpha})\varphi(\mathbf{r},\boldsymbol{\Omega},E) &\triangleq \boldsymbol{\Omega}\cdot\nabla\varphi(\mathbf{r},\boldsymbol{\Omega},E)+\Sigma_t(\mathbf{t};\mathbf{r},E)\,\varphi(\mathbf{r},\boldsymbol{\Omega},E) \\ &-\int_{4\pi} d\boldsymbol{\Omega}'\int_0^{E_f} dE'\,\Sigma_s(\mathbf{s};\mathbf{r},E'\to E,\boldsymbol{\Omega}'\to\boldsymbol{\Omega})\,\varphi(\mathbf{r},\boldsymbol{\Omega}',E') \\ &-\chi(\mathbf{p};\mathbf{r},E)\int_{4\pi} d\boldsymbol{\Omega}'\int_0^{E_f} dE'\,\nu\Sigma_f(\mathbf{f};\mathbf{r},E')\varphi(\mathbf{r},\boldsymbol{\Omega}',E') = Q(\mathbf{q};\mathbf{r},\boldsymbol{\Omega},E), \end{aligned} \tag{4.1}$$

subject to the customary vacuum boundary condition which specifies that there is no incoming flux of particles:

$$\varphi(\mathbf{r}_s, \boldsymbol{\Omega}, E) = 0, \;\; \mathbf{r}_s \in \partial V, \;\; \boldsymbol{\Omega} \cdot \mathbf{n} < 0, \;\; 0 < E < E_f, \tag{4.2}$$

where $\mathbf{n}$ denotes the unit outward normal vector at any point $\mathbf{r}_s \in \partial V$ on the body's outer surface ∂V, and E_f denotes the highest neutron energy in the medium.

The components of the vector $\boldsymbol{\alpha}$, which appears in the definition of the Boltzmann operator $L(\boldsymbol{\alpha})$, represent generically all of the model's imprecisely known scalar parameters, including macroscopic and/or microscopic neutron cross sections, fission spectrum parameters, forward and adjoint sources, and parameters that characterize the model responses (i.e., results of interest). Thus, the macroscopic cross-sections $\Sigma_t(\mathbf{t}; \mathbf{r}, E)$, $\Sigma_s(\mathbf{s}; \mathbf{r}, E' \to E, \boldsymbol{\Omega}' \to \boldsymbol{\Omega})$, $\nu\Sigma_f(\mathbf{f}; \mathbf{r}, E' \to E)$, the neutron fission spectrum $\chi(\mathbf{p}; \mathbf{r}, E' \to E)$, and the source $Q(\mathbf{q}; \mathbf{r}, \boldsymbol{\Omega}, E)$ generally depend not only on the spatial variable $\mathbf{r}$, on the energy variable E, and, possibly, solid angle $\boldsymbol{\Omega}$, but also on imperfectly known scalar-valued model parameters such as atomic number densities, microscopic cross-sections, and weighting functions. Specifically, the macroscopic total cross-section $\Sigma_t(\mathbf{t}; \mathbf{r}, E)$ is considered to depend on J_t imprecisely known scalar-valued model parameters denoted as t_i, $i = 1, \ldots, J_t$, which are considered to be the components of a vector of model parameters defined as $\mathbf{t} \triangleq [t_1, \ldots, t_{J_t}]^\dagger$. Similarly, the macroscopic scattering cross section $\Sigma_s(\mathbf{s}; E' \to E, \boldsymbol{\Omega}' \to \boldsymbol{\Omega})$ is considered to depend on J_s imprecisely known scalar-valued model parameters denoted as s_i, $i = 1, \ldots, J_s$, $\mathbf{s} \triangleq [s_1, \ldots, s_{J_s}]^\dagger$, while the effective macroscopic fission cross-section $\nu\Sigma_f(\mathbf{f}; \mathbf{r}, E')$ is considered to depend on J_f imprecisely known scalar-valued model parameters denoted as f_i, $i = 1, \ldots, J_f$, $\mathbf{f} \triangleq \left[f_1, \ldots, f_{J_f}\right]^\dagger$. Furthermore, the fission spectrum $\chi(\mathbf{p}; \mathbf{r}, E' \to E)$ is considered to depend on J_p imprecisely known scalar-valued parameters denoted as p_i, $i = 1, \ldots, J_p$, $\mathbf{p} \triangleq \left[p_1, \ldots, p_{J_p}\right]^\dagger$, while the source $Q(\mathbf{q}; \mathbf{r}, \boldsymbol{\Omega}, E)$ is considered to depend on J_q imprecisely known scalar-valued parameters denoted as q_i, $i = 1, \ldots, J_q$, $\mathbf{q} \triangleq \left[q_1, \ldots, q_{J_q}\right]^\dagger$.

The Hilbert space employed for defining the customary adjoint neutron transport operator will be denoted as H_B and is endowed with the following inner product, denoted as $\langle \varphi_1, \varphi_2 \rangle_B$, between two square-integrable functions $\varphi_1(\mathbf{r}, \boldsymbol{\Omega}, E) \in L_2(V \times \boldsymbol{\Omega} \times E) \in \mathrm{H}_B$ and $\varphi_2(\mathbf{r}, \boldsymbol{\Omega}, E) \in L_2(V \times \boldsymbol{\Omega} \times E) \in \mathrm{H}_B$:

$$\langle \varphi_1, \varphi_2 \rangle_B \triangleq \int dV \int_{4\pi} d\boldsymbol{\Omega} \int_0^{E_f} dE \, \varphi_1(\mathbf{r}, \boldsymbol{\Omega}, E) \varphi_2(\mathbf{r}, \boldsymbol{\Omega}, E). \tag{4.3}$$

In the Hilbert space H_B, the operator formally adjoint to $L(\boldsymbol{\alpha})$ will be denoted $L^+(\boldsymbol{\alpha})$ and is defined through the following relationship:

$$\langle \varphi^+, L(\boldsymbol{\alpha})\varphi \rangle_B = \langle \varphi, L^+(\boldsymbol{\alpha})\varphi^+ \rangle_B + P[\varphi, \varphi^+], \tag{4.4}$$

where the bilinear concomitant $P[\varphi, \varphi^+]$ is defined as follows:

$$P[\varphi,\varphi^{+}]\triangleq\int_{0}^{E_f}dE\left\{\int_{\boldsymbol{\Omega}\cdot\mathbf{n}<0}d\boldsymbol{\Omega}\int_{\partial V}|\boldsymbol{\Omega}\cdot\mathbf{n}|\varphi(\mathbf{r},\boldsymbol{\Omega},E)\varphi^{+}(\mathbf{r},\boldsymbol{\Omega},E)dA\right.$$
$$\left.-\int_{\boldsymbol{\Omega}\cdot\mathbf{n}>0}d\boldsymbol{\Omega}\int_{\partial V}\boldsymbol{\Omega}\cdot\mathbf{n}\varphi(\mathbf{r},\boldsymbol{\Omega},E)\varphi^{+}(\mathbf{r},\boldsymbol{\Omega},E)dA\right\}. \tag{4.5}$$

Carrying out the operations specified by the inner products in (4.4) yields the following expression for the adjoint operator $L^{+}(\boldsymbol{\alpha})$:

$$L^{+}(\boldsymbol{\alpha})\varphi^{+}\triangleq-\boldsymbol{\Omega}\cdot\nabla\varphi^{+}(\mathbf{r},\boldsymbol{\Omega},E)+\Sigma_t(\mathbf{t};\mathbf{r},E)\,\varphi^{+}(\mathbf{r},\boldsymbol{\Omega},E)$$
$$-\int_{4\pi}d\boldsymbol{\Omega}'\int_{0}^{E_f}dE'\,\Sigma_s(\mathbf{s};\mathbf{r},E\to E',\boldsymbol{\Omega}\to\boldsymbol{\Omega}')\,\varphi^{+}(\mathbf{r},\boldsymbol{\Omega}',E')$$
$$-\nu\Sigma_f(\mathbf{f};\mathbf{r},E)\int_{4\pi}d\boldsymbol{\Omega}'\int_{0}^{E_f}dE'\chi(\mathbf{p};\mathbf{r},E')\,\varphi^{+}(\mathbf{r},\boldsymbol{\Omega}',E'). \tag{4.6}$$

The adjoint flux, which will be denoted as $\varphi^{+}(\mathbf{r},\boldsymbol{\Omega},E)$, is the solution of the following "adjoint neutron transport" equation:

$$L^{+}(\boldsymbol{\alpha})\varphi^{+}=Q^{+}(\mathbf{k};\mathbf{r},\boldsymbol{\Omega},E), \tag{4.7}$$

where the source term $Q^{+}(\mathbf{k};\mathbf{r},\boldsymbol{\Omega},E)$ is related to the system's response (i.e., quantity of interest). When the forward flux is subject to the vacuum boundary condition given in (4.2), the adjoint flux $\varphi^{+}(\mathbf{r},\boldsymbol{\Omega},E)$ is customarily required to satisfy the following adjoint boundary condition:

$$\varphi^{+}(\mathbf{r}_s,\boldsymbol{\Omega},E)=0,\ \ \mathbf{r}_s\in\partial V,\ \ \boldsymbol{\Omega}\cdot\mathbf{n}>0. \tag{4.8}$$

The imposition of the boundary conditions given in (4.2) and (4.8) causes the bilinear concomitant defined in (4.5) to vanish, i.e., $P[\varphi,\varphi^{+}]\equiv 0$.

The source term $Q^{+}(\mathbf{k};\mathbf{r},\boldsymbol{\Omega},E)$ for the adjoint Boltzmann transport equation is considered to depend on J_k imprecisely known scalar-valued parameters denoted as k_i, $i=1,\ldots,J_k$, $\mathbf{k}\triangleq[k_1,\ldots,k_{J_k}]^{\dagger}$.

The model response considered in this work is a scalar-valued ratio of nonlinear functionals of the adjoint and forward fluxes, denoted as $R(\varphi,\varphi^{+};\boldsymbol{\alpha})$, having the following form:

$$R(\varphi, \varphi^+; \boldsymbol{\alpha}) \triangleq \frac{\int dV \int_{4\pi} d\boldsymbol{\Omega} \int_0^{E_f} dE\, G(\mathbf{d}; \varphi, \varphi^+)}{\int dV \int_{4\pi} d\boldsymbol{\Omega} \int_0^{E_f} dE\, H(\mathbf{d}; \varphi, \varphi^+)}, \tag{4.9}$$

where $G(\mathbf{d}; \varphi, \varphi^+)$ and $H(\mathbf{d}; \varphi, \varphi^+)$ are a suitably differentiable functions of their respective arguments, and where the vector $\mathbf{d} \triangleq [d_1, \ldots, d_{J_d}]^\dagger$ comprises J_d imprecisely known scalar-valued parameters d_i, $i = 1, \ldots, J_d$, which pertain only to the definitions of the functions $G(\mathbf{d}; \varphi, \varphi^+)$ and $H(\mathbf{d}; \varphi, \varphi^+)$. The mathematical definition considered in (4.9) generally represents the major quantities of interest in nuclear engineering (see, e.g., Stacey 2001), including multiplication factors, normalized power distributions, flux-averaged multi-group cross sections, normalized reaction rates, normalized detector responses, the well-known Rayleigh-Ritz functional, the Schwinger and Roussopoulos functionals, etc.

The response $R(\varphi, \varphi^+; \boldsymbol{\alpha})$ defined in (4.9) depends explicitly and/or implicitly, through the fluxes $\varphi^+(\mathbf{r}, \boldsymbol{\Omega}, E)$ and $\varphi(\mathbf{r}, \boldsymbol{\Omega}, E)$, on all of the imprecisely known model parameters. Therefore, it is convenient for subsequent mathematical derivations to consider that all of the imprecisely known scalar-valued model parameters are components of a "vector of imprecisely known model parameters," denoted as $\boldsymbol{\alpha}$ and defined as follows:

$$\begin{aligned} &\boldsymbol{\alpha} \triangleq [\alpha_1, \ldots, \alpha_{TP}]^\dagger \triangleq [\mathbf{t}, \mathbf{s}, \mathbf{f}, \mathbf{p}, \mathbf{q}, \mathbf{k}, \mathbf{d}]^\dagger, \\ &TP \triangleq J_t + J_s + J_f + J_p + J_q + J_k + J_d, \end{aligned} \tag{4.10}$$

where TP denotes the total number of imprecisely known scalar model parameters. Using the relations represented by (4.1), (4.2), (4.4), (4.7), and (4.8) yields the following well-known relation, which will be used in subsequent derivations:

$$\langle \varphi^+, Q \rangle_B = \langle \varphi, Q^+ \rangle_B. \tag{4.11}$$

The known nominal values of the model parameters will be denoted as $\boldsymbol{\alpha}^0 \triangleq [\alpha_1^0, \ldots, \alpha_{TP}^0]^\dagger$. Such nominal values are usually the known mean (expectation) values of the respective parameters. The nominal value of the flux is denoted as $\varphi^0(\mathbf{r}, \boldsymbol{\Omega}, E)$ and is obtained by solving (4.1) and (4.2) using the nominal parameter values $\boldsymbol{\alpha}^0$, i.e., $\varphi^0(\mathbf{r}, \boldsymbol{\Omega}, E)$ is the solution of

$$\left. \begin{aligned} L(\boldsymbol{\alpha}^0)\varphi^0(\mathbf{r}, \boldsymbol{\Omega}, E) = Q(\mathbf{q}^0; \mathbf{r}, \boldsymbol{\Omega}, E), \\ \varphi^0(\mathbf{r}_s, \boldsymbol{\Omega}, E) = 0, \ \ \mathbf{r}_s \in \partial V, \ \ \boldsymbol{\Omega} \cdot \mathbf{n} < 0, \ \ 0 < E < \infty \end{aligned} \right\} \tag{4.12}$$

Similarly, the nominal value of the adjoint flux, denoted as $\varphi^{+0}(\mathbf{r},\boldsymbol{\Omega},E)$, is obtained by solving (4.7) and (4.8) using the nominal parameter values $\boldsymbol{\alpha}^0$, i.e., $\varphi^{+0}(\mathbf{r},\boldsymbol{\Omega},E)$ is the solution of

$$\left.\begin{array}{r} L^{+}(\boldsymbol{\alpha}^0)\varphi^{+0}(\mathbf{r},\boldsymbol{\Omega},E) = Q^{+}\left(\mathbf{k}^0;\mathbf{r},\boldsymbol{\Omega},E\right), \\ \varphi^{+0}(\mathbf{r}_s,\boldsymbol{\Omega},E) = 0, \;\; \mathbf{r}_s \in \partial V, \;\; \boldsymbol{\Omega}\cdot\mathbf{n} > 0, \;\; 0 < E < \infty \end{array}\right\} \tag{4.13}$$

4.3 Application of the 1st-CASAM-L to Determine the First-Order Sensitivities of $R(\varphi,\varphi^+;\alpha)$

The total G-differential of the response defined in (4.9) to variations $\delta\boldsymbol{\alpha}\triangleq[\delta\alpha_1,\ldots,\delta\alpha_{TP}]^{\dagger}$ in the model parameters has the following expression:

$$\begin{aligned} &\delta R\left(\varphi^0,\varphi^{+0},\boldsymbol{\alpha}^0;\delta\varphi,\delta\varphi^+;\delta\boldsymbol{\alpha}\right) \\ &\triangleq \frac{d}{d\varepsilon}\left\{\frac{\int dV \int_{4\pi} d\Omega \int_0^{E_f} dE\, G\left[\mathbf{d}^0+\varepsilon\delta\mathbf{d};\varphi^0+\varepsilon\,\delta\varphi,\varphi^{+0}+\varepsilon\,\delta\varphi^+\right]}{\int dV \int_{4\pi} d\Omega \int_0^{E_f} dE\, H\left[\mathbf{d}^0+\varepsilon\delta\mathbf{d};\varphi^0+\varepsilon\,\delta\varphi,\varphi^{+0}+\varepsilon\,\delta\varphi^+\right]}\right\}_{\varepsilon=0} \\ &= \left\{\delta R\left(\varphi^0,\varphi^{+0},\boldsymbol{\alpha}^0;\delta\boldsymbol{\alpha}\right)\right\}_{dir} + \left\{\delta R\left(\varphi^0,\varphi^{+0},\boldsymbol{\alpha}^0;\delta\varphi,\delta\varphi^+\right)\right\}_{ind}, \end{aligned} \tag{4.14}$$

where the "direct-effect term" $\{\delta R(\varphi^0,\varphi^{+0},\boldsymbol{\alpha}^0;\delta\boldsymbol{\alpha})\}_{dir}$ is defined as follows:

$$\begin{aligned} \left\{\delta R\left(\varphi^0,\varphi^{+0},\boldsymbol{\alpha}^0;\delta\boldsymbol{\alpha}\right)\right\}_{dir} &\triangleq \frac{\sum_{i=1}^{J_d}\int dV \int_{4\pi} d\Omega \int_0^{E_f} dE \left\{\dfrac{\partial G(\mathbf{d};\varphi,\varphi^+)}{\partial d_i}\right\}\delta d_i}{\int dV \int_{4\pi} d\Omega \int_0^{E_f} dE\, H\left(\mathbf{d}^0;\varphi^0,\varphi^{+0}\right)} \\ &- \int dV \int_{4\pi} d\Omega \int_0^{E_f} dE\, G\left(\mathbf{d}^0;\varphi^0,\varphi^{+0}\right)\left[\int dV \int_{4\pi} d\Omega \int_0^{E_f} dE\, H\left(\mathbf{d}^0;\varphi^0,\varphi^{+0}\right)\right]^{-2} \\ &\times \sum_{i=1}^{J_d}\int dV \int_{4\pi} d\Omega \int_0^{E_f} dE \left\{\frac{\partial H(\mathbf{d};\varphi,\varphi^+)}{\partial d_i}\right\}\delta d_i, \end{aligned} \tag{4.15}$$

and where the "indirect-effect term" $\{\delta R(\varphi^0, \varphi^{+0}, \boldsymbol{\alpha}^0; \delta\varphi, \delta\varphi^+)\}_{ind}$ is defined below:

$$\{\delta R(\varphi^0, \varphi^{+0}, \boldsymbol{\alpha}^0; \delta\varphi, \delta\varphi^+)\}_{ind} \triangleq \frac{\int dV \int_{4\pi} d\boldsymbol{\Omega} \int_0^{E_f} dE \left\{ \frac{\partial G}{\partial \varphi} \delta\varphi(\mathbf{r}, \boldsymbol{\Omega}, E) + \frac{\partial G}{\partial \varphi^+} \delta\varphi^+(\mathbf{r}, \boldsymbol{\Omega}, E) \right\}}{\int dV \int_{4\pi} d\boldsymbol{\Omega} \int_0^{E_f} dE\, H(\mathbf{d}^0; \varphi^0, \varphi^{+0})}$$
$$- \int dV \int_{4\pi} d\boldsymbol{\Omega} \int_0^{E_f} dE\, G(\mathbf{d}^0; \varphi^0, \varphi^{+0}) \left[\int dV \int_{4\pi} d\boldsymbol{\Omega} \int_0^{E_f} dE\, H(\mathbf{d}^0; \varphi^0, \varphi^{+0}) \right]^{-2}$$
$$\times \int dV \int_{4\pi} d\boldsymbol{\Omega} \int_0^{E_f} dE \left\{ \frac{\partial H}{\partial \varphi} \delta\varphi(\mathbf{r}, \boldsymbol{\Omega}, E) + \frac{\partial H}{\partial \varphi^+} \delta\varphi^+(\mathbf{r}, \boldsymbol{\Omega}, E) \right\}. \tag{4.16}$$

All of the quantities appearing in (4.15) and (4.16) are to be evaluated at the "nominal values" $(\mathbf{d}^0; \varphi^0, \varphi^{+0})$. Since the nominal values $\varphi^0(\mathbf{r}, \boldsymbol{\Omega}, E)$ and $\varphi^{+0}(\mathbf{r}, \boldsymbol{\Omega}, E)$ of the forward and adjoint fluxes are known after having solved (4.1), (4.2), (4.7), and (4.8) using the nominal parameter values $\boldsymbol{\alpha}^0$, it follows that the direct-effect term defined in (4.15) can already be computed at this stage. On the other hand, the indirect-effect term defined in (4.16) can be computed only after having determined the flux variations $\delta\varphi(\mathbf{r}, \boldsymbol{\Omega}, E)$ and $\delta\varphi^+(\mathbf{r}, \boldsymbol{\Omega}, E)$, which are the solutions of the equations obtained by G-differentiating the original forward and adjoint transport equations and respective boundary conditions given in Eqs. (4.1), (4.2), (4.7), and (4.8). Performing these differentiations yields the following *First-Level Variational Sensitivity System* (1st-LVSS):

$$\begin{pmatrix} L(\boldsymbol{\alpha}) & 0 \\ 0 & L^+(\boldsymbol{\alpha}) \end{pmatrix} \begin{pmatrix} \delta\varphi(\mathbf{r}, \boldsymbol{\Omega}, E) \\ \delta\varphi^+(\mathbf{r}, \boldsymbol{\Omega}, E) \end{pmatrix} = \begin{pmatrix} Q_1^{(1)}(\boldsymbol{\alpha}, \varphi; \delta\boldsymbol{\alpha}) \\ Q_2^{(1)}(\boldsymbol{\alpha}, \varphi^+; \delta\boldsymbol{\alpha}) \end{pmatrix}, \tag{4.17}$$

together with the following boundary conditions:

$$\left. \begin{aligned} \delta\varphi(\mathbf{r}_s, \boldsymbol{\Omega}, E) = 0, \mathbf{r}_s \in \partial V,\ \ \boldsymbol{\Omega}\cdot\mathbf{n} < 0, 0 < E < E_f, \\ \delta\varphi^+(\mathbf{r}_s, \boldsymbol{\Omega}, E) = 0, \mathbf{r}_s \in \partial V,\ \ \boldsymbol{\Omega}\cdot\mathbf{n} > 0,\ 0 < E < E_f, \end{aligned} \right\} \tag{4.18}$$

where

$$
\begin{aligned}
Q_1^{(1)}(\boldsymbol{\alpha},\varphi;\delta\boldsymbol{\alpha}) &\triangleq \delta Q(\mathbf{q};\mathbf{r},\boldsymbol{\Omega},E) - \delta\Sigma_t(\mathbf{t};\mathbf{r},E)\varphi(\mathbf{r},\boldsymbol{\Omega},E) \\
&+ \int_{4\pi} d\boldsymbol{\Omega}' \int_0^{E_f} dE'\,\varphi(\mathbf{r},\boldsymbol{\Omega}',E')[\delta\Sigma_s(\mathbf{s};\mathbf{r},E'\to E,\boldsymbol{\Omega}'\to\boldsymbol{\Omega})] \\
&+ \delta\chi(\mathbf{p};\mathbf{r},E)\int_{4\pi} d\boldsymbol{\Omega}' \int_0^{E_f} dE'\,\varphi(\mathbf{r},\boldsymbol{\Omega}',E')\nu\Sigma_f(\mathbf{f};\mathbf{r},E') \\
&+ \chi(\mathbf{p};\mathbf{r},E)\int_{4\pi} d\boldsymbol{\Omega}' \int_0^{E_f} dE'\,\varphi(\mathbf{r},\boldsymbol{\Omega}',E')\delta\left[\nu\Sigma_f(\mathbf{f};\mathbf{r},E')\right],
\end{aligned}
\tag{4.19}
$$

$$
\begin{aligned}
Q_2^{(1)}(\boldsymbol{\alpha},\varphi^+;\delta\boldsymbol{\alpha}) &\triangleq \delta Q^+(\mathbf{k};\mathbf{r},\boldsymbol{\Omega},E) - \delta\Sigma_t(\mathbf{t};\mathbf{r},E)\varphi^+(\mathbf{r},\boldsymbol{\Omega},E) \\
&+ \int_{4\pi} d\boldsymbol{\Omega}' \int_0^{E_f} dE'\,\varphi^+(\mathbf{r},\boldsymbol{\Omega}',E')[\delta\Sigma_s(\mathbf{s};\mathbf{r},E\to E',\boldsymbol{\Omega}\to\boldsymbol{\Omega}')] \\
&+ \delta\left[\nu\Sigma_f(\mathbf{f};\mathbf{r},E)\right]\int_{4\pi} d\boldsymbol{\Omega}' \int_0^{E_f} dE'\chi(\mathbf{p};\mathbf{r},E')\varphi^+(\mathbf{r},\boldsymbol{\Omega}',E') \\
&+ \nu\Sigma_f(\mathbf{f};\mathbf{r},E)\int_{4\pi} d\boldsymbol{\Omega}' \int_0^{E_f} dE'\delta\chi(\mathbf{p};\mathbf{r},E')\,\varphi^+(\mathbf{r},\boldsymbol{\Omega}',E').
\end{aligned}
\tag{4.20}
$$

The following definitions were used in (4.19) and (4.20):

$$
\delta\Sigma_t(\mathbf{t};\mathbf{r},E) \triangleq \sum_{j=1}^{J_t} \frac{\partial\Sigma_t(\mathbf{t};\mathbf{r},E)}{\partial t_j}\,\delta t_j, \tag{4.21}
$$

$$
\delta\Sigma_s(\mathbf{s};\mathbf{r},E'\to E,\boldsymbol{\Omega}'\to\boldsymbol{\Omega}) \triangleq \sum_{j=1}^{J_s} \frac{\partial\Sigma_s(\mathbf{s};\mathbf{r},E'\to E,\boldsymbol{\Omega}'\to\boldsymbol{\Omega})}{\partial s_j}\,\delta s_j, \tag{4.22}
$$

$$
\delta\chi(\mathbf{p};\mathbf{r},E) \triangleq \sum_{j=1}^{J_p} \frac{\partial\chi(\mathbf{p};\mathbf{r},E)}{\partial p_j}\,\delta p_j, \tag{4.23}
$$

$$
\delta\left[\nu\Sigma_f(\mathbf{f};\mathbf{r},E)\right] \triangleq \sum_{j=1}^{J_f} \frac{\partial\left[\nu\Sigma_f(\mathbf{f};\mathbf{r},E)\right]}{\partial f_j}\,\delta f_j. \tag{4.24}
$$

The computationally expensive evaluation of the indirect-effect term using (4.16) can be avoided by expressing this indirect-effect term alternatively in terms of the solution of the First-Level Adjoint Sensitivity System (1st-LASS), which is constructed by introducing a Hilbert space denoted as H_1, comprising square-integrable vector-valued functions $\mathbf{u}^{(1)}(\mathbf{r},\boldsymbol{\Omega},E)\triangleq\left[u_1^{(1)}(\mathbf{r},\boldsymbol{\Omega},E),u_2^{(1)}\ (\mathbf{r},\boldsymbol{\Omega},E)\right]^{\dagger}$, $\mathbf{v}^{(1)}(\mathbf{r},\boldsymbol{\Omega},E)\triangleq\left[v_1^{(1)}(\mathbf{r},\boldsymbol{\Omega},E),v_2^{(1)}(\mathbf{r},\boldsymbol{\Omega},E)\right]^{\dagger}$, having square-integrable components $u_1^{(1)}(\mathbf{r},\boldsymbol{\Omega},E)$, $u_2^{(1)}(\mathbf{r},\boldsymbol{\Omega},E)$, $v_1^{(1)}(\mathbf{r},\boldsymbol{\Omega},E)$, and $v_2^{(1)}(\mathbf{r},\boldsymbol{\Omega},E)$, and endowed with the following inner product

$$\left\langle \mathbf{u}^{(1)},\mathbf{v}^{(1)}\right\rangle_{(1)}\triangleq\sum_{j=1}^{2}\int dV\int_{4\pi}d\Omega\int_{0}^{E_f}dE\,u_j^{(1)}(\mathbf{r},\boldsymbol{\Omega},E)v_j^{(1)}(\mathbf{r},\boldsymbol{\Omega},E). \tag{4.25}$$

The 1st-LASS is constructed using the inner product defined in (4.25) by following the same procedure as outlined in Chap. 2, namely: (1) introduce the first-level adjoint function $\boldsymbol{\psi}^{(1)}(\mathbf{r},\boldsymbol{\Omega},E)\triangleq\left[\psi_1^{(1)}(\mathbf{r},\boldsymbol{\Omega},E),\psi_2^{(1)}(\mathbf{r},\boldsymbol{\Omega},E)\right]^{\dagger}$ to construct the operator adjoint to the left side of (4.17); (2) require that the right side of the resulting adjoint equation represent the same functional as the indirect effect term defined in (4.16); and (3) use the boundary conditions provided in (4.18) and select the adjoint boundary conditions for $\boldsymbol{\psi}^{(1)}(\mathbf{r},\boldsymbol{\Omega},E)\triangleq\left[\psi_1^{(1)}(\mathbf{r},\boldsymbol{\Omega},E),\psi_2^{(1)}(\mathbf{r},\boldsymbol{\Omega},E)\right]^{\dagger}$ so as to cause the resulting bilinear concomitant to vanish. Performing these operations leads to the following *First-Level Adjoint Sensitivity System* (1st-LASS) for the *first-level adjoint function* $\boldsymbol{\psi}^{(1)}(\mathbf{r},\boldsymbol{\Omega},E)$:

$$\left.\begin{aligned}L^{+}(\boldsymbol{\alpha}^0)\psi_1^{(1)}(\mathbf{r},\boldsymbol{\Omega},E)&=\Theta_1^{(1)}(\mathbf{d};\varphi,\varphi^{+})_{(\mathbf{d}^0;\varphi^0,\varphi^{+0})},\\ L(\boldsymbol{\alpha}^0)\psi_2^{(1)}(\mathbf{r},\boldsymbol{\Omega},E)&=\Theta_2^{(1)}(\mathbf{d};\varphi,\varphi^{+})_{(\mathbf{d}^0;\varphi^0,\varphi^{+0})},\end{aligned}\right\} \tag{4.26}$$

$$\left.\begin{aligned}\psi_1^{(1)}(\mathbf{r}_s,\boldsymbol{\Omega},E)&=0,\ \ \mathbf{r}_s\in\partial V,\ \ \boldsymbol{\Omega}\cdot\mathbf{n}>0;\ \ 0<E<E_f,\\ \psi_2^{(1)}(\mathbf{r}_s,\boldsymbol{\Omega},E)&=0,\ \ \mathbf{r}_s\in\partial V,\ \ \boldsymbol{\Omega}\cdot\mathbf{n}<0;\ \ 0<E<E_f,\end{aligned}\right\} \tag{4.27}$$

where

$$\begin{aligned}\Theta_1^{(1)}(\mathbf{d};\varphi,\varphi^{+})\triangleq&\frac{\partial G}{\partial\varphi}\left[\int dV\int_{4\pi}d\Omega\int_{0}^{E_f}dE\,H(\mathbf{d};\varphi,\varphi^{+})\right]^{-1}\\ &-\frac{\partial H}{\partial\varphi}\int dV\int_{4\pi}d\Omega\int_{0}^{E_f}dE\,G(\mathbf{d};\varphi,\varphi^{+})\left[\int dV\int_{4\pi}d\Omega\int_{0}^{E_f}dE\,H(\mathbf{d};\varphi,\varphi^{+})\right]^{-2},\end{aligned} \tag{4.28}$$

$$\Theta_2^{(1)}(\mathbf{d};\varphi,\varphi^+)\triangleq\frac{\partial G}{\partial\varphi^+}\left[\int dV\int_{4\pi}d\boldsymbol{\Omega}\int_0^{E_f}dE\,H(\mathbf{d};\varphi,\varphi^+)\right]^{-1}$$

$$-\frac{\partial H}{\partial\varphi^+}\int dV\int_{4\pi}d\boldsymbol{\Omega}\int_0^{E_f}dE\,G(\mathbf{d};\varphi,\varphi^+)\left[\int dV\int_{4\pi}d\boldsymbol{\Omega}\int_0^{E_f}dE\,H(\mathbf{d};\varphi,\varphi^+)\right]^{-2}. \tag{4.29}$$

In terms of the first-level adjoint function $\boldsymbol{\psi}^{(1)}(\mathbf{r},\boldsymbol{\Omega},E)$, the resulting expression of the *total first-order response sensitivity* is as follows:

$$\left\{\delta R\left(\varphi,\varphi^+,\boldsymbol{\alpha};\delta\varphi,\delta\varphi^+;\delta\boldsymbol{\alpha}\right)\right\}_{\left(\mathbf{d}^0;\varphi^0,\varphi^{+0}\right)}=\left\{\delta R\left(\varphi^0,\varphi^{+0},\boldsymbol{\alpha}^0;\delta\boldsymbol{\alpha}\right)\right\}_{dir}$$
$$+\left\langle\boldsymbol{\psi}^{(1)}(\mathbf{r},\boldsymbol{\Omega},E),\ \mathbf{Q}^{(1)}\left(\boldsymbol{\alpha}^0;\varphi^0,\varphi^{+0};\delta\boldsymbol{\alpha}\right)\right\rangle_{(1)} \tag{4.30}$$
$$\triangleq\sum_{m_1=1}^{TP}\left\{\frac{\partial R\left(\varphi,\varphi^+,\boldsymbol{\alpha};\boldsymbol{\psi}^{(1)}\right)}{\partial\alpha_{m_1}}\right\}_{\left(\boldsymbol{\alpha}^0;\varphi^0,\varphi^{+0}\right)}\delta\alpha_{m_1}.$$

The *partial first-order response sensitivities*, denoted in (4.30) as $\partial R\left(\varphi,\varphi^+,\boldsymbol{\alpha};\boldsymbol{\psi}^{(1)}\right)/\partial\alpha_{m_1}, m_1=1,\ldots,TP$, of the response $R(\varphi,\varphi^+,\boldsymbol{\alpha})$ to a generic parameter α_{m_1} are obtained from (4.30) by identifying the quantities that multiply the various parameter variations $\delta\alpha_{m_1}$. This identification yields the following expressions for the respective partial sensitivities:

$$\text{For}\quad j=1,\ldots,J_t:\quad\frac{\partial R\left(\varphi,\varphi^+,\boldsymbol{\alpha};\boldsymbol{\psi}^{(1)}\right)}{\partial\alpha_j}\triangleq\frac{\partial R\left(\varphi,\varphi^+,\boldsymbol{\alpha};\boldsymbol{\psi}^{(1)}\right)}{\partial t_j}$$
$$=-\int dV\int_{4\pi}d\boldsymbol{\Omega}\int_0^{E_f}dE\left[\psi_1^{(1)}(\mathbf{r},\boldsymbol{\Omega},E)\varphi(\mathbf{r},\boldsymbol{\Omega},E)\right. \tag{4.31}$$
$$\left.+\psi_2^{(1)}(\mathbf{r},\boldsymbol{\Omega},E)\varphi^+(\mathbf{r},\boldsymbol{\Omega},E)\right]\frac{\partial\Sigma_t(\mathbf{t};\mathbf{r},E)}{\partial t_j};$$

$$\text{For}\quad j=1,\ldots,J_s:\ \frac{\partial R\left(\varphi,\varphi^+,\boldsymbol{\alpha};\boldsymbol{\psi}^{(1)}\right)}{\partial\alpha_{J_t+j}}\triangleq\frac{\partial R\left(\varphi,\varphi^+,\boldsymbol{\alpha};\boldsymbol{\psi}^{(1)}\right)}{\partial s_j}$$

$$=\int dV\int_{4\pi}d\boldsymbol{\Omega}\int_0^{E_f}dE\,\psi_1^{(1)}(\mathbf{r},\boldsymbol{\Omega},E)$$

$$\times\int_{4\pi}d\boldsymbol{\Omega}'\int_0^{E_f}dE'\,\frac{\partial\Sigma_s(\mathbf{s};\mathbf{r},E'\to E,\boldsymbol{\Omega}'\to\boldsymbol{\Omega})}{\partial s_j}\varphi(\mathbf{r},\boldsymbol{\Omega}',E') \tag{4.32}$$

$$+\int dV\int_{4\pi}d\boldsymbol{\Omega}\int_0^{E_f}dE\,\psi_2^{(1)}(\mathbf{r},\boldsymbol{\Omega},E)$$

$$\times\int_{4\pi}d\boldsymbol{\Omega}'\int_0^{E_f}dE'\,\frac{\partial\Sigma_s(\mathbf{s};\mathbf{r},E\to E',\boldsymbol{\Omega}\to\boldsymbol{\Omega}')}{\partial s_j}\varphi^+(\mathbf{r},\boldsymbol{\Omega}',E');$$

$$\text{For}\quad j=1,\ldots,J_f:\ \frac{\partial R\left(\varphi,\varphi^+,\boldsymbol{\alpha};\boldsymbol{\psi}^{(1)}\right)}{\partial\alpha_{J_t+J_s+j}}\triangleq\frac{\partial R\left(\varphi,\varphi^+,\boldsymbol{\alpha};\boldsymbol{\psi}^{(1)}\right)}{\partial f_j}$$

$$=\int dV\int_{4\pi}d\boldsymbol{\Omega}\int_0^{E_f}dE\psi_1^{(1)}(\mathbf{r},\boldsymbol{\Omega},E)$$

$$\times\chi(\mathbf{p};\mathbf{r},E)\int_{4\pi}d\boldsymbol{\Omega}'\int_0^{E_f}dE'\,\frac{\partial\left[\nu\Sigma_f(\mathbf{f};\mathbf{r},E')\right]}{\partial f_j}\varphi(\mathbf{r},\boldsymbol{\Omega}',E') \tag{4.33}$$

$$+\int dV\int_{4\pi}d\boldsymbol{\Omega}\int_0^{E_f}dE\psi_2^{(1)}(\mathbf{r},\boldsymbol{\Omega},E)$$

$$\times\frac{\partial\left[\nu\Sigma_f(\mathbf{f};\mathbf{r},E)\right]}{\partial f_j}\int_{4\pi}d\boldsymbol{\Omega}'\int_0^{E_f}dE'\chi(\mathbf{p};\mathbf{r},E')\varphi^+(\mathbf{r},\boldsymbol{\Omega}',E');$$

$$\text{For} \quad j = 1, \ldots, J_p: \quad \frac{\partial R(\varphi, \varphi^+, \boldsymbol{\alpha}; \boldsymbol{\psi}^{(1)})}{\partial \alpha_{J_t+J_s+J_f+j}} \triangleq \frac{\partial R(\varphi, \varphi^+, \boldsymbol{\alpha}; \boldsymbol{\psi}^{(1)})}{\partial p_j}$$

$$= \int dV \int_{4\pi} d\boldsymbol{\Omega} \int_0^{E_f} dE\, \psi_1^{(1)}(\mathbf{r}, \boldsymbol{\Omega}, E)$$

$$\times \frac{\partial \chi(\mathbf{p}; \mathbf{r}, E)}{\partial p_j} \int_{4\pi} d\boldsymbol{\Omega}' \int_0^{E_f} dE'\, \nu \Sigma_f(\mathbf{f}; \mathbf{r}, E') \varphi(\mathbf{r}, \boldsymbol{\Omega}', E') \tag{4.34}$$

$$+ \int dV \int_{4\pi} d\boldsymbol{\Omega} \int_0^{E_f} dE\, \psi_2^{(1)}(\mathbf{r}, \boldsymbol{\Omega}, E)$$

$$\times \left[\nu \Sigma_f(\mathbf{f}; \mathbf{r}, E)\right] \int_{4\pi} d\boldsymbol{\Omega}' \int_0^{E_f} dE' \frac{\partial \chi(\mathbf{p}; \mathbf{r}, E')}{\partial p_j} \varphi^+(\mathbf{r}, \boldsymbol{\Omega}', E');$$

$$\text{For} \quad j = 1, \ldots, J_q: \quad \frac{\partial R(\varphi, \varphi^+, \boldsymbol{\alpha}; \boldsymbol{\psi}^{(1)})}{\partial \alpha_{J_t+J_s+J_f+J_p+j}} \triangleq \frac{\partial R(\varphi, \varphi^+, \boldsymbol{\alpha}; \boldsymbol{\psi}^{(1)})}{\partial q_j}$$

$$= \int dV \int_{4\pi} d\boldsymbol{\Omega} \int_0^{E_f} dE \frac{\partial Q(\mathbf{q}; \mathbf{r}, \boldsymbol{\Omega}, E)}{\partial q_j} \psi_1^{(1)}(\mathbf{r}, \boldsymbol{\Omega}, E); \tag{4.35}$$

$$\text{For} \quad j = 1, \ldots, J_k: \quad \frac{\partial R(\varphi, \varphi^+, \boldsymbol{\alpha}; \boldsymbol{\psi}^{(1)})}{\partial \alpha_{J_t+J_s+J_f+J_p+J_q+j}} \triangleq \frac{\partial R(\varphi, \varphi^+, \boldsymbol{\alpha}; \boldsymbol{\psi}^{(1)})}{\partial k_j}$$

$$= \int dV \int_{4\pi} d\boldsymbol{\Omega} \int_0^{E_f} dE \frac{\partial Q^+(\mathbf{k}; \mathbf{r}, \boldsymbol{\Omega}, E)}{\partial k_j} \psi_2^{(1)}(\mathbf{r}, \boldsymbol{\Omega}, E); \tag{4.36}$$

$$\text{For} \quad j = 1, \ldots, J_d: \quad \frac{\partial R(\varphi, \varphi^+, \boldsymbol{\alpha}; \boldsymbol{\psi}^{(1)})}{\partial \alpha_{J_t+J_s+J_f+J_p+J_q+J_k+j}} \triangleq \frac{\partial R(\varphi, \varphi^+, \boldsymbol{\alpha}; \boldsymbol{\psi}^{(1)})}{\partial d_j}$$

$$= \left[\int dV \int_{4\pi} d\boldsymbol{\Omega} \int_0^{E_f} dE\, H(\mathbf{d}^0; \varphi^0, \varphi^{+0})\right]^{-1} \int dV \int_{4\pi} d\boldsymbol{\Omega} \int_0^{E_f} dE \left\{\frac{\partial G(\mathbf{d}; \varphi, \varphi^+)}{\partial d_j}\right\}$$

$$- \frac{\int dV \int_{4\pi} d\boldsymbol{\Omega} \int_0^{E_f} dE\, G(\mathbf{d}^0; \varphi^0, \varphi^{+0})}{\left[\int dV \int_{4\pi} d\boldsymbol{\Omega} \int_0^{E_f} dE\, H(\mathbf{d}^0; \varphi^0, \varphi^{+0})\right]^2} \int dV \int_{4\pi} d\boldsymbol{\Omega} \int_0^{E_f} dE \left\{\frac{\partial H(\mathbf{d}; \varphi, \varphi^+)}{\partial d_j}\right\}. \tag{4.37}$$

4.4 Application of the 2nd-CASAM-L to Determine the Second-Order Sensitivities of $R(\varphi, \varphi^{+}; \alpha)$

The second-order sensitivities of the response $R(\varphi, \varphi^{+}; \boldsymbol{\alpha})$ with respect to the model parameters are computed by applying the 2nd-CASAM-L, as detailed in Chap. 2, by considering each of the sensitivities obtained in Sect. 4.3 as a "new" response and determining, in turn, the "total first-order G-variation" of each of these "new" responses.

4.4.1 Determination of the Second-Order Sensitivities of the Form $\partial^2 R\left(\varphi, \varphi^{+}; \boldsymbol{\psi}^{(1)}; \boldsymbol{\psi}_j^{(2)}; \boldsymbol{\alpha}\right) / \partial t_j \partial \alpha_{m_2}$

The second-order sensitivities $\partial^2 R\left(\varphi, \varphi^{+}; \boldsymbol{\psi}^{(1)}; \boldsymbol{\psi}_j^{(2)}; \boldsymbol{\alpha}\right) / \left(\partial t_j\right)\left(\partial \alpha_{m_2}\right), j = 1, \ldots,$ J_t, $m_2 = 1, \ldots, TP$, involving the total cross sections, will ultimately depend on a second-level adjoint function, which will be denoted as $\boldsymbol{\psi}_j^{(2)}$, and are obtained by applying the procedure presented in Chap. 2 to the first-order sensitivity $\partial R(\varphi, \varphi^{+}, \boldsymbol{\alpha}; \boldsymbol{\psi}^{(1)})/\partial t_j$ provided in (4.31). Thus, applying the definition of the G-differential to (4.31) yields the following relation:

$$
\begin{aligned}
\delta\left[\frac{\partial R\left(\varphi, \varphi^{+}, \boldsymbol{\alpha}; \boldsymbol{\psi}^{(1)}\right)}{\partial t_j}\right] &= \left\{\delta\left[\frac{\partial R\left(\varphi, \varphi^{+}, \boldsymbol{\alpha}; \boldsymbol{\psi}^{(1)}\right)}{\partial t_j}\right]\right\}_{dir} \\
&+ \left\{\delta\left[\frac{\partial R\left(\varphi, \varphi^{+}, \boldsymbol{\alpha}; \boldsymbol{\psi}^{(1)}\right)}{\partial t_j}\right]\right\}_{ind}, \quad j = 1, \ldots, J_t,
\end{aligned} \tag{4.38}
$$

where the "direct-effect" and "indirect-effect" terms are defined, for each $j = 1, \ldots,$ J_t, as follows:

$$
\begin{aligned}
\left\{\delta\left[\frac{\partial R\left(\varphi, \varphi^{+}, \boldsymbol{\alpha}; \boldsymbol{\psi}^{(1)}\right)}{\partial t_j}\right]\right\}_{dir} &\triangleq -\int dV \int_{4\pi} d\boldsymbol{\Omega} \int_0^{E_f} dE \sum_{m_2=1}^{J_t} \frac{\partial^2 \Sigma_t(\mathbf{t}; \mathbf{r}, E)}{\partial t_j \partial t_{m_2}} \delta t_{m_2} \\
&\times \left[\psi_1^{(1)}(\mathbf{r}, \boldsymbol{\Omega}, E)\varphi(\mathbf{r}, \boldsymbol{\Omega}, E) + \psi_2^{(1)}(\mathbf{r}, \boldsymbol{\Omega}, E)\varphi^{+}(\mathbf{r}, \boldsymbol{\Omega}, E)\right],
\end{aligned} \tag{4.39}
$$

and

$$\left\{\delta\left[\frac{\partial R\left(\varphi,\varphi^{+},\boldsymbol{\alpha};\boldsymbol{\psi}^{(1)}\right)}{\partial t_j}\right]\right\}_{ind}\triangleq-\int dV\int_{4\pi}d\boldsymbol{\Omega}\int_0^{E_f}dE\,\frac{\partial\Sigma_t(\mathbf{t};\mathbf{r},E)}{\partial t_j}$$
$$\times\left[\delta\psi_1^{(1)}(\mathbf{r},\boldsymbol{\Omega},E)\varphi(\mathbf{r},\boldsymbol{\Omega},E)+\delta\varphi(\mathbf{r},\boldsymbol{\Omega},E)\psi_1^{(1)}(\mathbf{r},\boldsymbol{\Omega},E)\right]$$
$$-\int dV\int_{4\pi}d\boldsymbol{\Omega}\int_0^{E_f}dE\frac{\partial\Sigma_t(\mathbf{t};\mathbf{r},E)}{\partial t_j}\left[\delta\psi_2^{(1)}(\mathbf{r},\boldsymbol{\Omega},E)\varphi^{+}(\mathbf{r},\boldsymbol{\Omega},E)\right.$$
$$\left.+\delta\varphi^{+}(\mathbf{r},\boldsymbol{\Omega},E)\psi_2^{(1)}(\mathbf{r},\boldsymbol{\Omega},E)\right]. \tag{4.40}$$

The direct-effect term defined in (4.39) can be computed immediately. On the other hand, the indirect-effect term defined in (4.40) can be computed only after having obtained the following quantities: (1) the solution $\delta\mathbf{u}(\mathbf{r},\boldsymbol{\Omega},E)\triangleq[\delta\varphi(\mathbf{r},\boldsymbol{\Omega},E),\ \delta\varphi^{+}(\mathbf{r},\boldsymbol{\Omega},E)]$ of the 1st-LVSS defined by (4.17) and (4.18); and (2) the functions $\delta\psi_1^{(1)}(\mathbf{r},\boldsymbol{\Omega},E)$ and $\delta\psi_2^{(1)}(\mathbf{r},\boldsymbol{\Omega},E)$, which are the solutions of the G-differentiated 1st-LASS, namely:

$$L^{+}(\boldsymbol{\alpha})\delta\psi_1^{(1)}(\mathbf{r},\boldsymbol{\Omega},E)-\frac{\partial\Theta_1^{(1)}(\mathbf{d};\varphi,\varphi^{+})}{\partial\varphi}\delta\varphi-\frac{\partial\Theta_1^{(1)}(\mathbf{d};\varphi,\varphi^{+})}{\partial\varphi^{+}}\delta\varphi^{+}$$
$$=Q_1^{(2)}\left(\boldsymbol{\alpha},\boldsymbol{\psi}^{(1)};\delta\boldsymbol{\alpha}\right),$$
$$L(\boldsymbol{\alpha})\delta\psi_2^{(1)}(\mathbf{r},\boldsymbol{\Omega},E)-\frac{\partial\Theta_2^{(1)}(\mathbf{d};\varphi,\varphi^{+})}{\partial\varphi}\delta\varphi-\frac{\partial\Theta_2^{(1)}(\mathbf{d};\varphi,\varphi^{+})}{\partial\varphi^{+}}\delta\varphi^{+}$$
$$=Q_2^{(2)}\left(\boldsymbol{\alpha},\boldsymbol{\psi}^{(1)};\delta\boldsymbol{\alpha}\right), \tag{4.41}$$

subject to the boundary conditions below:

$$\delta\psi_1^{(1)}(\mathbf{r}_s,\boldsymbol{\Omega},E)=0,\ \ \mathbf{r}_s\in\partial V,\ \ \boldsymbol{\Omega}\cdot\mathbf{n}>0;0<E<E_f,$$
$$\delta\psi_2^{(1)}(\mathbf{r}_s,\boldsymbol{\Omega},E)=0,\ \ \mathbf{r}_s\in\partial V,\ \ \boldsymbol{\Omega}\cdot\mathbf{n}<0;0<E<E_f. \tag{4.42}$$

where

$$
\begin{aligned}
Q_1^{(2)}\left(\boldsymbol{\alpha}^0, \boldsymbol{\psi}^{(1)}; \delta\boldsymbol{\alpha}\right) &\triangleq \sum_{i=1}^{J_d} \frac{\partial \Theta_1^{(1)}(\mathbf{d}; \varphi, \varphi^+)}{\partial d_i} \delta d_i - \delta\Sigma_t(\mathbf{r}, E)\psi_1^{(1)}(\mathbf{r}, \boldsymbol{\Omega}, E) \\
&+ \int_{4\pi} d\Omega' \int_0^{E_f} dE' \left[\delta\Sigma_s(\mathbf{r}, E \to E', \boldsymbol{\Omega} \to \boldsymbol{\Omega}')\right]\psi_1^{(1)}(\mathbf{r}, \boldsymbol{\Omega}', E') \\
&+ \delta\left[\nu\Sigma_f(\mathbf{f}; \mathbf{r}, E)\right] \int_{4\pi} d\Omega' \int_0^{E_f} dE' \chi(\mathbf{p}; \mathbf{r}, E')\psi_1^{(1)}(\mathbf{r}, \boldsymbol{\Omega}', E') \\
&+ \nu\Sigma_f(\mathbf{f}; \mathbf{r}, E) \int_{4\pi} d\Omega' \int_0^{E_f} dE' \, \delta[\chi(\mathbf{p}; \mathbf{r}, E')]\psi_1^{(1)}(\mathbf{r}, \boldsymbol{\Omega}', E'),
\end{aligned} \tag{4.43}
$$

$$
\begin{aligned}
Q_2^{(2)}\left(\boldsymbol{\alpha}^0, \boldsymbol{\psi}^{(1)}; \delta\boldsymbol{\alpha}\right) &\triangleq \sum_{i=1}^{J_d} \frac{\partial \Theta_2^{(1)}(\mathbf{d}; \varphi, \varphi^+)}{\partial d_i} \delta d_i - \delta\Sigma_t(\mathbf{r}, E)\psi_2^{(1)}(\mathbf{r}, \boldsymbol{\Omega}, E) \\
&+ \int_{4\pi} d\Omega' \int_0^{E_f} dE' \left[\delta\Sigma_s(\mathbf{r}, E' \to E, \boldsymbol{\Omega}' \to \boldsymbol{\Omega})\right]\psi_2^{(1)}(\mathbf{r}, \boldsymbol{\Omega}', E') \\
&+ \chi(\mathbf{p}; \mathbf{r}, E) \int_{4\pi} d\Omega' \int_0^{E_f} dE' \, \delta\left[\nu\Sigma_f(\mathbf{f}; \mathbf{r}, E')\right]\psi_2^{(1)}(\mathbf{r}, \boldsymbol{\Omega}', E') \\
&+ \delta\chi(\mathbf{p}; \mathbf{r}, E) \int_{4\pi} d\Omega' \int_0^{E_f} dE' \, \nu\Sigma_f(\mathbf{f}; \mathbf{r}, E')\psi_2^{(1)}(\mathbf{r}, \boldsymbol{\Omega}', E') \; .
\end{aligned} \tag{4.44}
$$

The system comprising (4.41) and (4.42) is called the *Second-Level Variational Sensitivity System* (2nd-LVSS). It is evident from these equations that the evaluation of the function $\delta\boldsymbol{\psi}^{(1)}(\mathbf{r}, \boldsymbol{\Omega}, E)$ is at least as expensive computationally as determining $\delta\mathbf{u}(\mathbf{r}, \boldsymbol{\Omega}, E)$. To avoid the need for solving the 2nd-LVSS, the indirect-effect term defined in (4.40) will be expressed in terms of the solution of a 2nd-Level Adjoint Sensitivity System (2nd-LASS), which will be constructed by following the general principles presented in Chap. 2, which comprise the following sequence of steps for the present particular case:

1. Write the coupled 2nd-LVSS and 1st-LVSS, namely (4.41) and (4.17), in the following matrix form:

$$
\begin{pmatrix} \mathbf{A}_{11}^{(1)}(\boldsymbol{\alpha}) & \mathbf{F}_{12}^{(2)}(\boldsymbol{\alpha}) \\ \mathbf{0} & \mathbf{F}_{11}^{(1)}(\boldsymbol{\alpha}) \end{pmatrix} \begin{pmatrix} \delta\boldsymbol{\psi}^{(1)}(\mathbf{r}, \boldsymbol{\Omega}, E) \\ \delta\mathbf{u}(\mathbf{r}, \boldsymbol{\Omega}, E) \end{pmatrix} = \begin{pmatrix} \mathbf{Q}^{(2)}(\boldsymbol{\alpha}, \boldsymbol{\psi}^{(1)}; \delta\boldsymbol{\alpha}) \\ \mathbf{Q}^{(1)}(\boldsymbol{\alpha}; \varphi, \varphi^+; \delta\boldsymbol{\alpha}) \end{pmatrix} \tag{4.45}
$$

where the following definitions have been used:

$$\delta\boldsymbol{\psi}^{(1)} \triangleq \begin{pmatrix} \delta\psi_1^{(1)}(\mathbf{r},\boldsymbol{\Omega},E) \\ \delta\psi_2^{(1)}(\mathbf{r},\boldsymbol{\Omega},E) \end{pmatrix}; \ \mathbf{Q}^{(2)}\left(\boldsymbol{\alpha},\boldsymbol{\psi}^{(1)};\delta\boldsymbol{\alpha}\right) \triangleq \begin{pmatrix} Q_1^{(2)}\left(\boldsymbol{\alpha},\boldsymbol{\psi}^{(1)};\delta\boldsymbol{\alpha}\right) \\ Q_2^{(2)}\left(\boldsymbol{\alpha},\boldsymbol{\psi}^{(1)};\delta\boldsymbol{\alpha}\right) \end{pmatrix}; \tag{4.46}$$

$$\mathbf{F}_{12}^{(2)}(\boldsymbol{\alpha}) \triangleq \begin{pmatrix} -\dfrac{\partial\Theta_1^{(1)}(\mathbf{d};\varphi,\varphi^+)}{\partial\varphi} & -\dfrac{\partial\Theta_1^{(1)}(\mathbf{d};\varphi,\varphi^+)}{\partial\varphi^+} \\ -\dfrac{\partial\Theta_2^{(1)}(\mathbf{d};\varphi,\varphi^+)}{\partial\varphi} & -\dfrac{\partial\Theta_2^{(1)}(\mathbf{d};\varphi,\varphi^+)}{\partial\varphi^+} \end{pmatrix}. \tag{4.47}$$

2. Define an inner product $\langle \mathbf{u}^{(2)}(\mathbf{r},\boldsymbol{\Omega},E),\mathbf{v}^{(2)}(\mathbf{r},\boldsymbol{\Omega},E)\rangle_{(2)}$ of two vector-valued functions $\mathbf{u}^{(2)} \triangleq \left[\mathbf{u}_1^{(2)},\mathbf{u}_2^{(2)}\right]^\dagger$ and $\mathbf{v}^{(2)} \triangleq \left[\mathbf{v}_1^{(2)},\mathbf{v}_2^{(2)}\right]^\dagger$, where $\mathbf{u}_1^{(2)} \triangleq \left[u_{11}^{(2)}(\mathbf{r},\boldsymbol{\Omega},E),u_{12}^{(2)}(\mathbf{r},\boldsymbol{\Omega},E)\right]^\dagger$, $\mathbf{u}_2^{(2)} \triangleq \left[u_{21}^{(2)}(\mathbf{r},\boldsymbol{\Omega},E),u_{22}^{(2)}(\mathbf{r},\boldsymbol{\Omega},E)\right]^\dagger$, $\mathbf{v}_1^{(2)} \triangleq \left[v_{11}^{(2)}(\mathbf{r},\boldsymbol{\Omega},E),v_{12}^{(2)}(\mathbf{r},\boldsymbol{\Omega},E)\right]^\dagger$, $\mathbf{v}_2^{(2)} \triangleq \left[v_{21}^{(2)}(\mathbf{r},\boldsymbol{\Omega},E),v_{22}^{(2)}(\mathbf{r},\boldsymbol{\Omega},E)\right]^\dagger$, with square-integrable components $u_{ij}^{(2)}(\mathbf{r},\boldsymbol{\Omega},E), v_{ij}^{(2)}(\mathbf{r},\boldsymbol{\Omega},E), i,j = 1,2$, such that:

$$\begin{aligned} \left\langle \mathbf{u}^{(2)}(\mathbf{r},\boldsymbol{\Omega},E),\mathbf{v}^{(2)}(\mathbf{r},\boldsymbol{\Omega},E)\right\rangle_{(2)} &\triangleq \sum_{i=1}^{2}\int dV\int_{4\pi} d\boldsymbol{\Omega}\int_0^{E_f} dE\,\mathbf{u}_i^{(2)}\cdot\mathbf{v}_i^{(2)} \\ &= \sum_{i=1}^{2}\sum_{j=1}^{2}\int dV\int_{4\pi} d\boldsymbol{\Omega}\int_0^{E_f} dE\,u_{ij}^{(2)}(\mathbf{r},\boldsymbol{\Omega},E)v_{ij}^{(2)}(\mathbf{r},\boldsymbol{\Omega},E). \end{aligned} \tag{4.48}$$

3. Define a Hilbert space, denoted as $\mathrm{H}_{(2)}$, which is endowed with the inner product defined in (4.48). For a block-matrix-valued linear operator $\mathbf{L}^{(2)} \triangleq \begin{pmatrix} \mathbf{L}_{11}^{(2)} & \mathbf{L}_{12}^{(2)} \\ \mathbf{L}_{21}^{(2)} & \mathbf{L}_{22}^{(2)} \end{pmatrix}$, having component matrices $\mathbf{L}_{ij}^{(2)},\ i,j = 1,2$, define its formal adjoint operator, denoted as $\mathbf{A}^{(2)} \triangleq \begin{pmatrix} \mathbf{A}_{11}^{(2)} & \mathbf{A}_{12}^{(2)} \\ \mathbf{A}_{21}^{(2)} & \mathbf{A}_{22}^{(2)} \end{pmatrix}$, through the following relationship:

$$\left\langle \mathbf{v}^{(2)},\mathbf{L}^{(2)}\mathbf{u}^{(2)}\right\rangle_{(2)} = \left\langle \mathbf{u}^{(2)},\mathbf{A}^{(2)}\mathbf{v}^{(2)}\right\rangle_{(2)} + P^{(2)}\left[\mathbf{u}^{(2)},\mathbf{v}^{(2)}\right], \tag{4.49}$$

where $P^{(2)}[\mathbf{u}^{(2)},\mathbf{v}^{(2)}]$ denotes the corresponding bilinear concomitant on the phase-space boundary $(\partial V \times \partial\boldsymbol{\Omega} \times \partial E)$.

Apply the definition provided in (4.48) to form the inner product of (4.45) with a yet undefined function $\boldsymbol{\psi}_j^{(2)}(\mathbf{r},\boldsymbol{\Omega},E) \triangleq \left[\boldsymbol{\psi}_{1,j}^{(2)}(\mathbf{r},\boldsymbol{\Omega},E),\boldsymbol{\psi}_{2,j}^{(2)}(\mathbf{r},\boldsymbol{\Omega},E)\right]^\dagger$, where $\boldsymbol{\psi}_{1,j}^{(2)} \triangleq \left[\psi_{11,j}^{(2)},\psi_{12,j}^{(2)}\right]^\dagger$, $\boldsymbol{\psi}_{2,j}^{(2)} \triangleq \left[\psi_{21,j}^{(2)},\psi_{22,j}^{(2)}\right]^\dagger$, with square-integrable components $\psi_{mn,j}^{(2)}(\mathbf{r},\boldsymbol{\Omega},E) \in L_2(V \times \boldsymbol{\Omega} \times E),\ m,n = 1,2$, to obtain:

$$\left\langle \begin{pmatrix} \boldsymbol{\psi}_{1,j}^{(2)} \\ \boldsymbol{\psi}_{2,j}^{(2)} \end{pmatrix}^{\dagger} \begin{pmatrix} \mathbf{A}_{11}^{(1)}(\boldsymbol{\alpha}) & \mathbf{F}_{12}^{(2)}(\boldsymbol{\alpha}) \\ \mathbf{0} & \mathbf{F}_{11}^{(1)}(\boldsymbol{\alpha}) \end{pmatrix} \begin{pmatrix} \delta\boldsymbol{\psi}^{(1)} \\ \delta\mathbf{u} \end{pmatrix} \right\rangle_{(2)} = \left\langle \begin{pmatrix} \boldsymbol{\psi}_{1,j}^{(2)} \\ \boldsymbol{\psi}_{2,j}^{(2)} \end{pmatrix}^{\dagger} \begin{pmatrix} \mathbf{Q}^{(2)} \\ \mathbf{Q}^{(1)} \end{pmatrix} \right\rangle_{(2)}. \tag{4.50}$$

Use the relation shown in (4.49) to recast the left side of (4.50) in the following form:

$$\begin{aligned} &\left\langle \begin{pmatrix} \boldsymbol{\psi}_{1,j}^{(2)} \\ \boldsymbol{\psi}_{2,j}^{(2)} \end{pmatrix}^{\dagger} \begin{pmatrix} \mathbf{A}_{11}^{(1)}(\boldsymbol{\alpha}) & \mathbf{F}_{12}^{(2)}(\boldsymbol{\alpha}) \\ \mathbf{0} & \mathbf{F}_{11}^{(1)}(\boldsymbol{\alpha}) \end{pmatrix} \begin{pmatrix} \delta\boldsymbol{\psi}^{(1)} \\ \delta\mathbf{u} \end{pmatrix} \right\rangle_{(2)} = P^{(2)}\left[\boldsymbol{\psi}_j^{(2)}, \delta\boldsymbol{\psi}^{(1)}, \delta\mathbf{u}\right] \\ &+\left\langle \begin{pmatrix} \delta\boldsymbol{\psi}^{(1)} \\ \delta\mathbf{u} \end{pmatrix}^{\dagger} \begin{pmatrix} \left[\mathbf{A}_{11}^{(1)}(\boldsymbol{\alpha})\right]^{+} & \mathbf{0} \\ \left[\mathbf{F}_{12}^{(2)}(\boldsymbol{\alpha})\right]^{+} & \left[\mathbf{F}_{11}^{(1)}(\boldsymbol{\alpha})\right]^{+} \end{pmatrix} \begin{pmatrix} \boldsymbol{\psi}_{1,j}^{(2)} \\ \boldsymbol{\psi}_{2,j}^{(2)} \end{pmatrix} \right\rangle_{(2)}, \end{aligned} \tag{4.51}$$

where the symbol $[]^{+}$ indicates "formal adjoint" and where the bilinear concomitant $P^{(2)}\left[\boldsymbol{\psi}_j^{(2)}, \delta\boldsymbol{\psi}^{(1)}, \delta\mathbf{u}\right]$ is defined as follows:

$$P^{(2)}\left[\boldsymbol{\psi}_j^{(2)}, \delta\boldsymbol{\psi}^{(1)}, \delta\mathbf{u}\right] \triangleq \sum_{i=1}^{4} P_i^{(2)}\left[\boldsymbol{\psi}_j^{(2)}, \delta\boldsymbol{\psi}^{(1)}, \delta\mathbf{u}\right], \tag{4.52}$$

$$\begin{aligned} P_1^{(2)}\left[\boldsymbol{\psi}_j^{(2)}, \delta\boldsymbol{\psi}^{(1)}, \delta\mathbf{u}\right] &\triangleq \int_0^{E_f} dE \int_{\boldsymbol{\Omega}\cdot\mathbf{n}<0} d\boldsymbol{\Omega} \int_{\partial V} |\boldsymbol{\Omega}\cdot\mathbf{n}| \delta\varphi\, \psi_{21,j}^{(2)} dA \\ &- \int_0^{E_f} dE \int_{\boldsymbol{\Omega}\cdot\mathbf{n}>0} d\boldsymbol{\Omega} \int_{\partial V} \boldsymbol{\Omega}\cdot\mathbf{n}\; \delta\varphi(\mathbf{r},\boldsymbol{\Omega},E)\, \psi_{21,j}^{(2)}(\mathbf{r},\boldsymbol{\Omega},E) dA, \end{aligned} \tag{4.53}$$

$$\begin{aligned} P_2^{(2)}\left[\boldsymbol{\psi}_j^{(2)}, \delta\boldsymbol{\psi}^{(1)}, \delta\mathbf{u}\right] &\triangleq \int_0^{E_f} dE \int_{\boldsymbol{\Omega}\cdot\mathbf{n}<0} d\boldsymbol{\Omega} \int_{\partial V} |\boldsymbol{\Omega}\cdot\mathbf{n}| \delta\varphi^{+} \psi_{22,j}^{(2)} dA \\ &- \int_0^{E_f} dE \int_{\boldsymbol{\Omega}\cdot\mathbf{n}>0} d\boldsymbol{\Omega} \int_{\partial V} \boldsymbol{\Omega}\cdot\mathbf{n}\, \delta\varphi^{+}(\mathbf{r},\boldsymbol{\Omega},E) \psi_{22,j}^{(2)}(\mathbf{r},\boldsymbol{\Omega},E) dA, \end{aligned} \tag{4.54}$$

$$\begin{aligned} P_3^{(2)}\left[\boldsymbol{\psi}_j^{(2)}, \delta\boldsymbol{\psi}^{(1)}, \delta\mathbf{u}\right] &\triangleq \int_0^{E_f} dE \int_{\boldsymbol{\Omega}\cdot\mathbf{n}<0} d\boldsymbol{\Omega} \int_{\partial V} |\boldsymbol{\Omega}\cdot\mathbf{n}| \delta\psi_1^{(1)} \psi_{11,j}^{(2)} dA \\ &- \int_0^{E_f} dE \int_{\boldsymbol{\Omega}\cdot\mathbf{n}>0} d\boldsymbol{\Omega} \int_{\partial V} \boldsymbol{\Omega}\cdot\mathbf{n}\; \delta\psi_1^{(1)}(\mathbf{r},\boldsymbol{\Omega},E) \psi_{11,j}^{(2)}(\mathbf{r},\boldsymbol{\Omega},E) dA, \end{aligned} \tag{4.55}$$

$$P_4^{(2)}\left[\boldsymbol{\psi}_j^{(2)},\delta\boldsymbol{\psi}^{(1)},\delta\mathbf{u}\right]\triangleq\int_0^{E_f}dE\int_{\boldsymbol{\Omega}\cdot\mathbf{n}<0}d\boldsymbol{\Omega}\int_{\partial V}|\boldsymbol{\Omega}\cdot\mathbf{n}|\delta\psi_2^{(1)}\psi_{12,j}^{(2)}dA$$
$$-\int_0^{E_f}dE\int_{\boldsymbol{\Omega}\cdot\mathbf{n}>0}d\boldsymbol{\Omega}\int_{\partial V}\boldsymbol{\Omega}\cdot\mathbf{n}\,\delta\psi_2^{(1)}(\mathbf{r},\boldsymbol{\Omega},E)\psi_{12,j}^{(2)}(\mathbf{r},\boldsymbol{\Omega},E)dA. \tag{4.56}$$

4. Reduce the bilinear concomitant $P^{(2)}\left[\boldsymbol{\psi}_j^{(2)},\delta\boldsymbol{\psi}^{(1)},\delta\mathbf{u}\right]$ in (4.52) to zero by using the boundary conditions shown in (4.18) and (4.42), and by imposing on the components of the second-level adjoint function $\boldsymbol{\psi}_j^{(2)}(\mathbf{r},\boldsymbol{\Omega},E)$ the following boundary conditions:

$$\left.\begin{array}{l}\psi_{11,j}^{(2)}(\mathbf{r}_s,\boldsymbol{\Omega},E)=0,\ \ \mathbf{r}_s\in\partial V,\ \ \boldsymbol{\Omega}\cdot\mathbf{n}<0;0<E<E_f\\ \psi_{12,j}^{(2)}(\mathbf{r}_s,\boldsymbol{\Omega},E)=0,\ \ \mathbf{r}_s\in\partial V,\ \ \boldsymbol{\Omega}\cdot\mathbf{n}>0;0<E<E_f\\ \psi_{21,j}^{(2)}(\mathbf{r}_s,\boldsymbol{\Omega},E)=0,\ \ \mathbf{r}_s\in\partial V,\ \ \boldsymbol{\Omega}\cdot\mathbf{n}>0;\ 0<E<E_f\\ \psi_{22,j}^{(2)}(\mathbf{r}_s,\boldsymbol{\Omega},E)=0,\ \ \mathbf{r}_s\in\partial V,\ \ \boldsymbol{\Omega}\cdot\mathbf{n}<0;\ 0<E<E_f\end{array}\right\} \tag{4.57}$$

Identify the second term on the right side of (4.51) with the indirect-effect term defined in (4.40) and use (4.50) to obtain the following expression for the indirect-effect term:

$$\left\{\delta\left[\frac{\partial R(\varphi,\varphi^+,\boldsymbol{\alpha};\boldsymbol{\psi}^{(1)})}{\partial t_j}\right]\right\}_{ind}=\left\langle\begin{pmatrix}\boldsymbol{\psi}_{1,j}^{(2)}\\ \boldsymbol{\psi}_{2,j}^{(2)}\end{pmatrix}^{\dagger}\begin{pmatrix}\mathbf{Q}^{(2)}\left(\boldsymbol{\alpha},\boldsymbol{\psi}^{(1)};\delta\boldsymbol{\alpha}\right)\\ \mathbf{Q}^{(1)}\left(\boldsymbol{\alpha};\varphi,\varphi^+;\delta\boldsymbol{\alpha}\right)\end{pmatrix}\right\rangle_{(2)}, \tag{4.58}$$

where the components of the second-level adjoint function $\boldsymbol{\psi}_j^{(2)}(\mathbf{r},\boldsymbol{\Omega},E)$ are the solutions of the following system of equations:

$$\begin{pmatrix}\left[\mathbf{A}_{11}^{(1)}(\boldsymbol{\alpha})\right]^+ & \mathbf{0}\\ \left[\mathbf{F}_{12}^{(2)}(\boldsymbol{\alpha})\right]^+ & \left[\mathbf{F}_{11}^{(1)}(\boldsymbol{\alpha})\right]^+\end{pmatrix}\begin{pmatrix}\boldsymbol{\psi}_{1,j}^{(2)}\\ \boldsymbol{\psi}_{2,j}^{(2)}\end{pmatrix}=\begin{pmatrix}-\varphi(\mathbf{r},\boldsymbol{\Omega},E)\partial\Sigma_t(\mathbf{t};\mathbf{r},E)/\partial t_j\\ -\varphi^+(\mathbf{r},\boldsymbol{\Omega},E)\partial\Sigma_t(\mathbf{t};\mathbf{r},E)/\partial t_j\\ -\psi_1^{(1)}(\mathbf{r},\boldsymbol{\Omega},E)\partial\Sigma_t(\mathbf{t};\mathbf{r},E)/\partial t_j\\ -\psi_2^{(1)}(\mathbf{r},\boldsymbol{\Omega},E)\partial\Sigma_t(\mathbf{t};\mathbf{r},E)/\partial t_j\end{pmatrix}. \tag{4.59}$$

The operator Eq. (4.59) together with the boundary conditions given in (4.57) constitute the *Second-Level Adjoint Sensitivity System* (2nd-LASS) for the *second-level adjoint function* $\boldsymbol{\psi}_j^{(2)}(\mathbf{r},\boldsymbol{\Omega},E) \triangleq \left[\boldsymbol{\psi}_{1,j}^{(2)}(\mathbf{r},\boldsymbol{\Omega},E), \boldsymbol{\psi}_{2,j}^{(2)}(\mathbf{r},\boldsymbol{\Omega},E)\right]$, $j = 1, \ldots, J_t$. The components of $\boldsymbol{\psi}_j^{(2)}(\mathbf{r},\boldsymbol{\Omega},E)$ are obtained by performing two "forward transport" and two "adjoint transport" computations, for each of the imprecisely known scalar model parameters appearing in the definition of the total macroscopic cross section $\Sigma_t(\mathbf{t};\mathbf{r},E)$. Subsequently, the second-level adjoint function $\boldsymbol{\psi}_j^{(2)}(\mathbf{r},\boldsymbol{\Omega},E)$ is used in (4.40) to eliminate the dependence of the second-order sensitivities on the variations $\delta\mathbf{u}(\mathbf{r},\boldsymbol{\Omega},E)$ and $\delta\boldsymbol{\psi}^{(1)}(\mathbf{r},\boldsymbol{\Omega},E)$, and hence on parameter variations, thereby enabling the efficient and exact computation of the indirect-effect term. The second-order partial sensitivities $\partial^2 R\left(\varphi,\varphi^+;\boldsymbol{\psi}^{(1)};\boldsymbol{\psi}_j^{(2)};\boldsymbol{\alpha}\right)/\left(\partial t_j\right)\left(\partial\alpha_{m_2}\right)$, for $j = 1, \ldots, J_t$; $m_2 = 1, \ldots, TP$, can now be determined by identifying the quantities multiplying the parameter variations $\delta\alpha_{m_2}$ in expression resulting from the sum of the direct-effect term provided in (4.39) and the indirect-effect term obtained in (4.58). This procedure yields the following expressions:

$$
\begin{aligned}
&\text{For } j, m_2 = 1, \ldots, J_t : \frac{\partial^2 R\left(\varphi,\varphi^+;\boldsymbol{\psi}^{(1)};\boldsymbol{\psi}_j^{(2)};\boldsymbol{\alpha}\right)}{\partial t_j \partial t_{m_2}} \\
&= -\int dV \int_{4\pi} d\Omega \int_0^{E_f} dE \; \frac{\partial^2 \Sigma_t(\mathbf{t};\mathbf{r},E)}{\partial t_j \partial t_{m_2}} \\
&\times \left[\psi_1^{(1)}(\mathbf{r},\boldsymbol{\Omega},E)\varphi(\mathbf{r},\boldsymbol{\Omega},E) + \psi_2^{(1)}(\mathbf{r},\boldsymbol{\Omega},E)\varphi^+(\mathbf{r},\boldsymbol{\Omega},E)\right] \\
&- \int dV \int_{4\pi} d\Omega \int_0^{E_f} dE \; \frac{\partial \Sigma_t(\mathbf{t};\mathbf{r},E)}{\partial t_{m_2}} \left[\psi_{11,j}^{(2)}(\mathbf{r},\boldsymbol{\Omega},E)\psi_1^{(1)}(\mathbf{r},\boldsymbol{\Omega},E)\right. \\
&+ \psi_{12,j}^{(2)}(\mathbf{r},\boldsymbol{\Omega},E)\psi_2^{(1)}(\mathbf{r},\boldsymbol{\Omega},E) + \psi_{21,j}^{(2)}(\mathbf{r},\boldsymbol{\Omega},E)\varphi(\mathbf{r},\boldsymbol{\Omega},E) \\
&\left. + \psi_{22,j}^{(2)}(\mathbf{r},\boldsymbol{\Omega},E)\varphi^+(\mathbf{r},\boldsymbol{\Omega},E)\right];
\end{aligned}
\tag{4.60}
$$

For $j = 1, \ldots, J_t;\ m_2 = 1, \ldots, J_s:\ \dfrac{\partial^2 R\left(\varphi, \varphi^+; \boldsymbol{\psi}^{(1)}; \boldsymbol{\psi}_j^{(2)}; \boldsymbol{\alpha}\right)}{\partial t_j \partial s_{m_2}}$

$$\begin{aligned}
&= \int dV \int_{4\pi} d\boldsymbol{\Omega} \int_0^{E_f} dE \int_{4\pi} d\boldsymbol{\Omega}' \int_0^{E_f} dE' \frac{\partial \Sigma_s(\mathbf{s}; \mathbf{r}, E \to E', \boldsymbol{\Omega} \to \boldsymbol{\Omega}')}{\partial s_{m_2}} \\
&\times \left[\psi_{11,j}^{(2)}(\mathbf{r}, \boldsymbol{\Omega}, E)\psi_1^{(1)}(\mathbf{r}, \boldsymbol{\Omega}', E') + \psi_{22,j}^{(2)}(\mathbf{r}, \boldsymbol{\Omega}, E)\varphi^+(\mathbf{r}, \boldsymbol{\Omega}', E')\right] \\
&+ \int dV \int_{4\pi} d\boldsymbol{\Omega} \int_0^{E_f} dE \int_{4\pi} d\boldsymbol{\Omega}' \int_0^{E_f} dE' \frac{\partial \Sigma_s(\mathbf{s}; \mathbf{r}, E' \to E, \boldsymbol{\Omega}' \to \boldsymbol{\Omega})}{\partial s_{m_2}} \\
&\times \left[\psi_{12,j}^{(2)}(\mathbf{r}, \boldsymbol{\Omega}, E)\psi_2^{(1)}(\mathbf{r}, \boldsymbol{\Omega}', E') + \psi_{21,j}^{(2)}(\mathbf{r}, \boldsymbol{\Omega}, E)\varphi(\mathbf{r}, \boldsymbol{\Omega}', E')\right];
\end{aligned} \tag{4.61}$$

For $j = 1, \ldots, J_t;\ m_2 = 1, \ldots, J_f:\ \dfrac{\partial^2 R\left(\varphi, \varphi^+; \boldsymbol{\psi}^{(1)}; \boldsymbol{\psi}_j^{(2)}; \boldsymbol{\alpha}\right)}{\partial t_j \partial f_{m_2}}$

$$\begin{aligned}
&= \int dV \int_{4\pi} d\boldsymbol{\Omega} \int_0^{E_f} dE \int_{4\pi} d\boldsymbol{\Omega}' \int_0^{E_f} dE' \chi(\mathbf{p}; \mathbf{r}, E') \frac{\partial \left[\nu\Sigma_f(\mathbf{f}; \mathbf{r}, E)\right]}{\partial f_{m_2}} \\
&\times \left[\psi_{11,j}^{(2)}(\mathbf{r}, \boldsymbol{\Omega}, E)\psi_1^{(1)}(\mathbf{r}, \boldsymbol{\Omega}', E') + \psi_{22,j}^{(2)}(\mathbf{r}, \boldsymbol{\Omega}, E)\varphi^+(\mathbf{r}, \boldsymbol{\Omega}', E')\right] \\
&+ \int dV \int_{4\pi} d\boldsymbol{\Omega} \int_0^{E_f} dE \int_{4\pi} d\boldsymbol{\Omega}' \int_0^{E_f} dE' \chi(\mathbf{p}; \mathbf{r}, E) \frac{\partial \left[\nu\Sigma_f(\mathbf{f}; \mathbf{r}, E')\right]}{\partial f_{m_2}} \\
&\times \left[\psi_{12,j}^{(2)}(\mathbf{r}, \boldsymbol{\Omega}, E)\psi_2^{(1)}(\mathbf{r}, \boldsymbol{\Omega}', E') + \psi_{21,j}^{(2)}(\mathbf{r}, \boldsymbol{\Omega}, E)\varphi(\mathbf{r}, \boldsymbol{\Omega}', E')\right];
\end{aligned} \tag{4.62}$$

For $j = 1, \ldots, J_t;\ m_2 = 1, \ldots, J_p:\ \dfrac{\partial^2 R\left(\varphi, \varphi^+; \boldsymbol{\psi}^{(1)}; \boldsymbol{\psi}_j^{(2)}; \boldsymbol{\alpha}\right)}{\partial t_j \partial p_{m_2}}$

$$\begin{aligned}
&= \int dV \int_{4\pi} d\boldsymbol{\Omega} \int_0^{E_f} dE \int_{4\pi} d\boldsymbol{\Omega}' \int_0^{E_f} dE' \frac{\partial \chi(\mathbf{p}; \mathbf{r}, E')}{\partial p_{m_2}} \nu\Sigma_f(\mathbf{f}; \mathbf{r}, E) \\
&\times \left[\psi_{11,j}^{(2)}(\mathbf{r}, \boldsymbol{\Omega}, E)\psi_1^{(1)}(\mathbf{r}, \boldsymbol{\Omega}', E') + \psi_{22,j}^{(2)}(\mathbf{r}, \boldsymbol{\Omega}, E)\varphi^+(\mathbf{r}, \boldsymbol{\Omega}', E')\right] \\
&+ \int dV \int_{4\pi} d\boldsymbol{\Omega} \int_0^{E_f} dE \int_{4\pi} d\boldsymbol{\Omega}' \int_0^{E_f} dE' \frac{\partial \chi(\mathbf{p}; \mathbf{r}, E)}{\partial p_{m_2}} \nu\Sigma_f(\mathbf{f}; \mathbf{r}, E') \\
&\times \left[\psi_{12,j}^{(2)}(\mathbf{r}, \boldsymbol{\Omega}, E)\psi_2^{(1)}(\mathbf{r}, \boldsymbol{\Omega}', E') + \psi_{21,j}^{(2)}(\mathbf{r}, \boldsymbol{\Omega}, E)\varphi(\mathbf{r}, \boldsymbol{\Omega}', E')\right];
\end{aligned} \tag{4.63}$$

$$\text{For } j=1,\ldots,J_t;\ m_2=1,\ldots,J_q:\ \frac{\partial^2 R\left(\varphi,\varphi^+;\boldsymbol{\psi}^{(1)};\boldsymbol{\psi}_j^{(2)};\boldsymbol{\alpha}\right)}{\partial t_j \partial q_{m_2}}$$
$$=\int dV\int_{4\pi} d\boldsymbol{\Omega}\int_0^{E_f} dE\,\psi_{21,j}^{(2)}(\mathbf{r},\boldsymbol{\Omega},E)\frac{\partial Q(\mathbf{q};\mathbf{r},\boldsymbol{\Omega},E)}{\partial q_{m_2}}; \tag{4.64}$$

$$\text{For } j=1,\ldots,J_t;\ m_2=1,\ldots,J_k:\ \frac{\partial^2 R\left(\varphi,\varphi^+;\boldsymbol{\psi}^{(1)};\boldsymbol{\psi}_j^{(2)};\boldsymbol{\alpha}\right)}{\partial t_j \partial k_{m_2}}$$
$$=\int dV\int_{4\pi} d\boldsymbol{\Omega}\int_0^{E_f} dE\,\psi_{22,j}^{(2)}(\mathbf{r},\boldsymbol{\Omega},E)\frac{\partial Q^+(\mathbf{k};\mathbf{r},\boldsymbol{\Omega},E)}{\partial k_{m_2}}; \tag{4.65}$$

$$\text{For } j=1,\ldots,J_t;\ m_2=1,\ldots,J_d:\ \frac{\partial^2 R\left(\varphi,\varphi^+;\boldsymbol{\psi}^{(1)};\boldsymbol{\psi}_j^{(2)};\boldsymbol{\alpha}\right)}{\partial t_j \partial d_{m_2}}$$
$$=\int dV\int_{4\pi} d\boldsymbol{\Omega}\int_0^{E_f} dE\left[\psi_{11,j}^{(2)}\frac{\partial \Theta_1^{(1)}(\mathbf{d};\varphi,\varphi^+)}{\partial d_{m_2}}+\psi_{12,j}^{(2)}\frac{\partial \Theta_2^{(1)}(\mathbf{d};\varphi,\varphi^+)}{\partial d_{m_2}}\right]. \tag{4.66}$$

The components of the second-level adjoint function $\boldsymbol{\psi}_j^{(2)}(\mathbf{r},\boldsymbol{\Omega},E)$, which appear in (4.60)–(4.66), are the solutions of the following system of equations:

$$L(\boldsymbol{\alpha})\psi_{11,j}^{(2)}(\mathbf{r},\boldsymbol{\Omega},E)=-\varphi(\mathbf{r},\boldsymbol{\Omega},E)\frac{\partial\Sigma_t(\mathbf{t};\mathbf{r},E)}{\partial t_j},\quad j=1,\ldots,J_t, \tag{4.67}$$

$$L^+(\boldsymbol{\alpha})\psi_{12,j}^{(2)}(\mathbf{r},\boldsymbol{\Omega},E)=-\varphi^+(\mathbf{r},\boldsymbol{\Omega},E)\frac{\partial\Sigma_t(\mathbf{t};\mathbf{r},E)}{\partial t_j},\quad j=1,\ldots,J_t, \tag{4.68}$$

$$L^+(\boldsymbol{\alpha})\psi_{21,j}^{(2)}(\mathbf{r},\boldsymbol{\Omega},E)=\frac{\partial\Theta_1^{(1)}(\mathbf{d};\varphi,\varphi^+)}{\partial\varphi}\psi_{11}^{(2)}(\mathbf{r},\boldsymbol{\Omega},E)$$
$$+\frac{\partial\Theta_2^{(1)}(\mathbf{d};\varphi,\varphi^+)}{\partial\varphi}\psi_{12}^{(2)}(\mathbf{r},\boldsymbol{\Omega},E)-\psi_1^{(1)}(\mathbf{r},\boldsymbol{\Omega},E)\frac{\partial\Sigma_t(\mathbf{t};\mathbf{r},E)}{\partial t_j},j=1,\ldots,J_t, \tag{4.69}$$

$$L(\boldsymbol{\alpha})\psi_{22,j}^{(2)}(\mathbf{r},\boldsymbol{\Omega},E)=\frac{\partial\Theta_1^{(1)}(\mathbf{d};\varphi,\varphi^+)}{\partial\varphi^+}\psi_{11}^{(2)}(\mathbf{r},\boldsymbol{\Omega},E)$$
$$+\frac{\partial\Theta_2^{(1)}(\mathbf{d};\varphi,\varphi^+)}{\partial\varphi^+}\psi_{12}^{(2)}(\mathbf{r},\boldsymbol{\Omega},E)-\psi_2^{(1)}(\mathbf{r},\boldsymbol{\Omega},E)\frac{\partial\Sigma_t(\mathbf{t};\mathbf{r},E)}{\partial t_j},j=1,\ldots,J_t, \tag{4.70}$$

where the following definitions were used:

$$\frac{\partial \Theta_1^{(1)}(\mathbf{d};\varphi,\varphi^+)}{\partial \varphi} \triangleq \frac{\left(\frac{\partial^2 G}{\partial \varphi^2}\right)}{\int dV \int_{4\pi} d\boldsymbol{\Omega} \int_0^{E_f} H dE} - \left\{ \left(\frac{\partial G}{\partial \varphi}\right) \left[\int dV \int_{4\pi} d\boldsymbol{\Omega} \int_0^{E_f} \left(\frac{\partial H}{\partial \varphi}\right) dE \right] \right.$$
$$\left. + \left(\frac{\partial^2 H}{\partial \varphi^2}\right) \left(\int dV \int_{4\pi} d\boldsymbol{\Omega} \int_0^{E_f} G dE \right) + \left(\frac{\partial H}{\partial \varphi}\right) \left[\int dV \int_{4\pi} d\boldsymbol{\Omega} \int_0^{E_f} \left(\frac{\partial G}{\partial \varphi}\right) dE \right] \right\}$$
$$\times \left(\int dV \int_{4\pi} d\boldsymbol{\Omega} \int_0^{E_f} H dE \right)^{-2} + 2 \left(\frac{\partial H}{\partial \varphi}\right) \left(\int dV \int_{4\pi} d\boldsymbol{\Omega} \int_0^{E_f} G dE \right)$$
$$\times \left[\int dV \int_{4\pi} d\boldsymbol{\Omega} \int_0^{E_f} \left(\frac{\partial H}{\partial \varphi}\right) dE \right] \left(\int dV \int_{4\pi} d\boldsymbol{\Omega} \int_0^{E_f} H dE \right)^{-3}, \tag{4.71}$$

$$\frac{\partial \Theta_1^{(1)}(\mathbf{d};\varphi,\varphi^+)}{\partial \varphi^+} \triangleq \frac{\left(\frac{\partial^2 G}{\partial \varphi \, \partial \varphi^+}\right)}{\int dV \int_{4\pi} d\boldsymbol{\Omega} \int_0^{E_f} H dE} - \left\{ \left(\frac{\partial G}{\partial \varphi}\right) \left[\int dV \int_{4\pi} d\boldsymbol{\Omega} \int_0^{E_f} \left(\frac{\partial H}{\partial \varphi^+}\right) dE \right] \right.$$
$$\left. + \left(\frac{\partial^2 H}{\partial \varphi \, \partial \varphi^+}\right) \left(\int dV \int_{4\pi} d\boldsymbol{\Omega} \int_0^{E_f} G dE \right) + \left(\frac{\partial H}{\partial \varphi}\right) \left[\int dV \int_{4\pi} d\boldsymbol{\Omega} \int_0^{E_f} \left(\frac{\partial G}{\partial \varphi^+}\right) dE \right] \right\}$$
$$\times \left(\int dV \int_{4\pi} d\boldsymbol{\Omega} \int_0^{E_f} H dE \right)^{-2} + 2 \left(\frac{\partial H}{\partial \varphi}\right) \left(\int dV \int_{4\pi} d\boldsymbol{\Omega} \int_0^{E_f} G dE \right)$$
$$\times \left[\int dV \int_{4\pi} d\boldsymbol{\Omega} \int_0^{E_f} \left(\frac{\partial H}{\partial \varphi^+}\right) dE \right] \left(\int dV \int_{4\pi} d\boldsymbol{\Omega} \int_0^{E_f} H dE \right)^{-3}, \tag{4.72}$$

$$\frac{\partial \Theta_2^{(1)}(\mathbf{d};\varphi,\varphi^+)}{\partial \varphi} \triangleq \frac{\left(\frac{\partial^2 G}{\partial \varphi\,\partial \varphi^+}\right)}{\int dV \int_{4\pi} d\boldsymbol{\Omega} \int_0^{E_f} H dE} - \left\{ \left(\frac{\partial G}{\partial \varphi^+}\right)\left[\int dV \int_{4\pi} d\boldsymbol{\Omega} \int_0^{E_f} \left(\frac{\partial H}{\partial \varphi}\right) dE\right] \right.$$
$$\left. + \left(\frac{\partial^2 H}{\partial \varphi\,\partial \varphi^+}\right)\left(\int dV \int_{4\pi} d\boldsymbol{\Omega} \int_0^{E_f} G dE\right) + \left(\frac{\partial H}{\partial \varphi^+}\right)\left[\int dV \int_{4\pi} d\boldsymbol{\Omega} \int_0^{E_f} \left(\frac{\partial G}{\partial \varphi}\right) dE\right]\right\}$$
$$\times \left(\int dV \int_{4\pi} d\boldsymbol{\Omega} \int_0^{E_f} H dE\right)^{-2} + 2\left(\frac{\partial H}{\partial \varphi^+}\right)\left(\int dV \int_{4\pi} d\boldsymbol{\Omega} \int_0^{E_f} G dE\right)$$
$$\times \left[\int dV \int_{4\pi} d\boldsymbol{\Omega} \int_0^{E_f} \left(\frac{\partial H}{\partial \varphi}\right) dE\right]\left(\int dV \int_{4\pi} d\boldsymbol{\Omega} \int_0^{E_f} H dE\right)^{-3}, \tag{4.73}$$

$$\frac{\partial \Theta_2^{(1)}(\mathbf{d};\varphi,\varphi^+)}{\partial \varphi^+} \triangleq \frac{\left(\frac{\partial^2 G}{\partial \varphi^{+2}}\right)}{\int dV \int_{4\pi} d\boldsymbol{\Omega} \int_0^{E_f} H dE} - \left\{ \left(\frac{\partial G}{\partial \varphi^+}\right)\left[\int dV \int_{4\pi} d\boldsymbol{\Omega} \int_0^{E_f} \left(\frac{\partial H}{\partial \varphi^+}\right) dE\right] \right.$$
$$\left. + \left(\frac{\partial^2 H}{\partial \varphi^{+2}}\right)\left(\int dV \int_{4\pi} d\boldsymbol{\Omega} \int_0^{E_f} G dE\right) + \left(\frac{\partial H}{\partial \varphi^+}\right)\left[\int dV \int_{4\pi} d\boldsymbol{\Omega} \int_0^{E_f} \left(\frac{\partial G}{\partial \varphi^+}\right) dE\right]\right\}$$
$$\times \left(\int dV \int_{4\pi} d\boldsymbol{\Omega} \int_0^{E_f} H dE\right)^{-2} + 2\left(\frac{\partial H}{\partial \varphi^+}\right)\left(\int dV \int_{4\pi} d\boldsymbol{\Omega} \int_0^{E_f} G dE\right)$$
$$\times \left[\int dV \int_{4\pi} d\boldsymbol{\Omega} \int_0^{E_f} \left(\frac{\partial H}{\partial \varphi^+}\right) dE\right]\left(\int dV \int_{4\pi} d\boldsymbol{\Omega} \int_0^{E_f} H dE\right)^{-3}. \tag{4.74}$$

Equations (4.67)–(4.70), which are just the detailed form of (4.59), together with the boundary conditions given in (4.57), constitute the *Second-Level Adjoint Sensitivity System* (2nd-LASS) for the *second-level adjoint function* $\boldsymbol{\psi}_j^{(2)}(\mathbf{r},\boldsymbol{\Omega},E) \triangleq \left[\boldsymbol{\psi}_{1,j}^{(2)}(\mathbf{r},\boldsymbol{\Omega},E), \boldsymbol{\psi}_{2,j}^{(2)}(\mathbf{r},\boldsymbol{\Omega},E)\right]$, $j = 1, \ldots, J_t$. The components of $\boldsymbol{\psi}_j^{(2)}(\mathbf{r},\boldsymbol{\Omega},E)$ are computed by solving two "forward transport" and two "adjoint transport" computations, for each of the imprecisely known scalar model parameters appearing in the definition of the total macroscopic cross section $\Sigma_t(\mathbf{t};\mathbf{r},E)$.

4.4.2 Determination of the Second-Order Sensitivities of the Form $\partial^2 R\left(\varphi,\varphi^+;\boldsymbol{\psi}^{(1)};\boldsymbol{\theta}_j^{(2)};\boldsymbol{\alpha}\right)/\partial s_j \partial \alpha_{m_2}$

The second-order sensitivities $\partial^2 R\left(\varphi,\varphi^+;\boldsymbol{\psi}^{(1)};\boldsymbol{\theta}_j^{(2)};\boldsymbol{\alpha}\right)/(\partial s_j)(\partial \alpha_{m_2})$, for $j = 1, \ldots, J_s$ and $m_2 = 1, \ldots, TP$, are obtained by following the same procedure as detailed above in Sect. 4.4.1, but replacing the function $\boldsymbol{\psi}_j^{(2)}(\mathbf{r},\boldsymbol{\Omega},E)$ by a second-level adjoint function denoted as $\boldsymbol{\theta}_j^{(2)}(\mathbf{r},\boldsymbol{\Omega},E) \triangleq \left[\boldsymbol{\theta}_{1,j}^{(2)}(\mathbf{r},\boldsymbol{\Omega},E), \boldsymbol{\theta}_{2,j}^{(2)}(\mathbf{r},\boldsymbol{\Omega},E)\right]^\dagger$, where $\boldsymbol{\theta}_{1,j}^{(2)} \triangleq \left[\theta_{11,j}^{(2)}(\mathbf{r},\boldsymbol{\Omega},E), \theta_{12,j}^{(2)}(\mathbf{r},\boldsymbol{\Omega},E)\right]^\dagger$, $\boldsymbol{\theta}_{2,j}^{(2)} \triangleq \left[\theta_{21,j}^{(2)}(\mathbf{r},\boldsymbol{\Omega},E), \theta_{22,j}^{(2)}(\mathbf{r},\boldsymbol{\Omega},E)\right]^\dagger$, with square-integrable components $\theta_{mn,j}^{(2)}(\mathbf{r},\boldsymbol{\Omega},E)$, $m,n = 1,2$. The end results are as follows:

$$\text{For } j = 1, \ldots, J_s;\ m_2 = 1, \ldots, J_t:\ \frac{\partial^2 R\left(\varphi,\varphi^+;\boldsymbol{\psi}^{(1)};\boldsymbol{\theta}_j^{(2)};\boldsymbol{\alpha}\right)}{\partial s_j \partial t_{m_2}}$$

$$= -\int dV \int_{4\pi} d\Omega \int_0^{E_f} dE\, \frac{\partial \Sigma_t(\mathbf{t};\mathbf{r},E)}{\partial t_{m_2}} \left[\theta_{11,j}^{(2)}(\mathbf{r},\boldsymbol{\Omega},E)\psi_1^{(1)}(\mathbf{r},\boldsymbol{\Omega},E)\right. \tag{4.75}$$

$$+\theta_{12,j}^{(2)}(\mathbf{r},\boldsymbol{\Omega},E)\psi_2^{(1)}(\mathbf{r},\boldsymbol{\Omega},E) + \theta_{21,j}^{(2)}(\mathbf{r},\boldsymbol{\Omega},E)\varphi(\mathbf{r},\boldsymbol{\Omega},E)$$

$$\left. +\theta_{22,j}^{(2)}(\mathbf{r},\boldsymbol{\Omega},E)\varphi^+(\mathbf{r},\boldsymbol{\Omega},E)\right];$$

$$
\begin{aligned}
&\text{For } j=1,\ldots,J_s;\ m_2=1,\ldots,J_s: \ \frac{\partial^2 R\left(\varphi,\varphi^+;\boldsymbol{\psi}^{(1)};\boldsymbol{\theta}_j^{(2)};\boldsymbol{\alpha}\right)}{\partial s_j \partial s_{m_2}} \\
&= \int dV \int_{4\pi} d\boldsymbol{\Omega} \int_0^{E_f} dE \int_{4\pi} d\boldsymbol{\Omega}' \int_0^{E_f} dE' \frac{\partial^2 \Sigma_s(\mathbf{s};\mathbf{r},E'\to E,\boldsymbol{\Omega}'\to\boldsymbol{\Omega})}{\partial s_j \partial s_{m_2}} \\
&\times\left[\psi_1^{(1)}(\mathbf{r},\boldsymbol{\Omega},E)\varphi(\mathbf{r},\boldsymbol{\Omega}',E')\right] \\
&+ \int dV \int_{4\pi} d\boldsymbol{\Omega} \int_0^{E_f} dE \int_{4\pi} d\boldsymbol{\Omega}' \int_0^{E_f} dE' \frac{\partial^2 \Sigma_s(\mathbf{s};\mathbf{r},E\to E',\boldsymbol{\Omega}\to\boldsymbol{\Omega}')}{\partial s_j \partial s_{m_2}} \\
&\times\left[\psi_2^{(1)}(\mathbf{r},\boldsymbol{\Omega},E)\varphi^+(\mathbf{r},\boldsymbol{\Omega}',E')\right] \\
&+ \int dV \int_{4\pi} d\boldsymbol{\Omega} \int_0^{E_f} dE \int_{4\pi} d\boldsymbol{\Omega}' \int_0^{E_f} dE' \frac{\partial \Sigma_s(\mathbf{s};\mathbf{r},E'\to E,\boldsymbol{\Omega}'\to\boldsymbol{\Omega})}{\partial s_{m_2}} \\
&\times\left[\theta_{12,j}^{(2)}(\mathbf{r},\boldsymbol{\Omega},E)\psi_2^{(1)}(\mathbf{r},\boldsymbol{\Omega}',E')+\theta_{21,j}^{(2)}(\mathbf{r},\boldsymbol{\Omega},E)\varphi(\mathbf{r},\boldsymbol{\Omega}',E')\right] \\
&+ \int dV \int_{4\pi} d\boldsymbol{\Omega} \int_0^{E_f} dE \int_{4\pi} d\boldsymbol{\Omega}' \int_0^{E_f} dE' \frac{\partial \Sigma_s(\mathbf{s};\mathbf{r},E\to E',\boldsymbol{\Omega}\to\boldsymbol{\Omega}')}{\partial s_{m_2}} \\
&\times\left[\theta_{11,j}^{(2)}(\mathbf{r},\boldsymbol{\Omega},E)\psi_1^{(1)}(\mathbf{r},\boldsymbol{\Omega}',E')+\theta_{22,j}^{(2)}(\mathbf{r},\boldsymbol{\Omega},E)\varphi^+(\mathbf{r},\boldsymbol{\Omega}',E')\right];
\end{aligned}
\tag{4.76}
$$

$$
\begin{aligned}
&\text{For } j=1,\ldots,J_s;\ m_2=1,\ldots,J_f: \ \frac{\partial^2 R\left(\varphi,\varphi^+;\boldsymbol{\psi}^{(1)};\boldsymbol{\theta}_j^{(2)};\boldsymbol{\alpha}\right)}{\partial s_j \partial f_{m_2}} \\
&= \int dV \int_{4\pi} d\boldsymbol{\Omega} \int_0^{E_f} dE \int_{4\pi} d\boldsymbol{\Omega}' \int_0^{E_f} dE' \frac{\partial\left[\nu\Sigma_f(\mathbf{f};\mathbf{r},E)\right]}{\partial f_{m_2}}\chi(\mathbf{p};\mathbf{r},E') \\
&\times\left[\theta_{11,j}^{(2)}(\mathbf{r},\boldsymbol{\Omega},E)\psi_1^{(1)}(\mathbf{r},\boldsymbol{\Omega}',E')+\theta_{22,j}^{(2)}(\mathbf{r},\boldsymbol{\Omega},E)\varphi^+(\mathbf{r},\boldsymbol{\Omega}',E')\right] \\
&+ \int dV \int_{4\pi} d\boldsymbol{\Omega} \int_0^{E_f} dE \int_{4\pi} d\boldsymbol{\Omega}' \int_0^{E_f} dE' \frac{\partial\left[\nu\Sigma_f(\mathbf{f};\mathbf{r},E')\right]}{\partial f_{m_2}}\chi(\mathbf{p};\mathbf{r},E) \\
&\times\left[\theta_{12,j}^{(2)}(\mathbf{r},\boldsymbol{\Omega},E)\psi_2^{(1)}(\mathbf{r},\boldsymbol{\Omega}',E')+\theta_{21,j}^{(2)}(\mathbf{r},\boldsymbol{\Omega},E)\varphi(\mathbf{r},\boldsymbol{\Omega}',E')\right];
\end{aligned}
\tag{4.77}
$$

$$\text{For } j=1,\ldots,J_s;\ m_2=1,\ldots,J_p:\ \frac{\partial^2 R\left(\varphi,\varphi^+;\boldsymbol{\psi}^{(1)};\boldsymbol{\theta}_j^{(2)};\boldsymbol{\alpha}\right)}{\partial s_j \partial p_{m_2}}$$

$$= \int dV \int_{4\pi} d\boldsymbol{\Omega} \int_0^{E_f} dE \int_{4\pi} d\boldsymbol{\Omega}' \int_0^{E_f} dE' \frac{\partial \chi(\mathbf{p};\mathbf{r},E')}{\partial p_{m_2}} \nu\Sigma_f(\mathbf{f};\mathbf{r},E)$$
$$\times \left[\theta_{11,j}^{(2)}(\mathbf{r},\boldsymbol{\Omega},E)\psi_1^{(1)}(\mathbf{r},\boldsymbol{\Omega}',E') + \theta_{22,j}^{(2)}(\mathbf{r},\boldsymbol{\Omega},E)\varphi^+(\mathbf{r},\boldsymbol{\Omega}',E')\right] \tag{4.78}$$
$$+ \int dV \int_{4\pi} d\boldsymbol{\Omega} \int_0^{E_f} dE \int_{4\pi} d\boldsymbol{\Omega}' \int_0^{E_f} dE' \frac{\partial \chi(\mathbf{p};\mathbf{r},E)}{\partial p_{m_2}} \nu\Sigma_f(\mathbf{f};\mathbf{r},E')$$
$$\times \left[\theta_{12,j}^{(2)}(\mathbf{r},\boldsymbol{\Omega},E)\psi_2^{(1)}(\mathbf{r},\boldsymbol{\Omega}',E') + \theta_{21,j}^{(2)}(\mathbf{r},\boldsymbol{\Omega},E)\varphi(\mathbf{r},\boldsymbol{\Omega}',E')\right];$$

$$\text{For } j=1,\ldots,J_s;\ m_2=1,\ldots,J_q:\ \frac{\partial^2 R\left(\varphi,\varphi^+;\boldsymbol{\psi}^{(1)};\boldsymbol{\theta}_j^{(2)};\boldsymbol{\alpha}\right)}{\partial s_j \partial q_{m_2}}$$
$$= \int dV \int_{4\pi} d\boldsymbol{\Omega} \int_0^{E_f} dE\, \theta_{21,j}^{(2)}(\mathbf{r},\boldsymbol{\Omega},E) \frac{\partial Q(\mathbf{q};\mathbf{r},\boldsymbol{\Omega},E)}{\partial q_{m_2}}; \tag{4.79}$$

$$\text{For } j=1,\ldots,J_s;\ m_2=1,\ldots,J_k:\ \frac{\partial^2 R\left(\varphi,\varphi^+;\boldsymbol{\psi}^{(1)};\boldsymbol{\theta}_j^{(2)};\boldsymbol{\alpha}\right)}{\partial s_j \partial k_{m_2}}$$
$$= \int dV \int_{4\pi} d\boldsymbol{\Omega} \int_0^{E_f} dE\, \theta_{22,j}^{(2)}(\mathbf{r},\boldsymbol{\Omega},E) \frac{\partial Q^+(\mathbf{k};\mathbf{r},\boldsymbol{\Omega},E)}{\partial k_{m_2}}; \tag{4.80}$$

$$\text{For } j=1,\ldots,J_s;\ m_2=1,\ldots,J_d:\ \frac{\partial^2 R\left(\varphi,\varphi^+;\boldsymbol{\psi}^{(1)};\boldsymbol{\theta}_j^{(2)};\boldsymbol{\alpha}\right)}{\partial s_j \partial d_{m_2}}$$
$$= \int dV \int_{4\pi} d\boldsymbol{\Omega} \int_0^{E_f} dE \left[\theta_{11,j}^{(2)}(\mathbf{r},\boldsymbol{\Omega},E) \frac{\partial \Theta_1^{(1)}(\mathbf{d};\varphi,\varphi^+)}{\partial d_{m_2}}\right. \tag{4.81}$$
$$\left. + \theta_{12,j}^{(2)}(\mathbf{r},\boldsymbol{\Omega},E) \frac{\partial \Theta_2^{(1)}(\mathbf{d};\varphi,\varphi^+)}{\partial d_{m_2}}\right].$$

The components of the second-level adjoint function $\boldsymbol{\theta}_j^{(2)}(\mathbf{r},\boldsymbol{\Omega},E)$ are the solutions of the following 2nd-LASS, for each $j = 1, \ldots, J_s$:

$$L(\boldsymbol{\alpha})\,\theta_{11,j}^{(2)}(\mathbf{r},\boldsymbol{\Omega},E) = \int_{4\pi} d\boldsymbol{\Omega}' \int_0^{E_f} dE' \, \frac{\partial \Sigma_s(\mathbf{s};\mathbf{r},E' \to E,\boldsymbol{\Omega}' \to \boldsymbol{\Omega})}{\partial s_j} \varphi(\mathbf{r},\boldsymbol{\Omega}',E'), \tag{4.82}$$

$$L^+(\boldsymbol{\alpha})\theta_{12,j}^{(2)}(\mathbf{r},\boldsymbol{\Omega},E) = \int_{4\pi} d\boldsymbol{\Omega}' \int_0^{E_f} dE' \, \frac{\partial \Sigma_s(\mathbf{s};\mathbf{r},E \to E',\boldsymbol{\Omega} \to \boldsymbol{\Omega}')}{\partial s_j} \varphi^+(\mathbf{r},\boldsymbol{\Omega}',E'), \tag{4.83}$$

$$\begin{aligned} L^+(\boldsymbol{\alpha})\theta_{21,j}^{(2)}(\mathbf{r},\boldsymbol{\Omega},E) &= \int_{4\pi} d\boldsymbol{\Omega}' \int_0^{E_f} dE' \, \psi_1^{(1)}(\mathbf{r},\boldsymbol{\Omega}',E') \frac{\partial \Sigma_s(\mathbf{s};\mathbf{r},E \to E',\boldsymbol{\Omega} \to \boldsymbol{\Omega}')}{\partial s_j} \\ &+ \frac{\partial \Theta_1^{(1)}(\mathbf{d};\varphi,\varphi^+)}{\partial \varphi} \theta_{11,j}^{(2)}(\mathbf{r},\boldsymbol{\Omega},E) + \frac{\partial \Theta_2^{(1)}(\mathbf{d};\varphi,\varphi^+)}{\partial \varphi} \theta_{12,j}^{(2)}(\mathbf{r},\boldsymbol{\Omega},E), \end{aligned} \tag{4.84}$$

$$\begin{aligned} L(\boldsymbol{\alpha})\theta_{22,j}^{(2)}(\mathbf{r},\boldsymbol{\Omega},E) &= \int_{4\pi} d\boldsymbol{\Omega}' \int_0^{E_f} dE' \, \psi_2^{(1)}(\mathbf{r},\boldsymbol{\Omega}',E') \frac{\partial \Sigma_s(\mathbf{s};\mathbf{r},E' \to E,\boldsymbol{\Omega}' \to \boldsymbol{\Omega})}{\partial s_j} \\ &+ \frac{\partial \Theta_1^{(1)}(\mathbf{d};\varphi,\varphi^+)}{\partial \varphi^+} \theta_{11,j}^{(2)}(\mathbf{r},\boldsymbol{\Omega},E) + \frac{\partial \Theta_2^{(1)}(\mathbf{d};\varphi,\varphi^+)}{\partial \varphi^+} \theta_{12,j}^{(2)}(\mathbf{r},\boldsymbol{\Omega},E), \end{aligned} \tag{4.85}$$

subject to the following boundary conditions:

$$\left.\begin{aligned} &\theta_{11,j}^{(2)}(\mathbf{r}_s,\boldsymbol{\Omega},E) = 0, \;\; \mathbf{r}_s \in \partial V, \;\; \boldsymbol{\Omega}\cdot\mathbf{n} < 0; \;\; 0 < E < E_f \\ &\theta_{12,j}^{(2)}(\mathbf{r}_s,\boldsymbol{\Omega},E) = 0, \;\; \mathbf{r}_s \in \partial V, \;\; \boldsymbol{\Omega}\cdot\mathbf{n} > 0; 0 < E < E_f \\ &\theta_{21,j}^{(2)}(\mathbf{r}_s,\boldsymbol{\Omega},E) = 0, \;\; \mathbf{r}_s \in \partial V, \;\; \boldsymbol{\Omega}\cdot\mathbf{n} > 0; \;\; 0 < E < E_f \\ &\theta_{22,j}^{(2)}(\mathbf{r}_s,\boldsymbol{\Omega},E) = 0, \;\; \mathbf{r}_s \in \partial V, \;\; \boldsymbol{\Omega}\cdot\mathbf{n} < 0; 0 < E < E_f \end{aligned}\right\} \tag{4.86}$$

It is important to note that the forward and adjoint operators appearing on the left side of the 2nd-LASS defined by (4.82)–(4.85) for the second-level adjoint function $\boldsymbol{\theta}_j^{(2)}$ are the same operators as appearing on the left side of the 2nd-LASS defined by (4.67)–(4.70) for the second-level adjoint function $\boldsymbol{\psi}_j^{(2)}$, the forward operator being the same as on the left side of the original forward transport Eq. (4.1), while the adjoint operator is the same as that appearing in the original adjoint transport Eq. (4.6). Furthermore, the second-level adjoint function $\boldsymbol{\theta}_j^{(2)}$ is also subject to forward and adjoint vacuum boundary conditions. Only the source terms on the right sides of the respective forward, 1st-LASS and 2nd-LASS differ from each other. Therefore, the same "forward" and "adjoint" software packages can be used for solving numerically the various equations underlying the 1st-LASS and the 2nd-LASS. Furthermore, the second-order sensitivities shown in (4.75)–(4.81) have *formally* the same expressions as the second-order sensitivities shown in (4.60)–(4.66), respectively, except that the second-level adjoint function $\boldsymbol{\theta}_j^{(2)}$ in (4.75)–(4.81) plays the role of the second-level adjoint function $\boldsymbol{\psi}_j^{(2)}$ in (4.60)–(4.66). Thus, the software package used for computing the sensitivities shown in (4.60)–(4.66) can also be used for computing the sensitivities shown in (4.75)–(4.81).

The expressions of the second-order sensitivities computed using (4.75) must be identical to those computed using (4.61), which provides an independent path for the mutual verification of the solutions $\boldsymbol{\theta}_j^{(2)}$ and $\boldsymbol{\psi}_j^{(2)}$, of the respective 2nd-LASS.

4.4.3 Determination of the Second-Order Sensitivities of the Form $\partial^2 R\left(\varphi, \varphi^+; \boldsymbol{\psi}^{(1)}; \mathbf{u}_j^{(2)}; \boldsymbol{\alpha}\right)/\partial f_j \partial \alpha_{m_2}$

The second-order sensitivities $\partial^2 R\left(\varphi, \varphi^+; \boldsymbol{\psi}^{(1)}; \mathbf{u}_j^{(2)}; \boldsymbol{\alpha}\right)/(\partial f_j)(\partial \alpha_{m_2}), j = 1, \ldots,$ J_f, $m_2 = 1, \ldots, TP$, are obtained by following the same procedure as detailed in Sect. 4.4.1 but replacing the function $\boldsymbol{\psi}_j^{(2)}(\mathbf{r}, \boldsymbol{\Omega}, E)$ by another second-level adjoint function, which will be denoted as $\mathbf{u}_j^{(2)}(\mathbf{r}, \boldsymbol{\Omega}, E) \triangleq \left[\mathbf{u}_{1,j}^{(2)}(\mathbf{r}, \boldsymbol{\Omega}, E), \mathbf{u}_{2,j}^{(2)}(\mathbf{r}, \boldsymbol{\Omega}, E)\right]^\dagger$, where $\mathbf{u}_{1,j}^{(2)} \triangleq \left[u_{11,j}^{(2)}(\mathbf{r}, \boldsymbol{\Omega}, E), u_{12,j}^{(2)}(\mathbf{r}, \boldsymbol{\Omega}, E)\right]^\dagger$, $\mathbf{u}_{2,j}^{(2)} \triangleq \left[u_{21,j}^{(2)}(\mathbf{r}, \boldsymbol{\Omega}, E), u_{22,j}^{(2)}(\mathbf{r}, \boldsymbol{\Omega}, E)\right]^\dagger$, with square-integrable components $u_{mn,j}^{(2)}(\mathbf{r}, \boldsymbol{\Omega}, E),\ m, n = 1, 2$. The end results are as follows:

$$
\text{For } j = 1, \ldots, J_f;\ m_2 = 1, \ldots, J_t: \ \frac{\partial^2 R\left(\varphi, \varphi^+; \boldsymbol{\psi}^{(1)}; \mathbf{u}_j^{(2)}; \boldsymbol{\alpha}\right)}{\partial f_j \partial t_{m_2}}
$$

$$
\begin{aligned}
&= -\int dV \int_{4\pi} d\Omega \int_0^{E_f} dE \, \frac{\partial \Sigma_t(\mathbf{t}; \mathbf{r}, E)}{\partial t_{m_2}} \left[u_{11,j}^{(2)}(\mathbf{r}, \boldsymbol{\Omega}, E)\psi_1^{(1)}(\mathbf{r}, \boldsymbol{\Omega}, E)\right. \\
&\quad + u_{12,j}^{(2)}(\mathbf{r}, \boldsymbol{\Omega}, E)\psi_2^{(1)}(\mathbf{r}, \boldsymbol{\Omega}, E) + u_{21,j}^{(2)}(\mathbf{r}, \boldsymbol{\Omega}, E)\varphi(\mathbf{r}, \boldsymbol{\Omega}, E) \\
&\quad \left. + u_{22,j}^{(2)}(\mathbf{r}, \boldsymbol{\Omega}, E)\varphi^+(\mathbf{r}, \boldsymbol{\Omega}, E)\right];
\end{aligned} \tag{4.87}
$$

For $j = 1, \ldots, J_f;\ m_2 = 1, \ldots, J_s:\ \dfrac{\partial^2 R\left(\varphi, \varphi^+; \boldsymbol{\psi}^{(1)}; \mathbf{u}_j^{(2)}; \boldsymbol{\alpha}\right)}{\partial f_j \partial s_{m_2}}$

$$= \int dV \int_{4\pi} d\boldsymbol{\Omega} \int_0^{E_f} dE \int_{4\pi} d\boldsymbol{\Omega}' \int_0^{E_f} dE' \frac{\partial \Sigma_s(\mathbf{s}; \mathbf{r}, E \to E', \boldsymbol{\Omega} \to \boldsymbol{\Omega}')}{\partial s_{m_2}}$$
$$\times \left[u_{11,j}^{(2)}(\mathbf{r}, \boldsymbol{\Omega}, E)\psi_1^{(1)}(\mathbf{r}, \boldsymbol{\Omega}', E') + u_{22,j}^{(2)}(\mathbf{r}, \boldsymbol{\Omega}, E)\varphi^+(\mathbf{r}, \boldsymbol{\Omega}', E')\right] \tag{4.88}$$
$$+ \int dV \int_{4\pi} d\boldsymbol{\Omega} \int_0^{E_f} dE \int_{4\pi} d\boldsymbol{\Omega}' \int_0^{E_f} dE' \frac{\partial \Sigma_s(\mathbf{s}; \mathbf{r}, E' \to E, \boldsymbol{\Omega}' \to \boldsymbol{\Omega})}{\partial s_{m_2}}$$
$$\times \left[u_{12,j}^{(2)}(\mathbf{r}, \boldsymbol{\Omega}, E)\psi_2^{(1)}(\mathbf{r}, \boldsymbol{\Omega}', E') + u_{21,j}^{(2)}(\mathbf{r}, \boldsymbol{\Omega}, E)\, \varphi(\mathbf{r}, \boldsymbol{\Omega}', E')\right];$$

For $j = 1, \ldots, J_f;\ m_2 = 1, \ldots, J_f:\ \dfrac{\partial^2 R\left(\varphi, \varphi^+; \boldsymbol{\psi}^{(1)}; \mathbf{u}_j^{(2)}; \boldsymbol{\alpha}\right)}{\partial f_j \partial f_{m_2}}$

$$= \int dV \int_{4\pi} d\boldsymbol{\Omega} \int_0^{\infty} dE\, \psi_1^{(1)}(\mathbf{r}, \boldsymbol{\Omega}, E)\chi(E)$$
$$\times \int_{4\pi} d\boldsymbol{\Omega}' \int_0^{\infty} dE'\, \varphi(\mathbf{r}, \boldsymbol{\Omega}', E') \frac{\partial^2 \left[\nu \Sigma_f(\mathbf{f}; \mathbf{r}, E')\right]}{\partial f_j \partial f_{m_2}}$$
$$+ \int dV \int_{4\pi} d\boldsymbol{\Omega} \int_0^{E_f} dE \psi_2^{(1)}(\mathbf{r}, \boldsymbol{\Omega}, E) \frac{\partial^2 \left[\nu \Sigma_f(\mathbf{f}; \mathbf{r}, E)\right]}{\partial f_j \partial f_{m_2}}$$
$$\times \int_{4\pi} d\boldsymbol{\Omega}' \int_0^{E_f} dE' \varphi^+(\mathbf{r}, \boldsymbol{\Omega}', E')\chi(E') \tag{4.89}$$
$$+ \int dV \int_{4\pi} d\boldsymbol{\Omega} \int_0^{E_f} dE \int_{4\pi} d\boldsymbol{\Omega}' \int_0^{E_f} dE' \chi(E') \frac{\partial \left[\nu \Sigma_f(\mathbf{f}; \mathbf{r}, E)\right]}{\partial f_{m_2}}$$
$$\times \left[u_{11,j}^{(2)}(\mathbf{r}, \boldsymbol{\Omega}, E)\psi_1^{(1)}(\mathbf{r}, \boldsymbol{\Omega}', E') + u_{22,j}^{(2)}(\mathbf{r}, \boldsymbol{\Omega}, E)\varphi^+(\mathbf{r}, \boldsymbol{\Omega}', E')\right]$$
$$+ \int dV \int_{4\pi} d\boldsymbol{\Omega} \int_0^{E_f} dE \int_{4\pi} d\boldsymbol{\Omega}' \int_0^{E_f} dE' \chi(E) \frac{\partial \left[\nu \Sigma_f(\mathbf{f}; \mathbf{r}, E')\right]}{\partial f_{m_2}}$$
$$\times \left[u_{12,j}^{(2)}(\mathbf{r}, \boldsymbol{\Omega}, E)\psi_2^{(1)}(\mathbf{r}, \boldsymbol{\Omega}', E') + u_{21,j}^{(2)}(\mathbf{r}, \boldsymbol{\Omega}, E)\varphi(\mathbf{r}, \boldsymbol{\Omega}', E')\right];$$

$$
\begin{aligned}
&\text{For } j=1,\ldots,J_f;\ m_2=1,\ldots,J_p: \ \frac{\partial^2 R\left(\varphi,\varphi^+;\boldsymbol{\psi}^{(1)};\mathbf{u}_j^{(2)};\boldsymbol{\alpha}\right)}{\partial f_j \partial p_{m_2}} \\
&= \int dV \int_{4\pi} d\boldsymbol{\Omega} \int_0^{E_f} dE \psi_1^{(1)}(\mathbf{r},\boldsymbol{\Omega},E) \frac{\partial \chi(\mathbf{p};\mathbf{r},E)}{\partial p_{m_2}} \\
&\times \int_{4\pi} d\boldsymbol{\Omega}' \int_0^{E_f} dE' \, \varphi(\mathbf{r},\boldsymbol{\Omega}',E') \frac{\partial \left[\nu\Sigma_f(\mathbf{f};\mathbf{r},E')\right]}{\partial f_j} \\
&+ \int dV \int_{4\pi} d\boldsymbol{\Omega} \int_0^{E_f} dE \psi_2^{(1)}(\mathbf{r},\boldsymbol{\Omega},E) \frac{\partial \left[\nu\Sigma_f(\mathbf{f};\mathbf{r},E)\right]}{\partial f_j} \\
&\times \int_{4\pi} d\boldsymbol{\Omega}' \int_0^{E_f} dE' \, \varphi^+(\mathbf{r},\boldsymbol{\Omega}',E') \frac{\partial \chi(\mathbf{p};\mathbf{r},E')}{\partial p_{m_2}} \\
&+ \int dV \int_{4\pi} d\boldsymbol{\Omega} \int_0^{E_f} dE \int_{4\pi} d\boldsymbol{\Omega}' \int_0^{E_f} dE' \, \frac{\partial \chi(\mathbf{p};\mathbf{r},E')}{\partial p_{m_2}} \nu\Sigma_f(\mathbf{f};\mathbf{r},E) \\
&\times \left[u_{11,j}^{(2)}(\mathbf{r},\boldsymbol{\Omega},E)\psi_1^{(1)}(\mathbf{r},\boldsymbol{\Omega}',E') + u_{22,j}^{(2)}(\mathbf{r},\boldsymbol{\Omega},E)\varphi^+(\mathbf{r},\boldsymbol{\Omega}',E')\right] \\
&+ \int dV \int_{4\pi} d\boldsymbol{\Omega} \int_0^{E_f} dE \int_{4\pi} d\boldsymbol{\Omega}' \int_0^{E_f} dE' \, \frac{\partial \chi(\mathbf{p};\mathbf{r},E)}{\partial p_{m_2}} \nu\Sigma_f(\mathbf{f};\mathbf{r},E') \\
&\times \left[u_{12,j}^{(2)}(\mathbf{r},\boldsymbol{\Omega},E)\psi_2^{(1)}(\mathbf{r},\boldsymbol{\Omega}',E') + u_{21,j}^{(2)}(\mathbf{r},\boldsymbol{\Omega},E)\varphi(\mathbf{r},\boldsymbol{\Omega}',E')\right];
\end{aligned}
\tag{4.90}
$$

$$
\begin{aligned}
&\text{For } j=1,\ldots,J_f;\ m_2=1,\ldots,J_q: \frac{\partial^2 R\left(\varphi,\varphi^+;\boldsymbol{\psi}^{(1)};\mathbf{u}_j^{(2)};\boldsymbol{\alpha}\right)}{\partial f_j \partial q_{m_2}} \\
&= \int dV \int_{4\pi} d\boldsymbol{\Omega} \int_0^{E_f} dE \, u_{21,j}^{(2)}(\mathbf{r},\boldsymbol{\Omega},E) \frac{\partial Q(\mathbf{q};\mathbf{r},\boldsymbol{\Omega},E)}{\partial q_{m_2}};
\end{aligned}
\tag{4.91}
$$

$$
\begin{aligned}
&\text{For } j=1,\ldots,J_f;\ m_2=1,\ldots,J_k: \frac{\partial^2 R\left(\varphi,\varphi^+;\boldsymbol{\psi}^{(1)};\mathbf{u}_j^{(2)};\boldsymbol{\alpha}\right)}{\partial f_j \partial k_{m_2}} \\
&= \int dV \int_{4\pi} d\boldsymbol{\Omega} \int_0^{E_f} dE \, u_{22,j}^{(2)}(\mathbf{r},\boldsymbol{\Omega},E) \frac{\partial Q^+(\mathbf{k};\mathbf{r},\boldsymbol{\Omega},E)}{\partial k_{m_2}};
\end{aligned}
\tag{4.92}
$$

For $j = 1, \ldots, J_f;\ m_2 = 1, \ldots, J_d:\ \dfrac{\partial^2 R\left(\varphi, \varphi^+; \boldsymbol{\psi}^{(1)}; \mathbf{u}_j^{(2)}; \boldsymbol{\alpha}\right)}{\partial f_j \partial d_{m_2}}$

$$= \int dV \int_{4\pi} d\mathbf{\Omega} \int_0^{E_f} dE \left[u_{11,j}^{(2)}(\mathbf{r}, \mathbf{\Omega}, E) \frac{\partial \Theta_1^{(1)}(\mathbf{d}; \varphi, \varphi^+)}{\partial d_{m_2}} + u_{12,j}^{(2)}(\mathbf{r}, \mathbf{\Omega}, E) \frac{\partial \Theta_2^{(1)}(\mathbf{d}; \varphi, \varphi^+)}{\partial d_{m_2}} \right]. \tag{4.93}$$

The components of $\mathbf{u}_j^{(2)}$ are the solutions of the following 2nd-LASS, for $j = 1, \ldots, J_f$:

$$L(\boldsymbol{\alpha})\, u_{11,j}^{(2)}(\mathbf{r}, \mathbf{\Omega}, E) = \int_{4\pi} d\mathbf{\Omega}' \int_0^{E_f} dE'\, \chi(\mathbf{p}; \mathbf{r}, E' \to E) \varphi(\mathbf{r}, \mathbf{\Omega}', E') \frac{\partial \left[\nu \Sigma_f(\mathbf{f}; \mathbf{r}, E')\right]}{\partial f_j}, \tag{4.94}$$

$$L^+(\boldsymbol{\alpha}) u_{12,j}^{(2)}(\mathbf{r}, \mathbf{\Omega}, E) = \frac{\partial \left[\nu \Sigma_f(\mathbf{f}; \mathbf{r}, E)\right]}{\partial f_j} \int_{4\pi} d\mathbf{\Omega}' \int_0^{E_f} dE' \chi(\mathbf{p}; \mathbf{r}, E \to E') \varphi^+(\mathbf{r}, \mathbf{\Omega}', E'), \tag{4.95}$$

$$L^+(\boldsymbol{\alpha}) u_{21,j}^{(2)}(\mathbf{r}, \mathbf{\Omega}, E) = \frac{\partial \Theta_1^{(1)}(\mathbf{d}; \varphi, \varphi^+)}{\partial \varphi} u_{11,j}^{(2)}(\mathbf{r}, \mathbf{\Omega}, E) + \frac{\partial \Theta_2^{(1)}(\mathbf{d}; \varphi, \varphi^+)}{\partial \varphi} u_{12,j}^{(2)}(\mathbf{r}, \mathbf{\Omega}, E) + \frac{\partial \left[\nu \Sigma_f(\mathbf{f}; \mathbf{r}, E)\right]}{\partial f_j} \int_{4\pi} d\mathbf{\Omega}' \int_0^{\infty} dE'\, \chi(\mathbf{p}; \mathbf{r}, E \to E') \psi_1^{(1)}(\mathbf{r}, \mathbf{\Omega}', E'), \tag{4.96}$$

$$L(\boldsymbol{\alpha}) u_{22,j}^{(2)}(\mathbf{r}, \mathbf{\Omega}, E) = \frac{\partial \Theta_1^{(1)}(\mathbf{d}; \varphi, \varphi^+)}{\partial \varphi^+} u_{11,j}^{(2)}(\mathbf{r}, \mathbf{\Omega}, E) + \frac{\partial \Theta_2^{(1)}(\mathbf{d}; \varphi, \varphi^+)}{\partial \varphi^+} u_{12,j}^{(2)}(\mathbf{r}, \mathbf{\Omega}, E) + \int_{4\pi} d\mathbf{\Omega}' \int_0^{E_f} dE' \chi(\mathbf{p}; \mathbf{r}, E' \to E) \psi_2^{(1)}(\mathbf{r}, \mathbf{\Omega}', E') \frac{\partial \left[\nu \Sigma_f(\mathbf{f}; \mathbf{r}, E')\right]}{\partial f_j}, \tag{4.97}$$

subject to the boundary conditions below:

$$\left.\begin{aligned}
&u_{11,j}^{(2)}(\mathbf{r}_s,\boldsymbol{\Omega},E)=0,\quad \mathbf{r}_s\in\partial V,\quad \boldsymbol{\Omega}\cdot\mathbf{n}<0;0<E<E_f\\
&u_{12,j}^{(2)}(\mathbf{r}_s,\boldsymbol{\Omega},E)=0,\quad \mathbf{r}_s\in\partial V,\quad \boldsymbol{\Omega}\cdot\mathbf{n}>0;0<E<E_f\\
&u_{21,j}^{(2)}(\mathbf{r}_s,\boldsymbol{\Omega},E)=0,\quad \mathbf{r}_s\in\partial V,\quad \boldsymbol{\Omega}\cdot\mathbf{n}>0;0<E<E_f\\
&u_{22,j}^{(2)}(\mathbf{r}_s,\boldsymbol{\Omega},E)=0,\quad \mathbf{r}_s\in\partial V,\quad \boldsymbol{\Omega}\cdot\mathbf{n}<0;0<E<E_f
\end{aligned}\right\}\tag{4.98}$$

The forward and adjoint operators appearing on the left side of the 2nd-LASS defined by (4.94)–(4.97) for the second-level adjoint function $\mathbf{u}_j^{(2)}$ are the same operators as appearing on the left side of the corresponding 2nd-LASS for the second-level adjoint function $\boldsymbol{\theta}_j^{(2)}$ and $\boldsymbol{\psi}_j^{(2)}$, respectively. Furthermore, all of these second-level adjoint functions are subject to the same respective boundary conditions. Only the source terms on the right sides of the respective 2nd-LASS differ from each other. Therefore, the same "forward" and "adjoint" software packages can be used for solving numerically the various forward and adjoint equations underlying the 1st-LASS and the 2nd-LASS. Furthermore, most of the second-order sensitivities can also be computed using the same basic software package.

It is important to note that the expressions of the second-order sensitivities given by (4.87) must be identical to those given by (4.62); this identity provides an independent path for the mutual verification of the solutions $\mathbf{u}_j^{(2)}$ and $\boldsymbol{\psi}_j^{(2)}$. Also, the expressions of the second-order sensitivities provided in (4.88) must be identical to those provided in (4.77); this identity provides an independent path for the mutual verification of the solutions $\mathbf{u}_j^{(2)}$ and $\boldsymbol{\theta}_j^{(2)}$.

4.4.4 Determination of the Second-Order Sensitivities of the Form $\partial^2 R\left(\varphi,\varphi^+;\boldsymbol{\psi}^{(1)};\mathbf{w}_j^{(2)};\boldsymbol{\alpha}\right)/\partial p_j\partial\alpha_{m_2}$

The second-order sensitivities $\partial^2 R\left(\varphi,\varphi^+;\boldsymbol{\psi}^{(1)};\mathbf{w}_j^{(2)};\boldsymbol{\alpha}\right)/(\partial p_j)(\partial\alpha_{m_2})$, $j=1,\ldots,J_p$, $m_2=1,\ldots,TP$, are obtained by following the same procedure as in Sect. 4.4.1 but replacing the function $\boldsymbol{\psi}_j^{(2)}(\mathbf{r},\boldsymbol{\Omega},E)$ by a second-level adjoint function denoted as $\mathbf{w}_j^{(2)}(\mathbf{r},\boldsymbol{\Omega},E) \triangleq \left[\mathbf{w}_{1,j}^{(2)}(\mathbf{r},\boldsymbol{\Omega},E), \mathbf{w}_{2,j}^{(2)}(\mathbf{r},\boldsymbol{\Omega},E)\right]^{\dagger}$, where $\mathbf{w}_{1,j}^{(2)} \triangleq \left[w_{11,j}^{(2)}(\mathbf{r},\boldsymbol{\Omega},E), w_{12,j}^{(2)}(\mathbf{r},\boldsymbol{\Omega},E)\right]^{\dagger}$, $\mathbf{w}_{2,j}^{(2)} \triangleq \left[w_{21,j}^{(2)}(\mathbf{r},\boldsymbol{\Omega},E), w_{22,j}^{(2)}(\mathbf{r},\boldsymbol{\Omega},E)\right]^{\dagger}$, with square-integrable components $w_{mn,j}^{(2)}(\mathbf{r},\boldsymbol{\Omega},E)$, $m,n = 1,2$. The end results are as follows:

$$\text{For } j = 1,\ldots,J_p;\; m_2 = 1,\ldots,J_t: \quad \frac{\partial^2 R\left(\varphi,\varphi^+;\boldsymbol{\psi}^{(1)};\mathbf{w}_j^{(2)};\boldsymbol{\alpha}\right)}{\partial p_j \partial t_{m_2}}$$

$$\begin{aligned} &= -\int dV \int_{4\pi} d\Omega \int_0^{E_f} dE \left[w_{11,j}^{(2)}(\mathbf{r},\boldsymbol{\Omega},E)\psi_1^{(1)}(\mathbf{r},\boldsymbol{\Omega},E) \right. \\ &+ w_{12,j}^{(2)}(\mathbf{r},\boldsymbol{\Omega},E)\psi_2^{(1)}(\mathbf{r},\boldsymbol{\Omega},E) + w_{21,j}^{(2)}(\mathbf{r},\boldsymbol{\Omega},E)\varphi(\mathbf{r},\boldsymbol{\Omega},E) \\ &\left. + w_{22,j}^{(2)}(\mathbf{r},\boldsymbol{\Omega},E)\varphi^+(\mathbf{r},\boldsymbol{\Omega},E)\right] \frac{\partial \Sigma_t(\mathbf{t};\mathbf{r},E)}{\partial t_{m_2}}; \end{aligned} \tag{4.99}$$

$$\text{For } j = 1,\ldots,J_p;\; m_2 = 1,\ldots,J_s: \quad \frac{\partial^2 R\left(\varphi,\varphi^+;\boldsymbol{\psi}^{(1)};\mathbf{w}_j^{(2)};\boldsymbol{\alpha}\right)}{\partial p_j \partial s_{m_2}}$$

$$\begin{aligned} &= \int dV \int_{4\pi} d\Omega \int_0^{E_f} dE \int_{4\pi} d\Omega' \int_0^{E_f} dE' \frac{\partial \Sigma_s(\mathbf{s};\mathbf{r},E \to E',\boldsymbol{\Omega} \to \boldsymbol{\Omega}')}{\partial s_{m_2}} \\ &\times \left[w_{11,j}^{(2)}(\mathbf{r},\boldsymbol{\Omega},E)\psi_1^{(1)}(\mathbf{r},\boldsymbol{\Omega}',E') + w_{22,j}^{(2)}(\mathbf{r},\boldsymbol{\Omega},E)\varphi^+(\mathbf{r},\boldsymbol{\Omega}',E')\right] \\ &+ \int dV \int_{4\pi} d\Omega \int_0^{E_f} dE \int_{4\pi} d\Omega' \int_0^{E_f} dE' \frac{\partial \Sigma_s(\mathbf{s};\mathbf{r},E' \to E,\boldsymbol{\Omega}' \to \boldsymbol{\Omega})}{\partial s_{m_2}} \\ &\times \left[w_{12,j}^{(2)}(\mathbf{r},\boldsymbol{\Omega},E)\psi_2^{(1)}(\mathbf{r},\boldsymbol{\Omega}',E') + w_{21,j}^{(2)}(\mathbf{r},\boldsymbol{\Omega},E)\varphi(\mathbf{r},\boldsymbol{\Omega}',E')\right]; \end{aligned} \tag{4.100}$$

$$\text{For } j=1,\ldots,J_p;\ m_2=1,\ldots,J_f: \quad \frac{\partial^2 R\left(\varphi,\varphi^+;\boldsymbol{\psi}^{(1)};\mathbf{w}_j^{(2)};\boldsymbol{\alpha}\right)}{\partial p_j \partial f_{m_2}}$$

$$\begin{aligned}
&= \int dV \int_{4\pi} d\boldsymbol{\Omega} \int_0^{E_f} dE\, \psi_1^{(1)}(\mathbf{r},\boldsymbol{\Omega},E)\frac{\partial \chi(\mathbf{p};\mathbf{r},E)}{\partial p_j} \\
&\quad \times \int_{4\pi} d\boldsymbol{\Omega}' \int_0^{E_f} dE'\, \varphi(\mathbf{r},\boldsymbol{\Omega}',E')\frac{\partial\left[\nu\Sigma_f(\mathbf{f};\mathbf{r},E')\right]}{\partial f_{m_2}} \\
&+ \int dV \int_{4\pi} d\boldsymbol{\Omega} \int_0^{E_f} dE\, \psi_2^{(1)}(\mathbf{r},\boldsymbol{\Omega},E)\frac{\partial\left[\nu\Sigma_f(\mathbf{f};\mathbf{r},E)\right]}{\partial f_{m_2}} \\
&\quad \times \int_{4\pi} d\boldsymbol{\Omega}' \int_0^{E_f} dE'\, \varphi^+(\mathbf{r},\boldsymbol{\Omega}',E')\frac{\partial \chi(\mathbf{p};\mathbf{r},E')}{\partial p_j} \\
&+ \int dV \int_{4\pi} d\boldsymbol{\Omega} \int_0^{E_f} dE \int_{4\pi} d\boldsymbol{\Omega}' \int_0^{E_f} dE'\, \frac{\partial\left[\nu\Sigma_f(\mathbf{f};\mathbf{r},E)\right]}{\partial f_{m_2}}\chi(\mathbf{p};\mathbf{r},E') \\
&\quad \times \left[w_{11,j}^{(2)}(\mathbf{r},\boldsymbol{\Omega},E)\psi_1^{(1)}(\mathbf{r},\boldsymbol{\Omega}',E') + w_{22,j}^{(2)}(\mathbf{r},\boldsymbol{\Omega},E)\varphi^+(\mathbf{r},\boldsymbol{\Omega}',E')\right] \\
&+ \int dV \int_{4\pi} d\boldsymbol{\Omega} \int_0^{E_f} dE \int_{4\pi} d\boldsymbol{\Omega}' \int_0^{E_f} dE'\, \chi(\mathbf{p};\mathbf{r},E)\frac{\partial\left[\nu\Sigma_f(\mathbf{f};\mathbf{r},E')\right]}{\partial f_{m_2}} \\
&\quad \times \left[w_{12,j}^{(2)}(\mathbf{r},\boldsymbol{\Omega},E)\psi_2^{(1)}(\mathbf{r},\boldsymbol{\Omega}',E') + w_{21,j}^{(2)}(\mathbf{r},\boldsymbol{\Omega},E)\varphi(\mathbf{r},\boldsymbol{\Omega}',E')\right];
\end{aligned} \tag{4.101}$$

$$
\begin{aligned}
&\text{For } j=1,\ldots,J_p;\ m_2=1,\ldots,J_p: \quad \frac{\partial^2 R\left(\varphi,\varphi^+;\boldsymbol{\psi}^{(1)};\mathbf{w}_j^{(2)};\boldsymbol{\alpha}\right)}{\partial p_j \partial p_{m_2}} \\
&= \int dV \int_{4\pi} d\boldsymbol{\Omega} \int_0^{E_f} dE\, \psi_1^{(1)}(\mathbf{r},\boldsymbol{\Omega},E) \frac{\partial^2 \chi(\mathbf{p};\mathbf{r},E)}{\partial p_j \partial p_{m_2}} \\
&\quad \times \int_{4\pi} d\boldsymbol{\Omega}' \int_0^{E_f} dE'\, \varphi(\mathbf{r},\boldsymbol{\Omega}',E') \nu\Sigma_f(\mathbf{f};\mathbf{r},E') \\
&+ \int dV \int_{4\pi} d\boldsymbol{\Omega} \int_0^{E_f} dE\, \psi_2^{(1)}(\mathbf{r},\boldsymbol{\Omega},E) \nu\Sigma_f(\mathbf{f};\mathbf{r},E) \\
&\quad \times \int_{4\pi} d\boldsymbol{\Omega}' \int_0^{E_f} dE'\, \varphi^+(\mathbf{r},\boldsymbol{\Omega}',E') \frac{\partial^2 \chi(\mathbf{p};\mathbf{r},E')}{\partial p_j \partial p_{m_2}} \\
&+ \int dV \int_{4\pi} d\boldsymbol{\Omega} \int_0^{E_f} dE \int_{4\pi} d\boldsymbol{\Omega}' \int_0^{E_f} dE' \frac{\partial \chi(\mathbf{p};\mathbf{r},E')}{\partial p_{m_2}} \nu\Sigma_f(\mathbf{f};\mathbf{r},E) \\
&\quad \times \left[w_{11,j}^{(2)}(\mathbf{r},\boldsymbol{\Omega},E) \psi_1^{(1)}(\mathbf{r},\boldsymbol{\Omega}',E') + w_{22,j}^{(2)}(\mathbf{r},\boldsymbol{\Omega},E) \varphi^+(\mathbf{r},\boldsymbol{\Omega}',E') \right] \\
&+ \int dV \int_{4\pi} d\boldsymbol{\Omega} \int_0^{E_f} dE \int_{4\pi} d\boldsymbol{\Omega}' \int_0^{E_f} dE' \frac{\partial \chi(\mathbf{p};\mathbf{r},E)}{\partial p_{m_2}} \nu\Sigma_f(\mathbf{f};\mathbf{r},E') \\
&\quad \times \left[w_{12,j}^{(2)}(\mathbf{r},\boldsymbol{\Omega},E) \psi_2^{(1)}(\mathbf{r},\boldsymbol{\Omega}',E') + w_{21,j}^{(2)}(\mathbf{r},\boldsymbol{\Omega},E) \varphi(\mathbf{r},\boldsymbol{\Omega}',E') \right];
\end{aligned}
\tag{4.102}
$$

$$
\begin{aligned}
&\text{For } j=1,\ldots,J_p;\ m_2=1,\ldots,J_q: \quad \frac{\partial^2 R\left(\varphi,\varphi^+;\boldsymbol{\psi}^{(1)};\mathbf{w}_j^{(2)};\boldsymbol{\alpha}\right)}{\partial p_j \partial q_{m_2}} \\
&= \int dV \int_{4\pi} d\boldsymbol{\Omega} \int_0^{E_f} dE\, w_{21,j}^{(2)}(\mathbf{r},\boldsymbol{\Omega},E) \frac{\partial Q(\mathbf{q};\mathbf{r},\boldsymbol{\Omega},E)}{\partial q_{m_2}};
\end{aligned}
\tag{4.103}
$$

$$
\begin{aligned}
&\text{For } j=1,\ldots,J_p;\ m_2=1,\ldots,J_k: \quad \frac{\partial^2 R\left(\varphi,\varphi^+;\boldsymbol{\psi}^{(1)};\mathbf{w}_j^{(2)};\boldsymbol{\alpha}\right)}{\partial p_j \partial k_{m_2}} \\
&= \int dV \int_{4\pi} d\boldsymbol{\Omega} \int_0^{E_f} dE\, w_{22,j}^{(2)}(\mathbf{r},\boldsymbol{\Omega},E) \frac{\partial Q^+(\mathbf{k};\mathbf{r},\boldsymbol{\Omega},E)}{\partial k_{m_2}};
\end{aligned}
\tag{4.104}
$$

$$\text{For } j = 1, \ldots, J_p;\ m_2 = 1, \ldots, J_d : \frac{\partial^2 R\left(\varphi, \varphi^+; \boldsymbol{\psi}^{(1)}; \mathbf{w}_j^{(2)}; \boldsymbol{\alpha}\right)}{\partial p_j \partial d_{m_2}}$$

$$= \int dV \int_{4\pi} d\boldsymbol{\Omega} \int_0^{E_f} dE \left[w_{11,j}^{(2)}(\mathbf{r}, \boldsymbol{\Omega}, E) \frac{\partial \Theta_1^{(1)}(\mathbf{d}; \varphi, \varphi^+)}{\partial d_{m_2}} + w_{12,j}^{(2)}(\mathbf{r}, \boldsymbol{\Omega}, E) \frac{\partial \Theta_2^{(1)}(\mathbf{d}; \varphi, \varphi^+)}{\partial d_{m_2}} \right]. \tag{4.105}$$

The components of the second-level adjoint function $\mathbf{w}_j^{(2)}(\mathbf{r}, \boldsymbol{\Omega}, E)$ are the solutions of the following 2nd-LASS, for $j = 1, \ldots, J_p$:

$$L(\boldsymbol{\alpha})\, w_{11,j}^{(2)}(\mathbf{r}, \boldsymbol{\Omega}, E) = \int_{4\pi} d\boldsymbol{\Omega}' \int_0^{E_f} dE' \frac{\partial \chi(\mathbf{p}; \mathbf{r}, E' \to E)}{\partial p_j} \nu \Sigma_f(\mathbf{f}; \mathbf{r}, E') \varphi(\mathbf{r}, \boldsymbol{\Omega}', E'), \tag{4.106}$$

$$L^+(\boldsymbol{\alpha}) w_{12,j}^{(2)}(\mathbf{r}, \boldsymbol{\Omega}, E) = \nu \Sigma_f(\mathbf{f}; \mathbf{r}, E) \int_{4\pi} d\boldsymbol{\Omega}' \int_0^{E_f} dE' \frac{\partial \chi(\mathbf{p}; \mathbf{r}, E \to E')}{\partial p_j} \varphi^+(\mathbf{r}, \boldsymbol{\Omega}', E'), \tag{4.107}$$

$$L^+(\boldsymbol{\alpha}) w_{21,j}^{(2)}(\mathbf{r}, \boldsymbol{\Omega}, E) = \frac{\partial \Theta_1^{(1)}(\mathbf{d}; \varphi, \varphi^+)}{\partial \varphi} w_{11,j}^{(2)}(\mathbf{r}, \boldsymbol{\Omega}, E) + \frac{\partial \Theta_2^{(1)}(\mathbf{d}; \varphi, \varphi^+)}{\partial \varphi} w_{12,j}^{(2)}(\mathbf{r}, \boldsymbol{\Omega}, E) + \nu \Sigma_f(\mathbf{f}; \mathbf{r}, E) \int_{4\pi} d\boldsymbol{\Omega}' \int_0^{E_f} dE' \frac{\partial \chi(\mathbf{p}; \mathbf{r}, E \to E')}{\partial p_j} \psi_1^{(1)}(\mathbf{r}, \boldsymbol{\Omega}', E'), \tag{4.108}$$

$$L(\boldsymbol{\alpha}) w_{22,j}^{(2)}(\mathbf{r}, \boldsymbol{\Omega}, E) = \frac{\partial \Theta_1^{(1)}(\mathbf{d}; \varphi, \varphi^+)}{\partial \varphi^+} w_{11,j}^{(2)}(\mathbf{r}, \boldsymbol{\Omega}, E) + \frac{\partial \Theta_2^{(1)}(\mathbf{d}; \varphi, \varphi^+)}{\partial \varphi^+} w_{12,j}^{(2)}(\mathbf{r}, \boldsymbol{\Omega}, E) + \int_{4\pi} d\boldsymbol{\Omega}' \int_0^{E_f} dE' \left[\nu \Sigma_f(\mathbf{f}; \mathbf{r}, E') \right] \psi_2^{(1)}(\mathbf{r}, \boldsymbol{\Omega}', E') \frac{\partial \chi(\mathbf{p}; \mathbf{r}, E' \to E)}{\partial p_j}, \tag{4.109}$$

subject to the boundary conditions below:

$$\left.\begin{aligned} w_{11,j}^{(2)}(\mathbf{r}_s,\boldsymbol{\Omega},E)&=0, \quad \mathbf{r}_s\in\partial V, \quad \boldsymbol{\Omega}\cdot\mathbf{n}<0; 0<E<E_f \\ w_{12,j}^{(2)}(\mathbf{r}_s,\boldsymbol{\Omega},E)&=0, \quad \mathbf{r}_s\in\partial V, \quad \boldsymbol{\Omega}\cdot\mathbf{n}>0; 0<E<E_f \\ w_{21,j}^{(2)}(\mathbf{r}_s,\boldsymbol{\Omega},E)&=0, \quad \mathbf{r}_s\in\partial V, \quad \boldsymbol{\Omega}\cdot\mathbf{n}>0; 0<E<E_f \\ w_{22,j}^{(2)}(\mathbf{r}_s,\boldsymbol{\Omega},E)&=0, \quad \mathbf{r}_s\in\partial V, \quad \boldsymbol{\Omega}\cdot\mathbf{n}<0; 0<E<E_f \end{aligned}\right\} \tag{4.110}$$

As in the previous sections, the operators appearing on the left side of the 2nd-LASS defined by (4.106)–(4.109) for the second-level adjoint function $\mathbf{w}_j^{(2)}(\mathbf{r},\boldsymbol{\Omega},E)$ are the same operators as appearing on the respective left sides of the 2nd-LASS for the second-level adjoint function $\boldsymbol{\theta}_j^{(2)}$, $\boldsymbol{\psi}_j^{(2)}$, and $\mathbf{u}_j^{(2)}$; all of these the second-level adjoint functions are subject to similar boundary conditions. Only the source terms on the right sides of the respective 2nd-LASS differ from each other. Therefore, the same "forward" and "adjoint" software packages can be used for solving numerically the various forward and adjoint equations underlying the 1st-LASS and the 2nd-LASS.

Note that the expressions of the second-order sensitivities provided in (4.99) must be identical to those provided in (4.63); this identity provides an independent mutual verification of the second-level adjoint functions $\mathbf{w}_j^{(2)}$ and $\boldsymbol{\psi}_j^{(2)}$. Furthermore, the expressions of the second-order sensitivities provided in (4.100) must be identical to those provided in (4.78); this identity provides an independent mutual verification of the second-level adjoint functions $\mathbf{w}_j^{(2)}$ and $\boldsymbol{\theta}_j^{(2)}$. Finally, the expressions of the second-order sensitivities provided in (4.101) must be identical to those provided in (4.90); this identity provides an independent path for the mutual verification of the solutions $\mathbf{w}_j^{(2)}$ and $\mathbf{u}_j^{(2)}$.

4.4.5 Determination of the Second-Order Sensitivities of the Form $\partial^2 R\left(\varphi,\varphi^+;\boldsymbol{\psi}^{(1)};\mathbf{g}_j^{(2)};\boldsymbol{\alpha}\right)/\partial q_j \partial \alpha_{m_2}$

The second-order sensitivities $\partial^2 R\left(\varphi,\varphi^+;\boldsymbol{\psi}^{(1)};\mathbf{g}_j^{(2)};\boldsymbol{\alpha}\right)/(\partial q_j)(\partial \alpha_{m_2})$, $j=1,\ldots,J_q$, $m_2=1,\ldots,TP$, are obtained by following the same procedure as in Sect. 4.4.1, but replacing the function $\boldsymbol{\psi}_j^{(2)}(\mathbf{r},\boldsymbol{\Omega},E)$ by a second-level adjoint function denoted as $\mathbf{g}_j^{(2)}(\mathbf{r},\boldsymbol{\Omega},E)\triangleq\left[\mathbf{g}_{1,j}^{(2)},\mathbf{g}_{2,j}^{(2)}\right]^\dagger$, with $\mathbf{g}_{1,j}^{(2)}\triangleq\left[g_{11,j}^{(2)}(\mathbf{r},\boldsymbol{\Omega},E),g_{12,j}^{(2)}(\mathbf{r},\boldsymbol{\Omega},E)\right]^\dagger$, $\mathbf{g}_{2,j}^{(2)}\triangleq\left[g_{21,j}^{(2)}(\mathbf{r},\boldsymbol{\Omega},E),g_{22,j}^{(2)}(\mathbf{r},\boldsymbol{\Omega},E)\right]^\dagger$, having square-integrable components $g_{mn,j}^{(2)}(\mathbf{r},\boldsymbol{\Omega},E)$, $m,n=1,2$. The end results are as follows:

For $j = 1, \ldots, J_q;\ m_2 = 1, \ldots, J_t:\ \dfrac{\partial^2 R\left(\varphi, \varphi^+; \boldsymbol{\psi}^{(1)}; \mathbf{g}_j^{(2)}; \boldsymbol{\alpha}\right)}{\partial q_j \partial t_{m_2}}$

$$= -\int dV \int_{4\pi} d\boldsymbol{\Omega} \int_0^{E_f} dE\, \frac{\partial \Sigma_t(\mathbf{t}; \mathbf{r}, E)}{\partial t_{m_2}} \left[g_{11,j}^{(2)}(\mathbf{r}, \boldsymbol{\Omega}, E)\psi_1^{(1)}(\mathbf{r}, \boldsymbol{\Omega}, E) \right. \tag{4.111}$$
$$\left. + g_{21,j}^{(2)}(\mathbf{r}, \boldsymbol{\Omega}, E)\varphi(\mathbf{r}, \boldsymbol{\Omega}, E) + g_{22,j}^{(2)}(\mathbf{r}, \boldsymbol{\Omega}, E)\varphi^+(\mathbf{r}, \boldsymbol{\Omega}, E)\right];$$

For $j = 1, \ldots, J_q;\ m_2 = 1, \ldots, J_s:\ \dfrac{\partial^2 R\left(\varphi, \varphi^+; \boldsymbol{\psi}^{(1)}; \mathbf{g}_j^{(2)}; \boldsymbol{\alpha}\right)}{\partial q_j \partial s_{m_2}}$

$$= \int dV \int_{4\pi} d\boldsymbol{\Omega} \int_0^{E_f} dE \int_{4\pi} d\boldsymbol{\Omega}' \int_0^{E_f} dE'\, \frac{\partial \Sigma_s(\mathbf{s}; \mathbf{r}, E \to E', \boldsymbol{\Omega} \to \boldsymbol{\Omega}')}{\partial s_{m_2}}$$
$$\times \left[g_{11,j}^{(2)}(\mathbf{r}, \boldsymbol{\Omega}, E)\psi_1^{(1)}(\mathbf{r}, \boldsymbol{\Omega}', E') + g_{22,j}^{(2)}(\mathbf{r}, \boldsymbol{\Omega}, E)\varphi^+(\mathbf{r}, \boldsymbol{\Omega}', E')\right] \tag{4.112}$$
$$+ \int dV \int_{4\pi} d\boldsymbol{\Omega} \int_0^{E_f} dE\, g_{21,j}^{(2)}(\mathbf{r}, \boldsymbol{\Omega}, E)$$
$$\times \int_{4\pi} d\boldsymbol{\Omega}' \int_0^{E_f} dE'\, \varphi(\mathbf{r}, \boldsymbol{\Omega}', E') \frac{\partial \Sigma_s(\mathbf{s}; \mathbf{r}, E' \to E, \boldsymbol{\Omega}' \to \boldsymbol{\Omega})}{\partial s_{m_2}};$$

For $j = 1, \ldots, J_q;\ m_2 = 1, \ldots, J_f:\ \dfrac{\partial^2 R\left(\varphi, \varphi^+; \boldsymbol{\psi}^{(1)}; \mathbf{g}_j^{(2)}; \boldsymbol{\alpha}\right)}{\partial q_j \partial f_{m_2}}$

$$= \int dV \int_{4\pi} d\boldsymbol{\Omega} \int_0^{E_f} dE \int_{4\pi} d\boldsymbol{\Omega}' \int_0^{E_f} dE'\, \frac{\partial \left[\nu\Sigma_f(\mathbf{f}; \mathbf{r}, E)\right]}{\partial f_{m_2}} \chi(\mathbf{p}; \mathbf{r}, E')$$
$$\times \left[g_{11,j}^{(2)}(\mathbf{r}, \boldsymbol{\Omega}, E)\psi_1^{(1)}(\mathbf{r}, \boldsymbol{\Omega}', E') + g_{22,j}^{(2)}(\mathbf{r}, \boldsymbol{\Omega}, E)\varphi^+(\mathbf{r}, \boldsymbol{\Omega}', E')\right] \tag{4.113}$$
$$+ \int dV \int_{4\pi} d\boldsymbol{\Omega} \int_0^{E_f} dE\, g_{21,j}^{(2)}(\mathbf{r}, \boldsymbol{\Omega}, E)$$
$$\times \int_{4\pi} d\boldsymbol{\Omega}' \int_0^{E_f} dE'\, \chi(\mathbf{p}; \mathbf{r}, E)\varphi(\mathbf{r}, \boldsymbol{\Omega}', E') \frac{\partial \left[\nu\Sigma_f(\mathbf{f}; \mathbf{r}, E')\right]}{\partial f_{m_2}};$$

For $j = 1, \ldots, J_q;\ m_2 = 1, \ldots, J_p:\ \dfrac{\partial^2 R\left(\varphi, \varphi^+; \boldsymbol{\psi}^{(1)}; \mathbf{g}_j^{(2)}; \boldsymbol{\alpha}\right)}{\partial q_j \partial p_{m_2}}$

$$= \int dV \int_{4\pi} d\boldsymbol{\Omega} \int_0^{E_f} dE \int_{4\pi} d\boldsymbol{\Omega}' \int_0^{E_f} dE' \frac{\partial \chi(\mathbf{p}; \mathbf{r}, E')}{\partial p_{m_2}} \nu\Sigma_f(\mathbf{f}; \mathbf{r}, E)$$
$$\times \left[g_{11,j}^{(2)}(\mathbf{r}, \boldsymbol{\Omega}, E) \psi_1^{(1)}(\mathbf{r}, \boldsymbol{\Omega}', E') + g_{22,j}^{(2)}(\mathbf{r}, \boldsymbol{\Omega}, E) \varphi^+(\mathbf{r}, \boldsymbol{\Omega}', E') \right] \tag{4.114}$$
$$+ \int dV \int_{4\pi} d\boldsymbol{\Omega} \int_0^{E_f} dE\, g_{21,j}^{(2)}(\mathbf{r}, \boldsymbol{\Omega}, E)$$
$$\times \frac{\partial \chi(\mathbf{p}; \mathbf{r}, E)}{\partial p_{m_2}} \int_{4\pi} d\boldsymbol{\Omega}' \int_0^{E_f} dE'\, \varphi(\mathbf{r}, \boldsymbol{\Omega}', E') \nu\Sigma_f(\mathbf{f}; \mathbf{r}, E');$$

For $j = 1, \ldots, J_q;\ m_2 = 1, \ldots, J_q:\ \dfrac{\partial^2 R\left(\varphi, \varphi^+; \boldsymbol{\psi}^{(1)}; \mathbf{g}_j^{(2)}; \boldsymbol{\alpha}\right)}{\partial q_j \partial q_{m_2}}$

$$= \int dV \int_{4\pi} d\boldsymbol{\Omega} \int_0^{E_f} dE\, \psi_1^{(1)}(\mathbf{r}, \boldsymbol{\Omega}, E) \frac{\partial^2 Q(\mathbf{q}; \mathbf{r}, \boldsymbol{\Omega}, E)}{\partial q_j \partial q_{m_2}} \tag{4.115}$$
$$+ \int dV \int_{4\pi} d\boldsymbol{\Omega} \int_0^{E_f} dE\, g_{21,j}^{(2)}(\mathbf{r}, \boldsymbol{\Omega}, E) \frac{\partial Q(\mathbf{q}; \mathbf{r}, \boldsymbol{\Omega}, E)}{\partial q_{m_2}};$$

For $j = 1, \ldots, J_q;\ m_2 = 1, \ldots, J_k:\ \dfrac{\partial^2 R\left(\varphi, \varphi^+; \boldsymbol{\psi}^{(1)}; \mathbf{g}_j^{(2)}; \boldsymbol{\alpha}\right)}{\partial q_j \partial k_{m_2}}$

$$= \int dV \int_{4\pi} d\boldsymbol{\Omega} \int_0^{E_f} dE\, g_{22,j}^{(2)}(\mathbf{r}, \boldsymbol{\Omega}, E) \frac{\partial Q^+(\mathbf{k}; \mathbf{r}, \boldsymbol{\Omega}, E)}{\partial k_{m_2}}; \tag{4.116}$$

For $j = 1, \ldots, J_q;\ m_2 = 1, \ldots, J_d:\ \dfrac{\partial^2 R\left(\varphi, \varphi^+; \boldsymbol{\psi}^{(1)}; \mathbf{g}_j^{(2)}; \boldsymbol{\alpha}\right)}{\partial q_j \partial d_{m_2}}$

$$= \int dV \int_{4\pi} d\boldsymbol{\Omega} \int_0^{E_f} dE\, g_{11,j}^{(2)}(\mathbf{r}, \boldsymbol{\Omega}, E) \frac{\partial \Theta_1^{(1)}(\mathbf{d}; \varphi, \varphi^+)}{\partial d_{m_2}}. \tag{4.117}$$

The components of the second-level adjoint function $\mathbf{g}_j^{(2)}(\mathbf{r}, \boldsymbol{\Omega}, E)$ satisfy the following 2nd-LASS, for $j = 1, \ldots, J_q$:

$$g_{12,j}^{(2)}(\mathbf{r}, \boldsymbol{\Omega}, E) \equiv 0, \tag{4.118}$$

$$L(\boldsymbol{\alpha})\, g_{11,j}^{(2)}(\mathbf{r}, \boldsymbol{\Omega}, E) = \frac{\partial Q(\mathbf{q}; \mathbf{r}, \boldsymbol{\Omega}, E)}{\partial q_j}, \tag{4.119}$$

$$L^+(\boldsymbol{\alpha}) g_{21,j}^{(2)}(\mathbf{r}, \boldsymbol{\Omega}, E) = \frac{\partial \Theta_1^{(1)}(\mathbf{d}; \varphi, \varphi^+)}{\partial \varphi} g_{11,j}^{(2)}(\mathbf{r}, \boldsymbol{\Omega}, E), \tag{4.120}$$

$$L(\boldsymbol{\alpha}) g_{22,j}^{(2)}(\mathbf{r}, \boldsymbol{\Omega}, E) = \frac{\partial \Theta_1^{(1)}(\mathbf{d}; \varphi, \varphi^+)}{\partial \varphi^+} g_{11,j}^{(2)}(\mathbf{r}, \boldsymbol{\Omega}, E), \tag{4.121}$$

subject to the following boundary conditions:

$$\left.\begin{array}{l} g_{11,j}^{(2)}(\mathbf{r}_s, \boldsymbol{\Omega}, E) = 0, \ \ \mathbf{r}_s \in \partial V, \ \ \boldsymbol{\Omega} \cdot \mathbf{n} < 0; 0 < E < E_f \\ g_{21,j}^{(2)}(\mathbf{r}_s, \boldsymbol{\Omega}, E) = 0, \ \ \mathbf{r}_s \in \partial V, \ \ \boldsymbol{\Omega} \cdot \mathbf{n} > 0; 0 < E < E_f \\ g_{22,j}^{(2)}(\mathbf{r}_s, \boldsymbol{\Omega}, E) = 0, \ \ \mathbf{r}_s \in \partial V, \ \ \boldsymbol{\Omega} \cdot \mathbf{n} < 0; 0 < E < E_f \end{array}\right\} \tag{4.122}$$

The following important identities provide mutual verification paths for the second-level adjoint functions:

1. The expressions of the second-order sensitivities provided by (4.111) must be identical to those provided by (4.64); this identity provides an independent mutual verification of the second-level adjoint functions $\mathbf{g}_j^{(2)}$ and $\boldsymbol{\psi}_j^{(2)}$.
2. The expressions of the second-order sensitivities provided by (4.112) must be identical to those provided by (4.79); this identity provides an independent mutual verification of the second-level adjoint functions $\mathbf{g}_j^{(2)}$ and $\boldsymbol{\theta}_j^{(2)}$.
3. The expressions of the second-order sensitivities provided by (4.113) must be identical to those provided by (4.91); this identity provides an independent path for the mutual verification of the solutions $\mathbf{g}_j^{(2)}$ and $\mathbf{u}_j^{(2)}$.
4. The expressions of the second-order sensitivities provided by (4.114) must be identical to those provided by (4.103); this identity provides an independent path for the mutual verification of the solutions $\mathbf{g}_j^{(2)}$ and $\mathbf{w}_j^{(2)}$.

4.4.6 Determination of the Second-Order Sensitivities of the Form $\partial^2 R\left(\boldsymbol{\varphi}, \boldsymbol{\varphi}^+; \boldsymbol{\psi}^{(1)}; \boldsymbol{\gamma}_j^{(2)}; \boldsymbol{\alpha}\right) / \partial k_j \partial \alpha_{m_2}$

The second-order sensitivities $\partial^2 R\left(\varphi, \varphi^+; \boldsymbol{\psi}^{(1)}; \boldsymbol{\gamma}_j^{(2)}; \boldsymbol{\alpha}\right) / \left(\partial k_j\right)\left(\partial \alpha_{m_2}\right)$, $j = 1, \ldots, J_k, m_2 = 1, \ldots, TP$, are obtained by following the same procedure as detailed in Sect. 4.4.1, but replacing the function $\boldsymbol{\psi}_j^{(2)}(\mathbf{r}, \boldsymbol{\Omega}, E)$ by a second-level adjoint function denoted as $\boldsymbol{\gamma}_j^{(2)}(\mathbf{r}, \boldsymbol{\Omega}, E) \triangleq \left[\boldsymbol{\gamma}_{1,j}^{(2)}, \boldsymbol{\gamma}_{2,j}^{(2)}\right]^\dagger$, with $\boldsymbol{\gamma}_{1,j}^{(2)} \triangleq \left[\gamma_{11,j}^{(2)}(\mathbf{r}, \boldsymbol{\Omega}, E), \gamma_{12,j}(\mathbf{r}, \boldsymbol{\Omega}, E)\right]^\dagger$, $\boldsymbol{\gamma}_{2,j}^{(2)} \triangleq \left[\gamma_{21,j}^{(2)}(\mathbf{r}, \boldsymbol{\Omega}, E), \gamma_{22,j}^{(2)}(\mathbf{r}, \boldsymbol{\Omega}, E)\right]^\dagger$, having square-integrable components $\gamma_{mn,j}^{(2)}(\mathbf{r}, \boldsymbol{\Omega}, E)$, $m, n = 1, 2$. The end results are as follows:

$$\text{For } j = 1, \ldots, J_k;\ m_2 = 1, \ldots, J_t: \frac{\partial^2 R\left(\varphi, \varphi^+; \boldsymbol{\psi}^{(1)}; \boldsymbol{\gamma}_j^{(2)}; \boldsymbol{\alpha}\right)}{\partial k_j \partial t_{m_2}}$$

$$= -\int dV \int_{4\pi} d\Omega \int_0^{E_f} dE \frac{\partial \Sigma_t(\mathbf{t}; \mathbf{r}, E)}{\partial t_{m_2}} \left[\gamma_{12,j}^{(2)}(\mathbf{r}, \boldsymbol{\Omega}, E) \psi_2^{(1)}(\mathbf{r}, \boldsymbol{\Omega}, E)\right. \tag{4.123}$$

$$\left. + \gamma_{21,j}^{(2)}(\mathbf{r}, \boldsymbol{\Omega}, E) \varphi(\mathbf{r}, \boldsymbol{\Omega}, E) + \gamma_{22,j}^{(2)}(\mathbf{r}, \boldsymbol{\Omega}, E) \varphi^+(\mathbf{r}, \boldsymbol{\Omega}, E)\right];$$

$$\text{For } j = 1, \ldots, J_k;\ m_2 = 1, \ldots, J_s: \frac{\partial^2 R\left(\varphi, \varphi^+; \boldsymbol{\psi}^{(1)}; \boldsymbol{\gamma}_j^{(2)}; \boldsymbol{\alpha}\right)}{\partial k_j \partial s_{m_2}}$$

$$= \int dV \int_{4\pi} d\Omega \int_0^{E_f} dE \int_{4\pi} d\Omega' \int_0^{E_f} dE' \frac{\partial \Sigma_s(\mathbf{s}; \mathbf{r}, E' \to E, \boldsymbol{\Omega}' \to \boldsymbol{\Omega})}{\partial s_{m_2}}$$

$$\times \left[\gamma_{12,j}^{(2)}(\mathbf{r}, \boldsymbol{\Omega}, E) \psi_2^{(1)}(\mathbf{r}, \boldsymbol{\Omega}', E') + \gamma_{21,j}^{(2)}(\mathbf{r}, \boldsymbol{\Omega}, E) \varphi(\mathbf{r}, \boldsymbol{\Omega}', E')\right] \tag{4.124}$$

$$+ \int dV \int_{4\pi} d\Omega \int_0^{E_f} dE\, \gamma_{22,j}^{(2)}(\mathbf{r}, \boldsymbol{\Omega}, E)$$

$$\times \int_{4\pi} d\Omega' \int_0^{E_f} dE'\, \varphi^+(\mathbf{r}, \boldsymbol{\Omega}', E') \frac{\partial \Sigma_s(\mathbf{s}; \mathbf{r}, E \to E', \boldsymbol{\Omega} \to \boldsymbol{\Omega}')}{\partial s_{m_2}};$$

$$\text{For } j = 1, \ldots, J_k;\ m_2 = 1, \ldots, J_f:\ \frac{\partial^2 R\left(\varphi, \varphi^+; \boldsymbol{\psi}^{(1)}; \boldsymbol{\gamma}_j^{(2)}; \boldsymbol{\alpha}\right)}{\partial k_j \partial f_{m_2}}$$

$$= \int dV \int_{4\pi} d\boldsymbol{\Omega} \int_0^{E_f} dE \int_{4\pi} d\boldsymbol{\Omega}' \int_0^{E_f} dE' \frac{\partial \left[\nu\Sigma_f(\mathbf{f}; \mathbf{r}, E')\right]}{\partial f_{m_2}} \chi(\mathbf{p}; \mathbf{r}, E)$$

$$\times \left[\gamma_{12,j}^{(2)}(\mathbf{r}, \boldsymbol{\Omega}, E)\psi_2^{(1)}(\mathbf{r}, \boldsymbol{\Omega}', E') + \gamma_{21,j}^{(2)}(\mathbf{r}, \boldsymbol{\Omega}, E)\varphi(\mathbf{r}, \boldsymbol{\Omega}', E')\right] \tag{4.125}$$

$$+ \int dV \int_{4\pi} d\boldsymbol{\Omega} \int_0^{E_f} dE\, \gamma_{22,j}^{(2)}(\mathbf{r}, \boldsymbol{\Omega}, E) \frac{\partial \left[\nu\Sigma_f(\mathbf{f}; \mathbf{r}, E)\right]}{\partial f_{m_2}}$$

$$\times \int_{4\pi} d\boldsymbol{\Omega}' \int_0^{E_f} dE' \chi(\mathbf{p}; \mathbf{r}, E')\varphi^+(\mathbf{r}, \boldsymbol{\Omega}', E');$$

$$\text{For } j = 1, \ldots, J_k;\ m_2 = 1, \ldots, J_p:\ \frac{\partial^2 R\left(\varphi, \varphi^+; \boldsymbol{\psi}^{(1)}; \boldsymbol{\gamma}_j^{(2)}; \boldsymbol{\alpha}\right)}{\partial k_j \partial p_{m_2}}$$

$$= \int dV \int_{4\pi} d\boldsymbol{\Omega} \int_0^{E_f} dE \int_{4\pi} d\boldsymbol{\Omega}' \int_0^{E_f} dE' \frac{\partial \chi(\mathbf{p}; \mathbf{r}, E)}{\partial p_{m_2}} \nu\Sigma_f(\mathbf{f}; \mathbf{r}, E')$$

$$\times \left[\gamma_{12,j}^{(2)}(\mathbf{r}, \boldsymbol{\Omega}, E)\psi_2^{(1)}(\mathbf{r}, \boldsymbol{\Omega}', E') + \gamma_{21,j}^{(2)}(\mathbf{r}, \boldsymbol{\Omega}, E)\varphi(\mathbf{r}, \boldsymbol{\Omega}', E')\right] \tag{4.126}$$

$$+ \int dV \int_{4\pi} d\boldsymbol{\Omega} \int_0^{E_f} dE\, \nu\Sigma_f(\mathbf{f}; \mathbf{r}, E)\gamma_{22,j}^{(2)}(\mathbf{r}, \boldsymbol{\Omega}, E)$$

$$\times \int_{4\pi} d\boldsymbol{\Omega}' \int_0^{E_f} dE'\, \varphi^+(\mathbf{r}, \boldsymbol{\Omega}', E') \frac{\partial \chi(\mathbf{p}; \mathbf{r}, E')}{\partial p_{m_2}};$$

$$\text{For } j = 1, \ldots, J_k;\ m_2 = 1, \ldots, J_q:\ \frac{\partial^2 R\left(\varphi, \varphi^+; \boldsymbol{\psi}^{(1)}; \boldsymbol{\gamma}_j^{(2)}; \boldsymbol{\alpha}\right)}{\partial k_j \partial q_{m_2}}$$

$$= \int dV \int_{4\pi} d\boldsymbol{\Omega} \int_0^{E_f} dE\, \gamma_{21,j}^{(2)}(\mathbf{r}, \boldsymbol{\Omega}, E) \frac{\partial Q(\mathbf{q}; \mathbf{r}, \boldsymbol{\Omega}, E)}{\partial q_{m_2}}; \tag{4.127}$$

For $j = 1, \ldots, J_k;\ m_2 = 1, \ldots, J_k$: $\dfrac{\partial^2 R\left(\varphi, \varphi^+; \boldsymbol{\psi}^{(1)}; \boldsymbol{\gamma}_j^{(2)}; \boldsymbol{\alpha}\right)}{\partial k_j \partial k_{m_2}}$

$$= \int dV \int_{4\pi} d\boldsymbol{\Omega} \int_0^{E_f} dE\, \psi_2^{(1)}(\mathbf{r}, \boldsymbol{\Omega}, E) \frac{\partial^2 Q^+(\mathbf{k}; \mathbf{r}, \boldsymbol{\Omega}, E)}{\partial k_j \partial k_{m_2}} + \int dV \int_{4\pi} d\boldsymbol{\Omega} \int_0^{E_f} dE\, \gamma_{22,j}^{(2)}(\mathbf{r}, \boldsymbol{\Omega}, E) \frac{\partial Q^+(\mathbf{k}; \mathbf{r}, \boldsymbol{\Omega}, E)}{\partial k_{m_2}}; \tag{4.128}$$

For $j = 1, \ldots, J_k;\ m_2 = 1, \ldots, J_d$: $\dfrac{\partial^2 R\left(\varphi, \varphi^+; \boldsymbol{\psi}^{(1)}; \boldsymbol{\gamma}_j^{(2)}; \boldsymbol{\alpha}\right)}{\partial k_j \partial d_{m_2}}$

$$= \int dV \int_{4\pi} d\boldsymbol{\Omega} \int_0^{E_f} dE\, \gamma_{12,j}^{(2)}(\mathbf{r}, \boldsymbol{\Omega}, E) \frac{\partial \Theta_2^{(1)}(\mathbf{d}; \varphi, \varphi^+)}{\partial d_{m_2}}. \tag{4.129}$$

The components of the second-level adjoint function $\boldsymbol{\gamma}_j^{(2)}(\mathbf{r}, \boldsymbol{\Omega}, E) \triangleq \left[\boldsymbol{\gamma}_{1,j}^{(2)}(\mathbf{r}, \boldsymbol{\Omega}, E), \boldsymbol{\gamma}_{2,j}^{(2)}(\mathbf{r}, \boldsymbol{\Omega}, E)\right]^\dagger$ satisfy the following 2nd-LASS, for $j = 1, \ldots, J_k$:

$$L(\boldsymbol{\alpha})\gamma_{11,j}^{(2)}(\mathbf{r}, \boldsymbol{\Omega}, E) = 0, \tag{4.130}$$

$$L^+(\boldsymbol{\alpha})\gamma_{12,j}^{(2)}(\mathbf{r}, \boldsymbol{\Omega}, E) = \frac{\partial Q^+(\mathbf{k}; \mathbf{r}, \boldsymbol{\Omega}, E)}{\partial k_j}, \tag{4.131}$$

$$L^+(\boldsymbol{\alpha})\gamma_{21,j}^{(2)}(\mathbf{r}, \boldsymbol{\Omega}, E) = \frac{\partial \Theta_2^{(1)}(\mathbf{d}; \varphi, \varphi^+)}{\partial \varphi} \gamma_{12,j}^{(2)}(\mathbf{r}, \boldsymbol{\Omega}, E), \tag{4.132}$$

$$L(\boldsymbol{\alpha})\gamma_{22,j}^{(2)}(\mathbf{r}, \boldsymbol{\Omega}, E) = \frac{\partial \Theta_2^{(1)}(\mathbf{d}; \varphi, \varphi^+)}{\partial \varphi^+} \gamma_{12,j}^{(2)}(\mathbf{r}, \boldsymbol{\Omega}, E). \tag{4.133}$$

subject to the boundary conditions below:

$$\left.\begin{aligned} \gamma_{12,j}^{(2)}(\mathbf{r}_s, \boldsymbol{\Omega}, E) = 0, \ \ \mathbf{r}_s \in \partial V, \ \ \boldsymbol{\Omega} \cdot \mathbf{n} > 0; 0 < E < E_f \\ \gamma_{21,j}^{(2)}(\mathbf{r}_s, \boldsymbol{\Omega}, E) = 0, \ \ \mathbf{r}_s \in \partial V, \ \ \boldsymbol{\Omega} \cdot \mathbf{n} > 0; 0 < E < E_f \\ \gamma_{22,j}^{(2)}(\mathbf{r}_s, \boldsymbol{\Omega}, E) = 0, \ \ \mathbf{r}_s \in \partial V, \ \ \boldsymbol{\Omega} \cdot \mathbf{n} < 0; 0 < E < E_f \end{aligned}\right\} \tag{4.134}$$

The following important identities provide mutual verification paths for the second-level adjoint functions:

1. The expressions of the second-order sensitivities provided by (4.123) must be identical to those provided by (4.65); this identity provides an independent mutual verification of the second-level adjoint functions $\boldsymbol{\gamma}_j^{(2)}$ and $\boldsymbol{\psi}_j^{(2)}$.
2. The expressions of the second-order sensitivities provided by (4.124) must be identical to those provided by (4.80); this identity provides an independent mutual verification of the second-level adjoint functions $\boldsymbol{\gamma}_j^{(2)}$ and $\boldsymbol{\theta}_j^{(2)}$.
3. The expressions of the second-order sensitivities provided by (4.125) must be identical to those provided by (4.92); this identity provides an independent path for the mutual verification of the solutions $\boldsymbol{\gamma}_j^{(2)}$ and $\mathbf{u}_j^{(2)}$.
4. The expressions of the second-order sensitivities provided by (4.126) must be identical to those provided by (4.104); this identity provides an independent path for the mutual verification of the solutions $\boldsymbol{\gamma}_j^{(2)}$ and $\mathbf{w}_j^{(2)}$.
5. The expressions of the second-order sensitivities provided by (4.127) must be identical to those provided by (4.116); this identity provides an independent path for the mutual verification of the solutions $\boldsymbol{\gamma}_j^{(2)}$ and $\mathbf{g}_j^{(2)}$.

4.4.7 Determination of the Second-Order Sensitivities of the Form $\partial^2 R\left(\varphi,\varphi^+;\boldsymbol{\psi}^{(1)};\mathbf{h}_j^{(2)};\boldsymbol{\alpha}\right)/\partial d_j \partial \alpha_{m_2}$

The second-order sensitivities $\partial^2 R\left(\varphi,\varphi^+;\boldsymbol{\psi}^{(1)};\mathbf{h}_j^{(2)};\boldsymbol{\alpha}\right)/(\partial d_j)(\partial \alpha_{m_2})$, $j = 1, \ldots, J_d$, $m_2 = 1, \ldots, TP$, are obtained by following the same procedure as detailed in Sect. 4.4.1, but replacing the function $\boldsymbol{\psi}_j^{(2)}(\mathbf{r},\boldsymbol{\Omega},E)$ by a second-level adjoint function denoted as $\mathbf{h}_j^{(2)}(\mathbf{r},\boldsymbol{\Omega},E) \triangleq \left[\mathbf{h}_{1,j}^{(2)},\mathbf{h}_{2,j}^{(2)}\right]^\dagger$, with $\mathbf{h}_{1,j}^{(2)} \triangleq \left[h_{11,j}^{(2)}(\mathbf{r},\boldsymbol{\Omega},E), h_{12,j}^{(2)}(\mathbf{r},\boldsymbol{\Omega},E)\right]^\dagger$, $\mathbf{h}_{2,j}^{(2)} \triangleq \left[h_{21,j}^{(2)}(\mathbf{r},\boldsymbol{\Omega},E), h_{22,j}^{(2)}(\mathbf{r},\boldsymbol{\Omega},E)\right]^\dagger$, having square-integrable components $h_{mn,j}^{(2)}(\mathbf{r},\boldsymbol{\Omega},E)$, $m,n = 1,2$. The end results are as follows:

$$
\begin{aligned}
&\text{For } j = 1,\ldots,J_d;\ m_2 = 1,\ldots,J_t: \ \frac{\partial^2 R\left(\varphi,\varphi^+;\boldsymbol{\psi}^{(1)};\mathbf{h}_j^{(2)};\boldsymbol{\alpha}\right)}{\partial d_j \partial t_{m_2}} \\
&= -\int dV \int_{4\pi} d\boldsymbol{\Omega} \int_0^{E_f} dE \, \frac{\partial \Sigma_t(\mathbf{t};\mathbf{r},E)}{\partial t_{m_2}} \left[h_{21,j}^{(2)}(\mathbf{r},\boldsymbol{\Omega},E)\varphi(\mathbf{r},\boldsymbol{\Omega},E)\right. \\
&\quad \left. + h_{22,j}^{(2)}(\mathbf{r},\boldsymbol{\Omega},E)\varphi^+(\mathbf{r},\boldsymbol{\Omega},E)\right];
\end{aligned}
\tag{4.135}
$$

For $j = 1, \ldots, J_d;\ m_2 = 1, \ldots, J_s$: $\dfrac{\partial^2 R\left(\varphi, \varphi^+; \boldsymbol{\psi}^{(1)}; \mathbf{h}_j^{(2)}; \boldsymbol{\alpha}\right)}{\partial d_j \partial s_{m_2}}$

$$= \int dV \int_{4\pi} d\boldsymbol{\Omega} \int_0^{E_f} dE\, h_{21,j}^{(2)}(\mathbf{r}, \boldsymbol{\Omega}, E)$$

$$\times \int_{4\pi} d\boldsymbol{\Omega}' \int_0^{E_f} dE'\, \varphi(\mathbf{r}, \boldsymbol{\Omega}', E') \frac{\partial \Sigma_s(\mathbf{s}; \mathbf{r}, E' \to E, \boldsymbol{\Omega}' \to \boldsymbol{\Omega})}{\partial s_{m_2}} \tag{4.136}$$

$$+ \int dV \int_{4\pi} d\boldsymbol{\Omega} \int_0^{E_f} dE\, h_{22,j}^{(2)}(\mathbf{r}, \boldsymbol{\Omega}, E)$$

$$\times \int_{4\pi} d\boldsymbol{\Omega}' \int_0^{E_f} dE'\, \varphi^+(\mathbf{r}, \boldsymbol{\Omega}', E') \frac{\partial \Sigma_s(\mathbf{s}; \mathbf{r}, E \to E', \boldsymbol{\Omega} \to \boldsymbol{\Omega}')}{\partial s_{m_2}};$$

For $j = 1, \ldots, J_d;\ m_2 = 1, \ldots, J_q$: $\dfrac{\partial^2 R\left(\varphi, \varphi^+; \boldsymbol{\psi}^{(1)}; \mathbf{h}_j^{(2)}; \boldsymbol{\alpha}\right)}{\partial d_j \partial q_{m_2}}$

$$= \int dV \int_{4\pi} d\boldsymbol{\Omega} \int_0^{\infty} dE\, h_{21,j}^{(2)}(\mathbf{r}, \boldsymbol{\Omega}, E) \frac{\partial Q(\mathbf{q}; \mathbf{r}, \boldsymbol{\Omega}, E)}{\partial q_{m_2}}; \tag{4.137}$$

For $j = 1, \ldots, J_d;\ m_2 = 1, \ldots, J_k$: $\dfrac{\partial^2 R\left(\varphi, \varphi^+; \boldsymbol{\psi}^{(1)}; \mathbf{h}_j^{(2)}; \boldsymbol{\alpha}\right)}{\partial d_j \partial k_{m_2}}$

$$= \int dV \int_{4\pi} d\boldsymbol{\Omega} \int_0^{\infty} dE\, h_{22,j}^{(2)}(\mathbf{r}, \boldsymbol{\Omega}, E) \frac{\partial Q^+(\mathbf{k}; \mathbf{r}, \boldsymbol{\Omega}, E)}{\partial k_{m_2}}; \tag{4.138}$$

For $j = 1, \ldots, J_d;\ m_2 = 1, \ldots, J_f:\ \dfrac{\partial^2 R\left(\varphi, \varphi^+; \boldsymbol{\psi}^{(1)}; \mathbf{h}_j^{(2)}; \boldsymbol{\alpha}\right)}{\partial d_j \partial f_{m_2}}$

$$
\begin{aligned}
&= \int dV \int_{4\pi} d\boldsymbol{\Omega} \int_0^{E_f} dE\, h_{21,j}^{(2)}(\mathbf{r}, \boldsymbol{\Omega}, E) \\
&\times \int_{4\pi} d\boldsymbol{\Omega}' \int_0^{E_f} dE' \chi(\mathbf{p}; \mathbf{r}, E' \to E) \varphi(\mathbf{r}, \boldsymbol{\Omega}', E') \frac{\partial \left[\nu\Sigma_f(\mathbf{f}; \mathbf{r}, E')\right]}{\partial f_{m_2}} \\
&+ \int dV \int_{4\pi} d\boldsymbol{\Omega} \int_0^{E_f} dE\, h_{22,j}^{(2)}(\mathbf{r}, \boldsymbol{\Omega}, E) \frac{\partial \left[\nu\Sigma_f(\mathbf{f}; \mathbf{r}, E)\right]}{\partial f_{m_2}} \\
&\times \int_{4\pi} d\boldsymbol{\Omega}' \int_0^{E_f} dE' \chi(\mathbf{p}; \mathbf{r}, E \to E') \varphi^+(\mathbf{r}, \boldsymbol{\Omega}', E');
\end{aligned}
\tag{4.139}
$$

For $j = 1, \ldots, J_d;\ m_2 = 1, \ldots, J_p:\ \dfrac{\partial^2 R\left(\varphi, \varphi^+; \boldsymbol{\psi}^{(1)}; \mathbf{h}_j^{(2)}; \boldsymbol{\alpha}\right)}{\partial d_j \partial p_{m_2}}$

$$
\begin{aligned}
&= \int dV \int_{4\pi} d\boldsymbol{\Omega} \int_0^{E_f} dE\, h_{21,j}^{(2)}(\mathbf{r}, \boldsymbol{\Omega}, E) \frac{\partial \chi(\mathbf{p}; \mathbf{r}, E)}{\partial p_{m_2}} \\
&\times \int_{4\pi} d\boldsymbol{\Omega}' \int_0^{E_f} dE'\, \varphi(\mathbf{r}, \boldsymbol{\Omega}', E') \nu\Sigma_f(\mathbf{f}; \mathbf{r}, E') \\
&+ \int dV \int_{4\pi} d\boldsymbol{\Omega} \int_0^{E_f} dE\, h_{22,j}^{(2)}(\mathbf{r}, \boldsymbol{\Omega}, E) \nu\Sigma_f(\mathbf{f}; \mathbf{r}, E) \\
&\times \int_{4\pi} d\boldsymbol{\Omega}' \int_0^{E_f} dE'\, \varphi^+(\mathbf{r}, \boldsymbol{\Omega}', E') \frac{\partial \chi(\mathbf{p}; \mathbf{r}, E')}{\partial p_{m_2}};
\end{aligned}
\tag{4.140}
$$

For $j = 1, \ldots, J_d;\ m_2 = 1, \ldots, J_d:\ \dfrac{\partial^2 R\left(\varphi, \varphi^+; \boldsymbol{\psi}^{(1)}; \mathbf{h}_j^{(2)}; \boldsymbol{\alpha}\right)}{\partial d_j \partial d_{m_2}}$

$$
\begin{aligned}
&= \frac{\int dV \int_{4\pi} d\boldsymbol{\Omega} \int_0^{E_f} dE \dfrac{\partial^2 G(\mathbf{d}; \varphi, \varphi^+)}{\partial d_j \partial d_{m_2}}}{\int dV \int_{4\pi} d\boldsymbol{\Omega} \int_0^{E_f} dE\, H(\mathbf{d}; \varphi, \varphi^+)} - \left[\int dV \int_{4\pi} d\boldsymbol{\Omega} \int_0^{E_f} dE \frac{\partial G(\mathbf{d}; \varphi, \varphi^+)}{\partial d_j}\right] \\
&\times \left[\int dV \int_{4\pi} d\boldsymbol{\Omega} \int_0^{E_f} dE \frac{\partial H(\mathbf{d}; \varphi, \varphi^+)}{\partial d_{m_2}}\right]\left[\int dV \int_{4\pi} d\boldsymbol{\Omega} \int_0^{E_f} dE\, H(\mathbf{d}; \varphi, \varphi^+)\right]^{-2} \\
&- \left[\int dV \int_{4\pi} d\boldsymbol{\Omega} \int_0^{E_f} dE\, G(\mathbf{d}; \varphi, \varphi^+)\right]\left[\int dV \int_{4\pi} d\boldsymbol{\Omega} \int_0^{E_f} dE \frac{\partial^2 H(\mathbf{d}; \varphi, \varphi^+)}{\partial d_j \partial d_{m_2}}\right] \\
&\times \left[\int dV \int_{4\pi} d\boldsymbol{\Omega} \int_0^{E_f} dE\, H(\mathbf{d}; \varphi, \varphi^+)\right]^{-2} - \left[\int dV \int_{4\pi} d\boldsymbol{\Omega} \int_0^{E_f} dE \frac{\partial H(\mathbf{d}; \varphi, \varphi^+)}{\partial d_j}\right] \\
&\times \left[\int dV \int_{4\pi} d\boldsymbol{\Omega} \int_0^{E_f} dE \frac{\partial G(\mathbf{d}; \varphi, \varphi^+)}{\partial d_{m_2}}\right]\left[\int dV \int_{4\pi} d\boldsymbol{\Omega} \int_0^{E_f} dE\, H(\mathbf{d}; \varphi, \varphi^+)\right]^{-2} \\
&+ 2\frac{\left[\int dV \int_{4\pi} d\boldsymbol{\Omega} \int_0^{E_f} dE\, G(\mathbf{d}; \varphi, \varphi^+)\right]\left[\int dV \int_{4\pi} d\boldsymbol{\Omega} \int_0^{E_f} dE \dfrac{\partial H(\mathbf{d}; \varphi, \varphi^+)}{\partial d_j}\right]}{\left[\int dV \int_{4\pi} d\boldsymbol{\Omega} \int_0^{E_f} dE\, H(\mathbf{d}; \varphi, \varphi^+)\right]^3} \\
&\times \left[\int dV \int_{4\pi} d\boldsymbol{\Omega} \int_0^{E_f} dE \frac{\partial H(\mathbf{d}; \varphi, \varphi^+)}{\partial d_{m_2}}\right].
\end{aligned}
\tag{4.141}
$$

The components of the second-level adjoint function $\mathbf{h}_{1,j}^{(2)} \triangleq \left[h_{11,j}^{(2)}, h_{12,j}^{(2)}\right]^{\dagger}$ are $h_{11,j}^{(2)}(\mathbf{r}, \boldsymbol{\Omega}, E) \equiv 0,\ h_{12,j}^{(2)}(\mathbf{r}, \boldsymbol{\Omega}, E) \equiv 0, j = 1, \ldots, J_d,$ while the components of the second-level adjoint function $\mathbf{h}_{2,j}^{(2)} \triangleq \left[h_{21,j}^{(2)}(\mathbf{r}, \boldsymbol{\Omega}, E), h_{22,j}^{(2)}(\mathbf{r}, \boldsymbol{\Omega}, E)\right]^{\dagger}$ satisfy the following 2nd-LASS, for $j = 1, \ldots, J_d$:

$$
\begin{aligned}
L^{+}(\boldsymbol{\alpha})h_{21,j}^{(2)}(\mathbf{r},\boldsymbol{\Omega},E) &= \frac{\dfrac{\partial^2 G(\mathbf{d};\varphi,\varphi^{+})}{\partial\varphi\partial d_j}}{\displaystyle\int dV \int_{4\pi} d\boldsymbol{\Omega} \int_0^{E_f} dE\, H(\mathbf{d};\varphi,\varphi^{+})} \\
&- \frac{\left(\dfrac{\partial H(\mathbf{d};\varphi,\varphi^{+})}{\partial\varphi}\right)\displaystyle\int dV \int_{4\pi} d\boldsymbol{\Omega} \int_0^{E_f} dE\, \frac{\partial G(\mathbf{d};\varphi,\varphi^{+})}{\partial d_j}}{\left[\displaystyle\int dV \int_{4\pi} d\boldsymbol{\Omega} \int_0^{E_f} dE\, H(\mathbf{d};\varphi,\varphi^{+})\right]^2} \\
&- \frac{\left(\dfrac{\partial G(\mathbf{d};\varphi,\varphi^{+})}{\partial\varphi}\right)\displaystyle\int dV \int_{4\pi} d\boldsymbol{\Omega} \int_0^{E_f} dE\, \frac{\partial H(\mathbf{d};\varphi,\varphi^{+})}{\partial d_j}}{\left[\displaystyle\int dV \int_{4\pi} d\boldsymbol{\Omega} \int_0^{E_f} dE\, H(\mathbf{d};\varphi,\varphi^{+})\right]^2} \\
&- \frac{\left(\dfrac{\partial^2 H(\mathbf{d};\varphi,\varphi^{+})}{\partial\varphi\partial d_j}\right)\displaystyle\int dV \int_{4\pi} d\boldsymbol{\Omega} \int_0^{E_f} dE\, G(\mathbf{d};\varphi,\varphi^{+})}{\left[\displaystyle\int dV \int_{4\pi} d\boldsymbol{\Omega} \int_0^{E_f} dE\, H(\mathbf{d};\varphi,\varphi^{+})\right]^2} \\
&+2\frac{\partial H(\mathbf{d};\varphi,\varphi^{+})}{\partial\varphi}\left[\int dV \int_{4\pi} d\boldsymbol{\Omega} \int_0^{E_f} dE\, G(\mathbf{d};\varphi,\varphi^{+})\right] \\
&\times\left[\int dV \int_{4\pi} d\boldsymbol{\Omega} \int_0^{E_f} dE\, \frac{\partial H(\mathbf{d};\varphi,\varphi^{+})}{\partial d_j}\right]\left[\int dV \int_{4\pi} d\boldsymbol{\Omega} \int_0^{E_f} dE\, H(\mathbf{d};\varphi,\varphi^{+})\right]^{-3}.
\end{aligned}
\tag{4.142}
$$

$$
\begin{aligned}
L(\boldsymbol{\alpha})h_{22,j}^{(2)}(\mathbf{r},\boldsymbol{\Omega},E) = {} & \frac{\dfrac{\partial^2 G(\mathbf{d};\varphi,\varphi^+)}{\partial\varphi^+\partial d_j}}{\int dV \int_{4\pi} d\boldsymbol{\Omega} \int_0^{E_f} dE\, H(\mathbf{d};\varphi,\varphi^+)} \\
& - \frac{\left(\dfrac{\partial H(\mathbf{d};\varphi,\varphi^+)}{\partial\varphi^+}\right)\int dV \int_{4\pi} d\boldsymbol{\Omega} \int_0^{E_f} dE\, \dfrac{\partial G(\mathbf{d};\varphi,\varphi^+)}{\partial d_j}}{\left[\int dV \int_{4\pi} d\boldsymbol{\Omega} \int_0^{E_f} dE\, H(\mathbf{d};\varphi,\varphi^+)\right]^2} \\
& - \frac{\left(\dfrac{\partial G(\mathbf{d};\varphi,\varphi^+)}{\partial\varphi^+}\right)\int dV \int_{4\pi} d\boldsymbol{\Omega} \int_0^{E_f} dE\, \dfrac{\partial H(\mathbf{d};\varphi,\varphi^+)}{\partial d_j}}{\left[\int dV \int_{4\pi} d\boldsymbol{\Omega} \int_0^{E_f} dE\, H(\mathbf{d};\varphi,\varphi^+)\right]^2} \\
& - \frac{\left(\dfrac{\partial^2 H(\mathbf{d};\varphi,\varphi^+)}{\partial\varphi^+\partial d_j}\right)\int dV \int_{4\pi} d\boldsymbol{\Omega} \int_0^{E_f} dE\, G(\mathbf{d};\varphi,\varphi^+)}{\left[\int dV \int_{4\pi} d\boldsymbol{\Omega} \int_0^{E_f} dE\, H(\mathbf{d};\varphi,\varphi^+)\right]^2} \\
& + 2\frac{\partial H(\mathbf{d};\varphi,\varphi^+)}{\partial\varphi^+}\left[\int dV \int_{4\pi} d\boldsymbol{\Omega} \int_0^{E_f} dE\, G(\mathbf{d};\varphi,\varphi^+)\right] \\
& \times \left[\int dV \int_{4\pi} d\boldsymbol{\Omega} \int_0^{E_f} dE\, \frac{\partial H(\mathbf{d};\varphi,\varphi^+)}{\partial d_j}\right]\left[\int dV \int_{4\pi} d\boldsymbol{\Omega} \int_0^{E_f} dE\, H(\mathbf{d};\varphi,\varphi^+)\right]^{-3}.
\end{aligned}
\tag{4.143}
$$

subject to the boundary conditions below:

$$\left.\begin{aligned} h^{(2)}_{21,j}(\mathbf{r}_s, \boldsymbol{\Omega}, E) = 0, \quad \mathbf{r}_s \in \partial V, \quad \boldsymbol{\Omega} \cdot \mathbf{n} > 0; 0 < E < E_f \\ h^{(2)}_{22,j}(\mathbf{r}_s, \boldsymbol{\Omega}, E) = 0, \quad \mathbf{r}_s \in \partial V, \quad \boldsymbol{\Omega} \cdot \mathbf{n} < 0; 0 < E < E_f \end{aligned}\right\} \tag{4.144}$$

The following relations provide independent mutual verifications of the second-level adjoint functions:

1. The expressions of the second-order sensitivities provided by (4.135) must be identical to those provided by (4.66); this identity provides an independent mutual verification of the second-level adjoint functions $\mathbf{h}_j^{(2)}$ and $\boldsymbol{\psi}_j^{(2)}$.
2. The expressions of the second-order sensitivities provided by (4.136) must be identical to those provided by (4.81); this identity provides an independent mutual verification of the second-level adjoint functions $\mathbf{h}_j^{(2)}$ and $\boldsymbol{\theta}_j^{(2)}$.
3. The expressions of the second-order sensitivities provided by (4.139) must be identical to those provided by (4.93); this identity provides an independent path for the mutual verification of the solutions $\mathbf{h}_j^{(2)}$ and $\mathbf{u}_j^{(2)}$.
4. The expressions of the second-order sensitivities provided by (4.140) must be identical to those provided by (4.105); this identity provides an independent path for the mutual verification of the solutions $\mathbf{h}_j^{(2)}$ and $\mathbf{w}_j^{(2)}$.
5. The expressions of the second-order sensitivities provided by (4.137) must be identical to those provided by (4.117); this identity provides an independent path for the mutual verification of the solutions $\mathbf{h}_j^{(2)}$ and $\mathbf{g}_j^{(2)}$.
6. The expressions of the second-order sensitivities provided by (4.138) must be identical to those provided by (4.129); this identity provides an independent path for the mutual verification of the solutions $\mathbf{h}_j^{(2)}$ and $\boldsymbol{\gamma}_j^{(2)}$.

The results obtained for the first- and second-order sensitivities of the response $R(\varphi, \varphi^+; \boldsymbol{\alpha})$ defined in (4.9) can readily be particularized to "reaction rate" responses. For example, setting everywhere $G(\mathbf{d}; \varphi, \varphi^+) \equiv \Sigma_d(\mathbf{d}; \mathbf{r}, \boldsymbol{\Omega}, E)\, \varphi(\mathbf{r}, \boldsymbol{\Omega}, E)$, where $\Sigma_d(\mathbf{d}; \mathbf{r}, \boldsymbol{\Omega}, E)$ denotes a (macroscopic-cross-section-like) function that models the interaction of a detector with the incident particles, and also setting $H(\mathbf{d}; \varphi, \varphi^+) \equiv \left[\int dV \int_{4\pi} d\boldsymbol{\Omega} \int_0^{E_f} dE\right]^{-1}$ = constant provides the complete set of expressions for the first- and second-order sensitivities of a "detector response" or a reaction rate response to model parameters.

4.5 Second-Order Sensitivity Analysis of the Schwinger and Roussopoulos Functionals

The Schwinger (Levine and Schwinger 1949) and Roussopolos (1953) functionals have traditionally been used for computing more accurate values of integral quantities of the forward and adjoint particle fluxes that would otherwise have been obtained when using approximate values of the forward and adjoint particle fluxes in source-driven nuclear systems. The Roussopoulos functional, which will be denoted as $R(\varphi, \varphi^+; \boldsymbol{\alpha})$, is defined as follows:

$$\begin{aligned} R(\varphi, \varphi^+; \boldsymbol{\alpha}) &\triangleq \langle \varphi(\mathbf{r}, \boldsymbol{\Omega}, E), Q^+(\mathbf{k}; \mathbf{r}, \boldsymbol{\Omega}, E) \rangle_B \\ &- \langle \varphi^+(\mathbf{r}, \boldsymbol{\Omega}, E), L(\boldsymbol{\alpha})\varphi(\mathbf{r}, \boldsymbol{\Omega}, E) - Q(\mathbf{q}; \mathbf{r}, \boldsymbol{\Omega}, E) \rangle_B. \end{aligned} \tag{4.145}$$

The Roussopoulos functional, henceforth written in abbreviated form as the R-functional, has been used for computing reaction rates using approximate fluxes, since it provides more accurate values of the respective reaction rate than would have otherwise been obtained if the respective reaction rate had been computed directly using the approximate fluxes. This property becomes mathematically evident by inserting approximate trial functions of the form $\varphi = \varphi^0 + \Delta\varphi$ and $\varphi^+ = \varphi^{+0} + \Delta\varphi^+$ (where φ^0 and φ^{+0} denote the exact but unknown respective flux values, while $\Delta\varphi$ and $\Delta\varphi^+$ denote unknown deviations from the exact but unknown values φ^0 and φ^{+0}) into the expression of the R-functional $R(\varphi, \varphi^+; \boldsymbol{\alpha})$, yielding the following expression:

$$\begin{aligned} &R\left(\varphi^0 + \Delta\varphi, \varphi^{+0} + \Delta\varphi^+; \boldsymbol{\alpha}\right) \triangleq \left\langle \left(\varphi^0 + \Delta\varphi\right), Q^+\left(\mathbf{k}^0\right) \right\rangle_B \\ &- \left\langle \varphi^{+0} + \Delta\varphi^+, L\left(\boldsymbol{\alpha}^0\right)\left(\varphi^0 + \Delta\varphi\right) - Q\left(\mathbf{q}^0\right) \right\rangle_B = \left\langle \varphi^0, Q^+ \right\rangle_B \\ &- \left\{ P^{(1)}\left[\Delta\varphi, \varphi^+\right] \right\}_{\boldsymbol{\alpha} \equiv \boldsymbol{\alpha}^0} - \left\langle \Delta\varphi^+(\mathbf{r}, \boldsymbol{\Omega}, E), L\left(\boldsymbol{\alpha}^0\right)(\Delta\varphi) \right\rangle_B. \end{aligned} \tag{4.146}$$

Consequently, if the deviation $\Delta\varphi$ satisfies a "no incoming flux" boundary condition identical to (4.2), the bilinear concomitant in (4.146) vanishes, yielding the following expression:

$$\begin{aligned} &R\left(\varphi^0 + \Delta\varphi, \varphi^{+0} + \Delta\varphi^+; \boldsymbol{\alpha}\right) \\ &= \left\langle \varphi^{+0}, Q\left(\mathbf{q}^0\right) \right\rangle_B - \left\langle \Delta\varphi^+(\mathbf{r}, \boldsymbol{\Omega}, E), L\left(\boldsymbol{\alpha}^0\right)(\Delta\varphi) \right\rangle_B. \end{aligned} \tag{4.147}$$

Consequently, if the approximate forward flux satisfies the same vacuum boundary condition as satisfied by the exact forward flux, then the replacement of approximate forward and adjoint fluxes into the R-functional yields the exact reaction rate of interest within a "second-order error term" of the form $\langle \Delta\varphi^+(\mathbf{r}, \boldsymbol{\Omega}, E), L(\boldsymbol{\alpha}^0)$

$(\Delta\varphi)\rangle_B$, as indicated in (4.147). Notably, the relation expressed by (4.147) holds if and only if the model parameters are *not* allowed to vary. In other words, the relationship expressed by (4.147) remains valid if and only if the approximate flux satisfies the exact vacuum boundary conditions, and if and only if the model parameters are perfectly well known, and their perfectly well-known values are the respective nominal/mean values.

In most practical nuclear engineering problems, the forward and adjoint fluxes are normalized to some given constant value, e.g., the given total power and the given maximum flux, such that $\varphi(\mathbf{r}, \mathbf{\Omega}, E) \triangleq C\psi(\mathbf{r}, \mathbf{\Omega}, E)$ and $\varphi^+(\mathbf{r}, \mathbf{\Omega}, E) \triangleq C^+\psi^+(\mathbf{r}, \mathbf{\Omega}, E)$, where C and C^+ are imprecisely known normalization constants. In such cases, the R-functional $R(\varphi, \varphi^+; \boldsymbol{\alpha})$ is evidently sensitive to uncertainties in both C and C^+. The sensitivity of the R-functional to these flux normalizations can be reduced by requiring it to be stationary with respect to both C and C^+, which is achieved by effecting the following sequence of operations: (1) replacing the respective normalized flux expressions in (4.145); (2) setting to zero the derivatives with respect to C and C^+ of the resulting expression and computing the expressions of C and C^+ at the respective stationary point; and (3) substituting the resulting expressions of C and C^+ back into (4.145). This sequence of operations yields the following expression for the resulting particular case of the R-functional, designated as $R_p(\varphi, \varphi^+; \boldsymbol{\alpha})$, where the subscript "$p$" denotes "particular":

$$\begin{aligned} R_p(\varphi, \varphi^+; \boldsymbol{\alpha}) &= \frac{\langle \varphi(\mathbf{r}, \mathbf{\Omega}, E), Q^+(\mathbf{k}; \mathbf{r}, \mathbf{\Omega}, E)\rangle_B}{\langle \varphi^+(\mathbf{r}, \mathbf{\Omega}, E), L(\boldsymbol{\alpha})\varphi(\mathbf{r}, \mathbf{\Omega}, E)\rangle_B} \\ &\quad \times \langle \varphi^+(\mathbf{r}, \mathbf{\Omega}, E), Q(\mathbf{q}; \mathbf{r}, \mathbf{\Omega}, E)\rangle_B \triangleq S(\varphi, \varphi^+; \boldsymbol{\alpha}). \end{aligned} \tag{4.148}$$

The expression shown in (4.148) was derived independently by Schwinger (Levine and Schwinger 1949) and is referred in the literature as the *Schwinger functional*, which will be written in abbreviated form as the *S-functional* and will be denoted as $S(\varphi, \varphi^+; \boldsymbol{\alpha})$. By design, the nominal value of the S-functional coincides with the nominal value of the R-functional, both yielding the nominal value of the reaction rate of interest, i.e.,

$$S\left(\varphi^0, \varphi^{+0}; \boldsymbol{\alpha}^0\right) = R\left(\varphi^0, \varphi^{+0}; \boldsymbol{\alpha}^0\right) = \left\langle \varphi^0(\mathbf{r}, \mathbf{\Omega}, E), Q^+\left(\mathbf{k}^0; \mathbf{r}, \mathbf{\Omega}, E\right)\right\rangle_B. \tag{4.149}$$

Evaluating $S(\varphi, \varphi^+; \boldsymbol{\alpha})$ using approximate trial functions of the form $\varphi = \varphi^0 + \Delta\varphi$ and $\varphi^+ = \varphi^{+0} + \Delta\varphi^+$ (where φ^0 and φ^{+0} denote the exact but unknown respective flux values, while $\Delta\varphi$ and $\Delta\varphi^+$ denote unknown deviations from the exact but unknown values φ^0 and φ^{+0}) yields the following expression:

$$\begin{aligned} & S\left(\varphi^0 + \Delta\varphi, \varphi^{+0} + \Delta\varphi^+; \boldsymbol{\alpha}^0\right) \\ & \triangleq \frac{\left\langle \varphi^0 + \Delta\varphi, Q^+\left(\mathbf{k}^0\right)\right\rangle_B \left\langle \varphi^{+0} + \Delta\varphi^+, Q\left(\mathbf{q}^0\right)\right\rangle_B}{\left\langle \varphi^{+0} + \Delta\varphi^+, L\left(\boldsymbol{\alpha}^0\right)\left(\varphi^0 + \Delta\varphi\right)\right\rangle_B} \\ & = \left\langle \varphi^{+0}, Q\right\rangle_B - \left\{P\left[\Delta\varphi, \varphi^+\right]\right\}_{\boldsymbol{\alpha}\equiv\boldsymbol{\alpha}^0} + \left[\left\langle \Delta\varphi, Q^+\left(\mathbf{k}^0\right)\right\rangle + \left\{P\left[\Delta\varphi, \varphi^+\right]\right\}_{\boldsymbol{\alpha}\equiv\boldsymbol{\alpha}^0}\right] \\ & \quad \times \frac{\left[\left\langle \Delta\varphi^+, Q\left(\mathbf{q}^0\right)\right\rangle + \left\{P\left[\Delta\varphi, \varphi^+\right]\right\}_{\boldsymbol{\alpha}\equiv\boldsymbol{\alpha}^0}\right]}{\left\langle \varphi^{+0}, Q\left(\mathbf{q}^0\right)\right\rangle_B}. \end{aligned} \tag{4.150}$$

If the unknown deviation $\Delta\varphi$ satisfies a "no incoming flux" boundary condition identical to (4.2), the bilinear concomitant $\{P[\Delta\varphi, \varphi^+]\}_{\boldsymbol{\alpha}\equiv\boldsymbol{\alpha}^0}$ in (4.150) vanishes and the resulting relation simplifies to the following form:

$$\begin{aligned} & S\left(\varphi^0 + \Delta\varphi, \varphi^{+0} + \Delta\varphi^+; \boldsymbol{\alpha}^0\right) \\ & = \left\langle \varphi^{+0}, Q\left(\mathbf{q}^0\right)\right\rangle_B + \frac{\left\langle \Delta\varphi, Q^+\left(\mathbf{k}^0\right)\right\rangle \left\langle \Delta\varphi^+, Q\left(\mathbf{q}^0\right)\right\rangle}{\left\langle \varphi^{+0}, Q\left(\mathbf{q}^0\right)\right\rangle_B}. \end{aligned} \tag{4.151}$$

As indicated in (4.151), the S-functional is less sensitive than the R-functional to the normalizations of the exact (but unknown) forward and adjoint fluxes while yielding the exact reaction rate of interest within a "second-order error term" of the form $\langle \Delta\varphi, Q^+(\mathbf{k}^0)\rangle \langle \Delta\varphi^+, Q(\mathbf{q}^0)\rangle / \langle \varphi^{+0}, Q(\mathbf{q}^0)\rangle_B$ when using approximate trial functions of the form $\varphi = \varphi^0 + \Delta\varphi$ and $\varphi^+ = \varphi^{+0} + \Delta\varphi^+$. The relation expressed by (4.151) holds if the approximate forward flux satisfies the same vacuum boundary as the exact flux and the model parameters are perfectly well known (their perfectly well-known values are the corresponding nominal values).

In summary, the relations expressed by (4.151) and (4.147) indicate that: (1) when the flux is not normalized, the use of the R-functional provides approximate trial functions for the forward and adjoint fluxes that allow the subsequent computation of the exact reaction rate of interest within a "second-order error term"; (2) on the other hand, when the flux is normalized, the use of the S-functional provides approximate trial functions for the forward and adjoint fluxes that are less sensitive to their respective normalizations constants while yielding the exact reaction rate of interest, also within a "second-order error term." However, as also indicated by (4.151) and (4.147), the respective "second-order error terms" differ from each other.

The well-known results shown in (4.151) and (4.147), as well as the properties of the Schwinger and Roussopoulos functionals that have been revisited in this section, were derived under the assumption that the nuclear system's parameters are perfectly well known, which is never the case in practice. As can be inferred from the expressions in (4.151) and (4.147), additional error terms are expected to arise from parameter uncertainties, which will induce, in turn, intrinsic errors in the forward and adjoint fluxes. As will be shown in the remainder of this section, these additional error terms will not be "second-order error terms" in the parameter variations since the first-order sensitivities of these functionals to the system parameters do not generally vanish when computed at the parameters' nominal values.

The first variation of the S-functional *at fixed parameter values* is obtained, by definition, as follows:

$$
\begin{aligned}
\{\delta S(\varphi,\varphi^+;\boldsymbol{\alpha})\}_{\boldsymbol{\alpha}\equiv\boldsymbol{\alpha}^0} &\triangleq \left\{\frac{\partial S(\varphi,\varphi^+;\boldsymbol{\alpha})}{\partial\varphi}\delta\varphi\right\}_{\boldsymbol{\alpha}\equiv\boldsymbol{\alpha}^0} + \left\{\frac{\partial S(\varphi,\varphi^+;\boldsymbol{\alpha})}{\partial\varphi^+}\delta\varphi^+\right\}_{\boldsymbol{\alpha}\equiv\boldsymbol{\alpha}^0} \\
&= \left\{\frac{\langle\delta\varphi,Q^+\rangle_B\langle\varphi^+,Q\rangle_B}{\langle\varphi^+,L(\boldsymbol{\alpha}^0)\varphi\rangle_B} - \frac{\langle\varphi,Q^+\rangle_B\langle\varphi^+,Q\rangle_B\langle\varphi^+,L(\boldsymbol{\alpha}^0)\delta\varphi\rangle_B}{\left[\langle\varphi^+,L(\boldsymbol{\alpha}^0)\varphi\rangle_B\right]^2}\right\}_{\boldsymbol{\alpha}\equiv\boldsymbol{\alpha}^0} \\
&+ \left\{\frac{\langle\varphi,Q^+\rangle_B\langle\delta\varphi^+,Q\rangle_B}{\langle\varphi^+,L(\boldsymbol{\alpha}^0)\varphi\rangle_B} - \frac{\langle\varphi,Q^+\rangle_B\langle\varphi^+,Q\rangle_B\langle\delta\varphi^+,L(\boldsymbol{\alpha}^0)\varphi\rangle_B}{\left[\langle\varphi^+,L(\boldsymbol{\alpha}^0)\varphi\rangle_B\right]^2}\right\}_{\boldsymbol{\alpha}\equiv\boldsymbol{\alpha}^0}.
\end{aligned}
\tag{4.152}
$$

Using the relations between the forward and adjoint Boltzmann equations, the expression in (4.152) can be simplified to the following form:

$$
\begin{aligned}
\{\delta S(\varphi,\varphi^+;\boldsymbol{\alpha})\}_{\boldsymbol{\alpha}\equiv\boldsymbol{\alpha}^0} &= \left\{\langle\delta\varphi,Q^+\rangle - \langle\varphi^+,L(\boldsymbol{\alpha}^0)\delta\varphi\rangle_B\right\}_{\boldsymbol{\alpha}\equiv\boldsymbol{\alpha}^0} \\
&= -\left\{P^{(1)}[\delta\varphi,\varphi^+]\right\}_{\boldsymbol{\alpha}\equiv\boldsymbol{\alpha}^0}.
\end{aligned}
\tag{4.153}
$$

If the variation $\delta\varphi(\mathbf{r},\boldsymbol{\Omega},E)$, or, equivalently, if the approximate flux, does *not* satisfy the same vacuum boundary conditions as the forward flux, then the S-functional is *not* stationary with regard to arbitrary variations in the forward and adjoint fluxes *at fixed parameter values,* and therefore exhibits first-order errors in $\delta\varphi$ on the domain's boundary, as indicated by (4.153). However, if the approximate flux *does* satisfy the vacuum boundary condition, i.e., if

$$
\delta\varphi(\mathbf{r}_s,\boldsymbol{\Omega},E) = 0, \quad \mathbf{r}_s \in \partial V, \quad \boldsymbol{\Omega}\cdot\mathbf{n} < 0, \quad 0 < E < \infty,
\tag{4.154}
$$

then the bilinear concomitant vanishes in (4.153), i.e., $\left\{P^{(1)}[\delta\varphi,\varphi^+]\right\}_{\boldsymbol{\alpha}\equiv\boldsymbol{\alpha}^0} = 0$, and consequently the first variation of the S-functional vanishes with regard to arbitrary variations in the forward and adjoint fluxes at fixed parameter values, i.e., $\{\delta S(\varphi,\varphi^+;\boldsymbol{\alpha})\}_{\boldsymbol{\alpha}\equiv\boldsymbol{\alpha}^0} = 0$.

The first variation of the R-functional with regard to arbitrary variations in the forward and adjoint fluxes *at fixed parameter values* takes on the following form:

$$\begin{aligned}
\{\delta R(\varphi,\varphi^{+};\boldsymbol{\alpha})\}_{\boldsymbol{\alpha}\equiv\boldsymbol{\alpha}^{0}} &\triangleq \left\{\frac{\partial R(\varphi,\varphi^{+};\boldsymbol{\alpha})}{\partial\varphi}\delta\varphi\right\}_{\boldsymbol{\alpha}\equiv\boldsymbol{\alpha}^{0}} + \left\{\frac{\partial R(\varphi,\varphi^{+};\boldsymbol{\alpha})}{\partial\varphi^{+}}\delta\varphi^{+}\right\}_{\boldsymbol{\alpha}\equiv\boldsymbol{\alpha}^{0}} \\
&= \left\{\langle Q^{+},\delta\varphi\rangle_{B} - \langle\delta\varphi^{+},L\varphi - Q\rangle_{B} - \langle\varphi^{+},L\delta\varphi\rangle_{B}\right\}_{\boldsymbol{\alpha}\equiv\boldsymbol{\alpha}^{0}} \\
&= \left\{\langle\delta\varphi,Q^{+}\rangle_{B} - \langle\delta\varphi,L^{+}\varphi^{+}\rangle_{B} - P^{(1)}[\delta\varphi,\varphi^{+}]\right\}_{\boldsymbol{\alpha}\equiv\boldsymbol{\alpha}^{0}} \\
&= -\left\{P^{(1)}[\delta\varphi,\varphi^{+}]\right\}_{\boldsymbol{\alpha}\equiv\boldsymbol{\alpha}^{0}}.
\end{aligned} \tag{4.155}$$

If the variation $\delta\varphi(\mathbf{r},\boldsymbol{\Omega},E)$, or, equivalently, if the approximate flux, does *not* satisfy the same vacuum boundary conditions as the forward flux, then the R-functional is *not* stationary with regard to arbitrary variations in the forward and adjoint fluxes *at fixed parameter values,* and therefore exhibits the first-order errors in $\delta\varphi$ on the domain's boundary as shown in (4.155). However, if the approximate flux *does* satisfy the vacuum boundary condition shown in (4.154), then the bilinear concomitant vanishes in (4.155), i.e., $\left\{P^{(1)}[\delta\varphi,\varphi^{+}]\right\}_{\boldsymbol{\alpha}\equiv\boldsymbol{\alpha}^{0}} = 0$, and consequently $\{\delta R(\varphi,\varphi^{+};\boldsymbol{\alpha})\}_{\boldsymbol{\alpha}\equiv\boldsymbol{\alpha}^{0}} = 0$, which implies that the R-functional is stationary when the forward and adjoint fluxes are allowed to vary independently of each other and independently of the model parameters, and the unknown flux variations continue to satisfy the vacuum boundary conditions, but the model parameters are not allowed to vary.

4.5.1 Application of the 1st-CASAM-L to Determine the First-Order Sensitivities of the Roussopoulos and Schwinger Functionals to Imprecisely Known Model Parameters

The Roussopoulos and the Schwinger functionals depend explicitly and implicitly, through the fluxes $\varphi^{+}(\mathbf{r},\boldsymbol{\Omega},E)$ and $\varphi(\mathbf{r},\boldsymbol{\Omega},E)$, on all of the imprecisely known model parameters $\boldsymbol{\alpha}$ defined in (4.10). It is evident from the forward and adjoint transport equations, namely (4.1) and (4.7), that parameter variations $\delta\boldsymbol{\alpha}\triangleq[\delta\alpha_{1},\ldots,\delta\alpha_{TP}]^{\dagger}$ around the nominal parameter values $\boldsymbol{\alpha}^{0}\triangleq\left[\alpha_{1}^{0},\ldots,\alpha_{TP}^{0}\right]^{\dagger}$ will induce variations in the forward and adjoint fluxes; these variations will be denoted as $\delta\varphi(\mathbf{r},\boldsymbol{\Omega},E)$ and $\delta\varphi^{+}(\mathbf{r},\boldsymbol{\Omega},E)$, respectively. These flux variations are the solutions of the *First-Level Variational Sensitivity System* (1st-LVSS) defined in (4.17) and (4.18). The first-order total sensitivity of the S-functional to variations $\delta\boldsymbol{\alpha}$, $\delta\varphi$, $\delta\varphi^{+}$ is provided by its first-order Gateaux-differential, denoted as $\delta S(\varphi^{0},\varphi^{+0},\boldsymbol{\alpha}^{0};\delta\varphi,\delta\varphi^{+};\delta\boldsymbol{\alpha})$, which is computed by definition from (4.148) as follows:

$$
\begin{aligned}
&\delta S\left(\varphi^0,\varphi^{+0},\boldsymbol{\alpha}^0;\delta\varphi,\delta\varphi^+;\delta\boldsymbol{\alpha}\right)\\
&\triangleq\left\{\frac{d}{d\varepsilon}\left[\frac{\left\langle\varphi^0+\varepsilon\delta\varphi,Q^+\left(\boldsymbol{\alpha}^0+\varepsilon\delta\boldsymbol{\alpha}\right)\right\rangle_B\left\langle\varphi^{+0}+\varepsilon\delta\varphi^+,Q\left(\boldsymbol{\alpha}^0+\varepsilon\delta\boldsymbol{\alpha}\right)\right\rangle_B}{\left\langle\varphi^{+0}+\varepsilon\delta\varphi^+,L\left(\boldsymbol{\alpha}^0+\varepsilon\delta\boldsymbol{\alpha}\right)\left(\varphi^0+\varepsilon\delta\varphi\right)\right\rangle_B}\right]\right\}_{\varepsilon=0}\\
&=\frac{\left[\left\langle\delta\varphi,Q^+\left(\boldsymbol{\alpha}^0\right)\right\rangle_B+\left\langle\varphi^0,\delta Q^+\left(\boldsymbol{\alpha}^0;\delta\boldsymbol{\alpha}\right)\right\rangle_B\right]\left\langle\varphi^{+0},Q\left(\boldsymbol{\alpha}^0\right)\right\rangle_B}{\left\langle\varphi^{+0},L\left(\boldsymbol{\alpha}^0\right)\varphi^0\right\rangle_B}\\
&+\frac{\left\langle\varphi^0,Q^+\left(\boldsymbol{\alpha}^0\right)\right\rangle_B\left[\left\langle\delta\varphi^+,Q\left(\boldsymbol{\alpha}^0\right)\right\rangle_B+\left\langle\varphi^{+0},\delta Q\left(\boldsymbol{\alpha}^0;\delta\boldsymbol{\alpha}\right)\right\rangle_B\right]}{\left\langle\varphi^{+0},L\left(\boldsymbol{\alpha}^0\right)\varphi^0\right\rangle_B}\\
&-\frac{\left\langle\varphi^0,Q^+\left(\boldsymbol{\alpha}^0\right)\right\rangle_B\left\langle\varphi^{+0},Q\left(\boldsymbol{\alpha}^0\right)\right\rangle_B}{\left\langle\varphi^{+0},L\left(\boldsymbol{\alpha}^0\right)\varphi^0\right\rangle_B^2}\\
&\times\left[\left\langle\delta\varphi^+,L\left(\boldsymbol{\alpha}^0\right)\varphi^0\right\rangle_B+\left\langle\varphi^{+0},\delta L\left(\boldsymbol{\alpha}^0;\delta\boldsymbol{\alpha}\right)\varphi^0+L\left(\boldsymbol{\alpha}^0\right)\delta\varphi\right\rangle_B\right]\\
&=\left\langle\delta\varphi,Q^+\left(\mathbf{r},\boldsymbol{\Omega},E;\boldsymbol{\alpha}^0\right)\right\rangle_B+\left\langle\varphi^0(\mathbf{r},\boldsymbol{\Omega},E),\delta Q^+\right\rangle_B.
\end{aligned}
\tag{4.156}
$$

The last equality in (4.156) indicates that the first-order Gateaux-differential $\delta S(\varphi^0,\varphi^{+0},\boldsymbol{\alpha}^0;\delta\varphi,\delta\varphi^+;\delta\boldsymbol{\alpha})\equiv\delta S(\varphi^0,\boldsymbol{\alpha}^0;\delta\varphi;\delta\boldsymbol{\alpha})$ depends only on the variations in the parameters and the variations in the forward flux and adjoint source, but is independent of the variations in the adjoint flux. Furthermore, (4.156) also indicates that the first-order Gateaux-differential $\delta S\left(\varphi^0,\boldsymbol{\alpha}^0;\delta\varphi;\delta\boldsymbol{\alpha}\right)$ of the S-functional does not vanish in general. This means that the S-functional is *not* accurate to second order in variations $\delta\boldsymbol{\alpha}\triangleq[\delta\alpha_1,\ldots,\delta\alpha_{TP}]^\dagger$ around the nominal parameter values $\boldsymbol{\alpha}^0$. The first-order Gateaux-differential $\delta S\left(\varphi^0,\boldsymbol{\alpha}^0;\delta\varphi;\delta\boldsymbol{\alpha}\right)$ would vanish if and only if $\delta\boldsymbol{\alpha}\equiv 0$, i.e., if all model parameters were perfectly well known, which is never the case in practice.

Applying the definition of the first-order Gateaux-differential to the definition of the R-functional given in (4.145) leads to the following expression for the first-order Gateaux-differential, denoted as $\delta R\left(\varphi^0,\varphi^{+0},\boldsymbol{\alpha}^0;\delta\varphi,\delta\varphi^+;\delta\boldsymbol{\alpha}\right)$, of the R-functional:

$$
\begin{aligned}
&\delta R\left(\varphi^0,\varphi^{+0},\boldsymbol{\alpha}^0;\delta\varphi,\delta\varphi^+;\delta\boldsymbol{\alpha}\right)\\
&\triangleq\left\{\frac{d}{d\varepsilon}\left\langle Q^+\left(\boldsymbol{\alpha}^0+\varepsilon\delta\boldsymbol{\alpha}\right),\varphi^0+\varepsilon\delta\varphi\right\rangle_B\right\}_{\varepsilon=0}-\left\{\frac{d}{d\varepsilon}\left\langle\varphi^{+0}+\varepsilon\delta\varphi^+,\right.\right.\\
&\left.\left.L\left(\boldsymbol{\alpha}^0+\varepsilon\delta\boldsymbol{\alpha}\right)\left(\varphi^0+\varepsilon\delta\varphi\right)-Q\left(\boldsymbol{\alpha}^0+\varepsilon\delta\boldsymbol{\alpha}\right)\right\rangle_B\right\}_{\varepsilon=0}\\
&=\left\langle Q^+\left(\mathbf{r},\boldsymbol{\Omega},E;\boldsymbol{\alpha}^0\right),\delta\varphi\right\rangle_B+\left\langle\delta Q^+\left(\mathbf{r},\boldsymbol{\Omega},E;\boldsymbol{\alpha}^0;\delta\boldsymbol{\alpha}\right),\varphi^0\right\rangle_B.
\end{aligned}
\tag{4.157}
$$

The result obtained in (4.157) indicates that the first-order Gateaux-differential $\delta R\left(\varphi^0,\varphi^{+0},\boldsymbol{\alpha}^0;\delta\varphi,\delta\varphi^+;\delta\boldsymbol{\alpha}\right)\equiv\delta R\left(\varphi^0,\boldsymbol{\alpha}^0;\delta\varphi;\delta\boldsymbol{\alpha}\right)$ is independent of the variations in the adjoint flux, depending only on the variations in the parameters and the variations in the forward flux and adjoint source. Furthermore, (4.157) also indicates that the first-order Gateaux-differential $\delta R\left(\varphi^0,\boldsymbol{\alpha}^0;\delta\varphi;\delta\boldsymbol{\alpha}\right)$ does not vanish in general. This means that the R-functional is *not* accurate to second order in variations

$\delta\boldsymbol{\alpha} \triangleq [\delta\alpha_1, \ldots, \delta\alpha_{TP}]^\dagger$ around the nominal parameter values $\boldsymbol{\alpha}^0$. The only situation in which $\delta R(\varphi^0, \boldsymbol{\alpha}^0; \delta\varphi; \delta\boldsymbol{\alpha})$ would vanish is if $\delta\boldsymbol{\alpha} \equiv 0$, i.e., if all model parameters were perfectly well known, which is never the case in practice. Comparing the results obtained in (4.156) and (4.157) indicates that the first-order Gateaux-differentials of the Schwinger and Roussopoulos functionals take on identical expressions when they are evaluated at the nominal parameter (and flux) values, i.e.,

$$\delta S\left(\varphi^0, \boldsymbol{\alpha}^0; \delta\varphi; \delta\boldsymbol{\alpha}\right) = \delta R\left(\varphi^0, \boldsymbol{\alpha}^0; \delta\varphi; \delta\boldsymbol{\alpha}\right). \tag{4.158}$$

The result obtained in (4.158) indicates that the sensitivities of the Schwinger and the Roussopoulos functionals to imprecisely known parameters (such as cross sections, number densities, forward and adjoint sources), computed at the nominal parameter values, are equivalent to each other, and could, in principle, be computed once the flux variation $\delta\varphi(\mathbf{r}, \boldsymbol{\Omega}, E; \boldsymbol{\alpha}^0; \delta\boldsymbol{\alpha})$ becomes available after solving (half of) the *First-Level Variational Sensitivity System* (1st-LVSS). However, the direct evaluation of the flux variation $\delta\varphi(\mathbf{r}, \boldsymbol{\Omega}, E; \boldsymbol{\alpha}^0; \delta\boldsymbol{\alpha})$ is computationally expensive and can be avoided by using the 1st-CASAM to express the "indirect effect term" $\langle Q^+(\mathbf{r}, \boldsymbol{\Omega}, E; \boldsymbol{\alpha}^0), \delta\varphi(\mathbf{r}, \boldsymbol{\Omega}, E; \boldsymbol{\alpha}^0; \delta\boldsymbol{\alpha})\rangle_B$ alternatively through the following equality:

$$\left\langle Q^+\left(\mathbf{r}, \boldsymbol{\Omega}, E; \boldsymbol{\alpha}^0\right), \delta\varphi\right\rangle_B = \left\langle \varphi^+(\mathbf{r}, \boldsymbol{\Omega}, E), Q_1^{(1)}(\boldsymbol{\alpha}, \varphi; \delta\boldsymbol{\alpha})\right\rangle_B. \tag{4.159}$$

Replacing (4.159) in (4.156) and (4.157) eliminates in the latter equations the appearance of the functions $\delta\varphi$, $\delta\varphi^+$, $\delta\boldsymbol{\alpha}$, and yields the following expression for the *total first-order sensitivity* of the Schwinger and Roussopoulos functionals:

$$\begin{aligned}
&\delta S\left(\varphi^0, \boldsymbol{\alpha}^0; \delta\varphi; \delta\boldsymbol{\alpha}\right) = \delta R\left(\varphi^0, \boldsymbol{\alpha}^0; \delta\varphi; \delta\boldsymbol{\alpha}\right) \\
&= \left\{\langle \varphi(\mathbf{r}, \boldsymbol{\Omega}, E), \delta Q^+(\mathbf{k}; \mathbf{r}, \boldsymbol{\Omega}, E)\rangle_B\right\}_{(\boldsymbol{\alpha}^0; \varphi^0, \varphi^{+0})} \\
&+\left\{\langle \varphi^+(\mathbf{r}, \boldsymbol{\Omega}, E), \delta Q(\mathbf{q}; \mathbf{r}, \boldsymbol{\Omega}, E) - \delta L(\boldsymbol{\alpha})\varphi(\mathbf{r}, \boldsymbol{\Omega}, E)\rangle_B\right\}_{(\boldsymbol{\alpha}^0; \varphi^0, \varphi^{+0})} \\
&\triangleq \sum_{m_1=1}^{TP} \left\{\frac{\partial R(\varphi, \varphi^+, \boldsymbol{\alpha})}{\partial \alpha_{m_1}}\right\}_{(\boldsymbol{\alpha}^0; \varphi^0, \varphi^{+0})} \delta\alpha_{m_1}.
\end{aligned} \tag{4.160}$$

The *partial first-order response sensitivities* of the Schwinger and Roussopoulos functionals to a generic parameter α_{m_1} are obtained from (4.160) by identifying the quantities that multiply the various parameter variations $\delta\alpha_{m_1}$, which yields the following expressions:

$$\begin{aligned}
&\text{For } j = 1, \ldots, J_t: \ \frac{\partial S(\varphi, \varphi^+, \boldsymbol{\alpha})}{\partial t_j} = \frac{\partial R(\varphi, \varphi^+, \boldsymbol{\alpha})}{\partial t_j} \\
&= -\int dV \int_{4\pi} d\Omega \int_0^{E_f} dE \, \frac{\partial \Sigma_t(\mathbf{t}; \mathbf{r}, E)}{\partial t_j} \varphi^+(\mathbf{r}, \boldsymbol{\Omega}, E)\varphi(\mathbf{r}, \boldsymbol{\Omega}, E);
\end{aligned} \tag{4.161}$$

$$\text{For } j = 1, \ldots, J_s; \frac{\partial S(\varphi, \varphi^+, \boldsymbol{\alpha})}{\partial s_j} = \frac{\partial R(\varphi, \varphi^+, \boldsymbol{\alpha})}{\partial s_j}$$

$$= \int dV \int_{4\pi} d\boldsymbol{\Omega} \int_0^{E_f} dE\, \varphi^+(\mathbf{r}, \boldsymbol{\Omega}, E) \times \int_{4\pi} d\boldsymbol{\Omega}' \int_0^{E_f} dE' \frac{\partial \Sigma_s(\mathbf{s}; \mathbf{r}, E' \to E, \boldsymbol{\Omega}' \to \boldsymbol{\Omega})}{\partial s_j} \varphi(\mathbf{r}, \boldsymbol{\Omega}', E'); \tag{4.162}$$

$$\text{For } j = 1, \ldots, J_f : \frac{\partial S(\varphi, \varphi^+, \boldsymbol{\alpha})}{\partial f_j} = \frac{\partial R(\varphi, \varphi^+, \boldsymbol{\alpha})}{\partial f_j}$$

$$= \int dV \int_{4\pi} d\boldsymbol{\Omega} \int_0^{E_f} dE\, \varphi^+(\mathbf{r}, \boldsymbol{\Omega}, E) \times \int_{4\pi} d\boldsymbol{\Omega}' \int_0^{E_f} dE' \frac{\partial \left[\nu\Sigma_f(\mathbf{f}; \mathbf{r}, E')\right]}{\partial f_j} \chi(\mathbf{p}; \mathbf{r}, E' \to E) \varphi(\mathbf{r}, \boldsymbol{\Omega}', E'); \tag{4.163}$$

$$\text{For } j = 1, \ldots, J_p; \frac{\partial S(\varphi, \varphi^+, \boldsymbol{\alpha})}{\partial p_j} = \frac{\partial R(\varphi, \varphi^+, \boldsymbol{\alpha})}{\partial p_j}$$

$$= \int dV \int_{4\pi} d\boldsymbol{\Omega} \int_0^{E_f} dE\, \varphi^+(\mathbf{r}, \boldsymbol{\Omega}, E) \times \int_{4\pi} d\boldsymbol{\Omega}' \int_0^{E_f} dE' \frac{\partial \chi(\mathbf{p}; \mathbf{r}, E' \to E)}{\partial p_j} \nu\Sigma_f(\mathbf{f}; \mathbf{r}, E') \varphi(\mathbf{r}, \boldsymbol{\Omega}', E'); \tag{4.164}$$

$$\text{For } j = 1, \ldots, J_q : \frac{\partial S(\varphi, \varphi^+, \boldsymbol{\alpha})}{\partial q_j} = \frac{\partial R(\varphi, \varphi^+, \boldsymbol{\alpha})}{\partial q_j} = \int dV \int_{4\pi} d\boldsymbol{\Omega} \int_0^{E_f} dE \frac{\partial Q(\mathbf{q}; \mathbf{r}, \boldsymbol{\Omega}, E)}{\partial q_j} \varphi^+(\mathbf{r}, \boldsymbol{\Omega}, E); \tag{4.165}$$

$$\text{For } j = 1, \ldots, J_k : \frac{\partial S(\varphi, \varphi^+, \boldsymbol{\alpha})}{\partial k_j} = \frac{\partial R(\varphi, \varphi^+, \boldsymbol{\alpha})}{\partial k_j} = \int dV \int_{4\pi} d\boldsymbol{\Omega} \int_0^{E_f} dE \frac{\partial Q^+(\mathbf{k}; \mathbf{r}, \boldsymbol{\Omega}, E)}{\partial k_j} \varphi(\mathbf{r}, \boldsymbol{\Omega}, E). \tag{4.166}$$

The same model parameter may appear in the definitions of more than one macroscopic cross section. For example, the isotopic number density of some element, generically denoted as N_i, could be an imprecisely known model parameter which might appear in the definitions of the total, scattering, and/or fission macroscopic cross sections, as well as in the forward and/or adjoint source terms. In such a case, the sensitivity of the response to the generic model parameter N_i would be the sum of the corresponding partial sensitivities computed from the above expressions.

4.5.2 Application of the 2nd-CASAM-L to Determine the Second-Order Sensitivities of the Schwinger and Roussopoulos Functionals to Imprecisely Known Model Parameters

The expressions of the second-order sensitivities of the Schwinger and Roussopoulos functionals are derived by applying the 2nd-CASAM-L to the expressions of the corresponding first-order sensitivities obtained in (4.161)–(4.166). Since the first-order sensitivities of the Schwinger functional are identical to the corresponding first-order sensitivities of the Roussopoulos functional, the corresponding second-order sensitivities of these functionals will also be identical to each other. To avoid undue repetitions, only the final results are provided in this section, below.

4.5.2.1 Determination of the Second-Order Sensitivities of the Form $\partial^2 S\left(\boldsymbol{\varphi}, \boldsymbol{\varphi}^+; \boldsymbol{\psi}_j^{(2)}; \boldsymbol{\alpha}\right) / \partial t_j \partial \boldsymbol{\alpha}_{m_2}$

The second-order partial sensitivities $\partial^2 S\left(\varphi, \varphi^+; \boldsymbol{\psi}_j^{(2)}; \boldsymbol{\alpha}\right) / (\partial t_j)(\partial \alpha_{m_2})$ have the following expressions:

$$\text{For } j, m_2 = 1, \ldots, J_t : \frac{\partial^2 S\left(\varphi, \varphi^+; \boldsymbol{\psi}_j^{(2)}; \boldsymbol{\alpha}\right)}{\partial t_j \partial t_{m_2}} = \frac{\partial^2 R\left(\varphi, \varphi^+; \boldsymbol{\psi}_j^{(2)}; \boldsymbol{\alpha}\right)}{\partial t_j \partial t_{m_2}}$$

$$\begin{aligned} &= - \int dV \int_{4\pi} d\boldsymbol{\Omega} \int_0^{E_f} dE\, \varphi^+(\mathbf{r}, \boldsymbol{\Omega}, E) \varphi(\mathbf{r}, \boldsymbol{\Omega}, E) \frac{\partial^2 \Sigma_t(\mathbf{t}; \mathbf{r}, E)}{\partial t_j \partial t_{m_2}} \\ &- \int dV \int_{4\pi} d\boldsymbol{\Omega} \int_0^{E_f} dE \frac{\partial \Sigma_t(\mathbf{t}; \mathbf{r}, E)}{\partial t_{m_2}} \left[\psi_{1,j}^{(2)}(\mathbf{r}, \boldsymbol{\Omega}, E) \varphi(\mathbf{r}, \boldsymbol{\Omega}, E) \right. \\ &\qquad \left. + \psi_{2,j}^{(2)}(\mathbf{r}, \boldsymbol{\Omega}, E) \varphi^+(\mathbf{r}, \boldsymbol{\Omega}, E) \right]; \end{aligned} \tag{4.167}$$

$$\text{For } j=1,\ldots,J_t;\; m_2=1,\ldots,J_s:\; \frac{\partial^2 S\left(\varphi,\varphi^+;\boldsymbol{\psi}_j^{(2)};\boldsymbol{\alpha}\right)}{\partial t_j\partial s_{m_2}}$$

$$= \frac{\partial^2 R\left(\varphi,\varphi^+;\boldsymbol{\psi}_j^{(2)};\boldsymbol{\alpha}\right)}{\partial t_j\partial s_{m_2}} = \int dV \int_{4\pi} d\boldsymbol{\Omega} \int_0^{E_f} dE\, \psi_{1,j}^{(2)}(\mathbf{r},\boldsymbol{\Omega},E)$$

$$\times \int_{4\pi} d\boldsymbol{\Omega}' \int_0^{E_f} dE'\, \varphi(\mathbf{r},\boldsymbol{\Omega}',E') \frac{\partial \Sigma_s(\mathbf{s};\mathbf{r},E'\to E,\boldsymbol{\Omega}'\to\boldsymbol{\Omega})}{\partial s_{m_2}} \tag{4.168}$$

$$+ \int dV \int_{4\pi} d\boldsymbol{\Omega} \int_0^{E_f} dE\, \psi_{2,j}^{(2)}(\mathbf{r},\boldsymbol{\Omega},E)$$

$$\times \int_{4\pi} d\boldsymbol{\Omega}' \int_0^{E_f} dE'\, \varphi^+(\mathbf{r},\boldsymbol{\Omega}',E') \frac{\partial \Sigma_s(\mathbf{s};\mathbf{r},E\to E',\boldsymbol{\Omega}\to\boldsymbol{\Omega}')}{\partial s_{m_2}};$$

$$\text{For } j=1,\ldots,J_t;\; m_2=1,\ldots,J_f:\; \frac{\partial^2 S\left(\varphi,\varphi^+;\boldsymbol{\psi}_j^{(2)};\boldsymbol{\alpha}\right)}{\partial t_j\partial f_{m_2}}$$

$$= \frac{\partial^2 R\left(\varphi,\varphi^+;\boldsymbol{\psi}_j^{(2)};\boldsymbol{\alpha}\right)}{\partial t_j\partial f_{m_2}} = \int dV \int_{4\pi} d\boldsymbol{\Omega} \int_0^{E_f} dE\, \psi_{1,j}^{(2)}(\mathbf{r},\boldsymbol{\Omega},E)\chi(\mathbf{p};\mathbf{r},E)$$

$$\times \int_{4\pi} d\boldsymbol{\Omega}' \int_0^{E_f} dE'\, \varphi(\mathbf{r},\boldsymbol{\Omega}',E') \frac{\partial\left[\nu\Sigma_f(\mathbf{f};\mathbf{r},E')\right]}{\partial f_{m_2}} \tag{4.169}$$

$$+ \int dV \int_{4\pi} d\boldsymbol{\Omega} \int_0^{E_f} dE\, \psi_{2,j}^{(2)}(\mathbf{r},\boldsymbol{\Omega},E) \frac{\partial\left[\nu\Sigma_f(\mathbf{f};\mathbf{r},E)\right]}{\partial f_{m_2}}$$

$$\times \int_{4\pi} d\boldsymbol{\Omega}' \int_0^{E_f} dE' \chi(\mathbf{p};\mathbf{r},E')\varphi^+(\mathbf{r},\boldsymbol{\Omega}',E');$$

For $j = 1, \ldots, J_t;\ m_2 = 1, \ldots, J_p:\ \dfrac{\partial^2 S\left(\varphi, \varphi^+; \boldsymbol{\psi}_j^{(2)}; \boldsymbol{\alpha}\right)}{\partial t_j \partial p_{m_2}}$

$$= \frac{\partial^2 R\left(\varphi, \varphi^+; \boldsymbol{\psi}_j^{(2)}; \boldsymbol{\alpha}\right)}{\partial t_j \partial p_{m_2}} = \int dV \int_{4\pi} d\boldsymbol{\Omega} \int_0^{E_f} dE\, \psi_{1,j}^{(2)}(\mathbf{r}, \boldsymbol{\Omega}, E) \frac{\partial \chi(\mathbf{p}; \mathbf{r}, E)}{\partial p_{m_2}}$$

$$\times \int_{4\pi} d\boldsymbol{\Omega}' \int_0^{E_f} dE'\, \varphi(\mathbf{r}, \boldsymbol{\Omega}', E') \left[\nu\Sigma_f(\mathbf{f}; \mathbf{r}, E')\right] \tag{4.170}$$

$$+ \int dV \int_{4\pi} d\boldsymbol{\Omega} \int_0^{E_f} dE\, \psi_{2,j}^{(2)}(\mathbf{r}, \boldsymbol{\Omega}, E) \nu\Sigma_f(\mathbf{f}; \mathbf{r}, E)$$

$$\times \int_{4\pi} d\boldsymbol{\Omega}' \int_0^{E_f} dE'\, \varphi^+(\mathbf{r}, \boldsymbol{\Omega}', E') \frac{\partial \chi(\mathbf{p}; \mathbf{r}, E')}{\partial p_{m_2}};$$

For $j = 1, \ldots, J_t;\ m_2 = 1, \ldots, J_q:\ \dfrac{\partial^2 S\left(\varphi, \varphi^+; \boldsymbol{\psi}_j^{(2)}; \boldsymbol{\alpha}\right)}{\partial t_j \partial q_{m_2}}$

$$= \frac{\partial^2 R\left(\varphi, \varphi^+; \boldsymbol{\psi}_j^{(2)}; \boldsymbol{\alpha}\right)}{\partial t_j \partial q_{m_2}} = \int dV \int_{4\pi} d\boldsymbol{\Omega} \int_0^{E_f} dE\, \psi_{1,j}^{(2)}(\mathbf{r}, \boldsymbol{\Omega}, E) \frac{\partial Q(\mathbf{q}; \mathbf{r}, \boldsymbol{\Omega}, E)}{\partial q_{m_2}}; \tag{4.171}$$

For $j = 1, \ldots, J_t;\ m_2 = 1, \ldots, J_k:\ \dfrac{\partial^2 S\left(\varphi, \varphi^+; \boldsymbol{\psi}_j^{(2)}; \boldsymbol{\alpha}\right)}{\partial t_j \partial k_{m_2}}$

$$= \frac{\partial^2 R\left(\varphi, \varphi^+; \boldsymbol{\psi}_j^{(2)}; \boldsymbol{\alpha}\right)}{\partial t_j \partial k_{m_2}} = \int dV \int_{4\pi} d\boldsymbol{\Omega} \int_0^{E_f} dE\, \psi_{2,j}^{(2)}(\mathbf{r}, \boldsymbol{\Omega}, E) \frac{\partial Q^+(\mathbf{k}; \mathbf{r}, \boldsymbol{\Omega}, E)}{\partial k_{m_2}}. \tag{4.172}$$

The components of the second-level adjoint function $\boldsymbol{\psi}_j^{(2)}(\mathbf{r}, \boldsymbol{\Omega}, E)$, which appears in the expressions provided in (4.167)–(4.172), are the solutions of the following *Second-Level Adjoint Sensitivity System* (2nd-LASS), for $j = 1, \ldots, J_t$:

$$L^+(\boldsymbol{\alpha})\psi_{1,j}^{(2)}(\mathbf{r}, \boldsymbol{\Omega}, E) = -\varphi^+(\mathbf{r}, \boldsymbol{\Omega}, E) \frac{\partial \Sigma_t(\mathbf{t}; \mathbf{r}, E)}{\partial t_j}, \tag{4.173}$$

$$L(\boldsymbol{\alpha})\psi_{2,j}^{(2)}(\mathbf{r},\boldsymbol{\Omega},E) = -\varphi(\mathbf{r},\boldsymbol{\Omega},E)\frac{\partial\Sigma_t(\mathbf{t};\mathbf{r},E)}{\partial t_j}, \tag{4.174}$$

subject to the following boundary conditions:

$$\left.\begin{array}{l}\psi_{1,j}^{(1)}(\mathbf{r}_s,\boldsymbol{\Omega},E) = 0, \quad \mathbf{r}_s \in \partial V; \ \boldsymbol{\Omega}\cdot\mathbf{n} > 0; 0 < E < E_f \\ \psi_{2,j}^{(1)}(\mathbf{r}_s,\boldsymbol{\Omega},E) = 0, \quad \mathbf{r}_s \in \partial V; \ \boldsymbol{\Omega}\cdot\mathbf{n} < 0; 0 < E < E_f\end{array}\right\} \tag{4.175}$$

For each of the imprecisely known scalar model parameters appearing in the definition of the total macroscopic cross section $\Sigma_t(\mathbf{t};\mathbf{r},E)$, the 2nd-LASS is solved by using two computations, one involving the "forward transport solver" and one involving the "adjoint transport solver," for which the same software can be used as for solving the original forward and adjoint transport equations.

4.5.2.2 Determination of the Second-Order Sensitivities of the Form $\partial^2 S\left(\boldsymbol{\varphi},\boldsymbol{\varphi}^+;\boldsymbol{\theta}_j^{(2)};\boldsymbol{\alpha}\right)/\partial s_j \partial \boldsymbol{\alpha}_{m_2}$

The second-order sensitivities $\partial^2 S\left(\varphi,\varphi^+;\boldsymbol{\theta}_j^{(2)};\boldsymbol{\alpha}\right)/(\partial s_j)(\partial\alpha_{m_2})$, $j = 1, \ldots, J_s$, $m_2 = 1, \ldots, TP$, will ultimately depend on a second-level adjoint function denoted as $\boldsymbol{\theta}_j^{(2)}(\mathbf{r},\boldsymbol{\Omega},E) \triangleq \left[\theta_{1,j}^{(2)}(\mathbf{r},\boldsymbol{\Omega},E), \theta_{2,j}^{(2)}(\mathbf{r},\boldsymbol{\Omega},E)\right]^\dagger$, and have the following expressions:

$$\text{For } j = 1, \ldots, J_s; \ m_2 = 1, \ldots, J_t: \ \frac{\partial^2 S\left(\varphi,\varphi^+;\boldsymbol{\theta}_j^{(2)};\boldsymbol{\alpha}\right)}{\partial s_j \partial t_{m_2}}$$

$$= \frac{\partial^2 R\left(\varphi,\varphi^+;\boldsymbol{\theta}_j^{(2)};\boldsymbol{\alpha}\right)}{\partial s_j \partial t_{m_2}} = -\int dV \int_{4\pi} d\Omega \int_0^{E_f} dE \ \frac{\partial\Sigma_t(\mathbf{t};\mathbf{r},E)}{\partial t_{m_2}} \tag{4.176}$$

$$\times\left[\theta_{1,j}^{(2)}(\mathbf{r},\boldsymbol{\Omega},E)\varphi(\mathbf{r},\boldsymbol{\Omega},E) + \theta_{2,j}^{(2)}(\mathbf{r},\boldsymbol{\Omega},E)\varphi^+(\mathbf{r},\boldsymbol{\Omega},E)\right];$$

For $j = 1, \ldots, J_s$; $m_2 = 1, \ldots, J_s$: $\dfrac{\partial^2 S\left(\varphi, \varphi^+; \boldsymbol{\theta}_j^{(2)}; \boldsymbol{\alpha}\right)}{\partial s_j \partial s_{m_2}}$

$$
\begin{aligned}
&= \frac{\partial^2 R\left(\varphi, \varphi^+; \boldsymbol{\theta}_j^{(2)}; \boldsymbol{\alpha}\right)}{\partial s_j \partial s_{m_2}} = \int dV \int_{4\pi} d\boldsymbol{\Omega} \int_0^{E_f} dE\, \varphi^+(\mathbf{r}, \boldsymbol{\Omega}, E) \\
&\times \int_{4\pi} d\boldsymbol{\Omega}' \int_0^{E_f} dE'\, \varphi(\mathbf{r}, \boldsymbol{\Omega}', E') \frac{\partial^2 \Sigma_s(\mathbf{s}; \mathbf{r}, E' \to E, \boldsymbol{\Omega}' \to \boldsymbol{\Omega})}{\partial s_j \partial s_{m_2}} \\
&+ \int dV \int_{4\pi} d\boldsymbol{\Omega} \int_0^{E_f} dE\, \theta_{1,j}^{(2)}(\mathbf{r}, \boldsymbol{\Omega}, E) \qquad (4.177) \\
&\times \int_{4\pi} d\boldsymbol{\Omega}' \int_0^{E_f} dE'\, \varphi(\mathbf{r}, \boldsymbol{\Omega}', E') \frac{\partial \Sigma_s(\mathbf{s}; \mathbf{r}, E' \to E, \boldsymbol{\Omega}' \to \boldsymbol{\Omega})}{\partial s_{m_2}} \\
&+ \int dV \int_{4\pi} d\boldsymbol{\Omega} \int_0^{E_f} dE\, \theta_{2,j}^{(2)}(\mathbf{r}, \boldsymbol{\Omega}, E) \\
&\times \int_{4\pi} d\boldsymbol{\Omega}' \int_0^{E_f} dE'\, \varphi^+(\mathbf{r}, \boldsymbol{\Omega}', E') \frac{\partial \Sigma_s(\mathbf{s}; \mathbf{r}, E \to E', \boldsymbol{\Omega} \to \boldsymbol{\Omega}')}{\partial s_{m_2}};
\end{aligned}
$$

For $j = 1, \ldots, J_s$; $m_2 = 1, \ldots, J_f$: $\dfrac{\partial^2 S\left(\varphi, \varphi^+; \boldsymbol{\theta}_j^{(2)}; \boldsymbol{\alpha}\right)}{\partial s_j \partial f_{m_2}}$

$$
\begin{aligned}
&= \frac{\partial^2 R\left(\varphi, \varphi^+; \boldsymbol{\theta}_j^{(2)}; \boldsymbol{\alpha}\right)}{\partial s_j \partial f_{m_2}} = \int dV \int_{4\pi} d\boldsymbol{\Omega} \int_0^{E_f} dE\, \theta_{1,j}^{(2)}(\mathbf{r}, \boldsymbol{\Omega}, E) \chi(\mathbf{p}; \mathbf{r}, E) \\
&\times \int_{4\pi} d\boldsymbol{\Omega}' \int_0^{E_f} dE'\, \varphi(\mathbf{r}, \boldsymbol{\Omega}', E') \frac{\partial \left[\nu \Sigma_f(\mathbf{f}; \mathbf{r}, E')\right]}{\partial f_{m_2}} \qquad (4.178) \\
&+ \int dV \int_{4\pi} d\boldsymbol{\Omega} \int_0^{E_f} dE\, \theta_{2,j}^{(2)}(\mathbf{r}, \boldsymbol{\Omega}, E) \frac{\partial \left[\nu \Sigma_f(\mathbf{f}; \mathbf{r}, E)\right]}{\partial f_{m_2}} \\
&\times \int_{4\pi} d\boldsymbol{\Omega}' \int_0^{E_f} dE'\, \chi(\mathbf{p}; \mathbf{r}, E') \varphi^+(\mathbf{r}, \boldsymbol{\Omega}', E');
\end{aligned}
$$

$$\text{For } j=1,\ldots,J_s;\ m_2=1,\ldots,J_p:\ \frac{\partial^2 S\left(\varphi,\varphi^+;\boldsymbol{\theta}_j^{(2)};\boldsymbol{\alpha}\right)}{\partial s_j \partial p_{m_2}}$$

$$= \frac{\partial^2 R\left(\varphi,\varphi^+;\boldsymbol{\theta}_j^{(2)};\boldsymbol{\alpha}\right)}{\partial s_j \partial p_{m_2}} = \int dV \int_{4\pi} d\boldsymbol{\Omega} \int_0^{E_f} dE\, \theta_{1,j}^{(2)}(\mathbf{r},\boldsymbol{\Omega},E) \frac{\partial \chi(\mathbf{p};\mathbf{r},E)}{\partial p_{m_2}}$$

$$\times \int_{4\pi} d\boldsymbol{\Omega}' \int_0^{E_f} dE'\, \varphi(\mathbf{r},\boldsymbol{\Omega}',E')\nu\Sigma_f(\mathbf{f};\mathbf{r},E') \tag{4.179}$$

$$+ \int dV \int_{4\pi} d\boldsymbol{\Omega} \int_0^{E_f} dE\, \theta_{2,j}^{(2)}(\mathbf{r},\boldsymbol{\Omega},E)\nu\Sigma_f(\mathbf{f};\mathbf{r},E)$$

$$\times \int_{4\pi} d\boldsymbol{\Omega}' \int_0^{E_f} dE'\, \varphi^+(\mathbf{r},\boldsymbol{\Omega}',E') \frac{\partial \chi(\mathbf{p};\mathbf{r},E')}{\partial p_{m_2}};$$

$$\text{For } j=1,\ldots,J_s;\ m_2=1,\ldots,J_q:\ \frac{\partial^2 S\left(\varphi,\varphi^+;\boldsymbol{\theta}_j^{(2)};\boldsymbol{\alpha}\right)}{\partial s_j \partial q_{m_2}}$$

$$= \frac{\partial^2 R\left(\varphi,\varphi^+;\boldsymbol{\theta}_j^{(2)};\boldsymbol{\alpha}\right)}{\partial s_j \partial q_{m_2}} = \int dV \int_{4\pi} d\boldsymbol{\Omega} \int_0^{E_f} dE\, \theta_{1,j}^{(2)}(\mathbf{r},\boldsymbol{\Omega},E) \frac{\partial Q(\mathbf{q};\mathbf{r},\boldsymbol{\Omega},E)}{\partial q_{m_2}}; \tag{4.180}$$

$$\text{For } j=1,\ldots,J_s;\ m_2=1,\ldots,J_k:\ \frac{\partial^2 S\left(\varphi,\varphi^+;\boldsymbol{\theta}_j^{(2)};\boldsymbol{\alpha}\right)}{\partial s_j \partial k_{m_2}}$$

$$= \frac{\partial^2 R\left(\varphi,\varphi^+;\boldsymbol{\theta}_j^{(2)};\boldsymbol{\alpha}\right)}{\partial s_j \partial k_{m_2}} = \int dV \int_{4\pi} d\boldsymbol{\Omega} \int_0^{E_f} dE\, \theta_{2,j}^{(2)}(\mathbf{r},\boldsymbol{\Omega},E) \frac{\partial Q^+(\mathbf{k};\mathbf{r},\boldsymbol{\Omega},E)}{\partial k_{m_2}}. \tag{4.181}$$

The components of the second-level adjoint function $\boldsymbol{\theta}_j^{(2)}(\mathbf{r},\boldsymbol{\Omega},E) \triangleq \left[\theta_{1,j}^{(2)}(\mathbf{r},\boldsymbol{\Omega},E), \theta_{2,j}^{(2)}(\mathbf{r},\boldsymbol{\Omega},E)\right]^\dagger$, which appear in (4.176)–(4.181), are the solutions of the following 2nd-LASS, for $j = 1, \ldots, J_s$:

$$L^{+}(\boldsymbol{\alpha})\theta_{1,j}^{(2)}(\mathbf{r},\boldsymbol{\Omega},E) = \int_{4\pi} d\Omega' \int_{0}^{E_f} dE' \frac{\partial \Sigma_s(\mathbf{s};\mathbf{r},E\to E',\boldsymbol{\Omega}\to\boldsymbol{\Omega}')}{\partial s_j}\varphi^{+}(\mathbf{r},\boldsymbol{\Omega}',E'), \tag{4.182}$$

$$L(\boldsymbol{\alpha})\theta_{2,j}^{(2)}(\mathbf{r},\boldsymbol{\Omega},E) = \int_{4\pi} d\Omega' \int_{0}^{E_f} dE' \frac{\partial \Sigma_s(\mathbf{s};\mathbf{r},E'\to E,\boldsymbol{\Omega}'\to\boldsymbol{\Omega})}{\partial s_j}\varphi(\mathbf{r},\boldsymbol{\Omega}',E'), \tag{4.183}$$

satisfying the following boundary conditions:

$$\left.\begin{aligned} &\theta_{1,j}^{(2)}(\mathbf{r}_s,\boldsymbol{\Omega},E)=0,\quad \mathbf{r}_s\in\partial V,\quad \boldsymbol{\Omega}\cdot\mathbf{n}>0;0<E<E_f \\ &\theta_{2,j}^{(2)}(\mathbf{r}_s,\boldsymbol{\Omega},E)=0,\quad \mathbf{r}_s\in\partial V,\quad \boldsymbol{\Omega}\cdot\mathbf{n}<0;0<E<E_f \end{aligned}\right\} \tag{4.184}$$

The forward and adjoint operators appearing on the left side of the 2nd-LASS defined by (4.182) and (4.183) for the second-level adjoint function $\boldsymbol{\theta}_j^{(2)}$ are the same operators as those appearing on the left side of the 2nd-LASS for the second-level adjoint function $\boldsymbol{\psi}_j^{(2)}$. Furthermore, the forward and, respectively, adjoint functions are subject to the same forward and, respectively, adjoint (vacuum) boundary conditions. Only the source terms on the right sides of the respective 2nd-LASS differ from each other. Therefore, the same "forward" and "adjoint" software packages can be used for solving numerically the various equations underlying the 1st-LASS and the 2nd-LASS. Furthermore, the software package used for computing the sensitivities shown in (4.167)–(4.172) can also be used for computing the sensitivities shown in (4.176)–(4.181).

The expressions of the second-order sensitivities computed using (4.176) must be identical to those computed using (4.168), i.e.,

$$
\begin{aligned}
&\text{For } j=1,\ldots,J_s;\ k=1,\ldots,J_t: \ \frac{\partial^2 S\left(\varphi,\varphi^+;\boldsymbol{\theta}_j^{(2)};\boldsymbol{\alpha}\right)}{\partial s_j \partial t_k} \\
&= -\int dV \int\limits_{4\pi} d\boldsymbol{\Omega} \int\limits_0^{E_f} dE \, \frac{\partial \Sigma_t(\mathbf{t};\mathbf{r},E)}{\partial t_k} \\
&\times \left[\theta_{1,j}^{(2)}(\mathbf{r},\boldsymbol{\Omega},E)\varphi(\mathbf{r},\boldsymbol{\Omega},E) + \theta_{2,j}^{(2)}(\mathbf{r},\boldsymbol{\Omega},E)\varphi^+(\mathbf{r},\boldsymbol{\Omega},E)\right] \\
&= \frac{\partial^2 S\left(\varphi,\varphi^+;\boldsymbol{\psi}_j^{(2)};\boldsymbol{\alpha}\right)}{\partial t_k \partial s_j} = \int dV \int\limits_{4\pi} d\boldsymbol{\Omega} \int\limits_0^{E_f} dE \, \psi_{1,k}^{(2)}(\mathbf{r},\boldsymbol{\Omega},E) \\
&\times \int\limits_{4\pi} d\boldsymbol{\Omega}' \int\limits_0^{E_f} dE' \, \varphi(\mathbf{r},\boldsymbol{\Omega}',E') \frac{\partial \Sigma_s(\mathbf{s};\mathbf{r},E'\to E,\boldsymbol{\Omega}'\to\boldsymbol{\Omega})}{\partial s_j} \\
&+ \int dV \int\limits_{4\pi} d\boldsymbol{\Omega} \int\limits_0^{E_f} dE \, \psi_{2,k}^{(2)}(\mathbf{r},\boldsymbol{\Omega},E) \\
&\times \int\limits_{4\pi} d\boldsymbol{\Omega}' \int\limits_0^{E_f} dE' \, \varphi^+(\mathbf{r},\boldsymbol{\Omega}',E') \frac{\partial \Sigma_s(\mathbf{s};\mathbf{r},E\to E',\boldsymbol{\Omega}\to\boldsymbol{\Omega}')}{\partial s_j}.
\end{aligned}
\tag{4.185}
$$

The relation shown in (4.185) provides an independent path for the mutual verification of the solutions $\boldsymbol{\theta}_j^{(2)}$ and $\boldsymbol{\psi}_j^{(2)}$, of the respective 2nd-LASS.

4.5.2.3 Determination of the Second-Order Sensitivities of the Form $\partial^2 S\left(\varphi,\varphi^+;\mathbf{u}_j^{(2)};\boldsymbol{\alpha}\right)/\partial f_j \partial \alpha_{m_2}$

The second-order sensitivities $\partial^2 S\left(\varphi,\varphi^+;\mathbf{u}_j^{(2)};\boldsymbol{\alpha}\right)/(\partial f_j)(\partial \alpha_{m_2})$, $j=1,\ldots,J_f$, $m_2=1,\ldots,TP$, will ultimately depend on a second-level adjoint function, which will be denoted as $\mathbf{u}_j^{(2)}$, and have the following expressions:

$$
\begin{aligned}
&\text{For } j=1,\ldots,J_f;\ m_2=1,\ldots,J_t: \ \frac{\partial^2 S\left(\varphi,\varphi^+;\mathbf{u}_j^{(2)};\boldsymbol{\alpha}\right)}{\partial f_j \partial t_{m_2}} \\
&= \frac{\partial^2 R\left(\varphi,\varphi^+;\mathbf{u}_j^{(2)};\boldsymbol{\alpha}\right)}{\partial f_j \partial t_{m_2}} = -\int dV \int\limits_{4\pi} d\boldsymbol{\Omega} \int\limits_0^{E_f} dE \, \frac{\partial \Sigma_t(\mathbf{t};\mathbf{r},E)}{\partial t_{m_2}} \\
&\times \left[u_{1,j}^{(2)}(\mathbf{r},\boldsymbol{\Omega},E)\varphi(\mathbf{r},\boldsymbol{\Omega},E) + u_{2,j}^{(2)}(\mathbf{r},\boldsymbol{\Omega},E)\varphi^+(\mathbf{r},\boldsymbol{\Omega},E)\right];
\end{aligned}
\tag{4.186}
$$

For $j = 1, \ldots, J_f;\ m_2 = 1, \ldots, J_s:\ \dfrac{\partial^2 S\left(\varphi, \varphi^+; \mathbf{u}_j^{(2)}; \boldsymbol{\alpha}\right)}{\partial f_j \partial s_{m_2}}$

$$
\begin{aligned}
&= \frac{\partial^2 R\left(\varphi, \varphi^+; \mathbf{u}_j^{(2)}; \boldsymbol{\alpha}\right)}{\partial f_j \partial s_{m_2}} = \int dV \int_{4\pi} d\boldsymbol{\Omega} \int_0^{E_f} dE\, u_{1,j}^{(2)}(\mathbf{r}, \boldsymbol{\Omega}, E) \\
&\times \int_{4\pi} d\boldsymbol{\Omega}' \int_0^{E_f} dE'\, \varphi(\mathbf{r}, \boldsymbol{\Omega}', E') \frac{\partial \Sigma_s(\mathbf{s}; \mathbf{r}, E' \to E, \boldsymbol{\Omega}' \to \boldsymbol{\Omega})}{\partial s_{m_2}} \\
&+ \int dV \int_{4\pi} d\boldsymbol{\Omega} \int_0^{E_f} dE\, u_{2,j}^{(2)}(\mathbf{r}, \boldsymbol{\Omega}, E) \\
&\times \int_{4\pi} d\boldsymbol{\Omega}' \int_0^{E_f} dE'\, \varphi^+(\mathbf{r}, \boldsymbol{\Omega}', E') \frac{\partial \Sigma_s(\mathbf{s}; \mathbf{r}, E \to E', \boldsymbol{\Omega} \to \boldsymbol{\Omega}')}{\partial s_{m_2}};
\end{aligned}
\tag{4.187}
$$

For $j = 1, \ldots, J_f;\ m_2 = 1, \ldots, J_f:\ \dfrac{\partial^2 S\left(\varphi, \varphi^+; \mathbf{u}_j^{(2)}; \boldsymbol{\alpha}\right)}{\partial f_j \partial f_{m_2}}$

$$
\begin{aligned}
&= \frac{\partial^2 R\left(\varphi, \varphi^+; \mathbf{u}_j^{(2)}; \boldsymbol{\alpha}\right)}{\partial f_j \partial f_{m_2}} = \int dV \int_{4\pi} d\boldsymbol{\Omega} \int_0^{\infty} dE\, \varphi^+(\mathbf{r}, \boldsymbol{\Omega}, E) \chi(\mathbf{p}; \mathbf{r}, E) \\
&\times \int_{4\pi} d\boldsymbol{\Omega}' \int_0^{\infty} dE'\, \varphi(\mathbf{r}, \boldsymbol{\Omega}', E') \frac{\partial^2 \left[\nu \Sigma_f(\mathbf{f}; \mathbf{r}, E')\right]}{\partial f_j \partial f_{m_2}} \\
&+ \int dV \int_{4\pi} d\boldsymbol{\Omega} \int_0^{E_f} dE\, u_{1,j}^{(2)}(\mathbf{r}, \boldsymbol{\Omega}, E) \chi(\mathbf{p}; \mathbf{r}, E) \\
&\times \int_{4\pi} d\boldsymbol{\Omega}' \int_0^{E_f} dE'\, \varphi(\mathbf{r}, \boldsymbol{\Omega}', E') \frac{\partial \left[\nu \Sigma_f(\mathbf{f}; \mathbf{r}, E')\right]}{\partial f_{m_2}} \\
&+ \int dV \int_{4\pi} d\boldsymbol{\Omega} \int_0^{E_f} dE\, u_{2,j}^{(2)}(\mathbf{r}, \boldsymbol{\Omega}, E) \frac{\partial \left[\nu \Sigma_f(\mathbf{f}; \mathbf{r}, E)\right]}{\partial f_{m_2}} \\
&\times \int_{4\pi} d\boldsymbol{\Omega}' \int_0^{E_f} dE' \chi(\mathbf{p}; \mathbf{r}, E') \varphi^+(\mathbf{r}, \boldsymbol{\Omega}', E');
\end{aligned}
\tag{4.188}
$$

$$\text{For } j=1,\ldots,J_f;\ m_2=1,\ldots,J_p:\ \frac{\partial^2 S\left(\varphi,\varphi^+;\mathbf{u}_j^{(2)};\boldsymbol{\alpha}\right)}{\partial f_j \partial p_{m_2}}$$

$$= \frac{\partial^2 R\left(\varphi,\varphi^+;\mathbf{u}_j^{(2)};\boldsymbol{\alpha}\right)}{\partial f_j \partial p_{m_2}} = \int dV \int_{4\pi} d\boldsymbol{\Omega} \int_0^{E_f} dE\,\varphi^+(\mathbf{r},\boldsymbol{\Omega},E)\frac{\partial \chi(\mathbf{p};\mathbf{r},E)}{\partial p_{m_2}}$$

$$\times \int_{4\pi} d\boldsymbol{\Omega}' \int_0^{E_f} dE'\,\varphi(\mathbf{r},\boldsymbol{\Omega}',E')\frac{\partial\left[\nu\Sigma_f(\mathbf{f};\mathbf{r},E')\right]}{\partial f_j}$$

$$+ \int dV \int_{4\pi} d\boldsymbol{\Omega} \int_0^{E_f} dE\,u_{1,j}^{(2)}(\mathbf{r},\boldsymbol{\Omega},E)\frac{\partial \chi(\mathbf{p};\mathbf{r},E)}{\partial p_{m_2}} \tag{4.189}$$

$$\times \int_{4\pi} d\boldsymbol{\Omega}' \int_0^{E_f} dE'\,\varphi(\mathbf{r},\boldsymbol{\Omega}',E')\,\nu\Sigma_f(\mathbf{f};\mathbf{r},E')$$

$$+ \int dV \int_{4\pi} d\boldsymbol{\Omega} \int_0^{E_f} dE\,u_{2,j}^{(2)}(\mathbf{r},\boldsymbol{\Omega},E)\nu\Sigma_f(\mathbf{f};\mathbf{r},E)$$

$$\times \int_{4\pi} d\boldsymbol{\Omega}' \int_0^{\infty} dE'\,\varphi^+(\mathbf{r},\boldsymbol{\Omega}',E')\frac{\partial \chi(\mathbf{p};\mathbf{r},E')}{\partial p_{m_2}};$$

$$\text{For } j=1,\ldots,J_f;\ m_2=1,\ldots,J_q:\ \frac{\partial^2 S\left(\varphi,\varphi^+;\mathbf{u}_j^{(2)};\boldsymbol{\alpha}\right)}{\partial f_j \partial q_{m_2}}$$

$$= \frac{\partial^2 R\left(\varphi,\varphi^+;\mathbf{u}_j^{(2)};\boldsymbol{\alpha}\right)}{\partial f_j \partial q_{m_2}} = \int dV \int_{4\pi} d\boldsymbol{\Omega} \int_0^{E_f} dE\,u_{1,j}^{(2)}(\mathbf{r},\boldsymbol{\Omega},E)\frac{\partial Q(\mathbf{q};\mathbf{r},\boldsymbol{\Omega},E)}{\partial q_{m_2}}; \tag{4.190}$$

$$\text{For } j=1,\ldots,J_f;\ m_2=1,\ldots,J_k:\ \frac{\partial^2 S\left(\varphi,\varphi^+;\mathbf{u}_j^{(2)};\boldsymbol{\alpha}\right)}{\partial f_j \partial k_{m_2}}$$

$$= \frac{\partial^2 R\left(\varphi,\varphi^+;\mathbf{u}_j^{(2)};\boldsymbol{\alpha}\right)}{\partial f_j \partial k_{m_2}} = \int dV \int_{4\pi} d\boldsymbol{\Omega} \int_0^{E_f} dE\,u_{2,j}^{(2)}(\mathbf{r},\boldsymbol{\Omega},E)\frac{\partial Q^+(\mathbf{k};\mathbf{r},\boldsymbol{\Omega},E)}{\partial k_{m_2}}. \tag{4.191}$$

The components of the second-level adjoint function $\mathbf{u}_j^{(2)}(\mathbf{r},\boldsymbol{\Omega},E) \triangleq \left[u_{1,j}^{(2)}(\mathbf{r},\boldsymbol{\Omega},E), u_{2,j}^{(2)}(\mathbf{r},\boldsymbol{\Omega},E)\right]^\dagger$, which appear in the expressions provided in (4.186)–(4.191), are the solutions of the following 2nd-LASS, for $j = 1, \ldots, J_f$:

$$L^{+}(\boldsymbol{\alpha})u_{1,j}^{(2)}(\mathbf{r},\boldsymbol{\Omega},E)$$
$$=\frac{\partial\left[\nu\Sigma_f(\mathbf{f};\mathbf{r},E)\right]}{\partial f_j}\int_{4\pi}d\boldsymbol{\Omega}'\int_0^{E_f}dE'\chi(\mathbf{p};\mathbf{r},E\to E')\varphi^{+}(\mathbf{r},\boldsymbol{\Omega}',E'), \tag{4.192}$$

$$L(\boldsymbol{\alpha})\,u_{2,j}^{(2)}(\mathbf{r},\boldsymbol{\Omega},E)$$
$$=\int_{4\pi}d\boldsymbol{\Omega}'\int_0^{E_f}dE'\,\chi(\mathbf{p};\mathbf{r},E'\to E)\varphi(\mathbf{r},\boldsymbol{\Omega}',E')\frac{\partial\left[\nu\Sigma_f(\mathbf{f};\mathbf{r},E')\right]}{\partial f_j}, \tag{4.193}$$

subject to following boundary conditions:

$$\left.\begin{array}{l} u_{1,j}^{(2)}(\mathbf{r}_s,\boldsymbol{\Omega},E)=0,\ \ \mathbf{r}_s\in\partial V,\ \ \boldsymbol{\Omega}\cdot\mathbf{n}>0;0<E<E_f \\ u_{2,j}^{(2)}(\mathbf{r}_s,\boldsymbol{\Omega},E)=0,\ \ \mathbf{r}_s\in\partial V,\ \ \boldsymbol{\Omega}\cdot\mathbf{n}<0;0<E<E_f \end{array}\right\} \tag{4.194}$$

The expressions of the second-order sensitivities provided in (4.186) must be identical to those provided in (4.169), i.e.,

$$\text{For } j=1,\ldots,J_f;\ \ k=1,\ldots,J_t:\ \ \frac{\partial^2 S\left(\varphi,\varphi^{+};\mathbf{u}_j^{(2)};\boldsymbol{\alpha}\right)}{\partial f_j\partial t_k}$$
$$=-\int dV\int_{4\pi}d\boldsymbol{\Omega}\int_0^{E_f}dE\,\frac{\partial\Sigma_t(\mathbf{t};\mathbf{r},E)}{\partial t_k}$$
$$\times\left[u_{1,j}^{(2)}(\mathbf{r},\boldsymbol{\Omega},E)\varphi(\mathbf{r},\boldsymbol{\Omega},E)+u_{2,j}^{(2)}(\mathbf{r},\boldsymbol{\Omega},E)\varphi^{+}(\mathbf{r},\boldsymbol{\Omega},E)\right]$$
$$=\frac{\partial^2 S\left(\varphi,\varphi^{+};\boldsymbol{\psi}_j^{(2)};\boldsymbol{\alpha}\right)}{\partial t_k\partial f_j}=\int dV\int_{4\pi}d\boldsymbol{\Omega}\int_0^{E_f}dE\,\psi_{1,k}^{(2)}(\mathbf{r},\boldsymbol{\Omega},E)\chi(\mathbf{p};\mathbf{r},E)$$
$$\times\int_{4\pi}d\boldsymbol{\Omega}'\int_0^{E_f}dE'\,\varphi(\mathbf{r},\boldsymbol{\Omega}',E')\frac{\partial\left[\nu\Sigma_f(\mathbf{f};\mathbf{r},E')\right]}{\partial f_j}$$
$$+\int dV\int_{4\pi}d\boldsymbol{\Omega}\int_0^{E_f}dE\,\psi_{2,k}^{(2)}(\mathbf{r},\boldsymbol{\Omega},E)\frac{\partial\left[\nu\Sigma_f(\mathbf{f};\mathbf{r},E)\right]}{\partial f_j}$$
$$\times\int_{4\pi}d\boldsymbol{\Omega}'\int_0^{E_f}dE'\chi(\mathbf{p};\mathbf{r},E')\varphi^{+}(\mathbf{r},\boldsymbol{\Omega}',E'). \tag{4.195}$$

The relation shown in (4.195) provides an independent path for the mutual verification of the solutions $\mathbf{u}_j^{(2)}$ and $\boldsymbol{\psi}_j^{(2)}$.

Furthermore, the expressions of the second-order sensitivities provided in (4.187) must be identical to those provided in (4.178), i.e.,

$$
\begin{aligned}
&\text{For } j=1,\ldots,J_f;\ k=1,\ldots,J_s: \ \frac{\partial^2 S\left(\varphi,\varphi^+;\mathbf{u}_j^{(2)};\boldsymbol{\alpha}\right)}{\partial f_j \partial s_k} \\
&= \int dV \int_{4\pi} d\boldsymbol{\Omega} \int_0^{E_f} dE\, u_{1,j}^{(2)}(\mathbf{r},\boldsymbol{\Omega},E) \\
&\quad \times \int_{4\pi} d\boldsymbol{\Omega}' \int_0^{E_f} dE'\, \varphi(\mathbf{r},\boldsymbol{\Omega}',E') \frac{\partial \Sigma_s(\mathbf{s};\mathbf{r},E'\to E,\boldsymbol{\Omega}'\to\boldsymbol{\Omega})}{\partial s_k} \\
&+ \int dV \int_{4\pi} d\boldsymbol{\Omega} \int_0^{E_f} dE\, u_{2,j}^{(2)}(\mathbf{r},\boldsymbol{\Omega},E) \\
&\quad \times \int_{4\pi} d\boldsymbol{\Omega}' \int_0^{E_f} dE'\, \varphi^+(\mathbf{r},\boldsymbol{\Omega}',E') \frac{\partial \Sigma_s(\mathbf{s};\mathbf{r},E\to E',\boldsymbol{\Omega}\to\boldsymbol{\Omega}')}{\partial s_k} \\
&= \frac{\partial^2 S\left(\varphi,\varphi^+;\boldsymbol{\theta}_j^{(2)};\boldsymbol{\alpha}\right)}{\partial s_k \partial f_j} = \int dV \int_{4\pi} d\boldsymbol{\Omega} \int_0^{E_f} dE\, \theta_{1,k}^{(2)}(\mathbf{r},\boldsymbol{\Omega},E)\chi(\mathbf{p};\mathbf{r},E) \\
&\quad \times \int_{4\pi} d\boldsymbol{\Omega}' \int_0^{E_f} dE'\, \varphi(\mathbf{r},\boldsymbol{\Omega}',E') \frac{\partial\left[\nu\Sigma_f(\mathbf{f};\mathbf{r},E')\right]}{\partial f_j} \\
&+ \int dV \int_{4\pi} d\boldsymbol{\Omega} \int_0^{E_f} dE\, \theta_{2,k}^{(2)}(\mathbf{r},\boldsymbol{\Omega},E) \frac{\partial\left[\nu\Sigma_f(\mathbf{f};\mathbf{r},E)\right]}{\partial f_j} \\
&\quad \times \int_{4\pi} d\boldsymbol{\Omega}' \int_0^{E_f} dE'\, \chi(\mathbf{p};\mathbf{r},E')\varphi^+(\mathbf{r},\boldsymbol{\Omega}',E').
\end{aligned}
\tag{4.196}
$$

The relation shown in (4.196) provides an independent path for the mutual verification of the solutions $\mathbf{u}_j^{(2)}$ and $\boldsymbol{\theta}_j^{(2)}$.

4.5.2.4 Determination of the Second-Order Sensitivities of the Form $\partial^2 S\left(\boldsymbol{\varphi},\boldsymbol{\varphi}^+;\mathbf{w}_j^{(2)};\boldsymbol{\alpha}\right)/\partial p_j \partial \boldsymbol{\alpha}_{m_2}$

The second-order sensitivities $\partial^2 S\left(\varphi,\varphi^+;\mathbf{w}_j^{(2)};\boldsymbol{\alpha}\right)/(\partial p_j)(\partial \alpha_{m_2})$, $j = 1, \ldots, J_p$, $m_2 = 1, \ldots, TP$, will ultimately depend on a second-level adjoint function, which will be denoted as $\mathbf{w}_j^{(2)}(\mathbf{r},\boldsymbol{\Omega},E) \triangleq \left[w_{1,j}^{(2)}(\mathbf{r},\boldsymbol{\Omega},E), w_{2,j}^{(2)}(\mathbf{r},\boldsymbol{\Omega},E)\right]^\dagger$, and have the following expression:

$$
\begin{aligned}
&\text{For } j = 1, \ldots, J_p;\ m_2 = 1, \ldots, J_t: \ \frac{\partial^2 S\left(\varphi,\varphi^+;\mathbf{w}_j^{(2)};\boldsymbol{\alpha}\right)}{\partial p_j \partial t_{m_2}} \\
&= \frac{\partial^2 R\left(\varphi,\varphi^+;\mathbf{w}_j^{(2)};\boldsymbol{\alpha}\right)}{\partial p_j \partial t_{m_2}} = -\int dV \int_{4\pi} d\boldsymbol{\Omega} \int_0^{E_f} dE \frac{\partial \Sigma_t(\mathbf{t};\mathbf{r},E)}{\partial t_{m_2}} \\
&\times \left[w_{1,j}^{(2)}(\mathbf{r},\boldsymbol{\Omega},E)\varphi(\mathbf{r},\boldsymbol{\Omega},E) + w_{2,j}^{(2)}(\mathbf{r},\boldsymbol{\Omega},E)\varphi^+(\mathbf{r},\boldsymbol{\Omega},E)\right];
\end{aligned}
\tag{4.197}
$$

$$
\begin{aligned}
&\text{For } j = 1, \ldots, J_p;\ m_2 = 1, \ldots, J_s: \ \frac{\partial^2 S\left(\varphi,\varphi^+;\mathbf{w}_j^{(2)};\boldsymbol{\alpha}\right)}{\partial p_j \partial s_{m_2}} \\
&= \frac{\partial^2 R\left(\varphi,\varphi^+;\mathbf{w}_j^{(2)};\boldsymbol{\alpha}\right)}{\partial p_j \partial s_{m_2}} = \int dV \int_{4\pi} d\boldsymbol{\Omega} \int_0^{E_f} dE\, w_{1,j}^{(2)}(\mathbf{r},\boldsymbol{\Omega},E) \\
&\times \int_{4\pi} d\boldsymbol{\Omega}' \int_0^{E_f} dE'\, \varphi(\mathbf{r},\boldsymbol{\Omega}',E') \frac{\partial \Sigma_s(\mathbf{s};\mathbf{r},E' \to E, \boldsymbol{\Omega}' \to \boldsymbol{\Omega})}{\partial s_{m_2}} \\
&+ \int dV \int_{4\pi} d\boldsymbol{\Omega} \int_0^{E_f} dE\, w_{2,j}^{(2)}(\mathbf{r},\boldsymbol{\Omega},E) \\
&\times \int_{4\pi} d\boldsymbol{\Omega}' \int_0^{E_f} dE'\, \varphi^+(\mathbf{r},\boldsymbol{\Omega}',E') \frac{\partial \Sigma_s(\mathbf{s};\mathbf{r},E \to E', \boldsymbol{\Omega} \to \boldsymbol{\Omega}')}{\partial s_{m_2}};
\end{aligned}
\tag{4.198}
$$

For $j = 1, \ldots, J_p;\ m_2 = 1, \ldots, J_q$: $\dfrac{\partial^2 S\left(\varphi, \varphi^+; \mathbf{w}_j^{(2)}; \boldsymbol{\alpha}\right)}{\partial p_j \partial q_{m_2}}$

$$= \frac{\partial^2 R\left(\varphi, \varphi^+; \mathbf{w}_j^{(2)}; \boldsymbol{\alpha}\right)}{\partial p_j \partial q_{m_2}} = \int dV \int_{4\pi} d\boldsymbol{\Omega} \int_0^{E_f} dE\, w_{1,j}^{(2)}(\mathbf{r}, \boldsymbol{\Omega}, E) \frac{\partial Q(\mathbf{q}; \mathbf{r}, \boldsymbol{\Omega}, E)}{\partial q_{m_2}}; \tag{4.199}$$

For $j = 1, \ldots, J_p;\ m_2 = 1, \ldots, J_k$: $\dfrac{\partial^2 S\left(\varphi, \varphi^+; \mathbf{w}_j^{(2)}; \boldsymbol{\alpha}\right)}{\partial p_j \partial k_{m_2}}$

$$= \frac{\partial^2 R\left(\varphi, \varphi^+; \mathbf{w}_j^{(2)}; \boldsymbol{\alpha}\right)}{\partial p_j \partial k_{m_2}} = \int dV \int_{4\pi} d\boldsymbol{\Omega} \int_0^{E_f} dE\, w_{2,j}^{(2)}(\mathbf{r}, \boldsymbol{\Omega}, E) \frac{\partial Q^+(\mathbf{k}; \mathbf{r}, \boldsymbol{\Omega}, E)}{\partial k_{m_2}}; \tag{4.200}$$

For $j = 1, \ldots, J_p;\ m_2 = 1, \ldots, J_f$: $\dfrac{\partial^2 S\left(\varphi, \varphi^+; \mathbf{w}_j^{(2)}; \boldsymbol{\alpha}\right)}{\partial p_j \partial f_{m_2}}$

$$\begin{aligned} &= \frac{\partial^2 R\left(\varphi, \varphi^+; \mathbf{w}_j^{(2)}; \boldsymbol{\alpha}\right)}{\partial p_j \partial f_{m_2}} = \int dV \int_{4\pi} d\boldsymbol{\Omega} \int_0^{E_f} dE\, \varphi^+(\mathbf{r}, \boldsymbol{\Omega}, E) \frac{\partial \chi(\mathbf{p}; \mathbf{r}, E)}{\partial p_j} \\ &\quad \times \int_{4\pi} d\boldsymbol{\Omega}' \int_0^{E_f} dE'\, \varphi(\mathbf{r}, \boldsymbol{\Omega}', E') \frac{\partial \left[\nu\Sigma_f(\mathbf{f}; \mathbf{r}, E')\right]}{\partial f_{m_2}} \\ &+ \int dV \int_{4\pi} d\boldsymbol{\Omega} \int_0^{E_f} dE\, w_{1,j}^{(2)}(\mathbf{r}, \boldsymbol{\Omega}, E) \chi(\mathbf{p}; \mathbf{r}, E) \\ &\quad \times \int_{4\pi} d\boldsymbol{\Omega}' \int_0^{E_f} dE'\, \varphi(\mathbf{r}, \boldsymbol{\Omega}', E') \frac{\partial \left[\nu\Sigma_f(\mathbf{f}; \mathbf{r}, E')\right]}{\partial f_{m_2}} \\ &+ \int dV \int_{4\pi} d\boldsymbol{\Omega} \int_0^{E_f} dE\, w_{2,j}^{(2)}(\mathbf{r}, \boldsymbol{\Omega}, E) \frac{\partial \left[\nu\Sigma_f(\mathbf{f}; \mathbf{r}, E)\right]}{\partial f_{m_2}} \\ &\quad \times \int_{4\pi} d\boldsymbol{\Omega}' \int_0^{E_f} dE' \chi(\mathbf{p}; \mathbf{r}, E \to E') \varphi^+(\mathbf{r}, \boldsymbol{\Omega}', E'); \end{aligned} \tag{4.201}$$

For $j=1,\ldots,J_p;\ m_2=1,\ldots,J_p:\ \dfrac{\partial^2 S\left(\varphi,\varphi^+;\mathbf{w}_j^{(2)};\boldsymbol{\alpha}\right)}{\partial p_j \partial p_{m_2}}$

$$
\begin{aligned}
&= \frac{\partial^2 R\left(\varphi,\varphi^+;\mathbf{w}_j^{(2)};\boldsymbol{\alpha}\right)}{\partial p_j \partial p_{m_2}} = \int dV \int_{4\pi} d\boldsymbol{\Omega} \int_0^{E_f} dE\, \varphi^+(\mathbf{r},\boldsymbol{\Omega},E) \frac{\partial^2 \chi(\mathbf{p};\mathbf{r},E)}{\partial p_j \partial p_{m_2}} \\
&\times \int_{4\pi} d\boldsymbol{\Omega}' \int_0^{E_f} dE'\, \nu\Sigma_f(\mathbf{f};\mathbf{r},E')\varphi(\mathbf{r},\boldsymbol{\Omega}',E') + \int dV \int_{4\pi} d\boldsymbol{\Omega} \int_0^{E_f} dE\, w_{1,j}^{(2)}(\mathbf{r},\boldsymbol{\Omega},E) \\
&\times \frac{\partial \chi(\mathbf{p};\mathbf{r},E)}{\partial p_{m_2}} \int_{4\pi} d\boldsymbol{\Omega}' \int_0^{E_f} dE'\, \varphi(\mathbf{r},\boldsymbol{\Omega}',E')\, \nu\Sigma_f(\mathbf{f};\mathbf{r},E') \\
&+ \int dV \int_{4\pi} d\boldsymbol{\Omega} \int_0^{E_f} dE\, w_{2,j}^{(2)}(\mathbf{r},\boldsymbol{\Omega},E)\, \nu\Sigma_f(\mathbf{f};\mathbf{r},E) \\
&\times \int_{4\pi} d\boldsymbol{\Omega}' \int_0^{E_f} dE'\, \varphi^+(\mathbf{r},\boldsymbol{\Omega}',E') \frac{\partial \chi(\mathbf{p};\mathbf{r},E')}{\partial p_{m_2}};
\end{aligned}
\tag{4.202}
$$

For $j=1,\ldots,J_p;\ m_2=1,\ldots,J_q:\ \dfrac{\partial^2 S\left(\varphi,\varphi^+;\mathbf{w}_j^{(2)};\boldsymbol{\alpha}\right)}{\partial p_j \partial q_{m_2}}$

$$
= \frac{\partial^2 R\left(\varphi,\varphi^+;\mathbf{w}_j^{(2)};\boldsymbol{\alpha}\right)}{\partial p_j \partial q_{m_2}} = \int dV \int_{4\pi} d\boldsymbol{\Omega} \int_0^{E_f} dE\, w_{1,j}^{(2)}(\mathbf{r},\boldsymbol{\Omega},E) \frac{\partial Q(\mathbf{q};\mathbf{r},\boldsymbol{\Omega},E)}{\partial q_{m_2}};
\tag{4.203}
$$

For $j=1,\ldots,J_p;\ m_2=1,\ldots,J_k:\ \dfrac{\partial^2 S\left(\varphi,\varphi^+;\mathbf{w}_j^{(2)};\boldsymbol{\alpha}\right)}{\partial p_j \partial k_{m_2}}$

$$
= \frac{\partial^2 R\left(\varphi,\varphi^+;\mathbf{w}_j^{(2)};\boldsymbol{\alpha}\right)}{\partial p_j \partial k_{m_2}} = \int dV \int_{4\pi} d\boldsymbol{\Omega} \int_0^{E_f} dE\, w_{2,j}^{(2)}(\mathbf{r},\boldsymbol{\Omega},E) \frac{\partial Q^+(\mathbf{k};\mathbf{r},\boldsymbol{\Omega},E)}{\partial k_{m_2}}.
\tag{4.204}
$$

The components of the second-level adjoint function $\mathbf{w}_j^{(2)}(\mathbf{r}, \boldsymbol{\Omega}, E) \triangleq \left[w_{1,j}^{(2)}(\mathbf{r}, \boldsymbol{\Omega}, E), w_{2,j}^{(2)}(\mathbf{r}, \boldsymbol{\Omega}, E)\right]^{\dagger}$, which appear in (4.197)–(4.204), are the solutions of the following 2nd-LASS, for $j = 1, \ldots, J_p$:

$$L^{+}(\boldsymbol{\alpha}) w_{1,j}^{(2)}(\mathbf{r}, \boldsymbol{\Omega}, E) = \nu\Sigma_f(\mathbf{f}; \mathbf{r}, E) \int_{4\pi} d\Omega' \int_0^{E_f} dE' \frac{\partial \chi(\mathbf{p}; \mathbf{r}, E \to E')}{\partial p_j} \varphi^{+}(\mathbf{r}, \boldsymbol{\Omega}', E'), \tag{4.205}$$

$$L(\boldsymbol{\alpha})\, w_{2,j}^{(2)}(\mathbf{r}, \boldsymbol{\Omega}, E) = \int_{4\pi} d\Omega' \int_0^{E_f} dE' \frac{\partial \chi(\mathbf{p}; \mathbf{r}, E' \to E)}{\partial p_j} \nu\Sigma_f(\mathbf{f}; \mathbf{r}, E') \varphi(\mathbf{r}, \boldsymbol{\Omega}', E'), \tag{4.206}$$

subject to the following boundary:

$$\left. \begin{aligned} & w_{1,j}^{(2)}(\mathbf{r}_s, \boldsymbol{\Omega}, E) = 0, \quad \mathbf{r}_s \in \partial V, \quad \boldsymbol{\Omega} \cdot \mathbf{n} > 0; 0 < E < E_f \\ & w_{2,j}^{(2)}(\mathbf{r}_s, \boldsymbol{\Omega}, E) = 0, \quad \mathbf{r}_s \in \partial V, \quad \boldsymbol{\Omega} \cdot \mathbf{n} < 0; \; 0 < E < E_f \end{aligned} \right\} \tag{4.207}$$

As in the previous sections, the operators appearing on the left side of the 2nd-LASS defined by (4.205) and (4.206) for the second-level adjoint function $\mathbf{w}_j^{(2)}(\mathbf{r}, \boldsymbol{\Omega}, E)$ are the same operators as appearing on the respective left sides of the 2nd-LASS for the second-level adjoint function $\boldsymbol{\theta}_j^{(2)}$, $\boldsymbol{\psi}_j^{(2)}$, and $\mathbf{u}_j^{(2)}$; all of these second-level adjoint functions are subject to similar boundary conditions. Only the source terms on the right sides of the respective 2nd-LASS differ from each other. Therefore, the same "forward" and "adjoint" software packages can be used for solving numerically the various forward and adjoint equations underlying the 1st-LASS and the 2nd-LASS.

The expressions of the second-order sensitivities provided in (4.197) must be identical to those provided in (4.170), i.e.,

For $j = 1, \ldots, J_p;\ k = 1, \ldots, J_t : \dfrac{\partial^2 S\left(\varphi, \varphi^+; \mathbf{w}_j^{(2)}; \boldsymbol{\alpha}\right)}{\partial p_j \partial t_k}$

$$
\begin{aligned}
&= -\int dV \int_{4\pi} d\boldsymbol{\Omega} \int_0^{E_f} dE \frac{\partial \Sigma_t(\mathbf{t}; \mathbf{r}, E)}{\partial t_k} \\
&\times \left[w_{1,j}^{(2)}(\mathbf{r}, \boldsymbol{\Omega}, E)\varphi(\mathbf{r}, \boldsymbol{\Omega}, E) + w_{2,j}^{(2)}(\mathbf{r}, \boldsymbol{\Omega}, E)\varphi^+(\mathbf{r}, \boldsymbol{\Omega}, E) \right] \\
&= \frac{\partial^2 S\left(\varphi, \varphi^+; \boldsymbol{\psi}_j^{(2)}; \boldsymbol{\alpha}\right)}{\partial t_k \partial p_j} = \int dV \int_{4\pi} d\boldsymbol{\Omega} \int_0^{E_f} dE\, \psi_{1,k}^{(2)}(\mathbf{r}, \boldsymbol{\Omega}, E) \\
&\times \frac{\partial \chi(\mathbf{p}; \mathbf{r}, E)}{\partial p_j} \int_{4\pi} d\boldsymbol{\Omega}' \int_0^{E_f} dE'\, \varphi(\mathbf{r}, \boldsymbol{\Omega}', E') \left[\nu\Sigma_f(\mathbf{f}; \mathbf{r}, E') \right] \\
&+ \int dV \int_{4\pi} d\boldsymbol{\Omega} \int_0^{E_f} dE\, \psi_{2,k}^{(2)}(\mathbf{r}, \boldsymbol{\Omega}, E) \nu\Sigma_f(\mathbf{f}; \mathbf{r}, E) \\
&\times \int_{4\pi} d\boldsymbol{\Omega}' \int_0^{E_f} dE'\, \varphi^+(\mathbf{r}, \boldsymbol{\Omega}', E') \frac{\partial \chi(\mathbf{p}; \mathbf{r}, E')}{\partial p_j}.
\end{aligned} \tag{4.208}
$$

The relation expressed by (4.208) provides an independent mutual verification of the second-level adjoint functions $\mathbf{w}_j^{(2)}$ and $\boldsymbol{\psi}_j^{(2)}$.

Furthermore, the expressions of the second-order sensitivities provided in (4.198) must be identical to those provided in (4.179), i.e.,

$$
\text{For } j = 1, \ldots, J_p;\ k = 1, \ldots, J_s: \quad \frac{\partial^2 S\left(\varphi, \varphi^+; \mathbf{w}_j^{(2)}; \boldsymbol{\alpha}\right)}{\partial p_j \partial s_k}
$$

$$
\begin{aligned}
&= \int dV \int_{4\pi} d\boldsymbol{\Omega} \int_0^{E_f} dE\, w_{1,j}^{(2)}(\mathbf{r}, \boldsymbol{\Omega}, E) \\
&\times \int_{4\pi} d\boldsymbol{\Omega}' \int_0^{E_f} dE'\, \varphi(\mathbf{r}, \boldsymbol{\Omega}', E') \frac{\partial \Sigma_s(\mathbf{s}; \mathbf{r}, E' \to E, \boldsymbol{\Omega}' \to \boldsymbol{\Omega})}{\partial s_k} \\
&+ \int dV \int_{4\pi} d\boldsymbol{\Omega} \int_0^{E_f} dE\, w_{2,j}^{(2)}(\mathbf{r}, \boldsymbol{\Omega}, E) \\
&\times \int_{4\pi} d\boldsymbol{\Omega}' \int_0^{E_f} dE'\, \varphi^+(\mathbf{r}, \boldsymbol{\Omega}', E') \frac{\partial \Sigma_s(\mathbf{s}; \mathbf{r}, E \to E', \boldsymbol{\Omega} \to \boldsymbol{\Omega}')}{\partial s_k} \\
&= \frac{\partial^2 S\left(\varphi, \varphi^+; \boldsymbol{\theta}_j^{(2)}; \boldsymbol{\alpha}\right)}{\partial s_k \partial p_j} = \int dV \int_{4\pi} d\boldsymbol{\Omega} \int_0^{E_f} dE\, \theta_{1,k}^{(2)}(\mathbf{r}, \boldsymbol{\Omega}, E) \\
&\times \frac{\partial \chi(\mathbf{p}; \mathbf{r}, E)}{\partial p_j} \int_{4\pi} d\boldsymbol{\Omega}' \int_0^{E_f} dE'\, \varphi(\mathbf{r}, \boldsymbol{\Omega}', E') \nu\Sigma_f(\mathbf{f}; \mathbf{r}, E') \\
&+ \int dV \int_{4\pi} d\boldsymbol{\Omega} \int_0^{E_f} dE\, \theta_{2,k}^{(2)}(\mathbf{r}, \boldsymbol{\Omega}, E) \nu\Sigma_f(\mathbf{f}; \mathbf{r}, E) \\
&\times \int_{4\pi} d\boldsymbol{\Omega}' \int_0^{E_f} dE'\, \varphi^+(\mathbf{r}, \boldsymbol{\Omega}', E') \frac{\partial \chi(\mathbf{p}; \mathbf{r}, E \to E')}{\partial p_j}.
\end{aligned}
\tag{4.209}
$$

The relation expressed by (4.209) provides an independent mutual verification of the second-level adjoint functions $\mathbf{w}_j^{(2)}$ and $\boldsymbol{\theta}_j^{(2)}$.

Finally, the expressions of the second-order sensitivities provided in (4.201) must be identical to those provided in (4.189), i.e.,

$$\text{For } j=1,\ldots,J_p;\ k=1,\ldots,J_f: \frac{\partial^2 S\left(\varphi,\varphi^+;\mathbf{w}_j^{(2)};\boldsymbol{\alpha}\right)}{\partial p_j \partial f_k}$$

$$= \int dV \int_{4\pi} d\boldsymbol{\Omega} \int_0^{E_f} dE \varphi^+(\mathbf{r},\boldsymbol{\Omega},E) \frac{\partial \chi(\mathbf{p};\mathbf{r},E)}{\partial p_j}$$

$$\times \int_{4\pi} d\boldsymbol{\Omega}' \int_0^{E_f} dE' \varphi(\mathbf{r},\boldsymbol{\Omega}',E') \frac{\partial\left[\nu\Sigma_f(\mathbf{f};\mathbf{r},E')\right]}{\partial f_k}$$

$$+ \int dV \int_{4\pi} d\boldsymbol{\Omega} \int_0^{E_f} dE w_{1,j}^{(2)}(\mathbf{r},\boldsymbol{\Omega},E) \chi(\mathbf{p};\mathbf{r},E)$$

$$\times \int_{4\pi} d\boldsymbol{\Omega}' \int_0^{E_f} dE' \varphi(\mathbf{r},\boldsymbol{\Omega}',E') \frac{\partial\left[\nu\Sigma_f(\mathbf{f};\mathbf{r},E')\right]}{\partial f_k}$$

$$+ \int dV \int_{4\pi} d\boldsymbol{\Omega} \int_0^{E_f} dE w_{2,j}^{(2)}(\mathbf{r},\boldsymbol{\Omega},E) \frac{\partial\left[\nu\Sigma_f(\mathbf{f};\mathbf{r},E)\right]}{\partial f_k}$$

$$\times \int_{4\pi} d\boldsymbol{\Omega}' \int_0^{E_f} dE' \chi(\mathbf{p};\mathbf{r},E') \varphi^+(\mathbf{r},\boldsymbol{\Omega}',E') \tag{4.210}$$

$$= \frac{\partial^2 S\left(\varphi,\varphi^+;\mathbf{u}_j^{(2)};\boldsymbol{\alpha}\right)}{\partial f_k \partial p_j} = \int dV \int_{4\pi} d\boldsymbol{\Omega} \int_0^{E_f} dE \varphi^+(\mathbf{r},\boldsymbol{\Omega},E) \frac{\partial \chi(\mathbf{p};\mathbf{r},E)}{\partial p_j}$$

$$\times \int_{4\pi} d\boldsymbol{\Omega}' \int_0^{E_f} dE' \varphi(\mathbf{r},\boldsymbol{\Omega}',E') \frac{\partial\left[\nu\Sigma_f(\mathbf{f};\mathbf{r},E')\right]}{\partial f_k}$$

$$+ \int dV \int_{4\pi} d\boldsymbol{\Omega} \int_0^{E_f} dE u_{1,k}^{(2)}(\mathbf{r},\boldsymbol{\Omega},E) \frac{\partial \chi(\mathbf{p};\mathbf{r},E)}{\partial p_j}$$

$$\times \int_{4\pi} d\boldsymbol{\Omega}' \int_0^{E_f} dE' \varphi(\mathbf{r},\boldsymbol{\Omega}',E') \nu\Sigma_f(\mathbf{f};\mathbf{r},E')$$

$$+ \int dV \int_{4\pi} d\boldsymbol{\Omega} \int_0^{E_f} dE u_{2,k}^{(2)}(\mathbf{r},\boldsymbol{\Omega},E) \nu\Sigma_f(\mathbf{f};\mathbf{r},E)$$

$$\times \int_{4\pi} d\boldsymbol{\Omega}' \int_0^{\infty} dE' \varphi^+(\mathbf{r},\boldsymbol{\Omega}',E') \frac{\partial \chi(\mathbf{p};\mathbf{r},E')}{\partial p_j}.$$

The relation shown in (4.210) provides an independent path for the mutual verification of the solutions $\mathbf{w}_j^{(2)}$ and $\mathbf{u}_j^{(2)}$.

4.5.2.5 Determination of the Second-Order Sensitivities of the Form $\partial^2 S\left(\boldsymbol{\varphi}, \boldsymbol{\varphi}^+; \mathbf{g}_j^{(2)}; \boldsymbol{\alpha}\right)/\partial q_j \partial \alpha_{m_2}$

The second-order sensitivities $\partial^2 S\left(\varphi, \varphi^+; \mathbf{g}_j^{(2)}; \boldsymbol{\alpha}\right)/(\partial q_j)(\partial \alpha_{m_2})$, $j = 1, \ldots, J_q$, $m_2 = 1, \ldots, TP$, will ultimately depend on a second-level adjoint function, which will be denoted as $\mathbf{g}_j^{(2)}(\mathbf{r}, \boldsymbol{\Omega}, E) \triangleq \left[g_{1,j}^{(2)}(\mathbf{r}, \boldsymbol{\Omega}, E), g_{2,j}^{(2)}(\mathbf{r}, \boldsymbol{\Omega}, E)\right]^\dagger \equiv \left[0, g_{2,j}^{(2)}(\mathbf{r}, \boldsymbol{\Omega}, E)\right]^\dagger$, which turns out to have its first component identically zero, i.e., $g_{1,j}^{(2)}(\mathbf{r}, \boldsymbol{\Omega}, E) \equiv 0$, $j = 1, \ldots, J_q$. The expressions of $\partial^2 S\left(\varphi, \varphi^+; \mathbf{g}_j^{(2)}; \boldsymbol{\alpha}\right)/(\partial q_j)(\partial \alpha_{m_2})$ are provided below:

$$\text{For } j = 1, \ldots, J_q;\ m_2 = 1, \ldots, J_t: \frac{\partial^2 S\left(\varphi, \varphi^+; \mathbf{g}_j^{(2)}; \boldsymbol{\alpha}\right)}{\partial q_j \partial t_{m_2}} = \frac{\partial^2 R\left(\varphi, \varphi^+; \mathbf{g}_j^{(2)}; \boldsymbol{\alpha}\right)}{\partial q_j \partial t_{m_2}}$$
$$= -\int dV \int_{4\pi} d\boldsymbol{\Omega} \int_0^{E_f} dE\, \frac{\partial \Sigma_t(\mathbf{t}; \mathbf{r}, E)}{\partial t_{m_2}} g_{2,j}^{(2)}(\mathbf{r}, \boldsymbol{\Omega}, E) \varphi^+(\mathbf{r}, \boldsymbol{\Omega}, E); \tag{4.211}$$

$$\text{For } j = 1, \ldots, J_q;\ m_2 = 1, \ldots, J_s: \frac{\partial^2 S\left(\varphi, \varphi^+; \mathbf{g}_j^{(2)}; \boldsymbol{\alpha}\right)}{\partial q_j \partial s_{m_2}}$$
$$= \frac{\partial^2 R\left(\varphi, \varphi^+; \mathbf{g}_j^{(2)}; \boldsymbol{\alpha}\right)}{\partial q_j \partial s_{m_2}} = \int dV \int_{4\pi} d\boldsymbol{\Omega} \int_0^{E_f} dE\, g_{2,j}^{(2)}(\mathbf{r}, \boldsymbol{\Omega}, E) \tag{4.212}$$
$$\times \int_{4\pi} d\boldsymbol{\Omega}' \int_0^{E_f} dE'\, \varphi^+(\mathbf{r}, \boldsymbol{\Omega}', E') \frac{\partial \Sigma_s(\mathbf{s}; \mathbf{r}, E \to E', \boldsymbol{\Omega} \to \boldsymbol{\Omega}')}{\partial s_{m_2}};$$

For $j = 1, \ldots, J_q;\ m_2 = 1, \ldots, J_f:\ \dfrac{\partial^2 S\left(\varphi, \varphi^+; \mathbf{g}_j^{(2)}; \boldsymbol{\alpha}\right)}{\partial q_j \partial f_{m_2}}$

$$= \frac{\partial^2 R\left(\varphi, \varphi^+; \mathbf{g}_j^{(2)}; \boldsymbol{\alpha}\right)}{\partial q_j \partial f_{m_2}} = \int dV \int_{4\pi} d\boldsymbol{\Omega} \int_0^{E_f} dE\, g_{2,j}^{(2)}(\mathbf{r}, \boldsymbol{\Omega}, E) \frac{\partial\left[\nu\Sigma_f(\mathbf{f}; \mathbf{r}, E)\right]}{\partial f_{m_2}}$$
$$\times \int_{4\pi} d\boldsymbol{\Omega}' \int_0^{E_f} dE' \chi(\mathbf{p}; \mathbf{r}, E') \varphi^+(\mathbf{r}, \boldsymbol{\Omega}', E'); \tag{4.213}$$

For $j = 1, \ldots, J_q;\ m_2 = 1, \ldots, J_p:\ \dfrac{\partial^2 S\left(\varphi, \varphi^+; \mathbf{g}_j^{(2)}; \boldsymbol{\alpha}\right)}{\partial q_j \partial p_{m_2}}$

$$= \frac{\partial^2 R\left(\varphi, \varphi^+; \mathbf{g}_j^{(2)}; \boldsymbol{\alpha}\right)}{\partial q_j \partial p_{m_2}} = \int dV \int_{4\pi} d\boldsymbol{\Omega} \int_0^{E_f} dE\, g_{2,j}^{(2)}(\mathbf{r}, \boldsymbol{\Omega}, E) \nu\Sigma_f(\mathbf{f}; \mathbf{r}, E) \tag{4.214}$$
$$\times \int_{4\pi} d\boldsymbol{\Omega}' \int_0^{E_f} dE'\, \varphi^+(\mathbf{r}, \boldsymbol{\Omega}', E') \frac{\partial \chi(\mathbf{p}; \mathbf{r}, E')}{\partial p_{m_2}};$$

For $j = 1, \ldots, J_q;\ m_2 = 1, \ldots, J_q:\ \dfrac{\partial^2 S\left(\varphi, \varphi^+; \mathbf{g}_j^{(2)}; \boldsymbol{\alpha}\right)}{\partial q_j \partial q_{m_2}}$

$$= \frac{\partial^2 R\left(\varphi, \varphi^+; \mathbf{g}_j^{(2)}; \boldsymbol{\alpha}\right)}{\partial q_j \partial q_{m_2}} = \int dV \int_{4\pi} d\boldsymbol{\Omega} \int_0^{E_f} dE\, \varphi^+(\mathbf{r}, \boldsymbol{\Omega}, E) \frac{\partial^2 Q(\mathbf{q}; \mathbf{r}, \boldsymbol{\Omega}, E)}{\partial q_j \partial q_{m_2}}; \tag{4.215}$$

For $j = 1, \ldots, J_q;\ m_2 = 1, \ldots, J_k:\ \dfrac{\partial^2 S\left(\varphi, \varphi^+; \mathbf{g}_j^{(2)}; \boldsymbol{\alpha}\right)}{\partial q_j \partial k_{m_2}}$

$$= \frac{\partial^2 R\left(\varphi, \varphi^+; \mathbf{g}_j^{(2)}; \boldsymbol{\alpha}\right)}{\partial q_j \partial k_{m_2}} = \int dV \int_{4\pi} d\boldsymbol{\Omega} \int_0^{E_f} dE\, g_{2,j}^{(2)}(\mathbf{r}, \boldsymbol{\Omega}, E) \frac{\partial Q^+(\mathbf{k}; \mathbf{r}, \boldsymbol{\Omega}, E)}{\partial k_{m_2}}. \tag{4.216}$$

The non-zero component, $g_{2,j}^{(2)}(\mathbf{r}, \boldsymbol{\Omega}, E)$, of the second-level adjoint function $\mathbf{g}_j^{(2)}(\mathbf{r}, \boldsymbol{\Omega}, E)$, which appears in (4.211)–(4.216), is the solution of the following 2nd-LASS:

$$L(\boldsymbol{\alpha})\, g_{2,j}^{(2)}(\mathbf{r},\boldsymbol{\Omega},E)=\frac{\partial Q(\mathbf{q};\mathbf{r},\boldsymbol{\Omega},E)}{\partial q_j},\; j=1,\ldots,J_q, \tag{4.217}$$

subject to the following boundary condition:

$$g_{2,j}^{(2)}(\mathbf{r}_s,\boldsymbol{\Omega},E)=0,\;\; \mathbf{r}_s\in\partial V,\;\; \boldsymbol{\Omega}\cdot\mathbf{n}<0;\;\; 0<E<E_f. \tag{4.218}$$

The expressions of the second-order sensitivities provided in (4.211) must be identical to those provided in (4.171), i.e.,

$$\text{For } j=1,\ldots,J_q;\;\; k=1,\ldots,J_t:\;\; \frac{\partial^2 S\left(\varphi,\varphi^+;\mathbf{g}_j^{(2)};\boldsymbol{\alpha}\right)}{\partial q_j\partial t_k}$$
$$=-\int dV\int_{4\pi} d\boldsymbol{\Omega}\int_0^{E_f} dE\,\frac{\partial\Sigma_t(\mathbf{t};\mathbf{r},E)}{\partial t_k}\, g_{2,j}^{(2)}(\mathbf{r},\boldsymbol{\Omega},E)\varphi^+(\mathbf{r},\boldsymbol{\Omega},E)$$
$$=\frac{\partial^2 S\left(\varphi,\varphi^+;\boldsymbol{\psi}_j^{(2)};\boldsymbol{\alpha}\right)}{\partial t_k\partial q_j}=\int dV\int_{4\pi} d\boldsymbol{\Omega}\int_0^{E_f} dE\,\psi_{1,k}^{(2)}(\mathbf{r},\boldsymbol{\Omega},E)\frac{\partial Q(\mathbf{q};\mathbf{r},\boldsymbol{\Omega},E)}{\partial q_j}. \tag{4.219}$$

The relation expressed by (4.219) provides an independent mutual verification of the second-level adjoint functions $\mathbf{g}_j^{(2)}$ and $\boldsymbol{\psi}_j^{(2)}$.

The expressions of the second-order sensitivities provided in (4.212) must be identical to those provided in (4.180), i.e.,

$$\text{For } j=1,\ldots,J_q;\;\; k=1,\ldots,J_s:\;\; \frac{\partial^2 S\left(\varphi,\varphi^+;\mathbf{g}_j^{(2)};\boldsymbol{\alpha}\right)}{\partial q_j\partial s_k}$$
$$=\int dV\int_{4\pi} d\boldsymbol{\Omega}\int_0^{E_f} dE\, g_{2,j}^{(2)}(\mathbf{r},\boldsymbol{\Omega},E)$$
$$\times\int_{4\pi} d\boldsymbol{\Omega}'\int_0^{E_f} dE'\,\varphi^+(\mathbf{r},\boldsymbol{\Omega}',E')\frac{\partial\Sigma_s(\mathbf{s};\mathbf{r},E\to E',\boldsymbol{\Omega}\to\boldsymbol{\Omega}')}{\partial s_k}$$
$$=\frac{\partial^2 S\left(\varphi,\varphi^+;\boldsymbol{\theta}_j^{(2)};\boldsymbol{\alpha}\right)}{\partial s_k\partial q_j}=\int dV\int_{4\pi} d\boldsymbol{\Omega}\int_0^{E_f} dE\,\theta_{1,k}^{(2)}(\mathbf{r},\boldsymbol{\Omega},E)\frac{\partial Q(\mathbf{q};\mathbf{r},\boldsymbol{\Omega},E)}{\partial q_j}. \tag{4.220}$$

The relation expressed by (4.220) provides an independent mutual verification of the second-level adjoint functions $\mathbf{g}_j^{(2)}$ and $\boldsymbol{\theta}_j^{(2)}$.

The expressions of the second-order sensitivities provided in (4.213) must be identical to those provided in (4.190), i.e.,

$$\begin{aligned}
&\textit{For } j = 1, \ldots, J_q;\ k = 1, \ldots, J_f: \quad \frac{\partial^2 S\left(\varphi, \varphi^+; \mathbf{g}_j^{(2)}; \boldsymbol{\alpha}\right)}{\partial q_j \partial f_k} \\
&= \int dV \int_{4\pi} d\Omega \int_0^{E_f} dE\, g_{2,j}^{(2)}(\mathbf{r}, \boldsymbol{\Omega}, E) \frac{\partial\left[\nu\Sigma_f(\mathbf{f}; \mathbf{r}, E)\right]}{\partial f_k} \\
&\times \int_{4\pi} d\Omega' \int_0^{E_f} dE' \chi(\mathbf{p}; \mathbf{r}, E') \varphi^+(\mathbf{r}, \boldsymbol{\Omega}', E') = \frac{\partial^2 S\left(\varphi, \varphi^+; \mathbf{u}_j^{(2)}; \boldsymbol{\alpha}\right)}{\partial f_k \partial q_j} \\
&= \int dV \int_{4\pi} d\Omega \int_0^{E_f} dE\, u_{1,k}^{(2)}(\mathbf{r}, \boldsymbol{\Omega}, E) \frac{\partial Q(\mathbf{q}; \mathbf{r}, \boldsymbol{\Omega}, E)}{\partial q_j}.
\end{aligned} \tag{4.221}$$

The relation shown in (4.221) provides an independent path for the mutual verification of the solutions $\mathbf{g}_j^{(2)}$ and $\mathbf{u}_j^{(2)}$.

The expressions of the second-order sensitivities provided in (4.214) must be identical to those provided in (4.203), i.e.,

$$\begin{aligned}
&\text{For } j = 1, \ldots, J_q;\ k = 1, \ldots, J_p: \quad \frac{\partial^2 S\left(\varphi, \varphi^+; \mathbf{g}_j^{(2)}; \boldsymbol{\alpha}\right)}{\partial q_j \partial p_k} \\
&= \int dV \int_{4\pi} d\Omega \int_0^{E_f} dE\, g_{2,j}^{(2)}(\mathbf{r}, \boldsymbol{\Omega}, E) \nu\Sigma_f(\mathbf{f}; \mathbf{r}, E) \\
&\times \int_{4\pi} d\Omega' \int_0^{E_f} dE' \varphi^+(\mathbf{r}, \boldsymbol{\Omega}', E') \frac{\partial \chi(\mathbf{p}; \mathbf{r}, E')}{\partial p_k} = \frac{\partial^2 S\left(\varphi, \varphi^+; \mathbf{w}_j^{(2)}; \boldsymbol{\alpha}\right)}{\partial p_k \partial q_j} \\
&= \int dV \int_{4\pi} d\Omega \int_0^{E_f} dE\, w_{1,k}^{(2)}(\mathbf{r}, \boldsymbol{\Omega}, E) \frac{\partial Q(\mathbf{q}; \mathbf{r}, \boldsymbol{\Omega}, E)}{\partial q_j}.
\end{aligned} \tag{4.222}$$

The relation shown in (4.222) provides an independent path for the mutual verification of the solutions $\mathbf{g}_j^{(2)}$ and $\mathbf{w}_j^{(2)}$.

4.5.2.6 Determination of the Second-Order Sensitivities of the Form $\partial^2 S\left(\boldsymbol{\varphi},\boldsymbol{\varphi}^+;\boldsymbol{\gamma}_j^{(2)};\boldsymbol{\alpha}\right)/\partial k_j \partial \boldsymbol{\alpha}_{m_2}$

The second-order sensitivities $\partial^2 S\left(\varphi,\varphi^+;\boldsymbol{\gamma}_j^{(2)};\boldsymbol{\alpha}\right)/(\partial k_j)(\partial \alpha_{m_2})$, $j = 1, \ldots, J_k$, $m_2 = 1, \ldots, TP$, will ultimately depend on a second-level adjoint function, which will be denoted as $\boldsymbol{\gamma}_j^{(2)}(\mathbf{r},\boldsymbol{\Omega},E) \triangleq \left[\gamma_{1,j}^{(2)}(\mathbf{r},\boldsymbol{\Omega},E),\gamma_{2,j}^{(2)}(\mathbf{r},\boldsymbol{\Omega},E)\right]^\dagger \equiv \left[\gamma_{1,j}^{(2)}(\mathbf{r},\boldsymbol{\Omega},E),0\right]^\dagger$, which turns out to have its second component identically zero, i.e., $\gamma_{2,j}^{(2)}(\mathbf{r},\boldsymbol{\Omega},E) \equiv 0, j = 1, \ldots, J_k$. The expressions of $\partial^2 S\left(\varphi,\varphi^+;\boldsymbol{\gamma}_j^{(2)};\boldsymbol{\alpha}\right)/(\partial k_j)(\partial \alpha_{m_2})$ are provided below:

$$\text{For } j = 1,\ldots,J_k;\ m_2 = 1,\ldots,J_t: \ \frac{\partial^2 S\left(\varphi,\varphi^+;\boldsymbol{\gamma}_j^{(2)};\boldsymbol{\alpha}\right)}{\partial k_j \partial t_{m_2}} = \frac{\partial^2 R\left(\varphi,\varphi^+;\boldsymbol{\gamma}_j^{(2)};\boldsymbol{\alpha}\right)}{\partial k_j \partial t_{m_2}}$$
$$= -\int dV \int_{4\pi} d\Omega \int_0^{E_f} dE \ \frac{\partial \Sigma_t(\mathbf{t};\mathbf{r},E)}{\partial t_{m_2}} \gamma_{1,j}^{(2)}(\mathbf{r},\boldsymbol{\Omega},E)\varphi(\mathbf{r},\boldsymbol{\Omega},E); \tag{4.223}$$

$$\text{For } j = 1,\ldots,J_k;\ m_2 = 1,\ldots,J_s: \ \frac{\partial^2 S\left(\varphi,\varphi^+;\boldsymbol{\gamma}_j^{(2)};\boldsymbol{\alpha}\right)}{\partial k_j \partial s_{m_2}}$$
$$= \frac{\partial^2 R\left(\varphi,\varphi^+;\boldsymbol{\gamma}_j^{(2)};\boldsymbol{\alpha}\right)}{\partial k_j \partial s_{m_2}} = \int dV \int_{4\pi} d\Omega \int_0^{E_f} dE\, \gamma_{1,j}^{(2)}(\mathbf{r},\boldsymbol{\Omega},E) \tag{4.224}$$
$$\times \int_{4\pi} d\Omega' \int_0^{E_f} dE'\, \varphi(\mathbf{r},\boldsymbol{\Omega}',E') \frac{\partial \Sigma_s(\mathbf{s};\mathbf{r},E' \to E,\boldsymbol{\Omega}' \to \boldsymbol{\Omega})}{\partial s_{m_2}};$$

$$\text{For } j = 1,\ldots,J_k;\ m_2 = 1,\ldots,J_f: \ \frac{\partial^2 S\left(\varphi,\varphi^+;\boldsymbol{\gamma}_j^{(2)};\boldsymbol{\alpha}\right)}{\partial k_j \partial f_{m_2}}$$
$$= \frac{\partial^2 R\left(\varphi,\varphi^+;\boldsymbol{\gamma}_j^{(2)};\boldsymbol{\alpha}\right)}{\partial k_j \partial f_{m_2}} = \int dV \int_{4\pi} d\Omega \int_0^{E_f} dE\, \gamma_{1,j}^{(2)}(\mathbf{r},\boldsymbol{\Omega},E) \tag{4.225}$$
$$\times \int_{4\pi} d\Omega' \int_0^{E_f} dE'\, \chi(\mathbf{p};\mathbf{r},E' \to E)\varphi(\mathbf{r},\boldsymbol{\Omega}',E') \frac{\partial \left[\nu\Sigma_f(\mathbf{f};\mathbf{r},E')\right]}{\partial f_{m_2}};$$

For $j = 1, \ldots, J_k;\ m_2 = 1, \ldots, J_p$: $\dfrac{\partial^2 S\left(\varphi, \varphi^+; \boldsymbol{\gamma}_j^{(2)}; \boldsymbol{\alpha}\right)}{\partial k_j \partial p_{m_2}}$

$$= \frac{\partial^2 R\left(\varphi, \varphi^+; \boldsymbol{\gamma}_j^{(2)}; \boldsymbol{\alpha}\right)}{\partial k_j \partial p_{m_2}} = \int dV \int_{4\pi} d\boldsymbol{\Omega} \int_0^{E_f} dE\, \gamma_{1,j}^{(2)}(\mathbf{r}, \boldsymbol{\Omega}, E) \tag{4.226}$$

$$\times \int_{4\pi} d\boldsymbol{\Omega}' \int_0^{E_f} dE' \left[\nu\Sigma_f(\mathbf{f}; \mathbf{r}, E')\right] \varphi(\mathbf{r}, \boldsymbol{\Omega}', E') \frac{\partial \chi(\mathbf{p}; \mathbf{r}, E' \to E)}{\partial p_{m_2}};$$

For $j = 1, \ldots, J_k;\ m_2 = 1, \ldots, J_q$: $\dfrac{\partial^2 S\left(\varphi, \varphi^+; \boldsymbol{\gamma}_j^{(2)}; \boldsymbol{\alpha}\right)}{\partial k_j \partial q_{m_2}}$

$$= \frac{\partial^2 R\left(\varphi, \varphi^+; \boldsymbol{\gamma}_j^{(2)}; \boldsymbol{\alpha}\right)}{\partial k_j \partial q_{m_2}} = \int dV \int_{4\pi} d\boldsymbol{\Omega} \int_0^{E_f} dE\, \gamma_{1,j}^{(2)}(\mathbf{r}, \boldsymbol{\Omega}, E) \frac{\partial Q(\mathbf{q}; \mathbf{r}, \boldsymbol{\Omega}, E)}{\partial q_{m_2}}; \tag{4.227}$$

For $j = 1, \ldots, J_k;\ m_2 = 1, \ldots, J_k$: $\dfrac{\partial^2 S\left(\varphi, \varphi^+; \boldsymbol{\gamma}_j^{(2)}; \boldsymbol{\alpha}\right)}{\partial k_j \partial k_{m_2}}$

$$= \frac{\partial^2 R\left(\varphi, \varphi^+; \boldsymbol{\gamma}_j^{(2)}; \boldsymbol{\alpha}\right)}{\partial k_j \partial k_{m_2}} = \int dV \int_{4\pi} d\boldsymbol{\Omega} \int_0^{E_f} dE\, \varphi(\mathbf{r}, \boldsymbol{\Omega}, E) \frac{\partial^2 Q^+(\mathbf{k}; \mathbf{r}, \boldsymbol{\Omega}, E)}{\partial k_j \partial k_{m_2}}. \tag{4.228}$$

The non-zero component, $\gamma_{1,j}^{(2)}(\mathbf{r}, \boldsymbol{\Omega}, E)$, of the second-level adjoint function $\boldsymbol{\gamma}_j^{(2)}(\mathbf{r}, \boldsymbol{\Omega}, E)$, which appears in (4.211)–(4.216), is the solution of the following 2nd-LASS, for $j = 1, \ldots, J_k$:

$$L^+(\boldsymbol{\alpha}) \gamma_{1,j}^{(2)}(\mathbf{r}, \boldsymbol{\Omega}, E) = \frac{\partial Q^+(\mathbf{k}; \mathbf{r}, \boldsymbol{\Omega}, E)}{\partial k_j}, \tag{4.229}$$

subject to the following boundary condition:

$$\gamma_{1,j}^{(2)}(\mathbf{r}_s, \boldsymbol{\Omega}, E) = 0, \quad \mathbf{r}_s \in \partial V, \quad \boldsymbol{\Omega} \cdot \mathbf{n} > 0; 0 < E < E_f. \tag{4.230}$$

The expressions of the second-order sensitivities provided in (4.223) must be identical to those provided in (4.172), i.e.,

$$\text{For } j = 1, \ldots, J_k;\ m = 1, \ldots, J_t : \frac{\partial^2 S\left(\varphi, \varphi^+; \boldsymbol{\gamma}_j^{(2)}; \boldsymbol{\alpha}\right)}{\partial k_j \partial t_m}$$

$$= -\int dV \int_{4\pi} d\boldsymbol{\Omega} \int_0^{E_f} dE \; \frac{\partial \Sigma_t(\mathbf{t}; \mathbf{r}, E)}{\partial t_m} \gamma_{1,j}^{(2)}(\mathbf{r}, \boldsymbol{\Omega}, E) \varphi(\mathbf{r}, \boldsymbol{\Omega}, E)$$

$$= \frac{\partial^2 S\left(\varphi, \varphi^+; \boldsymbol{\psi}_j^{(2)}; \boldsymbol{\alpha}\right)}{\partial t_m \partial k_j} = \int dV \int_{4\pi} d\boldsymbol{\Omega} \int_0^{E_f} dE \, \psi_{2,m}^{(2)}(\mathbf{r}, \boldsymbol{\Omega}, E) \frac{\partial Q^+(\mathbf{k}; \mathbf{r}, \boldsymbol{\Omega}, E)}{\partial k_j}. \tag{4.231}$$

The relation expressed by (4.231) provides an independent mutual verification of the second-level adjoint functions $\boldsymbol{\gamma}_j^{(2)}$ and $\boldsymbol{\psi}_j^{(2)}$.

The expressions of the second-order sensitivities provided in (4.224) must be identical to those provided in (4.181), i.e., $j = 1, \ldots, J_k;\ m = 1, \ldots, J_s$:

$$\frac{\partial^2 S\left(\varphi, \varphi^+; \boldsymbol{\gamma}_j^{(2)}; \boldsymbol{\alpha}\right)}{\partial k_j \partial s_m} = \int dV \int_{4\pi} d\boldsymbol{\Omega} \int_0^{E_f} dE \, \gamma_{1,j}^{(2)}(\mathbf{r}, \boldsymbol{\Omega}, E)$$

$$\times \int_{4\pi} d\boldsymbol{\Omega}' \int_0^{E_f} dE' \, \varphi(\mathbf{r}, \boldsymbol{\Omega}', E') \frac{\partial \Sigma_s(\mathbf{s}; \mathbf{r}, E' \to E, \boldsymbol{\Omega}' \to \boldsymbol{\Omega})}{\partial s_m}$$

$$= \frac{\partial^2 S\left(\varphi, \varphi^+; \boldsymbol{\theta}_j^{(2)}; \boldsymbol{\alpha}\right)}{\partial s_m \partial k_j} = \int dV \int_{4\pi} d\boldsymbol{\Omega} \int_0^{E_f} dE \, \theta_{2,m}^{(2)}(\mathbf{r}, \boldsymbol{\Omega}, E) \frac{\partial Q^+(\mathbf{k}; \mathbf{r}, \boldsymbol{\Omega}, E)}{\partial k_j}. \tag{4.232}$$

The relation expressed by (4.232) provides an independent mutual verification of the second-level adjoint functions $\boldsymbol{\gamma}_j^{(2)}$ and $\boldsymbol{\theta}_j^{(2)}$.

The expressions of the second-order sensitivities provided in (4.225) must be identical to those provided in (4.195), i.e., for $j = 1, \ldots, J_k;\ m = 1, \ldots, J_f$:

$$\frac{\partial^2 S\left(\varphi,\varphi^+;\boldsymbol{\gamma}_j^{(2)};\boldsymbol{\alpha}\right)}{\partial k_j \partial f_m} = \int dV \int_{4\pi} d\boldsymbol{\Omega} \int_0^{E_f} dE\, \gamma_{1,j}^{(2)}(\mathbf{r},\boldsymbol{\Omega},E)$$
$$\times \int_{4\pi} d\boldsymbol{\Omega}' \int_0^{E_f} dE' \chi(\mathbf{p};\mathbf{r},E'\to E)\varphi(\mathbf{r},\boldsymbol{\Omega}',E')\frac{\partial\left[\nu\Sigma_f(\mathbf{f};\mathbf{r},E')\right]}{\partial f_m}$$
$$= \frac{\partial^2 S\left(\varphi,\varphi^+;\mathbf{u}_j^{(2)};\boldsymbol{\alpha}\right)}{\partial f_m \partial k_j} = \int dV \int_{4\pi} d\boldsymbol{\Omega} \int_0^{E_f} dE\, u_{2,m}^{(2)}(\mathbf{r},\boldsymbol{\Omega},E)\frac{\partial Q^+(\mathbf{k};\mathbf{r},\boldsymbol{\Omega},E)}{\partial k_j}. \tag{4.233}$$

The relation shown in (4.233) provides an independent path for the mutual verification of the solutions $\boldsymbol{\gamma}_j^{(2)}$ and $\mathbf{u}_j^{(2)}$.

The expressions of the second-order sensitivities provided in (4.226) must be identical to those provided in (4.204), i.e.,

$$\text{For } j=1,\ldots,J_k;\ m=1,\ldots,J_p:\ \frac{\partial^2 S\left(\varphi,\varphi^+;\boldsymbol{\gamma}_j^{(2)};\boldsymbol{\alpha}\right)}{\partial k_j \partial p_m}$$
$$= \int dV \int_{4\pi} d\boldsymbol{\Omega} \int_0^{E_f} dE\, \gamma_{1,j}^{(2)}(\mathbf{r},\boldsymbol{\Omega},E) \int_{4\pi} d\boldsymbol{\Omega}' \int_0^{E_f} dE' \left[\nu\Sigma_f(\mathbf{f};\mathbf{r},E')\right]$$
$$\times \frac{\partial \chi(\mathbf{p};\mathbf{r},E'\to E)}{\partial p_m}\varphi(\mathbf{r},\boldsymbol{\Omega}',E') = \frac{\partial^2 S\left(\varphi,\varphi^+;\mathbf{w}_j^{(2)};\boldsymbol{\alpha}\right)}{\partial p_m \partial k_j} \tag{4.234}$$
$$= \int dV \int_{4\pi} d\boldsymbol{\Omega} \int_0^{E_f} dE\, w_{2,m}^{(2)}(\mathbf{r},\boldsymbol{\Omega},E)\frac{\partial Q^+(\mathbf{k};\mathbf{r},\boldsymbol{\Omega},E)}{\partial k_j}.$$

The relation shown in (4.234) provides an independent path for the mutual verification of the solutions $\boldsymbol{\gamma}_j^{(2)}$ and $\mathbf{w}_j^{(2)}$.

The expressions of the second-order sensitivities provided in (4.227) must be identical to those provided in (4.216), i.e.,

For $j = 1, \ldots, J_k;\ m = 1, \ldots, J_q : \dfrac{\partial^2 S\left(\varphi, \varphi^+; \boldsymbol{\gamma}_j^{(2)}; \boldsymbol{\alpha}\right)}{\partial k_j \partial q_m}$

$$= \int dV \int_{4\pi} d\boldsymbol{\Omega} \int_0^{E_f} dE\, \gamma_{1,j}^{(2)}(\mathbf{r}, \boldsymbol{\Omega}, E) \frac{\partial Q(\mathbf{q}; \mathbf{r}, \boldsymbol{\Omega}, E)}{\partial q_m} = \frac{\partial^2 S\left(\varphi, \varphi^+; \mathbf{g}_j^{(2)}; \boldsymbol{\alpha}\right)}{\partial q_m \partial k_j}$$
$$= \int dV \int_{4\pi} d\boldsymbol{\Omega} \int_0^{E_f} dE\, g_{2,m}^{(2)}(\mathbf{r}, \boldsymbol{\Omega}, E) \frac{\partial Q^+(\mathbf{k}; \mathbf{r}, \boldsymbol{\Omega}, E)}{\partial k_j}. \tag{4.235}$$

The relation shown in (4.235) provides an independent path for the mutual verification of the solutions $\boldsymbol{\gamma}_j^{(2)}$ and $\mathbf{g}_j^{(2)}$.

4.6 Chapter Summary

This chapter has illustrated the application of the 2nd-CASAM-L to perform sensitivity analysis of ratios of generic functionals of the forward and adjoint neutron flux distributions in a multiplying subcritical system which comprises a non-fission neutron source. The forward and adjoint flux distributions in such a system are modeled by the forward and, respectively, adjoint time-independent integro-differential neutron transport Boltzmann equations (see, e.g., Davison 1957; Stacey 1974, 2001). This physical system has been chosen as a paradigm application of the 2nd-CASAM-L because the Boltzmann equation comprises both differential and integral operators, in six independent variables, as well as a large number of model parameters. Using the integro-differential Boltzmann equation makes it possible to illustrate simultaneously the construction of adjoint operators both for differential and for integral operators. Solving the integro-differential Boltzmann equation epitomizes "large-scale computations." The system response considered for the application of the 2nd-CASAM-L is a ratio of two generic nonlinear functionals of the forward and adjoint particle fluxes, while the model parameters include isotopic number densities, microscopic cross sections, fission spectrum parameters, neutron sources, and detector response functions. Such responses represent the most general functionals of fundamental importance in reactor physics. Notably, the modeling of multiplying subcritical system using the Boltzmann equation considered in this chapter greatly generalizes the modeling of the spatial transmission of particles through homogenized materials considered in Chap. 3. Of course, closed-form analytical expressions for the response sensitivities of various order, such as obtained in Chap. 3, were no longer possible to obtain for the complex subcritical multiplying physical system which was analyzed in this chapter.

As particular cases of the response considered in Sects. 4.2–4.4, the Roussopoulos and Schwinger functionals for computing reaction rates in a multiplying medium with an internal non-fission source were also analyzed in this chapter, in two distinct contexts, namely: (1) in the traditional, but unrealistic, context of postulating imprecise trial functions for the forward and adjoint particle fluxes while assuming that the system parameters are perfectly well known; and (2) in the realistic context of dealing with systems characterized by imprecisely known parameters (cross sections, number densities, forward and adjoint sources, etc.) which, in turn, cause variations in the particle fluxes.

The results presented in this work have highlighted the fact that the Roussopoulos and Schwinger functionals are both stationary if and only if the approximate forward flux satisfies the same vacuum boundary as the exact (but unknown) flux, while the model parameters are assumed to be perfectly well known, and their perfectly well-known values coincide with the respective nominal values. It was also shown that, although both the Roussopoulos and the Schwinger functionals yield the exact reaction rate of interest within a "second-order error term," the respective "second-order error terms" differ from each other when approximate expressions for the fluxes are used as trial functions.

In realistic practical contexts, however, the values of the system parameters are not perfectly well known. For such realistic contexts, it has been shown that the first-order sensitivities of the Schwinger functional are identical to the corresponding sensitivities of the Roussopoulos functional to the model's parameters, and these first-order sensitivities are *not* identically zero. Hence, neither the Roussopoulos nor the Schwinger functionals are accurate to second order in parameter variations/uncertainties, and consequently are not accurate to second-order variations in the flux when such flux variations are caused by imprecisely known model parameters. Even more importantly, though, the derivations presented in this work demonstrate that the sensitivities of the Schwinger functional to model parameters coincide, at the nominal parameter values, with the respective sensitivities of the Roussopoulos functional, implying that the two functionals are equivalent when computing reaction rates in nuclear systems characterized by imprecisely known parameters. For convenience, the expressions of the first- and second-order sensitivities of the Roussopoulos and Schwinger functionals were also provided. The same forward and adjoint operators appear on the left sides of the various 2nd-LASS for the various second-level adjoint functions. Only the source terms on the right sides of the respective 2nd-LASS differed from each other. Therefore, the "forward" and "adjoint" software packages which are used for solving the original forward and adjoint transport equations can also be used for solving numerically the various equations underlying the 1st-LASS and the 2nd-LASS. Furthermore, the computation of the second-order sensitivities involves the evaluations of integrals of the same form as those needed for computing the first-order sensitivities. Therefore, the same software can be used for computing both the first-order and second-order sensitivities. Each of the mixed second-order sensitivities can be computed using two distinct second-level adjoint functions, thus providing an inherent *solution verification* mechanism that enables and ensures the accuracy verification of the solutions of all of the 2nd-LASS.

The methodology and results presented in this chapter for the physical system modeled by the Boltzmann neutron transport equation provide the basis for the analysis of the OECD/NEA reactor physics benchmark (Valentine 2006) which will be presented in Volume 2 of this book.

Chapter 5
Concluding Remarks

Having underscored in Chap. 1 of this volume the need for determining not only first-order but also higher-order sensitivities of model responses to model parameters, Chap. 2 of this volume has presented the mathematical framework of the "*nth-Order Comprehensive Adjoint Sensitivity Analysis Methodology for Response-Coupled Forward and Adjoint Linear Systems*" (abbreviated as "nth-CASAM-L"), where the integer "n" denotes an arbitrarily high order. The nth-CASAM-L enables the most efficient computation of exactly determined expressions of arbitrarily high-order sensitivities of a generic system response which depends on both the forward and adjoint state functions. The qualifier "comprehensive" is employed in order to highlight the fact that the framework of the nth-CASAM-L comprises not only internal model parameters but also includes the system's uncertain boundaries and internal interfaces in the model's phase-space. The validity mathematical framework of the nth-CASAM-L has been established using mathematical induction.

The mathematical frameworks of the nth-CASAM-L for the particular cases: $n = 1, 2, 3, 4$, and 5, have been developed and presented in detail, since these particular cases would be the most frequently used in practical applications. Furthermore, these particular cases also highlight the fundamental conceptual breakthrough underlying the nth-CASAM, namely: *the nth-CASAM is formulated in linearly increasing higher-dimensional Hilbert spaces, as opposed to exponentially increasing parameter-dimensional spaces.* In particular, for a scalar-valued response associated with a model comprising *TP* model parameters, the nth-CASAM (for $n = 1$) requires 1 adjoint computation for computing exactly all of the first-order response sensitivities; this is in contradistinction to all other methods, which would require at least *TP* forward computations. All of the second-order sensitivities are computed exactly by the nth-CASAM (for $n = 2$) in at most *TP* computations, as opposed to needing $O(TP^2)$ computations (as would be required by all other methods) and so on. For every lower-order sensitivity of interest, the nth-CASAM computes the *TP* "next-higher-order" sensitivities using *a single adjoint computation performed in a linearly increasing higher-dimensional Hilbert space.* Very importantly, the nth-CASAM is also applicable to the computation of sensitivities

D. G. Cacuci, *The nth-Order Comprehensive Adjoint Sensitivity Analysis Methodology, Volume I*, https://doi.org/10.1007/978-3-030-96364-4_5

of *operator-valued responses*, which cannot be handled by current methods when the underlying problem comprises many parameters. The nth-CASAM can be applied to any model (deterministic, statistical, etc.). The higher-level adjoint functions are computed using the same forward and adjoint solvers (i.e., computer codes) as used for solving the original forward and adjoint systems, thereby requiring relatively minor additional software development for computing the various-order sensitivities. The larger the number of model parameters, the more efficient the nth-CASAM becomes for computing arbitrarily high-order response sensitivities.

The application of the nth-CASAM-L has been illustrated in Chap. 3 by considering two paradigm physical models. One model described particle transport in a homogenized medium characterized by imprecisely known parameters, internal and external boundaries in a spatial domain. The other model described the time-evolution—in temporal domain having imprecisely known starting and final-time instances—of the concentration of a substance obeying a conservation law in a medium characterized by imprecisely known parameters and initial state. These models have been selected for illustrating the application of the nth-CASAM-L because they involve many imprecisely known parameters, internal and external interfaces and imprecisely known initial and final time-instances, yet they admit exact closed-form expressions for all sensitivities, for convenient verification of all results. The paradigm models analyzed in Chap. 3 admit many relative sensitivities that have identical values. The precise and efficient computation of many relative sensitivities that have identical values does not pose any difficulties to the nth-CASAM-L methodology but is known to pose extreme challenges to conventional statistical methods which rely on sensitivities being readily "sortable" by having relative values that clearly differ from each other. Since the models analyzed in Chap. 3 involve numerous parameters yet are sufficiently simple to admit closed-form exact solutions for all quantities (forward and adjoint state functions, responses, sensitivities, etc.) of interest, these paradigm models can be used as a benchmarks for comparing the nth-CASAM-L to conventional statistical and/or finite-difference methods.

The information provided by the first-order sensitivities usually indicates which second-order sensitivities are important and which could be neglected. Therefore, it is useful to prioritize the computation of the second-order sensitivities by using the rankings of the relative magnitudes of the first-order sensitivities as a "priority indicator": the larger the magnitude of the relative first-order sensitivity, the higher the priority for computing the corresponding second-order sensitivities. Also, since vanishing first-order sensitivities may indicate critical points of the response in the phase-space of model parameters, it is also of interest to compute the second-order sensitivities that correspond to vanishing first-order sensitivities. In practice, only those second-order partial sensitivities which are deemed important would need to be computed.

Each successively higher-order sensitivities are computed using the same principles as used for computing the second-order sensitivities. The computation of the third-order sensitivities can be prioritized by using the magnitudes/importance of the first-order and second-order sensitivities as guiding indicators, and so on. It has also been shown that aggregating several parameters into a single one significantly reduces the number of adjoint computations needed for the computation of second- and higher-order sensitivities. Aggregating parameters is not necessary for the computation of the first-order sensitivities since the computation of all of the first-order sensitivities requires just a single adjoint-computation for solving the 1st-LASS.

Chapter 4 has illustrated the application of the 2nd-CASAM-L to perform sensitivity analysis of ratios of generic functionals of the forward and adjoint neutron flux distributions in a multiplying subcritical system which comprises a non-fission neutron source. The forward and adjoint flux distributions in such a system are modeled by the forward and, respectively, adjoint time-independent integro-differential neutron transport Boltzmann equations. This physical system has been chosen as a paradigm application of the 2nd-CASAM-L because the Boltzmann equation comprises both differential and integral operators, in six independent variables, as well as a large number of model parameters. Using the integro-differential Boltzmann equation makes it possible to illustrate simultaneously the construction of adjoint operators both for differential and for integral operators. Solving the integro-differential Boltzmann equation epitomizes "large-scale computations" involving many parameters (including isotopic number densities, microscopic cross sections, fission spectrum parameters, neutron sources, and detector response functions).

Particularly important model responses that involve both the forward and adjoint functions in linear systems are various forms of Lagrangian functionals, the Roussopoulos functional for computing reaction rates, the Raleigh quotient for computing eigenvalues and/or separation constants when solving partial differential equations, the Schwinger functional for "normalization-free" solutions, and many others. The application of the 2nd-CASAM to these functionals has also been presented in detail in Chap. 4 since these functionals play a fundamental role in various optimization and control procedures, derivation of numerical methods for solving differential, integral, and integro-differential equations. The material presented in Chap. 4 sets the stage for a large-scale application of the 4th-CASAM-L to perform a fourth-order sensitivity and uncertainty analysis of an OECD/NEA reactor physics benchmark modeled by the Boltzmann transport equation involving 21,976 uncertain parameters, of which 7477 have non-zero nominal values. The fourth-order sensitivity and uncertainty analysis of this OECD/NEA reactor physics benchmark will be presented in Volume 2 of this book.

For response-coupled forward/adjoint linear systems, the nth-CASAM-L is the only practically implementable methodology which overcomes, to the largest possible practical extent, the curse of dimensionality while providing exact expressions (i.e., free of methodologically introduced approximations) for computing efficiently sensitivities of all orders of model responses to model parameters. By enabling the practical computation of arbitrarily high-order response sensitivities to model parameters, the nth-CASAM-L makes it possible to compare the relative values of the sensitivities of various order, in order to assess which sensitivities are important and which may actually be neglected. Such comparisons enable investigations of the convergence of the (multivariate) Taylor series expansion of the response in terms of parameter variations, as well as investigations of the actual validity of expressions that are derived from Taylor expansion of the response (e.g., response variances/covariance, skewness, kurtosis) as functions of the model's parameters. The larger the number of model parameters, the more efficient the nth-CASAM-L becomes for computing arbitrarily high-order response sensitivities. The nth-CASAM-L also enables the direct and exact derivation (first on paper, then programing the exact expression thus obtained for numerical evaluation) of the expression of a specific high-order sensitivity of interest, which can subsequently be computed directly; such a direct computation is not possible with any statistical method.

Volume 3 of this book will present the "nth-Order Comprehensive Adjoint Sensitivity Analysis Methodology for Nonlinear Systems" (abbreviation: nth-CASAM-N) for the practical, efficient, and exact computation of arbitrarily high-order sensitivities of responses to model parameters for nonlinear systems. The nth-CASAM-L together with the nth-CASAM-N are expected to revolutionize all of the fields of activities which require response sensitivities, including the fields of uncertainty quantification, model validation, optimization, data assimilation, model calibration, sensor fusion, reduced-order modeling, inverse problems, and predictive modeling.

References

Amaldi E (1959) The Production and Slowing Down of Neutrons. In: Handbuch der Physik, Vol. 38, Part 2, 1-659, Springer Verlag, Berlin, Germany

Bellman RE (1957) Dynamic programming. Rand Corporation, Princeton University Press, ISBN 978-0-691-07951-6, USA. Republished: Bellman, RE (2003) Dynamic Programming. Courier Dover Publications, ISBN 978-0-486-42809-3, USA

Cacuci DG (1981a) Sensitivity theory for nonlinear systems: I. Nonlinear functional analysis approach. J. Math. Phys 22:2794-2802

Cacuci DG (1981b) Sensitivity theory for nonlinear systems: II. Extensions to additional classes of responses. J. Math. Phys. 22:2803-2812

Cacuci DG (2015) Second-order adjoint sensitivity analysis methodology for computing exactly and efficiently first- and second-order sensitivities in large-scale linear systems: I. Computational methodology. J. Comp. Phys. 284:687–699

Cacuci DG (2016) Second-Order Adjoint Sensitivity Analysis Methodology (2nd-ASAM) for Large-Scale Nonlinear Systems: I. Theory. Nucl. Sci. Eng. 184:16–30

Cacuci DG (2018) The second-order adjoint sensitivity analysis methodology. CRC Press, Taylor & Francis Group, Boca Raton, FL, USA

Cacuci DG (2019a) Second-order sensitivity and uncertainty analysis of the Roussopoulos functional for a subcritical multiplying nuclear system with source, Annals Nucl. Energy 126: 281–291. https://doi.org/10.1016/j.anucene.2018.11.022

Cacuci DG (2019b) The Roussopoulos and Schwinger functionals for nuclear systems involving imprecisely known fluxes and parameters: distinctions and equivalences. Nucl. Sci. Eng. https://doi.org/10.1080/00295639.2018.1564504

Cacuci DG (2019c) Second-order sensitivities of a general functional of the forward and adjoint fluxes in a multiplying nuclear system with source. Nucl. Eng. Des. 344:83–106. https://doi.org/10.1016/j.nucengdes.2019.01.007

Cacuci DG (2019d) Towards Overcoming the Curse of Dimensionality: The Third-Order Adjoint Method for Sensitivity Analysis of Response-Coupled Linear Forward/Adjoint Systems, with Applications to Uncertainty Quantification and Predictive Modeling. Energies, 12:4216. https://doi.org/10.3390/en12214216.

Cacuci DG (2020) Second-order adjoint sensitivity analysis of a general ratio of functionals of the forward and adjoint fluxes in a multiplying nuclear system with source, An. Nucl. En. 135: 106956. https://doi.org/10.1016/j.anucene.2019.106956

Cacuci DG (2021a) Fourth-order comprehensive adjoint sensitivity analysis of response-coupled linear forward/adjoint systems. I. Theoretical framework. Energies 14:3335. https://doi.org/10.3390/en14113335

D. G. Cacuci, *The nth-Order Comprehensive Adjoint Sensitivity Analysis Methodology, Volume I*, https://doi.org/10.1007/978-3-030-96364-4

Cacuci DG (2021b) High-Order Deterministic Sensitivity Analysis and Uncertainty Quantification: Review and New Developments. *Energies* 14:6715. https://doi.org/10.3390/en14206715

Cacuci DG, Ionescu-Bujor M, Navon, MI (2005) Sensitivity and uncertainty analysis: Applications to large scale systems, Volume 2, Chapman & Hall/CRC, Boca Raton, USA

Cacuci, D.G., Badea, A. F., Badea, M. C., and Peltz, J. J. (2016) Efficient Computation of Operator-Type Response Sensitivities for Uncertainty Quantification and Predictive Modeling: Illustrative Application to a Spent Nuclear Fuel Dissolver Model," Int. J. Num. Meth. in Fluids, Wiley Online Library (wileyonlinelibrary.com). https://doi.org/10.1002/fld.4258.

Davison B (1957) Neutron transport theory. Oxford University Press, Oxford, UK.

Levine H, Schwinger J (1949) On the theory of diffraction by an aperture in an infinite plane screen. Phys. Rev. 75:1423

Lewins J (1965) IMPORTANCE: The Adjoint Function, Pergamon Press Ltd., Oxford, UK, .

Luo Z, Wang X, Liu D. (2020) Prediction on the static response of structures with large-scale uncertain-but-bounded parameters based on the adjoint sensitivity analysis. Structural and Multidisciplinary Optimization, 61:123–139. https://doi.org/10.1007/s00158-019-02349-w

Práger T, Kelemen FD (2014) Adjoint methods and their application in earth sciences. In: Faragó I, Havasi Á, Zlatev Z (Eds.) Advanced numerical methods for complex environmental models: Needs and availability. Bentham Science Publishers, Oak Park, IL, USA. Chapter 4A, p. 203–275

Roussopolos P (1953) Methodes variationeles en theories des collisions. C. R Acad. Sci. 236:1858

Stacey WM (1974) Variational methods in nuclear reactor physics, Academic Press, Inc., New York, USA

Stacey WM (2001) Nuclear reactor physics. John Wiley & Sons, New York. USA.

Shultis JK, Faw RE (2000) Radiation Shielding. American Nuclear Society, La Grange Park, IL, USA.

Tukey JW (1957) The Propagation of Errors, Fluctuations and Tolerances. Technical Reports No. 10-12, Princeton University, Princeton, NJ, USA

Valentine TE (2006) Polyethylene-reflected plutonium metal sphere subcritical noise measurements, SUB-PU-METMIXED-001. International Handbook of Evaluated Criticality Safety Benchmark Experiments, NEA/NSC/DOC(95)03/I-IX, Organization for Economic Co-Operation and Development (OECD), Nuclear Energy Agency (NEA), Paris, France

Weiberg AM, Wigner EP (1958) The Physical Theory of Neutron Chain Reactors. University of Chicago Press, Chicago, IL, USA

Wigner E P (1945) Effect of small perturbations on pile period, Chicago Report CP-G-3048, Chicago, IL, USA

Williams ML (1986) Perturbation Theory for Nuclear Reactor Analysis. In: Ronen Y (ed) Handbook of Nuclear Reactor Calculations, CRC Press, Boca Raton, FL, USA, Volume 3, p 63-188.

Williams ML, Engle WW (1977) The Concept of Spatial Channel Theory Applied to Reactor Shielding Analysis. Nucl. Sci. Eng. 62:92.

Index

D. G. Cacuci, *The nth-Order Comprehensive Adjoint Sensitivity Analysis Methodology, Volume I*, https://doi.org/10.1007/978-3-030-96364-4